Life History Evolution

Life History Evolution

Traits, Interactions, and Applications

Edited by

Michal Segoli
Mitrani Department of Desert Ecology
Blaustein Institutes for Desert Research
Ben-Gurion University of the Negev
Midreshet Ben-Gurion
Israel

Eric Wajnberg
INRAE, Sophia Antipolis, France
INRIA, Project Hephaistos, Sophia Antipolis, France
Departamento de Entomologia e Acarologia
Universidade de São Paulo
Escola Superior de Agricultura "Luiz de Queiroz"
Piracicaba, São Paulo, Brazil

This edition first published 2025
© 2025 John Wiley & Sons Ltd

All rights reserved, including rights for text and data mining and training of artificial intelligence technologies or similar technologies. No part of this publication may be reproduced, stored in a retrieval system, or transmitted, in any form or by any means, electronic, mechanical, photocopying, recording or otherwise, except as permitted by law. Advice on how to obtain permission to reuse material from this title is available at http://www.wiley.com/go/permissions.

The right of Michal Segoli and Eric Wajnberg to be identified as the authors of the editorial material in this work has been asserted in accordance with law.

Registered Offices
John Wiley & Sons, Inc., 111 River Street, Hoboken, NJ 07030, USA
John Wiley & Sons Ltd, New Era House, 8 Oldlands Way, Bognor Regis, West Sussex, PO22 9NQ, UK

For details of our global editorial offices, customer services, and more information about Wiley products visit us at www.wiley.com.

Wiley also publishes its books in a variety of electronic formats and by print-on-demand. Some content that appears in standard print versions of this book may not be available in other formats.

Trademarks: Wiley and the Wiley logo are trademarks or registered trademarks of John Wiley & Sons, Inc. and/or its affiliates in the United States and other countries and may not be used without written permission. All other trademarks are the property of their respective owners. John Wiley & Sons, Inc. is not associated with any product or vendor mentioned in this book.

Limit of Liability/Disclaimer of Warranty
While the publisher and authors have used their best efforts in preparing this work, they make no representations or warranties with respect to the accuracy or completeness of the contents of this work and specifically disclaim all warranties, including without limitation any implied warranties of merchantability or fitness for a particular purpose. No warranty may be created or extended by sales representatives, written sales materials or promotional statements for this work. This work is sold with the understanding that the publisher is not engaged in rendering professional services. The advice and strategies contained herein may not be suitable for your situation. You should consult with a specialist where appropriate. The fact that an organization, website, or product is referred to in this work as a citation and/or potential source of further information does not mean that the publisher and authors endorse the information or services the organization, website, or product may provide or recommendations it may make. Further, readers should be aware that websites listed in this work may have changed or disappeared between when this work was written and when it is read. Neither the publisher nor authors shall be liable for any loss of profit or any other commercial damages, including but not limited to special, incidental, consequential, or other damages.

Library of Congress Cataloging-in-Publication Data
Names: Segoli, Michal, editor. | Wajnberg, E., editor.
Title: Life history evolution : traits, interactions, and applications / edited by Michal Segoli, Eric Wajnberg.
Description: Hoboken, NJ : Wiley, 2025. | Includes index.
Identifiers: LCCN 2024026594 (print) | LCCN 2024026595 (ebook) | ISBN 9781394185726 (hardback) | ISBN 9781394185733 (adobe pdf) | ISBN 9781394185740 (epub)
Subjects: LCSH: Evolution (Biology) | Life cycles (Biology)
Classification: LCC QH366.2 .L544 2025 (print) | LCC QH366.2 (ebook) | DDC 576.8–dc23/eng20240809
LC record available at https://lccn.loc.gov/2024026594
LC ebook record available at https://lccn.loc.gov/2024026595

Cover Design: Wiley
Cover Image: Courtesy of Alfred Daniel Johnson

Set in 9.5/12.5pt STIXTwoText by Straive, Pondicherry, India

Contents

List of Contributors *xv*
Foreword *xix*
Preface *xxiii*

Part I Traits *1*

1 Body Size and Timing of Maturation *3*
Toomas Tammaru and Tiit Teder
1.1 Introduction *3*
1.2 Part I *3*
1.2.1 Benefits of Large Body Size *3*
1.2.2 Costs of Large Body Size *6*
1.2.3 Absolute Timing of Maturation *7*
1.2.4 Classical Models of Age and Size at Maturation *7*
1.2.5 Adding Environmental Variability – Reaction Norms for Age and Size at Maturation *8*
1.2.5.1 Reaction Norms in the Classical Models *9*
1.2.5.2 Plastic Growth Rates *10*
1.2.5.3 Introducing Developmental Thresholds *10*
1.2.5.4 Probabilistic Reaction Norms *11*
1.2.5.5 Integrating Physiological Realism *12*
1.3 Part II *13*
1.3.1 Bivariate reaction Norms with a Positive Slope *13*
1.3.2 Trade-Off Between Age and Size at Maturation *15*
1.3.3 The Rates of Evolution – Do We Study the Right Thing? *16*
1.4 Conclusions *18*
Acknowledgements *18*
References *18*

2 Evolution of Ageing and Lifespan *29*
Alexei A. Maklakov
2.1 Introduction *29*
2.2 Evolutionary Theory of Ageing *30*
2.2.1 Mutation Accumulation and Antagonistic Pleiotropy *31*
2.2.2 'Disposable Soma' and the Developmental Theory of Ageing *34*
2.3 Asynchronous Ageing *38*
2.4 Sex Differences in Ageing *40*
2.5 Williams and Anti-Williams: Age, Density and Condition-Dependence of Mortality *42*
2.6 Concluding Remarks *43*
Acknowledgements *43*
References *43*

3	**Offspring Size and Life History Theory: What Do We Know?: What Do We Still Need to Learn?** *49*	
	Dustin J. Marshall	
3.1	Offspring Size Defined *49*	
3.2	The Knowns *49*	
3.2.1	What Does Offspring Size Affect? *49*	
3.2.2	Offspring Size and Theory *51*	
3.2.3	Transgenerational Plasticity in Offspring Size – The Link Between Generations *53*	
3.2.4	Summary of the Knowns of Offspring Size *54*	
3.3	The Unknowns *55*	
3.3.1	The Offspring Size and Number – How Do They Scale with Costs? *55*	
3.3.2	Why Does Offspring Size Affect Fitness? *56*	
3.3.3	Who Controls Offspring Size? *56*	
3.3.4	Summary of Unknowns in Offspring Size *56*	
	References *57*	
4	**The Evolution of Insect Egg Loads: The Balance of Time and Egg Limitation** *61*	
	Michal Segoli, Miriam Kishinevsky, and George E. Heimpel	
4.1	Trade-Offs Between Early and Late Components of Reproduction *61*	
4.2	Time *vs.* Egg Limitation in Insects *61*	
4.3	Egg Maturation Patterns *62*	
4.4	The Relative Importance of Egg and Time Limitation *62*	
4.4.1	Oviposition Opportunities *64*	
4.4.2	Sugar Availability *65*	
4.4.3	The Interaction Between Oviposition Opportunities and Sugar Availability *65*	
4.4.4	Environmental Stochasticity *66*	
4.4.5	The Cost of Producing an Egg *67*	
4.4.6	Interaction Between Environmental Stochasticity and the Cost of Producing an Egg *68*	
4.5	Additional Life History Strategies to Overcome the Risk of Egg Limitation *68*	
4.6	Conclusions and Future Directions *69*	
	Acknowledgements *69*	
	References *70*	
5	**Sex-Specific Life Histories** *77*	
	Hanna Kokko	
5.1	Introduction *77*	
5.2	Various Unidirectional Effects: Unguarded X, Mother's Curse and Toxic Y *78*	
5.3	Multi-directionality: Coevolution of Different Traits *79*	
5.3.1	An Example of a Model: How Easily Aggression Shortens or Prolongs Lifespan *81*	
5.3.2	How to Make Predictions for the Two Sexes? *83*	
5.3.3	Embrace the Knowledge Gaps, Work to Fill Them *84*	
5.3.4	Caveats: Biology Is Diverse! *88*	
5.3.5	One Genome, Two Sexes *89*	
5.4	Towards Progress *90*	
	Acknowledgements *91*	
	References *91*	
6	**Parental Care and Life History** *97*	
	Hope Klug and Michael B. Bonsall	
6.1	What Is Parental Care and How Does It Relate to Life History? *97*	
6.2	Distinguishing Between the Origin and the Maintenance of Parental Care *98*	
6.3	Life History and the Origin of Parental Care *98*

6.4	Life History and the Maintenance of Parental Care	*100*
6.5	Co-evolution Between Parental Care, Offspring Traits and Parental Traits	*101*
6.6	Sexual Selection, Life History and Sex Differences in Parental Care	*102*
6.7	Stochasticity, Environmental Variability, Life History and Parental Care	*104*
6.7.1	Theoretical Advances: The Evolution of Parental Care in Stochastic Environments	*104*
6.7.2	Linking Predictions of Parental Care in Stochastic Environments to Empirical Patterns and Analysis	*105*
6.8	Plasticity and the Evolution of Parental Care	*106*
6.8.1	Abiotic Environment	*107*
6.8.2	Social Environment	*107*
6.9	Final Conclusions and Future Directions	*108*
	Acknowledgements	*108*
	References	*108*

7 Sex Allocation *113*
Jun Abe and Stuart A. West

7.1	Introduction	*113*
7.2	Fisher's Theory	*113*
7.3	Interaction with Relatives	*115*
7.3.1	Local Resource Competition	*115*
7.3.2	Local Mate Competition	*115*
7.3.3	Local Sperm Competition in Simultaneous Hermaphrodites	*118*
7.3.4	Local Resource Enhancement	*119*
7.4	Environmental Condition	*120*
7.4.1	Facultative Sex Ratio Adjustment	*121*
7.4.2	Environmental Sex Determination	*124*
7.4.3	Sex Change in Sequential Hermaphrodites	*124*
7.5	Future Directions	*126*
	Acknowledgements	*126*
	References	*126*

8 Life History Evolution: Complex Life Cycles Across Animal Diversity *131*
Andreas Heyland, Konstantin Khalturin, and Vincent Laudet

8.1	Integration of Metamorphic Development Within the Life Cycle	*131*
8.2	The Regulation of Metamorphic Development by Hormones	*131*
8.3	Review of Metamorphic Mechanisms Across Taxa with Ecological and Evolutionary Considerations	*132*
8.3.1	Vertebrate Metamorphosis	*134*
8.3.1.1	Environmental Regulation of Vertebrate Metamorphosis	*135*
8.3.1.2	Evolutionary Patterns of Life History Diversity Within and Across Vertebrate Taxa	*137*
8.3.2	Insect Metamorphosis	*138*
8.3.2.1	Environmental Control of Insect Metamorphosis	*138*
8.3.2.2	Evolution of Insect Metamorphosis	*139*
8.3.3	Regulation of the Bentho-Planktonic Life Cycles in Invertebrates	*140*
8.3.3.1	Settlement Cues and Signals	*140*
8.3.3.2	Immune System-Related Signals	*141*
8.3.3.3	The Neuro-endocrine System	*141*
8.3.3.4	Hormonal Systems as Regulators of Metamorphic Development	*142*
8.3.3.5	Environmental Stressors Affecting Metamorphic Development and Settlement Among Marine Invertebrates	*143*
8.3.3.6	Evolution of Metamorphosis Among Marine Invertebrate Groups	*143*
8.4	Anthropogenic Environmental Impacts and Global Climate Change	*144*
	References	*145*

9 Social Living and Life History Evolution, with a Focus on Ageing and Longevity 155
Judith Korb and Volker Nehring
9.1 Introduction 155
9.2 Ultimate Causes of Long Reproductive Lifespans 156
9.2.1 Theoretical Models 156
9.2.2 Empirical Evidence 158
9.2.2.1 Vertebrates 158
9.2.2.2 Social Arthropods 159
9.2.2.3 Conclusion: Ultimate Causes 161
9.2.3 Links Between Sociality, Longevity and Other Life History Traits 161
9.3 Colony Life History in Obligatory Eusocial Insects 162
9.4 Proximate Mechanisms 163
9.4.1 Social Vertebrates 164
9.4.2 Social Arthropods 165
9.4.2.1 The TI-J-Part: Rewiring of the I-J-Fecundity Axis 165
9.4.2.2 The LiFe-Part, with a Focus on Lifespan 166
9.4.3 Conclusion: Proximate Mechanisms 166
9.5 Conclusion 166
Acknowledgements 167
References 167

10 Integrating Dispersal in Life History 175
Dries Bonte
10.1 Introduction 175
10.2 Dispersal as Part of the Life History 176
10.3 The Theory of Dispersal and Life Histories 177
10.3.1 Dispersal Evolution in Response to Other Life History Attributes 178
10.3.2 Joint Evolution of Dispersal and Other Life History Attributes 179
10.3.3 The Evolution of Dispersal-Life History Reaction Norms 179
10.4 Dispersal-Life History Co-variation in Nature 180
10.4.1 Empirical Evidence of Dispersal Syndromes Among Species 180
10.4.2 Phylogenetic Signals in Dispersal Syndromes 181
10.4.3 Micro-geographic Variation Among Populations 181
10.4.4 Condition-Dependence and Within-Population Variation 183
10.5 Concluding Remarks and Outlook: Why Should We Care? 184
References 185

11 The Evolution of Human Life Histories 191
Megan Arnot and Ruth Mace
11.1 Introduction 191
11.2 Life History Trade-Offs 191
11.3 The Life Histories of Great Apes 192
11.4 Variation in Human Life History 195
11.5 Menopause and the Post-reproductive Lifespan 197
11.5.1 The Phylogenetic Patterning of Menopause and the Post-reproductive Lifespan 197
11.5.2 An Evolutionary Approach to Menopause 197
11.5.2.1 Menopause and the Post-reproductive Lifespan as a By-product 198
11.5.2.2 Menopause and the Post-reproductive Lifespan as an Adaptation 200
11.5.3 On Studying the Evolution of Menopause 204
11.6 Final Remarks 205
References 205

Part II Interactions *213*

12 Life History Traits in the Context of Predator–Prey Interactions *215*
Joseph Travis
- 12.1 Introduction *215*
- 12.2 Types of Predation *216*
- 12.3 Theory for Predator-Driven Life History Evolution *217*
- 12.3.1 Overview *217*
- 12.3.2 Predictions from Theory *218*
- 12.4 Empirical Evidence *220*
- 12.5 Adaptive Plasticity in Life Histories *222*
- 12.6 Future Directions *224*
  - Acknowledgements *225*
  - References *225*

13 Life History Trait Evolution in the Context of Host–Parasite Interactions *229*
Alison B. Duncan, Giacomo Zilio, and Oliver Kaltz
- 13.1 Introduction *229*
- 13.2 Host Life History Evolution in Response to Parasites *230*
- 13.2.1 Evolutionary Shifts in Reproductive Effort and Phenology: Some Theory and Empirical Evidence *230*
- 13.2.2 Costs of Resistance and a Brief Perspective on Tolerance *234*
- 13.2.3 Scaling Up: Parasite Effects on Host Dispersal Evolution *235*
- 13.2.4 Scaling Down: Symbionts Mediating Host Life History Evolution *236*
- 13.2.5 Complexity: The Importance of Eco-evolutionary Feedbacks and Multispecies Interactions *237*
- 13.3 Parasite Life History Evolution in Response to Hosts: The Case of Virulence *238*
- 13.3.1 Why Consider Virulence Evolution from a Life History Perspective? *239*
- 13.3.2 Sources of Shortened Infection Lifespan: Host Mortality and Infection Clearance *239*
- 13.3.3 Russian Doll Effect: Breaking Down the Virulence-Transmission Trade-Off into its Components *241*
- 13.3.4 Where to Put the Reproductive Effort: Vertical *vs.* Horizontal Transmission *241*
- 13.3.5 Evolution of Virulence and Dispersal: Small and Big Worlds *243*
- 13.3.6 Very Small Worlds: Parasite Life History Evolution in Response to Multiple Infections *244*
- 13.4 Concluding Remarks *245*
  - References *246*

14 How Do Microbial Symbionts Shape the Life Histories of Multicellular Organisms? *255*
Elad Chiel and Yuval Gottlieb
- 14.1 Introduction *255*
- 14.2 Categories of Microbial Symbiosis *256*
- 14.3 How Microbial Symbionts Are Involved in Essential Biological Functions of Their Hosts? *256*
- 14.4 Nutritional Microbial Symbionts *257*
- 14.5 Reproductive Microbial Symbionts *258*
- 14.6 Defensive Microbial Endosymbionts *259*
- 14.7 Diapause and Microbial Symbionts *262*
- 14.8 Concluding Remarks *262*
  - References *262*

15 Ecological and Evolutionary Links Between Defences and Life History Traits in Plants *269*
Xoaquín Moreira and Luis Abdala-Roberts
- 15.1 Evolutionary Ecology of Plant Defences Against Herbivores *269*
- 15.2 Correlated Evolution of Plant Defences and Life History Traits *270*
- 15.2.1 Defences and Growth-Related Traits *270*
- 15.2.1.1 Plant-Based Allocation Constraints *270*

15.2.1.2 The Role of Herbivore Pressure *272*
15.2.2 Defences and Reproductive Traits *274*
15.2.2.1 Plant Defences and Pollination-Related Traits *274*
15.2.2.2 Plant Defences and Seed Dispersal-Related Traits *275*
15.3 Tripartite Views Shed Insight into the Evolution of Plant Life History Traits *276*
15.4 Challenges for Future Research *277*
References *278*

16 Are you in Synch?: How the Timing of Plant and Insect Life History Events Affects Pollination Interactions *285*
Tamar Keasar and Tzlil Labin
16.1 Generalisation in Pollination Networks *285*
16.2 What Drives Flowering Phenology? *286*
16.3 What Drives Pollinator Phenology? *286*
16.4 Do Interacting Plant–Insect Species Share Similar Reaction Norms to Temperature? *286*
16.5 Species-Level Phenological Asynchrony and Generalized Pollination: A Case Study *287*
16.5.1 *A. coronaria*'s Pollination Biology *287*
16.5.2 *A. coronaria*'s Flowering Phenology *290*
16.5.3 The Phenology of *A. coronaria*'s Insect Visitors *290*
16.6 Community-Level Phenology and Pollination Specialisation *291*
16.7 Concluding Remarks *292*
References *293*

17 Life Histories in the Context of Mutualism *297*
Renee M. Borges
17.1 Introduction *297*
17.1.1 Life History and Other Traits that Predispose Organisms to Enter into Mutualism *298*
17.1.2 Asymmetry in Partner Interaction and Consequences for Mutualism *298*
17.2 Mutualism Benefits and Life History Traits *298*
17.2.1 Protection *299*
17.2.1.1 Bacterial Co-constructed Countershading Mechanism in Squids *299*
17.2.1.2 Ant–Plant Protective Interactions *299*
17.2.2 Nutrition *301*
17.2.2.1 Zooxanthellae and Corals *301*
17.2.2.2 Plants, Rhizobia and Mycorrhizae *301*
17.2.3 Farming *302*
17.2.3.1 Fungus-Farming Ants *303*
17.2.3.2 Fungus-Farming Termites *303*
17.2.4 Movement *303*
17.2.4.1 Movement in Plants *304*
17.2.4.2 Phoretic Animal Movement *305*
17.3 Future Directions *306*
References *307*

Part III Applications *315*

18 Life History and Climate Change *317*
Juha Merilä and Lei Lv
18.1 Introduction *317*
18.2 Effects of ACC on Life History Strategies and Trade-Offs *318*
18.3 Phenology *319*
18.4 Body Size *320*

18.5	Reproductive Output and Success *321*	
18.6	Survival and Senescence *322*	
18.7	Population Demography and Extinction Risk *324*	
18.8	Genetic or Environmental Responses *325*	
18.9	Conclusions and Outlook *326*	
	Acknowledgements *326*	
	References *326*	

19 Environmental Pollution Effects on Life History *333*
Denis Réale, Loïc Quevarec, and Jean-Marc Bonzom

19.1	Introduction *333*
19.2	The Role of Life History Theories in Ecotoxicology *334*
19.3	The Acquisition/Allocation Principle and the Responses of Organisms to Pollution *335*
19.3.1	Responses at the Individual Organism Level *335*
19.3.2	Short-Term Mean Population-Level Responses *337*
19.3.3	Long-Term Responses of the Population *337*
19.3.4	The Cost of Adapting to Pollutants *338*
19.4	Literature Survey on Mechanisms Involved in the Life History Responses to Pollutants *339*
19.4.1	Studies Analysing the Effects of a Range of Concentrations of a Pollutant *339*
19.4.2	Multigenerational Studies, Common Garden Experiments and Long-Term Evolutionary Responses of Populations *346*
19.5	Case Studies *347*
19.5.1	Responses of *Caenorhabditis elegans* to Environmental Stressors *347*
19.5.2	Responses of *Caenorhabditis elegans* Exposed to Ionising Radiation *348*
19.6	Conclusion and Future Directions *348*
	References *350*

20 Life History Evolution on Expansion Fronts *357*
Elodie Vercken and Ben L. Phillips

20.1	What Are Expansion Fronts and Why Are They Hotspots for Rapid Evolution *357*
20.1.1	A Quick Introduction to Expansion Theory *357*
20.1.2	Evolutionary Processes *358*
20.1.2.1	Spatial Sorting *358*
20.1.2.2	*r*-Selection *359*
20.1.2.3	Stochasticity *361*
20.1.3	Loss of Ecological Interactions *362*
20.2	Trade-Offs Matter *363*
20.3	Other Types of Expansions, How Our Expectations Might Change *364*
20.3.1	Pushed *vs.* Pulled Expansions *364*
20.3.2	Interaction with the Environment *366*
20.4	An Applied Case Study: The Cane Toad *367*
20.5	Summary and Future Directions *368*
	References *368*

21 Adaptive Evolution of Life History Traits in Urban Environments *375*
Yuval Itescu, Maud Bernard-Verdier, and Jonathan M. Jeschke

21.1	Introduction *375*
21.2	Urban Drivers of Selection on Life History Traits *375*
21.2.1	The Spatial Dimension: Habitat Change and Spatial Fragmentation *376*
21.2.2	The Temporal Dimension: Altered Temporal Cycles *377*
21.2.3	The Abiotic Dimension: A Diversity of Novel Abiotic Stressors *377*
21.2.4	The Biotic Dimension: Novel Communities and Biotic Interactions

21.2.5 The Anthropogenic Inputs Dimension: Human Introduction of Artificial and Supplemental Resources *378*
21.2.6 Pan-urban Effects *379*
21.3 Studying Evolution in Urban Areas *379*
21.3.1 The Field of Urban Evolution *379*
21.3.2 Methodological Approaches to Study Evolution in Urban Areas *380*
21.4 Available Evidence of Adaptive Life History Evolution in Urban Areas *380*
21.4.1 The Spatial Dimension *380*
21.4.1.1 Dispersal in Fragmented Landscapes *380*
21.4.1.2 Finding a Mate in a Fragmented Urban Landscape *384*
21.4.1.3 Patch Isolation and the Island Syndrome Hypothesis *385*
21.4.1.4 Reduced Access to High-Quality Resources and Food Limitation *385*
21.4.2 The Temporal Dimension *386*
21.4.2.1 Adaptation to Changes in Seasonal Cycles *386*
21.4.2.2 Effects of Changes to the Diel Cycle *386*
21.4.3 The Abiotic Dimension *386*
21.4.3.1 Light Pollution *386*
21.4.3.2 Noise Pollution *387*
21.4.3.3 Chemical Pollution *387*
21.4.3.4 Elevated Temperatures and Thermal Stress *387*
21.4.4 The Biotic Dimension *388*
21.4.5 The Anthropogenic Inputs Dimension *388*
21.4.6 Pan-urban Effects *389*
21.5 Synthesis and Perspectives *389*
21.5.1 Patterns, Biases and Gaps in Research on Urban Life History Evolution *389*
21.5.1.1 Common Patterns and Observed Trade-Offs *390*
21.5.1.2 Variation in Selection Pressures in Urban Areas *391*
21.5.1.3 Cascading Effects and Indirect Selection via Biotic Interactions *391*
21.5.2 The Future of Life History Evolution in Cities: Research and Practice *391*
21.5.2.1 Island Biology in the City *391*
21.5.2.2 Urban Trait Syndrome and Convergent Evolution *391*
21.5.2.3 Understanding Life History Evolution in Cities Is of High Practical Relevance *392*
21.5.2.4 Integrative Approaches for Studying Life History Evolution in Cities Are Essential *392*
Acknowledgements *393*
References *393*

22 Life History and Biological Control *403*
Paul K. Abram and Ryan L. Paul
22.1 Introduction *403*
22.2 Selecting Among Interspecific Life History Variation *406*
22.2.1 Intrinsic Rate of Increase as an Aggregate Metric for Comparing the Life Histories of Biological Control Agents *407*
22.2.2 Using Life History Trait Syndromes to Explain Interspecific Patterns *407*
22.2.3 Case Study: Life Histories of Some of the Most Commonly Used Augmentative Biological Control Agents *408*
22.3 Managing or Manipulating Intraspecific Life History Variation *409*
22.3.1 Selecting Biological Control Agents from Among Naturally Occurring Intraspecific Life History Variation *410*
Leveraging Intraspecific Variation in Life History Traits for Breeding Programmes to Improve Biological Control
nt Performance *410*
g Unintentional Artificial Selection in Captive Rearing *411*
r Evolution of Life History Traits After Biological Control Releases *411*
Core Evolutionary Dynamics and Hybridization in Biological Control Agents of Tamarisk
Environmental Management and Agent Release Strategies *413*

22.4.1	Field Resource Supplementation	*413*
22.4.2	Using Life History Information to Inform Release Strategies	*414*
22.4.3	Life History Traits and the Resilience of Biological Control to Climate Change	*414*
22.4.4	Life History Traits and the Resilience of Biological Control to Insecticides	*415*
22.4.5	Case Study: The Interaction Between Environmental Conditions and Life Histories of *Anagrus* Parasitoids of Leafhopper Eggs	*415*
22.5	Future Directions and Conclusions	*416*
	Acknowledgements	*417*
	References	*418*

23 Life History and Exploitative Management of Fish and Wildlife *425*
Marco Festa-Bianchet

23.1	Introduction	*425*
23.2	Life History Traits, Density-Dependence and Sustainable Harvest	*425*
23.3	Contrasting Life Histories and Harvest Potential	*427*
23.4	Ecological Plasticity and Evolutionary Sources of Variability: A Few Ungulate Examples	*429*
23.5	How Can Knowledge of Life History Traits Improve Harvest Management?	*429*
23.6	Life History and Trophy Hunting	*431*
23.7	Life History and Compensatory Population Responses to Harvest	*431*
23.8	The Special Case of Sexually Selected Infanticide	*432*
23.9	Can Harvest Affect the Evolution of Life History Strategies?	*432*
23.10	Conclusion and Future Directions	*434*
	Acknowledgements	*434*
	References	*435*

24 Life History and the Control of Diseases *439*
Jessica E. Metcalf and Justin K. Sheen

24.1	Introduction	*439*
24.2	Life History Outcomes: A Classic Theoretical Scaffold to Illustrate Predictions	*440*
24.3	Levels of Selection	*445*
24.4	The Complexities of Variance and Covariation in Empirical Systems	*448*
24.4.1	Variance	*448*
24.4.2	Covariance	*450*
24.5	Frontiers in Life History Evolution and Pathogen Control	*451*
24.5.1	Emerging Pathogens	*451*
24.5.2	Biological Control Agents	*452*
24.5.3	Pathogen Control Efforts in the Light of Host–Pathogen Coevolution	*452*
24.6	Conclusions	*452*
	References	*453*

Index *457*

List of Contributors

Luis Abdala-Roberts
Departamento de Ecología Tropical
Campus de Ciencias Biológicas y Agropecuarias
Universidad Autónoma de Yucatán
Mérida
México

Jun Abe
Faculty of Science
Kanagawa University
Yokohama
Japan

Paul K. Abram
Agriculture and Agri-Food Canada
Agassiz, BC
Canada

Megan Arnot
Department of Anthropology
University College London
London
UK

Maud Bernard-Verdier
Leibniz Institute of Freshwater Ecology and Inland Fisheries (IGB)
Berlin
Germany

Institute of Biology
Freie Universität Berlin
Berlin
Germany

Centre d'Ecologie et des Sciences de la Conservation (CESCO)
Sorbonne Université
Paris
France

Michael B. Bonsall
Department of Biology
University of Oxford
Oxford
UK

Dries Bonte
Department of Biology
Ghent University
Gent
Belgium

Jean-Marc Bonzom
Institut de Radioprotection et de Sûreté Nucléaire (IRSN)
PSE-ENV/ SERPEN/LECO, Cadarache
Saint Paul Lez Durance
France

Renee M. Borges
Centre for Ecological Sciences
Indian Institute of Science
Bangalore
India

Elad Chiel
Department of Biology and Environment
University of Haifa-Oranim
Tivon
Israel

Alison B. Duncan
ISEM
University of Montpellier, CNRS, IRD, EPHE
Montpellier
France

List of Contributors

Marco Festa-Bianchet
Département de biologie
Université de Sherbrooke
Sherbrooke, Québec
Canada

Yuval Gottlieb
Koret School of Veterinary Medicine
The Hebrew University of Jerusalem
Rehovot
Israel

George E. Heimpel
Department of Entomology
University of Minnesota
St. Paul, MN
USA

Andreas Heyland
Integrative Biology
University of Guelph
Guelph, ON
Canada

Yuval Itescu
Leibniz Institute of Freshwater Ecology and Inland Fisheries (IGB)
Berlin
Germany

Institute of Biology
Freie Universität Berlin
Berlin
Germany

Department of Evolutionary and Environmental Biology
University of Haifa
Haifa
Israel

Jonathan M. Jeschke
Leibniz Institute of Freshwater Ecology and Inland Fisheries (IGB)
Berlin
Germany

Institute of Biology
Freie Universität Berlin
Berlin
Germany

Oliver Kaltz
ISEM
University of Montpellier, CNRS, IRD, EPHE
Montpellier
France

Tamar Keasar
Department of Biology and Environment
University of Haifa-Oranim
Tivon
Israel

Konstantin Khalturin
Institute of Cellular and Organismic Biology (ICOB)
Academia Sinica
Taipei
Taiwan

Miriam Kishinevsky
Department of Integrative Biology
University of Wisconsin – Madison
Madison, WI
USA

Hanna Kokko
Institute for Organismic and Molecular Evolution
Johannes Gutenberg University of Mainz
Mainz
Germany

Judith Korb
Evolutionary Biology & Ecology
University of Freiburg
Freiburg
Germany

Research Institute for the Environment and Livelihoods
Charles Darwin University
Casuarina Campus
NT0909 Darwin
Australia

Hope Klug
Biology, Geology, and Environmental Science
University of Tennessee at Chattanooga
Chattanooga, TN
USA

Tzlil Labin
Department of Evolutionary and Environmental Biology
University of Haifa
Haifa
Israel

Vincent Laudet
Institute of Cellular and Organismic Biology (ICOB)
Academia Sinica
Taipei
Taiwan

Marine Eco-Evo-Devo Unit
Okinawa Institute of Science and Technology (OIST)
Okinawa
Japan

Lei Lv
School of Environmental Science and Engineering
Southern University of Science and Technology
Shenzhen
China

Division of Ecology and Evolution, Research School of Biology
Australian National University
Canberra, ACT
Australia

Ruth Mace
Department of Anthropology
University College London
London
UK

Toulouse School of Economics
Institute for Advanced Study at Toulouse
Toulouse
France

Alexei A. Maklakov
School of Biological Sciences
University of East Anglia
Norwich
UK

Dustin J. Marshall
School of Biological Sciences
Monash University
Clayton, Victoria
Australia

Juha Merilä
Ecological Genetics Research Unit, Organismal and Evolutionary Biology Research Programme, Faculty Biological & Environmental Sciences
University of Helsinki
Helsinki
Finland

Area of Ecology and Biodiversity, School of Biological Sciences
The University of Hong Kong
Hong Kong
Hong Kong SAR

Jessica E. Metcalf
Department of Ecology and Evolutionary Biology
Princeton University
Princeton, NJ
USA

Xoaquín Moreira
Misión Biológica de Galicia (CSIC)
"Ecology and Evolution of Plant-Herbivore Interactions" Group
Pontevedra, Galicia
Spain

Volker Nehring
Evolutionary Biology & Ecology
University of Freiburg
Freiburg
Germany

Ryan L. Paul
Department of Horticulture
Oregon State University
Corvalis, OR
USA

Ben L. Phillips
School of Molecular and Life Sciences
Curtin University
Perth
Australia

Loïc Quevarec
Institut de Radioprotection et de Sûreté Nucléaire (IRSN)
PSE-ENV/ SERPEN/LECO, Cadarache
Saint Paul Lez Durance
France

Denis Réale
Département de sciences biologiques
Université du Québec à Montréal
Montréal, QC
Canada

Michal Segoli
Mitrani Department of Desert Ecology
Blaustein Institutes for Desert Research
Ben-Gurion University of the Negev
Midreshet Ben-Gurion
Israel

Stephen C. Stearns
Department of Ecology and Evolutionary Biology
Yale University
New Haven, CT
USA

Justin K. Sheen
Department of Ecology and Evolutionary Biology
Princeton University
Princeton, NJ
USA

Toomas Tammaru
Department of Zoology
Institute of Ecology and Earth Sciences
University of Tartu
Tartu
Estonia

Tiit Teder
Department of Zoology
Institute of Ecology and Earth Sciences
University of Tartu
Tartu
Estonia

Department of Ecology
Faculty of Environmental Sciences
Czech University of Life Sciences Prague
Prague-Suchdol
Czech Republic

Joseph Travis
Department of Biological Science
Florida State University
Tallahassee, FL
USA

Elodie Vercken
Biologie des Populations Introduites, UMR 1355
INRAE-CNRS-UniCA
Institut Sophia Agrobiotech
Sophia Antipolis
France

Eric Wajnberg
INRAE, Sophia Antipolis, France
INRIA, Project Hephaistos, Sophia Antipolis, France

Departamento de Entomologia e Acarologia
Universidade de São Paulo
Escola Superior de Agricultura "Luiz de Queiroz"
Piracicaba, São Paulo, Brazil

Stuart A. West
Department of Biology
Oxford University
Oxford
UK

Giacomo Zilio
CEFE
University of Montpellier, CNRS, EPHE, IRD
Montpellier
France

Foreword

Some issues raised in life history evolution were foreshadowed in the 19th century by Darwin and Weismann and in the early 20th century by Pearl. The impetus they received in the mid-20th century from Lack, Cole, MacArthur, Wilson, Williams, Gadgil, Charlesworth and others stimulated a body of research that took off in the last quarter of the 20th century. After initial interest in broad comparative patterns explained by analogies to population dynamics (r- and K-selection), the focus shifted to theory couched in terms of age- and size-specific birth and death rates, to experimental evolution in both field and laboratories and to phenotypic plasticity within generations as well as genetic responses across generations. By the early 1990s, the body of work had become significant enough to be summarized in several books. Since then, the insights and intellectual toolkit of life history evolution have penetrated evolutionary and behavioural ecology, and its applications now inform fisheries and game management, biological pest control, conservation biology, gerontology and evolutionary medicine.

Life history evolution explains the evolution of specific traits, but it also raises larger issues. The traits explained include age and size at maturity, the number of reproductive events per lifetime, the number of offspring per reproductive event (clutch or litter size), the rate of ageing, lifespan, sex allocation and all their plastic responses. The larger issues raised include how to formulate a theory of phenotypic evolution, how to connect micro- with macro-evolution, and – the original and continuing concern of life history evolution – how evolution designs phenotypes for reproductive success. At its interface with ecology, life history evolution plays a central role in research on eco-evolutionary feedbacks. At its interface with comparative biology, it is energized by molecular phylogenetics. Its insights into organismal design are now being woven together with evolutionary physiology and genomics. These enduring contributions to several fields have made life history evolution a very successful intervention.

The present book summarizes the state of play at the end of the first quarter of the 21st century. The hard work done by its editors has paid off in a volume that is comprehensive, stimulating and useful.

This foreword addresses three questions: (1) What did I learn? (2) What issues should be mentioned that are not reviewed here? and (3) What do we not yet know?

I learned that work on the maturation event and its reaction norm should now emphasize theoretical and empirical studies of the coevolution of life histories with physiology and development. In the evolution of ageing, one question that remains unanswered is whether the apparent immortality of some coelenterates might be explained by aspects of their development and maintenance not shared with bilaterians. In marine invertebrates, development rate is more sensitive to temperature than metabolic rate, which implies that small increases in temperature decrease the costs of offspring development, allowing parents to decrease the size of their offspring and increase their fecundity. Insect parasitoids contrast adaptations to spatial *vs.* temporal stochasticity, allowing an intriguing dissection of risk spreading. Could experimental evolution demonstrate that adaptation to spatial risk maximizes arithmetic mean lifetime reproductive success, whereas adaptation to temporal risk maximizes the geometric mean? Daniel Bernoulli would have been intrigued.

The strikingly long lifespans of termite, ant and naked mole rat queens raise the issue of whether degree of sociality is related to longevity, either as a correlate or as a determinant. Antagonistic pleiotropy between the reproductive performance of queens and the mortality of workers helps to explain the striking difference in their lifespans. The molecular mechanisms of ageing in social insects hold an important lesson: 'although [they] are highly conserved and shared among nematodes, insects and mammals, they can become uncoupled and re-wired if ultimate forces select for a long lifespan […] in social organisms'.

Viewing menopause as an adaptation is more defensible than I had thought. One question concerns the implications of stringent quality control through follicular atresia. To extend the reproductive lifespan while maintaining the ancestral level of quality control, one must start life with an ovary so large that it might not fit into a foetus.

The analysis of predator–prey interactions on life history evolution in both parties taught me that the costs of plasticity may be quite small. Here the coupling of theory with experiments done in the field has convincingly revealed alternative explanations for earlier conclusions.

The effects of plant defences on life history traits are quite context-dependent, varying with the annual-perennial, monoecious-dioecious, and herbaceous-woody distinctions, and often mediated by the tensions between pollination, dispersal and defence. I was surprised to learn that specialist pollinators are not better matched to the phenology of the plants they pollinate than are generalist pollinators, and that Grimes' description of life history patterns for plants – competitive, stress tolerant and ruderal – can be applied to corals and mycorrhizae as well.

Evidence for the impact of global warming on life history traits is abundant and on population growth is scanty. Poikilothermic organisms do appear to be getting smaller as the world warms, with implications for fish fecundity. Physical and chemical pollution can change the expression and shape the evolution of life history traits in ways that are specific to species and situations, with many different outcomes. That some experiments have found little or no effect of pollution at the population level suggests that variation in individual responses could be masking impacts on populations.

The eco-evolutionary dynamics of life history evolution on the expansion fronts of invading species reveal the distinction between pushed and pulled expansion fronts and the option of using genetic backburn to slow range expansion. Urban environments challenge organisms that evolved in more natural habitats, creating a mismatch problem like those humans faced as they domesticated themselves and created the diseases of civilization. Work on the biological control of arthropods and weeds has great potential to illuminate life history evolution in general by using experimental evolution to enhance performance and by showing how control agents evolve after they are released into new habitats.

In game and fisheries management, optimal harvesting strategies, based on the elasticity of population growth rates to changes in age-specific mortality, can be counter-intuitive. Protecting lactating females in chamois is not the best idea, and where there is sexually selected infanticide, as in African lions and European brown bears, killing a dominant male as a trophy can have large indirect impacts because the remaining males kill juveniles to bring females into oestrus more quickly.

Because life history traits are the central mediators of natural selection, it is not surprising that many other traits connect to evolution through them. The initial simplifying insights of the field fifty years ago have since been complicated in many productive ways. Whether one regrets the loss of simplicity or celebrates the richness of nuanced diversity is probably beside the point: complexity is what progress in our understanding has brought us. We should be grateful for the overarching structure that ties together so many details.

The use of life history thinking in applied fields such as fisheries management, pollution analysis, biological control and conservation ecology has great potential to qualify and even change basic theory, for there the economic incentives are great, funding can be long-term, and the motivation to dig into details is strong.

While this book is quite comprehensive, it does leave out some issues that may be worth addressing in the future. In 1980 I made the mistake of burying an eco-evolutionary analysis of the coevolution of seed dispersal, lifespan, reproductive effort and body size in forest trees in an obscure publication (Stearns and Crandall 1981). It suggests an explanation for why there are trees and why there are forests. It has not been followed up and should be revisited. The questions are important; the initial insights need qualification.

The insights of life history evolution also impact evolutionary medicine in several ways, among them are: (1) Trade-offs help to shape cancer evolution under chemotherapy along the familiar fast-slow axis, with important implications, and the analysis of phenotypic plasticity helps us to understand the role of stem cells in initiating cancer (Aktipis et al. 2013). (2) The endocrine system of vertebrates is structured to mediate life history trade-offs. Wang et al. (2019) show how the evolutionary design of organisms for reproductive success embeds principles of systems design into their physiology, thereby suggesting a general approach to evolutionary physiology. Similar work is needed on plants and insects. (3) There are interesting parallels between urban evolution and the mismatch theme in evolutionary medicine: can organisms in cities be models for human mismatch to the environments that we have created? That deserves systematic exploration.

Acknowledging ignorance is a first step towards future research. Here are three things about life histories that we do not yet understand: (1) Why are bilaterians not able to remodel their tissues as do some coelenterates and thereby extend their lifespan, perhaps indefinitely? (2) How do complex life cycles – like those found in algae, jellies and parasites with multiple hosts – evolve? What governs the number of stages and the duration of each? (3) How can we explain major phylogenetic events, such as internal fertilization, endothermy or clutch sizes fixed at one, and their consequences for the subsequent evolution of life history traits?

Stephen C. Stearns
May 2024

References

Aktipis, C.A., Boddy, A.M., Gatenby, R.A., Brown, J.S., and Maley, C. (2013). Life history trade-offs in cancer evolution. *Nature Reviews Cancer* 13: 883–892.

Stearns, S.C., and Crandall, R.E. (1981). Bet-hedging and persistence as adaptations of colonizers. In: *Evolution today. Proceedings of thew second international conference of systematic and evolutionary biology* (eds. Scudder, G.G.E., and Reveal, J.L.), 371–383. Hunt Institute for Biological Documents.

Wang, A., Luan, H.H., and Medzhitov. R. (2019). An evolutionary perspective on immunometabolism. *Science* 363: 140.

Preface

Life history traits are defined as those related to the timing and magnitude of major events in the life of an organism (*e.g.*, birth, maturation, reproduction and death). Life history theory aims to explain the huge variation in these trait values and their combinations that we see in nature. Hence, life history strategies are at the heart of evolutionary biology and ecology, illuminating how organisms are structured and how they allocate resources, behave and evolve, in response to different environmental pressures.

Throughout the last decade, the study of life history evolution has experienced some major advances: First, while classical life history theory relied on optimization approaches and had some major empirical successes, more advanced approaches involving frequency- and density-dependency, explicit population dynamics and evolutionary game theory, have further promoted our understanding in the field. Second, classical theory often emerged from observations of life history patterns in a specific group of animals (*e.g.*, clutch size in birds). Later work extended and further developed these principles to the study of various organisms, including vertebrates, invertebrates (in particular arthropods), plants, fungi, bacteria, and even phages. At the same time, concepts from life history evolution were used to promote our understanding of human evolutionary ecology. Third, classical life history theory often focused on a few fundamental life history traits, most commonly including body size, lifespan, investment in reproduction and sex ratio. Later works extended our view to consider additional traits such as parental care, life cycles, sociality and dispersal, either as life history traits in their own right or in the context of their interaction with major life history traits. Fourth, while classical life history theory often addressed the evolution of each organism independently, later advances extended this approach to consider life history evolution also in the context of interspecific and even community-level co-evolutionary interactions, including predator–prey, parasite–host, plant–herbivore and mutualistic interactions. Fifth, with the advances in molecular, genetic and physiological methods, we can better understand the underlying proximate mechanisms of life history patterns. Sixth, while theory often focuses on evolutionary responses to environmental conditions, we now have accumulating evidence that evolutionary processes may occur at a short enough time scale to allow eco-evolutionary feedback. Finally, there is an increasing awareness that life history evolution responds to anthropogenic-induced changes such as climate change, pollution and urbanization and that insights from these fields could be incorporated into how we manage natural resources globally.

Several foundational books have laid the groundwork for the analysis of life history evolution. However, a comprehensive, up-to-date volume summarizing recent advances in the field – as highlighted above – and presenting current examples and applied aspects was lacking. The current volume aims to fill this gap by presenting current ideas, analyses, and case studies in life history evolution, spanning a wide range of taxa (from bacteria to insects to humans), traits and applications.

This volume starts with a foreword by Stephen C. Stearns, Yale University, one of the leading scientists in this field and the author of several books on life history evolution theory. The volume is then organized into three parts. The first addresses different traits that are often considered, and a few that are less commonly considered, in the study of life history evolution. The second part focuses on life history evolution in the context of interspecific interactions, and the third part discusses anthropogenic impacts and possible applications. In the first part, the first three chapters present recent developments in the study of classical life history traits, namely, body size and the timing of maturation, lifespan and offspring size and number. The fourth chapter discusses optimized egg loads focusing on parasitoid insects. Two chapters then address the question of sex-specific life histories and parental care, and an additional chapter focuses on the complexity of life cycles. Finally, three additional chapters discuss life history in social groups, in relation to dispersal, and the particular case of humans focusing on menopause. The second part then starts with three chapters presenting recent advances in life history evolution in the context of predator–prey, host–parasite and host–endosymbiont interactions. Three more chapters discuss life history evolution in the context of plant–herbivore, plant–pollinator, and mutualistic interactions. Finally, the last part starts with two chapters dealing with the effects of human-induced stresses such as climate change and pollution on the evolution of life history traits of different organisms. Two more chapters address the case of species range expansions and urban environments, followed by three final chapters discussing applications in biological control programmes against crop pests, wildlife management and disease management.

Throughout the volume, the goal of each chapter is to present the main current ideas and developments in the field and to give the readers the main information required for further investigation. Each chapter ends with a brief description of remaining knowledge gaps, listing several open questions to be addressed in the future.

We hope that this volume will broaden the scope of previous publications by discussing life history evolution of both classical life history traits, and in relation to behavioural, morphological and physiological traits (Part I), in the context of specific interactions (Part II), and its applications for resource and disease management (Part III).

We express our gratitude to all authors for their excellent contributions and their collaboration throughout the edition of this volume. A special thanks to Stephen C. Stearns for writing the foreword section for this book. All chapters underwent a reviewing process to increase their relevance and foster links between them. In this respect, we thank the referees for their time and insightful comments. These include Carlos Barata, Javier Belles, Judith Bronstein, Julien Cote, Damian K. Dowling, Anja Felmy, Jessica Forrest, Jean-Michel Gaillard, Megan Greischar, Ian Hardy, George Heimpel, Ruth Hufbauer, Richard Karban, Boris Krasnov, Kevin Lafferty, Yael Lubin, Allen Moore, Michael Moore, Michael Poulsen, John Reynolds, Jay Rosenheim, Locke Rowe, Inon Scharf, Rebecca Sear and Elsa Youngsteadt.

Lastly, we also express our thanks to the editors of Wiley Publishing for their continuous help and guidance to the IIAS (Hebrew University, Jerusalem) which hosted both of us for five month during the initial phase, and to Fapesp (São Paulo State, Brazil) which hosted one of us (Eric Wajnberg) for eight months during the final phase of the preparation of this book (Process # 2022/10870-1).

Michal Segoli and Eric Wajnberg
May 2024

Part I

Traits

1

Body Size and Timing of Maturation

Toomas Tammaru[1] and Tiit Teder[1,2]

[1] *Department of Zoology, Institute of Ecology and Earth Sciences, University of Tartu, Tartu, Estonia*
[2] *Department of Ecology, Faculty of Environmental Sciences, Czech University of Life Sciences Prague, Prague-Suchdol, Czech Republic*

1.1 Introduction

Body size is perhaps the most apparent trait of any organism, at least from the perspective of visually oriented observers, including us, humans. However, the implications of body size reach far beyond the limits of visual appearance as basic physical principles determine where and how an organism of a certain size can or cannot live. Just as a few examples, the circulation of vital substances within the body is critically dependent on the size of the organism, as are the ways an organism can or cannot move around in its environment. From a more ecological perspective, size largely determines the range of items one can include in its diet, as well as the range of the consumers one should be afraid of. Proceeding to more derived traits, body size sets the limits for permissible colouration patterns and adequate behavioural repertoire, in the context of both inter- and intraspecific interactions.

Since the dawn of life history theory as a separate discipline, the evolutionary explanation of adult body size has been viewed in the framework of optimality analysis of age and size at maturation. It is indeed straightforward to ask when it pays to start investing in reproduction instead of continuing the investment in one's own somatic growth (Figure 1.1). Such an approach allows us to conveniently break down the question about the optimal body size into the analysis of costs and benefits of early and late maturation and costs and benefits of large and small size *per se*.

In the present chapter, we briefly review the knowledge about such costs and benefits, and how the optimal balance between them may depend on environmental conditions. We conclude that, while we can often provide satisfactory evolutionary explanations for the differences in size, *e.g.*, the differences between biological sexes, among populations or among related species, we may not know enough to explain absolute values. To be able to do so, we may need a tighter integration of the ecologically based optimality approach with the study of physiological and ontogenetic phenomena, often termed constraints in the research tradition of evolutionary ecology. Furthermore, it is crucial to learn how fast body size and associated traits evolve.

The chapter consists of two parts. We first provide an overview of past and current developments in the field (Part I), and thereafter focus on a few selected topics related to our research interests (Part II). We illustrate a number of key points with examples from our own work, which we happen to know best. More generally, due to the 'ontogenetic constraints' of the authors, our approach will be biased towards insects. This should not be a big problem as multicellular species other than insects form just a minority among the known biodiversity. We will nevertheless try not to ignore them.

1.2 Part I

1.2.1 Benefits of Large Body Size

For many organisms, the benefits of large size are straightforward to understand. A frequent fitness benefit of large size in females is the so-called fecundity advantage. This implies that larger-bodied females are expected to produce more offspring than smaller conspecifics (see Lim et al. (2014) for a meta-analysis, and Pincheira-Donoso and Hunt (2017) for a critical

Life History Evolution: Traits, Interactions, and Applications, First Edition. Edited by Michal Segoli and Eric Wajnberg.
© 2025 John Wiley & Sons Ltd. Published 2025 by John Wiley & Sons Ltd.

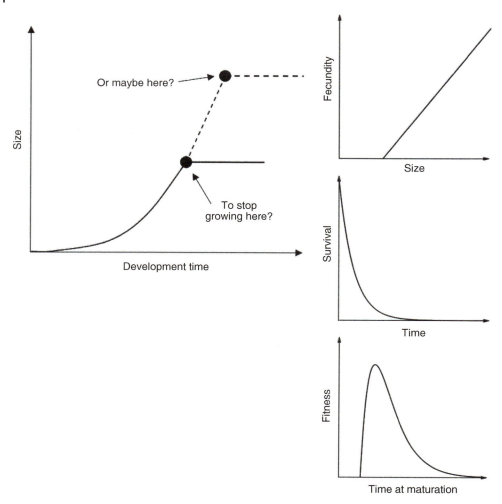

Figure 1.1 Left panel: conceptual setting for optimality analysis of age and size at maturation. We can ask when a juvenile should stop growing and switch to reproduction. Right panel: a simple model for optimal age and size at maturation for an organism with determinate growth. Size is assumed to increase proportionally with growing time (not shown). Fecundity is a linear function of size at maturation with a negative intercept. Probability of survival is a negative exponential function of time. Fitness is the product of survival to maturation and fecundity, the function attains its maximal value at some intermediate (optimal) value of time at maturation.

discussion of the concept of fecundity advantage). This can primarily result from large females having more resources to be converted into offspring or because they have better capabilities to obtain such resources. The pattern is clear and straightforward at least for many insects (Honěk 1993), some other invertebrates (reviewed in Roff 2002) and fish (Barneche et al. 2018).

The strength of the size–fecundity relationship naturally differs between different organisms and not in a random manner. One aspect to consider is the relative role of having more resources *vs.* the capability of obtaining them. This distinction can be illustrated by the contrast between capital- and income-breeding species (Drent and Daan 1980, Jönsson 1997). As applied to insects (Davis et al. 2016), the contrast is between females, which primarily depend on larvally derived resources (capital) in their reproduction, as opposed to females obtaining such resources through adult foraging (income). In capital-breeding insects, the correlation between body size and potential fecundity is often remarkably strong (*e.g.*, Tammaru et al. 1996b, 2002). This is because the mass of a newly eclosed female directly determines the amount of resources she can use for reproduction.

In females of capital-breeding insects, there is no need to move around in search of food. Amplified by the trade-off between fecundity and mobility, this has frequently led to the evolution of short adult life spans and overall behavioural simplification (Tammaru and Haukioja 1996, Davis et al. 2016). In such females (Figure 1.2), realized fecundity is largely

Figure 1.2 An ovipositing flightless female of the erebid moth *Orgyia antiqua*. The adult female does not move away from its pupal case and lives just for a few days. No costs of large adult size were found in this species, and realized fecundity is proportional to body mass (Tammaru et al. 2002). With such a reduced behavioural repertoire, there is little 'space' for costs of large adult size to operate. *Source:* © entomart.

determined by potential fecundity, as factors related to adult behaviour have little effect on the number of eggs laid. In contrast, in income-breeding insects with active long-living adults, the number of eggs laid should primarily depend on foraging success, efficiency in finding suitable oviposition substrates and survival rates. These fitness correlates may not need to depend strongly on body size, which can lead to substantial weakening of the fecundity advantage in income-breeding insects (Gotthard et al. 2007).

In an attempt to generalize from the comparison between capital- and income-breeding insects, we propose that the strong fecundity advantage should primarily be characteristic of animals in which females have simple behavioural repertoires. This suggestion is supported by the observations that the fecundity advantage of large female size may be weak, non-existent or even reversed in groups known for complex behavioural patterns, such as birds (Blums and Clark 2004, Magrath et al. 2009) and mammals (Isaac 2005), including humans (Valge et al. 2022). Plants represent the other extreme of behavioural complexity, and accordingly, there is typically a strong correlation between the size of a plant individual and its seed set (Shaanker et al. 1988, Weiner et al. 2009). Notably, however, in plants, the amount of resources readily available for reproduction and the ability to obtain further resources are both directly size-dependent.

Obviously, the fecundity advantage of large size is not the only source of the positive correlation between body size and fitness in females. Larger females may also produce offspring of better quality, partly through providing better-quality maternal care (Clutton-Brock et al. 1985, 1988; see also Chapter 6). In their meta-analysis, Lim et al. (2014) found evidence of a positive intraspecific correlation between maternal size and offspring size, broadly consistent across major taxa and environments. Such a relationship appears, however, not to be typical of insects (Fox and Czesak 2000). Large body size may also contribute to fitness through increased longevity in animals as different as flies (Reim et al. 2006) and primates (Blomquist et al. 2011; see also Chapter 2). Additionally, in some animal groups, males may actively prefer large females (Bonduriansky 2001).

For males, the advantages of large body size are frequently less clear, also being taxon-specific to a greater extent. This is primarily due to variations in the behavioural contexts of mate location and mate choice (see Choe and Crespi 1997, for the high diversity of mating systems in insects alone). Most typically, the higher reproductive output of larger males can be due to greater success in male-to-male competition or female choice (Shine 1989, Roff 1992, Andersson 1994, Dale et al. 2007). Opposite to the situation in females, we can therefore expect a stronger correlation between male size and fitness in species with more complex behavioural repertoires. In capital-breeding lepidopterans with simplified adult behaviours (Figure 1.2), for example, there may be no male-to-male interaction or female choice involved in the process of forming mating pairs (Tammaru et al. 1996a, van Dongen et al. 1998). This contrasts with the often much more complex behavioural interactions in butterflies, the most well-known income-breeding lepidopterans (Wiklund and Kaitala 1995, Kemp and Wiklund 2001), and various income-breeding moths (Phelan and Baker 1990, Svensson 1996). This difference may have left an imprint in the evolution of sex-specific body size: capital-breeding insects are generally characterized by strong female-biased sexual size dimorphism (SSD, females larger than males; Davis et al. 2016), while in butterflies, males can even be the larger sex (Wiklund and Forsberg 1991, Teder 2014). Such contrasts can also be made at higher taxonomic levels as insects in general tend to have female-biased SSD (Teder and Tammaru 2005) compared to the tendency towards males being larger than females in birds and mammals (Lindenfors et al. 2008). It appears likely that the opposite effect of behavioural complexity on fitness consequences of body size in the two sexes is largely responsible for this broad-scale pattern (see also Dale et al. 2007, Dial et al. 2008).

1.2.2 Costs of Large Body Size

The straightforward and often strong positive correlation between body size and major components of fitness raises the question about selective pressures counterbalancing the fitness advantage of large body size. Without such costs, we should expect a continuous evolutionary increase in body size. Such trends are sometimes indeed observed on the scale of tens of millions of years (known as Cope's rule: Kingsolver and Pfennig 2004, Hone and Benton 2005, Roy et al. 2024). Nevertheless, respective macro-evolutionary changes are still slower by many orders of magnitude than what could be expected on the basis of the nearly proportional relationship between body size and the number of eggs laid in capital-breeding insects, for example. In other words, the evolutionary stability of body size calls for an explanation (Hansen and Houle 2004, Rollinson and Rowe 2015).

The costs of large body size are usually much more challenging to see than the advantages. The available evidence of such costs was reviewed and synthesized by Blanckenhorn (2000). First, life history models usually assume that it takes more time to grow larger (see further in the chapter for a discussion of this assumption). This being the case, juveniles attempting to attain larger adult sizes are subjected to higher cumulative mortality risk and have lower chances of achieving maturity (see Teder and Kaasik 2023 for a consistent pattern). Because mortality is never zero, this principle cannot perhaps be questioned in its general form. As juvenile mortality rates are frequently dramatically high, for example, in fish (Moyle and Cech 2004) and herbivorous insects (Cornell and Hawkins 1995, Remmel et al. 2011), the survival costs of attaining large size must correspondingly be substantial. There is also empirical evidence of high predation rates causing evolution towards smaller body sizes (see, *e.g.*, Edley and Law 1988, Reznick et al. 1996, Stearns et al. 2000 for experimental studies, and the voluminous literature on selective effects of fishing, discussed further in the chapter; see also Chapters 12 and 23).

Besides the costs of attaining a large size, there may also be costs of being a large adult (Kozlowski 1991, Blanckenhorn 2000). The latter are more diverse, frequently taxon-specific to a considerable degree and also more elusive. Large adults may have low viability due to higher predation and parasitism rates or higher sensitivity to starvation. These costs may be mediated by increased detectability, reduced agility and/or higher energy requirements of larger adults. Also, larger males may have decreased mating success due to reduced mobility or higher energy requirements (Moya-Laraño et al. 2002). Blanckenhorn (2000) concluded that the costs of large size are generally poorly documented, and the evidence has remained fragmentary. The situation has hardly qualitatively changed since Blanckenhorn's (2000) review. As one reason, selective pressures against large size may be poorly known because respective costs can be manifested during infrequent selective events, such as episodes of extreme food scarcity. A technical complication is that size often correlates with body condition (nutritional status or similar), leading to the ambiguity of inferences based on phenotypic correlations of body size with fitness-related traits (Blanckenhorn 2000).

As an example, the question of explaining the evolutionary stability of body size is particularly challenging for capital-breeding moths in which realized fecundity can be considered proportional to female body size. Due to the extreme brevity and simplicity of their adult lives (Figure 1.2), there is hardly an opportunity for substantial costs of being a large adult to operate (Tammaru et al. 2002). This implies that the costs of large size should primarily be manifested during the larval development. However, folivorous lepidopteran larvae grow fast enough to double their masses in just two days (Tammaru and Esperk 2007). A simple calculation shows that no realistic constant mortality rates can counterbalance the considerable fecundity advantage of prolonging the growing period. Here the selective pressure against large size appears to primarily result from strongly positively size-dependent predation risk (Berger et al. 2006, Remmel and Tammaru 2009, Remmel et al. 2011; see also Chown and Gaston 2010). More generally, we believe that size-selective predation by insectivorous birds (and, in some systems, also invertebrate predators, Leduc et al. 2022) has a strong potential to explain the evolutionary stability of body size for a wide variety of insects (folivorous, and those with actively flying adults, at least; see also Teder et al. 2010).

In addition to attempts to measure selective pressures for/against large/small body size empirically, respective information can be obtained by comparing body sizes among populations subjected to different selection regimes. Specifically, such an approach allows us to address selective pressures that are hard to tackle in individual-based empirical studies, such as the idea that greater niche breadths favour larger individuals (Heaney 1978). A much-studied question in this context is the size of animals inhabiting islands compared to their mainland relatives (Lomolino 1985, Palkovacs et al. 2003, Benitez-Lopez et al. 2021). The correlative nature of such data (and problems with isolating adaptive genetic differences) poses limits to the causal interpretation of respective patterns. Nevertheless, the overall (but not universal; Meiri et al. 2008) trend towards the increased size in small vertebrate animals on islands is well consistent with the idea of the crucial importance of predation as a selective force against large body size.

Furthermore, with the increasing availability of reliable phylogenies, it has become feasible to study environmental determinants of optimal body size based on cross-species comparisons. The number of phylogeny-based studies on body size is rapidly increasing, with various macro-evolutionary trends beginning to be revealed (Baker et al. 2015, Cooney and Thomas 2020). While phylogenetic comparative studies providing evidence of costs/benefits of small/large sizes in particular taxa/environments are similarly proliferating (Olson et al. 2009, Collar et al. 2011, Riemer et al. 2018, Clarke 2021, He et al. 2023, Johnson et al. 2023, Seifert et al. 2023), it may still be premature to draw broad conclusions from this line of research.

1.2.3 Absolute Timing of Maturation

Most of the research discussed earlier focusses on body size while its counterpart, age at maturation, is primarily seen and analyzed as the reflection of the amount of time reserved for growth, with its very natural positive (allowing larger sizes to be attained) and negative (higher mortality) effects on fitness. Time is usually counted from the birth of the organism, with the fitness value of the absolute timing of maturation (in the morning or in the evening, today or in a week, this year or next year?) being often left out of consideration. Such an approach is justified for long-living organisms, where the time is measured in years. For such organisms, the absolute timing of the maturation decision should not matter as, in the general case, there are no *a priori* reasons to expect that one year would be better or worse than another in terms of the time point when maturity is reached.

The situation is very different for short-lived organisms in seasonal environments (*e.g.*, annual plants or insects), in which time is measured in weeks and days. Here the absolute timing of the maturation decision (calendar date) acquires crucial importance as postponing maturation (or any other landmark of the life cycle) by a week may have fatal consequences. Optimality models tailored for such organisms (Rowe and Ludwig 1991, Abrams et al. 1996) must be explicit with respect to absolute time. For temperate-zone insects, for example, it is crucial to reach the developmental stage adapted for hibernation before the onset of winter (Tauber et al. 1986, Teder 2020). Other determinants of the 'quality' of the absolute time point for insect maturation include factors such as attaining synchrony with the seasonal development of the food resource (young foliage and flowers; van Asch and Visser 2007, Teder 2020) or avoiding synchrony with predators (*e.g.*, Tiitsaar et al. 2013). A related but somewhat special case is the selection for protandry (earlier maturation of males compared to females) in seasonal environments. In particular, it has been shown that male fitness is maximized by maturation sometime before the seasonal peak of the availability of receptive females (Wiklund and Fagerström 1977, Nylin et al. 1993, Teder et al. 2021).

Another situation where absolute timing should matter is in the case of directional changes in population numbers. When the population is growing, it should pay to mature as early as possible to take maximal advantage of the growth, with the opposite being true in a declining population (Charlesworth 1980, Lande 1982). To adaptively adjust their life history decisions to population trends, the organisms should be able to use the information about the demographic situation in their population, which appears to be the case, *e.g.*, in rodents (Andreassen et al. 2021).

1.2.4 Classical Models of Age and Size at Maturation

Given sufficient time and genetic variability, quantitative traits are expected to have reached their optimal values, *i.e.*, the values at which fitness attains its maximum. Since the early developments of life history theory, numerous attempts have been made to explain observed body sizes in the framework of demographically based optimality models. In such models, demographic parameters are expressed as functions of 'maturation decision' (age and size at maturation), and the age maximizing (some measure of) fitness is determined (Figure 1.1.). Some of the early models, which can now be considered the classics of life history theory, are reviewed by Roff (1992, 2001, 2002) and Stearns (1992).

For organisms with indeterminate growth, exemplified by fish and lizards, the analysis can be formalized in terms of studying the effect of the timing of the maturation event on age-specific survival rates (l_x) and age-specific fecundities (m_x). Lotka–Euler equation relates these quantities to the instantaneous growth rate r, which can be used as the measure of fitness and assumed to be maximized by natural selection. With l_x being lower as age progresses, and m_x tending to increase in a trade-off manner, r is maximized at some intermediate age of maturation. Moreover, for vertebrates, it is reasonable to assume that the instantaneous juvenile mortality rate decreases with the increasing age of the mother at maturation. Using this type of approach, Stearns and Crandall (1981) could explain age and size at maturation in several species of lizards and salamanders with reasonable accuracy. Roff (1986), in turn, could predict age at maturation for a number of fish species and found a satisfactory correspondence with actual values.

For organisms with determinate growth, insects in particular, the assumptions of the model differ considerably. There is an abrupt switch from somatic growth to reproduction; the adult stage is frequently (though not always) short-lived, and the quality of the offspring is largely independent of the parameters of the mother. Also, in contrast to fish (Sogard 1997), mortality risk in insects is typically positively size-dependent (no chance to outgrow predators, see also earlier), and relative growth rates of the juveniles are frequently much higher than in many vertebrates. Using simple assumptions about larval growth and the size–fecundity relationship, the optimality model by Roff (1981) was able to predict the adult sizes of *Drosophila* flies.

Note that the discussed models have primarily been developed to explain age and size at maturation in females. Modelling optimal maturation decisions in males is more challenging as male fitness frequently depends on their success in male-to-male competition or in the face of female choice. In both cases, the fitness of each particular male is dependent on strategies adopted by other players in the setting. The strategies can change in frequency across the population and can evolve. An evolutionary analysis of such cases must therefore be based on a search for an evolutionarily stable strategy (ESS) by applying a game-theoretical approach (Maynard-Smith and Brown 1986). This contrasts the situation in females in which the fitness function of size (when determined by, *e.g.*, the fecundity advantage of size and mortality caused by generalist predators) needs not to depend on decisions made by other individuals in the population. Additionally, the adaptive landscape of male size is more strongly case specific than that of females, which may explain the relative scarcity of optimality models of male size. Nevertheless, such models exist and have been applied in the contexts of explaining alternative reproductive tactics (primarily in fish; Gross 1985, Engqvist and Taborsky 2016, Morris et al. 2016) and SSD (Hedrick and Temeles 1989, Badyaev 2002, Mollet et al. 2023).

While the success of species-specific models is encouraging, the ability of a model to explain observed body size can hardly be treated as solid evidence of an evolutionary explanation having been found. Even with a reasonable prediction, we may not have correctly identified selective forces shaping the values of age and size at maturation; one can construct various models that describe the observations equally well. More specifically, species-specific models have relied on some parameters estimated with intolerably high uncertainty. For mobile insects in particular, mortality rates in the wild are extremely hard to measure. This is something the *Drosophila* model by Roff (1981) also suffers from. These analyses inevitably had to rely on just robust estimates of both larval and adult mortality.

More recently, perhaps acknowledging such difficulties, the focus of research efforts appears to have moved away from empirically based species-specific optimality models. In other words, the attempts to directly answer the underlying question 'Why is this animal of the size it is?' appear to be decreasing in frequency. Indeed, this would imply solving the hard task of quantifying stabilizing selection, *i.e.*, measuring the full set of selective forces both in favour and against large body size. Nevertheless, of more recent attempts of this kind, we acknowledge the remarkable work of Fairbairn (2007) on water striders. For these insects, various selective forces on body size were assessed, and it was concluded that fecundity selection favours longer abdomens in females, this being balanced by a negative relationship between female size and reproductive longevity. Once again, the picture was less clear for males.

Instead of providing species-specific quantitative predictions, the primary focus of theoretical work in the field has shifted to qualitative analysis. It has been asked how the models' output changes when the models' assumptions are modified. Some of the predictions of such analyses are rather intuitive (*e.g.*, grow larger when there is a strong positive effect of body size on fecundity), while some may not be, such as those relating optimal maturation decisions to intricate patterns of mortality (Abrams and Rowe 1996, Marty et al. 2011). What is also not intuitive at first glance is the dependence of the results on the measure of fitness used in the optimality models (Roff 1992, Berrigan and Koella 1994, Roff 2001, Breck et al. 2020), even at the qualitative level. In addition to r, the instantaneous rate of increase, R_0, the lifetime reproductive success of an individual, is widely used. A crucial difference between these measures is that R_0 is not affected by the timing of reproduction, whereas reproduction early in individual life contributes more to the value of r. Such sensitivity calls for caution in choosing the appropriate fitness measure and interpreting the results, with spatio-temporal dynamics of the population having a pivotal role (Brommer 2000).

1.2.5 Adding Environmental Variability – Reaction Norms for Age and Size at Maturation

Both age and size at maturation are phenotypically plastic traits. To begin with the perhaps most intuitive example, organisms subjected to severe nutritional stress rarely attain sizes characteristic of their more fortunate conspecifics. A natural next step in the research on optimal age and size at maturation is to describe and explain the dependence of the maturation

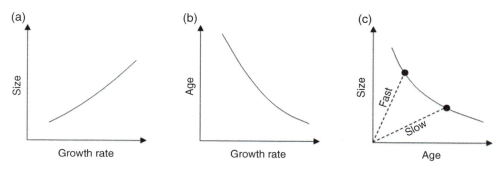

Figure 1.3 Reaction norms for age and size at maturation. (a) The dependence of size at maturation on environmental quality, expressed in terms of growth rate. (b) The dependence of age at maturation on environmental quality. (c) The bivariate reaction norm integrating (a) and (b), environmental quality can be read from the image as the slope of the growth trajectories.

decision on environmental parameters. Such a relationship is a biological phenomenon of interest *per se*, but studying plasticity also provides a powerful tool for understanding selective pressures acting on both age and size at maturation in general.

Plasticity can be visualized using the concept of reaction norm. A reaction norm can be presented as a graph showing the dependence of the phenotypic value of the focal trait on some environmental variable (Figure 1.3a,b). As both age and size at maturation typically change when the maturation decision responds to a change in the environment, it pays to modify the graphical presentation to capture the response in both these variables simultaneously (Stearns 1983, 1992, Schlichting and Pigliucci 1998). In the resulting graph, the plane is formed by axes for age and body size, with each point on this plane representing a theoretical maturation decision (Figure 1.3c). A line drawn through realized maturation decisions can be interpreted as a bivariate reaction norm, displaying the response of both size and age at maturation to changes in some environmental variable. The values of the environmental variable cannot be directly read from the graph. In the typical case, however, the environmental variable in focus is the one which determines individual growth rates. Growth rate, often interpretable as the quality of the environment, is reflected by the slope of the line drawn from the origin to a particular point on the reaction norm.

1.2.5.1 Reaction Norms in the Classical Models

Classical models for age and size at maturation (see earlier) can be and were extended to address the question about the optimal shape of reaction norms for age and size at maturation. Clearly, a new element that needed to be introduced was the variation in environmental variables. In particular, environment-specific differences in resource availability entered the models as variation in individual growth rates, with an additional set of assumptions characterizing mortality schedules. In particular, Stearns and Koella (1986) assumed that resource availability correlates with mortality and that this effect may differ for adults and juveniles. Exploring the behaviour of the model in the parameter space, Stearns and Koella (1986) were able to derive various optimal shapes of the reaction norms for different sets of parameters and relationships (Figure 1.4). The optimal reaction norms have, however, one common feature: optimal age at maturation invariably increases with decreasing growth rate. In contrast, optimal size at maturation could either decrease or increase. Further in the chapter, we will denote such reaction norms as having a negative and a positive slope, respectively. In the case of a negative slope, there is a negative correlation between age and size at maturation (longer development periods are associated with smaller final sizes), and a positive correlation in the case of a positive slope (longer development periods are associated with larger sizes).

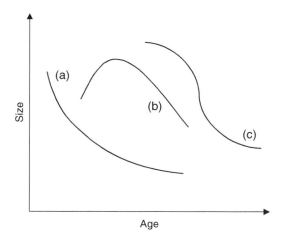

Figure 1.4 A schematic presentation of different shapes of bivariate reaction norms for age and size at maturation (Figure 1.3), which have been found to be optimal under different assumptions: (a) L-shaped; (b) dome-shaped; and (c) sigmoid. Redrawn from Perrin and Rubin (1990) and Stearns (1992).

Besides the overall slope, the shape of the reaction norm may have, of course, additional parameters, so that we can talk about L-shaped and sigmoid reaction norms (Figure 1.4). In particular, Stearns and Koella (1986) predicted an L-shaped reaction norm for the case in which there was no dependence of mortality rate on growth rate or when adult mortality increased as growth rate decreased. However, when juvenile mortality increased with decreasing growth rate, more complex shapes appeared.

Kozlowski and Wiegert (1987) explored a similar set of assumptions but used a different definition of fitness (R_0) that further increased the diversity of the predicted shapes of reaction norms. While the L-shaped reaction norms were still common among their predictions, there were also qualitative differences from the results of Stearns and Koella (1986). In particular, they found that the decision to mature can also be postponed with an increasing growth rate. In turn, Perrin and Rubin (1990) explored the conditions that could lead to dome-shaped reaction norms (Figure 1.4), those combining sections with a negative and positive slope. Dome-shaped reaction norms were found to be optimal when a negative relationship between growth and survival rates was assumed.

Later elaborations of such models have primarily contributed by adding ecological realism, such as considering the spatial structure of the environment and density-dependent regulation of the populations (Kawecki 1993, Kawecki and Stearns 1993, Burd et al. 2006, Marty et al. 2011). As an overall summary, these works show that the spatio-temporal structure of the populations may have an important role in shaping selective pressures on age and size at maturation, as do the competitive relationships among the individuals. In such cases, an ESS approach should be applied to derive optimal shapes of reaction norms. Although adding complexity adds realism, this comes at some cost of the loss in generality as ecological realism is inevitably case specific.

1.2.5.2 Plastic Growth Rates

The classical models predicting age and size at maturation assumed that environmental factors such as food availability directly determine juvenile growth rates. Under this assumption, growth rate can essentially be treated as an environmental variable (Figure 1.3). Clearly, such assumptions disregard the possibility of adaptive variation in growth rates *per se*, i.e. the possibility that the juveniles may choose to grow slower than physiologically possible. Inspired by accumulating empirical evidence of plasticity in growth rates (*e.g.*, Nylin et al. 1989, Lima and Dill 1990), Abrams et al. (1996) constructed models for time and size at maturation in which the assumption of environmental determination of growth rates was relaxed. These authors analyzed two situations where selective pressure for faster development is expected to arise. First, they considered life history evolution in a seasonal environment, postulating an optimum for the absolute timing of maturation and a fitness penalty for deviating from the optimal value. The severity of the seasonal time stress was varied by manipulating season length. In the second model, the selection for faster development was generated by increased juvenile mortality. In both cases, growth rates were assumed to be flexible, with faster growth being costly in terms of higher mortality risk. Inevitably, optimization models with two variables (here, time at maturation and growth rate) are more complex, and the two-dimensional setting should yield a diverse array of predictions. This was indeed the case. Quite intuitively, in the model for seasonal time stress, the authors concluded that the most probable responses to relaxed time stress would include increased size, decreased growth rate, and longer development time. However, different outcomes are possible depending on details of the assumptions. In particular, an optimal change may consist of a simultaneous increase in growth rate and size both in response to increasing or decreasing time stress. In the second model, the most likely response to increased mortality risk was smaller adult size, but no definitive conclusions could be drawn for growth rate and development time. An analysis of the models showed that the model outcomes are critically dependent on mathematical details such as second derivatives of the underlying relationships. This may be considered discouraging as respective empirical relationships are rarely known to such an extent.

The model by Abrams et al. (1996) has inspired a great deal of empirical research. Manipulatively increased time stress has been shown to almost invariably lead to shorter developmental periods and lower body size, with mixed evidence with respect to plastic changes in growth rate (Dmitriew 2011; see also further in the chapter). In some contrast, Abrams et al. (1996) seem not to have given comparable momentum to theoretical work in the field (however, see Abrams and Rowe 1996, Yearsley et al. 2004, Morris et al. 2016).

1.2.5.3 Introducing Developmental Thresholds

Day and Rowe (2002) highlighted the discrepancy between the high diversity of shapes of reaction norms for age and size at maturation predicted by optimality models, and the fact that the empirically recorded reaction norms were much less diverse, with the L-shaped ones (Figure 1.4) clearly dominating. They argued that while L-shaped reaction norms appear among the predictions of the classical models, the assumptions leading to such predictions are quite specific and are unlikely to be generally met. In their work, they proceeded from the idea that such a general phenomenon should more likely be based on generally applicable principles.

The model of Day and Rowe (2002) proposed to explain the L-shaped reaction norms within a minimalistic demographic optimality framework, which is well consistent with the authors' strive for maximal generality. The model assumes just a simple growth function, constant instantaneous mortality and a monotonous increase of fecundity with size. As a distinctive feature, the authors added the concept of overhead threshold to postulate the minimal size that has to be reached by the growing organism to attain positive potential fecundity. For maturation sizes exceeding the threshold, fecundity is a function of the difference between the actual size and the threshold size, and in the simplest case, proportional to it. The analysis showed that an L-shaped reaction norm is an almost universal prediction derived from the overhead threshold model.

The key assumption of the overhead threshold model is quite straightforward. It is indeed intuitive to assume that, for any given body design, some minimum size must be attained to produce a viable and functional organism. Ultimately, this is because the laws of physics set the limits for body size under which each particular body design can be functional. These are just the resources exceeding the viability limit that can be invested into reproduction, which results in a size–fecundity relationship with a negative intercept (Figure 1.5). Not only are the assumptions of this model easy to capture but also the predictions. In the framework of the overhead threshold model, the principle leading to the L-shaped reaction norms can readily be verbally expressed: 'Grow as long you reach the threshold value, as below that you have zero fitness. Having reached the threshold, continue growing because, initially, you gain a lot in terms of the relative increase in your fecundity. Note that such gains will diminish quickly, and in the case of slow growth, be satisfied with a smaller final size.'

The overhead threshold model has received wide support primarily from studies on arthropod maturation (*e.g.*, Juliano et al. 2004, Plaistow et al. 2004, Etile and Despland 2008, Teder et al. 2008). Studying damselflies, Nilsson-Örtman and Rowe (2021), provided empirical confirmation for a distinctive prediction of the model. The prediction was that, while food reductions at sizes below the threshold should delay maturation, those occurring above the threshold should lead to an earlier onset of maturity. Nevertheless, despite the overall enthusiasm, some system-specific empirically based criticism of the overhead threshold model has also been presented (Harney et al. 2013, Helm et al. 2017).

1.2.5.4 Probabilistic Reaction Norms

The classical models for age and size at maturation are deterministic, meaning there is only one outcome for any combination of the input parameters. While simple deterministic models provide indispensable tools for proving and illustrating phenomena at the most general level, their generality comes at the cost of biological realism. It is characteristic of biological

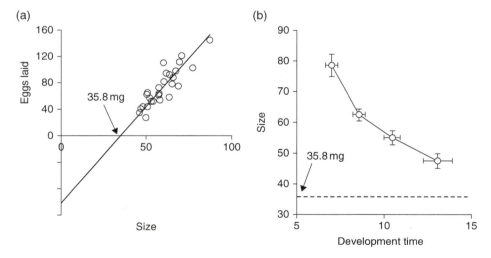

Figure 1.5 (a) The dependence of fecundity on adult body size (expressed as live mass of pupae) in the geometrid moth *Epirrita autumnata* (Adapted from Tammaru et al. 1996b). (b) Reaction norm for age and size at maturation (±95% CI) in response to manipulation of larval food availability (four treatments, food availability decreasing from left to right; time is measured in days since the beginning of final larval instar) in *E. autumnata* (Adapted from Tammaru 1998). In concert, these two relationships show a good correspondence with the overhead threshold model by Day and Rowe (2002). There is a critical body size corresponding to zero fecundity, which is close to the horizontal asymptote for the bivariate reaction norm.

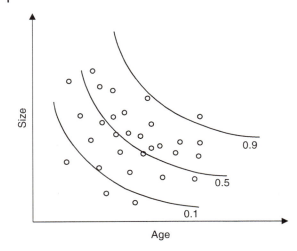

Figure 1.6 A probabilistic reaction norm. Dots indicate particular hypothetic maturation decisions. The lines delimit the areas where the probability of maturing is within the indicated range.

systems to include stochastic variation: two identical organisms in similar conditions are unlikely to reach maturity at the same point in time, primarily due to unmeasured and perhaps unmeasurable differences in their micro-environments throughout their ontogenies. If applied at the population level, genetic differences among the individuals further contribute to the variation in the reaction norms. To cope with such variability, we need stochastic models for age and size at maturation. To meet the challenge, Heino et al. (2002) introduced the concept of probabilistic maturation reaction norms (PMRN). A PMRN is basically a function that describes how the probability of becoming mature during a specified time period is related to the age and size of the organism. As the maturation status of an individual is a binary variable (did mature or did not), the relationship can be conveniently modelled in the framework of logistic regression (or respective generalized linear model [GLM]). PMRNs can be visualized in the conventional time *vs.* size plane so that the areas with different probabilities of becoming mature are mapped (Figure 1.6).

Besides accounting for and illustrating the variability in maturation decisions *per se*, in whatever context this may be of interest, the PMRN approach has various other advantages. A practical advantage of the approach is that a PMRN can be constructed based on cross-sectional data as long as the sizes, ages and maturation times of the sampled organisms can be determined. The PMRN approach is also free of the bias that mortality can introduce into studies that record maturation reaction norms more traditionally, *i.e.*, through manipulating growth rates and recording treatment-specific average ages and sizes at maturation (*e.g.*, Gebhardt and Stearns 1988, Tammaru 1998). A major advantage is also that the GLM approach allows for the inclusion of covariates. This means that the decision to mature should not necessarily be modelled as dependent on the size and age of the organism only. Other variables characterizing the organisms (*e.g.*, some condition index) or their environment (*e.g.*, temperature) can also be included as predictors. Such an approach also provides a powerful tool for separating genetically and environmentally based differences when, for example, reaction norms are compared across populations (Dieckmann and Heino 2007). A potential drawback is, however, that, like any parametric method, a PMRN is sensitive to the assumptions of the models having been met.

The original PMRN models were designed to study fish populations where the approach is still primarily used (Heino et al. 2015, Niu et al. 2023). In consistency with annual reproduction events, typical of fish, time was treated as a discrete variable, with the probability of maturation in a particular year serving as the response variable. Further works have adjusted the approach for the case of continuous time (van Dooren et al. 2005, Kuparinen et al. 2008), and for approaching different questions such as predator-induced plasticity (Beckerman et al. 2010), the evolvability of maturation thresholds (Harney et al. 2013) and non-genetic inheritance (Plaistow et al. 2015). More recently, Legault and Kingsolver (2020) applied the probabilistic approach to describe maturation in an insect (*Manduca sexta*), separately modelling stochasticity in growth and in the maturation decision.

1.2.5.5 Integrating Physiological Realism

Understanding the evolution of ontogenetic growth (including ontogenetic switches such as maturation) is challenging because it necessarily requires integrating processes occurring at different levels of biological organization, from molecules to ecosystems. Not only are the processes themselves so different that it is hard to find a common denominator to analyze them jointly, but the same appears to apply to the approaches the biologists working in different fields are used to.

Evolutionary ecology aims to explain growth and maturation patterns in the demographic optimality framework, accounting for the effects of different ontogenetic decisions on natality and mortality. Under this approach, fitness, the quantity assumed to be maximized, is derived from such demographic parameters in one way or another. However, not all predictions based on demographic optimality can be realized, as various physiological or developmental constraints preclude some of them. Such constraints have not been commonly integrated into the models of age and size at maturation. To what extent this is a problem must depend on the type of question asked. When we are interested in the qualitative effect of some selective force on the shape of the reaction norm, a purely demographically based model may well be sufficient. However, it cannot suffice if we take the maximalist approach and wish to provide a quantitative evolutionary explanation for the observed body sizes. The need for considering constraints, or to include more physiological realism, has repeatedly been put forward (Reznick 1990, Bernardo 1993, Ricklefs and Wikelski 2002, Berner and Blanckenhorn 2007, De Block et al. 2008, Flatt and Heyland 2011), but we still appear to lack a generally accepted protocol for how should this be done.

One way to integrate physiology into optimality models is by assuming a particular mathematical form of the growth curve. Many classical models rely on the assumption of the von Bertalanffy growth curve, primarily for modelling convenience. Such a convention has, however, been criticized, partly because the shape of the von Bertalanffy curve itself results from the ontogenetic dynamics of the allocation of resources to growth *vs.* reproduction, leading thus to some circularity in the causation (Day and Taylor 1997, Czarnoleski and Kozlowski 1998). Ideally, constraints should not be integrated into the models in a binary way ('must be' *vs.* no limits; possible *vs.* impossible). Instead of seeing the constraints as absolute, they should be quantified on a continuous scale. We see two principal ways this could be done. First, physiological constraints could be quantified in terms of fitness costs associated with approaching some physiological limits. Under such an approach, there should be no fundamental difference between selective pressures and constraints: developmental/physiological constraints may be seen as internally (ontogenetically, physiologically) based selective pressures (Schwenk and Wagner 2004), and we treat them differently just because these selective pressures are rooted outside the applied explanatory framework (demographic, in our case). Second, one may attempt to predict the evolutionary time needed for the constraint to relax under selection pressure. A constraint can then be considered absolute within time frames well below this limit (10 million years, *e.g.*) but not when evolutionary processes at a longer time scale are considered.

While we are unaware of any example of a truly satisfactory way of integrating physiological constraints into optimality models for age and size at maturation, we see a system that may well constitute a promising candidate for developing life history models, which are quantitatively explicit about constraints. This system is the growth and maturation of insects, with distinct larval instars as the key element. Growing insect larvae periodically renew their exoskeletons, including mouthparts and parts of their respiratory (tracheal) system, with the active growth period between two successive moults being called an instar (Sehnal 1985). Typically, insects go through three to seven instars during their larval development (Esperk et al. 2007).

While there is undoubtedly more than one reason why insect larvae need to moult, evidence is accumulating in favour of oxygen limitation as the main driver of the moulting process (Greenlee and Harrison 2004, Callier and Nijhout 2011). A larva attempting to grow 'too much' within an instar will face oxygen shortage. Accordingly, various lines of evidence suggest substantial constraints on the within-instar growth increments (Tammaru 1998, Kivelä et al. 2016). As the physics behind this constraint is straightforward, the constraint could be quantifiable in the first sense (see earlier), *i.e.*, fitness consequences of oxygen shortage should be measurable. Also, as the number of insect species is vast, and obtaining relevant empirical data is relatively straightforward, the constraint should be readily quantifiable in the second sense (see earlier) as well, *i.e.*, the degree of phylogenetic conservatism in respective phenomena can be traced based on cross-species comparative analyses (Tammaru et al. 2015, Kivelä et al. 2020). Mechanistic knowledge will hardly limit the integration of demographic optimality and physiology as there is respectable knowledge of the proximate mechanisms of insect growth and maturation (Shingleton 2011, Nijhout et al. 2014, Shingleton and Vea 2023). Additionally, while the patterns of within-instar growth are supposed to be constrained, the number of instars appears to be flexible in many insects (Esperk et al. 2007). This may allow for elucidating contrasts to be built between constrained and unconstrained ways to prolong growth schedules.

1.3 Part II

1.3.1 Bivariate reaction Norms with a Positive Slope

Both empirical evidence and many theoretical models agree that the bivariate reaction norms for age and size at maturation are most frequently L-shaped (see Figure 1.4), or, more generally, they have a negative slope in the size *vs.* time parameter plane. We thus suggest that a systematic approach to studying the shapes of such reaction norms should treat reaction norms with a negative slope as the typical or 'default' case. Research attention should then focus on asking in which systems and why reaction norms with a positive slope occur. In other words, we may ask when and why it sometimes pays to mature earlier and at a smaller size than 'usual' (Figure 1.7). Intuitively, such a decision should primarily pay off when a 'disaster' can be expected to happen, though the disasters can be of a different nature, as discussed further in the chapter.

Accordingly, Teder et al. (2014) performed a meta-analysis of the prevalence of negatively *vs.* positively sloped reaction norms, based on case studies reporting responses of insect larvae to variable quality or quantity of food. More than 200 relevant studies could be retrieved, even if the vast majority of them had been conducted in contexts other than evolutionary

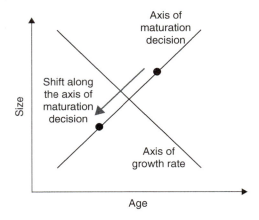

Figure 1.7 A schematic presentation of possible plastic and evolutionary changes in reaction norms for age and size at maturation (redrawn from Tammaru et al. 2000, 2010). The shifts can occur along the axis of maturation decision (earlier maturation at lower mass, grey arrow) or along the axis of growth rate (orthogonal to the grey arrow, not shown). We suggest an approach in which the ubiquitous plastic shifts along the axis of growth rate are treated as the default, while plastic shifts along the axis of maturation decision call for an evolutionary explanation. It is the opposite for genetic changes: evolutionary shifts along the axis of maturation decision are the default, while shifts involving growth rate changes deserve attention.

ecology. As an overall conclusion, the authors found that negative correlations between size and development time (substitute for age at maturation) in response to variations in diet quality and quantity were ubiquitous, confirming the a priori expectation. The cases of positive correlations were largely limited to insects feeding on discrete food items, primarily parasitoids. Indeed, after a discrete food item is finished (and the larva cannot locate a new one), nothing will be gained from prolonging the immature development. Besides parasitoids, various other mass-provisioning insects, such as many solitary bees and dung beetles, might be expected to display positive correlations between size and development time. By contrast, with virtually no meaningful exceptions, herbivorous and predatory insects are characterized by reaction norms with a negative slope. This is consistent with the idea that predictable resource depletion, a scenario selecting for positively sloped reaction norms, is not frequent for these animals (Speight et al. 2008, Price et al. 2011).

Some animals can sense the presence of predators and may adjust their growth decisions correspondingly (see Chapter 12). The theory generally predicts that, in such cases, juveniles should slow down growth in an attempt to remain unnoticed by the predators (Abrams and Rowe 1996) for which indeed evidence exists (Dmitriew 2011). Alternatively, it may occasionally pay to mature early at a smaller size when a cue of high mortality risk is received, especially when the adults are bound to leave the dangerous larval environment. A reaction norm with a positive slope will be the result. Such an ecological setting is characteristic of various insects, and the corresponding response has sometimes indeed been recorded (Peckarsky et al. 2001, Lee et al. 2021). Nevertheless, such a response does not appear to be the rule for this group of animals (Johansson and Stoks 2005). On the other hand, in a more general form, this principle may have a rather broad applicability. Indeed, early maturation in humans having experienced harsh conditions in early childhood, frequently at the expense of somatic growth, has also been interpreted as an adaptive escape from adverse environmental conditions (Belsky et al. 1991).

In insects, contacts between conspecific larvae frequently lead to early cessation of feeding at lower masses, resulting thus in a positively sloped reaction norm. Numerous case studies on various species are consistent with these observations (Goulson and Cory 1995, Tammaru et al. 2000, Välimäki et al. 2013). It has been proposed that the early pupation of crowded larvae may be a response to anticipation of food shortage or an adaptation to prepare the adults to leave the overcrowded environment (Haukioja et al. 1988). Neither of these explanations seems to be generally applicable, partly because such crowding responses also occur in insects unlikely to experience resource depletion. It appears more plausible that crowded conditions are perceived as a cue of elevated mortality risk, either because the larvae mistake the conspecifics for potential predators or because aggregations of larvae are more likely to attract their natural enemies (Vellau and Tammaru 2012).

We are unaware of reviews of the slopes of the reaction norms comparable to that by Teder et al. (2014) for insects. However, Teder et al. (2014) retrieved some data on other ectothermic animals, arachnids and amphibians in particular, which also largely conform with the conclusions of the prevalence of negatively sloped diet-induced reaction norms. Other than in the context of diet variations, positively sloped reaction norms have however been recorded in systems in which other types of disasters can be expected. For example, many species of anurans respond to the deterioration of their larval environment (pond drying) by increasing their developmental rates at the expense of reduced size at metamorphosis (Denver 1997, Richter-Boix et al. 2011), with mixed evidence for a comparable response in age and size at maturation in aquatic insect larvae (Juliano and Stoffregen 1994, De Block and Stoks 2005, Jannot et al. 2008). Similarly, insect larvae may respond to cues of approaching winter by maturing earlier and at a smaller size (Johansson et al. 2001, Gotthard 2008), as may do desert plants in response to water stress (Aronson et al. 1992).

The dominance of 'simple' negatively sloped reaction norms in herbivorous insects (Teder et al. 2014) might be partly explained by 'ecological simplicity' of these animals, as compared to fish, for example. Herbivorous insects frequently live in relative isolation, both as individuals and as populations. This is due to a typically high degree of resource specialization, and top-down regulation by generalist predators. Both intra- and interspecific interference is rarely an issue. The ecology of fish differs from that of such insects in various essential ways. This is primarily due to significant population- and community-level processes, which may substantially interfere with selective pressures on age and size at maturation. Accordingly, fish-oriented models have incorporated phenomena such as the cross-environment correlation between growth rate and mortality risk, density-dependent growth and the effect of predation rate on food availability in juveniles (Abrams and Rowe 1996, Marty et al. 2011). Naturally, all of this increases the complexity of the models, making a wide array of different predictions possible. Moreover, the indeterminate growth of these animals further complicates the picture, as there remains a possibility to compensate for small sizes at maturation by growing later in life. Consistently, for fish, there is empirical evidence for reaction norms for age and size at maturation, which are not monotonically decreasing, with the dome-shaped ones being a frequent outcome (see Figure 1.4; Alm 1959, Perrin and Rubin 1990, Jonsson et al. 2013). Nevertheless, reaction norms with an overall positive slope appear still to be rare also in fish (Marty et al. 2011).

In sharp contrast to reaction norms in response to food levels, positive slopes dominate in reaction norms in response to varying temperatures. In ectotherms, higher temperatures within non-stressful limits invariably lead to faster development, and in most (but not all) cases, juveniles experiencing higher temperatures develop into smaller adults. The respective relationship, called the temperature–size rule, has given rise to much debate (Atkinson 1994, Berrigan and Charnov 1994, Atkinson and Sibly 1997). It has not been immediately clear what the physiological basis of such a relationship may be, and if and how it can be adaptive. The wide variety of both ideas and available evidence is too voluminous to be covered here, and we advise the interested reader to consult recent reviews on the subject (Audzijonyte et al. 2019, Verberk et al. 2021). Briefly, current opinions seem to converge on the key role of oxygen supply (which is both size- and temperature-dependent) as either proximate or ultimate (selection to avoid hypoxia) explanation behind the temperature-dependence of body size (but see Einum et al. 2021, Atkinson et al. 2022). Nevertheless, such physiology-based relationships can be modified by ecological factors such as seasonal time stress or predictably temperature-dependent food availability. For example, latitudinal variation in body size of a damselfly is best explained by the combined effects of temperature and time stress caused by variations in the length of the growing season (Hassall 2013).

1.3.2 Trade-Off Between Age and Size at Maturation

Classical models of age and size at maturation rely on the assumption of a positive genetic correlation between these two traits, implying a trade-off between adult size and survival to maturation. In other words, the models assume that organisms should extend their growth periods to attain larger sizes. This is certainly the case when growth rates are environmentally determined, with the organisms maximizing their growth effort within the physiologically attainable limits for given resource levels (not an unrealistic scenario in some systems; Ayres and MacLean 1987, Kause et al. 1999). However, the possibility for adaptive plasticity in growth rates could mitigate or even eliminate the trade-off between age and size. Theoretical models relying on the assumption of plastic growth rates have been reviewed earlier in this chapter, with the extent and importance of such phenomena in nature remaining to be discussed.

The evidence of adaptive plasticity in growth rates is overwhelming (see reviews by Arendt 1997, Dmitriew 2011). While the effect of *e.g.*, food levels on growth is trivial, growth rates have also been found to respond to seasonal time constraints, cues of increased predation risk and those of habitat deterioration. Nevertheless, plasticity in growth rates has not always been found when it would have been expected (Dmitriew 2011), such as in the responses of odonates to cues of high-time stress (De Block and Stocks 2004). There is also ample evidence of differences in growth rates among populations, often tested under a common-garden design to reveal genetic differences. Most commonly, differences in growth rates have been found among populations originating from different latitudes (Conover and Present 1990, Laugen et al. 2003, Stoks et al. 2012, Blanckenhorn et al. 2018, Kojima et al. 2020; but see Meister et al. 2017).

There is no doubt that the broad evidence of within-species variability in growth rate justifies the view that the trait should be regarded as a life history trait of its own right, and a triangular relationship among size, age and growth rate should be accounted for in life history studies (Abrams et al. 1996, Davidowitz and Nijhout 2004). Nevertheless, the available evidence should hardly be used to discredit the basic idea of a trade-off between age and size at maturation. In particular, in our meta-analytical studies on proximate determinants of SSD in insects, we have found convincing evidence that, overall, the larger body size of females is associated with longer developmental periods compared to conspecific males (Teder 2014, Teder et al. 2021). The same pattern is suggested by our empirical studies on the determination of SSD (Tammaru et al. 2010, Sõber et al. 2019) and among-population differences in body size (Vellau et al. 2013, Meister et al. 2017): whenever there is a 'reason' to grow larger, this is primarily achieved by a longer developmental period, not by an increase in growth rates. Thus, it seems that a shift along the maturation decision axis (Figure 1.7) should be treated as the default for evolutionary changes (at least in insects). In contrast, deviations from the default, *i.e.*, shifts including a component of growth rate, call for case-specific evolutionary explanations. In a way, the situation for evolutionary changes is thus the opposite of that for plastic changes discussed earlier (Figure 1.7).

Consistently, selection experiments on *Drosophila* have found clear evidence that flies selected for prolonged development are larger at adult emergence, and flies selected for large size have longer developmental periods (reviewed in Flatt 2020), with similar trends in rats and mice (Millar and Hickling 1991). Positive genetic correlations between age and size at maturation – a genetic phenomenon underlying the micro-evolutionary trade-off between age and size – have also been widely observed in other systems (Roff 2000). For an evolutionary change in a natural environment, fishing-induced selection on fish stocks has frequently resulted in early maturation at smaller body sizes (Jørgensen et al. 2007, Kendall et al. 2009, Heino et al. 2015; see also Chapter 23). However, the possibility that there has been simultaneous selection on both these traits adds some ambiguity to the interpretation in this particular context. Also, the generally larger size of insular rodents compared to their mainland conspecifics is associated with later maturation (Adler and Levins 1994).

As a methodological note, we would like to stress that defining and measuring growth rates is not necessarily straightforward and needs attention. One aspect is that, given the curvilinearity of any realistic growth trajectory, the conclusions about growth rates are highly sensitive to mathematical details, *i.e.*, the way how exactly growth rates are calculated, or linearized before doing so (Figure 1.8; see also Berner and Blanckenhorn 2007, Tammaru et al. 2010, Meister et al. 2017). In the case of complex growth patterns, characteristic of arthropods, for example, the growth curves cannot be linearized by any transformations. In our work, we have strived to approximations of instantaneous growth rates, *i.e.*, the derivative of body mass with respect to time, based on 24-hour mass increments (Tammaru and Esperk 2007, Meister et al. 2017). We admit that such a procedure is not feasible in all systems, but, in any case, researchers should be careful about what exactly they mean when talking about growth rates. Unfortunately, this is not always the case.

1.3.3 The Rates of Evolution – Do We Study the Right Thing?

It is a usual characteristic of the optimality approach to avoid the question about the rate of evolutionary change: it is just assumed that the focal variables have had sufficient time to reach the evolutionary equilibrium. Different approaches are needed to answer the question of how fast this happens.

Selection experiments attempting to bring about a genetic change in body size have mostly been successful. This is consistent with the frequent observation of moderately high heritabilities of body size (Mousseau and Roff 1987, Roff 2001). Just as some examples, in *Drosophila*, the response of about 15% in body mass can readily be achieved by ten generations of artificial selection on size (Hillesheim and Stearns 1991, Partridge and Fowler 1993, Partridge et al. 1999; see also Teuschl et al. 2007). Significant evolution of body size (about 1.5 times increase in body mass) has incidentally happened in another model insect, the moth *M. sexta*, in the course of maintaining a laboratory population over 30 years (D'Amico et al. 2001; see Chown and Gaston 2010 for other insect examples).

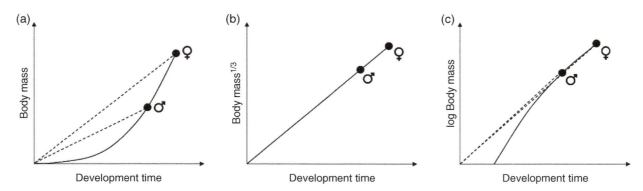

Figure 1.8 Sensitivity of the measures of growth rate to mathematical details. Males and females follow the same (cubic) growth trajectory, but females stop growing later than males, attaining larger sizes (Adapted from Tammaru and Esperk 2007). If the (integral) growth rate is calculated by dividing mass by development period, we will get higher growth rates for females (a). The growth rates of the sexes do not differ for the linearized growth trajectory (cube-root transformed mass) (b), but the analysis would return slightly higher male growth rates for logarithmically transformed masses (c).

Grainger and Levine (2022) provided an analysis of rapid evolutionary changes in body size in natural conditions. As an example, the body size of a rotifer decreased by 14% just in two weeks of applying a temperature treatment (Walczyńska et al. 2017). Substantial evidence of rapid changes in body size in natural conditions, though not in response to a natural selective factor, comes from fish populations subjected to size-selective fishing (see also Chapter 23). Here, the magnitude of the changes has reached up to 20–30% for both age and size at maturation (Hutchings and Baum 2005, Jørgensen et al. 2007) over the less than 100-year-long observation period. In this context, separating genetic changes from plastic ones is frequently challenging, but, overall, fishing-induced evolution in age and size at maturation can hardly be questioned (Swain et al. 2007, Heino et al. 2015; see also Chapter 23). Studies on *Homo floresiensis* (Diniz-Filho et al. 2019) and Sicilian elephant (Baleka et al. 2021) provide charismatic examples of rapid body size evolution at somewhat longer time scales. Works explicitly documenting (rapid) evolution and evolvability of the bivariate reaction norms for age and size at maturation are understandably more scarce, but nevertheless present (*e.g.*, Wild et al. 1994, Hutchings 2011, Harney et al. 2013).

It is clear that body size can evolve rapidly and sometimes actually does so. However, the question about the realized rates of body size evolution on the macro-evolutionary time scales appears to be a different one. Studying a fish species, Reznick et al. (1997) concluded that the macro-evolutionarily attested rates of body size evolution are by many orders of magnitude lower than what can be seen on the micro-evolutionary time scale. The review of Uyeda et al. (2011) provides broad-scale support for this observation. In particular, across various taxa, the rates of body size evolution observed over time scale longer than about a million years have mostly been notably low. This is consistent with the general idea of prevailing evolutionary stasis in various morphological traits (Eldredge et al. 2005, Estes and Arnold 2007).

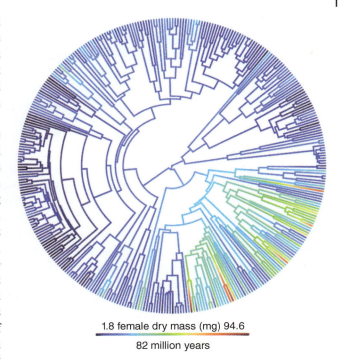

Figure 1.9 The reconstructed evolution of body size for northern European representatives of the moth family Geometridae (Adapted from Ude et al. 2024). Body sizes are coded from blue (small) to red (large), the depth of the tree (radius of the circle) is 82 million years. The picture suggests high phylogenetic conservatism of body sizes.

The overall evolutionary conservatism of body size is reflected in the fact that body size can just very rarely be used to separate sister species (see Gaston and Chown 2013 for discussion). This has been known to taxonomists for centuries and shows that the phenomenon is not limited to taxa, which could have been subjected to formal phylogenetic analysis so far. As a particular example, we present here a recent reconstruction of body size evolution in geometrid moths (Figure 1.9). We see that body size is a highly conservative trait in this group of insects. The sizes do not appear to have changed much during tens of millions of years. Subfamily-level groups have largely retained their ancestral sizes, and with a few (notable) exceptions, congeneric species that diverged less than 10 million years ago are virtually indistinguishable in size. While body size can evolve fast, we do not see frequent signs of this having happened in the evolutionary past. Persistent size differences among major clades of Geometridae cannot readily be explained by clade-specific differences in ecologically based selective pressures. Indeed, the family is highly homogeneous ecologically, and the potential of ecological traits to account for interspecific differences in body size is not high in geometrid moths in general (Seifert et al. 2023, Foerster et al. 2024).

Explaining different pictures of the evolutionary process emerging from micro- and macro-evolutionary observations is a field of active scientific debate (Houle and Rossoni 2022, Rolland et al. 2023). In particular, various ideas have been proposed to reconcile the discordant estimates of the rate of evolution obtained from different time spans. While there are different sources of potential methodological artefacts (Harmon et al. 2021), biologically meaningful phenomena are likely to be involved. A prominent one relies on the assumption that, in widely distributed species, conflicting micro-evolutionary trends in different parts of the range may considerably slow down the evolution of the species as a whole (Futuyma 2010). Also, selection pressures may vary in time so that relative stasis observed at the macroscale may be the outcome of integration over numerous microscale changes which may cancel each other (*e.g.*, Bergland et al. 2014).

However, for body size in particular, the discrepancy between the micro- and macro-evolutionary estimates of the rate of evolution may result from the ephemeral character of rapid changes in this trait. It may well be the case that the long-term evolutionary potential of the populations selected to deviate rapidly from their lineage-specific optimal body size may not be high. In particular, a trait as central as body size interacts with so many physiological and ecological traits and processes that being of a 'wrong size' would be eventually counter-selected in one way or another. This would be consistent with the idea that high phenotypic integration hinders responses to selection (Wagner and Altenberg 1996, Pigliucci 2003). Just a few genes with a major effect may be responsible for the rapid change in body size but awaiting complementary changes in all coadapted genes takes a long time. Before this happens, the affected populations will be vulnerable, they may return to their 'right sizes' or face extinction (see Olsen et al. 2004 for a potential example).

If this is the case, it may bear two cautionary messages. First, we may provocatively ask whether experimental studies on the evolution of body size have any relevance at all with respect to explaining the 'real' evolution of this trait (see Di Martino and Liow 2021, for an example). Can the evolutionary changes observed over a few generations be extrapolated to time scales longer by orders of magnitude? Second, should we be worried about the fate of the populations forced to evolve quickly by, *e.g.*, recent human-induced environmental changes? In any case, however, the high evolutionary conservatism suggests that the maximalist approach to understanding the evolution of body size, implying the ability to explain the absolute values of organisms for this trait, should not ignore the phylogenetic context.

1.4 Conclusions

There is a respectable amount of knowledge and understanding about selective forces that have shaped the body size of multicellular organisms, and the responses of age and size at maturation to environmental conditions. However, we may still be unable to provide a satisfactory answer to perhaps childishly naïve but nevertheless the fundamental question, 'Why is each animal of the size it is?' There are definitely many ways to move forward. The one we would like to highlight specifically is to pay explicit attention to quantifying selective pressures rooted in physiological and ontogenetic processes, frequently but perhaps misleadingly referred to as constraints. A related avenue is the wider application of phylogenetic comparative methods to study environmental correlates of different body sizes and the features of respective reaction norms across species. This could also shed light on the applicability of the results of short-term studies to explain the evolution of body size at the scale of millions of years, something that appears frequently to have happened at a surprisingly low pace.

Acknowledgements

The authors were supported by the Estonian Research Council grant PRG741, and the Internal Grant Agency of the Faculty of Environmental Sciences, Czech University of Life Sciences Prague (Grant no. 42900/1322/3208). We thank our colleagues John T. Clarke, Robert B. Davis, Toomas Esperk, Peeter Hõrak, Ants Kaasik, Leonard Opare, Marina Semtšenko and Kadri Ude for their help and advice. Figure 1.9 is based on the work of Erki Õunap and Kadri Ude. The editors and an anonymous reviewer provided valuable suggestions.

References

Abrams, P.A., Leimar, O., Nylin, S., and Wiklund, C. (1996). The effect of flexible growth rates on optimal sizes and development times in a seasonal environment. *The American Naturalist* 147: 381–395.

Abrams, P.A., and Rowe, L. (1996). The effects of predation on the age and size of maturity of prey. *Evolution* 50: 1052–1061.

Adler, G.H., and Levins, R. (1994). The island syndrome in rodent populations. *Quarterly Review of Biology* 69: 473–490.

Alm, G. (1959). Connection between maturity, size and age in fishes. *Report of the Institute of Freshwater Research, Drottningholm* 40: 1–145.

Andersson, M. (1994). *Sexual selection*. Princeton University Press.

Andreassen, H.P., Sundell, J., Ecke, F., Halle, S., Haapakoski, M., Henttonen, H., Huitu, O., Jacob, J., Johnsen, K., Koskela, E., Luque-Larena, J.J., Lecomte, N., Leirs, H., Mariën, J., Neby, M., Rätti, O., Sievert, T., Singleton, G.R., van Cann, J., Broecke, B.V., and Ylönen, H. (2021). Population cycles and outbreaks of small rodents: ten essential questions we still need to solve. *Oecologia* 195: 601–622.

Arendt, J.D. (1997). Adaptive intrinsic growth rates: an integration across taxa. *Quarterly Review of Biology* 72: 149–177.

Aronson, J., Kigel, J., Shmida, A., and Klein, J. (1992). Adaptive phenology of desert and Mediterranean populations of annual plants grown with and without water stress. *Oecologia* 89: 17–26.

Atkinson, D. (1994). Temperature and organism size – a biological law for ectotherms? *Advances in Ecological Research* 25: 1–58.

Atkinson, D., Leighton, G., and Berenbrink, M. (2022). Controversial roles of oxygen in organismal responses to climate warming. *Biological Bulletin* 243: 207–219.

Atkinson, D., and Sibly, R.M. (1997). Why are organisms usually bigger in colder environments? Making sense of a life history puzzle. *Trends in Ecology & Evolution* 12: 235–239.

Audzijonyte, A., Barneche, D.R., Baudron, A.R., Belmaker, J., Clark, T.D., Marshall, C.T., Morrongiello, J.R., and van Rijn, I. (2019). Is oxygen limitation in warming waters a valid mechanism to explain decreased body sizes in aquatic ectotherms? *Global Ecology and Biogeography* 28: 64–77.

Ayres, M.P., and Maclean, S.F. (1987). Development of birch leaves and the growth energetics of *Epirrita autumnata* (Geometridae). *Ecology* 68: 558–568.

Badyaev, A.V. (2002). Growing apart: an ontogenetic perspective on the evolution of sexual size dimorphism. *Trends in Ecology & Evolution* 17: 369–378.

Baker, J., Meade, A., Pagel, M., and Venditti, C. (2015). Adaptive evolution toward larger size in mammals. *Proceedings of the National Academy of Sciences of the United States of America* 112: 5093–5098.

Baleka, S., Herridge, V.L., Catalano, G., Lister, A.M., Dickinson, M.R., Di Patti, C., Barlow, A., Penkman, K.E.H., Hofreiter, M., and Paijmans J.L.A (2021). Estimating the dwarfing rate of an extinct Sicilian elephant. *Current Biology* 31: 3606–3612.

Barneche, D.R., Robertson, D.R., White, C.R., and Marshall, D.J. (2018). Fish reproductive-energy output increases disproportionately with body size. *Science* 360: 642–644.

Beckerman, A.P., Rodgers, G.M., and Dennis, S.R. (2010). The reaction norm of size and age at maturity under multiple predator risk. *Journal of Animal Ecology* 79: 1069–1076.

Belsky, J., Steinberg, L., and Draper, P. (1991). Childhood experience, interpersonal development, and reproductive strategy: an evolutionary theory of socialization. *Child Development* 62: 647–670.

Benitez-Lopez, A., Santini, L., Gallego-Zamorano, J., Milá, B., Walkden, P., Huijbregts, M.A.J., and Tobias, J.A. (2021). The island rule explains consistent patterns of body size evolution in terrestrial vertebrates. *Nature Ecology & Evolution* 5: 768–786.

Berger, D., Walters, R., and Gotthard, K. (2006). What keeps insects small? – Size dependent predation on two species of butterfly larvae. *Evolutionary Ecology* 20: 575–589.

Bergland, A.O., Behrman, E.L., O'Brien, K.R., Schmidt, P.S., and Petrov, D.A. (2014). Genomic evidence of rapid and stable adaptive oscillations over seasonal time scales in *Drosophila*. *PLoS Genetics* 10(11): e1004775.

Bernardo, J. (1993). Determinants of maturation in animals. *Trends in Ecology & Evolution* 8: 166–173.

Berner, D., and Blanckenhorn, W.U. (2007). An ontogenetic perspective on the relationship between age and size at maturity. *Functional Ecology* 21: 505–512.

Berrigan, D., and Charnov, E.L. (1994). Reaction norms for age and size at maturity in response to temperature: a puzzle for life historians. *Oikos* 70: 474–478.

Berrigan, D., and Koella, J.C. (1994). The evolution of reaction norms: simple models for age and size at maturity. *Journal of Evolutionary Biology* 7: 549–566.

Blanckenhorn, W.U. (2000). The evolution of body size: what keeps organisms small? *Quarterly Review of Biology* 75: 385–407.

Blanckenhorn, W.U., Bauerfeind, S.S., Berger, D., Davidowitz, G., Fox, C.W., Guillaume, F., Nakamura, S., Nishimura, K., and Sasaki, H. (2018). Life history traits, but not body size, vary systematically along latitudinal gradients on three continents in the widespread yellow dung fly. *Ecography* 41: 2080–2091.

Blomquist, G.E., Sade, D.S., and Berard, J.D. (2011). Rank-related fitness differences and their demographic pathways in semi-free-ranging rhesus macaques (*Macaca mulatta*). *International Journal of Primatology* 32: 193–208.

Blums, P., and Clark, R.G. (2004). Correlates of lifetime reproductive success in three species of European ducks. *Oecologia* 140: 61–67.

Bonduriansky, R. (2001). The evolution of male mate choice in insects: a synthesis of ideas and evidence. *Biological Reviews* 76: 305–339.

Breck, J.E., Simon, C.P., Rutherford, E.S., Low, B.S., Lamberson, P.J., and Rogers, M.W. (2020). The geometry of reaction norms yields insights on classical fitness functions for Great Lakes salmon. *PLoS ONE* 15(3): e0228990.

Brommer, J.E. (2000). The evolution of fitness in life-history theory. *Biological Reviews* 75: 377–404.

Burd, M., Read, J., Sanson, G.D., and Jaffre, T. (2006). Age-size plasticity for reproduction in monocarpic plants. *Ecology* 87: 2755–2764.

Callier, V., and Nijhout, H.F. (2011). Control of body size by oxygen supply reveals size-dependent and size-independent mechanisms of molting and metamorphosis. *Proceedings of the National Academy of Sciences of the United States of America* 108: 14664–14669.

Charlesworth, B. (1980). *Evolution in age-structured populations.* Cambridge University Press.

Choe, J.C., and Crespi, B.J. (1997). *The evolution of mating systems in insects and arachnids.* Cambridge University Press.

Chown, S.L., and Gaston, K.J. (2010). Body size variation in insects: a macroecological perspective. *Biological Reviews* 85: 139–169.

Clarke, J.T. (2021). Evidence for general size-by-habitat rules in actinopterygian fishes across nine scales of observation. *Ecology Letters* 24: 1569–1581.

Clutton-Brock, T.H., Albon, S.D., and Guiness, F.E. (1985). Parental investment and sex differences in juvenile mortality in birds and mammals. *Nature* 313: 131–133.

Clutton-Brock, T.H., Albon, S.D., and Guiness, F.E. (1988). Reproductive success in male and female deer. In: *Reproductive success* (ed. Clutton-Brock, T.H.), 325–343. University of Chicago Press.

Collar, D.C., Schulte, J.A. II, and Losos, J.B. (2011). Evolution of extreme body size disparity in monitor lizards (Varanus). *Evolution* 65: 2664–2680.

Conover, D.O., and Present, T.M.C. (1990). Countergradient variation in growth rate: compensation for length of the growing season among Atlantic silversides from different latitudes. *Oecologia* 83: 316–324.

Cooney, C.R., and Thomas, G.H. (2020). Heterogeneous relationships between rates of speciation and body size evolution across vertebrate clades. *Nature Ecology & Evolution* 5: 101–110.

Cornell, H.V., and Hawkins, B.A. (1995). Survival patterns and mortality sources of herbivorous insects: some demographic-trends. *The American Naturalist* 145: 563–593.

Czarnoleski, M., and Kozlowski, J. (1998). Do Bertalanffy's growth curves result from optimal resource allocation? *Ecology Letters* 1: 5–7.

D'Amico, L.J., Davidowitz, G., and Nijhout, H.F. (2001). The developmental and physiological basis of body size evolution in an insect. *Proceedings of the Royal Society B: Biological Sciences* 268: 1589–1593.

Dale, J., Dunn, P.O., Figuerola, J., Lislevand, T., Szekely, T., and Whittingham, L.A. (2007). Sexual selection explains Rensch's rule of allometry for sexual size dimorphism. *Proceedings of the Royal Society B: Biological Sciences* 1628: 2971–2979.

Davidowitz, G., and Nijhout, H.F. (2004). The physiological basis of reaction norms: the interaction among growth rate, the duration of growth and body size. *Integrative and Comparative Biology* 44: 443–449.

Davis, R.B., Javoiš, J., Kaasik, A., Õunap, E., and Tammaru, T. (2016). An ordination of life histories using morphological proxies: capital *vs.* income breeding in insects. *Ecology* 97: 2112–2124.

Day, T., and Rowe, L. (2002). Developmental thresholds and the evolution of reaction norms for age and size at life-history transitions. *The American Naturalist* 159: 338–350.

Day, T., and Taylor, P.D. (1997). Von Bertalanffy's growth equation should not be used to model age and size at maturity. *The American Naturalist* 149: 381–393.

De Block, M., and Stoks, R. (2004). Life-history variation in relation to time constraints in a damselfly. *Oecologia* 140: 68–75.

De Block, M., and Stoks, R. (2005). Pond drying and hatching date shape the tradeoff between age and size at emergence in a damselfly. *Oikos* 108: 485–494.

De Block, M., Slos, S., Johansson, F., and Stoks, R. (2008). Integrating life history and physiology to understand latitudinal size variation in a damselfly. *Ecography* 31: 115–123.

Denver, R.J. (1997). Proximate mechanisms of phenotypic plasticity in amphibian metamorphosis. *American Zoologist* 37: 172–184.

Di Martino, E., and Liow L.H. (2021). Trait-fitness associations do not predict within-species phenotypic evolution over 2 million years. *Proceedings of the Royal Society B: Biological Sciences* 288: 20202047.

Dial, K.P., Greene, E., and Irschick, D. (2008). Allometry of behavior. *Trends in Ecology & Evolution* 23: 394–401

Dieckmann, U., and Heino, M. (2007). Probabilistic maturation reaction norms: their history, strengths, and limitations. *Marine Ecology Progress Series* 335: 253–269.

Diniz-Filho, J.A.E., Jardim, L., Rangel, T.F., Holden, P.B., Edwards, N.R., Hortal, J., Santos, A.M.C., and Raia, P. (2019). Quantitative genetics of body size evolution on islands: an individual-based simulation approach. *Biology Letters* 15: 20190481.

Dmitriew, C.M. (2011). The evolution of growth trajectories: what limits growth rate? *Biological Reviews* 86: 97–116.

Drent, R.H., and Daan, S. (1980). The prudent parent: energetic adjustments in avian breeding. *Ardea* 68: 225–252.

Edley, M.T., and Law, R. (1988). Evolution of life histories and yields in experimental populations of *Daphnia magna*. *Biological Journal of the Linnean Society* 34: 309–326.

Einum, S., Bech, C., and Kielland, Ø.N. (2021). Quantitative mismatch between empirical temperature-size rule slopes and predictions based on oxygen limitation. *Scientific Reports* 11: 23594.

Eldredge, N., Thompson, J.N., Brakefield, P.M., Gavrilets, S., Jablonski, D., Jackson, J.B.C., Lenski, R.E., Lieberman, B.S., McPeek, M.A., and Miller, W. (2005). The dynamics of evolutionary stasis. *Paleobiology* 31: 133–145.

Engqvist, L., and Taborsky, M. (2016). The evolution of genetic and conditional alternative reproductive tactics. *Proceedings of the Royal Society B: Biological Sciences* 283: 20152945.

Esperk, T., Tammaru, T., and Nylin, S. (2007). Intraspecific variability in number of larval instars in insects. *Journal of Economic Entomology* 100: 627–645.

Estes, S., and Arnold, S.J. (2007). Resolving the paradox of stasis: models with stabilizing selection explain evolutionary divergence on all timescales. *The American Naturalist* 169: 227–244.

Etile, E., and Despland, E. (2008). Developmental variation in the forest tent caterpillar: life history consequences of a threshold size for pupation. *Oikos* 117: 135–143.

Fairbairn, D.J. (2007). Sexual dimorphism in the water strider, *Aquarius remigis*: a case study of adaptation in response to sexually antagonistic selection. In: *Sex, size and gender roles: evolutionary studies of sexual size dimorphism* (eds. Fairbairn, D.J., Blanckenhorn, W.U., and Szekely, T.), 97–105. Oxford University Press.

Flatt, T. (2020). Life-history evolution and the genetics of fitness components in *Drosophila melanogaster*. *Genetics* 214: 3–48.

Flatt, T., and Heyland, A. (2011). *Mechanisms of life-history evolution: the genetics and physiology of life history traits and trade-off*. Oxford University Press.

Foerster, S. I. A., Clarke, J. T., Õunap, E., Teder, T., & Tammaru, T. (2024). A comparative study of body size evolution in moths: Evidence of correlated evolution with feeding and phenology-related traits. *Journal of Evolutionary Biology* 37: 891–904.

Fox, C.W., and Czesak, M.E. (2000). Evolutionary ecology of progeny size in arthropods. *Annual Review of Entomology* 45: 341–369.

Futuyma, D.J. (2010). Evolutionary constraint and ecological consequences. *Evolution* 64: 1865–1884.

Gaston, K.J., and Chown, S.L. (2013). Macroecological patterns in insect body size. In: *Animal body size: linking pattern and process across space, time, and taxonomic group* (eds. Smith, F.A., and Lyons, S.K.), 13–61. The University of Chicago Press.

Gebhardt, M.D., and Stearns, S.C. (1988). Reaction norms for developmental time and weight at eclosion in *Drosophila mercatorum*. *Journal of Evolutionary Biology* 1: 335–354.

Gotthard, K. (2008). Adaptive growth decisions in butterflies. *BioScience* 58: 222–230.

Gotthard, K., Berger, D., and Walters, R. (2007). What keeps insects small? Time limitation during oviposition reduces the fecundity benefit of female size in a butterfly. *The American Naturalist* 169: 768–779.

Goulson, D., and Cory, J.S. (1995). Responses of *Mamestra brassicae* (Lepidoptera, Noctuidae) to crowding: interactions with disease resistance, colour phase and growth. *Oecologia* 104: 416–423.

Grainger, T.N., and Levine, J.M. (2022). Rapid evolution of life-history traits in response to warming, predation and competition: a meta-analysis. *Ecology Letters* 25: 541–554.

Greenlee, K.J., and Harrison, J.F. (2004). Development of respiratory function in the American locust *Schistocerca americana* II. Within-instar effects. *Journal of Experimental Biology* 207: 509–517.

Gross, M.R. (1985). Disruptive selection for alternative life histories in salmon. *Nature* 313: 47–48.

Hansen, T.F., and Houle, D. (2004). Evolvability, stabilizing selection, and the problem of stasis. In: *Phenotypic integration* (eds. Pigliucci, M., and Preston, K.), 130–150. Oxford University Press.

Harmon, L.J., Pennell, M.W., Henao-Diaz, L.F., Rolland, J., Sipley, B.N., and Uyeda, J.C. (2021). Causes and consequences of apparent timescaling across all estimated evolutionary rates. *Annual Review of Ecology, Evolution, and Systematics* 52: 587–609.

Harney, E., van Dooren, T.J.M., Paterson, S., and Plaistow, S.J. (2013). How to measure maturation: a comparison of probabilistic methods used to test for genotypic variation and plasticity in the decision to mature. *Evolution* 67: 525–538.

Hassall, C. (2013). Time stress and temperature explain continental variation in damselfly body size. *Ecography* 36: 894–903.

Haukioja, E., Pakarinen, E., Niemelä, P., and Iso-Iivari, L. (1988). Crowding-triggered phenotypic responses alleviate consequences of crowding in *Epirrita autumnata* (Lep., Geometridae). *Oecologia* 75: 549–558.

He, J., Tu, J., Yu, J., and Jiang, H. (2023). A global assessment of Bergmann's rule in mammals and birds. *Global Change Biology* 29: 5199–5210.

Heaney, L.R. (1978). Island area and body size of insular mammals: evidence from the tri-colored squirrel (*Callosciurus prevosti*) of Southeast Asia. *Evolution* 32: 29–44.

Hedrick, A.V., and Temeles, E.J. (1989). The evolution of sexual dimorphism in animals: hypotheses and tests. *Trends in Ecology & Evolution* 4: 136–138.

Heino, M., Dieckmann, U., and Godø, O.R. (2002). Measuring probabilistic reaction norms for age and size at maturation. *Evolution* 56: 669–678.

Heino, M., Pauli, B.D., and Dieckmann, U. (2015). Fisheries-induced evolution. *Annual Review of Ecology, Evolution, and Systematics* 46: 461–480.

Helm, B.R., Rinehart, J.P., Yocum, G.D., and Bowsher, J.A. (2017). Metamorphosis is induced by food absence rather than a critical weight in the solitary bee, *Osmia lignaria*. *Proceedings of the National Academy of Sciences of the United States of America* 114: 10924–10929.

Hillesheim, E., and Stearns, S.C. (1991). The responses of *Drosophila melanogaster* to artificial selection on body weight and its phenotypic plasticity in two larval food environments. *Evolution* 45: 1909–1923.

Hone, M.J., and Benton, D.W. (2005). The evolution of large size: how does Cope's Rule work? *Trends in Ecology & Evolution* 20, 4–6.

Honěk, A. (1993). Intraspecific variation in body size and fecundity in insects: a general relationship. *Oikos* 66: 483–492.

Houle, D., and Rossoni, D.M. (2022). Complexity, evolvability, and the process of adaptation. *Annual Review of Ecology, Evolution, and Systematics* 53: 137–159.

Hutchings, J.A. (2011). Old wine in new bottles: reaction norms in salmonid fishes. *Heredity* 106: 421–437.

Hutchings, J.A., and Baum, J.K. (2005). Measuring marine fish biodiversity: temporal changes in abundance, life history and demography. *Philosophical Transactions of the Royal Society B: Biological Sciences* 360: 315–338.

Isaac, J.L. (2005). Potential causes and life-history consequences of sexual size dimorphism in mammals. *Mammal Review* 35: 101–115.

Jannot, J.E., Wissinger, S.A., and Lucas, J.R. (2008). Diet and a developmental time constraint alter life-history trade-offs in a caddis fly (Trichoptera: Limnephilidae). *Biological Journal of the Linnean Society* 95: 495–504.

Johansson, F., and Stoks, R. (2005). Adaptive plasticity in response to predators in dragonfly larvae and other aquatic insects. In: *Insect evolutionary ecology*, (eds. Fellowes, M.D.E., Holloway, G.J., and Rolff, J.), 343–366. CABI Publishing.

Johansson, F., Stoks, R., Rowe, L., and De Block, M. (2001). Life history plasticity in a damselfly: effects of combined time and biotic constraints. *Ecology* 82: 1857–1869.

Johnson, J.V., Finn, C., Guirguis, J., Goodyear, L.E.B., Harvey, L.P., Magee, R., Ron, S., and Pincheira-Donoso, D. (2023). What drives the evolution of body size in ectotherms? A global analysis across the amphibian tree of life. *Global Ecology and Biogeography* 32: 1311–1322.

Jönsson, K. I. (1997). Capital and income breeding as alternative tactics of resource use in reproduction. *Oikos* 78: 57–66.

Jonsson, B., Jonsson, N., and Finstad, A.G. (2013). Effects of temperature and food quality on age and size at maturity in ectotherms: an experimental test with Atlantic salmon. *Journal of Animal Ecology* 82: 201–210.

Jørgensen, C., Enberg, K., Dunlop, E.S., Arlinghaus, R., Boukal, D.S., Brander, K. Ernande, B., Gårdmark, A., Johnston, F., Matsumura, S., Pardoe, H., Raab, K., Silva, A., Vainikka, A., Dieckmann, U., Heino, M., and Rijnsdorp, A.D. (2007). Ecology: managing evolving fish stocks. *Science* 318: 1247–1248.

Juliano, S.A., Olson, J.R., Murrell, E.G., and Hatle, J.D. (2004). Plasticity and canalization of insect reproduction: testing alternative models of life history transitions. *Ecology* 85: 2986–2996.

Juliano, S.A., and Stoffregen, T.L. (1994). Effects of habitat drying on size at and time to metamorphosis in the tree hole mosquito *Aedes triseriatus*. *Oecologia* 97: 369–376.

Kause, A., Saloniemi, I., Haukioja, E., and Hanhimäki, S. (1999). How to become large quickly: quantitative genetics of growth and foraging in a flush feeding lepidopteran larva. *Journal of Evolutionary Biology* 12: 471–482.

Kawecki, T.J. (1993). Age and size at maturity in a patchy environment: fitness maximization versus evolutionary stability. *Oikos* 66: 309–317.

Kawecki, T.J., and Stearns, S.C. (1993). The evolution of life histories in spatially heterogeneous environments: optimal reaction norms revisited. *Evolutionary Ecology* 7: 155–174.

Kemp, D.J., and Wiklund, C. (2001). Fighting without weaponry: a review of male-male contest competition in butterflies. *Behavioral Ecology and Sociobiology* 49: 429–442.

Kendall, N.W., Hard, J.J., and Quinn, T.P. (2009). Quantifying six decades of fishery selection for size and age at maturity in sockeye salmon. *Evolutionary Applications* 2: 523–536.

Kingsolver, J.G., and Pfennig, D.W. (2004). Individual-level selection as a cause of Cope's rule of phyletic size increase. *Evolution* 58: 1608–1612.

Kivelä, S.M., Davis, R.B., Esperk, T., Gotthard, K., Mutanen, M., Valdma, D., and Tammaru, T. (2020). Comparative analysis of larval growth in Lepidoptera reveals instar-level constraints. *Functional Ecology* 34: 1391–1403.

Kivelä, S.M., Friberg, M., Wiklund, C., Leimar, O., and Gotthard, K. (2016). Towards a mechanistic understanding of insect life history evolution: oxygen-dependent induction of moulting explains moulting sizes. *Biological Journal of the Linnean Society* 117: 586–600.

Kojima, W., Nakakura, T., Fukuda, A., Lin, C.P., Harada, M., Hashimoto, Y., Kawachi, A., Suhama, S., and Yamamoto, R. (2020). Latitudinal cline of larval growth rate and its proximate mechanisms in a rhinoceros beetle. *Functional Ecology* 34: 1577–1588.

Kozlowski, J. (1991). Optimal energy allocation models: an alternative to the concepts of reproductive effort and cost of reproduction. *Acta Oecologica* 12: 11–33.

Kozlowski, J., and Wiegert, R.G. (1987). Optimal age and size at maturity in annuals and perennials with determinate growth. *Evolutionary Ecology* 1: 231–244.

Kuparinen, A., O'Hara, R., and Merilä, J. (2008). Probabilistic models for continuous ontogenetic transition processes. *PLoS ONE* 3(11): e3677.

Lande, R. (1982). Elements of a quantitative genetic model of life history evolution. In: *Evolution and genetics of life histories* (eds. Dingle, H., and Hegman, J.P.), 21–29. Springer.

Laugen, A.T., Laurila, A., Räsänen, K., and Merilä, J. (2003). Latitudinal countergradient variation in the common frog (*Rana temporaria*) development rates: evidence for local adaptation. *Journal of Evolutionary Biology* 16: 996–1005.

Leduc, S., Rosenberg, T., Johnson, A.D., and Segoli, M. (2022). Nest provisioning with parasitized caterpillars by female potter wasps: costs and potential mechanisms. *Animal Behaviour* 188: 99–109.

Lee, Z.A., Baranowski, A.K., and Preisser, E.L. (2021). Auditory predator cues affect monarch (*Danaus plexippus*; Lepidoptera: Nymphalidae) development time and pupal weight. *Acta Oecologica* 111: 103740.

Legault, G., and Kingsolver, J.G. (2020). A stochastic model for predicting age and mass at maturity of insects. *The American Naturalist* 196: 227–240.

Lim, J.N., Senior, A.M., and Nakagawa, S. (2014). Heterogeneity in individual quality and reproductive trade-offs within species. *Evolution* 68: 2306–2318.

Lima, S.L., and Dill, L.M. (1990). Behavioral decisions made under the risk of predation: a review and prospectus. *Canadian Journal of Zoology* 68: 619–640.

Lindenfors, P., Gittleman, J.L., and Jones, K.E. (2008). Sexual size dimorphism in mammals. In: *Sex, size and gender roles: evolutionary studies of sexual size dimorphism* (eds. Fairbairn, D., Szekely, T., and Blanckenhorn, W.U.), 16–26. Oxford University Press.

Lomolino, M.V. (1985). Body size of mammals on islands: the island rule reexamined. *The American Naturalist* 125: 310–316.

Magrath, M.J.L., Santema, P., Bouwman, K.M., Brinkhuizen, D.M., Griffith, S.C., and Langmore, N.E. (2009). Seasonal decline in reproductive performance varies with colony size in the fairy martin, *Petrochelidon ariel*. *Behavioral Ecology and Sociobiology* 63: 661–672.

Marty, L., Dieckmann, U., Rochet, M.-J., and Ernande, B. (2011). Impact of environmental covariation in growth and mortality on evolving maturation reaction norms. *The American Naturalist* 177: E98–E118.

Maynard-Smith, J., and Brown, R.L. (1986). Competition and body size. *Theoretical Population Biology* 30: 166–179.

Meiri, S., Cooper, N., and Purvis, A. (2008). The Island rule: made to be broken? *Proceedings of the Royal Society B: Biological Sciences* 275: 141–148.

Meister, H., Esperk, T., Välimäki, P., and Tammaru, T. (2017). Evaluating the role and measures of juvenile growth rate: latitudinal variation in insect life histories. *Oikos* 126: 1726–1737.

Millar, J.S., and Hickling G.J. (1991). Body size and the evolution of mammalian life histories. *Functional Ecology* 5: 588–593.

Mollet, F.M., Enberg, K., Boukal, D.S., Rijnsdorp, A.D., and Dieckmann, U. (2023). An evolutionary explanation of female-biased sexual size dimorphism in North Sea plaice, *Pleuronectes platessa* L. *Ecology and Evolution* 13: e8070.

Morris, M.R., Friebertshauser, R.J., Rios-Cardenas, O., Liotta, M.N., and Abbott, J.K. (2016). The potential for disruptive selection on growth rates across genetically influenced alternative reproductive tactics. *Evolutionary Ecology* 30: 519–533.

Mousseau, T.A., and Roff, D.A. (1987). Natural selection and the heritability of fitness components. *Heredity* 59: 181–197.

Moya-Laraño, J., Halaj, J., and Wise, DH (2002). Climbing to reach females: Romeo should be small. *Evolution* 56: 420–425.

Moyle, P.B., and Cech, J.J. Jr (2004). *An introduction to ichthyology*. Pearson Prentice Hall.

Nijhout, H.F., Riddiford, L.M., Mirth, C., Shingleton, A.W., Suzuki, Y., and Callier, V. (2014). The developmental control of size in insects. *Wiley Interdisciplinary Reviews: Developmental Biology* 3: 113–134.

Nilsson-Örtman, V., and Rowe, L. (2021). The evolution of developmental thresholds and reaction norms for age and size at maturity. *Proceedings of the National Academy of Sciences of the United States of America* 118: e2017185118.

Niu, J., Huss, M., Vasemägi, A., and Gårdmark, A. (2023). Decades of warming alters maturation and reproductive investment in fish. *Ecosphere* 14: e4381.

Nylin, S., Wickman, P.-O., and Wiklund, C. (1989). Seasonal plasticity in growth and development of the speckled wood butterfly, *Pararge aegeria* (Satyrinae). *Biological Journal of the Linnean Society* 38: 155–171.

Nylin, S., Wiklund, C., Wickman, P.-O., and Garcia-Barros, E. (1993). Absence of trade-offs between sexual size dimorphism and early male emergence in a butterfly. *Ecology* 74: 1414–1427.

Olsen, E.M., Heino, M., Lilly, G.R., Morgan, M.J., Brattey, J., Ernande, B., and Dieckmann, U. (2004). Maturation trends indicative of rapid evolution preceded the collapse of northern cod. *Nature* 428: 932–935.

Olson, V.A., Davies, R.G., Orme, C.D.L., Thomas, G.H., Meiri, S., Blackburn, T. M., Gaston, K.J., Owens, I.P.F., and Bennett, P.M. (2009). Global biogeography and ecology of body size in birds. *Ecology Letters* 12: 249–259.

Palkovacs, E.P., Marschner, M., Ciofi, C., Gerlach, J., and Caccone, A. (2003). Are the native giant tortoises from the Seychelles really extinct? A genetic perspective based on mtDNA and microsatellite data. *Molecular Ecology* 12: 1403–1413.

Partridge, L., and Fowler, K. (1993). Responses and correlated responses to artificial selection on thorax length in *Drosophila melanogaster*. *Evolution* 47: 213–226.

Partridge, L., Langelan, R., Fowler, K., Zwaan, B., and French, V. (1999). Correlated responses to selection on body size in *Drosophila melanogaster*. *Genetics Research* 74: 43–54.

Peckarsky, B.L., Taylor, B.W., McIntosh, A.R., McPeek, M.A., and Lytle, D.A. (2001). Variation in mayfly size at metamorphosis as a developmental response to risk of predation. *Ecology* 82: 740–757.

Perrin, N., and Rubin, J.F. (1990). On dome-shaped norms of reaction for size-to-age at maturity in fishes. *Functional Ecology* 4: 53–57.

Phelan, P., and Baker, T.C. (1990). Comparative study of courtship in twelve phycitine moths (Lepidoptera: Pyralidae). *Journal of Insect Behavior* 3: 303e326.

Pigliucci, M. (2003). Phenotypic integration: studying the ecology and evolution of complex phenotypes. *Ecology Letters* 6: 265–272.

Pincheira-Donoso, D., and Hunt, J. (2017). Fecundity selection theory: concepts and evidence. *Biological Reviews* 92: 341–356.

Plaistow, S.J., Lapsley, C.T., Beckerman, A.P., and Benton, T.G. (2004). Age and size at maturity: sex, environmental variability and developmental thresholds. *Proceedings of the Royal Society B: Biological Sciences* 271: 919–924.

Plaistow, S.J., Shirley, C., Collin, H., Cornell, S.J., and Harney, E.D. (2015). Offspring provisioning explains clone-specific maternal age effects on life history and life span in the water flea, *Daphnia pulex*. *The American Naturalist* 186: 376–389.

Price, P.W., Denno, R.F., Eubanks, M.D., Finke, D.L., and Kaplan, I. (2011). *Insect ecology: behavior, populations and communities*. Cambridge University Press.

Reim, C., Teuschl, Y., and Blanckenhorn, W.U. (2006). Size-dependent effects of temperature and food stress on energy reserves and starvation resistance in yellow dung flies. *Evolutionary Ecology Research* 8: 1215–1234.

Remmel, T., Davison, J., and Tammaru, T. (2011). Quantifying predation on folivorous insect larvae: the perspective of life-history evolution. *Biological Journal of the Linnean Society* 104: 1–18.

Remmel, T., and Tammaru, T. (2009). Size-dependent predation risk in tree-feeding insects with different colouration strategies: a field experiment. *Journal of Animal Ecology* 78: 973–980.

Reznick, D.N. (1990). Plasticity in age and size at maturity in male guppies (*Poecilia reticulata*): an experimental evaluation of alternative models of development. *Journal of Evolutionary Biology* 3: 185–203.

Reznick, D.N., Butler IV, M.J., Rodd, F.H., and Ross, P. (1996). Life-history evolution in guppies (*Poecilia reticulata*). 6. Differential mortality as a mechanism for natural selection. *Evolution* 50: 1651–1660.

Reznick, D.N., Shaw, F.H., Rodd, F.H., and Shaw, R.G. (1997). Evaluation of the rate of evolution in natural populations of guppies (*Poecilia reticulata*). *Science* 275: 1934–1937.

Richter-Boix, A., Tejedo, M., and Rezende, E.L. (2011). Evolution and plasticity of anuran larval development in response to desiccation. A comparative analysis. *Ecology and Evolution* 1: 15–25.

Ricklefs, R.E., and Wikelski, M. (2002). The physiology/life-history nexus. *Trends in Ecology & Evolution* 17: 462–468.

Riemer, K., Guralnick, R.P., and White, E.P. (2018). No general relationship between mass and temperature in endothermic species. *eLife* 7: e27166.

Roff, D.A. (1981). On being the right size. *The American Naturalist* 118: 405–422.

Roff, D.A. (1986). Predicting body size with life-history models. *BioScience* 36: 316–323.

Roff, D.A. (1992). *The evolution of life histories*. Theory and analysis. Chapman and Hall.

Roff, D.A. (2000). Trade-offs between growth and reproduction: analysis of the quantitative genetic evidence. *Journal of Evolutionary Biology* 13: 434–445.

Roff, D.A. (2001). Age and size at maturity. In: *Evolutionary ecology: concepts and case studies* (eds. Fox, C.W., Roff, D.A., and Fairbairn, D.J.), 99–112. Oxford Academic.

Roff, D.A. (2002). *Life history evolution*. Sinauer Associates.

Rolland J., Henao-Diaz, L.F., Doebeli, M., Germain, R., Harmon, L.J., Knowles, L.L., Liow, L.H., Mank, J.E., Machac, A., Otto, S.P., Pennell, M., Salamin, N., Silvestro, D., Sugawara, M., Uyeda, J., Wagner, C.E., and Schluter, D. (2023). Conceptual and empirical bridges between micro- and macroevolution. *Nature Ecology & Evolution* 7: 1181–1193.

Rollinson, N., and Rowe, L. (2015). Persistent directional selection on body size and a resolution to the paradox of stasis. *Evolution* 69: 2441–2451.

Rowe, L., and Ludwig, D. (1991). Size and timing of metamorphosis in complex life histories: time constraints and variation. *Ecology* 72: 413–427.

Roy, S., Brännström, Å., and Dieckmann, U. (2024). Ecological determinants of Cope's rule and its inverse. *Communications Biology* 7: 38.

Schlichting, C.D., and Pigliucci, M. (1998). *Phenotypic evolution: a reaction norm perspective*. Sinauer Associates.

Schwenk, K., and Wagner, G.P. (2004). The relativism of constraints on phenotypic evolution. In: *Phenotypic integration: studying the ecology and evolution of complex phenotypes* (eds. Pigliucci, M., and Preston, K.), 390–408. Oxford University Press.

Sehnal, F. (1985). Growth and life cycles. In: *Comprehensive insect physiology, biochemistry and pharmacology*, Vol. 2: Postembryonic development (eds. Kerkut, G.A., and Gilbert, L.I.), 1–86. Pergamon Press.

Seifert, C.L., Strutzenberger, P., and Fiedler, K. (2023). How do host plant use and seasonal life cycle relate to insect body size: a case study on European geometrid moths (Lepidoptera: Geometridae). *Journal of Evolutionary Biology* 36: 743–752.

Shaanker, R.U., Ganeshaiah, K.N., and Bawa, K.S. (1988). Parent-offspring conflict, sibling rivalry, and brood size patterns in plants. *Annual Review of Ecology and Systematics* 19: 177–205.

Shine, R. (1989). Ecological causes for the evolution of sexual size dimorphism: a review of the evidence. *Quarterly Review of Biology* 64: 419–461.

Shingleton, A.W. (2011). Evolution and the regulation of growth and body size. In: *Mechanisms of life-history evolution: the genetics and physiology of life history traits and trade-off* (eds. Flatt, T., and Heyland, A.), 43–55. Oxford University Press.

Shingleton, A.W., and Vea, I.M. (2023). Sex-specific regulation of development, growth and metabolism. *Seminars in Cell & Developmental Biology* 138: 117–127.

Sõber, V., Sandre, S.-L., Esperk, T., Teder, T., and Tammaru, T. (2019). Ontogeny of sexual size dimorphism revisited: females grow for a longer time and also faster. *PLoS ONE* 14(4): e0215317.

Sogard, S. (1997). Size-selective mortality in the juvenile stage of teleost fishes: a review. *Bulletin of Marine Science* 60: 1129–1157.

Speight, M.R., Hunter, M.D., and Watt, A.D. (2008). *Ecology of insects: concepts and applications*. Wiley-Blackwell.

Stearns, S.C. (1983). A natural experiment in life-history evolution: field data on the evolution of mosquitofish, *Gambusia affinis*, to Hawaii. *Evolution* 37: 601–617.

Stearns, S.C. (1992). *The evolution of life-histories*. Oxford University Press.

Stearns, S.C., Ackermann, M., Doebeli, M., and Kaiser, M. (2000). Experimental evolution of aging, growth, and reproduction in fruitflies. *Proceedings of the National Academy of Sciences of the United States of America* 97: 3309–3313.

Stearns, S.C., and Crandall, R.E. (1981). Quantitative predictions of delayed maturity. *Evolution* 35: 455–463.

Stearns, S.C., and Koella, J.C. (1986). The evolution of phenotypic plasticity in life-history traits: predictions of reaction norms for age and size at maturity. *Evolution* 40: 893–913.

Stoks, R., Swillen, I., and De Block, M. (2012). Behaviour and physiology shape the growth accelerations associated with predation risk, high temperatures and southern latitudes in *Ischnura* damselfly larvae. *Journal of Animal Ecology* 81: 1034–1040.

Svensson, M. (1996). Sexual selection in moths: the role of chemical communication. *Biological Reviews* 71: 113–135.

Swain, D.P., Sinclair, A.F., and Hanson, J.M. (2007). Evolutionary response to size-selective mortality in an exploited fish population. *Proceedings of the Royal Society B: Biological Sciences* 274: 1015–1022.

Tammaru, T. (1998). Determination of adult size in a folivorous moth: constraints at instar level? *Ecological Entomology* 23: 80–89.

Tammaru, T., and Esperk, T. (2007). Growth allometry of immature insects: larvae do not grow exponentially. *Functional Ecology* 21: 1099–1105.

Tammaru, T., Esperk, T., and Castellanos, I. (2002). No evidence for costs of being large in females of *Orgyia* spp. (Lepidoptera, Lymantriidae): larger is always better. *Oecologia* 133: 430–438.

Tammaru, T., Esperk, T., Ivanov, V., and Teder, T. (2010). Proximate sources of sexual size dimorphism in insects: locating constraints on larval growth schedules. *Evolutionary Ecology* 24: 161–175.

Tammaru, T., and Haukioja, E. (1996). Capital breeders and income breeders among Lepidoptera – consequences to population dynamics. *Oikos* 77: 561–564.

Tammaru, T., Kaitaniemi, P., and Ruohomäki, K. (1996b). Realized fecundity in *Epirrita autumnata* (Lepidoptera: Geometridae): relation to body size and consequences to population dynamics. *Oikos* 77: 407–416.

Tammaru, T., Ruohomäki, K., and Montola, M. (2000). Crowding-induced plasticity in *Epirrita autumnata* (Lepidoptera: Geometridae): weak evidence of specific modifications in reaction norms. *Oikos* 90: 171–181.

Tammaru, T., Ruohomäki, K., and Saikkonen, K. (1996a). Components of male fitness in relation to body size in *Epirrita autumnata* (Lepidoptera, Geometridae). *Ecological Entomology* 21: 185–192.

Tammaru, T., Vellau, H., Esperk, T., and Teder, T. (2015). Searching for constraints by cross-species comparison: reaction norms for age and size at maturity in insects. *Biological Journal of the Linnean Society* 114: 296–307.

Tauber, M.J., Tauber, C.A., and Masaki, S. (1986). *Seasonal adaptations of insects*. Oxford University Press.

Teder, T. (2014). Sexual size dimorphism requires a corresponding sex difference in development time: a meta-analysis in insects. *Functional Ecology* 28: 479–486.

Teder, T. (2020). Phenological responses to climate warming in temperate moths and butterflies: species traits predict future changes in voltinism. *Oikos* 129: 1051–1060.

Teder, T., Esperk, T., Remmel, T., Sang, A., and Tammaru, T. (2010). Counterintuitive size patterns in bivoltine moths: late-season larvae grow larger despite lower food quality. *Oecologia* 162: 117–125.

Teder, T., and Kaasik, A. (2023). Early-life food stress hits females harder than males in insects: a meta-analysis of sex differences in environmental sensitivity. *Ecology Letters* 26: 1419–1431.

Teder, T., Kaasik, A., Taits, K., and Tammaru, T. (2021). Why do males emerge before females? Sexual size dimorphism drives sexual bimaturism in insects. *Biological Reviews* 96: 2461–2475.

Teder, T., and Tammaru, T. (2005). Sexual size dimorphism within species increases with body size in insects. *Oikos* 108: 321–334.

Teder, T., Tammaru, T., and Esperk, T. (2008). Dependence of phenotypic variance in body size on environmental quality. *The American Naturalist* 172: 223–232.

Teder, T., Vellau, H., and Tammaru, T. (2014). Age and size at maturity: a quantitative review of diet-induced reaction norms in insects. *Evolution* 68: 3217–3228.

Teuschl, Y., Reim, C., and Blanckenhorn, W.U. (2007). Correlated responses to artificial body size selection in growth, development, phenotypic plasticity and juvenile viability in yellow dung flies. *Journal of Evolutionary Biology* 20: 87–103.

Tiitsaar, A., Kaasik, A., and Teder, T. (2013). The effects of seasonally variable dragonfly predation on butterfly assemblages. *Ecology* 94: 200–207.

Ude, K., Õunap, E., Kaasik, A., Davis, R.B., Javoiš, J., Foerster, S.I.A., and Tammaru, T. (2024) Evolution of wing shape in geometrid moths: phylogenetic effects dominate over ecology. *Journal of Evolutionary Biology* 37: 526–537.

Uyeda, J.C., Hansen T.F., Arnold S.J., and Pienaar J. (2011). The million year wait for macroevolutionary bursts. *Proceedings of the National Academy of Sciences of the United States of America* 108: 15908–15913.

Valge, M., Meitern, R., and Hõrak, P. (2022). Sexually antagonistic selection on educational attainment and body size in Estonian children. *Annals of the New York Academy of Sciences* 1516: 271–285.

Välimäki, P., Kivelä, S.M., and Mäenpää, M.I. (2013). Temperature- and density-dependence of diapause induction and its life history correlates in the geometrid moth *Chiasmia clathrata* (Lepidoptera: Geometridae). *Evolutionary Ecology* 27: 1217–1233.

van Asch, M., and Visser, M.E. (2007). Phenology of forest caterpillars and their host trees: the importance of synchrony. *Annual Review of Entomology* 52: 37–55.

van Dongen, S., Matthysen, E., Sprengers, E., and Dhondt, A.A. (1998). Mate selection by male winter moths *Operophtera brumata* (Lepidoptera, Geometridae): adaptive male choice or female control? *Behaviour* 135: 29–42.

van Dooren, T.J.M., Tully, T., and Ferrière, R. (2005). The analysis of reaction norms for age and size at MATURITY using maturation rate models. *Evolution* 59: 500–506.

Vellau, H., Leppik, E., Frerot, B., and Tammaru, T. (2013). Detecting a difference in reaction norms for size and time at maturation: pheromone strains of the European corn borer (*Ostrinia nubilalis*: Lepidoptera, Crambidae). *Evolutionary Ecology Research* 15: 589–599.

Vellau, H., and Tammaru, T. (2012). Larval crowding leads to unusual reaction norms for size and time at maturity in a geometrid moth (Lepidoptera: Geometridae). *European Journal of Entomology* 109: 181–186.

Verberk, W.C.E.P., Atkinson, D., Hoefnagel, K.N., Hirst, A.G., Horne, C.R., and Siepel, H. (2021). Shrinking body sizes in response to warming: explanations for the temperature-size rule with special emphasis on the role of oxygen. *Biological Reviews* 96: 247–268.

Wagner, G.P., and Altenberg, L. (1996). Complex adaptations and the evolution of evolvability. *Evolution* 50: 967–976.

Walczyńska, A., Franch-Gras, L., and Serra, M. (2017). Empirical evidence for fast temperature-dependent body size evolution in rotifers. *Hydrobiologia* 796: 191–200.

Weiner, J., Campbell, L. G., Pino, J., and Echarte, L. (2009). The allometry of reproduction within plant populations. *Journal of Ecology* 97: 1220–1233.

Wiklund, C., and Fagerström, T. (1977). Why do males emerge before females? A hypothesis to explain the incidence of protandry in butterflies. *Oecologia* 31: 153–158.

Wiklund, C., and Forsberg, J. (1991). Sexual size dimorphism in relation to female polygamy and protandry in butterflies: a comparative study of Swedish Pieridae and Satyridae. *Oikos* 60: 373–381.

Wiklund, C., and Kaitala, A. (1995). Sexual selection for large male size in a polyandrous butterfly: the effect of body size on male versus female reproductive success in *Pieris napi*. *Behavioral Ecology* 6: 6–13.

Wild, V., Simianer, H., Gjøen, H.-M., and Gjerde, B. (1994). Genetic parameters and genotype-environment interaction for early sexual maturity in Atlantic salmon (*Salmo salar*). *Aquaculture* 128: 51–65.

Yearsley, J.M., Kyriazakis, I., and Gordon, I.J. (2004). Delayed costs of growth and compensatory growth rates. *Functional Ecology* 18: 563–570.

2

Evolution of Ageing and Lifespan

Alexei A. Maklakov

School of Biological Sciences, University of East Anglia, Norwich, UK

> *It is indeed remarkable that after a seemingly miraculous feat of morphogenesis, a complex metazoan should be unable to perform the much simpler task of merely maintaining what is already formed.*
>
> (George Williams 1957)
>
> *…most species suffer some cancer of old age, whether old occurs at 80 weeks or 80 years…*
>
> (Peto et al. 1975)
>
> *When 900 years old you reach, look as good you will not.*
>
> (Yoda, Return of the Jedi 1983)

2.1 Introduction

Striped field mice (*Rhabdomys pumilio*) usually live no longer than two months in nature and can survive for a maximum of four years in captivity, yet a different mouse-sized rodent of southern Africa, a naked mole-rat (*Heterocephalus glaberrodents*), can live for 17 years in the wild, and over 28 years in captivity (Buffenstein 2008). Size is of importance here because lifespan often scales with body mass. Nevertheless, accounting for size makes naked mole-rats look even more impressive – their lifespan is five times what is expected for a mammal of their size.

This anecdotal example illustrates several principles that are supported by diverse empirical studies of ageing in animal populations in nature and in the laboratory. First, there is large diversity of lifespans and ageing rates across species and populations (Jones et al. 2014), and evolutionary biology of ageing seeks to explain the origin and maintenance of this variation.

Second, lifespan in natural populations is a combination of environmental (often called extrinsic) factors, such as predators, pathogens, intraspecific competition and accidents, and intrinsic physiological deterioration with advancing age, commonly referred to as ageing or senescence. In laboratories, or other forms of captivity, animals often live much longer in the absence of natural hazards (Tidière et al. 2016), and the effect of ageing on organismal performance can become more pronounced. Nevertheless, intrinsic ageing can interact with environmentally driven mortality to determine lifespan, for example, by making the organisms more susceptible to predation, infectious disease, or abiotic stress.

Third, there is substantial difference in lifespan and ageing among individuals within populations that is caused in part by environmental effects, including gene-by-environment interactions. For example, non-breeding naked mole-rats live only around four years in the wild, four times less than their breeding counterparts. This is likely driven by the costs of venturing outside the protective environment of their colony increasing the chance of predation and accidents (Buffenstein 2008). However, the difference between breeders and non-breeders essentially disappears in captivity, where non-breeders can survive for much longer (see earlier). This suggests that lifespan is shaped by the interaction between social status and environment in this species.

Why do organisms age? Ageing, or senescence – a decline in performance with age, which leads to reduced reproduction and increased probability of death – reduces Darwinian fitness, and one could anticipate that natural selection will act against the evolution of ageing. It is sometimes suggested that ageing is inevitable because of gradual accumulation of

Life History Evolution: Traits, Interactions, and Applications, First Edition. Edited by Michal Segoli and Eric Wajnberg.
© 2025 John Wiley & Sons Ltd. Published 2025 by John Wiley & Sons Ltd.

genetic damage in cells. However, living organisms possess means of self-maintenance and regeneration, damaged cells can be repaired or replaced, and some organisms may not age at all (Jones et al. 2014). Perhaps the sheer diversity of lifespans and ageing rates across and within species already provides an indication that ageing is malleable and subject to the force of natural selection. Early evolutionary thinking by Weismann (1889) focused on naïve group-selection approaches where ageing evolves for the benefit of the species, although this author seems to abandon his group-selectionist stance in later years (see Rose 1991).

A different way to deal with the problem of senescence, and its negative effects on Darwinian fitness, was to suggest that ageing does not occur in nature. Indeed, one of the main co-founders of the evolutionary theory of ageing (ETA), Sir Peter Medawar, suggested that: 'Whether animals can, or cannot, reveal an innate deterioration is almost literally a domestic problem; the fact is that under the exactions of natural life they do not do so. They simply do not live that long' (Medawar 1952).

This observation is incorrect because ageing is prevalent in natural populations and does have the predicted detrimental effect on fitness, the extent of which varies across species (Nussey et al. 2013). Medawar (1952)'s mistake can be attributed to lack of good data on ageing in wild animals, which only began to accumulate in the 1970s. Nevertheless, the effect of this historical observation on human mind is so strong that references to 'non-ageing' populations in nature appear in biogerontological literature until this date and coexist with the large volume of work documenting senescence in vital rates in a broad array of species across taxa (Nussey et al. 2013).

It is interesting, from the perspective of the history of science, that the largely erroneous observation that ageing almost does not occur in nature led to the development of the population genetic theory of ageing that did withstand the test of time. The devil is, as always, in the details: mutations that cause ageing can accumulate, or be selected for, in a population not only when selection is completely absent as implied by the naïve 'no-ageing-in-the-wild' idea, but when it weakens with age.

2.2 Evolutionary Theory of Ageing

The first glimpses of the contemporary ETA can be seen in the works of Fisher (1930) and Haldane (1941). Fisher (1930) developed the concepts of intrinsic rate of increase of the population (r, Fisher called it the Malthusian parameter), which corresponds to the average individual fitness in a population, and individual reproductive value (Fisher 1930), which measures contribution of specific age classes to future generations of a population growing at rate r. Fisher (1930) also hinted that effectiveness of selection at different ages can mould what he referred to as 'natural death' (Charlesworth 2000). Haldane (1941) pointed out that Huntington's disease, caused by a deleterious dominant allele with very high penetrance, can be present at a relatively high frequency in a population because of its late age of onset (35.5 years), and that in the past '… men and women seldom lived much beyond forty, so postponement of onset beyond this age had no selective advantage'. These ideas were developed by Medawar (1952) who formulated the first explicit theory of the evolution of ageing based on population genetics principles, and further by Williams (1957), Hamilton (1966) and Charlesworth (1980).

The ETA maintains that ageing evolves because the force of natural selection declines with increasing age. More specifically, selection against mortality declines with age because young individuals have more reproductive potential than old individuals (Hamilton 1966, Moorad et al. 2019) (Figure 2.1). Therefore, there is a greater fitness cost to early mortality. On the other hand, selection for age-specific fecundity declines with age because the fraction of individuals belonging to a specific age decreases with increasing age. The force of age-specific selection is maximized and constant throughout the development and prior to reproduction but declines over time and converges with zero at the last age of reproduction (Hamilton 1966, Rose 1991, Moorad 2014, Moorad et al. 2019) (Figure 2.1). This decline in the force of selection has been aptly termed a 'selection shadow' (Fabian and Flatt 2011, Flatt and Partridge 2018), the part of organismal life history where selection is totally or partially 'blind', and this 'blindness' has profound implications for the evolution of age-specific life histories.

The diminished force of selection with age can result in accumulation of deleterious alleles whose effects on performance are concentrated in late-life (*mutation accumulation* (MA); Medawar 1952) and in selection for alleles whose effects are beneficial in early-life but detrimental in late-life (*antagonistic pleiotropy* [AP]; Williams 1957). These population genetic forces shape the evolution of age-specific physiology that results in senescence of cells, tissues, and organisms. Evolutionary physiological theories of ageing, such as the 'disposable soma' theory (DST) of ageing, consider the effect of natural selection on organismal physiology and explicitly link proximate mechanisms to age-specific life histories and vital rates. The DST is based on the idea of resource allocation between key life history traits and directly integrates the notion of diminished force of selection with age with a specific proximate mechanism – resource allocation – that underlies the evolution of ageing

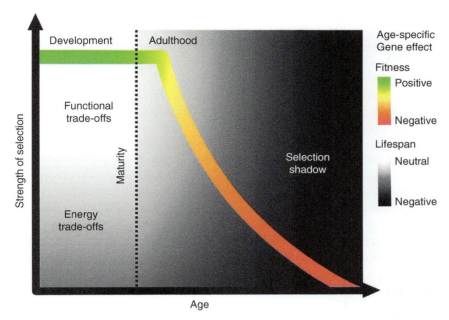

Figure 2.1 The strength of age-specific selection is maximized during early life – specifically, during development and early adulthood until the age of first reproduction – but declines after the start of reproduction and reaches zero at the age of last reproduction (Hamilton 1966). This figure also illustrates some possible age-specific effects of alleles on fitness and lifespan across the life course. For example, an antagonistically pleiotropic (AP) allele can have positive effect on fitness during development and early adulthood when selection if maximal but can become progressively deleterious with increasing age. This can happen because the physiological requirements of an adult organism differ from those of a developing organism or a young adult. This can also happen because the detrimental effects take time to accumulate and only become apparent at later ages. However, the effects of such an allele on lifespan can vary with age depending on whether the trade-off between lifespan (somatic maintenance) and other fitness-related traits (*e.g.,* development, growth, and reproduction) is based on energy allocation or function. The negative allelic effect on lifespan can result from competitive energy allocation between development, growth, and reproduction on the one hand, and somatic maintenance on the other hand, resulting in energy trade-offs as suggested by the 'disposable soma' theory (DST). However, functional trade-offs resulting from suboptimal regulation of gene expression in late life can lead to suboptimal physiological function with increasing age, as proposed by the developmental theory of ageing (DTA, also known as 'early-life inertia'). Under functional trade-offs, optimizing gene expression in adulthood improves both fitness and lifespan, without early-life costs (*Source:* Maklakov and Chapman 2019 with permission).

(Kirkwood 1977, Kirkwood and Holliday 1979). More recently, a different evolutionary physiological theory, the developmental theory of ageing (DTA), has been developed and is increasingly tested in empirical studies (de Magalhaes and Church 2005, Carlsson et al. 2021). This proximate theory focuses on suboptimal regulation of age-specific physiology due to diminishing selection with age. Historically, there has been a lot of confusion between the ultimate population genetic theories (MA and AP) and the proximate physiological theory DST, and it is quite common to see in the literature that there are three main evolutionary theories of ageing. However, as has been noted repeatedly by many authors, DST is a proximate physiological theory that can only evolve via alleles with classical antagonistically pleiotropic effects. For example, an allele that increases investment into immediate reproduction at the cost of cellular maintenance leading to increased probability of death in later ages is an antagonistically pleiotropic allele acting via resource allocation mechanism, as envisaged by the DST. While it is possible to consider AP alleles that do not act via resource allocation mechanism, it is impossible to consider the evolution of ageing via resource allocation without alleles that have clear antagonistically pleiotropic properties. Thus, DST is a special case of AP. Here I will follow the recently proposed hierarchical model of the ETA (Lemaitre et al. 2024), focusing first on the higher-level ultimate population genetic theories of ageing (MA and AP), before considering the proximate physiological models (DST and DTA).

2.2.1 Mutation Accumulation and Antagonistic Pleiotropy

Medawar (1952) famously referred to ageing as 'an unsolved problem in biology' in his inaugural lecture in 1951 and then proceeded to provide us with the foundation of the ETA, which stands until this date. Medawar (1952) considered a hypothetical population of theoretically immortal organisms with a constant rate of death due to extrinsic mortality hazards, such as abiotic stress, starvation, predation, competition, pathogens or accidents. His insight was that because individuals gradually disappear from the population due to extrinsic mortality, the contribution of each age-class to future generations decreases with increasing age. Medawar (1952)'s most well-acknowledged contribution was the realization that ageing can then evolve via accumulation of late-acting deleterious germline mutations. This process is commonly referred to today as 'mutation accumulation' (MA) and it has been progressively formalized and developed further (Charlesworth 1980, 2000, 2001, Wachter et al. 2013, 2014).

The second route for the evolution of ageing is based on the observation that alleles that increase early-life performance at the cost of late-life performance may be beneficial and can be selected for and spread to fixation in a population. Such genes are called antagonistically pleiotropic, and the theory is generally known as 'antagonistic pleiotropy' (AP, Williams 1957). Interestingly, Medawar (1946) mentioned this explicitly in 1946 by saying: 'It is by no means difficult to imagine a genetic endowment which can favour young animals only at the expense of their elders; or rather, at their own expense when they grow old'. It is the second part of this sentence that incontrovertibly points to the genes with antagonistically pleiotropic effects that can contribute to the evolution of ageing. Later, Medawar (1952) was even more explicit by noting that '... even small advantage conferred early in the life of the individual may outweigh a catastrophic disadvantage withheld until later'.

The AP was later championed by Williams (1957) who is generally credited with developing it into a fully-fledged theory. Later mathematical analysis by Charlesworth (1980) and Rose (1985) provided formal integration of AP with population genetic theory and showed that while some AP alleles are predicted to go to fixation, AP is also capable of maintaining genetic polymorphism (see Rose 1991).

There is a considerable body of literature dedicated to testing the roles of MA and AP in the evolution of ageing. Early studies often used quantitative genetic approach to characterize the contribution of MA and AP alleles to age-specific genetic variance. Specifically, studies tested the prediction that dominance variance, as well as homozygote variance and, consequently, inbreeding depression should increase with age under MA but not under AP (Hughes and Charlesworth 1994, Charlesworth and Hughes 1996). This is because deleterious alleles with late-life effects on fitness are expected to accumulate at higher equilibrium frequencies. Indeed, inbreeding depression increased with age in *Drosophila melanogaster* fruit flies, which was interpreted as general support for MA-type alleles in the evolution of ageing. However, later re-appraisal of these ideas found that evolution of ageing under AP can also result in increase in dominance variance and inbreeding depression with age (Moorad and Promislow 2009). Therefore, while quantitative genetic studies of age-specific genetic variance components and inbreeding depression provide invaluable support for the ETA and help us to understand the roles of genetic and environmental factors in ageing (Moorad and Promislow 2009), they cannot help to quantify the relative importance of MA and AP in the evolution of senescence.

There are more direct ways of looking for the contribution of MA alleles to rates of ageing. For example, recent analyses of data from UK Biobank suggest that single nucleotide polymorphisms (SNPs) associated with late-onset diseases occur at higher frequencies than SNPs associated with earlier-onset diseases (Donertas et al. 2021), in line with the prediction that selection on germline mutations with late-acting effects is reduced. In a different study using mammalian transcriptomes across five different species, genes with relatively high levels of expression in late adulthood had lower levels of conservation, thus harbouring more deleterious alleles compared to genes expressed in young adulthood (Turan et al. 2019). These findings are in line with MA-derived prediction that deleterious genes whose effects on performance are concentrated in late life are under weaker natural selection.

Initial models of MA assumed mutations whose effects were confined to a particular age, or perhaps span a range of adjacent age classes (Charlesworth 2001). Under such scenario, there would be no correlation between mutational effects on early-life performance (when the effects are presumed to be neutral) and late-life performance (where the effects are deleterious). This assumption, however, could result in the evolution of the so-called 'walls of death' with infinite mortality rates (Wachter et al. 2013, 2014) and lead to senescence at ever younger ages (Aubier and Galipaud 2023) because fixation of late-acting deleterious mutations promotes fixation of other deleterious mutations. This contradicts the empirical findings suggesting that 'walls of death' are uncommon, and mutations are commonly 'positively pleiotropic', *i.e.*, they affect all age classes in the same direction resulting in positive pleiotropy across ages. One potential solution to both aforementioned problems lies in the assumption that the effect of deleterious mutations on performance can become progressively worse with age (Wachter et al. 2013, 2014, Maklakov et al. 2015). Indeed, recent work in *D. melanogaster* identified specific alleles whose deleterious effects on performance increased when flies became older (Brengdahl et al. 2020, 2023). Interestingly, a recent study looking into the evolution of human lifespan suggests positive genetic correlations between early-life fitness and late-life mortality, at least in males, which is in line with MA model of ageing evolution (Moorad and Walling 2017).

AP enjoyed much stronger focus in terms of rigorous and diverse empirical tests and there is perhaps more direct support for the role of AP type of alleles in the evolution of ageing. The first line of evidence comes from classical experimental evolution studies in the laboratory where animals are limited to acquiring fitness during set periods across their life course, commonly in early- and late adulthood, where 'early' and 'late' are defined arbitrarily according to the life cycle of the species. The first studies of this kind were performed using *D. melanogaster*, where selection for delayed reproduction resulted in flies that lived longer and, sometimes, reproduced better late in life at the cost of reduced reproduction in early life (Luckinbill et al. 1984, Rose 1984, Partridge et al. 1999). These results have been later confirmed in other species (reviewed in McHugh and Burke 2022) suggesting that selection for improved late-life performance commonly results in reduced early-life performance, and *vice versa*, in line with AP.

The data from genome-wide association studies (GWAS) in human datasets commonly provides support for both MA and AP alleles in ageing. For example, there is an excess of alleles with AP properties among genes linked to ageing and late-life onset of disease in humans (Rodriguez et al. 2017, Donertas et al. 2021).

Perhaps the most direct illustration of AP in ageing comes from the studies of long-lived mutants. Mutations in the *daf-2* gene encoding insulin/IGF-1 receptor in nutrient-sensing insulin/insulin-like signalling (IIS) pathway in *Caenorhabditis elegans* nematode worms can dramatically increase survival, with some alleles more than doubling median and maximum lifespan (Gems et al. 1998, Zhao et al. 2021). However, many of these alleles also have negative pleiotropic effects on other traits, such as development, movement, and fecundity, which would lead to reduced fitness in more natural conditions. Indeed, when at least some long-lived *daf-2* mutants compete against wild-type counterparts in laboratory evolution experiments, they quickly go to extinction especially when facing fluctuating food environments (Chen et al. 2007). These *daf-2* alleles, as well as variants of some other genes in the IIS pathway, such as *age-1*, provide us with excellent examples of AP in action. There are several studies of the so-called 'longevity mutants' in other species, including baker's yeast (*Saccharomyces cerevisiae*), fruit flies (*D. melanogaster*) and house mice (*Mus musculus*). Overall, competition experiments suggest that long-lived mutants tend to be outcompeted by wild-type counterparts, and the differences become more pronounced in challenging environments (Briga and Verhulst 2015), providing further experimental support for AP.

Specifically, *age-1*, a gene encoding a catalytic subunit of the PI3 kinase in the IIS pathway, is a superb illustration of the importance of gene-by-environment interactions when considering 'the cost of long life' in evolution of ageing. The first long-lived mutant ever discovered (Friedman and Johnson 1988), *C. elegans* bearing *age-1* (*hx546*) allele, was not only long-lived by also resistant to a variety of ecological stresses, such bacterial pathogens and temperature fluctuations. Why does this allele not replace a wild type? It turns out that these mutants are not doing very well under starvation, and they are quickly outcompeted by the wild-type worms in conditions that simulate fluctuating access to food (Walker et al. 2000). However, things quickly become more complicated when we start considering different environmental factors. Remember that *age-1* (*hx546*) mutants are resistant to various abiotic stresses, including heat stress. When starvation and heat stress were combined in a single experiment, the long-lived mutants outcompeted the wild-type animals (Savory et al. 2014).

In nature, animals face a wide range of changing environments, and it is key to account for this complexity if we are to understand how ageing evolves. This suggests that future tests of putative antagonistically pleiotropic alleles require more sophisticated experimental approaches. Ideally, we should aim to test the performance of specific alleles in natural populations. However, this approach may not always be feasible for a range of logistical, ethical and conservation concerns. The alternative approaches may include various scenarios of semi-natural conditions or laboratory microcosms that simulate natural environments but prevent the spread of experimentally introduced alleles in natural populations. Overall, there is broad support for both MA and AP types of alleles in the evolution of ageing using a variety of approaches including quantitative genetics, experimental laboratory evolution, cross-species phylogenetic comparisons, GWAS, age-specific trade-offs in natural populations and detailed studies of specific alleles with putative MA or AP properties (Rose 1991, Hughes and Reynolds 2005, Lemaitre et al. 2015, Rodriguez et al. 2017, Flatt and Partridge 2018, Donertas et al. 2021).

Moreover, our understanding of the evolution of ageing would be incomplete without linking age-specific allelic effects to the physiology of the organism. A holistic approach to the biology of ageing would allow us to establish how the decline in the force of natural selection with age shapes organismal physiology, and, ultimately, the evolution of vital rates, age-specific survival and reproduction. Such understanding will help in bridging the gap between the evolutionary ecology of ageing and applied biogerontology (also known as 'geroscience'). Currently, the study of the evolutionary origin and ecology of ageing in natural populations on one hand, and mechanistic and translational research aimed at understanding the molecular basis of senescence and improving healthy ageing in humans, on the other hand, often proceed in isolation. Both fields stand to gain considerably from closer integration. Linking natural selection with physiology can help distinguish between different evolutionary models of ageing, while considering how evolutionary forces shape life history trade-offs can help identify hidden costs of improved late-life performance. Evolutionary physiological models of senescence (see further in the chapter) provide the direct route for the integration between evolutionary and mechanistic approaches to biology of ageing.

2.2.2 'Disposable Soma' and the Developmental Theory of Ageing

While we have a good understanding of the theoretical population genetics of ageing (see earlier), the ETA is not complete without considering the age-specific effects of genes on age-specific physiology. One of the earliest models to consider the proximate mechanisms behind the evolution of ageing is the DST (Kirkwood 1977, Kirkwood and Holliday 1979). The DST, in its current form, refers to competitive allocation of limited resources between key life history traits, *i.e.*, growth, survival and reproduction (Kirkwood and Rose 1991, Kirkwood 2005). Somatic cells accumulate molecular damage with time, and this damage has to be identified and repaired. Somatic maintenance and repair under DST is understood broadly and includes very different processes such as intracellular DNA and protein repair and cellular renewal. This resource allocation model suggests that diversion of resources to costly somatic maintenance to improve survival will reduce resources available for reproduction, and *vice versa*. Because natural selection attempts to maximize fitness rather than longevity, allocation of resources to somatic maintenance will not allow perfect repair and renewal leading to accumulation of somatic damage with age. This proximate explanation is fully in line with AP theory. Consider an allele that causes more resources to be allocated to early reproduction at the expense of somatic maintenance leading to reduced reproduction and increased probability of death later in life. Such an AP allele can be beneficial in a population where early reproduction provides strong selective advantage – for example, in a rapidly growing population using an ephemeral resource – in which case this allele will be strongly favoured by selection and can rapidly go to fixation leading to the evolution of faster ageing in this population.

There is considerable correlative evidence in support of the DST theory, although there is paucity of studies directly demonstrating the DST mechanism in action. First, laboratory selection for improved late performance resulting in reduced early-life reproduction is often interpreted as evidence for resource allocation between current reproduction and future survival (Kirkwood and Rose 1991). However, while these results are indeed consistent with DST, they do not provide direct evidence in the absence of demonstrated allocation of resources, and other studies documented laboratory evolution of reproduction uncorrelated with the evolution of lifespan (Zajitschek et al. 2016, 2019).

Perhaps the strongest support for the DST comes from studies of natural populations. First, it is well documented that investment in early-life fitness components correlates with reduced late-life performance (Lemaitre et al. 2015), as shown across 24 vertebrates (mostly birds, mammals and two reptiles). Second, brood size manipulation studies in birds provide the best experimental support for the so-called 'cost of reproduction' in nature. Landmark studies in Collared Flycatchers (*Ficedula albicollis*) on the island of Gotland in Sweden (Gustafsson and Sutherland 1988, Gustafsson and Part 1990), and in European Jackdaws (*Coloeus monedula*) in Groningen, Netherlands (Boonekamp et al. 2015) show that experimental enlargement of broods increases parental mortality with age, although, apparently, the birds can cope with artificially increased broods for one year, and several years of experimentation can be necessary to show actuarial senescence (Santos and Nakagawa 2012, Boonekamp et al. 2015). One interpretation of these results is that when birds are compelled to invest more than they bargained for to reproduce, they have fewer resources to allocate to somatic maintenance resulting in faster accumulation of damage leading to more rapid senescence. However, as discussed previously (Partridge and Andrews 1985, Boonekamp et al. 2015) experimentally increased reproduction could have a disproportionate effect on late-life mortality without accelerating ageing, for example, by increasing the exposure of parents to pathogens or predators.

Overall, it seems likely that at least some of the AP alleles that contribute to the evolution of ageing will be involved in governing resource allocation trade-offs envisaged by the DST. However, there is a large scope for future work in this area, because the relative importance of resource allocation trade-offs in ageing is not yet established.

There is a quite wide range of studies in different organisms focusing on the relationship between reproduction and survival whose results are inconsistent with the primary role of resource allocation in the evolution of ageing and lifespan (reviewed in Partridge et al. 2005, Flatt 2011, Flatt and Heyland 2011, Flatt and Partridge 2018). For example, reduced nutrient intake without malnutrition, known as 'dietary restriction', is commonly associated with reduced reproduction and increased lifespan, suggesting that animals allocate resources from reproduction to somatic maintenance when food is in short supply (Kirkwood and Shanley 2005). However, adding just one essential amino acid, *i.e.*, methionine, to dietary-restricted *D. melanogaster* restores their reproduction to normal levels without affecting their lifespan (Grandison et al. 2009) contradicting the resource allocation scenario. Similarly, experimental laboratory evolution studies in invertebrates suggest that the evolution of long life is not obligatorily associated with reduced reproduction. In fact, sometimes lifespans can rapidly evolve without any effect on reproduction (see earlier), and sometimes the evolution of long life is associated with improved reproductive performance (see further in the chapter the section entitled Williams and anti-Williams: age, density and condition-dependence of mortality). These studies are in line with MA theory and suggest that populations harbour standing age-specific genetic variation for fitness.

Eusocial organisms also provide an interesting angle to look at the survival-reproduction trade-off (Korb and Heinze 2021; see also Chapter 9). As discussed earlier, there is no difference in lifespan between reproductive and non-reproductive colony members in a eusocial naked mole-rat in the absence of extrinsic mortality hazards. In eusocial insects, reproducing queens produce copious numbers of eggs throughout their lives and outlive non-reproductive workers sometimes by the order of magnitude. The queen of the black garden ant (*Lasius niger*) can live for over 28 years compared with one or two years only for workers (males only survive for a few weeks; Parker et al. 2004). However, high levels of relatedness within these societies can have profound effect on how we measure fitness, and it is possible that such eusocial colonies should be viewed as a 'superorganism', where queens represent the metaphorical 'germline', while workers represent the 'soma' (Kramer et al. 2016; see also Chapter 9). More broadly, the division of labour within the colony is predicted to result in different ageing rates in different types of individuals (Pen and Flatt 2021). Nevertheless, on the level of an individual organism, increased reproduction does not always correspond to reduced survival.

Perhaps the clearest example of uncoupling of survival-reproduction trade-off comes from ingenious experimental work by Dillin et al. (2002) using *C. elegans*. Earlier work recognized that many life history traits in *C. elegans*, including development time, body size, reproduction and lifespan, are shaped by insulin/IGF-1 nutrient-sensing signalling pathway (IIS). This pathway is evolutionarily conserved and has strong effect on life history traits across the animal kingdom (Gems and Partridge 2001, Kenyon 2001; see also Chapters 8 and 9). Specifically, as discussed earlier, *daf-2* mutants are often exceptionally long-lived, but all suffer from more or less obvious fitness detriments (Gems et al. 1998, Chen et al. 2007). However, Dillin et al. (2002) used RNA interference (RNAi), which allows age-specific knockdown of gene expression in *C. elegans*. This elegant approach enables researchers to reduce the expression of the gene of interest at different stages across the life cycle. Therefore, this method opens doors for direct testing of the key tenet of the ETA, namely that the force of natural selection declines with age after reproductive maturity resulting in the evolution of ageing.

Early-life *daf-2* RNAi starting from the egg stage until death results in a substantial lifespan extension but also reduces early reproduction (Figure 2.2). Such RNAi treatment produces a phenotype that is similar to a partial-loss-of-function mutant allele, and the outcome of this treatment is in line with the DST assuming that the trade-off is maintained by resource allocation from early reproduction to improved somatic maintenance. Similar, albeit smaller, reductions in early-life fecundity are observed when *daf-2* is knocked down at different developmental stages (Figure 2.2). It is quite clear that reduced *daf-2* expression leading to reduction in insulin signalling improves somatic maintenance and increases lifespan but interferes with organismal function in a way that reduces early-life reproduction, as predicted by the DST. However, when the authors allowed the animals to develop normally and knocked down *daf-2* only on the first day of adulthood when young *C. elegans* are reproductively fully mature, they managed to achieve the exact same level of lifespan extension without any reduction in early-life reproduction whatsoever (Figure 2.2). Thus, while reproduction and survival in *C. elegans* are regulated by *daf-2*, this genetic trade-off is not based on resource allocation, *i.e.*, lifespan extension does not require reduction in reproductive performance.

Such clear examples of uncoupling of survival-reproduction trade-offs that contradict resource allocation principle led to the development of a new physiological evolutionary theory: the developmental theory of ageing (DTA). As often happens in science, it required, initially, taking a good look at the foundations of the ETA.

When Williams (1957) outlined the AP theory of ageing, he provided an abstract example of an AP allele involved in calcium deposition in a vertebrate. This hypothetical allele contributes positively to bone calcification during development and thus to fitness. However, later in life, this allele contributes to calcium deposition in blood vessels increasing the probability of arterial calcification in adulthood, thereby reducing fitness. This example of an AP allele does not involve resource allocation but depends purely on gene function and the age-specific allelic effect on fitness. Because the strength of selection for age-specific regulation of the expression of this allele declines with age (Hamilton 1966), ageing can evolve via AP alleles simply because of their suboptimal expression in late life. Dillin et al. (2002)'s findings of age-specific fitness effects of *daf-2* in *C. elegans* fit this scenario perfectly.

The DTA (also known as 'early-life inertia') maintains that ageing is caused by increasingly suboptimal regulation of gene function with advancing age because of the reduction in the force of natural selection (Figure 2.3). The first explicit discussion of this evolutionary physiological theory was provided by de Magalhaes and Church (2005), who suggested that ageing can result as 'an unintended outcome of development' because selection optimizes development and reproduction rather than late-life performance, and directly placed these ideas under the umbrella of the AP theory. They also described two

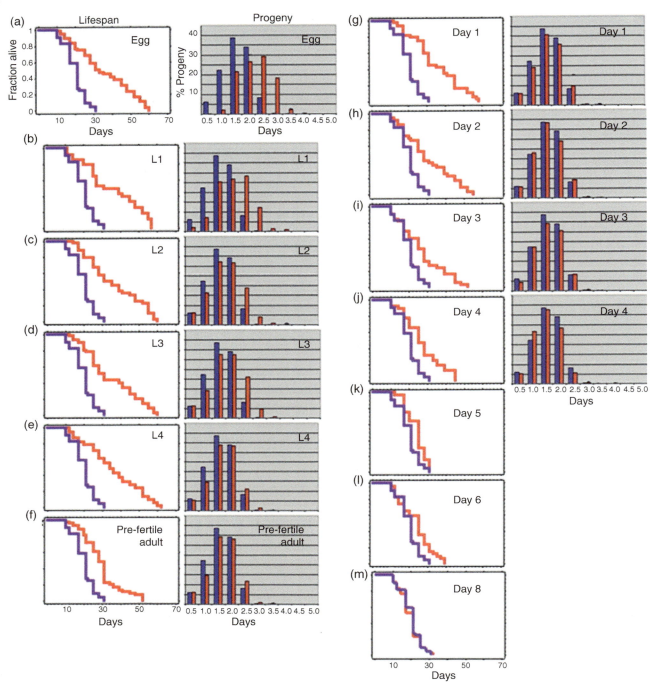

Figure 2.2 Dillin et al. (2002) provide the most striking example of uncoupling increased lifespan from reduced reproduction. *daf-2* is a receptor in evolutionarily conserved insulin/IGF-1 signalling pathway that shapes lifespan and age-specific reproduction in *Caenorhabditis elegans*. Downregulation of *daf-2* expression (*daf-2* knockdown) using RNA interference (RNAi) at different stages across the life cycle affects lifespan and age-specific reproduction. X-axis is age in days, while y-axis is either a fraction of animals alive, or percent of total progeny produced. Blue lines and bars represent wild-type animals that were fed bacteria carrying empty RNAi vector as sham control, while red lines and bars represent experimental animals that were fed on bacteria expressing *daf-2* dsRNA. When *daf-2* expression was downregulated from early-life onwards (from egg or different larval stages, L1–L4, panels (a) to (e)), lifespan strongly increased but the reproductive schedule also shifted, with long-lived *daf-2* RNAi animals reproducing less in early-life, which would likely be detrimental to their fitness in a growing population. However, when *daf-2* was downregulated in early adulthood (*e.g.*, days of adulthood 1 and 2 as depicted on panels (g) and (h)), the same level of lifespan extension was achieved compared with egg and larval stages, but without concomitant reduction or delay in early reproduction. (*Source:* Reproduced with permission from Dillin et al. 2002 / American Association for the Advancement of Science - AAAS.

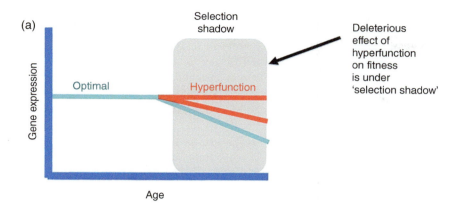

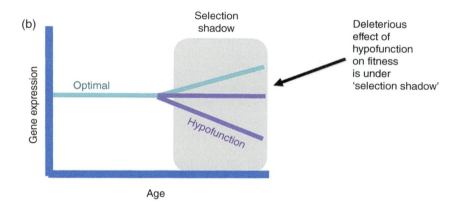

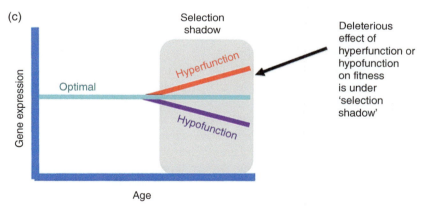

Figure 2.3 Illustration of the DTA, or 'early-life inertia', from Lemaitre et al. (2024), illustrating how decline in the force of selection can lead to misregulation of gene expression with increasing age. This figure depicts several scenarios where gene expression is either higher or lower than optimal for a given age. The 'selection shadow' refers to late life when selection is weak (see Figure 2.1). In the first scenario (a), constant expression of a gene at a given level can lead to deleterious effects on fitness in late ages, but natural selection is too weak (under 'selection shadow') to result in the evolution of a modifier gene that will downregulate late-life expression of the focal gene (optimal level of expression is shown in turquoise, suboptimal level is in red). This scenario is commonly referred to as 'hyperfunction'. Second (b), it could be beneficial to increase gene expression with age, but selection in late life is too weak, so expression evolves to be below the optimal level in late life, either because it remains constant, or perhaps even declines with age (this scenario commonly referred to as 'hypofunction'). Finally (c), it could be optimal to maintain gene expression at constant level with age but weak selection in late life results in mis-regulation and either over- (hyperfunction) or under- (hypofunction) expression in old age. The key take-home message is that decline in the force of selection with age results in mis-regulation of gene expression, which in turn results in suboptimal biological function (*Source:* Lemaitre et al. 2024 / PLOS / CC BY 4.0).

different mechanistic routes by which optimization of early-life performance can harm late-life performance and cause ageing: first, some developmental mechanisms can 'fade' with age, and second, continued action of other developmental mechanisms can become detrimental (as in Williams (1957)'s hypothetical example, see earlier). The latter mechanism, also known as 'hyperfunction', received a lot of attention in the literature and Blagosklonny (2006, 2012) discussed the hypothesis that over-expression of target-of-rapamycin (TOR) signalling pathway is a primary contributor to ageing across taxa (Gems 2022) for the detailed description of the history and the evidence in relation to the 'hyperfunction theory of ageing'). Nevertheless, many genes that regulate lifespan are under-expressed in adulthood suggesting 'hypofunction' (Maklakov and Chapman 2019) is a very common phenomenon as well.

There is good experimental evidence for the DTA beyond the initial findings (Ezcurra et al. 2018, Lind et al. 2019, Lind et al. 2021), but it is largely limited to the *C. elegans* system and there is great scope for future empirical work here. One important caveat is that much of the experimental work has been done in stable and benign laboratory conditions (but see Carlsson et al. 2021), while genotype-by-age-by-environment interactions are likely to shape the evolution of ageing and lifespan in nature (Wilson et al. 2008).

The DTA-like processes have been increasingly discussed and studied in the last decade and different authors used different terminology to describe them (*e.g.*, quasi-programme, programmatic ageing, early-life inertia, just to name a few) (de Magalhaes 2012, Carlsson et al. 2021, Gems 2022). Additionally, the relationship between ultimate population genetic theories of ageing (MA and AP) and different versions of evolutionary physiological theories (DST, DTA, 'hyperfunction') became very unclear. To rectify this problem, Lemaitre and colleagues (Lemaitre et al. 2024) attempted to provide a unified view of the ETA, which includes a hierarchy of evolutionary models of ageing from ultimate population genetic theories to proximate physiological theories, and specifically developed different scenarios for the DTA theory operating via either 'hyperfunction' or 'hypofunction' routes (see Figure 2.3). One of the outcomes of this approach was a realization that the DTA/'early-life inertia' can operate under both MA and AP types of alleles (see Aubier and Galipaud 2023).

While the initial discussion of the DTA focused exclusively on AP-like scenarios where alleles have positive early-life effects at the cost of late-life performance, it is not a necessary prerequisite for the evolution of ageing under the DTA. An age-specific optimization of selection approach shows that ageing evolves in a non-ageing population via DTA also in the absence of AP alleles. A deleterious MA allele that exceeds optimal function in older individuals (*i.e.*, causes 'hyperfunction' in old age) while falling short of optimal function in young individuals can still segregate in the population under mutation-selection balance and contribute to the evolution of ageing (Lemaitre et al. 2024).

2.3 Asynchronous Ageing

Early evolutionary theory suggested that reduction in the force of selection with age will cause a synchronous evolution of ageing in all traits. Specifically, Williams (1957) and Maynard Smith (1959) suggested that all traits will deteriorate at a similar rate. Williams (1957) included this as one of the predictions for the evolution of ageing via AP that senescence should always be a generalized deterioration and that selection will operate in a way that harmonizes the rate of ageing across different traits. Williams (1957)'s argument for synchrony was based on selection operating to decrease senescence in those traits that deteriorate faster than other traits in an organism, and the simultaneous accumulation of late-acting deleterious alleles under relaxed selection in traits that senesce slower. More recent ideas suggested that selection should operate specifically against senescence in traits that are more closely related to fitness. This hypothesis was put forward to explain the empirical observations of asynchronous ageing (Boonekamp et al. 2018, Rodriguez-Munoz et al. 2019). Indeed, there is substantial evidence for asynchrony in ageing rates across different traits (Herndon et al. 2002, Hayward et al. 2015, Gaillard and Lemaitre 2017) suggesting that early theoretical arguments in favour of synchronous ageing were at the very least insufficient to describe the natural variation in trait-specific ageing (Moorad and Ravindran 2022). For example, Hayward et al. (2015) provide a striking example of the variation in rates of ageing across 20 different traits in wild population of Soay sheep (*Ovis aries*; Figure 2.4). Recent theoretical work indeed suggests that both predictions, *i.e.*, synchronous ageing, and faster ageing in traits that are less important to fitness, are not supported by the formal evolutionary analysis (Moorad and Ravindran 2022).

The new evolutionary theory maintains that traits that are under stronger selection in early-life will show more rapid senescence (Moorad and Ravindran 2022). This is because the decline in the force of selection with age – the general feature of the evolution of ageing – happens faster for the traits that are under stronger selection. This opens large scope for empirical work, and this is one of the areas of research in the evolutionary biology of ageing where gaps in our knowledge are particularly pronounced.

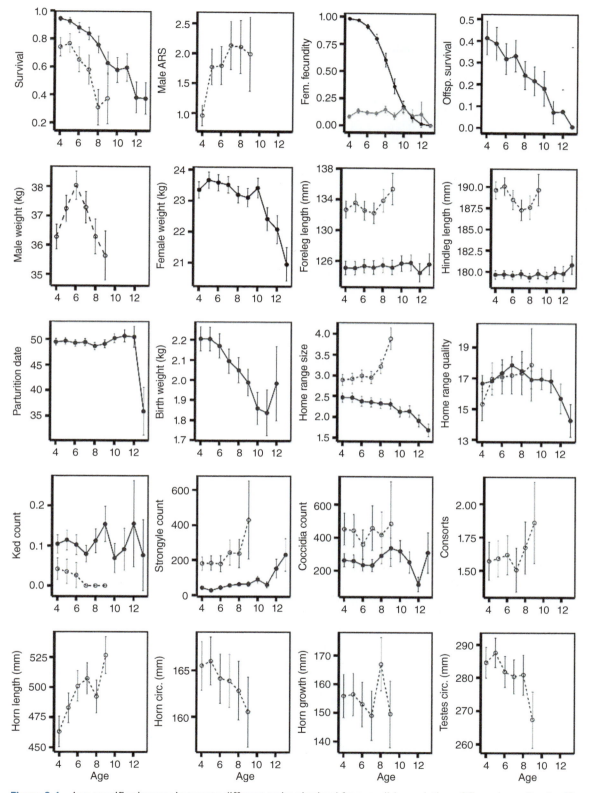

Figure 2.4 Age-specific changes in twenty different traits obtained from a wild population of Soay sheep *O. aries*. The data is shown as mean values ± SE across different ages (the estimates are from generalized linear mixed models rather than raw values, see Hayward et al. (2015) for details). Solid lines and symbols represent females, while broken lines and open symbols represent males. Note that in the 'Fem. Fecundity' panel, the probability of a female giving birth at a given age is plotted in black, while the probability of her twinning is plotted in grey. Male ARS refers to male annual reproductive success. Male and female weight predictions are shown in different panels to make sex-specific trends clearly visible because males are much heavier than females (*Source:* Hayward et al. / 2015 with permission of ELSEVIER.). The picture shows the Soay sheep (*O. aries*) female and lamb from free-living population at St Kilda archipelago. *Source:* Photo by Arpat Ozgul.

Figure 2.4 (Continued)

2.4 Sex Differences in Ageing

Sex differences in lifespan and ageing are common across the animal kingdom (Promislow 1992, Liker and Szekely 2005, Austad and Fischer 2016; see also Chapter 5). There is a wide diversity of sex-specific lifespans across taxa. For example, in mammals, females commonly outlive males (Promislow 1992, Lemaitre et al. 2020), while the opposite is observed in birds (Liker and Szekely 2005). Nevertheless, in some mammals, males live longer than females (Lemaitre et al. 2020) and many bird species show 'mammalian' female lifespan advantage (Liker and Szekely 2005), while in many species, there is no sex difference in lifespan altogether.

This is an exciting and growing field of theoretical and empirical research and there is much left to learn about the evolution of sex-specific life histories. Males and females commonly have different routes to successful reproduction (Parker 1979), which can lead to sex-specific selection across different stages of the course of life (Williams 1957, Trivers 1972, 1985, Bonduriansky et al. 2008, Maklakov and Lummaa 2013). Evolutionary theory states that selection will optimize fitness rather than lifespan and because the relationship between fitness and lifespan can differ between the sexes (Trivers 1985, Bonduriansky et al. 2008, Scharer et al. 2012), it follows that selection can produce sex differences in lifespan.

Early evolutionary hypotheses focused on the role of sexual selection in males in generating male-biased mortality in mammals (Williams 1957, Promislow 1992), but these ideas have been since developed and broadened to encompass a wide variety of sex-specific patterns of ageing. There is strong empirical evidence to suggest that sex differences in lifespan are shaped by the interaction between sex-specific selection and local environmental conditions (Lemaitre et al. 2020). For example, studies in model organisms in laboratory conditions and wild populations in nature suggest that sex differences in lifespan may depend on food availability and social environment in a given population (Festa-Bianchet et al. 2006, Nakagawa et al. 2012, Zajitschek et al. 2013). The relationship between diet- and sex-specific lifespans provides a particularly good example of the complexity of causes behind sexual dimorphism in lifespan. Macro-nutrient food composition expressed as protein:carbohydrate (P:C) ratio directly affects reproduction and lifespan in animals. Studies in insects (Maklakov et al. 2008) and mice (Solon-Biet et al. 2015) show that different P:C ratios can be necessary to optimize reproduction and lifespan in the two sexes. Therefore, which sex lives longer could depend directly on the type of macro-nutrients available in the environment.

Perhaps one of the most direct ways of testing the hypothesis that sex-specific selection can generate sex differences in lifespan is to impose sex-specific selection experimentally. In a dioicous nematode *Caenorhabditis remanei*, males live longer than females. However, experimental increase in adult male mortality for 20 generations led to the reduction in male, but not female, lifespan, and, consequently, the evolution of sexual monomorphism in this trait (Chen and Maklakov 2014; Figure 2.5). On the other hand, when increase in adult male mortality was condition-dependent (*i.e.*, favouring survival of fast-moving males in the context of mate search), it led to the evolution of greater dimorphism in lifespan (Figure 2.5). This shows that sex-specific selection can rapidly generate dimorphism or monomorphism in lifespan provided there is standing sex-specific genetic variation for this trait.

The aforementioned example suggests that there could be substantial levels of sex-specific genetic variation for lifespan in the populations. This is not always the case as sexes share most of their genes and, therefore, the intersexual genetic correlation (r_{MF}) can constrain evolution in the face of sex-specific selection resulting in intralocus sexual conflict (Lande 1980, Bonduriansky and Chenoweth 2009). For example, the evolution of optimal dietary choice can be constrained by shared genetic architecture and the sexes may not be able to reach their optimal dietary intake even in the presence of sufficient types of nutrients in their environment, which will affect sex-specific lifespan and ageing. Several studies showed that nutrient intake can be indeed suboptimal in both sexes but the role of intralocus sexual conflict in generating this outcome requires further work (Reddiex et al. 2013, Hawkes et al. 2022). Hawkes et al. (2022) provide perhaps the clearest example to date that intralocus conflict can indeed constrain adaptive evolution of sex-specific nutrient intake and shape sex differences in lifespan. They examined male and female responses to varying P:C ratios in the decorated cricket *Gryllodes sigillatus* to show that P:C intake has opposing effect on lifespan in the presence of a strong r_{MF} for nutrient intake. Moreover, while both sexes were displaced from their nutritional optima, females were affected more strongly, suggesting that intralocus conflict over nutrient intake can contribute to faster ageing and shorter lifespan in females of this species.

Beyond sex-specific selection, there are several mechanisms based on asymmetrical genetic inheritance that can putatively contribute to sex differences in lifespan and ageing (see also Chapter 5). One of the earlier models in this regard was so-called 'unguarded X', *i.e.*, the hypothesis that the heterogametic sex is hampered by recessive deleterious mutations that are located on unpaired (hence, 'unguarded') X (or Z) sex chromosome (Trivers 1985). This idea aligns well with classical evolutionary genetics and assumes that most deleterious mutations maintained in the population via mutation-selection balance are recessive (Lynch et al. 1999). Empirical evidence in support of this idea is mixed (Brengdahl et al. 2018, Sultanova et al. 2020, Cayuela et al. 2022). While the logic behind this hypothesis is very straightforward, recent population genetic model parametrized using data from the empirical studies, suggests that 'unguarded X' can potentially explain only a small proportion of the observed sex difference in lifespan in animals (Connallon et al. 2022).

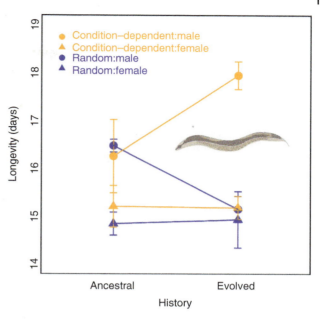

Figure 2.5 Experimental evolution of sexual dimorphism and monomorphism in lifespan in *C. remanei* nematodes. Males live longer than females in this species. Experimentally increased extrinsic male mortality in adulthood (by removing 80% of the males in each population and replacing them with the same number of virgin young males derived from the offspring cohort of this population) resulted in the evolution of shorter-lived males (blue circles and triangles represent mean longevities for males and females respectively, ±SE) after 20 generations. Because there was no effect on female longevity, sexual dimorphism in lifespan effectively disappeared after 20 generations of experimental evolution. On the other hand, when adult male mortality was condition-dependent (*i.e.*, surviving males were selected based on their success in mate search behaviour, see Chen and Maklakov (2014) for details), the same rate of extrinsic mortality resulted in the evolution of longer-lived males (orange circles). Because female longevity did not evolve in this experimental regime (orange triangles), the overall result was the increase in sexual dimorphism in lifespan under condition-dependent mortality. Longevity data comes from four replicate populations evolving under random or condition-dependent mortality, measured contemporaneously (ancestral populations were cryopreserved during the experiment) in ancestral and evolved populations. Overall, this experiment underscores the power of sex-specific selection to increase or decrease sexual dimorphism in lifespan. Graphic of female *C. remanei* by Zahida Sultanova. *Source:* Chen and Maklakov, 2014 / with permission of ELSEVIER.

A more recent hypothesis that aims to explain sex differences in lifespan-based asymmetrical inheritance of sex chromosomes is the 'toxic Y'. This idea originates from the same observation as the 'unguarded X', *i.e.*, that the heterogametic XY (or ZW) sex can suffer from shorter lifespan, but gives a different proximate explanation, namely, that accumulation of repetitive DNA on the Y chromosome can lead to higher mortality in males (Brown et al. 2020). The hypothesis was directly tested in *D. melanogaster* by generating a range of experimental Y chromosomes that differ in the amount of heterochromatic repetitive DNA. Remarkably, this had no effect on sex differences in organismal physiology and did not affect sexual dimorphism in lifespan (Delanoue et al. 2023). Instead, *D. melanogaster* lifespan seems to be directly affected by phenotypic sex rather than by sex chromosomes: when XY males are feminized by expressing the sex-determination *transformer^F* (*tra^F*) gene, they live longer than non-feminized males, while XX females masculinized by *tra^F* knockout live shorter lives than non-masculinized females (Delanoue et al. 2023).

Another form of asymmetric genetic inheritance that has been extensively studied in relation to sex differences in lifespan and ageing is the maternal inheritance of mitochondrial DNA (mDNA). In most metazoans, offspring inherit their mtDNA from females, while males do not pass their mtDNA to the next generation. This reduces selection against mtDNA mutations that harm males but not females (Frank and Hurst 1996), in a process that was later named 'the Mother's Curse' (Gemmell et al. 2004). Additionally, sexually antagonistic mtDNA alleles that benefit females, but harm males, can be positively selected (Innocenti et al. 2011). There is a wide range of empirical studies testing the predictions derived from the Mother's Curse. Comparative studies showing that females tend to outlive males in many taxa are in line with all asymmetric inheritance theories, including Mother's Curse, and, therefore, not unique. Furthermore, in many taxa males live longer than females (Tower 2006, Maklakov and Lummaa 2013). Several experimental studies suggested that mtDNA mutations can indeed have deleterious effects on male lifespan and ageing (Camus et al. 2012), but overall the evidence in favour of the critical role of the Mother's Curse in the evolution of sex differences in senescence remains relatively weak (Dowling and Adrian 2019). While the logic of 'sex-specific selective sieve' (Innocenti et al. 2011) resulting from uniparental transmission of mtDNA is hard to dispute, the Mother's Curse can be resolved via selection acting on nuclear genes to compensate for male-harming effects of mtDNA mutations (Connallon et al. 2018). Nevertheless, mito-nuclear co-evolution is still predicted to result in substantial 'male mitochondrial load', such that the role of Mother's Curse in the evolution of sex differences in ageing certainly requires further study.

2.5 Williams and Anti-Williams: Age, Density and Condition-Dependence of Mortality

Williams (1957) predicted that increased rate of adult mortality should result in the evolution of increased rate of senescence. This prediction created an ongoing confusion among modellers, demographers, and experimentalists (both organismal and molecular biologists) when discussing the ways of testing the ETA (see Caswell and Shyu 2017 and references therein; Moorad et al. 2019, da Silva 2020, Day and Abrams 2020, de Vries et al. 2023). It must be emphasized that this is just one of the predictions that Williams (1957) postulated (see Gaillard and Lemaitre 2017) for an in-depth look at the rest of Williams (1957)'s predictions and how well they have done in tests), but it is the one that attracted the lion's share of attention. The main confusion comes from the fact that Williams (1957)'s formulation was too broad and too brief to account for different ecologically relevant scenarios. It is widely appreciated today that increase in extrinsic (also called environmental or ecological) age-independent mortality does not necessarily result in the evolution of increased rates of senescence (reviewed in Moorad et al. 2020, de Vries et al. 2023).

Ageing evolves because selection on survival and fecundity decline with age (Hamilton 1966). Essentially, increased extrinsic mortality does not influence selection in the absence of density dependence because reduced probability of survival and reduced population growth rate cancel each other out in an exponentially growing population (Caswell and Shyu 2017, Day and Abrams 2020). Therefore, increased extrinsic mortality does not affect the evolution of ageing under these specific conditions. However, when density dependence regulates populations via reducing juvenile recruitment (this could happen either indirectly via reduced fecundity or directly via reduced chances of recruitment even if fecundity is unaffected), Williams (1957)'s prediction holds. It is important to note that Williams (1957) explicitly wrote about extrinsic 'adult mortality', and thus his prediction arguably refers to age-dependent mortality scenario. Interestingly, de Vries et al. (2023) make an interesting point that the prevalence of studies in support of Williams (1957)'s original prediction indirectly suggests that populations are often subject to density dependence via reduced juvenile recruitment. It is certainly the case that the studies simulating increased extrinsic mortality in laboratory conditions and finding Williams-like outcomes (Stearns et al. 2000, Chen and Maklakov 2012) were conducted in a way that imposed density dependence via reduced juvenile recruitment.

When density dependence regulates populations via increased mortality in old age, increased extrinsic mortality leads to the evolution of slower senescence, something that de Vries et al. (2023) aptly named 'anti-Williams'. This is because higher extrinsic mortality reduces the proportion of deaths in old age that are due to density dependence (Abrams 1993). However, density dependence affecting primarily old individuals is not the only way in which higher extrinsic mortality can lead to the evolution of slower senescence.

Trinidadian guppies (*Poecilia reticulata*) inhabit a system of rivers running parallel to the mountains. In lower parts of the rivers, guppies coexist with a number of predator fish species, while in upper parts the predators are generally absent (Reznick et al. 2004; see also Chapter 12). This creates a natural experiment where replicate guppy populations evolve in either high-predation or low-predation environments. In line with the life history theory, guppies in high-predation sites mature earlier and invest more in reproduction than guppies from low-predation site. However, in an example of 'anti-Williams' ageing in a natural population, high-predation guppies also live longer and have longer reproductive lifespan than their low-predation counterparts. This intriguing finding can be explained by the theory developed primarily by Abrams (1993) and Williams and Day (2003). This theory focuses on the interaction between extrinsic mortality and age-specific physiological deterioration. When extrinsic mortality is condition-dependent, for example, preferentially removes low-fitness individuals from the population, increase in such mortality concomitantly increases selection against physiological deterioration across ages, thereby increasing selection against senescence. Therefore, high rates of condition-dependent mortality can lead to the evolution of slower senescence. This hypothesis was later tested experimentally in the laboratory setting using *C. remanei* nematodes. When different rates of extrinsic mortality were imposed haphazardly on adult worms while maintaining density-dependent regulation of population size via offspring recruitment (Chen and Maklakov 2012), high-mortality populations evolved shorter intrinsic lifespans as predicted by classic theory while taking age-specific density dependence into account (Abrams 1993). However, when exact same rates of extrinsic mortality were condition-dependent (based on the ability to survive acute heat stress), the evolutionary outcome was the opposite, *i.e.*, high-mortality populations evolved longer intrinsic lifespans. This underscores the role of the environment in the evolution of ageing and lifespan.

2.6 Concluding Remarks

The study of the biology of ageing is undergoing an unprecedented renaissance in recent years. Several factors contribute to this. First, we have strong theoretical foundation in the form of ETA that allows us to make specific predictions when testing the evolution and expression of ageing in different environments. Nevertheless, the effect of the environment on ageing is only partly understood, while it is likely to affect both expression (Kawasaki et al. 2008) and the evolution (Williams et al. 2006, Cotto and Ronce 2014) of ageing. Second, some aspects of the classical theory have been challenged by new discoveries and prompted new theoretical and empirical research. This work is ongoing and will likely lead to many interesting findings in the coming years. Third, molecular biology of ageing has made rapid progress in the last couple of decades, and efforts are being made to reconcile our advances in the mechanistic understanding of ageing with ETA into a coherent integrative framework that combines genetic and physiological theories of ageing. Fourth, there is strong societal interest in translational research on ageing, leading to the rapid development of biogerontology (also known as 'geroscience'). There is much to be done to fully integrate evolutionary biology and biogerontology, but there has been a lot of progress in this regard in recent years, which will likely benefit both fields of research.

Acknowledgements

I thank my colleagues who kindly read the entire chapter and made valuable comments that certainly made it better: Tracey Chapman, Damian Dowling, Urban Friberg, Jean-Michel Gaillard, Jean-François Lemaître, Jacob Moorad and the editors of the volume. The remaining mistakes are mine.

References

Abrams, P.A. (1993). Does increased mortality favor the evolution of more rapid senescence? *Evolution* 47(3): 877–887.
Aubier, T.G., and Galipaud, M. (2023). Senescence evolution under the catastrophic accumulation of deleterious mutations. *Evolution Letters* 8: 212–221.
Austad, S.N., and Fischer, K.E. (2016). Sex differences in lifespan. *Cell Metabolism* 23(6): 1022–1033.
Blagosklonny, M.V. (2006). Aging and immortality – Quasi-programmed senescence and its pharmacologic inhibition. *Cell Cycle* 5(18): 2087–2102.

Blagosklonny, M.V. (2012). Answering the ultimate question "What is the proximal cause of aging?" *Aging* 4(12): 861–877.

Bonduriansky, R., and Chenoweth, S.F. (2009). Intralocus sexual conflict. *Trends in Ecology & Evolution* 24: 280–288.

Bonduriansky, R., Maklakov, A., Zajitschek, F., and Brooks, R. (2008). Sexual selection, sexual conflict and the evolution of ageing and lifespan. *Functional Ecology* 22: 443–453.

Boonekamp, J.J., Mulder, E., and Verhulst, S. (2018). Canalisation in the wild: effects of developmental conditions on physiological traits are inversely linked to their association with fitness. *Ecology Letters* 21(6): 857–864.

Boonekamp, J.J., Salomons, H.M., Bouwhuis, S., Dijkstra, C., and Verhulst, S. (2015). Reproductive effort accelerates actuarial senescence in wild birds: an experimental study. *Ecology Letters* 18(3): 315–315.

Brengdahl, M.I., Kimber, C.M., Elias, P., Thompson, J., and Friberg, U. (2020). Deleterious mutations show increasing negative effects with age in *Drosophila melanogaster*. *BMC Biology* 18: 128.

Brengdahl, M., Kimber, C.M., Maguire-Baxter, J., and Friberg, U. (2018). Sex differences in life span: females homozygous for the X chromosome do not suffer the shorter life span predicted by the unguarded X hypothesis. *Evolution* 72(3): 568–577.

Brengdahl, M.I., Kimber, C.M., Shenoi, V.N., Dumea, M., Mital, A., and Friberg, U. (2023). Age-specific effects of deletions: implications for aging theories. *Evolution* 77(1): 254–263.

Briga, M., and Verhulst, S. (2015). What can long-lived mutants tell us about mechanisms causing aging and lifespan variation in natural environments? *Experimental Gerontology* 71: 21–26.

Brown, E.J., Nguyen, A.H., and Bachtrog, D. (2020). The Y chromosome may contribute to sex-specific ageing in *Drosophila*. *Nature Ecology & Evolution* 4(6): 853–962.

Buffenstein, R. (2008). Negligible senescence in the longest living rodent, the naked mole-rat: insights from a successfully aging species. *Journal of Comparative Physiology B: Biochemical Systems and Environmental Physiology* 178(4): 439–445.

Camus, M.F., Clancy, D.J., and Dowling, D.K. (2012). Mitochondria, maternal inheritance, and male aging. *Current Biology* 22(18): 1717–1721.

Carlsson, H., Ivimey-Cook, E., Duxbury, E.M.L., Edden, N., Sales, K., and Maklakov, A.A. (2021). Ageing as "early-life inertia": disentangling life-history trade-offs along a lifetime of an individual. *Evolution Letters* 5(5): 551–564.

Caswell, H., and Shyu, E. (2017). Senescence, selection gradients and mortality. In: *The evolution of senescence in the tree of life* (eds. Shefferson, R.P., Jones, O.R., and Salguero-Gomez, R.), 56–82. Cambridge University Press.

Cayuela, H., Lemaitre, J.F., Lena, J.P., Ronget, V., Martinez-Solano, I., Muths, E., Pilliod, D.S., Schmidt, B.R., Sanchez-Montes, G., Gutierrez-Rodriguez, J., Pyke, G., Grossenbacher, K., Lenzi, O., Bosch, J., Beard, K.H., Woolbright, L.L., Lambert, B.A., Green, D.M., Jreidini, N., Garwood, J.M., Fisher, R.N., Matthews, K., Dudgeon, D., Lau, A., Speybroeck, J., Homan, R., Jehle, R., Baskale, E., Mori, E., Arntzen, J.W., Joly, P., Stiles, R.M., Lannoo, M.J., Maerz, J.C., Lowe, W.H., Valenzuela-Sanchez, A., Christiansen, D.G., Angelini, C., Thirion, J.M., Merila J., Colli, G.R., Vasconcellos, M.M., Boas, T.C.V., Arantes, I.D., Levionnois, P., Reinke, B.A., Vieira, C., Marais, G.A.B., Gaillard, J.M., and Miller, D.A.W. (2022). Sex-related differences in aging rate are associated with sex chromosome system in amphibians. *Evolution* 76(2): 346–356.

Charlesworth, B. (1980). *Evolution in age-structured populations*. Cambridge University Press.

Charlesworth, B. (2000). Fisher, Medawar, Hamilton and the evolution of aging. *Genetics* 156(3): 927–931.

Charlesworth, B. (2001). Patterns of age-specific means and genetic variances of mortality rates predicted by the mutation-accumulation theory of ageing. *Journal of Theoretical Biology* 210(1): 47–65.

Charlesworth, B., and Hughes, K.A. (1996). Age-specific inbreeding depression and components of genetic variance in relation to the evolution of senescence. *Proceedings of the National Academy of Sciences of the United States of America* 93(12): 6140–6145.

Chen, H.-Y., and Maklakov, A.A. (2012). Longer lifespan evolves under high rates of condition dependent mortality. *Current Biology* 22: 2140–2143.

Chen, H.-Y., and Maklakov, A.A. (2014). Condition dependence of male mortality drives the evolution of sex differences in longevity. *Current Biology* 24(20): 2423–2427.

Chen, J.J., Senturk, D., Wang, J.L., Muller, H.G., Carey, J.R., Caswell, H., and Caswell-Chen, E.P. (2007). A demographic analysis of the fitness cost of extended longevity in *Caenorhabditis elegans*. *Journals of Gerontology Series B: Biological Sciences and Medical Sciences* 62(2): 126–135.

Connallon, T., Beasley, I.J., McDonough, Y., and Ruzicka, F. (2022). How much does the unguarded X contribute to sex differences in life span? *Evolution Letters* 6(4): 319–329.

Connallon, T., Camus, M.F., Morrow, E.H., and Dowling, D.K. (2018). Coadaptation of mitochondrial and nuclear genes, and the cost of mother's curse. *Proceedings of the Royal Society B: Biological Sciences* 285(1871): 20172257.

Cotto, O., and Ronce, O. (2014). Maladaptation as a source of senescence in habitats variable in space and time. *Evolution* 68(9): 2481–2493.

da Silva, J. (2020). Williams' intuition about extrinsic mortality was correct. *Trends in Ecology & Evolution* 35(5): 378–379.

Day, T., and Abrams, P.A. (2020). Density dependence, senescence, and Williams' hypothesis. *Trends in Ecology & Evolution* 35(4): 300–302.

de Magalhaes, J.P. (2012). Programmatic features of aging originating in development: aging mechanisms beyond molecular damage? *FASEB Journal* 26(12): 4821–4826.

de Magalhaes, J.P., and Church, G.M. (2005). Genomes optimize reproduction: aging as a consequence of the developmental program. *Physiology* 20: 252–259.

de Vries, C., Galipaud, M., and Kokko, H. (2023). Extrinsic mortality and senescence: a guide for the perplexed. *Peer Community Journal* 3: e29.

Delanoue, R., Clot, C., Leray, C., Pihl, T., and Hudry, B. (2023). Y chromosome toxicity does not contribute to sex-specific differences in longevity. *Nature Ecology & Evolution* 7(8): 1245–1256.

Dillin, A., Crawford, D.K., and Kenyon, C. (2002). Timing requirements for insulin/IGF-1 signaling in *C. elegans*. *Science* 298(5594): 830–834.

Donertas, H.M., Fabian, D.K., Fuentealba, M., Partridge, L., and Thornton, J.M. (2021). Common genetic associations between age-related diseases. *Nature Aging* 1(4): 400–412.

Dowling, D.K., and Adrian, R.E. (2019). Challenges and prospects for testing the mother's curse hypothesis. *Integrative and Comparative Biology* 59(4): 875–889.

Ezcurra, M., Benedetto, A., Sornda, T., Gilliat, A.F., Au, C., Zhang, Q., van Schelt, S., Petrache, A.L., Wang, H., de la Guardia, Y., Bar-Nun, S., Tyler, E., Wakelam, M.J., Gems, D. (2018). *C. elegans* eats its own intestine to make yolk leading to multiple senescent pathologies. *Current Biology* 28: 2544–2556.

Fabian, D., and Flatt, T. (2011). The evolution of aging. *Nature Education Knowledge* 3(10): 9.

Festa-Bianchet, M., Coulson, T., Gaillard, J.M., Hogg, J.T., and Pelletier, F. (2006). Stochastic predation events and population persistence in bighorn sheep. *Proceedings of the Royal Society B: Biological Sciences* 273(1593): 1537–1543.

Fisher, R.A. (1930). *The genetical theory of natural selection*. Oxford University Press.

Flatt, T. (2011). Survival costs of reproduction in *Drosophila*. *Experimental Gerontology* 46(5): 369–375.

Flatt, T., and Heyland, A. (2011). *Mechanisms of life history evolution: the genetics and physiology of life history traits and trade-offs*. Oxford University Press.

Flatt, T., and Partridge, L. (2018). Horizons in the evolution of aging. *BMC Biology* 16: 93.

Frank, S.A., and Hurst, L.D. (1996). Mitochondria and male disease. *Nature* 383(6597): 224–224.

Friedman, D.B., and Johnson, T.E. (1988). A mutation in the afe-1 gene in *Caenorhabditis elegans* lengthens life and reduces hermaphrodite fertility. *Genetics* 118(1): 75–86.

Gaillard, J.M., and Lemaitre, J.F. (2017). The Williams' legacy: a critical reappraisal of his nine predictions about the evolution of senescence. *Evolution* 71(12): 2768–2785.

Gemmell, N.J., Metcalf, V.J., and Allendorf, F.W. (2004). Mother's curse: the effect of mtDNA on individual fitness and population viability. *Trends in Ecology & Evolution* 19(5): 238–244.

Gems, D. (2022). The hyperfunction theory: an emerging paradigm for the biology of aging. *Ageing Research Reviews* 74: 18.

Gems, D., and Partridge, L. (2001). Insulin/IGF signalling and ageing: seeing the bigger picture. *Current Opinion in Genetics & Development* 11(3): 287–292.

Gems, D., Sutton, A.J., Sundermeyer, M.L., Albert, P.S., King, K.V., Edgley, M.L., Larsen, P.L., and Riddle, D.L. (1998). Two pleiotropic classes of daf-2 mutation affect larval arrest, adult behavior, reproduction and longevity in *Caenorhabditis elegans*. *Genetics* 150(1): 129–155.

Grandison, R.C., Piper, M.D.W., and Partridge, L. (2009). Amino-acid imbalance explains extension of lifespan by dietary restriction in *Drosophila*. *Nature* 462(7276): 1061–1164.

Gustafsson, L., and Part, T. (1990). Acceleration of senescence in the collared flycatcher *Ficedula albicollis* by reproductive costs. *Nature* 347(6290): 279–281.

Gustafsson, L., and Sutherland, W.J. (1988). The costs of reproduction in the collared flycatcher, *Ficedula albicollis*. *Nature* 335(6193): 813–815.

Haldane, J.B.S. (1941). *New paths in genetics*. Allen and Unwin, Ltd.

Hamilton, W.D. (1966). The moulding of senescence by natural selection. *Journal of Theoretical Biology* 12: 12–45.

Hawkes, M., Lane, S.M., Rapkin, J., Jensen, K., House, C.M., Sakaluk, S.K., and Hunt, J. (2022). Intralocus sexual conflict over optimal nutrient intake and the evolution of sex differences in life span and reproduction. *Functional Ecology* 36(4): 865–881.

Hayward, A.D., Moorad, J., Regan, C.E., Berenos, C., Pilkington, J.G., Pemberton, J.M., and Nussey, D.H. (2015). Asynchrony of senescence among phenotypic traits in a wild mammal population. *Experimental Gerontology* 71: 56–68.

Herndon, L.A., Schmeissner, P.J., Dudaronek, J.M., Brown, P.A., Listner, K.M., Sakano, Y., Paupard, M.C., Hall, D.H., and Driscoll, M. (2002). Stochastic and genetic factors influence tissue-specific decline in ageing *C. elegans*. *Nature* 419(6909): 808–814.

Hughes, K.A., and Charlesworth, B. (1994). A genetic analysis of senescence in *Drosophila*. *Nature* 367(6458): 64–66.

Hughes, K.A., and Reynolds, R.M. (2005). Evolutionary and mechanistic theories of aging. *Annual Review of Entomology* 50: 421–445.

Innocenti, P., Morrow, E.H., and Dowling, D.K. (2011). Experimental evidence supports a sex-specific selective sieve in mitochondrial genome evolution. *Science* 332(6031): 845–848.

Jones, O.R., Scheuerlein, A., Salguero-Gomez, R., Camarda, C.G., Schaible, R., Casper, B.B., Dahlgren, J.P., Ehrlen, J., Garcia, M.B., Menges, E.S., Quintana-Ascencio, P.F., Caswell, H., Baudisch, A., and Vaupel, J.W. (2014). Diversity of ageing across the tree of life. *Nature* 505(7482): 169–173.

Kawasaki, N., Brassil, C.E., Brooks, R.C., and Bonduriansky, R. (2008). Environmental effects on the expression of life span and aging: an extreme contrast between wild and captive cohorts of *Telostylinus angusticollis* (Diptera : Neriidae). *The American Naturalist* 172(3): 346–357.

Kenyon, C. (2001). A conserved regulatory system for aging. *Cell* 105(2): 165–168.

Kirkwood, T.B.L. (1977). Evolution of aging. *Nature* 270(5635): 301–304.

Kirkwood, T.B.L. (2005). Understanding the odd science of aging. *Cell* 120(4): 437–447.

Kirkwood, T.B.L., and Holliday, R. (1979). Evolution of aging and longevity. *Proceedings of the Royal Society B: Biological Sciences* 205(1161): 531–546.

Kirkwood, T.B.L., and Rose, M.R. (1991). Evolution of senescence – late survival sacrificed for reproduction. *Philosophical Transactions of the Royal Society B: Biological Sciences* 332(1262): 15–24.

Kirkwood, T.B.L., and Shanley, D.P. (2005). Food restriction, evolution and ageing. *Mechanisms of Ageing and Development* 126(9): 1011–1016.

Korb, J., and Heinze, J. (2021). Ageing and sociality: why, when and how does sociality change ageing patterns? *Philosophical Transactions of the Royal Society B: Biological Sciences* 376(1823): 8.

Kramer, B.H., van Doorn, G.S., Weissing, F.J., and Pen, I. (2016). Lifespan divergence between social insect castes: challenges and opportunities for evolutionary theories of aging. *Current Opinion in Insect Science* 16: 76–80.

Lande, R. (1980). Sexual dimorphism, sexual selection, and adaptation in polygenic characters. *Evolution* 34(2): 292–305.

Lemaitre, J.-F., Berger, V., Bonenfant, C., Douhard, M., Gamelon, M., Plard, F., and Gaillard, J.M. (2015). Early-late life trade-offs and the evolution of ageing in the wild. *Proceedings of the Royal Society B: Biological Sciences* 282(1806): 20150209.

Lemaitre, J.-F., Moorad, J., Gaillard, J.M., Maklakov, A.A., and Nussey, D.H. (2024). A unified framework for evolutionary genetic and physiological theories of aging. *PLoS Biology* 22(2): e3002513.

Lemaitre, J.-F., Ronget, V., Tidiere, M., Allaine, D., Berger, V., Cohas, A., Colchero, F., Conde, D.A., Garratt, M., Liker, A., Marais, G.A.B., Scheuerlein, A., Székely, T., and Gaillard, J.M. (2020). Sex differences in adult lifespan and aging rates of mortality across wild mammals. *Proceedings of the National Academy of Sciences of the United States of America* 117(15): 8546–8553.

Liker, A., and Szekely, T. (2005). Mortality costs of sexual selection and parental care in natural populations of birds. *Evolution* 59(4): 890–897.

Lind, M.I., Ravindran, S., Sekajova, Z., Carlsson, H., Hinas, A., and Maklakov, A.A. (2019) Experimentally reduced insulin/IGF-1 signaling in adulthood extends lifespan of parents and improves Darwinian fitness of their offspring, *Evolution Letters*, 3(2): 207–216.

Lind, M.I., Carlsson, H., Duxbury, E.M.L., Ivimey-Cook, E., and Maklakov, A.A. (2021). Cost-free lifespan extension via optimization of gene expression in adulthood aligns with the developmental theory of ageing. *Proceedings of the Royal Society B: Biological Sciences* 288(1944): 20201728.

Luckinbill, L.S., Arking, R., Clare, M.J., Cirocco, W.C., and Buck, S.A. (1984). Selection for delayed senescence in *Drosophila melanogaster*. *Evolution* 38(5): 996–1003.

Lynch, M., Blanchard, J., Houle, D., Kibota, T., Schultz, S., Vassilieva, L., and Willis, J. (1999). Perspective: spontaneous deleterious mutation. *Evolution* 53(3): 645–663.

Maklakov, A.A., and Chapman, T. (2019). Evolution of ageing as a tangle of trade-offs: energy versus function. *Proceedings of the Royal Society B: Biological Sciences* 286(1911): 20191604.

Maklakov, A.A., and Lummaa, V. (2013). Evolution of sex differences in lifespan and aging: causes and constraints. *BioEssays* 35(8): 717–724.

Maklakov, A.A., Rowe, L., and Friberg, U. (2015). Why organisms age: evolution of senescence under positive pleiotropy? *BioEssays* 37(7): 802–807.

Maklakov, A.A., Simpson, S.J., Zajitschek, F., Hall, M.D., Dessmann, J., Clissold, F., Raubenheimer, D., Bondurianksy, R., and Brooks, R.C. (2008). Sex-specific fitness effects of nutrient intake on reproduction and lifespan. *Current Biology* 18(14): 1062–1066.

Maynard Smith, J. (1959). The rate of ageing in *Drosophila subobscura*. In: *CIBA foundation symposium – the lifespan of animals (colloquia on ageing)* (eds. Wolstenholme, G.E.W., and O'Conner, M.), 269–301. Churchill.

McHugh, K.M., and Burke, M.K. (2022). From microbes to mammals: the experimental evolution of aging and longevity across species. *Evolution* 76(4): 692–707.

Medawar, P.B. (1946). Old age and natural death. *Modern Quarterly* 1: 30–56.

Medawar, P.B. (1952). *An unresolved problem of biology*. H.K. Lewis.

Moorad, J.A. (2014). Individual fitness and phenotypic selection in age-structured populations with constant growth rates. *Ecology* 95(4): 1087–1095.

Moorad, J.A., and Promislow, D.E.L. (2009). What can genetic variation tell us about the evolution of senescence? *Proceedings of the Royal Society B: Biological Sciences* 276(1665): 2271–2278.

Moorad, J., Promislow, D., and Silvertown, J. (2019). Evolutionary ecology of senescence and a reassessment of Williams' 'extrinsic mortality' hypothesis. *Trends in Ecology & Evolution* 34(6): 519–530.

Moorad, J., Promislow, D., and Silvertown, J. (2020). George C. Williams' problematic model of selection and senescence: time to move on. *Trends in Ecology & Evolution* 35(4): 303–305.

Moorad, J.A., and Ravindran, S. (2022). Natural selection and the evolution of asynchronous aging. *The American Naturalist* 199(4): 551–563.

Moorad, J.A., and Walling, C.A. (2017). Measuring selection for genes that promote long life in a historical human population. *Nature Ecology & Evolution* 1(11): 1773–1781.

Nakagawa, S., Lagisz, M., Hector, K.L., and Spencer, H.G. (2012). Comparative and meta-analytic insights into life extension via dietary restriction. *Aging Cell* 11(3): 401–409.

Nussey, D.H., Froy, H., Lemaitre, J.-F., Gaillard, J.M., and Austad, S.N. (2013). Senescence in natural populations of animals: widespread evidence and its implications for bio-gerontology. *Ageing Research Reviews* 12(1): 214–225.

Parker, G.A. (1979). Sexual selection and sexual conflict. In: *Sexual selection and reproductive competition in insects* (eds. Blum, M.S., and Blum, N.A.), 123–166. Academic Press.

Parker, J.D., Parker, K.M., Sohal, B.H., Sohal, R.S., and Keller, L. (2004). Decreased expression of Cu-Zn superoxide dismutase 1 in ants with extreme lifespan. *Proceedings of the National Academy of Sciences of the United States of America* 101(10): 3486–3489.

Partridge, L., and Andrews, R. (1985). The effect of reproductive activity on the longevity of male *Drosophila melanogaster* is not caused by a acceleration and aging. *Journal of Insect Physiology* 31(5): 393–395.

Partridge, L., Gems, D., and Withers, D.J. (2005). Sex and death: what is the connection? *Cell* 120(4): 461–472.

Partridge, L., Prowse, N., and Pignatelli, P. (1999). Another set of responses and correlated responses to selection on age at reproduction in *Drosophila melanogaster*. *Proceedings of the Royal Society B: Biological Sciences* 266(1416): 255–261.

Pen, I., and Flatt, T. (2021). Asymmetry, division of labour and the evolution of ageing in multicellular organisms. *Philosophical Transactions of the Royal Society B: Biological Sciences* 376: 20190729.

Peto R, Roe FJ, Lee PN, Levy L, Clack J. (1975.) Cancer and ageing in mice and men. *British Journal of Cancer* 32(4): 411–26.

Promislow, D.E.L. (1992). Costs of sexual selection in natural populations of mammals. *Proceedings of the Royal Society B: Biological Sciences* 247(1320): 203–210.

Reddiex, A.J., Gosden, T.P., Bondurianksy, R., and Chenoweth, S.F. (2013). Sex-specific fitness consequences of nutrient intake and the evolvability of diet preferences. *The American Naturalist* 182(1): 91–102.

Reznick, D.N., Bryant, M.J., Roff, D., Ghalambor, C.K., and Ghalambor, D.E. (2004). Effect of extrinsic mortality on the evolution of senescence in guppies. *Nature* 431(7012): 1095–1099.

Rodriguez, J.A., Marigorta, U.M., Hughes, D.A., Spataro, N., Bosch, E., and Navarro, A. (2017). Antagonistic pleiotropy and mutation accumulation influence human senescence and disease. *Nature Ecology & Evolution* 1(3): 0055.

Rodriguez-Munoz, R., Boonekamp, J.J., Liu, X.P., Skicko, I., Haugland Pedersen, S., Fisher, D.N., Hopwood, P., and Tregenza, T. (2019). Comparing individual and population measures of senescence across 10 years in a wild insect population. *Evolution* 73(2): 293–302.

Rose, M.R. (1984). Laboratory evolution of posponed senescence in *Drosophila melanogaster*. *Evolution* 38(5): 1004–1010.

Rose, M.R. (1985). Life-history evolution with antagonistic pleiotropy and overalpping generations. *Theoretical Population Biology* 28(3): 342–358.

Rose, M.R. (1991). *Evolutionary biology of aging*. 1st ed. Oxford University Press.

Santos, E.S.A., and Nakagawa, S. (2012). The costs of parental care: a meta-analysis of the trade-off between parental effort and survival in birds. *Journal of Evolutionary Biology* 25(9): 1911–1917.

Savory, F.R., Benton, T.G., Varma, V., Hope, I.A., and Sait, S.M. (2014). Stressful environments can indirectly select for increased longevity. *Ecology and Evolution* 4(7): 1176–1185.

Scharer, L., Rowe, L., and Arnqvist, G. (2012). Anisogamy, chance and the evolution of sex roles. *Trends in Ecology & Evolution* 27(5): 260–264.

Solon-Biet, S.M., Walters, K.A., Simanainen, U.K., McMahon, A.C., Ruohonen, K., Ballard, J.W.O., Raubenheimer, D., Handelsman, J., Le Couteur, D.G., and Simpson, S.J. (2015). Macronutrient balance, reproductive function, and lifespan in aging mice. *Proceedings of the National Academy of Sciences of the United States of America* 112(11): 3481–3486.

Stearns, S.C., Ackermann, M., Doebeli, M., and Kaiser, M. (2000). Experimental evolution of aging, growth, and reproduction in fruitflies. *Proceedings of the National Academy of Sciences of the United States of America* 97(7): 3309–3313.

Sultanova, Z., Garcia-Roa, R., and Carazo, P. (2020). Condition-dependent mortality exacerbates male (but not female) reproductive senescence and the potential for sexual conflict. *Journal of Evolutionary Biology* 33(8): 1086–1096.

Tidière, M., Gaillard, J.M., Berger, V., Müller, D.W.H., Bingaman Lackey, L., Gimenez, O., Clauss, M., and Lemaître, J.-F. (2016). Comparative analyses of longevity and senescence reveal variable survival benefits of living in zoos across mammals. *Scientific Reports* 6(1): 36361.

Tower, J. (2006). Sex-specific regulation of aging and apoptosis. *Mechanisms of Ageing and Development* 127: 705–718.

Trivers, R. (1972). Parental investment and sexual selection. In: *Sexual selection and the descent of man 1871–1971* (ed. Campbell, B.), 136–179. Aldine.

Trivers, R. (1985). *Social evolution*. Benjamin-Cummings Publishing.

Turan, Z.G., Parvizi, P., Donertas, H.M., Tung, J., Khaitovich, P., and Somel, M. (2019). Molecular footprint of Medawar's mutation accumulation process in mammalian aging. *Aging Cell* 18(4): 12.

Wachter, K.W., Evans, S.N., and Steinsaltz, D. (2013). The age-specific force of natural selection and biodemographic walls of death. *Proceedings of the National Academy of Sciences of the United States of America* 110(25): 10141–10146.

Wachter, K.W., Steinsaltz, D., and Evans, S.N. (2014). Evolutionary shaping of demographic schedules. *Proceedings of the National Academy of Sciences of the United States of America* 111: 10846–10853.

Walker, D.W., McColl, G., Jenkins, N.L., Harris, J., and Lithgow, G.J. (2000). Natural selection – Evolution of lifespan in *C. elegans*. *Nature* 405(6784): 296–297.

Weismann, A. (1889). *Essays upon heredity and kindred biological problems*. Clarendon Press.

Williams, G.C. (1957). Pleiotropy, natural selection, and the evolution of senescence. *Evolution* 11(4): 398–411.

Williams, P.D., and Day, T. (2003). Antagonistic pleiotropy, mortality source interactions, and the evolutionary theory of senescence. *Evolution* 57(7): 1478–1488.

Williams, P.D., Day, T., Fletcher, Q., and Rowe, L. (2006). The shaping of senescence in the wild. *Trends in Ecology & Evolution* 21(8): 458–463.

Wilson, A.J., Charmantier, A., and Hadfield, J.D. (2008). Evolutionary genetics of ageing in the wild: empirical patterns and future perspectives. *Functional Ecology* 22(3): 431–442.

Yoda. (1983). Star Wars: Episode VI - Return of the Jedi.

Zajitschek, F., Georgolopoulos, G., Vourlou, A., Ericsson, M., Zajitschek, S.R.K., Friberg, U., and Maklakov, A.A. (2019). Evolution under dietary restriction decouples survival from fecundity in *Drosophila melanogaster* females. *Journals of Gerontology A: Biological Sciences and Medical Sciences* 74(10): 1542–1548.

Zajitschek, F., Zajitschek, S.R.K., Canton, C., Georgolopoulos, G., Friberg, U., and Maklakov, A.A. (2016). Evolution under dietary restriction increases male reproductive performance without survival cost. *Proceedings of the Royal Society B: Biological Sciences* 283(1825): 20152726.

Zajitschek, F., Zajitschek, S.R.K., Friberg, U., and Maklakov, A.A. (2013). Interactive effects of sex, social environment, dietary restriction, and methionine on survival and reproduction in fruit flies. *Age* 35(4): 1193–1204.

Zhao, Y., Zhang, B., Marcu, I., Athar, F., Wang, H.Y., Galimov, E.R., Chapman, H., and Gems, D. (2021). Mutation of daf-2 extends lifespan via tissue-specific effectors that suppress distinct life-limiting pathologies. *Aging Cell* 20(3): e13324.

3

Offspring Size and Life History Theory: What Do We Know?

What Do We Still Need to Learn?

Dustin J. Marshall

School of Biological Sciences, Monash University, Clayton, Victoria, Australia

3.1 Offspring Size Defined

Before discussing offspring size, we need to define it. The sorts of distinctions we make about what is, and is not, offspring size determine the scope of our discussions and the forces that are relevant to the ecology and evolution of offspring size. It is relatively straightforward to define offspring size in species that produce discrete propagules (*e.g.*, seeds or eggs), where each propagule is released from the parent and completes development independently of the parent. In such instances, offspring mass probably describes offspring size very well. However, in species where offspring are retained and provisioned on or in the parent for some or all of their development, offspring size becomes more nebulous. We could describe offspring size as the size of the egg or ovicell for these species, but in many instances, this will not capture the subsequent size of offspring that are released. For example, ovicell size would not describe the total investment mothers make in offspring in humans, nor would it describe the size of offspring that are released from the mother. This complicating effect of post-ovulatory provisioning, which occurs from bryozoans to fish and even to mammals, means that a different measure of offspring size is necessary. Similarly, some amphibians, insects and marine invertebrates lay 'trophic' or 'nurse' eggs (Spight 1976, Kudo and Nakahira 2005, Dugas et al. 2016) whereby mothers augment their investment in offspring with unfertilized eggs that are consumed while the offspring develop – an investment that affects the size of offspring but only after the release of the propagule from the mother. Perhaps most obviously, mammals and birds provision their offspring with milk or collected food after they have been released from the mother, and this provisioning can more than double the size of the provisioned offspring. Should we consider the size of the neonate or hatching as offspring size, or should we only count the size of weaned young or fledglings? No discrete definition will ever be perfect for describing the biology of a continuum. But, for our purposes, I will define offspring size as the size of offspring when released from the parent and gains external resources that are not products of the maternal reproductive tract. Under this definition, the mass of mammalian neonates or avian eggs represents offspring size in mammals and birds, while the size of offspring once they have consumed all trophic eggs is appropriate for species with trophic eggs. The egg of a livebearing fish does not represent offspring size but the fry, larvae or juvenile released from the mother does. Accordingly, egg size captures offspring size in oviparous species of all kinds, whereas neonate size captures offspring size in viviparous species.

3.2 The Knowns

3.2.1 What Does Offspring Size Affect?

Within any one species, larger offspring tend to perform better than smaller offspring. By performing better, I mean larger offspring have higher survival, more rapid growth, or greater reproductive output, all of which imply that offspring size affects fitness. Offspring size can affect performance in every life history stage (Bernardo 1996, Marshall et al. 2018). The list of studies showing relationship between offspring size and some element of offspring performance grows constantly, so attempting to quantify and list these effects exhaustively is quixotic. Larger offspring can have higher hatching success or survival (Bagenal 1969), and can be more resistant to predation (Palmer 1990), starvation (Berkeley et al. 2004), competition (Allen et al. 2008), or stressors of various kinds including anthropogenic pollutants (Marshall 2008, Johnson 2022).

Larger offspring tend to take longer to develop, presumably because they have more cells/tissues that need to be created or differentiated (Gillooly et al. 2002). Hence, assuming the same rate of biological work, larger eggs will take longer to develop than smaller eggs. Larger eggs might actually have lower capacity for developmental work relative to their size because metabolic rate tends to scale with offspring size with an exponent of less than 1 (Pettersen et al. 2015, 2018). Consequently, larger eggs have relatively lower metabolic rates with which to power development, thereby making the relationship between offspring size and development time even steeper.

While the intraspecific relationship between offspring size and development time is relatively well resolved, interspecific relationships seem less well resolved. In part, the relationship between offspring size and development depends on whether larvae feed or not. Across most marine invertebrates, larger eggs take longer to hatch but spend less time in the larval phase (Levitan 2000). Church et al. (2019) argued that there is no relationship between offspring size and development time among species in insects specifically, contradicting earlier compilations across the same group (Maino et al. 2017). While among-species comparisons should not be used to infer within-species patterns more generally, the conflicting conclusions of the two studies are worth discussing, if only to encourage further research in this specific group. To me, the discrepancy arises because Church et al. (2019) concludes there is no relationship between offspring size based on a phylogenetically controlled analysis that finds a marginally non-significant effect. Meanwhile, a regression that excludes phylogeny finds a positive relationship between offspring size and development for the same dataset. What conclusion should we make? It is important to note that phylogenetically controlled analyses are necessary for accounting for a lack of evolutionary independence among related groups in order to conclude whether there is an association between two traits that is independent of shared ancestry alone (Harvey and Pagel 1991). I am in favour of using phylogenetically controlled analyses for describing macro-evolutionary patterns, therefore. However, in the particular analysis by Church et al. (2019), there is strong covariance between clade and the predictor, egg size. Specifically, most of the species with larger eggs are polyneopterans, such that the power to detect effects of egg size on developmental time over and above the phylogenetically associated effect is greatly diminished. In this case, based on this analysis, I would therefore conclude that there is a strong tendency for larger eggs to take longer to develop, but more data for species with larger eggs outside of the Polyneopteran clade are required. I predict that, with more data, a strong, positive relationship between egg size and development time in insects, even while accounting for phylogenetic (non)independence will be found. Happily, the database of egg sizes provided by Church et al. (2019) provides a rich resource of egg sizes. We just need more estimates of development time in insects.

Offspring size effects extend well into ontogeny. Larger offspring often access more resources initially than their smaller conspecifics, perhaps because the size of their resource-capturing structures (*e.g.*, mouths, leaves, roots, filters) co-varies with offspring size (Marshall et al. 2018). Offspring size can even define the trophic mode of offspring, such that tadpoles from smaller eggs might be herbivorous while conspecifics from larger eggs might be omnivorous (Martin and Pfennig 2010). Offspring size can affect offspring dispersal, via direct mass effects (*e.g.*, smaller seeds might disperse further; Parciak 2002), indirect effects on development time (*e.g.*, larger eggs take longer to hatch so they drift in currents for longer), or via offspring size-mediated developmental dimorphisms (*e.g.*, small eggs hatch as free-swimming feeding larvae while larger eggs develop directly into benthic juveniles, Krug 1998). Offspring size effects can persist throughout the life history, affecting the fecundity and even longevity of those offspring as adults (Kesselring et al. 2012). These long-lasting effects might emerge from the fact that small size/growth advantages early in life can translate into major differences later if size affects resource access or phenology. In some instances, offspring size can be the single most important intrinsic driver of offspring performance and may be the reason why maternal effects typically dwarf genetic effects in quantitative genetic studies of offspring phenotype and performance (Bernardo 1996).

It is important to note that larger offspring do not always perform better than smaller offspring of the same species. In some environments for some species, larger offspring can even have lower performance than smaller offspring. For example, tadpoles from larger eggs do better in cooler temperatures but worse in warmer temperatures (Kaplan 1992). More generally, the effects of offspring size are context-dependent, *i.e.*, increased offspring size may carry significant benefits under some circumstances but not under others. For example, it can be much better to be a larger offspring when competing against conspecifics of the same age but offspring size has no benefit to offspring when they compete against established adults (Allen et al. 2008). A range of factors have been shown to alter the offspring size–performance relationship, which alters the fitness returns of any particular offspring size and, as we will explore later, this leads to plasticity in size of offspring that parents produce.

The ubiquity and strength of offspring size effects inevitably mean that variation in offspring size has ecological consequences. Offspring size mediates density-dependence (albeit in context-specific ways) and competitive interactions among species (Stanton 1984, Cameron et al. 2017). Offspring size can drive population dynamics, with cohorts of larger, better-quality offspring showing higher recruitment success than cohorts of smaller offspring (Benton et al. 2005, Beckerman et al. 2006). Consequently, temporal patterns in offspring size can drive temporal patterns in the recruitment of individuals into adult populations (Marshall 2021). Offspring size in resident species can even affect subsequent community assembly – presumably because offspring size affects the feeding niche of one species, it will also, therefore, shape the resource environment available to other species, thus changing the trajectories of communities (Davis and Marshall 2014). Similarly, relative offspring size can mediate the sign of density-dependent interactions in *Bugula neritina* from competition to facilitation, for example, in some instances, larger offspring can facilitate the performance of smaller offspring because they can change local food regimes (Cameron et al. 2017).

From an applied perspective, because larger offspring often perform better, the role of offspring size is increasingly appreciated in altering the dynamics of exploited species (Marshall et al. 2010; see also Chapter 23). For example, larger mothers tend to produce larger offspring within any one species, so the removal of larger mothers might disproportionately affect the replenishment of a harvested species (Barneche et al. 2018). Similarly, because benign environments tend to favour the production of smaller offspring, the release of hatchery-reared offspring into wild populations might alter the distribution of offspring sizes in the wild, thereby altering recruitment success (Heath et al. 2003). That offspring size is so variable, affects so many components of offspring fitness, and can alter ecological dynamics, it is perhaps unsurprising that biologists have long sought to understand the evolutionary and ecological drivers of variation in offspring size.

3.2.2 Offspring Size and Theory

We have seen that larger offspring tend to perform better than smaller offspring and that, all else being equal, larger offspring are of higher 'quality'. Why then would parents produce smaller, lower-quality offspring? Answering this question requires an understanding of the theory of offspring size, some of the reasons are intuitively appealing but others are more complicated.

For the moment, let us assume that selection for offspring size acts solely on mothers, such that any strategy that increases maternal fitness will be selected for (we will explore why this may or may not be true later). Two aspects of offspring size therefore determine maternal fitness – the offspring size–fitness relationship and the offspring size–fecundity relationship. As we have seen, the offspring size–fitness relationship varies but is typically positive: larger offspring tend to have higher fitness than smaller offspring. The reason why mothers might still be favoured by producing offspring that are not as large as possible, however, comes from the second component – offspring size–fecundity. Theory has long assumed that mothers have a finite supply of resources, such that they can either produce a few larger offspring or more numerous, smaller offspring (Lack 1947, Vance 1973a,b, Smith and Fretwell 1974). In other words, the theory assumes a trade-off between offspring size and fecundity. From these two fitness components (offspring performance and number), we can see that two different species could produce very different offspring sizes with the same amount of resources (and hence very different fecundities) and have equivalent maternal fitness (Figure 3.1). So, the theory holds that the size of offspring that is optimal (*i.e.*, maximizes maternal fitness) depends principally on the relationship between size and performance.

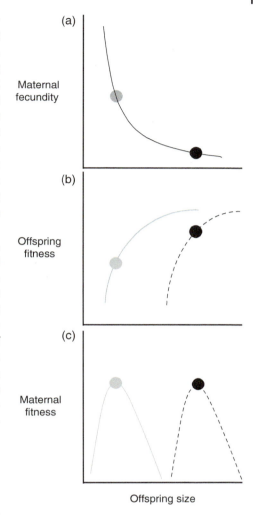

Figure 3.1 Schematic representation showing two hypothetical species that produce offspring of different sizes (shown in grey and black). Panel (a) shows the maternal fecundity and offspring size are both on the same trade-off between size and number (indicated in the black line). Panel (b) shows the species-specific relationships between offspring size and fitness (indicated by the grey line and black dashed line for each species). Panel (c) shows the species-specific relationships between offspring size and maternal fitness (which is the product of fecundity and offspring fitness). Note that each species has a different optimal offspring size where maternal fitness is maximized. The relationship between offspring size and fitness is thought to be the chief driver of selection on offspring size, while maternal fitness is typically the parameter that is thought to be maximized. Note that in neither case is offspring fitness maximized relative to maternal fitness, illustrating the conflict between the provisioning strategy that maximizes the fitness of offspring and parents.

Generally speaking, when increasing offspring size yields a large increase in fitness, then all else being equal, larger offspring will be favoured. When increasing offspring size yields little fitness return, then *ceteris paribus*, smaller offspring will be favoured (Smith and Fretwell 1974).

The theory base of offspring size goes back to Lack (1947) who first considered how offspring size increased offspring performance but traded off with offspring number, and that maternal fitness was maximized by balancing these two effects. The first formal theory was published by Vance (1973a,b) which focused on marine invertebrates specifically but contained an offspring size–performance relationship and size–number trade-off. The following year, Smith and Fretwell (1974) published a more broadly known theory that contained a more generic offspring size–number relationship. These theoretical works formed the foundation of offspring size optimality models, which continue to be used to understand offspring size variation today. These approaches will always be useful but they inevitably exclude some key elements, namely the role of dispersal and interactions. Later, I will explore how including these elements might change theoretical predictions.

So far, we have focused on the offspring size–fitness relationship but we should also consider the size–fecundity relationship. The offspring size–number trade-off was long assumed to be fixed and invariant, but more recently there have been theoretical explorations of how this might change. For example, Sakai and Harada (2001, 2004) assumed provisioning larger offspring was more energetically wasteful than provisioning smaller offspring, particularly for smaller mothers with less resources. Hence, the costs of producing larger offspring were disproportionately greater for smaller mothers and they would therefore be favoured by selection to produce smaller offspring than larger mothers, even if the offspring size–fitness relationship was identical for both maternal size classes. These ideas have since been extended (Filin 2015) and illustrate that both the size–number trade-off and the size–fitness relationship shape selection on offspring size and both should be considered.

We often assume that offspring size and number trade off in a straightforward fashion but the necessary empirical evidence for this assumption is actually lacking. Importantly, estimating the observed relationship between offspring size and number within species does not tell us whether size and number actually trade off (van Noordwijk and de Jong 1986). As eloquently illustrated by Reznick et al. (2000), observing or failing to observe a trade-off in phenotypic states provides no indication of whether these two phenotypes involve a trade-off in processes (Garland et al. 2022). To be clear, phenotypic states here refer to covariances between two traits, whereas processes refer to mechanistic links between those traits. The theory makes assumptions only about processes, not states, so while it is tempting to regard a negative relationship between offspring size and number as evidence of a process trade-off (as many authors did), we really should not (Stearns 1992). Thus, a major gap in our knowledge is the process of how offspring size and number trade off against each other (see Unknowns, further in the chapter, but also Lim et al. 2014).

Optimal offspring size theory explores the fitness returns of different offspring size 'strategies' based on the simplifying assumption that offspring do not interact with each other and habitats are uniformly distributed. Subsequent theory introduced more complexity, using game theory to explore how frequency-dependence, dispersal and habitat limitation might alter selection on offspring size (Parker and Begon 1986, McGinley et al. 1987). For example, producing smaller offspring might result in poorer offspring performance *per capita*, but it allows mothers to produce many more offspring. This, in turn, increases the size of the dispersal kernel and the probability of finding and colonizing rare, good-quality habitats (McGinley et al. 1987). This fecundity benefit might shift the balance in favour of producing smaller offspring even if they perform poorly on average. Conversely, as mothers increase in size, they may benefit from producing larger offspring, not because those offspring are better necessarily, but producing larger offspring will reduce fecundity, and could reduce the negative effects of density-dependence. Today, it is clear that all of these factors can apply more or less depending on the species and environment, and anything that affects the offspring size–fitness relationship will affect selection on offspring size accordingly.

The web of selection acting on offspring size is complex but theory can be remarkably successful in predicting how offspring size should vary with different environmental conditions (Einum and Fleming 2000, Marshall and Keough 2008). For example, theory that focuses on how temperature affects the costs of development, whereby small increases in temperature alter developmental costs (according to the differential sensitivity of metabolic *vs.* developmental rates), successfully predicts how temperature affects offspring size across a range of species, from invertebrates to fish, *i.e.*, when developmental costs decrease with temperature, warmer mothers tend to produce smaller offspring and *vice versa* (Pettersen et al. 2019). This covariance between maternal environment and the size of offspring that mothers produce is known more generally as transgenerational plasticity and further in the chapter I explore this phenomenon in more detail.

3.2.3 Transgenerational Plasticity in Offspring Size – The Link Between Generations

Parents provisioning their offspring must balance two competing forces acting on offspring size and number: as a consequence, they should invest as little as is optimal in each offspring in order to maximize fecundity. The fitness return of any one offspring investment strategy (here 'strategy' refers to both the size and the number of offspring that are produced) depends on the environment that offspring experiences (Burgess and Marshall 2014). In some instances, parents can directly determine the key environmental conditions that offspring experience, for example, parents can choose specific oviposition sites, or create favourable natal habitats such as nests (Dolia et al. 2023). In other instances, parents cannot alter the environment offspring experience but parents also experience that environment themselves and can therefore predict it, for example, environmental temperature in aquatic habitats (Burgess and Marshall 2011) or colony size in honey bees (Amiri et al. 2020). Finally, parents might not experience the offspring environment if there is a significant delay between egg laying and offspring emergence but they can anticipate the environment offspring will experience because of predictable patterns in environmental variation (*e.g.*, seasons). Thus, through a range of mechanisms, parents have information about the environment their offspring are likely to experience and can provision their offspring accordingly – a process known as transgenerational plasticity or adaptive parental effects. For example, parents laying eggs in benign environments (that are conducive to survival and growth) might reduce per-offspring investment because offspring do not require extensive resources to thrive, allowing parents to maximize fecundity (Stamps 2006). The converse is true when parents lay offspring in harsh environments (Fox et al. 1997). The list of examples of transgenerational plasticity in offspring size in response to different environmental conditions continues to grow (Yin et al. 2019), but one key issue – the capacity of parents to predict the offspring environments – tends to be overlooked persistently (Uller et al. 2013, Burgess and Marshall 2014).

Transgenerational plasticity in offspring size should minimize the mismatch between the optimal offspring size and the size of offspring that are produced, but transgenerational plasticity is imperfect. Parents are unlikely to be able to change the size of their offspring instantaneously and once provisioned, it may be physiologically impossible to reverse or alter this provisioning even when the offspring environment changes. Hence mismatches between optimal and actual offspring size are inevitable when there are lags between offspring provisioning and offspring release into temporally variable environments: the degree of mismatch will depend on the relative temporal resolution of each. In more formal terms, the capacity of parents to optimally provision their offspring via adaptive transgenerational plasticity depends on the relatively temporal auto-correlation of the environmental variation (Burgess and Marshall 2014). If environments vary predictably over the timescale at which parents assess the environment and provision their offspring, then transgenerational plasticity is possible and favoured by selection. However, if environments vary too rapidly or unpredictably over the biological window that parents have, then adaptive transgenerational plasticity is impossible and will not be favoured by selection. For example, at a field site where I do experiments, the temperature of seawater on any one day is highly predictive of the temperature up to six weeks into the future (Marshall 2021). In my study species, the cheilostome bryozoan *B. neritina* (see Figure 3.2), larvae are provisioned from fertilized ovicells over the period of about one week and so in this case the environment is sufficiently predictable over the appropriate temporal scale for parents to provision their offspring optimally. Hence there is a strong (negative) relationship between offspring size, environmental temperature and offspring performance in this system (Marshall 2021). Too few studies assess environmental predictability in a formal sense, particularly at the scales that are appropriate to the study organism but the tools for doing so are readily available and I strongly recommend them (Marshall and Burgess 2015). In the absence of such assessments, many experiments seeking to study transgenerational plasticity in offspring size risk being ignorant of whether their environmental manipulations are biologically

Figure 3.2 *Bugula neritina* is a colonial marine invertebrate that broods its offspring in chambers known as ovicells before releasing them. The larvae depend entirely on maternally derived resources while swimming and searching for an appropriate settlement site before permanently attaching to a surface, undergoing metamorphosis and beginning to feed. This species is a model organism for studying offspring size effects in the marine environment.

relevant or meaningful – such experiments may manipulate environments in ways that parents cannot perceive as worth responding to. As a consequence, the ubiquity and strength of transgenerational plasticity in offspring remain unclear – there are too many naïve studies that find no evidence for adaptive transgenerational plasticity (Uller et al. 2013), and I suspect this is because many of these studies had no chance of finding it.

Nevertheless, there appear to be some consistent patterns in offspring size and its relationship with the environment within species. As mentioned earlier, higher temperatures tend to favour the production of smaller offspring (Pettersen et al. 2019). Because development rate is more temperature sensitive than metabolic rate in most organisms Marshall et al. (2021), small increase in temperature decreases the costs of offspring development, such that parents can decrease the size of their offspring thereby increasing their fecundity (Pettersen et al. 2019). Whether these differences in developmental costs that drive offspring size–temperature relationships within species also drive the same patterns that we observe across many species is unclear (Marshall et al. 2018). Changing food environments also generates changes in offspring size. In a range of species, lower food availability tends to induce increased investment in offspring size (Fox and Czesak 2000). In such instances, parents probably seek to buffer their offspring from the negative effects of reduced food availability by increasing their offspring's nutritional reserves. Food quality has similar effects on offspring provisioning. For example, butterflies laying eggs on higher-quality host plants tend to make smaller offspring than those laying eggs on lower-quality host plants (Swanson et al. 2016).

While I have focused mostly on adaptive transgenerational plasticity, it must be noted that changes in offspring size do not always benefit offspring. Assuming that selection acts to maximize maternal fitness, some environments favour provisioning strategies that do not reflect any optima for that round of reproduction. For example, experimentally halving the size of *B. neritina* mothers (they are modular, colonial organisms) results in those mothers dramatically reducing the size of offspring that they produce in the short term (Marshall and Keough 2004). This reduction occurs despite the offspring environment likely remaining the same. Halved mothers reduce the size of their offspring so as to minimize their allocation to a reproductive activity that has already been initiated, and thereby increase their allocation to growth. In this scenario, selection probably favours colonies of a certain size and so mothers are trade-off current reproduction in favour of somatic growth to increase fitness in the longer term. This example illustrates that parents trade off not only allocation to the size and number of offspring but also to current *vs.* future reproduction and growth. This complex web of trade-offs means that any change in parental environment or phenotype will alter the costs and benefits of any one provisioning strategy. For iteroparous organisms, it is therefore important to view any allocation strategy in this broader context. From a practical perspective, this means that offspring size can act as a phenotypic link across generations. For example, stress during the parental generation could mean that mothers produce smaller, lower-quality offspring that perform less well, even if the offspring themselves are not exposed to that stressor (Stamps 2006). Thus, while my primary focus on adaptive transgenerational plasticity has been its role as a buffer for offspring, insulating them from environmental variation, it can also act as a conduit, transmitting stressor effects across generations (Guillaume et al. 2016).

In addition to external environmental cues inducing changes in offspring size, there are also correlates of offspring size that relate to the parental phenotypes themselves. For example, larger individuals tend to produce larger offspring than smaller conspecifics within a range of organisms (Lim et al. 2014, Barneche et al. 2018). Age also affects offspring size independently of adult size, albeit in complex ways – sometimes older individuals produce smaller offspring, and sometimes they produce larger. It is worth noting, however, that disentangling size, age and condition is not always straightforward, particularly for field populations (Jones et al. 2014). Unfortunately, this biological uncertainty translates to uncertainty in the management of exploited populations (Marshall et al. 2010; see also Chapter 23). If larger/older mothers produce larger offspring that have consistently higher chances of performing well, then any management strategy that fails to protect larger mothers will disproportionately reduce the replenishment of exploited populations. However, most fisheries management fails to formally account for such effects because the links between maternal size, offspring size, and offspring fitness are unclear (Marshall et al. 2021). Nevertheless, I would argue that, in the absence of definitive evidence, a more conservative approach would be to assume that offspring from larger individuals, given they are typically larger, do have the best chance of recruiting and we should therefore manage adult populations accordingly (Marshall et al. 2021).

3.2.4 Summary of the Knowns of Offspring Size

Standard theory assumes that parents face a trade-off between the size and number of offspring that they produce for a given amount of resources. Bigger offspring tend to perform better than smaller offspring, in terms of survival, growth and reproduction. However, the relationship between offspring size and performance changes with the environment that offspring experience and so we should expect and observe a relationship between offspring size and various environmental conditions and parental phenotypes. Despite decades of research in this field, however, key uncertainties regarding offspring size remain. In the subsequent section, I highlight what I believe are the outstanding problems in the field.

3.3 The Unknowns

3.3.1 The Offspring Size and Number – How Do They Scale with Costs?

The offspring size–number trade-off is central to offspring size theory and has deep intuitive appeal, but our empirical understanding of this issue is surprisingly limited. As discussed earlier, the observation (or not) of a relationship between the size and number of offspring provides little insight into the process of allocation to size *vs.* number. A recent study (Furness et al. 2022) sought to deal with this issue by accounting for maternal body size in amphibians, which was assumed to be a proxy of resource availability. This approach is an improvement on studies that fail to account for body size but it still does not really capture the process of allocation to size *vs.* number – instead, it shows the evolutionary outcomes of these allocation processes and describes these phenotypic endpoints. More generally, we need a better understanding of exactly how the costs of production scale with the size and number of offspring that are being produced (Garland et al. 2022). Theory assumes strictly that both these costs each scale at exactly 1, for example, a doubling of offspring size or doubling fecundity, doubles the total costs of production. However, there are good reasons to believe the costs of fecundity and the costs of specific offspring sizes might scale non-linearly.

First, if offspring consume significant resources while being provisioned, then any reproduction strategy that extends the provisioning phase will increase the costs of that strategy (Sakai and Harada 2001). Hence, if producing more offspring in a given reproductive bout reduces the rate at which the entire brood is resourced and extends the brooding period, then the costs per offspring will disproportionally increase (Filin 2015). While theory speculates about this, empirical tests are exceedingly rare, though there are intriguing attempts in plants that have explored this issue (Sakai and Harada 2001). Costs might also deviate from scaling at 1:1 with fecundity because mothers might reuse reproductive structures over time (Filin 2015). To explain this, imagine that mothers must first grow reproductive structures such as follicles for creating eggs, these structures could then be reused in later rounds of reproduction. For that first reproductive bout, mothers would therefore pay higher costs for given fecundity than for later bouts, thereby generating a systematic difference in the relationship between fecundity and costs of fecundity between reproductive bouts. Again, while some theory deals with the ancillary costs of reproductive structures (Filin 2015), there is very little empiricism on this topic. Furthermore, if overall whole-organism metabolism scales with body mass at less than 1, and reproductive tissue accounts for a significant proportion of body mass (both of which apply in many organisms), then higher fecundities should impose relatively lower, but absolutely higher, metabolic loads on reproductive females. Again, this seems likely but largely untested empirically. Finally, there are alternative approaches to understanding how offspring size and number functionally trade-off against each other, namely: quantitative genetics. By explicitly estimating the genetic covariance between offspring size and number, one can quantify how much scope for independent evolution in both traits can occur, or whether there are quantitative constraints (*sensu* Futuyma 2010) that prevent mothers from producing offspring that are both large and numerous. Examples of such approaches are rare (see Rollinson and Rowe 2015 for a comprehensive meta-analysis of all that are available) but extremely valuable. For example, Johnson (2022) examined the heritability of offspring size and body size in fish. Similarly, estimating how fecundity changes with artificial selection on egg size can provide valuable insights into how these two traits co-vary from an evolutionary standpoint (Schwarzkopf et al. 1999).

Similar to the assumption that fecundity costs scale proportionately to fecundity, theory has also assumed that offspring costs scale proportionately to offspring size. We should be sceptical about the robustness of this assumption. First, almost nothing in nature is strictly linear so this would be a rare exception. Second, just like fecundity, if making a larger offspring means taking longer to produce a batch, then the costs of producing those offspring will rise due to higher overhead costs of reproduction (Sakai and Harada 2001, 2004). Again, theory has speculated about this, but there are very few empirical tests. So, we are left with the precarious situation where the key components of offspring size theory have not been adequately tested empirically and we have good reasons to suspect that standard assumptions are wrong. This seems like important territory for empiricists to explore, as changing these assumptions would likely yield very different predictions and reshape our understanding of offspring size strategies.

3.3.2 Why Does Offspring Size Affect Fitness?

The mechanisms by which offspring size affects offspring performance are remarkably unclear in most, if not all organisms (Marshall et al. 2018). Larger offspring might simply have more resources than smaller offspring such that they can use these additional resources for more fitness-enhancing functions. Alternatively, larger offspring might be able to convert their additional resources to larger or better resource acquisition structures (leaves, roots, mouths) than smaller offspring, giving them a resource acquisition advantage. For example, Sinervo et al. (1992) showed elegantly how offspring size alters interference competitive abilities in lizards. Offspring size effects could be driven by size alone. While studies are rare, there are examples of larger offspring more able to access refuges from predators (Palmer 1990). Finally, larger offspring might simply be more metabolically efficient, because larger conspecifics should have lower mass-specific metabolic rates, and they may be able to accumulate energy reserves more effectively (Pettersen et al. 2018, 2022). None of these effects are mutually exclusive but evidence for any of them is patchy and more studies are needed.

3.3.3 Who Controls Offspring Size?

Perhaps the biggest problem in studying the evolutionary ecology of offspring size is assigning fitness (Wolf and Brodie 1998, Wolf and Wade 2001). Classic life history theory, as it relates to offspring size, focuses exclusively on fitness returns to parents, usually mothers. Hence, throughout all of our discussions here, we have focused on the balance between size and fecundity and producing offspring of a size that maximizes maternal fitness. But, there is another theory, based on quantitative genetics, which is equally relevant and important that assigns fitness solely to the organism which has the traits, phenotypes and genotypes of interest, in this case, the offspring. Under this world view, we should think solely about how selection maximizes offspring fitness, without considering mothers at all. These two divergent views are simply ends of the same continuum and ultimately pose a bigger question: whose phenotype is offspring size? On the one hand, offspring size, like any other trait, is simply the trait of the offspring – it is after all the generation in which the trait manifests. On the other hand, it is hard to argue that offspring alone determine their size. There is a vast literature showing that mothers that have no genetic differences on average (and therefore neither do their offspring) produce very different-sized offspring depending on the maternal environment (Fox and Czesak 2000, Guinnee et al. 2004). These studies strongly suggest that offspring size is a plastic trait, under maternal control. But this does not mean offspring size has no heritable, genetic underpinning (Rollinson and Rowe 2015). Of course, it must – many studies demonstrate genetic component to offspring size (Miles and Wayne 2009, Amiri et al. 2020). So, I and others would argue that offspring size is an odd trait in that it affects the fitness of both mothers and offspring and its phenotypic value is set partly by offspring but mostly by mothers (Trivers 1974, Bernardo 1996). What does this mean for theory and how should we talk about the fitness consequences of different offspring size strategies? Wolf and Wade (2001) provide a wonderful paper on how to assign fitness to offspring and parents. They provide an elegant theoretical framework that describes the various issues, and how to resolve them. To very coarsely summarize this work, if offspring size is mostly maternally determined, then it is appropriate to attribute the fitness consequences of offspring size to mothers, in accordance with life history theory. On the other hand, if offspring size is determined by the genotype of offspring, then a quantitative genetics approach is appropriate. I suspect that the former is more likely than the latter given the transgenerational plasticity studies cited earlier, but it is remarkable how poorly tested and poorly parameterized the models of Wolf and Wade (2001) are in an empirical context. This represents a fundamental knowledge gap in the field that has persisted ever since this paper, which I believe represents a priority for future research. The problem (and perhaps the reason for this problem's persistence) is that estimating the necessary parameters requires a multigenerational quantitative genetics approach – a daunting but necessary undertaking. I urge those interested in this issue to read Wolf and Wade (2001) as well as Hadfield (2013) and consider taking on this challenge.

3.3.4 Summary of Unknowns in Offspring Size

Despite decades of research, we still have a remarkably poor understanding of why larger offspring tend to perform better than smaller offspring, whether larger offspring cost proportionately more to make than smaller offspring, and how fecundity costs scale with the number of offspring. While we have long assumed that we should attribute fitness consequences of different offspring size strategies to mothers, there is actually very little evidence for this assumption. In other words, we lack a mechanistic understanding of why offspring size affects fitness, and how much different offspring strategies cost. At a higher conceptual level, we do not really know which generation determines the size that offspring are, nor how selection acts to shape offspring size, despite making strong assumptions on these issues. To me, this is all very exciting because it means we have much more to do: these are empirical questions that are as daunting as they are rewarding.

References

Allen, R.M., Buckley, Y.M., and Marshall, D.J. (2008). Offspring size plasticity in response to intraspecific competition: an adaptive maternal effect across life-history stages. *The American Naturalist* 171: 225–237.

Amiri, E., Le, K., Melendez, C.V., Strand, M.K., Tarpy, D.R., and Rueppell, O. (2020). Egg-size plasticity in *Apis mellifera*: honey bee queens alter egg size in response to both genetic and environmental factors. *Journal of Evolutionary Biology* 33: 534–543.

Bagenal, T.B. (1969). Relationship between egg size and fry survival in brown trout *Salmo trutta* L. *Journal of Fish Biology* 1: 349–353.

Barneche, D.R., Robertson, D.R., White, C.R., and Marshall, D.J. (2018). Fish reproductive-energy output increases disproportionately with body size. *Science* 360: 642–645.

Beckerman, A.P., Benton, T.G., Lapsley, C.T., and Koesters, N. (2006). How effective are maternal effects at having effects? *Proceedings of the Royal Society B: Biological Sciences* 273: 485–493.

Benton, T.G., Plaistow, S.J., Beckerman, A.P., Lapsley, C.T., and Littlejohns, S. (2005). Changes in maternal investment in eggs can affect population dynamics. *Proceedings of the Royal Society B: Biological Sciences* 272: 1351–1356.

Berkeley, S.A., Chapman, C., and Sogard, S.M. (2004). Maternal age as a determinant of larval growth and survival in a marine fish, *Sebastes melanops*. *Ecology* 85: 1258–1264.

Bernardo, J. (1996). The particular maternal effect of propagule size, especially egg size: patterns models, quality of evidence and interpretations. *American Zoologist* 36: 216–236.

Burgess, S.C., and Marshall, D.J. (2011). Temperature-induced maternal effects and environmental predictability. *Journal of Experimental Biology* 214: 2329–2336.

Burgess, S.C., and Marshall, D J. (2014). Adaptive parental effects: the importance of estimating environmental predictability and offspring fitness appropriately. *Oikos* 123: 769–776.

Cameron, H., Monro, K., and Marshall, D.J. (2017). Should mothers provision their offspring equally? A manipulative field test. *Ecology Letters* 20: 1025–1033.

Church, S.H., Donoughe, S., de Medeiros, B.A.S., and Extavour, C.G. (2019). Insect egg size and shape evolve with ecology but not developmental rate. *Nature* 571: 58–62.

Davis, K., and Marshall, D.J. (2014). Offspring size in a resident species affects community assembly. *Journal of Animal Ecology* 83: 322–331.

Dolia, J., Das, A., and Kelkar, N. (2023). House-warming: wild king cobra nests have thermal regimes that positively affect hatching success and hatchling size. *Journal of Thermal Biology* 112: 103468.

Dugas, M.B., Moore, M.P., Martin, R.A., Richards-Zawacki, C.L., and Sprehn, C.G. (2016). The pay-offs of maternal care increase as offspring develop, favouring extended provisioning in an egg-feeding frog. *Journal of Evolutionary Biology* 29: 1977–1985.

Einum, S., and Fleming, I.A. (2000). Highly fecund mothers sacrifice offspring survival to maximize fitness. *Nature* 405: 565–567.

Filin, I. (2015). The relation between maternal phenotype and offspring size, explained by overhead material costs of reproduction. *Journal of Theoretical Biology* 364: 168–178.

Fox, C.W., and Czesak, M.E. (2000). Evolutionary ecology of progeny size in arthropods. *Annual Review of Entomology* 45: 341–369.

Fox, C.W., Thakar, M.S., and Mosseau, T.A. (1997). Egg size plasticity in a seed beetle: an adaptive maternal effect. *The American Naturalist* 149: 149–163.

Furness, A.I., Venditti, C., and Capellini, I. (2022). Terrestrial reproduction and parental care drive rapid evolution in the trade-off between offspring size and number across amphibians. *PLoS Biology* 20(1): e3001495.

Futuyma, D.J. (2010). Evolutionary constraint and ecological consequences. *Evolution* 64: 1865–1884.

Garland, T., Downs, C.J., and Ives, A.R. (2022). Trade-offs (and constraints) in organismal biology. *Physiological and Biochemical Zoology* 95: 82–112.

Gillooly, J.F., Charnov, E.L., West, G.B., Savage, V.M., and Brown, J.H. (2002). Effects of size and temperature on developmental time. *Nature* 417: 70–73.

Guillaume, A.S., Monro, K., and Marshall, D.J. (2016). Transgenerational plasticity and environmental stress: do paternal effects act as a conduit or a buffer? *Functional Ecology* 30: 1175–1184.

Guinnee, M.A., West, S.A., and Little, T.J. (2004). Testing small clutch size models with *Daphnia*. *The American Naturalist* 163: 880–887.

Hadfield, J.D. (2013). The quantitative genetic theory of parental effects. In: *The evolution of parental care* (eds. Royle, N.J., Smiseth, P.T., and Kolliker, M.), 267–284. Oxford University Press.

Harvey, P.H., and Pagel, M.D. (1991). *The comparative method in evolutionary biology*. Oxford University Press.

Heath, D.D., Heath, J.W., Bryden, C.A., Johnson, R.M., and Fox, C.W. (2003). Rapid evolution of egg size in captive salmon. *Science* 299: 1738–1740.

Johnson, D.W. (2022). Selection on offspring size and contemporary evolution under ocean acidification. *Nature Climate Change* 12: 757–760.

Jones, O.R., Scheuerlein, A., Salguero-Gómez, R., Camarda, C.G., Schaible, R., Casper, B.B., Dahlgren, J.P., Ehrlén, J., García, M.B., Menges, E.S., Quintana-Ascencio, P.F., Caswell, H., Baudisch, A., and Vaupel, J.W. (2014). Diversity of ageing across the tree of life. *Nature* 505: 169–173.

Kaplan, R.H. (1992). Greater maternal investment can decrease offspring surivival in the frog *Bombina orientalis*. *Ecology* 73: 280–288.

Kesselring, H., Wheatley, R., and Marshall, D.J. (2012). Initial offspring size mediates trade-off between fecundity and longevity in the field. *Marine Ecology Progress Series* 465: 129–136.

Krug, P.J. (1998). Poecilogony in an estuarine opisthobranch: planktotrophy, lecithotrophy, and mixed clutches in a population of the ascoglossan *Alderia modesta*. *Marine Biology* 132: 483–494.

Kudo, S., and Nakahira, T. (2005). Trophic-egg production in a subsocial bug: adaptive plasticity in response to resource conditions. *Oikos* 111: 459–464.

Lack, D. (1947). The significance of clutch size. *Ibis* 89: 302–352.

Levitan, D.R. (2000). Optimal egg size in marine invertebrates: theory and phylogenetic analysis of the critical relationship between egg size and development time in echinoids. *The American Naturalist* 156: 175–192.

Lim, J.N., Senior, A.M., and Nakagawa, S. (2014). Heterogeneity in individual quality and reproductive trade-offs within species. *Evolution* 68: 2306–2318.

Maino, J.L., Pirtle, E.I., and Kearney, M.R. (2017). The effect of egg size on hatch time and metabolic rate: theoretical and empirical insights on developing insect embryos. *Functional Ecology* 31: 227–234.

Marshall, D.J. (2008). Transgenerational plasticity in the sea: a context-dependent maternal effect across life-history stages. *Ecology* 89: 418–427.

Marshall, D.J. (2021). Temperature-mediated variation in selection on offspring size: a multi-cohort field study. *Functional Ecology* 35: 2219–2228.

Marshall, D.J., Bode, M., Mangel, M., Arlinghaus, R., and Dick, E.J. (2021). Reproductive hyperallometry and managing the world's fisheries. *Proceedings of the National Academy of Sciences of the United States of America* 118(34): e2100695118.

Marshall, D.J., and Burgess, S.C. (2015). Deconstructing environmental predictability: seasonality, environmental colour and the biogeography of marine life histories. *Ecology Letters* 18: 174–181.

Marshall, D.J., Heppell, S.S., Munch, S.B., and Warner, R.R. (2010). The relationship between maternal phenotype and offspring quality: do older mothers really produce the best offspring? *Ecology* 91: 2862–2873.

Marshall, D.J., and Keough, M.J. (2004). When the going gets rough: effect of maternal size manipulation on offspring quality. *Marine Ecology Progress Series* 272: 301–305.

Marshall, D.J., and Keough, M.J. (2008). The relationship between offspring size and performance in the sea. *The American Naturalist* 171: 214–224.

Marshall, D.J., Pettersen, A.K., and Cameron, H. (2018). A global synthesis of offspring size variation, its eco-evolutionary causes and consequences. *Functional Ecology* 32: 1436–1446.

Martin, R.A., and Pfennig, D.W. (2010). Maternal investment influences expression of resource polymorphism in amphibians: implications for the evolution of novel resource-use phenotypes. *PLoS ONE* 5(2): e9117.

McGinley, M.A., Temme, D.H., and Geber, M.A. (1987). Parental investment in offspring in variable environments: theoretical and empirical considerations. *The American Naturalist* 130: 370–398.

Miles, C.M., and Wayne, M.L. (2009). Life history trade-offs and response to selection on egg size in the polychaete worm *Hydroides elegans*. *Genetica* 135: 289–298.

Palmer, A.R. (1990). Predator size, prey size and the scaling of vulnerability: hatchling gastropods *vs.* barnacles. *Ecology* 71: 759–775.

Parciak, W. (2002). Environmental variation in seed number, size and dispersal of a fleshy-fruited plant. *Ecology* 83: 780–793.

Parker, G.A., and Begon, M. (1986). Optimal egg size and clutch size – effects of environment and maternal phenotype. *The American Naturalist* 128: 573–592.

Pettersen, A.K., Schuster, L., and Metcalfe, N.B. (2022). The evolution of offspring size: a metabolic scaling perspective. *Integrative and Comparative Biology* 62: 1492–1502.

Pettersen, A.K., White, C.R., Bryson-Richardson, R.J., and Marshall, D.J. (2018). Does the cost of development scale allometrically with offspring size? *Functional Ecology* 32: 762–772.

Pettersen, A.K., White, C.R., Bryson-Richardson, R.J., and Marshall, D.J. (2019). Linking life-history theory and metabolic theory explains the offspring size-temperature relationship. *Ecology Letters* 22: 518–526.

Pettersen, A.K., White, C.R., and Marshall, D.J. (2015). Why does offspring size affect performance? Integrating metabolic scaling with life-history theory. *Proceedings of the Royal Society B: Biological Sciences* 282: 20151946.

Reznick, D., Nunney, L., and Tessier, A. (2000). Big houses, big cars, superfleas and the costs of reproduction. *Trends in Ecology & Evolution* 15: 421–425.

Rollinson, N., and Rowe, L. (2015). Persistent directional selection on body size and a resolution to the paradox of stasis. *Evolution* 69: 2441–2451.

Sakai, S., and Harada, Y. (2001). Why do large mothers produce large offspring? Theory and a test. *The American Naturalist* 157: 348–359.

Sakai, S., and Harada, Y. (2004). Size-number trade-off and optimal offspring size for offspring produced sequentially using a fixed amount of reserves. *Journal of Theoretical Biology* 226: 253–264.

Schwarzkopf, L., Blows, M.W., and Caley, M.J. (1999). Life-history consequences of divergent selection on egg size in *Drosophila melanogaster*. *The American Naturalist* 29: 333–340.

Sinervo, B., Doughty, P., Huey, R.B., and Zamudio, K. (1992). Allometric engineering: a causal analysis of natural selection on offspring size. *Science* 258: 1927–1930.

Smith, C.C., and Fretwell, S.D. (1974). The optimal balance between size and number of offspring. *The American Naturalist* 108: 499–506.

Spight, T.M. (1976). Hatching size and the distribution of nurse eggs among prosobranch embryos. *Biological Bulletin* 150: 491–499.

Stamps, J.A. (2006). The silver spoon effect and habitat selection by natal dispersers. *Ecology Letters* 9: 1179–1185.

Stanton, M.L. (1984). Seed variation in wild radish: effect of seed size on components of seedling and adult fitness. *Ecology* 65: 1105–1112.

Stearns, S.C. (1992). *The evolution of life histories*. Oxford University Press.

Swanson, E.M., Espeset, A., Mikati, I., Bolduc, I., Kulhanek, R., Whiter, W.A., Kenzie, S., and Snell-Rood, E.C. (2016). Nutrition shapes life-history evolution across species. *Proceedings of the Royal Society B: Biological Sciences* 283: 20152764.

Trivers, R.L. (1974). Parent-offspring conflict. *American Zoologist* 14: 249–264.

Uller, T., Nakagawa, S., and English, S. (2013). Weak evidence for anticipatory parental effects in plants and animals. *Journal of Evolutionary Biology*: 26: 2161–2170.

van Noordwijk, A.J., and de Jong, G. (1986). Acquisition and allocation of resources: their Iifluence on variation in life history tactics. *The American Naturalist* 128: 137–142.

Vance, R.R. (1973a). On reproductive strategies in marine benthic invertebrates. *The American Naturalist* 107: 339–352.

Vance, R.R. (1973b). More on reproductive strategies in marine benthic invertebrates. *The American Naturalist* 107: 353–361.

Wolf, J.B., and Brodie, E.D. (1998). The coadaptation of parental and offspring characters. *Evolution* 52: 299–308.

Wolf, J.B., and Wade, M.J. (2001). On the assignment of fitness to parents and offspring: whose fitness is it and when does it matter? *Journal of Evolutionary Biology* 14: 347–356.

Yin, J., Zhou, M., Lin, Z., Li, Q.Q., and Zhang, Y.-Y. (2019). Transgenerational effects benefit offspring across diverse environments: a meta-analysis in plants and animals. *Ecology Letters* 22: 1976–1986.

4

The Evolution of Insect Egg Loads

The Balance of Time and Egg Limitation

Michal Segoli[1], Miriam Kishinevsky[2], and George E. Heimpel[3]

[1] *Mitrani Department of Desert Ecology, Blaustein Institutes for Desert Research, Ben-Gurion University of the Negev, Midreshet Ben-Gurion, Israel*
[2] *Department of Integrative Biology, University of Wisconsin - Madison, Madison, WI, USA*
[3] *Department of Entomology, University of Minnesota, St. Paul, MN, USA*

4.1 Trade-Offs Between Early and Late Components of Reproduction

Reproduction is composed of a series of sequential events, which may differ tremendously between different plant and animal groups. For example, animal-pollinated flowering plants first invest in gamete production (pollen and ovules) and in structures to attract pollinators (*e.g.*, petals, scent and nectar), and later in the production of seeds and structures that allow their dispersal. Birds typically first invest in nest construction, then in egg-laying and incubation, and finally in guarding and food-provisioning for the nestlings. Insects typically first invest in egg production, then in the location or construction of suitable oviposition sites, then in egg-laying, and in some cases in guarding and/or food-provisioning of the developing offspring.

Despite these variations, all organisms share a common life history challenge, namely that investment in early components of reproduction is likely to come at the expense of later functions in the reproductive sequence (Rosenheim et al. 2010, 2014). However, information on resource availability and other constraints during later reproductive events is not always available at the early stages of reproduction. For example, at the time of nest construction and egg-laying, birds often cannot fully predict resource availability for feeding nestlings later in the season. Hence, they may not be able to estimate how many offspring they will be able to feed (Mock and Forbes 1995). Similarly, at the time of ovule and flower production, plants often cannot fully predict the amount of resources that will be available to them later for seed production, or how much pollination they will receive (Burd 1995, Burd et al. 2009). Hence, a major question in the evolution of life history strategies is how organisms balance their investment in different sequential reproductive components under uncertainty.

4.2 Time *vs.* Egg Limitation in Insects

The trade-off between early and late investment in reproduction has been studied extensively in insects that lay their eggs in discrete units of resource, and in particular in parasitoids (insects that have an immature life stage that develops on or within a single host individual, ultimately killing it; Godfray 1994; see Figure 4.1). The reproductive success of female parasitoids is considered to be limited mainly by the number of eggs that they can produce, or by the number of suitable hosts that they can find during their lifetime. Importantly, as in the examples mentioned earlier, early investment in egg production is likely to come at the expense of the time and energy that will be required later for locating high-quality hosts. Therefore, parasitoids and other animals that distribute their eggs among oviposition sites should balance their investment in egg production *vs.* their investment in energy reserves in order to maximize their lifetime reproductive success (Iwasa et al. 1984, Mangel 1987, 1989a, 1989b).

Egg load, or the number of mature oocytes available for oviposition, is at the heart of this life history trade-off. Investment in egg number may trade off with egg size (see Chapter 3 on the number and size of offspring), with the lifespan of the

Figure 4.1 A parasitoid female wasp of the genus *Acroclisoides* ovipositing into hemipteran eggs. *Source:* Photo by Alfred Daniel J.

Egg limitation   Time limitation

Figure 4.2 Illustration of egg limitation and time limitation in parasitoids. Parasitoid illustrations represent a braconid wasp ovipositing into a caterpillar. The white ovals within the parasitoid abdomen represent eggs, and the caterpillars represent suitable hosts. *Source:* Figure created by L. Segoli.

organism (see Chapter 2 on the evolution of ageing), as well as with other life history functions (see other chapters in this book). Hypothetically, under a perfect balance, a female is expected to produce the exact number of eggs that she will be able to lay over her lifetime. However, due to environmental stochasticity, females often cannot anticipate the exact number and quality of oviposition sites they are going to encounter during their lifetime, or even during a single foraging bout. Hence, individual females may experience episodes during which they have sufficient eggs, but not enough time to locate additional oviposition sites or hosts (*i.e.*, time limitation), and episodes during which oviposition sites or energy reserves to locate them are available, but eggs have already been depleted, either permanently or temporarily (*i.e.*, egg limitation) (Figure 4.2). While the risk of egg limitation is likely to lead to selection for higher fecundity, the risk of time limitation is likely to lead to selection for higher investment in energy reserves. This raises the following question: What is the relative importance of these two opposing risks in shaping insect reproductive strategies? Before addressing this question, we discuss some of the main patterns of egg maturation in insects.

4.3 Egg Maturation Patterns

Egg maturation patterns may vary between insect species, populations, and individuals. The first source of variation is the number of mature eggs carried by the female upon emergence (*i.e.*, initial egg load). While some insects emerge with all of their eggs already mature ('pro-ovigeny'), others continue producing eggs throughout their life ('synovigeny'). Notably, these two reproductive strategies are not dichotomous, but rather represent a continuum determined by the relative proportion of eggs that are mature upon emergence out of the total number of eggs produced throughout the insect's life (coined the 'ovigeny index') (Jervis et al. 2001). Second, following emergence, egg loads may be further influenced by oviposition rate. This, in turn, may depend on oviposition opportunities in the environment, but also on oviposition decisions made by the female (Skinner 1985). The third factor of importance is egg maturation rate, which again may depend on oviposition opportunities (Wu and Heimpel 2007, Dieckhoff et al. 2014), and the female's age and physiological condition (Papaj 2000, Keinan et al. 2017). Finally, in some species, females may resorb eggs ('oosorption') and convert their nutrients into energy reserves (Bell and Bohm 1975). Importantly, while in strictly pro-ovigenic species egg depletion is irreversible, synovigenic females can at least partially compensate for oviposition by maturing more eggs (Dieckhoff et al. 2014). Nevertheless, synovigenic females may still be prone to the risk of egg limitation either in an absolute sense (*e.g.*, towards the end of their lives) or during a certain point in their lives (*e.g.*, towards the end of the day) (Boggs 1997, Heimpel et al. 1997, 1998, Legaspi and Legaspi 2004, 2008). Thus, synovigenic and pro-ovigenic species are both potentially subject to trade-offs between egg and time limitation.

4.4 The Relative Importance of Egg and Time Limitation

The question of the relative importance of these two opposing risks initiated a long-lasting debate in the literature, with many assuming that egg limitation is of minor importance (reviewed in Rosenheim et al. 2008, Rosenheim 2011). In support of this view, some authors argued that under a realistic level of stochasticity in reproductive opportunities, parasitoids and other insects that deposit eggs in discrete units of resource should produce a surplus of eggs, making the occurrence of actual egg limitation negligible, to the point where it could be ignored (Sevenster et al. 1998, Ellers et al. 2000). This is consistent with the view that selection should maximize the rate of oviposition (Charnov and Skinner 1984, 1985, Visser et al. 1992, Godfray 1994) and, more generally with classical foraging theory, which was based on the assumption that organisms behave in a way that maximizes the harvest rate of resources from their environment (Stephens and Krebs 1986). This view was also based in part on field observations showing that parasitoids often use low-quality hosts such as species of insects with low suitability for parasitoid development (Heimpel et al. 2003), a behaviour consistent with time limitation (Heimpel and Casas 2008). For example, Janssen (1989) found that two species of *Drosophila* parasitoids accepted a host species of very low suitability in the field, and argued that this behaviour is consistent with time limitation. Based upon these arguments, the notion that parasitoids are mostly time-limited dominated the literature until the mid-1980s (reviewed in Rosenheim et al. 2008, Rosenheim 2011), and still persists to some extent (Wajnberg 2006, de Bruijn et al. 2018, Chen et al. 2021, Ode et al. 2022).

In opposition to this view, the claim was made that egg limitation should not be ignored (Iwasa et al. 1984, Mangel 1987, 1989a, 1989b, Getz and Mills 1996, Heimpel and Collier 1996, Shea et al. 1996, Heimpel et al. 1998, Schreiber 2006, 2007, Kon and Schreiber 2009, Okuyama 2015). The rationale is that, if an organism is never limited by a certain resource (*e.g.*, eggs), it should be selected to invest less in that resource, and more in other functions. Optimization models considering the trade-offs between egg number and size, or between egg number and adult survival, supported this notion, showing that, under optimal investment, a certain proportion of females in the population (even if small) is likely to become egg-limited (Rosenheim 1996). Moreover, models suggested that even if the actual proportion of egg-limited females is low, egg production can still represent a considerable contribution to the total costs of oviposition in terms of foregone future fitness returns (Rosenheim 1999a, b, Rosenheim et al. 2000). Furthermore, the ecological importance of egg limitation is expected to be higher even in cases where the actual proportion of egg-limited females is low. This is because females who become egg-limited are normally those who exploit exceptionally rich habitats and thus make disproportionately large contributions to the next generation (Rosenheim 2011).

In addition to these theoretical arguments, multiple studies have provided evidence for the actual occurrence of egg limitation in the field (reviewed in Heimpel and Rosenheim 1998). One line of evidence comes from dissections of the ovaries of field-caught females and counts of the number of mature eggs that are present. Studies using this method have demonstrated that a low to moderate proportion of females can actually become egg-limited under field conditions (Ode 1994, Heimpel et al. 1996, Ellers et al. 1998, Heimpel and Rosenheim 1998, Cronin and Strong 1999, Casas et al. 2000, Segoli and Rosenheim 2013a, Phillips and Kean 2017). For example, in two small pro-ovigenic parasitoids of the genus *Anagrus* that attack eggs of grape leafhoppers, *ca.* 9–13% of females collected soon after their natural death in the field were depleted of eggs (Segoli and Rosenheim 2013a). In the synovigenic parasitoid *Gonatocerus ashmeadi* that parasitizes eggs of the glassy-winged sharpshooter, *Homalodisca vitripennis*, 17% of recently dead females collected in a citrus orchard had no eggs in their ovaries (Irvin et al. 2014). Similarly, in the pro-ovigenic gall-forming midge, *Rhopalomyia californica*, the egg supplies of *ca.* 17% of females collected from the field were completely exhausted (Rosenheim et al. 2008). In contrast, no evidence for egg depletion was found in the synovigenic ichneumonid parasitoid, *Diadegma insulare*, when attacking the diamondback moth *Plutella xylostella* under field conditions (Lee and Heimpel 2008). Similar results were found for the soybean aphid parasitoid *Binodoxys communis*, although egg loads dropped slightly with increasing host density, suggesting that egg limitation could occur during host outbreaks (Dieckhoff et al. 2014).

A second line of evidence for egg limitation comes from projected estimates of the probability of becoming egg-limited based on empirical data on longevity, egg maturation rates, and oviposition opportunities in the field (Weisser et al. 1997, Ellers et al. 1998, Heimpel and Rosenheim 1998, Casas et al. 2000). For example, in the aphelinid parasitoid *Aphytis aonidiae*, empirical data suggested that the encounter rate with its host, the San Jose scale, can be higher than the egg maturation rate, and hence, it was estimated that most females are likely to become egg-limited by the end of each foraging day (Heimpel et al. 1998, Rosenheim et al. 2000). Driessen and Hemerik (1992) explored how the risk of egg limitation would change for the parasitoid *Leptopilina clavipes*, that attacks *Drosophila* larvae, across different scenarios of habitat quality. They predicted that under the conditions found in the field, *ca.* 13% of the females in the population would become egg-limited.

Another form of indirect evidence is provided via documentation of behavioural responses of females when they become closer to depleting their eggs, or towards the end of their lives, with females generally becoming choosier towards their host when experiencing a high risk of egg limitation, and less choosy when they face high risk of time limitation (Minkenberg et al. 1992, Roitberg et al. 1993, Fletcher et al. 1994, Prokopy et al. 1994, Heimpel and Rosenheim 1995, Heimpel et al. 1996,

Sirot and Bernstein 1996, van Randen and Roitberg 1996, Heimpel and Rosenheim 1998, Mangel and Heimpel 1998, Babendreier and Hoffmeister 2002, Bezemer and Mills 2003, Kishinevsky and Keasar 2015). For example, females of the parasitoid *Aphytis lingnanensis* laid fewer eggs per host (indicating higher choosiness) when they were close to depleting their eggs (Rosenheim and Rosen 1991), and females of the parasitoid *Leptopilina heterotoma* were more likely to parasitize low-quality hosts following exposure to dropping barometric pressure (Roitberg et al. 1993). Low pressure is associated with thunderstorms, which are known to cause considerable mortality in small insects, so this result suggests lower choosiness as the perception of time limitation increases, which is consistent with the theory. Evidence is not limited to parasitoids. In tephritid fruit flies within the genera *Anastrepha* and *Bactrocera*, females more often reject lower-quality host plants, and lay fewer eggs per fruit under conditions leading to egg limitation (Diaz-Fleischer and Aluja 2003, Xu et al. 2012). In the geometrid moth species *Scotopteryx chenopodiata*, females that are injured or deprived of food show higher oviposition rates and shorter latency before ovipositing on low-quality host plants (Javois and Tammaru 2004). In fig wasps, females exhibit a higher tendency to accept a fig for oviposition with increasing age (Yadav and Borges 2018). The occurrence of such a variety of behavioural responses to risks of both egg and time limitation in a wide range of organisms suggests that the two corresponding opposing selection forces are important in shaping insect reproductive strategies.

The accumulation of both theoretical and empirical evidence for the occurrence and importance of egg limitation in insects has led to what could be considered a paradigm shift, acknowledging the importance of both time and egg limitation, and highlighting the balance between the two corresponding opposing selection forces. Ultimately, this shift came from the recognition that lifetime reproductive success – rather than oviposition rate – is the appropriate currency to maximize in evolutionary models of oviposition (Houston et al. 1988). The resolution of this debate allowed the field to move forward, shifting the focus from the question of which risk is more important, to questions regarding the effects of different environmental factors and phylogenetic constraints on this balance and how associated trade-offs affect the evolution of insect reproductive strategies. In the next sections, we will explore predictions and evidence for the importance of several such factors, including oviposition opportunities, food (sugar) availability, environmental stochasticity and the cost of producing eggs, as well as the interactions between these factors.

4.4.1 Oviposition Opportunities

Perhaps the simplest prediction stemming from models is that high availability of oviposition opportunities in the environment should increase the risk of egg depletion, thereby selecting for higher investment in egg supply (Rosenheim 1996, Sevenster et al. 1998, Ellers 2000, Rosenheim 2011, Segoli and Wajnberg 2020). Some empirical evidence supports the first part of this prediction. For example, in *Microctonus hyperodae*, a parasitoid that was introduced to New Zealand to control an invasive weevil, field-collected females exhibited lower egg loads and approached egg limitation late in the season when the host population was at its highest densities (Phillips and Kean 2017). Similarly, in the parasitoids *Anagrus erythroneurae* and *Anagrus daanei*, which attack eggs of grape leafhoppers, the estimated number of eggs laid by females in the field (calculated as the estimated initial egg load per female minus egg load upon their death) was higher in vineyards with denser leafhopper populations than in vineyards with low leafhopper densities (Segoli and Rosenheim 2013a). In contrast, however, in the parasitoid *A. aonidiae*, no association was found between egg loads of field-collected females and local host density (Heimpel and Rosenheim 1998).

Absent or weak relationships between host availability and egg limitation can be explained by behavioural and physiological responses to navigate the trade-off between egg and time limitation. These may include rapid egg maturation following oviposition (Rivero-Lynch and Godfray 1997, Wu and Heimpel 2007, Bodin et al. 2009, Casas et al. 2009, Dieckhoff and Heimpel 2010, Dieckhoff et al. 2014), and state-dependent oviposition behaviour, with females becoming choosier when they approach egg limitation (see previous section). In some parasitoids, adult females are able to feed on host individuals. For these species a switch from oviposition to host feeding as egg load declines not only reduces oviposition rate but also contributes to increased egg maturation, thereby attenuating a negative relationship between egg load and host availability (Heimpel and Collier 1996, Wu and Heimpel 2007). Such plastic responses are consistent with the second part of the aforementioned theoretical prediction of higher egg production in environments with many oviposition opportunities, as a means to oppose the risk of egg limitation.

Evidence for evolutionary adaptations to the risk of egg limitation at the population level includes the *Drosophila* parasitoid *Asobara tabida*, where females from southern regions of Europe, which tend to support higher host populations, were shown to have higher initial egg loads than those from northern regions (Kraaijeveld and Vanderwel 1994). In another

Drosophila parasitoid, *Leptopilina boulardi*, females from humid environments, where again host densities tend to be higher, exhibit higher initial egg loads than females from dry environments (Moiroux et al. 2010). In the flour moth parasitoid *Venturia canescens*, females from grain storage facilities, where hosts tend to aggregate, emerge with higher egg loads at the expense of energy reserves and flight ability, compared to females from natural habitats (Pelosse et al. 2007, Amat et al. 2017). Finally, in the parasitoid *A. daanei*, females from vineyards, where the leafhopper host reaches high densities, emerge with more and smaller eggs than those from wild grapes in riparian natural habitats, suggesting local adaptation in egg loads in response to host densities (Segoli and Rosenheim 2013a, 2013b).

At the species level, it has long been suggested that parasitoid species attacking hosts of an earlier developmental stage have higher fecundities than those attacking hosts of later stages (Price 1973, 1974, Jervis et al. 2012). This is consistent with the interpretation that parasitoids can evolutionarily adjust their egg loads to host availability, as hosts of early developmental stages naturally occur at higher densities in the environment, and hence provide more oviposition opportunities (Godfray 1994, Mayhew 2016). In addition, parasitoid species with higher fecundity and shorter lifespan were found to be associated with agricultural habitats, whereas the opposite suite of traits was found to be more dominant in natural areas (Kishinevsky and Keasar 2022). This can again be explained by the differences between the two habitats – agricultural plots tend to have high densities of hosts and are more highly disturbed by insecticides, harvesting or other human activities (Kishinevsky and Ives 2022), possibly selecting for species with fast reproduction and short lifespan.

4.4.2 Sugar Availability

Many insects rely on sugar resources in the form of floral nectar or hemipteran honeydew during the adult stage, which may substantially increase their lifespan (Waage 1983, Lee et al. 2004, Chen and Fadamiro 2006, Wäckers et al. 2006, Wäckers et al. 2008, Zhu et al. 2013, Wang et al. 2014, Lahiri et al. 2017, Kishinevsky et al. 2018), whereas it is generally assumed not to contribute significantly to egg production (Rivero and Casas 1999, Giron and Casas 2003, Jervis et al. 2008, Visser and Ellers 2008, Balzan and Wäckers 2013). Moreover, sugar feeding can fuel flight and could thus provide opportunities to locate more oviposition sites (Forsse et al. 1992, Wanner et al. 2006, Fahrner et al. 2014, Heimpel 2019). Studies have demonstrated that the availability of sugar resources may be limited in the field (Heimpel and Jervis 2005, Segoli and Rosenheim 2013c, Kishinevsky et al. 2018, Kishinevsky and Keasar 2021), potentially affecting the balance between the relative risks of egg and time limitation.

So, how should insects adjust their reproductive allocations according to sugar availability in the environment? A model of insect life history evolution, based on Monte Carlo simulations coupled with genetic algorithms, predicted that optimal investment in egg number should increase with the availability of life-extending sugar sources (Segoli and Wajnberg 2020). This is likely because the additional energy obtained from feeding reduces the need for initial investment in energy reserves and allows adult females to survive longer to lay more eggs during their lifetime.

Several studies provide evidence for a potential positive effect of food availability on the number of eggs laid by females in the field (Lee and Heimpel 2008, Segoli and Rosenheim 2013c, Tena et al. 2015) and in the laboratory (Harvey et al. 2001, Eliopoulos et al. 2003, Zhang et al. 2011). In addition, there is evidence that oosorption can be greatly enhanced by sugar deprivation. For example, egg loads of sugar-deprived *Aphytis melinus* that were also deprived of hosts dropped from *ca.* 10 to *ca.* 1 over a 36-hour period, while those of sugar-fed individuals did not drop at all (Heimpel et al. 1997). Similar results have been reported for other insect species (Bell and Bohm 1975, Dieckhoff and Heimpel 2010). Thus, sugar meals can have an important impact on egg loads even without directly supporting egg maturation. Despite this accumulated theoretical and empirical evidence for the importance of sugar feeding in mediating insect reproductive strategies, we are unaware of any evidence for responses in egg loads to sugar availability in the environment at the population or species level.

4.4.3 The Interaction Between Oviposition Opportunities and Sugar Availability

Segoli and Wajnberg (2020) further explored theoretically the interaction between oviposition opportunities (in terms of host density) and sugar availability in their effect on the evolution of insect reproductive allocations. Their model predicted that under conditions of high food availability, the optimal egg load will increase with host density, at the expense of energy reserves. However, when sugar availability was limited, egg load was predicted to increase primarily at the expense of egg size (Figure 4.3). This is probably because females cannot compromise their energy reserves when no other food sources are available. To our knowledge, these predictions have never been experimentally tested. However, they are consistent with empirical evidence from the species *A. daanei*. In this species, females from vineyards – where host populations can reach

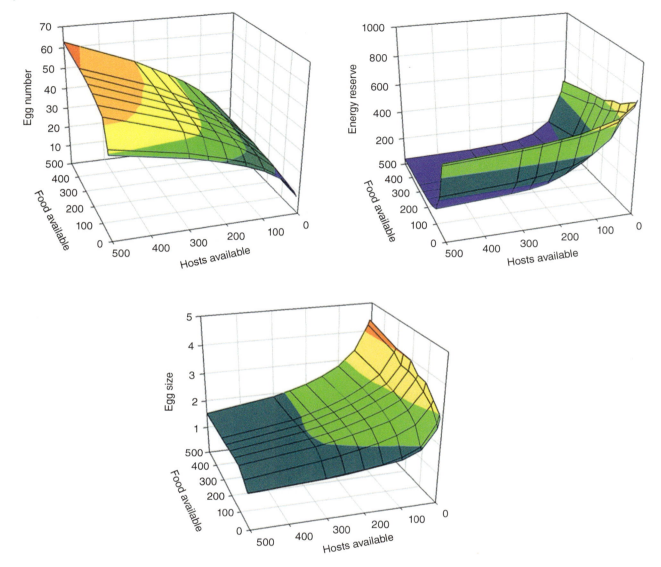

Figure 4.3 Results from a model by Segoli and Wajnberg (2020), predicting the optimal life history traits of a simulated parasitoid female in an environment in which both host and food items are present in varying densities. The results are consistent with empirical data on life history traits of the parasitoid *Anagrus daanei* vineyards and riparian habitats in California (see text).

high densities – seem to emerge with more eggs at the expense of egg size (Segoli and Rosenheim 2013b), but not at the expense of lifespan (Segoli et al. 2018). This may be explained by the low availability of sugar sources such as floral nectar in the sampled vineyards in California, USA (Segoli and Rosenheim 2013c), possibly leading to selection for the maintenance of high initial energy reserves (see also Chapter 22, on life history and biological control).

4.4.4 Environmental Stochasticity

Stochasticity in oviposition opportunities may stem from variation in survival prospects or variation in the availability and distribution of oviposition sites (*e.g.*, hosts) (Ellers et al. 2000). While the effect of mean number of oviposition opportunities on the occurrence of egg limitation and investment in egg production is expected to be relatively simple (see previous section), the effect of stochasticity in oviposition opportunities on egg limitation is more difficult to anticipate. Several models predicted that increased spatial stochasticity in oviposition opportunities should lead to increased investment in egg supplies (allowing females to exploit rich habitats), and consequently to a decrease in the proportion of females becoming egg-limited (Rosenheim 1996, Sevenster et al. 1998, Ellers 2000). This is consistent with Liebig's law of the minimum (Brock 2002), suggesting that, in stochastic environments, over-investment in securing the cheapest resource (in this case assumed to be the egg, as opposed to investment in finding a host) is expected (Ellers 2000, Rosenheim et al. 2010, Rosenheim 2011).

While temporal stochasticity has often been hypothesized to have an effect similar to that of spatial stochasticity (Godfray 1994, van Baalen 2000), a model by Rosenheim (2011) predicted that temporally stochastic environments do not necessarily favour increased fecundity. Rather, they should favour either increased (when eggs are least expensive), decreased (when eggs are most expensive) or intermediate maxima in fecundity (when egg costs are intermediate). The reason that temporal variation acts differently from spatial variation is probably because it imposes particularly stiff penalties on strategies that perform very poorly during generations when opportunities for reproduction are unusually rare. In addition, the model predicted that, even in cases where (either spatial or temporal) environmental stochasticity reduces the likelihood of egg limitation, it could still increase the ecological importance of egg limitation (*i.e.*, the proportion of future reproduction that is foregone due to females becoming egg-limited).

Unfortunately, evidence for the effect of spatial *vs.* temporal stochasticity on the evolution of insect reproductive strategies is lacking. The higher fecundity of parasitoids from agricultural compared to natural habitats could be potentially attributed to higher stochasticity in such environments (Kishinevsky and Ives 2022). However, this cannot be easily disentangled from the effect of higher host density that also characterizes agricultural environments. Evidence from other biological systems may include a comparative analysis of the reproductive strategies in angiosperm plants, which suggests that plants faced with greater spatial unpredictability in pollen availability produce a larger number of ovules per flower (Burd et al. 2009).

4.4.5 The Cost of Producing an Egg

While some models assumed a trade-off between egg number and size (Rosenheim 1996, Segoli and Wajnberg 2020), the minimal investment required to produce a single egg may be to a large extent phylogenetically constrained. Different insect taxa produce eggs that vary tremendously in size and nutritional content, with some species producing eggs that are rich in yolk and containing sufficient nutrients for the completion of embryonic development, while others produce relatively small eggs that are poor in nutrients (Fox and Czesak 2000). For example, some species of parasitoids produce yolk-deficient hydropic eggs (that inflate after they are injected into the body of the host), and embryonic development occurs using nutrients absorbed directly from the host tissues (Thompson and Hagen 1999). Importantly, the cost of producing a single egg may influence the balance between the risks of time and egg limitation.

Not surprisingly, models predict that the overall investment in reproduction, as well as the proportion of egg-limited females, increases with the cost of producing a single egg (Rosenheim 2011). However, empirical evidence is lacking. To address this question, we have gathered information on egg size and on the actual occurrence of egg depletion in the field for different parasitoid species. Data were gathered from the literature (Cronin and Strong 1993, Ode 1994, Visser 1994, Ellers et al. 1998, Heimpel and Rosenheim 1998, Phillips et al. 1998, Casas et al. 2000, Daane and Costello 2000, Ellers et al. 2001, Phillips and Baird 2001, Pexton and Mayhew 2002, Heimpel and Casas 2008, Lee and Heimpel 2008, Irvin and Hoddle 2009, Cummins et al. 2011, Segoli and Rosenheim 2013a, 2013b, Irvin et al. 2014) and augmented by unpublished data from multiple contributors (T.M. Blackburn, J.T. Cronin, J. Ellers, M.S. Hoddle, A.R. Kraaijeveld, J.C. Lee, C.B. Phillips, J.A. Rosenheim, M. Visser, R.A. Wharton). For each species, we estimated egg size as the mean ratio between egg length and parasitoid body length. In addition, we gathered estimates of the proportion of females that exhausted their egg supply in the field. Average values were used when several sources reported on the same species. The association between the estimated proportion of egg-limited females and the estimated relative egg size was analyzed while accounting for parasitoid phylogeny. The taxonomic tree was built following Blaimer et al. (2023) with a standardized branch length. The function cor_phylo () in the package phyr (Li et al. 2020) in R version 4.1.2 (R Core Team 2021) was used for the analysis. The results suggest a positive association between the relative size of the egg and the probability of a female becoming egg-limited per species, as predicted by theory (Figure 4.4; $r = 0.671$, $p = 0.0023$ when not accounting for parasitoid phylogeny, and $r = 0.641$, $p = 0.0132$, when accounting for phylogeny). As far as we know, this is the first empirical evidence of such an association. Information on the proportion of egg-limited females for more insect species is needed to confirm the generality of this pattern.

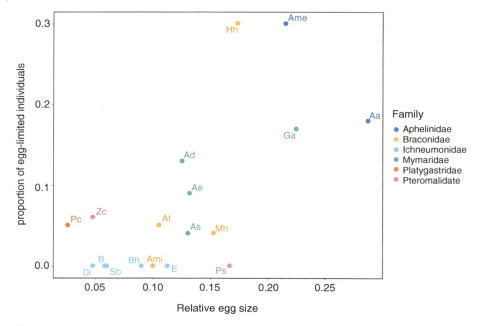

Figure 4.4 Relationship between relative egg size (egg length/body length) and the proportion of individuals collected from the field reported having no eggs. Each dot represents one parasitoid species: *Anagrus daanei* (Ad); *Anagrus erythroneura* (Ae); *Anagrus sophiae* (As); *Aphaereta minuta* (Ami); *Aphytis aonidiae* (Aa); *Aphytis melinus* (Ame); *Asobara tabida* (At); *Banchus sp.* (B); *Barytarbes honestus* (Bh); *Diadegma insulare* (Di); *Habrobracon hebetor* (Hh); *Euryproctus* sp. (E); *Gonatocerus ashmeadi* (Ga); *Microctonus hyperodae* (Mh); *Platygaster californica* (Pc); *Pteromalus sequester* (Ps); *Sympherta burra* (Sb); *Zatropis capitis* (Zc). Colour codes represent different parasitoid families.

4.4.6 Interaction Between Environmental Stochasticity and the Cost of Producing an Egg

The common prediction that egg loads should increase with environmental stochasticity is based on the assumption that the cost of producing a single egg is smaller than the cost of locating an oviposition site to lay a single egg. Under this assumption, it is worthwhile for a female to produce a surplus of eggs to be able to exploit exceptionally rich habitats. While this assumption is probably often correct, it should not be assumed to be true under all conditions. Cases in which the egg is expensive to produce (see previous section) and oviposition sites are relatively easy to locate may result in reduced fecundity as environmental stochasticity increases (Ellers 2000, Rosenheim 2011). However, to our knowledge, there is no empirical evidence for the interaction between the cost of egg production and environmental stochasticity in their effect on egg load dynamics and evolution.

4.5 Additional Life History Strategies to Overcome the Risk of Egg Limitation

In some situations, insects may be highly prone to egg limitation. For example, hosts of large size may accommodate many developing parasitoids. A parasitoid female ovipositing in such a host may be selected to produce a large brood in order to optimally utilize it (Skinner 1985, Mayhew and Glaizot 2001), potentially leading to egg depletion. Gregarious parasitoids (where multiple individual parasitoids develop on the same host individual) tend to have higher egg loads than closely related solitary species (where a single parasitoid individual develops on each host; Heimpel 2000), thereby decreasing this risk. However, in extreme cases, females may have difficulties producing enough eggs to optimally utilize the body of a single host. One potential evolutionary mechanism to overcome such a limitation is polyembryonic development. Polyembryony is a mode of development where multiple genetically identical embryos develop from a single egg via embryonic division (Strand 2003). Obligate polyembryony is extremely rare in the animal kingdom, yet it is surprisingly common in some groups of parasitoids (Godfray 1994, Craig et al. 1997). While several advantages have been proposed for the evolution of polyembryony (*e.g.*, reduced genetic conflict among siblings and the ability to adjust offspring numbers to host quality), overcoming egg limitation may be a major selective force (Segoli et al. 2010, Keasar and Wajnberg 2019). For example, parasitoid species of the genus *Copidosoma*, which are all polyembryonic, attack moth eggs and their offspring emerge from the moth caterpillar that can be a few orders of magnitude larger than the parasitoid. In fact, the number of wasps that can complete their development within a single host could reach several hundred or even over a thousand individuals (Strand 2003). Hence, a single parasitoid female probably would not have been able to produce and carry the number of eggs required to effectively exploit a single host, nor would the host egg be able to accommodate the deposition of so many parasitoid eggs (Segoli et al. 2010). Another potential mechanism is the exploitation of a single large host by multiple females. For example, in species of the genus *Melittobia*, the host is jointly exploited by the founding female and by her daughters (Matthews et al. 2009). In the genus *Sclerodermus* multiple females may also cooperate in exploiting a single host (Tang et al. 2014).

4.6 Conclusions and Future Directions

The realization that time and egg limitation both play an important role in shaping insect lifetime reproductive success and reproductive strategies was crucial for the field to move forward. However, despite important theoretical advancements, many predictions remained untested. The actual occurrence of egg limitation in the field has been demonstrated in a handful of species, but this information offers a snapshot in time rather than illustrating a dynamic process. Some evidence for plastic and evolutionary responses to impending egg limitation exist, and they support the hypothesis that the risk of egg limitation can be an important selective force. However, evidence for evolutionary responses to food availability, habitat stochasticity, egg size and the interactions between these factors is lacking.

Based on the knowledge gaps that we have identified, we propose several future directions. First, we suggest the development of an egg limitation metric that encompasses not only the availability of eggs but also the opportunity to deposit them in a certain environment (see Figure 4.2). This will provide a more accurate estimate of the relative risks in a continuous rather than in an absolute way allowing a more precise testing of hypotheses. For example, a female with few eggs in an environment with few oviposition opportunities will be estimated to be less prone to the risk of egg limitation than a female with few eggs in an environment with many oviposition opportunities. Such a metric could also potentially incorporate factors such as clutch size, egg maturation rate, and mortality rate in the environment.

Second, we suggest evolutionary experiments in which insects are reared on different combinations of high and low host and sugar availability over many generations and then subjected to assays assessing egg limitation and/or physiological and behavioural responses to impending egg limitation according to theoretical predictions. Perhaps more challenging, but still likely to be rewarding, is to run selection under constant *vs.* stochastic conditions, and/or in spatially *vs.* temporally stochastic environments. Such experiments could potentially be conducted using species with 'expensive' eggs *vs.* those with relatively 'cheap' eggs or species of varying clutch sizes. In all cases, studies would benefit from considering multiple life history traits and trade-offs simultaneously rather than one at a time.

Third, we suggest considering eco-evolutionary feedback between insect reproductive traits and resource availability in their environment. For example, female egg loads could be affected by host availability, but, at the same time, could cause changes in host density in the environment, resulting in a feedback loop. This can be tested by simulation models or by running experiments in which host density is kept constant *vs.* being allowed to fluctuate in response to parasitoid reproduction.

Finally, as reflected in this chapter, the study of insect egg loads is heavily based on knowledge from parasitoid wasps. However, insights may be relevant to many other insects as well as other arthropod species. We suggest that the field could benefit from expanding the taxonomic scope to include species representing diverse groups, habitats and life histories.

Acknowledgements

We acknowledge support from the Israel Institute for Advanced Studies at the Hebrew University, Israel. We thank Tim M. Blackburn, James T. Cronin, Jacintha Ellers, Michael S. Hoddle, Alex R. Kraaijeveld, Jena C. Lee, Craig B. Phillips, Jay A. Rosenheim, Marcel Visser and Robert A. Wharton for contributing unpublished data.

References

Amat, I., van Alphen, J.J.M., Kacelnik, A., Desouhant, E., and Bernstein, C. (2017). Adaptations to different habitats in sexual and asexual populations of parasitoid wasps: a meta-analysis. *PeerJ* 5: e3699.

Babendreier, D., and Hoffmeister, T.S. (2002). Superparasitism in the solitary ectoparasitoid *Aptesis nigrocincta*: the influence of egg load and host encounter rate. *Entomologia Experimentalis et Applicata* 105: 63–69.

Balzan, M.V., and Wäckers, F.L. (2013). Flowers to selectively enhance the fitness of a host-feeding parasitoid: adult feeding by *Tuta absoluta* and its parasitoid *Necremnus artynes*. *Biological Control* 67: 21–31.

Bell, W.J., and Bohm, M.K. (1975). Oosorption in insects. *Biological Reviews of the Cambridge Philosophical Society* 50: 373–396.

Bezemer, T.M., and Mills, N.J. (2003). Clutch size decisions of a gregarious parasitoid under laboratory and field conditions. *Animal Behaviour* 66: 1119–1128.

Blaimer, B.B., Santos, B.F., Cruaud, A., Gates, M.W., Kula, R.R., Mikó, I., Rasplus, J.-Y., Smith, D.R., Talamas, E.J., Brady, S.G., and Buffington, M.L. (2023). Key innovations and the diversification of Hymenoptera. *Nature Communications* 14: 1212.

Bodin, A., Jaloux, B., Delbecque, J.P., Vannier, F., Monge, J.P., and Mondy, N. (2009). Reproduction in a variable environment: how does *Eupelmus vuilleti*, a parasitoid wasp, adjust oogenesis to host availability? *Journal of Insect Physiology* 55: 643–648.

Boggs, C.L. (1997). Reproductive allocation from reserves and income in butterfly species with differing adult diets. *Ecology* 78: 181–191.

Brock, W.H. (2002). *Justus Von Liebig: the chemical gatekeeper*. Cambridge University Press.

Burd, M. (1995). Ovule packaging in stochastic pollination and fertilization environments. *Evolution* 49: 100–109.

Burd, M., Ashman, T.L., Campbell, D.R., Dudash, M.R., Johnston, M.O., Knight, T.M., Mazer, S.J., Mitchell, R J., Steets, J.A., and Vamosi, J.C. (2009). Ovule number per flower in a world of unpredictable pollination. *American Journal of Botany* 96: 1159–1167.

Casas, J., Nisbet, R.M., Swarbrick, S., and Murdoch, W.W. (2000). Eggload dynamics and oviposition rate in a wild population of a parasitic wasp. *Journal of Animal Ecology* 69: 185–193.

Casas, J., Vannier, F., Mandon, N., Delbecque, J.P., Giron, D., and Monge, J.P. (2009). Mitigation of egg limitation in parasitoids: immediate hormonal response and enhanced oogenesis after host use. *Ecology* 90: 537–545.

Charnov, E.L., and Skinner, S.W. (1984). Evolution of host selection and clutch size in parasitoid wasps. *Florida Entomologist* 67: 5–21.

Charnov, E.L., and Skinner, S.W. (1985). Complementary approaches to the understanding of parasitoid oviposition decisions. *Environmental Entomology* 14: 383–391.

Chen, L., and Fadamiro, H.Y. (2006). Comparing the effects of five naturally occurring monosaccharide and oligosaccharide sugars on longevity and carbohydrate nutrient levels of a parasitic phorid fly, *Pseudacteon tricuspis*. *Physiological Entomology* 31: 46–56.

Chen, W.B., Weng, Q.F., Nie, R., Zhang, H.Z., Jing, X.Y., Wang, M.Q., Li, Y.Y., Mao, J.J., and Zhang, L.S. (2021). Optimizing photoperiod, exposure time, and host-to-parasitoid ratio for mass-rearing of *Telenomus remus*, an egg parasitoid of *Spodoptera frugiperda*, on *Spodoptera litura* eggs. *Insects* 12: 1050.

Craig, S.F., Slobodkin, L.B., Wray, G.A., and Biermann, C.H. (1997). The 'paradox' of polyembryony: a review of the cases and a hypothesis for its evolution. *Evolutionary Ecology* 11: 127–143.

Cronin, J.T., and Strong, D.R. (1993). Superparasitism and mutual interference in the egg parasitoid *Anagrus delicatus* (Hymenoptera, Mymaridae). *Ecological Entomology* 18: 293–302.

Cronin, J.T., and Strong, D.R. (1999). Dispersal-dependent oviposition and the aggregation of parasitism. *The American Naturalist* 154: 23–36.

Cummins, H.M., Wharton, R.A., and Colvin, A.M. (2011). Eggs and egg loads of field-collected ctenoplematinae (Hymenoptera: Ichneumonidae): evidence for phylogenetic constraints and life-history trade-offs. *Annals of the Entomological Society of America* 104: 465–475.

Daane, K.M., and Costello, M.J. (2000). Variegated and western grape leafhoppers. In: *Raisin production manual* (ed. Christensen, P.L), 173–181. University of California, Agricultural and Natural Resources.

de Bruijn, J.A.C., Vet, L.E.M., and Smid, H.M. (2018). Costs of persisting unreliable memory: reduced foraging efficiency for free-flying parasitic wasps in a wind tunnel. *Frontiers in Ecology and Evolution* 6: 160.

Diaz-Fleischer, F., and Aluja, M. (2003). Behavioural plasticity in relation to egg and time limitation: the case of two fly species in the genus *Anastrepha* (Diptera: Tephritidae). *Oikos* 100: 125–133.

Dieckhoff, C., and Heimpel, G.E. (2010). Determinants of egg load in the soybean aphid parasitoid *Binodoxys communis*. *Entomologia Experimentalis et Applicata* 136: 254–261.

Dieckhoff, C., Theobald, J.C., Wackers, F.L., and Heimpel, G.E. (2014). Egg load dynamics and the risk of egg and time limitation experienced by an aphid parasitoid in the field. *Ecology and Evolution* 4: 1739–1750.

Driessen, G., and Hemerik, L. (1992). The time and egg budget of *Leptopilina clavipes*, a parasitoid of larval *Drosophila*. *Ecological Entomology* 17: 17–27.

Eliopoulos, P.A., Harvey, J.A., Athanassiou, C.G., and Stathas, G.J. (2003). Effect of biotic and abiotic factors on reproductive parameters of the synovigenic endoparasitoid *Venturia canescens*. *Physiological Entomology* 28: 268–275.

Ellers, J. (2000). Searching for more general weight conjectures, using the symmetric group as an example. *Journal of Algebra* 225: 602–629.

Ellers, J., Bax, M., and van Alphen, J.J.M. (2001). Seasonal changes in female size and its relation to reproduction in the parasitoid *Asobara tabida*. *Oikos* 92: 309–314.

Ellers, J., Sevenster, J G., and Driessen, G. (2000). Egg load evolution in parasitoids. *The American Naturalist* 156: 650–665.

Ellers, J., van Alphen, J.J.M., and Sevenster, J.G. (1998). A field study of size-fitness relationships in the parasitoid *Asobara tabida*. *Journal of Animal Ecology* 67: 318–324.

Fahrner, S.J., Lelito, J.P., Blaedow, K., Heimpel, G.E., and Aukema, B.H. (2014). Factors affecting the flight capacity of *Tetrastichus planipennisi* (Hymenoptera: Eulophidae), a classical biological control agent of *Agrilus planipennis* (Coleoptera: Buprestidae). *Environmental Entomology* 43: 1603–1612.

Fletcher, J.P., Hughes, J.P., and Harvey, I.F. (1994). Life expectancy and egg load affect oviposition decisions of a solitary parasitoid. *Proceedings of the Royal Society B: Biological Sciences* 258: 163–167.

Forsse, E., Smith, S.M., and Bourchier, R.S. (1992). Flight initiation in the egg parasitoid *Trichogramma minutum* – effects of ambient temperature, mates, food, and host eggs. *Entomologia Experimentalis et Applicata* 62: 147–154.

Fox, C.W., and Czesak, M.E. (2000). Evolutionary ecology of progeny size in arthropods. *Annual Review of Entomology* 45: 341–369.

Getz, W.M., and Mills N.J. (1996). Host-parasitoid coexistence and egg-limited encounter rates. *The American Naturalist* 148: 333–347.

Giron, D., and Casas J. (2003). Lipogenesis in an adult parasitic wasp. *Journal of Insect Physiology* 49: 141–147.

Godfray, H.C.J. (1994). *Parasitoids: behavioral and evolutionary ecology*. Princeton University Press.

Harvey, J.A., Harvey, I.F., and Thompson, D.J. (2001). Lifetime reproductive success in the solitary endoparasitoid, *Venturia canescens*. *Journal of Insect Behavior* 14: 573–593.

Heimpel, G.E. (2000). Effects of parasitoid clutch size on host parasitoid population dynamics. In: *Parasitoid population biology* (eds. Hochberg, M.E., and Ives, A.R.), 27–40. Princeton University Press.

Heimpel, G.E. (2019). Linking parasitoid nectar feeding and dispersal in conservation biological control. *Biological Control* 132: 36–41.

Heimpel, G.E., and Casas, J. (2008). Parasitoid foraging and oviposition behaviour in the field. In: *Behavioral ecology of insect parasitoids – from theoretical approaches to field applications* (eds. Wajnberg, E., Bernstein, C., and van Alphen, J.), 52–70. Blackwell.

Heimpel, G.E., and Collier, T.R. (1996). The evolution of host-feeding behaviour in insect parasitoids. *Biological Reviews of the Cambridge Philosophical Society* 71: 373–400.

Heimpel, G.E., and Jervis, M.A. (2005). Does floral nectar improve biological control by parasitoids? In: *Plant-provided food and plant-carnivore mutualism* (eds. Wäckers, F., van Rijn, P.C.J., and Bruin, J.), 267–304. Cambridge University Press.

Heimpel, G.E., Mangel, M., and Rosenheim, J.A. (1998). Effects of time limitation and egg limitation on lifetime reproductive success of a parasitoid in the field. *The American Naturalist* 152: 273–289.

Heimpel, G.E., Neuhauser, C., and Hoogendoorn, M. (2003). Effects of parasitoid fecundity and host resistance on indirect interactions among hosts sharing a parasitoid. *Ecology Letters* 6: 556–566.

Heimpel, G.E., and Rosenheim, J.A. (1995). Dynamic host feeding by the parasitoid *Aphytis melinus* – the balance between current and future reproduction. *Journal of Animal Ecology* 64: 153–167.

Heimpel, G.E., and Rosenheim, J.A. (1998). Egg limitation in parasitoids: a review of the evidence and a case study. *Biological Control* 11: 160–168.

Heimpel, G.E., Rosenheim, J.A., and Kattari, D. (1997). Adult feeding and lifetime reproductive success in the parasitoid *Aphytis melinus*. *Entomologia Experimentalis et Applicata* 83: 305–315.

Heimpel, G.E., Rosenheim, J.A., and Mangel, M. (1996). Egg limitation, host quality, and dynamic behavior by a parasitoid in the field. *Ecology* 77: 2410–2420.

Houston, A., Clark, C., McNamara, J., and Mangel, M. (1988). Dynamic models in behavioral and evolutionary ecology. *Nature* 332: 29–34.

Irvin, N.A., Espinosa, J.S., and Hoddle, M.S. (2014). Maximum realised lifetime parasitism and occurrence of time limitation in *Gonatocerus ashmeadi* (Hymenoptera: Mymaridae) foraging in citrus orchards. *Biocontrol Science and Technology* 24: 662–679.

Irvin, N.A., and Hoddle, M.S. (2009). Egg maturation, oosorption, and wing wear in *Gonatocerus ashmeadi* (Hymenoptera: Mymaridae), an egg parasitoid of the glassy-winged sharpshooter, *Homalodisca vitripennis* (Hemiptera: Cicadellidae). *Biological Control* 48: 125–132.

Iwasa, Y., Suzuki, Y., and Matsuda, H. (1984). Theory of oviposition strategy of parasitoids. 1. Effect of mortality and limited egg number. *Theoretical Population Biology* 26: 205–227.

Janssen, A. (1989). Optimal host selection by *Drosophila* parasitoids in the field. *Functional Ecology* 3: 469–479.

Javois, J., and Tammaru, T. (2004). Reproductive decisions are sensitive to cues of life expectancy: the case of a moth. *Animal Behaviour* 68: 249–255.

Jervis, M.A., Ellers, J., and Harvey, J.A. (2008). Resource acquisition, allocation, and utilization in parasitoid reproductive strategies. *Annual Review of Entomology* 53: 361–385.

Jervis, M.A., Heimpel, G.E., Ferns, P.N., Harvey, J.A., and Kidd, N.A.C. (2001). Life-history strategies in parasitoid wasps: a comparative analysis of 'ovigeny'. *Journal of Animal Ecology* 70: 442–458.

Jervis, M.A., Moe, A., and Heimpel, G.E. (2012). The evolution of parasitoid fecundity: a paradigm under scrutiny. *Ecology Letters* 15: 357–364.

Keasar, T., and Wajnberg, E. (2019). Evolutionary constraints on polyembryony in parasitic wasps: a simulation model. *Oikos* 128: 347–359.

Keinan, Y., Kishinevsky, M., and Keasar, T. (2017). Intraspecific variability in egg maturation patterns and associated life-history trade-offs in a polyembryonic parasitoid wasp. *Ecological Entomology* 42: 587–594.

Kishinevsky, M., Cohen, N., Chiel, E., Wajnberg, E., and Keasar, T. (2018). Sugar feeding of parasitoids in an agroecosystem: effects of community composition, habitat and vegetation. *Insect Conservation and Diversity* 11: 50–57.

Kishinevsky, M., and Ives, A.R. (2022). The success of a habitat specialist biological control agent in the face of disturbance. *Ecosphere* 13: e4050.

Kishinevsky, M., and Keasar, T. (2015). State-dependent host acceptance in the parasitoid *Copidosoma koehleri*: the effect of intervals between host encounters. *Behavioral Ecology and Sociobiology* 69: 543–549.

Kishinevsky, M., and Keasar, T. (2021). Sugar feeding by parasitoids inside and around vineyards varies with season and weed management practice. *Agriculture Ecosystems & Environment* 307: 107229.

Kishinevsky, M., and Keasar, T. (2022). Trait-based characterisation of parasitoid wasp communities in natural and agricultural areas. *Ecological Entomology* 47: 657–667.

Kon, R., and Schreiber, S J. (2009). Multiparasitoid-host interactions with egg-limited encounter rates. *SIAM Journal on Applied Mathematics* 69: 959–976.

Kraaijeveld, A.R., and Vanderwel, N.N. (1994). Geographic variation in reproductive success of the parasitoid *Asobara tabida* in larvae of several *Drosophila* species. *Ecological Entomology* 19: 221–229.

Lahiri, S., Orr, D., Cardoza, Y.J., and Sorenson, C. (2017). Longevity and fecundity of the egg parasitoid *Telenomus podisi* provided with different carbohydrate diets. *Entomologia Experimentalis et Applicata* 162: 178–187.

Lee, J.C., and Heimpel, G.E. (2008). Floral resources impact longevity and oviposition rate of a parasitoid in the field. *Journal of Animal Ecology* 77: 565–572.

Lee, J.C., Heimpel, G.E., and Leibee, G.L. (2004). Comparing floral nectar and aphid honeydew diets on the longevity and nutrient levels of a parasitoid wasp. *Entomologia Experimentalis et Applicata* 111: 189–199.

Legaspi, J C., and Legaspi, B.C. (2004). Does a polyphagous predator prefer prey species that confer reproductive advantage? Case study of *Podisus maculiventris*. *Environmental Entomology* 33: 1401–1409.

Legaspi, J.C., and Legaspi, B.C. (2008). Ovigeny in selected generalist predators. *Florida Entomologist* 91: 133–135.

Li, D., Dinnage, R., Nell, L.A., Helmus, M.R., and Ives, A.R. (2020). phyr: an R package for phylogenetic species-distribution modelling in ecological communities. *Methods in Ecology and Evolution* 11: 1455–1463.

Mangel, M. (1987). Oviposition site selection and clutch size in insects. *Journal of Mathematical Biology* 25: 1–22.

Mangel, M. (1989a). Evolution of host selection in parasitoids – does the state of the parasitoid matter. *The American Naturalist* 133: 688–705.

Mangel, M. (1989b). An evolutionary interpretation of the motivation to oviposit. *Journal of Evolutionary Biology* 2: 157–172.

Mangel, M., and Heimpel, G.E. (1998). Reproductive senescence and dynamic oviposition behaviour in insects. *Evolutionary Ecology* 12: 871–879.

Matthews, R.W., Gonzalez, J.M., Matthews, J.R., and Deyrup, L.D. (2009). Biology of the parasitoid *Melittobia* (Hymenoptera: Eulophidae). *Annual Review of Entomology* 54: 251–266.

Mayhew, P.J. (2016). Comparing parasitoid life histories. *Entomologia Experimentalis et Applicata* 159: 147–162.

Mayhew, P.J., and Glaizot, O. (2001). Integrating theory of clutch size and body size evolution for parasitoids. *Oikos* 92: 372–376.

Minkenberg, O.P.J.M., Tatar, M., and Rosenheim, J.A. (1992). Egg load as a major source of variability in insect foraging and oviposition behavior. *Oikos* 65: 134–142.

Mock, D.W., and Forbes, L.S. (1995). The evolution of parental optimism. *Trends in Ecology & Evolution* 10: 130–134.

Moiroux, J., Le Lann, C., Seyahooei, M.A., Vernon, P., Pierre, J.S., van Baaren, J., and van Alphen, J.J.M. (2010). Local adaptations of life-history traits of a *Drosophila* parasitoid, *Leptopilina boulardi*: does climate drive evolution? *Ecological Entomology* 35: 727–736.

Ode, P.J. (1994). Female-biased sex allocation in the outcrossing parasitic wasp, Bracon hebetor Say (Hymenoptera: Braconidae). PhD thesis, University of Wisconsin.

Ode, P.J., Vyas, D.K., and Harvey, J.A. (2022). Extrinsic inter- and intraspecific competition in parasitoid wasps. *Annual Review of Entomology* 67: 305–328.

Okuyama, T. (2015). Egg limitation in host-parasitoid dynamics: an individual-based perspective. *Theoretical Ecology* 8: 327–331.

Papaj, D.R. (2000). Ovarian dynamics and host use. *Annual Review of Entomology* 45: 423–448.

Pelosse, P., Bernstein, C., and Desouhant, E. (2007). Differential energy allocation as an adaptation to different habitats in the parasitic wasp *Venturia canescens*. *Evolutionary Ecology* 21: 669–685.

Pexton, J.J., and Mayhew, P.J. (2002). Siblicide and life-history evolution in parasitoids. *Behavioral Ecology* 13: 690–695.

Phillips, C.B., and Baird, D.B. (2001). Geographic variation in egg load of *Microctonus hyperodae* Ioan (Hymenoptera: Braconidae) and its implications for biological control success. *Biocontrol Science and Technology* 11: 371–380.

Phillips, C.B., and Kean, J.M. (2017). Response of parasitoid egg load to host dynamics and implications for egg load evolution. *Journal of Evolutionary Biology* 30: 1313–1324.

Phillips, C.B., Proffitt, J.R., and Goldson, S.L. (1998). Potential to enhance the efficacy of *Microctonus hyperodae* Loan. Proceedings of the 51st New Zealand plant protection conference, pp. 16–22.

Price, P.W. (1973). Reproductive strategies in parasitoid wasps *The American Naturalist* 107: 684–693.

Price, P.W. (1974). Streategies for egg-production. *Evolution* 28: 76–84.

Prokopy, R.J., Roitberg, B.D., and Vargas, R.I. (1994). Effects of egg load on finding and acceptance of host Fruit in *Ceratitis capitata* flies. *Physiological Entomology* 19: 124–132.

R Core Team (2021). *R: a language and environment for statistical computing*. R Foundation for Statistical Computing. https://www.r-project.org/.

Rivero, A., and Casas, J. (1999). Incorporating physiology into parasitoid behavioral ecology: the allocation of nutritional resources. *Researches on Population Ecology* 41: 39–45.

Rivero-Lynch, A.P., and Godfray, H.C.J. (1997). The dynamics of egg production, oviposition and resorption in a parasitoid wasp. *Functional Ecology* 11: 184–188.

Roitberg, B.D., Sircom, J., Roitberg, C.A., van Alphen, J.J.M., and Mangel, M. (1993). Life expectancy and reproduction. *Nature* 364: 108–108.

Rosenheim, J.A. (1996). An evolutionary argument for egg limitation. *Evolution* 50: 2089–2094.

Rosenheim, J.A. (1999a). Characterizing the cost of oviposition in insects: a dynamic model. *Evolutionary Ecology* 13: 141–165.

Rosenheim, J.A. (1999b). The relative contributions of time and eggs to the cost of reproduction. *Evolution* 53: 376–385.

Rosenheim, J.A. (2011). Stochasticity in reproductive opportunity and the evolution of egg limitation in insects. *Evolution* 65: 2300–2312.

Rosenheim, J.A., Alon, U., and Shinar, G. (2010). Evolutionary balancing of fitness-limiting factors. *The American Naturalist* 175: 662–674.

Rosenheim, J.A., Heimpel, G.E., and Mangel, M. (2000). Egg maturation, egg resorption and the costliness of transient egg limitation in insects. *Proceedings of the Royal Society B: Biological Sciences* 267: 1565–1573.

Rosenheim, J.A., Jepsen, S.J., Matthews, C.E., Smith, D.S., and Rosenheim, M.R. (2008). Time limitation, egg limitation, the cost of oviposition, and lifetime reproduction by an insect in nature. *The American Naturalist* 172: 486–496.

Rosenheim, J.A., and Rosen, D. (1991). Foraging and oviposition decisions in the parasitoid *Aphytis lingnanensis* – distinguishing the influences of egg load and experience. *Journal of Animal Ecology* 60: 873–893.

Rosenheim, J.A., Williams, N.M., and Schreiber, S.J. (2014). Parental optimism versus parental pessimism in plants: how common should we expect pollen limitation to be? *The American Naturalist* 184: 75–90.

Schreiber, S.J. (2006). Host-parasitoid dynamics of a generalized Thompson model. *Journal of Mathematical Biology* 52: 719–732.

Schreiber, S.J. (2007). Periodicity, persistence, and collapse in host–parasitoid systems with egg limitation. *Journal of Biological Dynamics* 1: 273–287.

Segoli, M., Harari, A.R., Rosenheim, J.A., Bouskila, A., and Keasar, T. (2010). The evolution of polyembryony in parasitoid wasps. *Journal of Evolutionary Biology* 23: 1807–1819.

Segoli, M., and Rosenheim J.A. (2013a). Limits to the reproductive success of two insect parasitoid species in the field. *Ecology* 94: 2498–2504.

Segoli, M., and Rosenheim, J.A. (2013b). The link between host density and egg production in a parasitoid insect: comparison between agricultural and natural habitats. *Functional Ecology* 27: 1224–1232.

Segoli, M., and Rosenheim, J.A. (2013c). Spatial and temporal variation in sugar availability for insect parasitoids in agricultural fields and consequences for reproductive success. *Biological Control* 67: 163–169.

Segoli, M., Sun, S.C., Nava, D.E., and Rosenheim, J.A. (2018). Factors shaping life history traits of two proovigenic parasitoids. *Integrative Zoology* 13: 297–306.

Segoli, M., and Wajnberg, E. (2020). The combined effect of host and food availability on optimized parasitoid life-history traits based on a three-dimensional trade-off surface. *Journal of Evolutionary Biology* 33: 850–857.

Sevenster, J.G., Ellers, J., and Driessen, G. (1998). An evolutionary argument for time limitation. *Evolution* 52: 1241–1244.

Shea, K., Nisbet, R.M., Murdoch, W.W., and Yoo, H.J.S. (1996). The effect of egg limitation on stability in insect host-parasitoid population models. *Journal of Animal Ecology* 65: 743–755.

Sirot, E., and Bernstein, C. (1996). Time sharing between host searching and food searching in parasitoids: state-dependent optimal strategies. *Behavioral Ecology* 7: 189–194.

Skinner, S.W. (1985). Clutch size as an optimal foraging problem for insects. *Behavioral Ecology and Sociobiology* 17: 231–238.

Stephens, D.W., and Krebs, J.R. (1986). *Foraging theory*. Princeton University Press.

Strand, M. (2003). Polyembryony. In: *Encyclopedia of insects* (eds. Carde, M.R., and Resch, V.), 928–932. Academic Press.

Tang, X.Y., Meng, L., Kapranas, A., Xu, F.Y., Hardy, I.C.W., and Li, B.P. (2014). Mutually beneficial host exploitation and ultra-biased sex ratios in quasisocial parasitoids. *Nature Communications* 5: 4942

Tena, A., Pekas, A., Cano, D., Wäckers, F.L., and Urbaneja, A. (2015). Sugar provisioning maximizes the biocontrol service of parasitoids. *Journal of Applied Ecology* 52: 795–804.

Thompson, S.N., and Hagen, K.S. (1999). Nutrition of entomophagous insects and other arthropods. In: *Handbook of biological control* (eds. Bellows, T.S., and Fisher, T.W.), 594–652. Academic press.

van Baalen, M. (2000). The evolution of parasitoid egg load. In: *Parasitoid population biology* (eds. Hochberg, M.E., and Ives, A.R.), 103–120. Princeton University Press.

van Randen, E.J., and Roitberg, B.D. (1996). The effect of egg load on superparasitism by the snowberry fly. *Entomologia Experimentalis et Applicata* 79: 241–245.

Visser, M.E. (1994). The importance of being large – the relationship between size and fitness in females of the parasitoid *Aphaereta minuta* (Hymenoptera, Braconidae). *Journal of Animal Ecology* 63: 963–978.

Visser, B., and Ellers, J. (2008). Lack of lipogenesis in parasitoids: a review of physiological mechanisms and evolutionary implications. *Journal of Insect Physiology* 54: 1315–1322.

Visser, M.E., van Alphen, J.J.M., and Hemerik, L. (1992). Adaptive superparasitism and patch time allocation in solitary parasitoids – an ESS model. *Journal of Animal Ecology* 61: 93–101.

Waage, J.K. (1983). Aggregation in field parasitoid populations – foraging time allocation by a population of *Diadegma* (Hymenoptera, Ichneumonidae). *Ecological Entomology* 8: 447–453.

Wäckers, F.L., Lee, J.C., Heimpel, G.E., Winkler, K., and Wagenaar, R. (2006). Hymenopteran parasitoids synthesize 'honeydew-specific' oligosaccharides. *Functional Ecology* 20: 790–798.

Wäckers, F.L., van Rijn, P.C.J., and Heimpel, G.E. (2008). Honeydew as a food source for natural enemies: making the best of a bad meal? *Biological Control* 45: 176–184.

Wajnberg, E. (2006). Time allocation strategies in insect parasitoids: from ultimate predictions to proximate behavioral mechanisms. *Behavioral Ecology and Sociobiology* 60: 589–611.

Wang, W., Lu, S.L., Liu, W.X., Cheng, L.S., Zhang, Y.B., and Wan, F.H. (2014). Effects of five naturally occurring sugars on the longevity, oogenesis, and nutrient accumulation pattern in adult females of the synovigenic parasitoid *Neochrysocharis formosa* (Hymenoptera: Eulophidae). *Neotropical Entomology* 43: 564–573.

Wanner, H., Gu, H., and Dorn, S. (2006). Nutritional value of floral nectar sources for flight in the parasitoid wasp, *Cotesia glomerata*. *Physiological Entomology* 31: 127–133.

Weisser, W.W., Volkl, W., and Hassell, M.P. (1997). The importance of adverse weather conditions for behaviour and population ecology of an aphid parasitoid. *Journal of Animal Ecology* 66: 386–400.

Wu, Z.S., and Heimpel, G.E. (2007). Dynamic egg maturation strategies in an aphid parasitoid. *Physiological Entomology* 32: 143–149.

Xu, L.L., Zhou, C.M., Xiao, Y., Zhang, P.F., Tang, Y., and Xu, Y.J. (2012). Insect oviposition plasticity in response to host availability: the case of the tephritid fruit fly *Bactrocera dorsalis*. *Ecological Entomology* 37: 446–452.

Yadav, P., and Borges, R.M. (2018). Why resource history matters: age and oviposition history affect oviposition behaviour in exploiters of a mutualism. *Ecological Entomology* 43: 473–482.

Zhang, Y.B., Liu, W.X., Wang, W., Wan, F.H., and Li, Q. (2011). Lifetime gains and patterns of accumulation and mobilization of nutrients in females of the synovigenic parasitoid, *Diglyphus isaea* Walker (Hymenoptera: Eulophidae), as a function of diet. *Journal of Insect Physiology* 57: 1045–1052.

Zhu, P.Y., Gurr, G.M., Lu, Z.X., Heong, K,. Chen, G.H., Zheng, X.S., Xu, H.X., and Yang, Y.J. (2013). Laboratory screening supports the selection of sesame (*Sesamum indicum*) to enhance *Anagrus* spp. parasitoids (Hymenoptera: Mymaridae) of rice planthoppers. *Biological Control* 64: 83–89.

5

Sex-Specific Life Histories

Hanna Kokko

Institute for Organismic and Molecular Evolution, Johannes Gutenberg University of Mainz, Mainz, Germany

5.1 Introduction

Killer whales (*Orcinus orca*) look like majestic, top-of-the-food chain animals, and one could be excused to think that the mightiest predator in the ocean is a mature male orca. But that would be wrong. Adult male orcas do very badly should their mother die: their mortality increases massively (for males > 30 years, by more than eightfold) in the year following their mother's death (Foster et al. 2012). Males only travel away from their natal group to mate, to thereafter return to their mother (who keeps feeding their adult sons, Wright et al. 2016). Luckily for them, a mother's death during their own lifespan is a relatively rare occurrence; females of this species have far longer lifespans than males (Foster et al. 2012).

This is an example of a dramatic sex difference in lifespan. Why did it evolve? More generally, why do mammalian males as a whole die at a younger age than females do, while a typical bird species will show the opposite pattern (Pipoly et al. 2015)?

An analogous set of 'why' questions exists with respect to reproduction. For reproductive traits, however, some of the 'why' questions boil down to a simple definitional issue. Males produce the smaller gametes, while females produce the larger ones. Therefore, while one can ask a range of sensible 'why' questions about the asymmetry in gamete morphology in the first place (in other words, why did anisogamy evolve; Lessells et al. 2009, Lehtonen and Parker 2014), there is no particular exciting answer to the question of why the producers of small gametes are labelled males. It is simply the convention biologists use to avoid a mess where everything that sounds vaguely male-like to us humans might be used to make a non-human individual be labelled a male.

It is worth reflecting on this fact a little. It would be massively impractical to equate maleness with 'that subpopulation that dies younger'. Although that may seem far-fetched, many other asymmetries in contemporary humans come with impracticalities that similarly prevent them from being good general definitions of an individual's sex. Sex chromosomes, for example, would leave sex undefined in organisms that lack sex chromosomes (or in species where researchers simply have not yet conducted any studies of its sex determination). For non-humans, we likewise cannot talk about 'gender' as it relates to self-reported identities (Tannenbaum et al. 2019, de Vries and Lehtonen 2023).

With respect to biological sex, the purpose of labels is to aid communication: to set a stage where we can proceed past terminology and ask why traits combine in particular ways. Teaching experience shows that students new to biology might assume that Y chromosomes, highly mobile and explorative adults (or their gametes), larger-bodied individuals, aggressive behaviours or (if one is a fungus) cells that donate nuclei to other cells as opposed to receiving them, are all indicative of some sort of 'maleness'. In the case of, for example, fungi, the term 'male' is indeed being used to describe the asymmetry of donating *vs.* receiving nuclei in cells (Niewenhuis and Aanen 2012).

In scientific terminology, however, definitional clarity can help to see where patterns of co-variation are not automatically given but require an explanation. If one strips everything else from the definition of a male and simply states that males produce sperm (or pollen) while females produce eggs (or ovules), the question comes into focus: any observed sex differences, for example, sexual size dimorphism, sex differences in basic metabolic rate or the mean age of reproduction, should not be assumed to be part of a 'maleness' or a 'femaleness' syndrome. The field should strive to understand precisely why the co-variation of a trait with the sex of the individual arises. In other words, why should a subpopulation that follows a 'live fast, die young' schedule associate with them producing microgametes (*i.e.*, being male) rather than macrogametes (being female)? Or is this even true? The short summary is that it appears to be often true but not always.

Below, I will first talk about the causalities that have their origins in sex determination and the associated genetic architecture. I call this a unidirectional route because changes in sex determination and, for example, the size of the sex chromosomes, are usually not argued to coevolve with the organism's life history traits; there is little scope for coevolution. Thereafter, I will turn to the question of the coevolution between components of life history traits and the two sexes. This is a bi- or possibly multi-directional set of questions because there are potentially many traits, expressed in the two sexes (and possibly at different stages of ontogeny), that all can potentially trade-off with each other or experience selection in directions that align with each other or be antagonistic, with positive or negative feedback.

5.2 Various Unidirectional Effects: Unguarded X, Mother's Curse and Toxic Y

There are some exciting ideas that might help explain, in an indirect manner, some of the variation in male and female lifespans. These relate to the fact that there is great diversity in sex determination mechanisms (Bachtrog et al. 2014). Thus, it might not be the maleness or femaleness *per se*, but the underlying genetic architecture, which happens to also create males and females, that creates lifespan differences. Males may be the heterogametic sex (XY systems, *e.g.*, mammals) or the homogametic one (ZW, *e.g.*, birds), or sex may be associated with different ploidy (*e.g.*, under haplodiploidy, males are haploid, developing from unfertilized eggs, while females are diploid), or there may be no genetic sex determination at all, with, for example, the temperature of the developing egg instead determining the sex of the offspring.

The fact that sex determination mechanisms vary allows, in principle, detecting effects of the fact that deleterious alleles are expressed in the sex that has only a single copy of the X (or Z) chromosome (the unguarded X hypothesis), which could shorten the lifespan of the heterogametic sex. Likewise, it is possible to test the idea that the accumulation of repetitive DNA on the Y (or W) chromosome could be responsible for faster male senescence, as silencing the repeats may become ineffective in ageing males. The diversity of sex determination mechanisms then offers comparisons where the effect of males' responses to the mating system and other aspects of the ecology of each species (*e.g.*, sexual dimorphism in body size) can vary largely independently of the genetic architecture of males and females.

Additionally, there is also the potential that 'mother's curse', the fact that mitochondria are under no selection to make males perform well (except in the very rare cases where mitochondria break from their usual uniparental inheritance via females; Allison et al. 2021), might shorten male lifespan (Maklakov and Lummaa 2013; see Chapter 2, this volume, for further explanation). This last factor presents a larger challenge in any comparative analysis, as there is little variation on which sex suffers from the mother's curse: it is almost always the male, thus it is difficult to disentangle mitochondria-related trouble from all other effects that may be detrimental to male lifespan.

Even so, the mother's curse has been tested comparatively in mammals. Although there is no variation in which sex is the dead-end for mitochondria in this taxon, one can assume that between-species variation in mtDNA substitution rates would be reflected in sexually dimorphic life histories if mother's curse plays a strong role in shaping life history traits. There was, however, little support for this across 104 species of mammals (Cayuela et al. 2023).

The unguarded X hypothesis has, in a meta-analysis spanning several taxa, gained rather clear support (Xirocostas et al. 2020), and in amphibians, ageing rates (but not lifespan) behaved as expected based on the unguarded X hypothesis. However, a recent parameterization of a population genetic model casts some doubt on the interpretation that the rather stark lifespan differences — 7.1% and 20.9% longer lifespans, on average, for the homogametic sex, depending on whether it is the female or the male, respectively, that has the unguarded sex chromosome (Xirocostas et al. 2020) — really can result from the proposed mechanism alone (Connallon et al. 2022). The current state of affairs reflects difficulties in determining the true cause of sex differences in lifespan or ageing rate. If a significant proportion of data comes from a bird-mammal comparison, then the XY *vs.* ZW comparison lacks statistical power. There are numerous relevant differences between these taxa: a typical mammal has more stark parenting role differentiation, and a different type of sexual competition, than the avian-style phenotypic sex differences. The latter often feature plumage dimorphism, song and, for example, differences in arrival timing from migration, but less intense sex difference in parenting roles than that of a typical mammal. All these and additional features, such as dosage compensation (which, again, differs between birds and mammals), may matter greatly for the final result (Pipoly et al. 2015).

Comparative analyses can of course be improved to include more variables, and intriguing new data suggests that the size of the chromosomes may matter (Sultanova et al. 2023). All else being equal, a larger chromosome contains more genes that may impact lifespan, and for the toxic Y hypothesis, a large Y chromosome presents a large target for deleterious mutations. For mammals, taking chromosome size into account provided thus new support for both the unguarded X and toxic Y hypotheses, and mammalian males in species with particularly small Y chromosomes live relatively longer.

I am not aware of a study that has attempted to put equal weight on chromosomes and on the numerous behavioural variables that can conceivably impact the outcome. This is also not the only possible way to investigate the matter. In *Drosophila*, for example, the experimental creation of females that express recessive alleles on the X to the same degree as males has yielded no evidence supporting the unguarded X hypothesis (Brengdahl et al. 2018), while experiments that create homozygosity through inbreeding offer some support (Sultanova et al. 2018). For the toxic Y hypothesis, CRISPR-Cas9 has been used to create flies that allow direct tests of the predictions (again, in *Drosophila*; Delanoue et al. 2023) with no support for the hypothesis. Being phenotypically male or female was found to be responsible for the lifespan differences (Delanoue et al. 2023), rather than the effects of the Y chromosome specifically. It is interesting that this single-species study, which was able to probe deeply into the mechanisms, gives in its gist a similar answer to the question as does the mathematical analysis of Connallon et al. (2022): sex-specific selection may create sexually divergent life histories more easily, or more strongly, than details of the genetic architecture.

5.3 Multi-directionality: Coevolution of Different Traits

Life history theory could be defined as the question of understanding why 'Darwinian demons' cannot exist. The demon, a brainchild of Williams (1966), is an organism that combines all the life history traits that can conceivably be thought to be favoured by selection: it matures immediately upon being born, has infinite fecundity every time point that it is alive, and it is also immortal. Obviously, there are not enough carbon atoms in the universe to make enough copies of the demon. Therefore, organisms are expected to face limitations in all the mentioned life history traits and be subject to trade-offs between them. Natural selection cannot optimize everything simultaneously, and instead, one expects organisms to evolve, in some sense, the best compromise solution.

Males and females of the same species might find their best solutions to differ. However, before turning to the theoretical underpinnings of the male-female difference, it is necessary to first know the basics about the history of ideas in an interspecific (rather than intersexual) context. *r-K* selection in life history theory refers to a set of ideas that originated with MacArthur and Wilson (1967) and were elaborated on by Pianka (1970). In this framework, species differ in which life history traits are optimized (at the expense of other traits). The letter *r* refers to the intrinsic growth rate and *K* to the carrying capacity in the model of logistic growth:

$$\frac{1}{N}\frac{dN}{dt} = r\left(1 - \frac{N}{K}\right) \tag{5.1}$$

Here, N is the current population size, $\frac{dN}{dt}$ describes its change over time, and $\frac{1}{N}\frac{dN}{dt}$ thus describes the *per capita* growth. The key idea is that some species are at the '*r*-selected' end of a continuum, where success requires maximizing r. This is achieved when producing very many offspring over a short period of time (few of them are expected to survive). Simultaneously, the organism will forego other goals, such as achieving a long life or a large body size. The '*K*-selected' species optimize differently. *K* refers to carrying capacity: Equation (5.1) describes a non-growing population, $\frac{dN}{dt} = 0$, when $N = K$ (this special value of N is often denoted N^*). The mental image of a *K*-selected species involves a population that already occupies all the possible breeding spots, and this makes it difficult for juveniles to recruit into the breeding population. A parent consequently needs to make their offspring well equipped to succeed in this competition. Accordingly, they only produce a few of them at a time (Pianka 1970; review in Jeschke et al. 2019). Note that by focusing on offspring production by mothers, studies in the *r-K* framework were mostly interested in females. The female bias in life history studies has persisted ever since (Archer et al. 2022).

Importantly, the *r-K* contrast was also thought to co-vary with other patterns, such as faster development, earlier maturity and smaller body size when *r*-selected (Pianka 1970), creating 'syndromes' of co-varying traits (Stearns 1977). The *r-K* framework is intuitive, but in the decades that followed its conception, the 'syndromes' it predicts received a very critical look, especially by theoreticians (Stearns 1977, Roff 2002; reviewed in Jeschke et al. 2019). Although it has not stopped inspiring research on both theoretical and empirical fronts (Engen and Saether 2016 and Saether et al. 2016, 2021 provide intriguing examples), today's efforts to structure interspecific variation in life histories tend to talk about a fast-slow axis (*e.g.*, Bielby et al. 2007, Allen et al. 2017, Tarka et al. 2018, Bakewell et al. 2020, Cayuela et al. 2020, Ferreira et al. 2023), instead of a *r-K* contrast – though microbial research offers examples of very recent uses of the *r-K* framework (Wei and Zhang 2019, Marshall et al. 2023), a topic to which we shall return at the end of this chapter.

Why did the field turn away from '*r*' and '*K*' as endpoints of a continuum? The key issue is the following: Just because one can imagine a set of co-varying traits, for example, a set that appears to make sense for a '*r*-selected' population, one has not yet pinpointed the reasons *why* evolutionary theory should predict the co-occurrence of such a set (Stearns 1977, Roff 2002, Jeschke et al. 2019). As Stearns (1977) put it, 'the logic of comparison is weaker than the logic of prediction'. In other words, it is all too easy to tabulate traits and imagine a 'typical' case at each end of a continuum. The real work is to use evolutionary principles to see whether, and under what conditions, selection really will shape multiple traits into the purported sets of co-varying syndromes (Stearns 1977, Roff 2002).

Should one accept the *r-K* endpoints too easily, one may have succumbed to the so-called availability heuristic (Esgate et al. 2005), a mental shortcut that can lead to a biased view of the world by overemphasizing cases that come to mind very easily. Pianka (1970) himself offered insects as typical *r*-selected organisms and vertebrates as *K*-selected organisms. Should one nod in agreement, one has fallen victim to the availability heuristic. A more open-minded look at the question is to ask if the same organism might also combine trait values from either end of the *r-K* spectrum, breaking the proposed syndrome. For example, fast development was supposedly an *r*-selected trait (Pianka 1970), while a large body size was assumed to result from *K*-selection. But is it really not possible to find a case where an organism that is selected to be large for its adult life is also selected to develop as rapidly as possible away from its juvenile size? When a large body size confers protection from many predators, this syndrome-breaking combination appears wholly plausible: juveniles are then vulnerable for as long as they are still small (see also Chapter 12, this volume).

A more general point is that succeeding in terms of *r* does not automatically imply less success in terms of *K* (Jeschke et al. 2019, Marshall et al. 2023). The carrying capacity of a population combines properties of the environment, such as resource availability, with the traits of the species, and pitting it against *r* as an alternative to invest in is theoretically awkward. Interestingly, the question can be defined better: are there situations where a genotype not only replaces others (the criterion for its evolutionary success) but also leads the population to have the highest population density that is achievable given the ecological setting? The answer is yes (Charlesworth 1980), but it comes with caveats: there should be a limited supply of one resource type that becomes limiting. An example of such a resource is breeding sites in territorial birds, in which case evolution maximizes the number of individuals competing for available sites but not, for example, the mean production of offspring from the sites (Kokko et al. 2001). The availability of sites is, in this case, the one crucial resource that limits population growth. When the ecological situation is more complex, one should not expect evolution to provide solutions that maximize the number of individuals at equilibrium or population growth rate. Pertinent for the topic of this chapter, sexual reproduction can cause strong deviations from the naïve expectation that either growth rates or carrying capacities are maximized. If there is sexual conflict, males may evolve behaviours or genotypes that are detrimental to females, and this makes the population grow less well (Kokko 2021).

It is thus a healthy development that modern attempts to place species along a continuum often use the words 'slow' or 'fast' to classify life histories, instead of claiming them to be *r*- or *K*-selected. Firstly, this separates the observations from the discussion of the properties of a specific population growth model (the logistic), which may or may not fit a particular species well. Secondly, a focus on a fast-slow axis, instead of asking whether species are *r*- or *K*-selected, avoids using the word 'selection'. This hopefully helps to remember that finding a pattern, *i.e.*, that species differ from each other, does not mean that one has simultaneously completed the task of understanding the process of selection that led to the pattern.

Recently, the study of slow and fast life histories has been complemented by a framework called POLS for 'pace-of-life syndromes' (Réale et al. 2010), where the quest is to identify sets of co-occurring life history traits, together with differences in physiological traits (*e.g.*, metabolic rates) as well as behavioural traits, for example, high or low exploration rate or aggression. This relatively recent development is more relevant for the study of sex-specific life histories than the *r-K* concept ever was, for the simple reason that the *r-K* ideas have typically not been extended to make claims about the relative position of males and females along the purported continuum. The POLS framework has recently experienced this expansion (Hämäläinen et al. 2018, Immonen et al. 2018, Tarka et al. 2018, Moschilla et al. 2019; note that Réale et al. 2010, too, mentioned a few cases where male and female traits needed to be listed separately, but sex specificity was not their main focus). However, the fact that *r-K* selection did not experience sex-specific scrutiny does not make its history irrelevant to understand issues in the present, sex-specific context. The reason is that there are strong parallels between the *r-K* features that have attracted criticism and the way the POLS has developed.

A key lesson arising from the *r-K* history is that observations, even if systematically collected, are not yet (evolutionary) explanations. Progress on both fronts is necessary. Here, one can ask whether the POLS framework arose because observations strongly suggested the existence of syndromes. The answer is clearly no: the defining paper (Réale et al. 2010) is a collection of arguments, ideas and examples, instead of having arisen from, for example, a principal component analysis suggestive of syndrome-like patterns either within or between species. The order of events (idea first, tests later, *e.g.*, Royauté et al. 2018) is, of course, not unusual in science, and Réale et al. (2010) also, laudably, already discussed known exceptions to the patterns they proposed. But it is interesting to reflect on the hypothetical temporal sequence where the data collection effort would have happened first. Given that Royauté et al.'s (2018) dataset shows meagre support for the POLS (see also Moiron et al. 2020), science might have proceeded to an entirely different set of ideas regarding trait co-variation had there not been strong prior views regarding the expected patterns (Réale et al. 2010). Turning back to the real temporal timeline, a very recent analysis has also shown that the slow-fast axis is far better supported in an interspecific context than in within-species comparisons of individuals (van de Walle et al. 2023).

In addition to the situation of an 'ideas-first' instead of 'data-first' history, there is a problem of the ideas having arisen with very little modelling effort (see Mathot and Frankenhuis 2018 for documentation of the scarcity of modelling effort on the topic). This means that the dangers exposed by the *r-K* selection debate, *i.e.*, the availability heuristic and the mistaken feeling that having identified variation is equivalent to having explained its evolutionary origin, might apply to the present approaches too. In the following sections, I hope to convince the reader that this is indeed the case and end with some guidance for the future.

5.3.1 An Example of a Model: How Easily Aggression Shortens or Prolongs Lifespan

The reason why formal modelling can help is that it helps steer clear of the cognitive biases, such as the availability heuristic, by forcing a researcher to make explicit what the assumption structures are. This also allows subsequent criticism to be more precise (see, *e.g.*, the discussion that followed a model for the evolutionary origin of animal personalities, Wolf et al. 2007, where it was pointed out that results change towards less maintenance of individual differences if an animal has time to play the relevant behavioural games many times before it dies; McElreath et al. 2007). Because modelling effort in this field has been scant (Mathot and Frankenhuis 2018), I will provide a new example that focuses on the role of aggression, which is the sixth trait in the table of Réale et al. (2010), assumed to link with a short life and other 'fast' POL traits. The example will show how easily ecology can make aggressiveness co-vary with either long or short lives in a territorial bird. My small model below provides a mathematical treatment of the arguments provided by Laskowski et al. 2020: aggression may prolong life if it allows superior access to resources.

Aggression is known to play a crucial role in the life of the Gouldian finch *Erythrura gouldiae*. The real-life system of this Australian finch species is rather complex, with black-headed and red-headed morphs (Figure 5.1). The colour locus is on the Z chromosome (Toomey et al. 2018). Genomically, the Gouldian finch polymorphism shows signs of balancing selection (Kim et al. 2019), very likely driven by the ecological finding that the bird's ecology satisfies criteria of the so-called Hawk–Dove game (Kokko et al. 2014). In this game, aggressive types are favoured if rare, but aggression starts to 'backfire' if common, as high rates of fighting (with other redheads) appear to trade-off with parenting duties (Pryke and Griffith 2009). The species is highly mobile, and both red-headed and black-headed individuals need to locate nest holes at the beginning of each breeding season. Red-headed individuals are able to occupy nest holes when competing with black-headed ones, but once surrounded by many other redheads, they do not put them to good use; their breeding success suffers in a negatively frequency-dependent manner. For a full model explaining why neither colour can reach fixation, see Kokko et al. (2014). Here I am presenting a much more stylized model that boils the argument down to the bare essentials: if intuition suggests that the redheads are the morph with the 'fast' life, modelling can confirm this prediction to be true in some ecological settings but not in others.

Figure 5.1 The black-headed and the red-headed morph of the Gouldian finch, *Erythrura gouldiae*. *Source:* Photo by Sarah Pryke.

Consider just two types of individuals that are maintained asexually (as in the original stylized Hawk–Dove game), with differing levels of aggressiveness. I assume that a proportion p_nest of adults can find a nest site, but the aggressive (red-headed) morph is somewhat better at doing so. Thus, if its proportion at the beginning of the breeding season is x_0, its proportion among nest-site holders is $x_\text{breed} = \alpha/(\alpha x_0 + (1 - x_0))$ (e.g., $\alpha = 1.2$ predicts that each redhead individual has a 20% elevated probability of becoming the owner of a site when competing with a blackhead). All blackheads that managed to get a breeding site – despite being in an unfavourable position to do so – breed successfully, while some of the redheads are assumed to fail. I assume that the failure probability increases with the proportion of reds, with two options that the model can be run with; the relevant proportion may be the overall proportion of redheads ($p_\text{fail} = 1 - e^{-x_0/2}$) or their proportion among nest-site holders ($p_\text{fail} = 1 - e^{-x_\text{breed}/2}$). The results are similar (both qualitatively and quantitatively) between these options, and, for brevity, I will only give results for the case where the proportion of reds among the breeders predicts the breeding failures of redheads.

Let us consider two scenarios regarding annual survival. In both cases, the baseline survival of redheads is 99% of that of blackheads, i.e., we agree that their aggression links physiologically with a tendency to live shorter lives.

In scenario (a), survival of adults scales as 0.6, 0.5 and 0.5 for non-breeders, failed breeders and successful breeders (who are assumed to fledge two same-sex chicks as the parent), respectively (for redheads, these numbers are then multiplied by 0.99). Since failed breeders and successful breeders survive equally poorly relative to non-breeders, this scenario focuses on the costs of breeding and assumes that they cannot be escaped by those who attempt to breed but fail. Following the demography of the system over many years (Figure 5.2a) reveals that the system stabilizes at approximately two-thirds blackheads and one-third redheads (rather similar to the situation in nature, despite the model ignoring the details of how the colour alleles are inherited). In this assumption set, redheads can be computed to have a somewhat shorter lifespan than blackheads (2.04 years *vs.* 2.14 years, Figure 5.2a).

In another scenario set (b), we retain that redheads are predisposed to a fast life (the 99% survival relative to blackheads is still there). Moreover, we assume that competition for breeding sites is equally strong as before, but in this version of the model, we make a difference between the breeders that fail and those that succeed. The rationale is that successful breeders have to spend more parenting effort, and therefore they survive less well than failed breeders. Failed breeders may also survive better than non-breeders if the ecological setting has the nest holes situated in habitats that also offer the best protection from other challenges such as predation and the best foraging. If the above argumentation makes failed breeders the

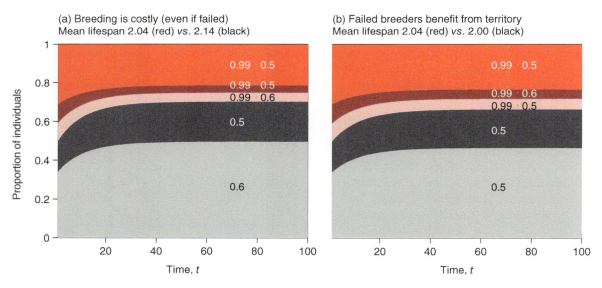

Figure 5.2 The results of the model of Gouldian finch morphs differing in aggressiveness over 100 generations (x-axis) when starting with a proportion of 0.5 of red-headed individuals. The colours indicate the proportions of (from top to bottom) successfully breeding redheads, failed redhead breeders and non-breeding redheads, followed by breeding and non-breeding blackheads. The numbers within each coloured zone indicate the subsequent survival of each type of individual (see text for detail), in some cases involving a multiplicative form where, for example, 0.99·0.5 combines the poorer background survival of redheads (99% of that of blackheads) with the baseline survival of breeders. The assumption structure is different for (a), where non-breeding blackheads are the best surviving category, than for (b), where redheads that gained a territory but failed to breed survive better than any other category.

part of the population with the best survival, the relative ranking of successful breeders and non-breeders is not yet clear. In the example (Figure 5.2b), I simply make them survive equally well (0.5). As a whole, redheads are now predicted to be the longer-lived morph (Figure 5.3b: 2.04 *vs.* 2.00 years on average). Why? They are better than blackheads at occupying nest holes, but they also more often end up as failed breeders – and breeding failures, in this assumption set, spare them costs of breeding.

The stylized nature of this example is intentional, and it skips many aspects of real Gouldian finches (*e.g.*, aggressive 'reds' are more shy in the context of novel object inspection, Williams et al. 2012). The reason the example should nevertheless be thought-provoking is that it shows how very subtle changes in ecological assumptions, with neither scenario (a) nor (b) *a priori* more likely than the other, can create ecologies in which aggression is favoured *despite* it causing a shorter lifespan (Figure 5.2a) and others where it is favoured (Figure 5.2b), in part *because* it prolongs lifespan. Aggression can do so by giving access to survival-enhancing resources (see also McElreath et al. 2007, Réale et al. 2010, and Chang et al. 2024), with the full effect being masked for those individuals who succeed in breeding while exposed for those who do not breed successfully in a given year.

5.3.2 How to Make Predictions for the Two Sexes?

The above sections were a rather long detour away from explicit comparisons of the two sexes. Let us now get back to the issue of whether males and females follow different syndromes. Within-species comparisons are commonplace in the POLS framework (Réale et al. 2010, van de Walle et al. 2023), but here, van de Walle et al.'s (2023) summary is 'neither slow nor fast, just diverse', as they did not find good alignment of individuals along a slow-fast axis within any species they examined. These authors, however, did not try to compare two sexes; their within-species data had a female focus. If one tried to do so, would the two sexes arrange themselves neatly along a fast-slow axis?

The explicit sex-specific predictions (Hämäläinen et al. 2018, Immonen et al. 2018, Tarka et al. 2018, Moschilla et al. 2019) emphasize, correctly, that it is unlikely that one sex is always the faster one. According to Tarka et al. (2018), polygyny should make males the 'faster' sex, while polyandry should predict the opposite, with females being 'faster'. It has also been predicted that situations where most resources are occupied create selection to be at the slow end of the POLS (with consequences for mating systems too, Arnqvist et al. 2022), a prediction strongly reminiscent of K-selection.

Are the newer predictions equally vulnerable to criticism as the classic r-K predictions were? Let us discuss the polygyny–polyandry comparison first. Here, Tarka et al. (2018) remind us that monogamy is also possible, and it should lie between polygyny and polyandry so that monogamy should be associated with the smallest sex difference in the pace of life. At first sight, this sequence from polygyny via monogamy to polyandry offers a pleasing symmetry, from male-faster via no difference to a female-faster system. However, the symmetry offered by the words (the endings -gyny and -andro are literal opposites) may mislead. Polyandry is not, in reality, the opposite of polygyny. Polyandry typically offers multiple mating opportunities for males too, and as discussed by Holman and Kokko (2013), the formally correct term would then be 'polygynandry' (why this word is not more often employed is unclear – perhaps it is simply too much of a mouthful). Words can mislead one into thinking that a prediction is on more solid ground than it really is, just like summarizing a complex set of causalities with one variable in an equation ('K') can tempt one to assign it an exaggerated causal role.

What about monogamy, the case predicted to yield little sexual difference in life histories (Tarka et al. 2018)? The word itself evokes no asymmetry between the mating chances of the two sexes. But there is a difference between the symmetry in the number of mates involved in monogamy (one mate for either sex) and the real-life parenting roles under monogamy. In monogamous mammals, for example, the female is still the only parent that is able to perform gestation or lactation. Indeed, Promislow (1992), in a comparative analysis, found mortality to be frequently female-biased in monogamous mammals. Monogamy is similarly associated with female-biased susceptibility to parasites in a meta-analysis of diverse taxa (Wittman and Cox 2021). Humans offer particularly good data on how much pregnancy and childbirth impact female life histories (see also Chapter 11 in this volume). We may be used to thinking that human females are the longer-lived sex, but this familiar observation may only have arisen since the demographic transition (Bolund et al. 2016), which has reduced the number of times that women are exposed to the costs of reproduction.

Above, I also mentioned the prediction that high resource competition (with most resources occupied) should make species and their mating systems approach the 'slow' end of the spectrum (Arnqvist et al. 2022). This prediction was tested, admirably, by using the same experimental setup for 12 different seed beetle species, ending up with much more support for the idea. Arnqvist et al. (2022) particularly emphasize the role of ecology: some species rely on annually fluctuating resources (seeds of annual plants) that are difficult for the beetles to utilize fully (low infestation in the field), while others use a more stable supply of seeds from trees, and this leads to strong intraspecific competition. The contrast between unstable and stable resource availability, then, is argued to align species with fast and slow paces of life, respectively, with effects on mating dynamics too (Arnqvist et al. 2022).

The stability argument (though without a two-sexes angle) was present in the old *r-K* framework too (Pianka 1970). Do the *r-K* criticisms apply to the modern test in this context too? Is it possible that the field is testing ideas that are not based on any formal theory? The situation is somewhat akin to the phrase 'absence of evidence is not evidence of absence', but since here we have empirical evidence (Arnqvist et al. 2022), the correct target of criticism is the absence of theory. A clunky, but in the present context, more accurate version of the above phrase is that the absence of theoretical work solidifying the predictions does not yet constitute proof that the predictions are wrong. They simply lack a solid theoretical footing.

In many settings, the ephemerality of resources is known to have eco-evolutionary consequences (Butterworth et al. 2023), but the effects can vary. For example, in the field of senescence evolution (a topic that is related to the pace of life), a counterintuitive effect arises: extrinsic mortality rates do not at all impact selection to delay senescence if populations experience exponential growth that is occasionally interrupted by catastrophic events (*e.g.*, the disappearance of local habitat patches) that indiscriminately kill individuals (de Vries et al. 2023). If, however, juveniles can only recruit into a population when an adult breeder has died, extrinsic mortality begins to matter for the age of onset of senescence (de Vries et al. 2023). This latter scenario has a certain *K*-selected flavour about it, but it does not straightforwardly associate the slowest life histories with the most stable environments. Instead, the slow end emerges as an evolutionary prediction when juveniles are the life stage that suffers disproportionately from high population density, and additionally, extrinsic mortality impacting all population members must be low. The lesson is the by now familiar one: careful theoretical scrutiny can offer more precise conditions for when a prediction is fulfilled than a too hastily compiled alignment of words based on intuition alone.

5.3.3 Embrace the Knowledge Gaps, Work to Fill Them

Thus far, I have not dared make too many solid predictions about male and female life histories. The issue comes back to the problem mentioned at the beginning of this chapter: there are many traits that sound 'male-like', but their co-variation is not assured, nor has it been explained, just because one can easily conjure an image of a male individual that features all the relevant traits in a combination that 'feels right'. Published lists of potentially co-occurring traits include a fast growth rate, heavy investment into early reproduction (leading to short lifespan), weak immune responses, being aggressive, bold and explorative, while also exhibiting low levels of parental care (Tarka et al. 2018). Initially, it is easy to nod in agreement when reading this list: all these traits sound very male-like, predicting investment in combat rather than care, and living short lives as a consequence. Yet, why precisely should avoiding parental care duties be associated with a live-fast-die-young life history? Specifically, if parenting is an investment in offspring survival at the expense of a parent's own survival, should lack of investment not predict a longer, not a shorter lifespan?

These kinds of questions are not trivial, and they form the reason to build models from first principles (Parker 2014). Ever since the 1970s, a strong theme in our understanding of sex differences has been the economic realization that, given a fixed budget, an organism can create more gametes per time unit when the gametes are small (Parker et al. 1972). A small caveat is in place: the prediction should acknowledge the tacit assumption that the total budget for gamete production should not depend on the sex of the parent. In reality, this independence is unlikely to be true across all cases; the female budget may be considerably larger (Hayward and Gillooly 2011). Still, there is a good reason why models tend to assume symmetry. If we are interested in a 'null model' where we assume no other differences between the sexes than what is imposed by their definition, then it makes sense not to resort to any budgetary differences, as these would require causal explanations themselves (de Vries and Lehtonen 2023).

Thus, let us consider a model where the only assumption about males (in addition to their definition, *i.e.*, that their gametes are small) is that they produce more gametes than females do. It follows that males have a greater risk that their gametes fail to become a zygote. Males are easily under stronger selection than females to make sure that their gametes outcompete those of others (Parker 1979, Janicke et al. 2016). The phrase 'to make sure' has to be interpreted properly in this context. Succeeding in fertilizing females is typically a zero-sum game, which means that one male's gain is another male's loss: paternity can be strived for, but not everyone can possibly succeed in outcompeting everyone else. This in no way prevents the mating system from requiring competitive investment for there to be any chances for a particular male.

Given that male gametes are more numerous than female gametes, it can be shown (Lehtonen 2022) that the male relationship between the number of matings and fitness easily becomes steeper than the same relationship for females. For the exact mathematics and different scenarios, see Lehtonen (2022), but for the present purpose, intuition is sufficiently aided by noting that it is easier for an individual with fewer gametes to begin with (*i.e.*, a female) to end up in a situation where all gametes have been found and are now able to begin development. From this point onwards, new matings do not improve reproductive success, unless one assumes taxon-specific mechanisms, such as nutritious nuptial gifts. The relationship between the number of matings and reproductive success is the famous Bateman gradient, which was first discovered empirically (Bateman 1948, with an improved replication by Davies et al. 2023) and later became a building block of many mathematical arguments behind sex differences (*e.g.*, Arnold and Duvall 1994, Jones 2009, Kokko et al. 2012, Lehtonen et al. 2016, Lehtonen 2022).

Even so, a sex difference in the Bateman gradient is not yet proof that males should be predestined to follow a 'live fast, die young' strategy. The Bateman gradient only quantifies how much (if at all) fitness improves if an extra mating happens. It does not comment on whether it is difficult to achieve any extra matings (Kokko et al. 2012). Only if this difficulty exists, can we expect strong selection on traits that improve mating success (sexual selection). These then might trade-off with other components of fitness, such as the ability to provide parental care or survival.

Given some variation in sexually selected trait values, some individuals will consistently outperform others of the same sex, leading to large variance in reproductive success. The next step is to understand how strongly current reproductive success (which I will write as r) depends on the relevant trait value, which I call 'competitiveness' to cover various taxon-specific possibilities. In Figure 5.3, this is modelled as $r_m \propto c_m^\alpha$ for males, where the symbol $\propto$ stands for 'proportional to', and c measures an individual's allocation in traits that help their own gametes outcompete those of others, *i.e.*, level of competitiveness. For females, $r_f \propto c_f^\beta$, and the sexual asymmetry is introduced into the model by assuming $\alpha > \beta$. This is visible in a lower slope of the female curve in Figure 5.3a (present either as a decelerating shape of the curve while the male curve is slightly accelerating, left panel, or as a steeper linear slope if using log-scaled plots, right panel).

One important but often overlooked point: even if $\alpha > \beta$, it remains true (assuming $\alpha > \beta > 0$) that both sexes are selected for increased competitiveness. This has biological consequences. For example, consider migrating birds. It is easy to imagine why males are selected to arrive early if this helps them outcompete same-sex competitors when the first-arriving bird can claim a high-quality territory. But the exact same reasoning should also apply to females if late females arrive only to find the best territory already occupied (by a male and an earlier female who already paired with him), which then limits late females' reproductive success. Therefore, having identified a plausible benefit for males who arrive early does not yet explain why males often arrive earlier than females. The job is only half done, if one does not also seek to understand why precisely an analogous process of selection on females should be weaker (Kokko et al. 2006 give some answers, for example, male-biased adult sex ratios may provide the necessary asymmetry).

In general, positive slopes (Figure 5.3a) appear to predict ever-increasing competitiveness in both sexes. But as predictions go, 'everlasting improvement' is a rather meaningless statement. Can we ever hope to match predictions with data if the prediction for a trait value is 'as large as possible'? This problem simply highlights that Figure 5.3a is not yet a complete model. The graphs are not wrong, but they depict only one component of fitness. Completing the model requires realizing that competitiveness is an investment; attempts to increase competitiveness are unlikely to come for free.

In general, life history theory can only make predictions for traits that trade off with something else. Typically, this is done by assuming that achieving an increase in one component of fitness requires reducing the investment in some other resource. Early life history theory typically talks about energy as the unit of investment, but individual situations may require looking into, for example, oxidative stress difficulties of optimizing a gene regulatory network for early and late life simultaneously (Maklakov and Chapman 2019; see also Chapter 2, this volume), or cell-level mechanisms that create trade-offs between fast growth and lifespan (even in unicellular organisms; Nakaoka and Wakamoto 2017). Regardless of the mechanism, the principle is the same: if achieving something requires sacrificing something else, then the value at which selection stops favouring further increases depends on how strongly a fitness component changes with the levels of investment.

To see how this plays out in our present context, let us assume that the competitiveness values for males and females, c_m and c_f, come with an unavoidable trade-off: higher c means higher mortality, which we denote μ_m or μ_f in a sex-specific manner. To keep the model as simple as possible, I do not model here any within-sex variation in 'quality' or 'condition' (for a model including such effects, see Hooper et al. 2018), or the fact that sexes may also differ in how long it takes them to become mature or to reach a level of competitiveness that allows actual reproduction to take place. Instead, for simplicity, I assume that individuals reap the reproductive benefits of competitiveness every moment that they are alive, *i.e.*, I do not model juveniles separately from adults.

To create the examples, I assume that the mortality cost increases disproportionately when investment into competitiveness increases; very large c_f or c_m values penalize viability a lot. I assume that the 'rules' are the same for either sex, *i.e.*, neither has greater vulnerability in any *a priori* sense: the functions are identical, $\mu_f(c_f) = (1 + c_f^2)/10$ and

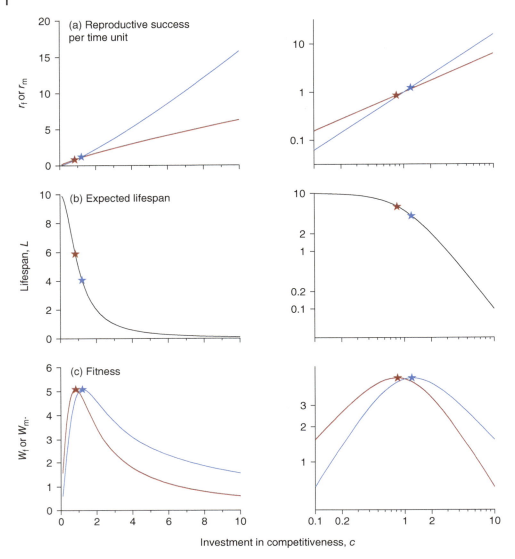

Figure 5.3 A simple model predicting male (blue lines and symbols) and female (red lines and symbols) life histories for a model where males use $\alpha = 1.2$ while females use $\beta = 0.8$. The result is (a) more male than female reproduction per unit time, which is explicable as there are fewer males in the population at any point than there are females, as shown in (b): males live for 4 units of time while females live for 6 6 (L denotes expected lifespan, see text for the equation for L_m and L_f). Both sexes achieve the same lifetime fitness (c) but they achieve this at different values of competitiveness (x-axis). See text for model details. The stars indicate the best values for fitness; thus, the stars are located at the peak in (c) but do not necessarily maximize reproductive rate (a) or lifespan (b). Left panels show the results using linear scales; right panels show the same data on logarithmic scales, which help show the symmetries of the situation.

$\mu_m(c_m) = (1 + c_m^2)/10$, where the division by 10 implements the assumed baseline mortality hazard of 0.1 (for a tutorial of this type of mortality arguments see Kokko 2024). Lifespans only evolve to be different from each other if c_f differs from c_m. The expected lifespans are $L_f = 10/(1 + c_f^2)$ and $L_m = 10/(1 + c_m^2)$. These values are depicted in Figure 5.3b.

Given our simplifying assumption that individuals have constant competitiveness all their lives, male fitness is a product of reproductive success (per unit time) and lifespan:

$$W_m = \frac{c_m^\alpha}{C_M} \frac{1}{(1 + c_m^2)} \tag{5.2a}$$

where C_M reflects the collective competitive pressure created by all males in the population. C_M is a function of an average male's investment in two ways: competition is stronger if males on average use a high c_m, but there is also an indirect effect that large c_M shortens male lifespans, which diminishes the number of individuals contributing to C_M. One could spend some time thinking about the exact ways to model C_M, with $C_M = \sum_i c_{i,m}^\alpha$ (summation over all males alive at present), but luckily it turns out that this does not matter, as long as our focal individual only contributes very little to C_M, which is true if the population is not very small. We can then simply treat C_M as a constant that scales the vertical axis of Figure 5.3c.

For females, we have:

$$W_f = \frac{c_f^\beta}{C_F} \frac{1}{(1 + c_f^2)} \tag{5.2b}$$

where C_F reflects the competitive pressure created by all females in the population for the relevant resources that are needed for reproduction. C_F and C_M are equally unimportant: when we seek for the best value of c_f or c_m, the division by C_F or C_M represent constants and do not change the location of the optimum, i.e., c_f^* or c_m^*. (For those who like to think of derivatives: when one computes where the relevant derivatives are zero, it does not matter whether this zero value of $\frac{\partial W_f}{\partial c_f}$ is being divided by a large or a small C_F). That C_M and C_F just represent a vertical shifting of the optimization problem is clearest on a log scale:

$$\begin{cases} \ln W_m = \alpha \ln c_m - \ln C_M - \ln(1 + c_m^2) \\ \ln W_f = \beta \ln c_f - \ln C_F - \ln(1 + c_f^2) \end{cases} \tag{5.3}$$

Still, there is an unavoidable caveat that needs to be discussed. I am assuming a large population so that each individual's own choices of c_m and c_f cannot perturb the total of all competitive investments of all same-sex individuals significantly from its current mean. When local mating groups are very small, a different type of analysis is needed, but that would deviate from the point I want to make here: how minimal the conditions are to create strong differences in lifespan between the sexes.

For both males and females, increasing investment in outcompeting others pays off in the context of reproduction, but this comes with a counteracting reduction of lifespan. The investment pays off as long as the cost slope (reduction in lifespan, Figure 5.3a) is shallower than the slope indicating reproductive gains (Figure 5.3b). Since I made the plausible assumption that massive investment leads to massively high mortality, the slope of the lifespan becomes steeper at high c_f or c_m, and at some value of c_f (or c_m) the slopes are equivalent in magnitude but opposite in sign. Further investment in greater competitiveness is from this point onwards no longer selected for. Mathematically, this happens at $c_m^* = \sqrt{\frac{\alpha}{2-\alpha}}$, $c_f^* = \sqrt{\frac{\beta}{2-\beta}}$ (and, indeed, is not dependent on C_M and C_F). This corresponds to the peaks of the W_f and W_m curves in Figure 5.3c.

The $\alpha > \beta$ assumption translates into a steeper positive slope on a log scale, and this implies that males are selected to invest more in outcompeting other males than females are to outcompete other females. Males, consequently, can be expected to evolve shorter lifespans too. This also implies a female-biased sex ratio among adults if the primary sex ratio is unbiased.

Thus, rather minimal assumptions are sufficient to create a situation where one of the sexes invests in a 'live fast, die young' strategy (higher c, leading to a shorter lifespan L but with more reproductive success per unit time spent alive), relative to the other sex. All that is required is a difference between α and β i.e., that being more competitive than another same-sex individual boosts relative reproductive success more in one sex than in another. The biological interpretation is that there can be asymmetries in how much doing something risky – anything that shortens lifespan – pays off in terms of current reproductive opportunities. The fact that male variance in reproductive success often exceeds that of females (Janicke et al. 2016) creates an *a priori* prediction that males might often be the 'fast' sex, especially when the mating system predicts strong sexual selection. Symptoms of this may then depend on ecology, for example, males may be more susceptible to parasitism (Moore and Wilson 2002), predation (Christe et al. 2006, Okada et al. 2021; but see Worthington and Swallow 2010), or food limitation (especially during growth to be the larger sex, if the species has male-biased sexual dimorphism; Martin et al. 2007).

When making strong directional predictions, however, the lessons of the Gouldian finch example above should be kept in mind. Ecology can matter. A trait that appears potentially lifespan-shortening (aggression) does not necessarily have that effect when it also confers priority access to resources (Laskowski et al. 2021 and Chang et al. 2024), and the frequency with which an organism can invest in offspring production can impact its placement on a fast-slow axis. Consider male Nazca boobies *Sula granti* (a species of seabird). For most measured traits, male boobies age more slowly than females, in a system where female shortage creates situations where not all males can breed every year (Tompkins and Anderson 2019). This difference in ageing in an otherwise monomorphic bird is in agreement with the idea that taking sabbatical years from breeding prolongs lifespan relative to breeding every year (Jouventin and Dobson 2002, Shaw and Levin 2013). Intriguingly, in male boobies showed earlier age-related declines than females did is, the probability to breed at all. This task is clearly more challenging for males to complete than for females, given that male boobies, thanks to their longer lifespan, are in surplus and face tougher competition for breeding opportunities.

In general, parenting is often costly, and experimental evaluation of the magnitude of this effect has been crucial to our understanding of reproductive effort, for example, to understand why birds lay smaller clutches than the number of young they can raise in any given year. For example, in a classic study, Daan et al. (1996) provided evidence that kestrel (*Falco tinnunculus*) parents, who are experimentally forced to raise more young, have elevated mortality in the subsequent year. Recently, these effects have been explored in multi-year settings, which allows detecting effects that extend to quantifying senescence. In a beautiful study of jackdaws, *Corvus monedula*, birds were manipulated to raise enlarged or reduced broods, and the latter delayed senescence relative to the former, with 24% increase in lifespan on average (Boonekamp et al. 2020).

Jackdaws are so biparental that Boonekamp's (2020) study (sensibly) did not focus on sex differences, but, as a whole, the pieces of the puzzle point in the direction where mating competition is nowhere near the only dangerous activity for a male bird. In an analysis of 194 species, male birds' mortality costs were shown to be highly sensitive to both mating competition and their roles as paternal care providers (Liker and Székely 2005). It thus remains possible that avoiding care indirectly increases mortality, as it frees a male to return to engage in male-male competition. This can be very costly, and the resulting life history may be fast. Lekking birds indeed show that males disappear from populations sooner than females, leading to mammal-like female-biased adult sex ratios (Donald 2007). However, the key then is the mortality cost of developing or expressing the competitive traits, not the avoidance of care *per se*. Therefore, estimating if males really should be the faster-paced sex than females will require evaluating just how dangerous different types of breeding activities are and to what extent efficient mating competition requires absence of commitment to providing other types of care. When estimated, these trade-offs are not always strict (Safari et al. 2019), which then also feeds back into impacting the evolution of sex-specific care strategies themselves.

5.3.4 Caveats: Biology Is Diverse!

While clear in principle, the above model lacks nuance. Even ungulates, which in some sense occupy a very sexually stereotypical niche with male combat and female care of young, are not maximally fast; instead, males show some restraint early in their lives, and their reproductive effort peaks in what is called prime age (Mysterud et al. 2004). Sometimes costs of sexually selected traits are only paid during the mating season (Kraus et al. 2008), other times at a very different part of the life cycle. In dragonflies, for example, survival costs of adult ornamentation appear to be paid already in the juvenile stage (Moore 2023). Also, as the booby example above showed, males are not always the shorter-lived sex. Male crickets live longer and age more slowly than females (Archer et al. 2012, Hawkes et al. 2022). In birds, adult sex ratios are typically male-biased (indicative of higher male than female survival), except in lekking birds (Donald 2007). However, to represent the exception to an exception, in lekking manakins, it takes so long for a male to become an alpha male that selection switches to favouring very long-lived males (several times longer lifespan than that of females; McDonald 1993). Finally, senescence of the germline and the soma may occur at different rates: cockerels may end up behaviourally dominant but with poor sperm, with unfortunate consequences for female fertility if they only mate with dominant males (Dean et al. 2010).

All these factors combine into a diverse set of patterns that the model above simplifies away. Especially, all differences between juvenile performance (viability, growth) and adult traits (including their age-dependence after maturity has been reached) were absent in it. Thus, while the model did include a trade-off between early and late life (a central theme in our understanding of why senescence evolves in the first place; Johnson and Hixon 2011, Lemaitre et al. 2015, Mérot et al. 2020; see also Chapter 2, this volume), its handling of the trade-off between mating effort and lifespan was rather simplistic (for real-life examples, see Hunt et al. 2006, Maklakov et al. 2009, 2010, Preston et al. 2011, Lizé et al. 2014 and Lukas and Clutton-Brock 2014; but see also Festa-Bianchet 2012 for a case with less clear trade-offs). It is clear that temporal aspects can create lifespan expectations that deviate from the straightforward assumption that males die earlier. A drastic but thought-provoking example is that sex-changing fish may be either protandrous (small, young males turning to females once they have grown to be large enough) or protogynous (the opposite), largely depending on which sex benefits more from being large (Todd et al. 2016). In sequential hermaphrodites, it is difficult to even talk about a male or a female lifespan. Still, they can be inspiring: their male as well as female function might benefit from being large and/or socially dominant, and the steepness of this relationship is important for determining which function they opt for during different stages of ontogeny.

5.3.5 One Genome, Two Sexes

If one sex is selected to follow a different pace of life than the other, the temporal aspects of the trade-off exist with another significant complication: the two sexes develop from largely the same genome, which may make it difficult for both sexes to schedule their lives optimally and independently of selection on the other sex. Obviously, this does not prevent sexual dimorphism from evolving, including sexually divergent life histories. The manakin or killer whale examples above are not even that extreme, when remembering the extremely long lifespans of certain ant queens compared with those of the males they mated with before founding a colony (Heinze 2016, Kramer et al. 2022; see also Chapter 9, this volume). Still, sexual conflict can persist in an unresolved state for a very large number of generations (Ruzicka et al. 2019), and it can impact how the two sexes respond to selection on life history traits. Experimental evolution has revealed, in a seed beetle system, that upward selection on male lifespan decreases male fitness (Berg and Maklakov 2012), a result that is counterintuitive only at first sight. Males may very well be better off investing in a 'live fast, die young' type of life history. Highly intriguingly, the same treatments, *i.e.*, selection for long-lived males, increased relative female fitness.

Another exciting example of correlated evolution between the sexes is an experimental study of broad-horned flour beetles *Gnatocerus cornutus* (Okada et al. 2021). Males of this species have exaggerated, sexually selected traits (mandibles). Large mandibles trade-off with locomotion, increasing vulnerability to predation (the relevant anti-predatory behaviour is running away rather than fighting the predator). Female fitness was indirectly improved in experimental evolution treatments where intense predation led to small-mandibled males. The experiment thus revealed that even though males were the only sex morphologically expressing the exaggerated mandibles, both sexes paid fitness costs for the trait being present in the population.

Genetic correlations may make the two sexes respond similarly in terms of lifespan not only in experimental evolution but also in nature (Macartney and Bonduriansky 2022). This is thought-provoking because concordant lifespan responses do not necessarily imply that fitness, too, is concordant between the sexes. After all, whenever there are reasons to evolve to be at different points along a fast-slow axis, the longest lives are not necessarily the fittest ones. Genetic correlations have even been argued to underlie the fact that female humans live so long past menopause. The argument is that males can at least potentially benefit from long lives if paired with younger mates (Tuljapurkar et al. 2007), and if this selects for long-lived males, female lifespan might increase as a correlated response (indirect selection). However, for humans specifically, the explanation should also be pitted against the fact that humans are a social species with extended families, where grandmothers can significantly boost their daughter's reproductive success (Chapman et al. 2019; see also Chapter 11 in this volume). To return to the killer whale conundrum at the very beginning of this chapter: whale species with menopause (a small minority of all whales) have females that are extremely long-lived for their body size, but they do not have exceptionally long reproductively active lifespans for their body size (Ellis et al. 2024). This, together with females living far longer than males, suggests that selection has actively promoted female lifespan in some whale species via chances to help younger kin (Johnstone and Cant 2010). Indirect response to selection on males appears unlikely in these cases.

Sexual conflict can sometimes be (at least partially) resolved in intriguing ways. Consider, for example, the Atlantic salmon *Salmo salar* (Barson et al. 2015, Czorlich et al. 2018), a species where juveniles migrate from freshwater spawning grounds to the sea and then return to breed once maturity has been reached. Age at maturity is strongly dependent on a gene (VHLL3) whose expression, interestingly, is associated with size and age at maturity in humans too (Barson et al. 2015). If a fish is homozygous for the 'early' allele, it will return to breed at an earlier age, which presumably has both costs and benefits: the cost is that there is less time for growth in the sea, and the benefit is that the first breeding occurs without waiting for one more year (during which time death can strike). The balance of costs and benefits is unlikely to be identical for the two sexes. Highly interestingly, heterozygous individuals behave in a very sex-specific manner: more like 'early' fish if male, and more like 'late' fish if female, conceivably because large size in female fish generally predicts high fecundity, while the correlation between size and success is not as strong in males (Barson et al. 2015).

The above examples mainly focused on intra-sexual conflict, where the same locus is expressed in the two sexes and experiences selection in opposing directions. Intersexual conflict over life history traits is another possibility. Males may interact with females in a manner that is suboptimal for the latter (or *vice versa*). Toxic ejaculates are a famous example (Alonzo and Pizzari 2010, Edward et al. 2015). Here, males evolve traits that actively shift the female life history towards the faster end of the spectrum. This benefits a male in situations where he is unlikely to be the sire of eggs that the female lays much later in life. Finally, while costs are often mediated by matings, not only inter-sexual relations matter. In carrion flies and seed beetles, being housed with males is detrimental to female lifespan, while being housed with either males or females is detrimental for males, relative to males kept alone (Maklakov and Bonduriansky 2009).

5.4 Towards Progress

Above, I have repeatedly warned against lumping traits too quickly into expected co-variation patterns and repeated the call by Mathot and Frankenhuis (2018) for more explicit theory on life history variation across and within species. I will end this chapter with one more look at the *r-K* concept, as there are thought-provoking developments worth knowing about — regardless of the framework one has chosen to structure one's thoughts.

Mallet (2012) pointed out that the emphasis on '*K*' may be a historical accident. A linearly decreasing population growth rate, which standard ecology textbooks write as $\frac{1}{N}\frac{dN}{dt} = r(1-N/K)$ (Eq. (5.1)) or $\frac{dN}{dt} = rN(1-N/K)$, could equally well be formulated as $\frac{1}{N}\frac{dN}{dt} = r - \alpha N$. Intriguingly, this was the formulation used in all early texts on this topic (Verhulst 1838, Pearl and Reed 1920, Lotka 1925, Volterra 1927). The equilibrium population size is then described as $N^* = r/\alpha$ (instead of $N^* = K$). This notation, had the field stuck to it, makes it easier to appreciate that the equilibrium population size is an emergent property that combines the species' intrinsic life history characteristics (that determine *r*) with their modifications as density increases (captured by the coefficient α). Mallet (2012) argues that many ecological debates could have been cleared earlier, and some confusions might never have arisen at all, had the field kept teaching students the insights that follow from $N^* = r/\alpha$.

In particular, $N^* = r/\alpha$ predicts that all else being equal, a higher intrinsic growth rate *r* increases the carrying capacity. This is the diagonal opposite to the intuition of Pianka (1970) and many others, that organisms should be able to do well at only one end of the '*r-K* axis'.

The ability of a high *r* to improve the carrying capacity makes sense as an *a priori* prediction. Let us imagine an organism that evolves to be more efficient at using resources as a whole. Should we then not expect it to be able to grow better at all population sizes, whether conspecifics are currently rare or common? Quoting Marshall et al. (2023), 'discussions over which population parameter is maximised by evolution, be it *r* or *K*, are somewhat misplaced'.

Marshall et al. (2023) argue that it is instead healthy to ask explicitly what processes are necessary to break the *a priori* expected positive co-variation of the intrinsic growth rate *r* and the carrying capacity *K*. In their meta-analysis of 36 studies of microbial populations, the relationship was positive in only four studies. Negative relationships can be expected if α, which describes how steeply growth rates diminish when the population becomes more dense, is higher for those populations that have, perhaps via changes in metabolism, managed to increase their intrinsic growth rate *r*. If genotypes differ in their ability to cope with sparse and dense competitor densities, such that it is difficult to perform well in both contexts (Wright et al. 2019), then the relationship between α and *r* produces patterns that align with both '*r-K*' and the fast-slow frameworks. The important insight is that the necessary changes from a null model are now presented explicitly, making the required empirical tests clearer too.

The consequences for sex-specific life histories should be clear. Null models, to which minimal sexual asymmetries are added (de Vries and Lehtonen 2023), are invaluable for making sure we understand how broad patterns can arise. Thereafter, much can be different between taxa, be it the genetic architecture of important traits, genetic co-variation, or ecological settings that modulate the sex-specific (yet coevolving) responses to biotic and abiotic challenges. A healthy dialogue between empiricists and theoreticians will fuel understanding of how we understand both the broad patterns and why the exceptions, either rare ones or fairly common ones, arise where they do.

Acknowledgements

I thank the two editors as well as Jean-Michel Gaillard and Tom Keaney for very helpful feedback on this work, Willem Frankenhuis for invaluable discussions, Sarah Pryke for permission to use her photograph of Gouldian finches and for our past discussions of this system, and the Alexander von Humboldt Foundation for funding.

References

Allen, W.L., Street, S.E., and Capellini, I. (2017). Fast life history traits promote invasion success in amphibians and reptiles. *Ecology Letters* 20: 222–230.

Allison, T.M., Radzvilavicius, A.L., and Dowling, D.K. (2021). Selection for biparental inheritance of mitochondria under hybridization and mitonuclear fitness interactions. *Proceedings of the Royal Society B: Biological Sciences* 288: 20211600.

Alonzo, S.H., and Pizzari, T. (2010). Male fecundity stimulation: conflict and cooperation within and between the sexes: model analysis and coevolutionary dynamics. *The American Naturalist* 175: 174–185.

Archer, C.R., Paniw, M., Vega-Trejo, R., and Sepil, I. (2022). A sex skew in life-history research: the problem of missing males. *Proceedings of the Royal Society B: Biological Sciences* 289: 20221117.

Archer, C.R., Zajitschek, F., Sakaluk, S.K., Royle, N.J., and Hunt, J. (2012). Sexual selection affects the evolution of lifespan and ageing in the decorated cricket *Gryllodes sigillatus*. *Evolution* 66: 3088–3100.

Arnold, S.J., and Duvall, D. (1994). Animal mating systems: a synthesis based on selection theory. *The American Naturalist* 143: 317–348.

Arnqvist, G., Rönn, J., Watson, C., Goenaga, J., and Immonen, E. (2022). Concerted evolution of metabolic rate, economics of mating, ecology, and pace of life across seed beetles. *Proceedings of the National Academy of Sciences of the United States of America* 119: e2205564119.

Bachtrog, D., Mank, J.E., Peichel, C.L., Kirkpatrick, M., Otto, S.P., Ashman, T.-L., Hahn, M.W., Kitano, J., Mayrose, I., Ming, R., Perrin, N., Ross, L., Valenzuela, N., Vamosi, J.C., and The Tree of Sex Consortium (2014). Sex determination: why so many ways of doing it? *PLoS Biology* 12(7): e1001899.

Bakewell, A.T., Davis, K.E., Freckleton, R.P., Isaac, N.J.B., and Mayhew, P.J. (2020). Comparing life histories across taxonomic groups in multiple dimensions: how mammal-like are insects? *The American Naturalist* 195: 70–81.

Barson, N.J., Aykanat, T., Hindar, K., Baranski, M., Bolstad, G.H., Fiske, P., Jacq, C., Jensen, A.J., Johnston, S.E., Karlsson, S., Kent, M., Moen, T., Niemelä, E., Nome, T., Naesje, T.F., Orell, P., Romakkaniemi, A., Saegrov, H., Urdal, K., Erkinaro, J., Lien, S., and Primmer, C.R. (2015). Sex-dependent dominance at a single locus maintains variation in age at maturity in salmon. *Nature* 528: 405–408.

Bateman, A.J. (1948). Intrasexual selection in *Drosophila*. *Heredity* 2: 349–368.

Berg, E.C., and Maklakov, A.A. (2012). Sexes suffer from suboptimal lifespan because of genetic conflict in a seed beetle. *Proceedings of the Royal Society B: Biological Sciences* 279: 4296–4302.

Bielby, J., Mace, G.M., Bininda-Emonds, O.R.P., Cardillo, M., Gittleman, J.L., Jones, K.E., Orme, C.D.L., and Purvis, A. (2007). The fast-slow continuum in mammalian life history: an empirical reevaluation. *The American Naturalist* 169: 748–757.

Bolund, E., Lummaa, V., Smith, K.R., Hanson, H.A., and Maklakov, A.A. (2016). Reduced costs of reproduction in females mediate a shift from a male-biased to a female-biased lifespan in humans. *Scientific Reports* 6: 24672.

Boonekamp, J.J., Bauch, C., and Verhulst, S. (2020). Experimentally increased brood size accelerates actuarial senescence and increases subsequent reproductive effort in a wild bird population. *Journal of Animal Ecology* 89: 1395–1407.

Brengdahl, M., Kimber, C.M., Maguire-Baxter, J., and Friberg, U. (2018). Sex differences in life span: females homozygous for the X chromosome do not suffer the shorter life span predicted by the unguarded X hypothesis. *Evolution* 72: 568–577.

Butterworth, N.J., Benbow, M.E., and Barton, P.S. (2023). The ephemeral resource patch concept. *Biological Reviews* 98: 697–726.

Cayuela, H., Gaillard, J.-M., Vieira, C., Ronget, V., Gippet, J.M.W., Garcia, T.C., Marais, G.A.B., and Lemaitre, J.-F. (2023). Sex differences in adult lifespan and aging rate across mammals: a test of the 'Mother Curse hypothesis'. *Mechanisms of Ageing and Development* 212: 111799.

Cayuela, H., Lemaitre, J.-F., Bonnaire, E., Pichenot, J., and Schmidt, B.R. (2020). Population position along the fast-slow life-history continuum predicts intraspecific variation in actuarial senescence. *Journal of Animal Ecology* 89: 1069–1079.

Chang, C.-c., Moiron, M., Sánchez-Tójar, A., Niemelä, P.T., and Laskowski, K.L. (2024). What is the meta-analytic evidence for life-history trade-offs at the genetic level? *Ecology Letters* 27: e14354.

Chapman, S.N., Pettay, J.E., Lummaa, V., and Lahdenperä, M. (2019). Limits to fitness benefits of prolonged post-reproductive lifespan in women. *Current Biology* 29: 645–650.

Charlesworth, B. (1980). *Evolution in age-structured populations*. Cambridge University Press.

Christe, P., Keller, L., and Roulin, A. (2006). The predation cost of being a male: implications for sex-specific rates of ageing. *Oikos* 114: 381–394.

Connallon, T., Beasley, I.J., McDonough, Y., and Ruzicka, F. (2022). How much does the unguarded X contribute to sex differences in life span? *Evolution Letters* 6: 319–329.

Czorlich, Y., Aykanat, T., Erkinaro, J., Orell, P., and Primmer, C.P. (2018). Rapid sex-specific evolution of age at maturity is shaped by genetic architecture in Atlantic salmon. *Nature Ecology & Evolution* 2: 1800–1807.

Daan, S., Deerenberg, C., and Dijkstra, C. (1996). Increased daily work precipitates natural death in the kestrel. *Journal of Animal Ecology* 65: 539–544.

Davies, N., Janicke, T., and Morrow, E.H. (2023). Evidence for stronger sexual selection in males than in females using an adapted method of Bateman's classic study of *Drosophila melanogaster*. *Evolution* 77: 2420–2430.

de Vries, C., Galipaud, M., and Kokko, H. (2023). Extrinsic mortality and senescence: a guide for the perplexed. *Peer Community Journal* 3: e29.

de Vries, C., and Lehtonen, J. (2023). Sex-specific assumptions and their importance in models of sexual selection. *Trends in Ecology & Evolution* 38: 927–935.

Dean, R., Cornwallis, C.K., Løvlie, H., Wodley, K., Richardson, D.S., and Pizzari, T. (2010). Male reproductive senescence causes potential for sexual conflict over mating. *Current Biology* 20: 1193–1196.

Delanoue, R., Clot, C., Leray, C., Pihl, T., and Hudry, B. (2023). Y chromosome toxicity does not contribute to sex-specific differences in longevity. *Nature Ecology & Evolution* 7: 1245–1256.

Donald, P.F. (2007). Adult sex ratios in wild bird populations. *Ibis* 149: 671–692.

Edward, D.A., Stockley, P., and Hosken, D.J. (2015). Sexual conflict and sperm competition. *Cold Spring Harbour Perspectives Biology* 7: a017707.

Ellis, S., Franks, D.W., Kronborg Nielsen, M.L., Weiss, M.N., and Croft, D.P. (2024). The evolution of menopause in toothed whales. *Nature* 627: 579–585.

Engen, S., and Saether, B.-E. (2016). r- and K-selection in fluctuating populations is determined by the evolutionary trade-off between two fitness measures: growth rate and lifetime reproductive success. *Evolution* 71: 167–173.

Esgate, A., Groome, D., Baker, K., Heathcote, D., and Kemp, R. (2005). *An introduction to applied cognitive psychology*. Psychology Press.

Ferreira, M.S., Dickman, C.R., Fisher, D.O., Figueiredo, M.d.S.L., and Vieira, M.V. (2023). Marsupial position on life-history continua and the potential contribution of life-history trait to population growth. *Proceedings of the Royal Society B: Biological Sciences* 290: 20231316.

Festa-Bianchet, M. (2012). The cost of trying: weak interspecific correlations among life-history components in male ungulates. *Canadian Journal of Zoology* 90: 1072–1085.

Foster, E.A., Franks, D.W., Mazzi, S., Darden, S.K., Balcomb, K.C., Ford, J.K.B., and Croft, D.P. (2012). Adaptive prolonged postreproductive life span in killer whales. *Science* 337: 1313.

Hämäläinen, A., Immonen, E., Tarka, M., and Schuett, W. (2018). Evolution of sex-specific pace-of-life syndromes: causes and consequences. *Behavioral Ecology and Sociobiology* 72: 50.

Hawkes, M., Lane, S.M., Rapkin, J., Jensen, K., House, C.M., Sakaluk, S.K., and Hunt, J. (2022). Intralocus sexual conflict over optimal nutrient intake and the evolution of sex differences in life span and reproduction. *Functional Ecology* 36: 865–881.

Hayward, A., and Gillooly, J.F. (2011). The cost of sex: quantifying energetic investment in gamete production by males and females. *PLoS ONE* 6(1): e16557.

Heinze, J. (2016). The male has done his work – the male may go. *Current Opinion in Insect Science* 16: 22–27.

Holman, L., and Kokko, H. (2013). The consequences of polyandry for population viability, extinction risk and conservation. *Philosophical Transactions of the Royal Society, B: Biological Sciences* 368: 20120053.

Hooper, A.K., Lehtonen, J., Schwanz, L.E., and Bondurianksy, R. (2018). Sexual competition and the evolution of condition-dependent ageing. *Evolution Letters* 2: 37–48.

Hunt, J., Jennions, M.D., Spyrou, N., and Brooks, R. (2006). Artificial selection on male longevity influences age-dependent reproductive effort in the black field cricket *Teleogryllus commodus*. *The American Naturalist* 168: E72–E86.

Immonen, E., Hämäläinen, A., Schuett, W., and Tarka, M. (2018). Evolution of sex-specific pace-of-life syndromes: genetic architecture and physiological mechanisms. *Behavioral Ecology and Sociobiology* 72: 60.

Janicke, T., Häderer, I.K., Lajeunesse, M.J., and Anthes, N. (2016). Darwinian sex roles confirmed across the animal kingdom. *Science Advances* 2: e1500983.

Jeschke, J.M., Gabriel, W., and Kokko, H. (2019). r-strategists/K-strategists. In: *Encyclopedia of ecology*. 2nd ed. (eds. Jørgensen, S.E., and Fath, B.D.), 193–201. Elsevier.

Johnson, D.W., and Hixon, M.A. (2011). Sexual and lifetime selection on body size in a marine fish: the importance of life-history trade-offs. *Journal of Evolutionary Biology* 24: 1653–1663.

Johnstone, R.A., and Cant, M.A. (2010). The evolution of menopause in cetaceans and humans: the role of demography. *Proceedings of the Royal Society B: Biological Sciences* 277: 3765–3771.

Jones, A.G. (2009). On the opportunity for sexual selection, the Bateman gradient and the maximum intensity of sexual selection. *Evolution* 63: 1673–1684.

Jouventin, P., and Dobson, F.S. (2002). Why breed every other year? The case of albatrosses. *Proceedings of the Royal Society B: Biological Sciences* 269: 1955–1961.

Kim, K.-W., Jackson, B.C., Zhang, H., Toews, D.P.L., Taylor, S.A., Greig, E.I., Lovette, I.J., Liu, M.M., Davison, A., Griffith, S.C., Zeng, K., and Burke, T. (2019). Genetics and evidence for balancing selection of a sex-linked colour polymorphism in a songbird. *Nature Communications* 10: 1852.

Kokko, H. (2021). The stagnation paradox: the ever-improving but (more or less) stationary population fitness. *Proceedings of the Royal Society B: Biological Sciences* 288: 20212145.

Kokko, H. (2024). Who is afraid of modelling time as a continuous variable? *Methods in Ecology and Evolution* 15: 1736–1756.

Kokko, H., Griffith, S.C., and Pryke, S.R. (2014). The Hawk–Dove game in a sexually reproducing species explains a colourful polymorphism of an endangered bird. *Proceedings of the Royal Society B: Biological Sciences* 281: 20141794.

Kokko, H., Gunnarsson, T.G., Morrell, L.J., and Gill, J.A. (2006). Why do female migratory birds arrive later than males? *Journal of Animal Ecology* 75: 1293–1303.

Kokko, H., Klug, H., and Jennions, M.D. (2012). Unifying cornerstones of sexual selection: operational sex ratio, Bateman gradient, and the scope for competitive investment. *Ecology Letters* 15: 1340–1351.

Kokko, H., Sutherland, W.J., and Johnstone, R.A. (2001). The logic of territory choice: implications for conservation and source-sink dynamics. *The American Naturalist* 157: 459–463.

Kramer, B.H., van Doorn, G.S., Arani, B.M.S., and Pen, I. (2022). Eusociality and the evolution of aging in superorganisms. *The American Naturalist* 200: 63–80.

Kraus, C., Eberle, M., and Kappeler, P.M. (2008). The costs of risky male behaviour: sex differences in seasonal survival in a small sexually monomorphic primate. *Proceedings of the Royal Society B: Biological Sciences* 275: 1635–1644.

Lehtonen, J. (2022). Bateman gradients from first principles. *Nature Communications* 13: 3591.

Lehtonen, J., and Parker, G.A. (2014). Gamete competition, gamete limitation, and the evolution of the two sexes. *Molecular Human Reproduction* 20: 1161–1168.

Lehtonen, J., Parker, G.A., and Schärer, L. (2016). Why anisogamy drives ancestral sex roles. *Evolution* 70: 1129–1135.

Lemaitre, J.-F., Berger, V., Bonenfant, C., Douhard, M., Gamelon, M., Plard, F., and Gaillard, J.-M. (2015). Early-late life trade-offs and the evolution of ageing in the wild. *Proceedings of the Royal Society B: Biological Sciences* 282: 20150209.

Lessells, C.M., Snook, R.R., and Hosken, D.J. (2009). The evolutionary origin and maintenance of sperm: selection for a small, motile gamete mating type. In: *Sperm biology: an evolutionary perspective* (eds. Birkhead, T.R., Hosken, D.J., and Pitnick, S.), 43–67. Academic Press.

Liker, A. and Székely, T. (2005). Mortality costs of sexual selection and parental care in natural populations of birds. *Evolution* 59: 890–897.

Lizé, A., Price, T.A.R., Heys, C., Lewis, Z., and Hurst, G.D.D. (2014). Extreme cost of rivalry in a monandrous species: male-male interactions result in failure to acquire mates and reduced longevity. *Proceedings of the Royal Society B: Biological Sciences* 281: 20140631.

Lotka, A.J. (1925). *Elements of physical biology*. Williams & Wilkins.

Lukas, D., and Clutton-Brock, T. (2014). Costs of mating competition limit male lifetime breeding success in polygynous mammals. *Proceedings of the Royal Society B: Biological Sciences* 281: 20140418.

MacArthur, R.H., and Wilson, E.O. (1967). *The theory of island biogeography*. Princeton University Press.

Macartney, E.L., and Bondurianski, R. (2022). Does female resistance to mating select for live-fast die-young strategies in males? A comparative analysis in the genus *Drosophila*. *Journal of Evolutionary Biology* 35: 192–200.

Maklakov, A.A., and Bondurianski, R. (2009). Sex differences in survival costs of homosexual and heterosexual interactions: evidence from a fly and a beetle. *Animal Behaviour* 77: 1375–1379.

Maklakov, A.A., Bondurianski, R., and Brooks, R.C. (2009). Sex differences, sexual selection, and ageing: an experimental evolution approach. *Evolution* 63: 2941–2503.

Maklakov, A.A., Cayetano, L., Brooks, R.C., and Bondurianski, R. (2010). The roles of life-history selection and sexual selection in the adaptive evolution of mating behavior in a beetle. *Evolution* 64: 1273–1282.

Maklakov, A.A., and Chapman, T. (2019). Evolution of ageing as a tangle of trade-offs: energy versus function. *Proceedings of the Royal Society B: Biological Sciences* 286: 20191604.

Maklakov, A.A., and Lummaa, V. (2013). Evolution of sex differences in lifespan and aging: causes and constraints. *BioEssays* 35: 717–724.

Mallet, J. (2012). The struggle for existence: how the notion of carrying capacity, K, obscures the links between demography, Darwinian evolution, and speciation. *Evolutionary Ecology Research* 14: 627–665.

Marshall, D.J., Cameron, H.E., and Loreau, M. (2023). Relationships between intrinsic population growth rate, carrying capacity and metabolism in microbial populations. *The ISME Journal* 17: 2140–2143.

Martín, C.A., Alonso, J.C., Alonso, J.A., Palacin, C., Magaña, M., and Martín, B. (2007). Sex-biased juvenile survival in a bird with extreme size dimorphism, the great bustard *Otis tarda*. *Journal of Avian Biology* 38: 335–346.

Mathot, K.J., and Frankenhuis, W.E. (2018). Models of pace-of-life syndromes (POLS): a systematic review. *Behavioral Ecology and Sociobiology* 72: 41.

McDonald, D.B. (1993). Demographic consequences of sexual selection in the long-tailed manakin. *Behavioral Ecology* 4: 297–309.

McElreath, R., Luttbeg, B., Fogarty, S.P., Brodin, T., and Sih, A. (2007). Evolution of animal personalities. *Nature* 450: E5.

Mérot, C., Llaurens, V., Normandeau, E., Bernatchez, L., and Wellenreuther, M. (2020). Balancing selection via life-history trade-offs maintains an inversion polymorphism in a seaweed fly. *Nature Communications* 11: 670.

Moiron, M., Laskowski, K.L., and Niemelä, P.T. (2020). Individual differences in behaviour explain variation in survival: a meta-analysis. *Ecology Letters* 23: 399–408.

Moore, M.P. (2023). Ornamented species incur higher male mortality in the larval stage. *Biology Letters* 19: 20230108.

Moore, S.L., and Wilson, K. (2002). Parasites as a viability cost of sexual selection in natural populations of mammals. *Science* 297: 2015–2018.

Moschilla, J.A., Tomkins, J.L. and Simmons, L.W. (2019). Sex-specific pace-of-life syndromes. *Behavioral Ecology* 30: 1096–1105.

Mysterud, A., Langvatn, R., and Stenseth, N.C. (2004). Patterns of reproductive effort in male ungulates. *Journal of Zoology* 264: 209–215.

Nakaoka, H., and Wakamoto, Y. (2017). Aging, mortality, and the fast growth trade-off of *Schizosaccharomyces pombe*. *PLoS Biology* 15(6): e2001109.

Nieuwenhuis, B.P.S., and Aanen, D.K. (2012). Sexual selection in fungi. *Journal of Evolutionary Biology* 25: 2397–2411.

Okada, K., Katsuki, M., Sharma, M.D., Kyiose, K., Seko, T., Okada, Y., Wilson, A.J., and Hosken, D.J. (2021). Natural selection increases female fitness by reversing the exaggeration of a male sexually selected trait. *Nature Communications* 12: 3420.

Parker, G.A. (1979). Sexual selection and sexual conflict. In: *Sexual selection and reproductive competition in insects* (eds. Blum, M.S., and Blum, N.A.), 123–166. Academic Press.

Parker, G.A. (2014). The sexual cascade and the rise of pre-ejaculatory (Darwinian) sexual selection, sex roles, and sexual conflict. *Cold Spring Harbour Perspectives in Biology* 6: a017509.

Parker, G.A., Baker, R.R., and Smith, V.G.F. (1972). The origin and evolution of gamete dimorphism and the male-female phenomenon. *Journal of Theoretical Biology* 36: 181–198.

Pearl, R., and Reed, L.J. (1920). On the rate of growth of the population of the United States since 1790 and its mathematical representation. *Proceedings of the National Academy of Sciences of the United States of America* 6: 275–288.

Pianka, E.R. (1970). On r- and K-selection. *The American Naturalist* 104: 592–597.

Pipoly, I., Bokony, V., Kirkpatrick, M., Donald, P.F., Székely, T., and Liker, A. (2015). The genetic sex-determination system predicts adult sex ratios in tetrapods. *Nature* 527: 91–64.

Preston, B.T., Jalme, M.S., Hingrat, Y., Lacroix, F., and Sorci, G. (2011). Sexually extravagant males age more rapidly. *Ecology Letters* 14: 1017–1024.

Promislow, D.E.L. (1992). Costs of sexual selection in natural populations of mammals. *Proceedings of the Royal Society B: Biological Sciences* 247: 203–210.

Pryke, S.R., and Griffith, S.C. (2009). Socially mediated trade-offs between aggression and parental effort in competing color morphs. *The American Naturalist* 174: 455–464.

Réale, D., Garant, D., Humphries, M.M., Bergeron, P., Careau, V., and Montiglio, P.-O. (2010). Personality and the emergence of the pace-of-life syndrome concept at the population level. *Philosophical Transactions of the Royal Society, B: Biological Sciences* 365: 4051–4063.

Roff, D.A. (2002). *Life history evolution*. Sinauer.

Royauté, R., Berdal, M.A., Garrison, C.R., and Dochtermann, N.A. (2018). Paceless life? A meta-analysis of the pace-of-life syndrome hypothesis. *Behavioral Ecology and Sociobiology* 72: 64.

Ruzicka, F., Hill, M.S., Pennell, T.M., Flis, I., Ingleby, F.C., Mott, R., Fowler, K., Morrow, E.H., and Reuter, M. (2019). Genome-wide sexually antagonistic variants reveal long-standing constraints on sexual dimorphism in fruit flies. *PLoS Biology* 17(4): e3000244.

Saether, B.-E., Engen, S., Gustafsson, L., Grøtan, V., and Vriend, S.J.G. (2021). Density-dependent adaptive topography in a small passerine bird, the collared flycatcher. *The American Naturalist* 197: 93–110.

Saether, B.E., Grotan, V., Engen, S., Coulson, T., Grant, P.R., Visser, M.E., Brommer, J.E., Grant, R.B., Gustafsson, L., Hatchwell, B.J., Jerstad, K., Karell, P., Pietiäinen, H., Roulin, A., Røstad, O.W., and Weimerskirch, H. (2016). Demographic routes to variability and regulation in bird populations. *Nature Communications* 7: 12001.

Safari, I., Goymann, W., and Kokko, H. (2019). Male-only care and cuckoldry in black coucals: does parenting hamper sex life? *Proceedings of the Royal Society B: Biological Sciences* 286: 20182789.

Shaw, A.K., and Levin, S.A. (2013). The evolution of intermittent breeding. *Journal of Mathematical Biology* 66: 685–703.

Stearns, S.C. (1977). The evolution of life history traits: a critique of the theory and a review of the data. *Annual Review of Ecology and Systematics* 8: 145–171.

Sultanova, Z., Andic, M., and Carazo, P. (2018). The "unguarded-X" and the genetic architecture of lifespan: inbreeding results in a potentially maladaptive sex-specific reduction of female lifespan in *Drosophila melanogaster*. *Evolution* 72: 540–552.

Sultanova, Z., Downing, P.A. and Carazo, P. (2023). Genetic sex determination, sex chromosome size and sex-specific lifespans across tetrapods. *Journal of Evolutionary Biology* 36: 480–494.

Tannenbaum, C., Ellis, R.P., Eyssel, F., Zou, J., and Schiebinger, L. (2019). Sex and gender analysis improves science and engineering. *Nature* 575: 137–146.

Tarka, M., Guenther, A., Niemelä, P.T., Nakagawa, S., and Noble, D.W.A. (2018). Sex differences in life history, behavior, and physiology along a slow-fast continuum: a meta-analysis. *Behavioral Ecology and Sociobiology* 72: 132.

Todd, E.V., Liu, H., Muncaster, S., and Gemmell, N.J. (2016). Bending genders: the biology of natural sex change in fish. *Sexual Development* 10: 223–241.

Tompkins, E.M., and Anderson, D.J. (2019). Sex-specific patterns of senescence in Nazca boobies linked to mating system. *Journal of Animal Ecology* 88: 986–1000.

Toomey, M.B., Marques, C.I., Andrade, P., Araújo, P.M., Sabatino, S., Gazda, M.A., Adonso, S., Lopes, R.J., Corbo, J.C., and Carneiro, M. (2018). A non-coding region near Follistatin controls head colour polymorphism in the Gouldian finch. *Proceedings of the Royal Society B: Biological Sciences* 285: 20181788.

Tuljapurkar, S.D., Puleston, C.O., and Gurven, M.D. (2007). Why men matter: mating patterns drive evolution of human lifespan. *PLoS ONE* 2(8): e785.

van de Walle, J., Fay, R., Gaillard, J.-M., Pelletier, F., Hamel, S., Gamelon, M., Barbraud, C., Blanchet, F.G., Blumstein, D.T., Charmantier, A., Delord, K., Larue, B., Martin, J., Mills, J.A., Milot, E., Mayer, F.M., Rotella, J., Saether, B.-E., Teplitsky, C., van de Pol, M., van Vuren, D.H., Visser, M.E., Wells, C.P., Yarrall, J., and Jenouvrier, S. (2023). Individual life histories: neither slow nor fast, just diverse. *Proceedings of the Royal Society B: Biological Sciences* 290: 20230511.

Verhulst, P.-F. (1838). Notice sur la loi que la population suit dans son accroissement. *Correspondance Mathématique et Physique* 10: 113–121.

Volterra, V. (1927). Variazioni e fluttuazioni del numero d'individui in specie animali conviventi. *Regio Comitato Talassografico Italiano, Memoria* 131: 1–142.

Wei, X., and Zhang, J. (2019). Environment-dependent pleiotropic effects of mutations on the maximum growth rate r and carrying capacity K of population growth. *PLoS Biology* 17(1): e3000121.

Williams, C.G. (1966). *Adaptation and natural selection: a critique of some current thought*. Princeton University Press.

Williams, L.J., King, A.J., and Mettke-Hofmann, C. (2012). Colourful characters: head colour reflects personality in a social bird, the Gouldian finch, *Erythrura gouldiae*. *Animal Behaviour* 84: 159–165.

Wittman, T.N., and Cox, R.M. (2021). The evolution of monogamy is associated with reversals from male to female bias in the survival cost of parasitism. *Proceedings of the Royal Society B: Biological Sciences* 288: 20210421.

Wolf, M., van Doorn, G.S., Leimar, O., and Weissing, F.J. (2007). Life-history trade-offs favour the evolution of animal personalities. *Nature* 447: 581–585.

Worthington, A.M., and Swallow, J.G. (2010). Gender differences in survival and antipredatory behavior in stalk-eyed flies. *Behavioral Ecology* 21: 759–766.

Wright, B.M., Stredulinsky, E.H., Ellis, G.M., and Ford, J.K.B. (2016). Kin-directed food sharing promotes lifetime natal philopatry of both sexes in a population of fish-eating killer whales, *Orcinus orca*. *Animal Behaviour* 115: 81–95.

Wright, J., Bolstad, G.H., Araya-Ajoy, Y.G., and Dingemanse, N.J. (2019). Life-history evolution under fluctuating density-dependent selection and the adaptive alignment of pace-of-life syndromes. *Biological Reviews* 94: 230–247.

Xirocostas, Z.A., Everingham, S.E., and Moles, A.T. (2020). The sex with the reduced sex chromosome dies earlier: a comparison across the tree of life. *Biology Letters* 16: 20190867.

6

Parental Care and Life History

Hope Klug[1] and Michael B. Bonsall[2]

[1] Biology, Geology, and Environmental Science, University of Tennessee at Chattanooga, Chattanooga, TN, USA
[2] Department of Biology, University of Oxford, Oxford, UK

6.1 What Is Parental Care and How Does It Relate to Life History?

Life history traits such as size at birth, growth rate, age and size at maturity, lifespan and the size of offspring are traits that are directly linked with fitness. As such, selection is mediated through phenotypic variation in these traits (Stearns 1992). Parental care is most broadly defined as any parental trait that is likely to increase the fitness of offspring and that is likely to have originated and is currently maintained for this function (Smiseth et al. 2012). More specifically, parental care is a parental behaviour that (1) occurs post-fertilization (or after the production of daughter cells if reproduction is asexual), (2) is directed at offspring and (3) appears likely to increase the offsprings' lifetime reproductive success (Bonsall and Klug 2011). In some cases, parental care is considered a life history strategy or trait. In other cases, parental care is considered a strategy or trait that influences and is influenced by life history. This distinction (*i.e.,* whether or not parental care itself is considered a life history trait) is largely one of semantics. In this chapter, for simplicity, when we refer to life history traits, we are referring to basic life history traits that are the principal components of fitness, such as stage-specific rates of maturation, mortality and mating (Stearns 1992) and thus distinguish between parental care and other life history traits.

Parental care in nature is highly diverse and ranges from simple preparation of a rearing environment or guarding of offspring during the egg stage to more complex forms of care such as provisioning, grooming and support of offspring during intraspecific interactions as juveniles or adults (reviewed in Balshine 2012 and Klug and Bonsall 2014). For example, in both invertebrates and vertebrates, parents in a range of species: (1) protect offspring from predators during the egg and/or juvenile stages; (2) prevent desiccation of eggs; (3) oxygenate eggs; (4) remove offspring waste during egg and juvenile stages; (5) provision young through lactation, the preparation of food, and/or the capture of prey; (6) reduce offspring energetic expenditure through transport behaviour or thermoregulation; (7) engage in teaching of young and/or provide behavioural support during intraspecific interactions; (8) increase offspring immune function; (9) adaptively manipulate offspring developmental rate; and (10) protect offspring from parasites, parasitoids, and disease (reviewed in Alonso-Alvarez and Velando 2012 and Klug and Bonsall 2014). Providing parental care is costly to parents, such that it reduces current reproductive success due to energetic expenditure, increased predation risk and susceptibility to disease (reviewed in Clutton-Brock 1991 and Alonso-Alvarez and Velando 2012). In addition, when parents are providing care, they are frequently unable to acquire new matings (Kokko and Jennions 2008), although this is not always the case. In some species, however, females prefer males that provide parental care, in which case care can increase (rather than decrease) current mating and reproductive success (Lindström et al. 2006, Stiver and Alonzo 2009, Alonzo 2012). Given its effects on offspring and parental fitness, parental care is strongly under natural selection, and there can be a complex evolutionary relationship between parental care and basic life history traits (Figure 6.1).

In this chapter, we provide a detailed overview of how parental care and other life history traits interact from both ecological and evolutionary perspectives. We first discuss the life history conditions that are most likely to lead to the origin of parental care across environmental contexts. We then discuss the basic life history traits (*i.e.,* stage-specific rates of maturation, mortality, etc.) that can mediate the maintenance and elaboration of parental care once it is present in a population,

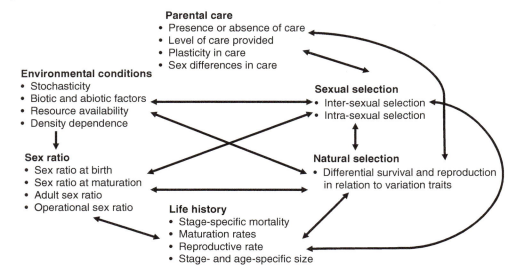

Figure 6.1 The complex links between life history, parental care and environmental conditions. Environmental conditions will directly influence sex ratio as well as the natural and sexual selection that individuals in a population experience. Due to the social interactions and resource use that are linked with survival and reproduction, sexual and natural selection can also influence abiotic and biotic environmental conditions. Life history will affect and be affected by sex ratio. Parental care will be directly influenced by natural and sexual selection, as well as life history, and environmental conditions and sex ratio (*e.g.,* primary, adult, OSR) will also influence care through the effects of sex ratio on life history and natural and sexual selection.

and we provide an overview of our understanding of how parental care and other life history traits co-evolve once parental care is present in a population. Males and females frequently differ in life history traits, and we then discuss how such differences can lead to asymmetry in parental care, as well as the links between basic life history traits, parental care and sexual selection. In addition, we discuss the links between stochasticity, phenotypic plasticity, life history and parental care. Finally, we provide an overview of areas of research that warrant further attention.

6.2 Distinguishing Between the Origin and the Maintenance of Parental Care

The conditions that favour the origin of parental care will often not be the same as those conditions that shape the maintenance of parental care (discussed in Klug et al. 2012, Royle et al. 2016). When we consider the factors that lead to the origin of a trait, such as parental care, we are interested in the demographic, ecological and evolutionary factors that lead to the origin of that trait when that trait was previously absent or rare in the population. After parental care originates and begins to spread in a population, the relative frequency of that trait will increase, and density-dependent processes will become important. As such, the selection acting on parental care is likely to be affected by a different set of genetic, physiological and ecological factors than those that influence the origin of the trait (Klug et al. 2012, Head et al. 2014). As a result, understanding the evolution of parental care in relation to given life history strategies requires consideration of the distinct processes that can influence the origin *vs.* maintenance of traits. Throughout this chapter, we are mindful, when relevant, of the distinction between the origin and the maintenance and elaboration of parental care when reviewing both theoretical and empirical research on the topic.

6.3 Life History and the Origin of Parental Care

Life history conditions strongly influence whether parental care will originate in a system (Table 6.1). In particular, theoretical work that utilizes a stage-structured mathematical approach suggests that stage-specific rates of mortality and maturation will determine whether a rare mutant that exhibits parental care can invade a resident population that does not

Table 6.1 General life history conditions that are most likely to lead to the evolutionary origin and maintenance of parental care. See text for relevant citations.

Life history trait	Conditions favouring the origin of care	Conditions favouring the maintenance of care	Exceptions
Offspring mortality	High mortality in the absence of care	High egg mortality if adult survival is density-dependent; moderate and high egg mortalities if egg survival is density-dependent	If care increases time spent in a relatively safe stage, the origin of care will also be favoured at low egg death rates
Adult mortality	High parental mortality when egg and adult survival are density-independent	Low adult mortality if egg survival is density-dependent; no effect of adult mortality if adult survival is density-dependent	The origin of care will be favoured most strongly at lower adult mortalities if (1) egg survival is density-dependent and (2) adult mortality is density-dependent and adult density is high
Egg maturation rate	Low egg maturation rates	High egg maturation rates if egg survival is density-dependent; low egg maturation rates if adult survival is density-dependent	If care extends into the juvenile stage, the origin of care can be favoured at high egg maturation rates
Duration of juvenile stage	Short juvenile stage	Short juvenile stage	

exhibit parental care (Klug and Bonsall 2010, Klug et al. 2013a). Offspring need, which is frequently measured as offspring mortality in the absence of parental care, can strongly influence whether the origin of parental care is selected for. Specifically, when offspring survive poorly in the absence of care due to harsh environmental conditions (*e.g.*, intense predation, limited resources, disease), theoretical work suggests that there will be strong natural selection favouring the origin of parental care (Webb et al. 2002, Klug and Bonsall 2010) (Table 6.1). Empirical work supports this prediction. For example, parental care is more common in invertebrates when offspring mortality in the absence of care is high (reviewed in Clutton-Brock 1991 and Wong et al. 2013). Theoretical work also suggests that the origin of parental care is more likely to be favoured when parental mortality is relatively high *vs.* low, although parents must survive long enough to be able to provide care (Klug and Bonsall 2010) (Table 6.1). Such a pattern potentially occurs because individuals will be selected to invest more heavily into current reproductive success when there is reduced potential for future reproduction due to high adult mortality. Empirically, well-developed parental care tends to be more common in fishes that have relatively short lifespans, whereas the absence of parental care in fishes is correlated with relatively long lifespans (Winemiller and Rose 1992). However, in some cases, we might also expect parental care to originate when adult mortality is relatively low, and theoretical work suggests that offspring mortality and density-dependent processes can interact to influence the origin of parental care. Specifically, parental care is most likely to originate at lower adult mortalities if (1) egg survival is density-dependent (*vs.* density-independent), a condition that is likely to be associated with a greater need for care, and (2) adult mortality is density-dependent and adult density is relatively high, a condition under which there will be reduced potential for future reproduction (Reyes et al. 2016) (Table 6.1). Together, such theoretical work (Klug and Bonsall 2010, Reyes et al. 2016) therefore suggests that care can originate evolutionarily across a range of adult mortalities. Indeed, in nature, parental care has originated in animals that have diverse adult survival strategies (patterns of care reviewed in Clutton-Brock 1991 and Royle et al. 2012). In general, though, relatively few empirical studies have explored the link between adult survival and the evolutionary origin of parental care. To the best of our knowledge, no empirical studies have examined how density-dependent processes interact with adult survival to influence the origin of parental care, a topic that warrants further empirical attention.

Similarly, the duration of pre-adult stages and the rate of maturation are expected to influence the origin of parental care (Table 6.1). When parental care is provided only during the egg stage, theoretical work suggests that care is more likely to originate when egg maturation rate is relatively low (*i.e.*, when eggs develop slowly) (Klug and Bonsall 2010). Such a pattern occurs because fitness is maximized by making the egg stage relatively safe when eggs develop slowly and therefore spend a relatively long amount of time in that stage. In contrast, when care is provided to both eggs and juveniles, the origin of parental care will be most strongly selected for at greater egg maturation rates (*i.e.*, when eggs develop quickly) (Klug and Bonsall 2010). Under such conditions, both the egg and juvenile stages are relatively safe, and, as such, fitness is maximized when eggs mature, leave the egg stage and enter the relatively safe juvenile stage quickly. In general, when both the egg and juvenile stages are relatively safe, faster egg maturation and care will be favoured because there is some risk of dying at any point in time that an individual spends in the egg stage, so individuals will be selected to mature quickly. Mathematical modelling also suggests that the relative, rather than absolute, mortality of the egg *vs.* juvenile stages will influence the origin of parental care. Natural selection will, all else equal, favour spending relatively short periods of time in stages that have high mortality (Shine 1978, Remeš and Martin 2002, Warkentin 2011, Klug and Bonsall 2014). Empirical research supports this prediction on an ecological time scale. For example, many amphibians, fishes, birds, reptiles and invertebrates can adjust hatching time in response to perceived mortality risk associated with predation, disease and other factors (Remeš and Martin 2002, Warkentin 2011). However, animals are likely constrained in their ability to adjust development, hatching, and metamorphosis (Warkentin 2011), and natural selection will in some cases favour longer developmental times in order to allow offspring to grow larger. As a result, in some species, natural selection has favoured parental control of offspring developmental rates (Shine 1978, Klug and Bonsall 2014). Shine (1978) noted that there is a positive relationship between egg size and the occurrence of parental care in many species of invertebrates, fishes, reptiles and amphibians. Shine (1978) hypothesized that such a pattern occurs because parents make the egg stage relatively safe by providing parental care and then increase the time offspring spend in the egg stage by producing large eggs (safe-harbour hypothesis; Shine 1978, 1989; but see Nussbaum and Schultz 1989). This, then, is expected to decrease the time offspring spend in a relatively dangerous stage (Shine 1978, 1989). There is mixed empirical support for this prediction. Using a comparative approach, research focused on insects found no relationship between egg sizes, which is typically correlated with the duration of the egg stage, and the mode of parental care provided (Gilbert and Manica 2010). In contrast, egg size is correlated with the evolution of parental care in frogs, although the most likely evolutionary trajectory in frogs is for large eggs to evolve first and then parental care to originate next (*i.e.*, the most likely evolutionary sequence is from small to large eggs followed by the origin of care) (Summers et al. 2006). This finding is inconsistent with the theoretical predictions of Shine (1978), as the findings of Summers et al. (2006) suggest that larger eggs, which take longer to develop, preceded the evolution of parental care (whereas Shine (1978) suggested parental care preceded the evolution of larger eggs). Nonetheless, such empirical research highlights the important role that egg size and developmental rate have played in the evolution of parental care. More recent theoretical research found that parental manipulation of offspring developmental rate that decreases time spent in relatively dangerous stage(s) can in-of-itself be a benefit of parental care that can directly favour the evolutionary origin of parental care (Klug and Bonsall 2014). Empirically, few studies have examined parental manipulation of offspring developmental rate, but research on the topic suggests that spider and burying beetle parents can adaptively manipulate offspring developmental rate (Li 2002, Smiseth et al. 2003).

Further, when parents can manipulate offspring developmental rate, theoretical research suggests that parental care has the potential to originate over a broader range of life history conditions in comparison to the case in which parents cannot manipulate offspring developmental rate (Klug and Bonsall 2014). When the only benefit of parental care is increased offspring survival, parental care is expected to originate only when offspring mortality in the absence of care is relatively high (*i.e.*, when offspring need care the most; Klug and Bonsall 2010). However, when parental care increases time spent in a relatively safe egg stage, parental care is expected to originate even when offspring mortality is low (Klug and Bonsall 2014). Such a pattern occurs because parental manipulation of offspring developmental rate can function as a benefit of care. When parental manipulation of offspring developmental rate occurs, the benefits of providing care can outweigh the costs, even when offspring survive relatively well without care (Klug and Bonsall 2014). This theoretical research illustrates that life history, and in particular offspring developmental rate and mortality, can strongly influence the evolutionary origin of parental care. Finally, stochasticity can also influence the evolutionary origin of parental care in relation to life history. The links between stochastic processes, life history and parental care are discussed below.

6.4 Life History and the Maintenance of Parental Care

Once parental care originates evolutionarily, individuals who exhibit care are expected to increase in frequency in the population. As the relative frequency of individuals who provide care increases, density-dependence among individuals who provide care will influence selection on parental care and other life history trait values (Klug et al. 2012), and theoretical research suggests that whether egg *vs.* adult survival is density-dependent will influence selection for the maintenance of parental care (*i.e.*, we expect different patterns of care depending on whether density-dependent processes act on the egg stage *vs.* the adult stage; Reyes et al. 2016). This theoretical research suggests that parental care will be maintained when egg mortality is relatively high (*i.e.*, when offspring are likely to die in the absence of care; Reyes et al. 2016), which is consistent with the pattern associated with the origin of parental care (Table 6.1). However, the maintenance of parental care will be most strongly selected for at high egg death rates when adult survival is density-dependent. When egg survival is density-dependent, however, selection for the maintenance of parental care will be similar at moderate and high egg mortality levels (Reyes et al. 2016) (Table 6.1). In contrast to the conditions that favour the origin of care, the maintenance of parental care will be most strongly selected for when adult death rate is relatively low if egg survival is density-dependent (Reyes et al. 2016) (Table 6.1). When adult survival is density-dependent, however, adult mortality is not expected to influence the maintenance of parental care (Reyes et al. 2016) (Table 6.1). Moreover, in contrast to the conditions that favour the origin of parental care, theoretical research suggests that the maintenance of care will be most strongly selected for at relatively high egg maturation rates when egg survival is density-dependent (Reyes et al. 2016). Consistent with the conditions that favour the origin of care, theoretical research predicts that the maintenance of care will be most strongly selected for at relatively low egg maturation rates if adult mortality is density-dependent (Reyes et al. 2016) (Table 6.1). In general, the maintenance of egg-only care will be most strongly selected for when the duration of the juvenile stage is relatively short (Reyes et al. 2016) (Table 6.1). Collectively, this theoretical work suggests that the conditions that favour the maintenance of parental care will depend on whether density-dependent processes influence egg *vs.* adult survival. These results further highlight that basic aspects of ecology (*i.e.*, density-dependence acting at different life history stages) can influence the conditions that favour the maintenance of parental care.

To the best of our knowledge, empirical research has not explicitly tested the predictions outlined in the previous paragraph. However, predictions made by Reyes et al. (2016) suggest both egg and adult mortality influence the maintenance of parental care. Consistent with this general prediction, parental care is maintained across evolutionary time in many animals in which there is high offspring need (*i.e.,* high offspring mortality in the absence of care due to harsh environmental conditions, a need for provisioning or thermoregulation, offspring dependence on guarding, etc.; reviewed in Clutton-Brock 1991 and Royle et al. 2012). Similarly, adult mortality risk can influence parental care once it is present in a species. In the bird species including the white-breasted nuthatch (*Sitta carolinensis*) and the red-breasted nuthatch (*Sitta canadensis*), parents responded to a simulated adult predator by increasing the length of time between nest visits and aborting visits to the nest more frequently (Ghalambor and Martin 2000). Similarly, empirical research reveals that maturation rate influences parental care. For instance, females of the egg-carrying spitting spider (*Scytodes pallida*) adjust hatching time of eggs in response to the threat of predation (Li 2002). Specifically, eggs that are carried by females in the presence of egg predators hatch sooner than those not carried and those that are carried in the absence of predators, and the chemical cues of predators are sufficient to induce parental adjustment of offspring hatching time (Li 2002). Likewise, parental care in burying beetles (*Nicrophorus vespilloides*) reduces the duration of the larval stage (Smiseth et al. 2003).

The above research illustrates the link between life history and parental care once care has originated in a system. In addition, once parental care originates, we would expect co-evolutionary feedback between parental care and other life history traits, which is the focus of the next section.

6.5 Co-evolution Between Parental Care, Offspring Traits and Parental Traits

Co-evolutionary dynamics will ultimately shape patterns of parental care and other life history traits once care is present in a system (Figure 6.1). The selection acting on both life history traits and parental care will be shaped by among-individual interactions (reviewed in Moore et al. 1997, Kölliker et al. 2012). Specifically, the evolution of life history traits and parental care will depend on both an individual's own genotype (*i.e.*, their direct genetic effects) and the genotypes that contribute to the family environment (*i.e.*, their indirect genetic effects) (Moore et al. 1997, Kölliker et al. 2012). Indirect genetic effects are present when the genotype of one individual affects gene expression in another individual, as mediated by the social interaction between the two individuals (Moore et al. 1997, Kölliker et al. 2012). When parental care is provided, one or both parents are socially interacting with their offspring. Given this, both parents and offspring have the potential to influence gene expression in the other individual, which in turn will influence the selection acting on traits. Such indirect genetic effects are therefore expected to lead to co-adaptation between parent and offspring traits (reviewed in Kölliker et al. 2012). In particular, such co-adaptation is expected to strongly influence the level of parental care provided and offspring needs (Kölliker et al. 2012). Specifically, offspring need is expected to co-evolve with care, and offspring often become more dependent on care through evolutionary time. Consistent with this prediction, in species in which parental care has recently evolved, offspring tend to survive relatively well in the absence of care. For example, in the flagfish (*Jordanella floridae*), in which paternal care has relatively recently evolved, most (~90% of) eggs survive when care is absent and predators are also absent (Klug et al. 2005). Flagfish males improve the survival of their eggs when predators are present but decrease egg survival when alone with eggs (Klug et al. 2005). In species in which care has a long evolutionary history, such as mammals, offspring fail to survive in the absence of parental care (Clutton-Brock 1991, Balshine 2012). In general, we would expect offspring to become more dependent on care – and hence offspring mortality in the absence of care to increase – across evolutionary time.

In addition, there can be co-evolution between parental care and other parental and/or offspring traits. In mammals, increased litter size is likely to follow the evolution of paternal care (Stockley and Hobson 2016). Similarly, egg size and parental care often co-evolve in fishes (Kolm and Ahnesjö 2005). In frogs, phylogenetic analyses revealed that the evolution of large egg size often preceded the evolution of parental care (Summers et al. 2006), and egg size is a predictor of parental care in bony fishes (Iglesias-Rios et al. 2022). While some earlier research suggested that large egg size preceded the evolution of parental care (*i.e.*, the most likely evolutionary transition was the origin of large eggs followed by the evolutionary origin of care; Shine 1978; but see Nussbaum and Schultz 1989), more recent research suggests that the relationship between egg size and parental care is the result of complex co-evolutionary processes in which either larger egg size or parental care can precede the other (Kolm and Ahnesjö 2005, Royle et al. 2016). In the burying beetle (*N. vespilloides*), the strong positive effect of parental care on offspring growth can mask the relatively weak benefits of larger offspring size (Monteith et al. 2012), illustrating that parental care can interact with egg size to influence offspring survival.

6.6 Sexual Selection, Life History and Sex Differences in Parental Care

Males and females differ in basic life history traits, and, at the most fundamental level, females produce larger gametes than males (see also Chapter 5, this volume). This anisogamy can lead to a range of basic life history differences between males and females, including differences in the number of gametes produced, such that males produce many small gametes and females produce fewer larger gametes (Bateman 1948, Smith and Fretwell 1974). Anisogamy can also influence differences in sex roles and mating strategies between males and females (Bateman 1948). These differences between males and females in basic life history traits can also influence sex-specific patterns of parental care. If we consider species that provide some form of parental care, maternal care is most common in mammals and invertebrates, biparental care is most common in birds and paternal care is most common in fishes (Royle et al. 2012; see Figure 6.2). With regard to the origin of parental care,

Figure 6.2 Parental care is diverse in nature, and one or both sexes provide care in many species. While maternal care is more common across species than paternal care, male care occurs in some animals and is linked closely to life history. For example, (a) sand goby (*Pomatoschistus minutus*) males guard, fan and clean their eggs within a nest, and (b) giant water bug (*Belostoma lutarium*) males provide parental care to eggs that are laid on their back by females. In both species, offspring survival is dependent on male care, and offspring will not survive in the absence of care.

theoretical research suggests that life history will influence sex-specific patterns of care (Klug et al. 2013a). Paternal and biparental care are most likely to originate when (1) male egg maturation rate is relatively low (*vs.* high), (2) female egg maturation rate is relatively high (*vs.* low), (3) male and female eggs have high (*vs.* low) mortality in the absence of care, (4) male and female adults have high (*vs.* low) mortality and (5) male and female juvenile survival is high (*vs.* low) (Klug et al. 2013a). These conditions also hold for maternal care, except that both male and female egg maturation rates are expected to be high in this case. This theoretical work suggests that the basic life history conditions that give rise to the origin of maternal, paternal and biparental care will be similar and are influenced by sex-specific rates of mortality and maturation. However, this research also suggests that sex differences in maturation rate can be a key factor that impacts the origin of care, as sex-specific maturation rates influence the number of males *vs.* females who can mate and potentially provide care, which in turn influences the sex-specific fitness associated with providing care. Empirically, little work has focused on how sex differences in maturation rate influence the origin of different forms of care. However, in the black guillemot (*Cepphus grylle*), a species with biparental care, male chicks hatch on average one day sooner than females (Cook and Monaghan 2004). The link between basic life history and sex-specific patterns of parental care is an area of research that warrants further empirical attention.

Life history can also influence transitions between different forms of parental care once care has evolved. Theoretical research suggests that transitions from paternal to maternal or biparental care will be most likely to occur when (1) baseline egg death rate is low, (2) males have fast and females have slow egg maturation rates, (3) adult and juvenile mortality is high in females and (4) adult and juvenile mortality is low in males. In all other cases, transitions from maternal to paternal or biparental care will be most likely to occur (Klug et al. 2013b). This theoretical research highlights that sex differences in basic life history can influence transitions among different forms of parental care once care is present in a system. Such a pattern occurs because sex differences in basic life history influence parental residual reproductive value through their effects on survival to reproduction, which in turn creates sex differences in the net fitness costs and benefits of providing care (Klug et al. 2013b). Empirically, to the best of our knowledge, no research has explored the link between sex differences in life history and sex-specific patterns of care. However, transitions among different forms of care are common, particularly in anurans and fish (Reynolds et al. 2002, Mank et al. 2005).

In addition to the direct effects of basic life history on parental care, life history can also affect the adult sex ratio (ratio of adult males to females, ASR) and operational sex ratio (ratio of males to females currently prepared to mate in a given location, OSR), which can then in turn influence sexual selection, sex roles and parental care (Kokko and Jennions 2012, Fromhage and Jennions 2016, Jennions and Fromhage 2016) (Figure 6.1). For example, sex differences in egg, juvenile and adult survival can lead to a biased ASR and/or OSR and/or differences between ASR and OSR. Specifically, if males are less likely to survive the pre-adult stage, all else equal, we would expect a female-biased ASR and OSR (discussed in Klug et al. 2022). If the ASR is male-biased due to biases in the sex ratio at maturation, we would expect a positive relationship between ASR and the proportion of male care provided (Kokko and Jennions 2012, Fromhage and Jennions 2016). Similarly, if the ASR becomes more male-biased due to similar changes in male adult mortality in both the time-in state (*i.e.*, the mating pool) and the time-out state (*i.e.*, the state in which individuals process the consequences of mating), the OSR is expected to become more male-biased, and the proportion of male parental care is expected to increase in a population as ASR increases (Fromhage and Jennions 2016). In contrast, if the ASR becomes more male-biased due to sex-biased adult mortality in only the time-in state, the OSR will become more male-biased as more females than males die, but the proportion of male parental care will remain invariant across ASRs (Fromhage and Jennions 2016). These results highlight that the source of variation in sex ratio biases – which can be due to sex differences in mortality during different life history stages and states – can directly influence population-level parameters (OSR, ASR) and sex-specific patterns of parental care (Kokko and Jennions 2012, Fromhage and Jennions 2016, Jennions and Fromhage 2016) (Figure 6.1).

In addition, and as mentioned above, sexual selection for paternal care can offset the net costs of care for males and directly lead to the evolution of male-only care (Alonzo 2012). In such cases, there will not necessarily be a trade-off between current and future reproduction (Stiver and Alonzo 2009), which is consistent with some empirical research demonstrating female mate preferences based on males providing care. For example, in the sand goby (*Pomatoschistus minutus*), a species with exclusive male care, females show mate preferences for males showing higher levels of male care (Lindström et al. 2006).

Empirically, a range of other life history characteristics also influence sex-specific patterns of parental care, and sex-specific patterns of care also influence offspring life history. For instance, external fertilization is more likely to lead to paternal care in fishes (Figure 6.2), whereas internal fertilization is more likely to lead to maternal care (Mank et al. 2005, Benun Sutton and Wilson 2019). Such a pattern likely occurs because females are physically associated with offspring, whereas males are not when fertilization occurs within the female's body. In birds, there is co-variation between parental care, life history and environmental conditions. Higher-elevation avian species have lower fecundity but provide more parental care and male provisioning is increased at higher elevations (Badyaev and Ghalambor 2001). In mammals, there are shorter lactation periods in species in which males transport offspring (West and Capellini 2016). Similarly, ecology and life history interact to drive diversity in parental care in insects. For instance, feeding behaviour, environmental conditions, the presence of predators and social structure can interact to influence patterns of care across insect species (Wong et al. 2013).

6.7 Stochasticity, Environmental Variability, Life History and Parental Care

Environmental variability influences life history traits (Hastings and Caswell 1979) such as optimal offspring size and the evolution of strategies where a range of offspring phenotypes is produced to deal with future environmental variability (so-called bet-hedging strategies). However, the role of environmental variation and change in the evolution of parental care is less well understood. Verbal arguments suggest a potential role for environmental fluctuations in the evolution of care. For example, when environmental conditions are harsh and competition for resources is intense such that there is high variability in resource acquisition, the benefits to care are likely to be high (Clutton-Brock 1991, Wilson 2000). By contrast, alternative arguments suggest that, under the assumption of r- and K-selection theories, parental care is more likely to be associated with constant and stable environments (Stearns 1976). Similarly, empirical studies have highlighted that parental decisions (such as provisioning) and hence the benefits and costs of providing care are influenced by the level of variability in a given environment (McGinley et al. 1987, Dziminski et al. 2009, Low et al. 2012). In this section, we review both theoretical and empirical advances in our understanding of the evolution of parental care in fluctuating (stochastic) environments.

6.7.1 Theoretical Advances: The Evolution of Parental Care in Stochastic Environments

Theoretical work developed by Carlisle (1982) investigated parental care allocations in variable environments. Using an optimality approach to construct 'curves' of fitness against level of care, Carlisle (1982) showed that when environmental variation is predictable, more care should be provided to successive offspring as environmental conditions improve (and less so in deteriorating environments). In non-uniform (non-constant and unstable) environments where the risks of parent mortality increase, levels of care should be expected to increase. However, if increasing the costs of care increases the risk of parent mortality in these non-constant environments, we would expect an evolutionary transition to less care (Carlisle 1982). The dynamic nature of the evolution of parental care in stochastic environments requires much more scrutiny.

To begin to address this, using a stage-structured life history framework, we have considered how environmental variability influences the evolutionary dynamics of parental care (Bonsall and Klug 2011). By considering this in a very general manner, allowing individuals to experience a fixed number of different environments (that affect the various costs and benefits of parental care differently) and deriving appropriate fitness measures for variable environments (Lewontin and Cohen 1969), we show that environmental variability can influence the evolution of care. Stochastic environments can reduce both the fitness gains and selection for parental care, particularly when the costs of parental care operate across multiple life history traits (such as parental survival and reproductive rate). Generally, this is consistent with the idea that variable environments will often be associated with a decreased need for parental care (Stearns 1976), as there is expected greater investment in offspring number, increased reproductive effort, early maturity and time to first reproduction. By contrast, if the only cost of care is to increase parental death rate, any environmental variability that then affects adult death rate increases the fitness associated with care. Unpredictable environments can then increase the fitness benefits of parental care (Clutton-Brock 1991), with low adult survival favouring the evolution of parental care (Klug and Bonsall 2010). The interaction between the predictability of the environment, the mode of reproductive parity (*e.g.*, single *vs.* multiple broods) and the evolution of parental care remains to be more fully elucidated. Iteroparity (multiple broods) and investment in a small number of high-quality offspring have been argued to favour the evolution of parental care (Wilson 2000). Alternatively, investment by semelparous (single brood) organisms into parental care should be favoured due to the low costs of care in terms of future reproduction (Tallamy and Brown 1999). However, in terms of lifetime reproductive success, if adult survival of semelparous organisms is low and the nature of the environment is unpredictable, more nuanced approaches are required to understand the interaction between life history and the evolution of parental care in stochastic environments.

6.7.2 Linking Predictions of Parental Care in Stochastic Environments to Empirical Patterns and Analysis

Empirically, the evolutionary transitions to different forms (paternal, maternal or biparental care) of parental care (Klug et al. 2013b, Azad et al. 2022) are likely to be influenced by environmental variability. Theory predicts that these evolutionary transitions and maintenance of parental care (compared to its origin) depend on the relative investment patterns of males and females (Klug et al. 2013b): anisogamous investment patterns are more likely to favour transitions to increased maternal care. Paternal care is expected when males have low survival and/or the survival of male offspring is low (Figure 6.2). Offspring needs (low offspring survival without care) can also influence the evolution of paternal care (Klug et al. 2013b).

Notwithstanding the dearth of investigations on the life history drivers of the evolutionary transitions of parental care in stochastic environments, the availability of phylogenies has led to numerous studies on birds, insects and fish species, exploring the correlates associated with these evolutionary transitions to different forms of parental care (Székely and Reynolds 1995, Goodwin et al. 1998, Reynolds et al. 2002). The amazingly diverse cichlid fishes have various forms of parental care with wide interspecific variation in which sex stays with the offspring. Comparative work has shown that uniparental care (female-only, male-only) has arisen in this group of fish at least 20 times (Goodwin et al. 1998). The diversity of cichlid life histories has been attributed to environmental variation: large-scale climatic and geological fluctuations have influenced wet–dry transitions, generating widespread habitat variability and hence rapid adaptive radiations (Ivory et al. 2016). As such, forms of care and helping within the cichlids are likely to be shaped by environmental cues (Kasper et al. 2017).

In a comparative study on the evolutionary transitions in parental care among shorebirds (Székely and Reynolds 1995), paternal care was shown to be more common than maternal care. In one clade (the Charadriidae – plovers, dotterels, lapwings), the evolution of male care had several independent origins, whereas in another clade (the Scolopacidae – sandpipers), male care is an ancestral state with a contemporaneous shift to increased levels of biparental care. While male and female shorebirds tend to have similar patterns of survival (Méndez et al. 2018), environmental differences in habitats (harsh *vs.* benign) can affect male and female reproductive success, offspring abandonment and the patterns of parental care (paternal, maternal and biparental) in these birds (Reynolds and Székely 1997).

Provisioning of care between males and females is also influenced by environmental stochasticity. Using space-for-time approaches, Trivers (1974) and Vincze et al. (2017) have shown how temperature variation influences patterns of biparental care across different species of plovers (*Charadrius* spp.). As temperatures increase and become more variable, paternal care is more common as males increase their effort relative to females during egg incubation. Under changing environments, due to climate warming, increases in both mean temperature and variation of extremes (Pörtner et al. 2022), patterns of parental care can be expected to be heavily disrupted.

Parent-offspring conflicts are tightly linked to the evolution of parental care. The evolutionary conflict that arises can occur as parents value offspring equally while offspring value their own survival more than current or future siblings (Trivers 1974). Resolution of this has focused on conflicts over resource provisioning (*e.g.*, Parker and Macnair 1978, Godfray 1995a, 1995b) and the degree of information transmitted between parents and offspring (*e.g.*, Godfray and Parker 1991). However, the premise that parents equally value offspring might be infelicitous, and the influence of life histories in variable environments influences parental effort and care decisions. Environmental fluctuations coupled with the reproductive value of offspring have important implications for the evolution of parental care. Parents might not value all offspring equally; reproductive value varies across offspring – with older offspring actually having greater value than young offspring. Similarly, sexual selection might lead to some offspring having greater reproductive value than others. If environmental fluctuations affect offspring survival, this has implications for patterns of parental investment (Low 1978). With low reproductive value of young offspring, parental care provided at early life stages should be limited (Low 1978). The evolution of life histories is then critically linked to environmental uncertainties. Within marsupials, for example, delayed implantation (and the associated limited costs with early foetus production and/or abortion) is most likely favoured due to unpredictable environmental conditions (Low 1978).

Parental care occurs in social environments where individuals interact: parents interact with offspring, females and males interact to determine level of care and caring parents interact with other individuals in a social (related) and non-social way. As such, cooperative breeding systems, in which there is spatial and temporal variation in habitats, breeding opportunities and life history decisions, provide ways to understand the evolution of behaviours. For instance, the superb starling *Lamprotornis superbus* is a cooperatively breeding passerine, endemic to savannah in east Africa that forms spatially subdivided complex groups (Rubenstein 2011). This starling experiences unpredictable environments (within and between-year patterns of rainfall have high variance) and this environmental variability influences territory quality, resource abundance and hence overall reproductive success – which is low (~13% of all eggs laid are fledged). However, variance in this reproductive success (fitness) is strongly positively correlated with social groups being larger and having more nest helpers (and hence more carers) (Rubenstein 2011). Greater levels of parental care influence offspring survival in unpredictable environments (Bonsall and Klug 2011).

Finally, population ecological processes matter. Population density fluctuations (driven by ecological processes of competition and/or predation) can influence the environment under which offspring are reared. Increases in con- or hetero-specific individuals will alter the conditions for parental care (Carlisle 1982). Competitive environments, coupled with external environmental fluctuations, might further influence the evolution of parental care (Potticary et al. 2023). Climate change has altered the phenological and seasonal activity of many species (Both and Visser 2005, Visser and Both 2005, Forrest 2016, Lemoine 2021, Buckley 2022). For example, differences in the intensity of within- and between-species competition in burying beetles have been attributed to changes in seasonal activity and phenological overlaps between species. Disrupting these population ecological processes of competition across different and/or changing environments affects the levels of resource availability (the between-species effects) with implications for the differential survival of male and female beetles (the within-species effects). This has consequences for the provisioning of parental care by burying beetles (Potticary et al. 2023). Similarly, variation in levels of predation can influence the parental care outcomes. For instance, the effects of environmental variation and predation interact to affect clutch survival in glass frogs (Lehtinen et al. 2014). Clutch mortality due to predation is observed to be significantly higher in the absence of male frog attendance and in the rainy season, compared to levels of predation when the males are in attendance and in the dry season (Lehtinen et al. 2014). Using a manipulative experiment to test this observation, Lehtinen et al. (2014) investigated the role of paternal egg attendance (by removing males) on clutch survival in a glass frog, *Hyalinobatrachium orientale*, in different seasons. Males were more likely to attend to eggs during dry seasons, and clutches where males were removed had significantly higher risk of mortality (Lehtinen et al. 2014). Importantly, seasonal environments provide unique opportunities to evaluate the role of environmental variation in the evolution of parental care.

Critical in this is the mismatch between the frequency of the environmental fluctuations and the generation time of the animal, its breeding cycle and the timing of offspring provisioning. Rapidly changing but predictable environments will have different implications for the evolution of care than slowly changing but unpredictable environments. Understanding the role of changing environments in the evolution of parental care, particularly under global warming scenarios, requires urgent empirical and theoretical attention.

6.8 Plasticity and the Evolution of Parental Care

Phenotypic plasticity is the capacity of single genotypes to produce a range of phenotypes (Stearns 1982). Parental care strategies are expected to be affected by environmental drivers. However, it remains relatively poorly explored how life history variation interacts with environmental variability to affect plasticity in parental care. Understanding these environmental effects often necessitates understanding how individual responses (in providing parental care) vary amongst different environments. These sorts of 'reaction norms' provide a way in which to quantify shifts in parental care behaviours (*e.g.*, care *vs.* no care) and/or parental care strategies (*e.g.*, biparental, maternal and/or paternal care).

Under some specific assumptions, reaction norms can allow gene × environment interactions associated with behaviours to be investigated. They are our main conceptual tools for understanding the ecological and evolutionary patterns associated with the plasticity of care.

For the framework constructed around life history variation and fitness, we can use analogous approaches to trait-based reaction norms. For instance, by deriving measures of parental care in different environments, we can use a 'behavioural reaction norm' approach (Royle et al. 2014) to disentangle the consequences of parental care strategies in different environments (Figure 6.3).

The slope of the behavioural response curve with respect to different environments (Figure 6.3) provides a way to quantify plasticity in care. By considering how different cost-benefit ratios vary across environments and/or how life history variations contribute to care – through fitness differences between care and no-care strategies – we can quantify how plasticity in parental care and/or parental care strategies might evolve.

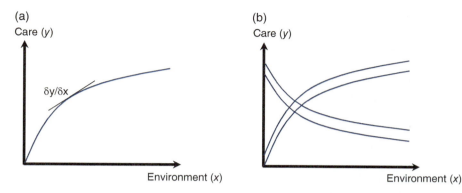

Figure 6.3 Schematic illustration of reaction norms for parental care based on behavioural traits (*Source:* adapted from Royle et al. 2014). (a) Variation in care in response to varying environments, where the slope of the relationship (at a point) provides a measure of the plasticity in care. From our framework, values for the care trait could be determined from fitness differences between care and no-care strategies. (b) Quantifying different trade-offs in traits, life histories and/or different parental care strategies (*e.g.*, paternal, maternal, biparental) (different lines) allows variation in plasticity (variation in slopes at a given level of environment (*x*) and care (*y*)) to be quantified.

6.8.1 Abiotic Environment

Parental care is expected to evolve to mitigate environmental unpredictability (Wilson 2000). We have shown that parental care is favoured when it increases egg survival (implicitly in harsh environments) (Klug and Bonsall 2010). However, the evolution of plasticity of care in abiotic environments has a focus on performance across the different environmentally generated phenotypes.

Bet-hedging is a stochastic response to this sort of environmental variability. Through the stochastic variation in the production of phenotypes, parents that spread reproductive events across time and space could maximise their lifetime reproductive success through differential offspring survival in different environments. Bet-hedging strategies introduce plasticity concepts into Shine's (1978) safe-stage hypothesis and into our ideas about developmental drivers and parental care (Klug et al. 2013b), allowing for a broader suite of mechanisms to drive the evolution of parental care. How life history traits and trade-offs interact with unpredictable environments to influence the evolution of plasticity in parental care is yet to be fully understood.

Recently, Westrick et al. (2023) introduced a more integrative framework for understanding plasticity and the evolution of parental care. They argue that plasticity, particularly in parental care, cannot be fully resolved without understanding plasticity within and between individuals. They further argue that it is essential to understand how mechanisms associated with individual variation drive behavioural evolution. Behaviours are the products of trade-offs in life histories, and, as noted above, this sort of variation amongst individuals in different environments is realized through reaction norms. However, plasticity within individuals allows adjustment in behaviours to local temporal or spatial ecological conditions (Westrick et al. 2023), so simple correlative approaches might be limited in unravelling these amongst and between individual mechanisms driving the evolution of plasticity in parental care.

6.8.2 Social Environment

The plasticity of care in social environments is a phenomenon associated with the evolution of different types of caring strategies. The social environment is the most likely to be of importance in offspring provisioning. A mate that does (or does not) bring resources affects decisions on whether a parent forages for food or defends offspring. The interaction between offspring and parents is most likely to influence the evolution and co-evolution of parental care strategies (Royle et al. 2014).

Focusing on burying beetles, Moss and Moore (2021) investigated how benign and harsh environments influenced the provision of offspring care by females and males. Even in harsh environments, sexual conflicts between parents over the provision of care persisted, with socially plastic responses. Females provided equivalent care, irrespective of the environmental conditions. However, males deserted earlier and provisioned offspring less in harsh environments (Moss and Moore 2021).

Beyond the plasticity of biparental care, equally important in the social context of plasticity in parental care is the provision by helpers that are usually individuals that have forgone reproduction or are post-reproduction. Understanding the provision of helping necessitates understanding patterns of relatedness between parents, helpers and offspring and the demographic structuring of breeding populations. Recently, Roper et al. (2023) show that inclusive fitness measures for selection predict that age-specific patterns of survival may not be constant and can be greater than zero in post-reproductive age classes (*e.g.*, grandmothers survive longer than expected; see also Chapter 11, this volume). The fact that patterns of relatedness and age structure influence life histories has important implications for the evolution of plasticity in parental care. For instance, plastic responses to unpredictable environments might decline with age if age-dependent survival declines monotonically. Houslay et al. (2020) have shown, in burying beetle systems, that younger mothers were more likely to show plasticity in levels of parental care. However, age-dependent survival patterns are complex (Roper et al. 2023) and plastic responses in care might be favoured in older age classes.

Obviously, there is much still to be understood about plasticity in parental care. For instance, formulating which care-related life history traits will be most variable in unpredictable environments and how these might influence the costs and benefits of parental care requires much rigorous theoretical work. Furthermore, a more robust understanding of the interaction between life histories and environmental effects requires further theoretical and empirical work.

6.9 Final Conclusions and Future Directions

The link between life history and parental care is complex and mediated by a range of factors and selective pressures (Figure 6.1). Basic life history will influence the origin and maintenance of parental care, and life history will influence and be influenced by differences between males and females, stochasticity and plasticity. We would expect basic life history and parental care to co-evolve.

In the future, research on this topic would be enhanced by paying greater attention to the following questions: (1) Stochasticity and care: Do stochastic environments make the sequence of parental care between the sexes more, or less, labile? Are males more likely to reduce care in stochastic environments compared to females (or *vice versa*)? (2) Environmental variability and care: How is parental care for offspring in different life stages influenced by environmental variability? Do parents really value all offspring equally in variable *vs.* stable environments? How can plasticity mediate the maintenance of care in a changing environment? Are patterns of biparental care more likely to be disrupted as environments fluctuate? (3) Climate change and care: Can parental care mediate the effects of climate change? Or, on the flip side, will parental care become maladaptive in some species as phenologies shift in response to climate change? (4) Phylogenetic relationships between care and life history: Based on phylogenetic analyses, what relationships exist between the origin and the loss of parental care in relation to basic life history across various mating systems?

Acknowledgements

This material is based upon work supported by the National Science Foundation under Grant no. DEB 1552721 (to H.K.).

References

Alonso-Alvarez, C., and Velando, A. (2012). Benefits and costs of parental care. In: *The evolution of parental care* (eds. Royle N.J., Smiseth P.T., and Kölliker, M.), 40–61. Oxford University Press.

Alonzo, S.H. (2012). Sexual selection favours male parental care, when females can choose. *Proceedings of the Royal Society B: Biological Sciences* 279: 1784–1790.

Azad, T., Alonzo, S.H., Bonsall, M.B., and Klug, H. (2022). Life history, mating dynamics and the origin of parental care. *Journal of Evolutionary Biology* 35: 379–390.

Badyaev, A.V., and Ghalambor, C.K. (2001). Evolution of life histories along elevational gradients: trade-off between parental care and fecundity. *Ecology* 82: 2948–2960.

Balshine, S. (2012). Patterns of parental care in vertebrates. In: *The evolution of parental care* (eds. Royle N.J., Smiseth P.T., and Kölliker, M.), 62–80. Oxford University Press.

Bateman, A.J. (1948). Intrasexual selection in *Drosophila*. *Heredity* 2: 349–368.

Benun Sutton, F., and Wilson, A.B. (2019). Where are all the moms? External fertilization predicts the rise of male parental care in bony fishes. *Evolution* 73: 2451–2460.

Bonsall, M.B., and Klug, H. (2011). The evolution of parental care in stochastic environments. *Journal of Evolutionary Biology* 24: 645–655.

Both, C., and Visser, M.E. (2005). The effect of climate change on the correlation between avian life-history traits. *Global Change Biology* 11: 1606–1613.

Buckley, L.B. (2022). Temperature-sensitive development shapes insect phenological responses to climate change. *Current Opinion in Insect Science* 52: 100897.

Carlisle, T.R. (1982). Brood success in variable environments: implications for parental care allocation. *Animal Behaviour* 30: 824–836.

Clutton-Brock, T.H. (1991). *The evolution of parental care*. Princeton University Press.

Cook, M.I., and Monaghan, P. (2004). Sex differences in embryo development periods and effects on avian hatching patterns. *Behavioral Ecology* 15: 205–209.

Dziminski, M.A., Vercoe, P.E., and Roberts, J.D. (2009). Variable offspring provisioning and fitness: a direct test in the field. *Functional Ecology* 23: 164–171.

Forrest, J.R. (2016). Complex responses of insect phenology to climate change. *Current Opinion in Insect Science* 17: 49–54.

Fromhage, L., and Jennions, M.D. (2016). Coevolution of parental investment and sexually selected traits drives sex-role divergence. *Nature Communications* 7: 12517.

Ghalambor, C.K., and Martin, T.E. (2000). Parental investment strategies in two species of nuthatch vary with stage-specific predation risk and reproductive effort. *Animal Behaviour* 60: 263–267.

Gilbert, J.D., and Manica, A. (2010). Parental care trade-offs and life-history relationships in insects. *The American Naturalist* 176: 212–226.

Godfray, H.C.J. (1995a). Signaling of need between parents and young: parent-offspring conflict and sibling rivalry. *The American Naturalist* 146: 1–24.

Godfray, H.C.J. (1995b). Evolutionary theory of parent-offspring conflict. *Nature* 376: 133–138.

Godfray, H.C.J., and Parker, G.A. (1991). Clutch size, fecundity and parent-offspring conflict. *Philosophical Transactions of the Royal Society, B: Biological Sciences* 332: 67–79.

Goodwin, N.B., Balshine-Earn, S., and Reynolds, J.D. (1998). Evolutionary transitions in parental care in cichlid fish. *Proceedings of the Royal Society B: Biological Sciences* 265: 2265–2272.

Hastings, A., and Caswell, H. (1979). Role of environmental variability in the evolution of life history strategies. *Proceedings of the National Academy of Sciences of the United States of America* 76: 4700–4703.

Head, M.L., Hinde, C.A., Moore, A.J., and Royle, N.J. (2014). Correlated evolution in parental care in females but not males in response to selection on paternity assurance behaviour. *Ecology Letters* 17: 803–810.

Houslay, T.M., Kitchener, P.A., and Royle, N.J. (2020). Are older parents less flexible? Testing age-dependent plasticity in *Nicrophorus vespilloides* burying beetles. *Animal Behaviour* 162: 79–86.

Iglesias-Rios, R., Lobón-Cerviá, J., do Amaral, C.R.L., Garber, R., and Mazzoni, R. (2022). Egg size is a good predictor of parental care behaviour among bony fishes. *Ecology of Freshwater Fish* 31: 492–498.

Ivory, S.J., Blome, M.W., King, J.W., McGlue, M.M., Cole, J.E., and Cohen, A.S. (2016). Environmental change explains cichlid adaptive radiation at Lake Malawi over the past 1.2 million years. *Proceedings of the National Academy of Sciences of the United States of America* 113: 11895–11900.

Jennions, M., and Fromhage, L. (2016). Not all sex ratios are equal: the fisher condition, parental care and sexual selection. *Philosophical Transactions of the Royal Society, B: Biological Sciences* 372: 20160312.

Kasper, C., Vierbuchen, M., Ernst, U., Fischer, S., Radersma, R., Raulo, A., Cunha-Saraiva, F., Wu, M., Mobley, K.B., and Taborsky, B. (2017). Genetics and developmental biology of cooperation. *Molecular Ecology* 26: 4364–4377.

Klug, H., Alonzo, S.H., and Bonsall, M.B. (2012). Theoretical foundations of parental. In: *The evolution of parental care* (eds. Royle N.J., Smiseth P.T., and Kölliker, M.), 40–61. Oxford University Press.

Klug, H., and Bonsall, M.B. (2010). Life history and the evolution of parental care. *Evolution* 64: 823–835.

Klug, H., and Bonsall, M.B. (2014). What are the benefits of parental care? The importance of parental effects on developmental rate. *Ecology and Evolution* 4: 2330–2351.

Klug, H., Bonsall, M.B., and Alonzo, S.H. (2013a). The origin of parental care in relation to male and female life history. *Ecology and Evolution* 3: 779–791.

Klug, H., Bonsall, M.B., and Alonzo, S.H. (2013b). Sex differences in life history drive evolutionary transitions among maternal, paternal, and bi-parental care. *Ecology and Evolution* 3: 792–806.

Klug, H., Chin, A., and St Mary, C.M. (2005). The net effects of guarding on egg survivorship in the flagfish, *Jordanella floridae*. *Animal Behaviour* 69: 661–668.

Klug, H., Langley, C., and Reyes, E. (2022). Cascading effects of pre-adult survival on sexual selection. *Royal Society Open Science* 9: 211973.

Kokko, H., and Jennions, M.D. (2008). Parental investment, sexual selection and sex ratios. *Journal of Evolutionary Biology* 21: 919–948.

Kokko, H., and Jennions, M.D. (2012). Sex differences in parental care. In: *The evolution of parental care* (eds. Royle N.J., Smiseth P.T., and Kölliker, M.), 40–61. Oxford University Press.

Kölliker, M., Royle, N.J., Smiseth, P.T., and Kölliker, M. (2012). Parent–offspring co-adaptation. In: *The evolution of parental care* (eds. Royle, N.J., Smiseth, P.T., and Kölliker, M.), 40–61. Oxford University Press.

Kolm, N., and Ahnesjö, I. (2005). Do egg size and parental care coevolve in fishes? *Journal of Fish Biology* 66: 1499–1515.

Lehtinen, R.M., Green, S.E., and Pringle, J.L. (2014). Impacts of paternal care and seasonal change on offspring survival: a multiseason experimental study of a Caribbean frog. *Ethology* 120: 400–409.

Lemoine, N.P. (2021). Phenology dictates the impact of climate change on geographic distributions of six co-occurring North American grasshoppers. *Ecology and Evolution* 11: 18575–18590.

Lewontin, R.C., and Cohen, D. (1969). On population growth in a randomly varying environment. *Proceedings of the National Academy of Sciences of the United States of America* 62: 1056–1060.

Li, D. (2002). Hatching responses of subsocial spitting spiders to predation risk. *Proceedings of the Royal Society B: Biological Sciences* 269: 2155–2161.

Lindström, K., St. Mary, C.M., and Pampoulie, C. (2006). Sexual selection for male parental care in the sand goby, *Pomatoschistus minutus*. *Behavioral Ecology and Sociobiology* 60: 46–51.

Low, B.S. (1978). Environmental uncertainty and the parental strategies of marsupials and placentals. *The American Naturalist* 112: 197–213.

Low, M., Makan, T., and Castro, I. (2012). Food availability and offspring demand influence sex-specific patterns and repeatability of parental provisioning. *Behavioral Ecology* 23: 25–34.

Mank, J.E., Promislow, D.E.L., and Avise, J.C. (2005). Phylogenetic perspectives in the evolution of parental care in ray-finned fishes. *Evolution* 59: 1570–1578.

McGinley, M.A., Temme, D.H., and Geber, M.A. (1987). Parental investment in offspring in variable environments: theoretical and empirical considerations. *The American Naturalist* 130: 370–398.

Méndez, V., Alves, J.A., Gill, J.A., and Gunnarsson, T.G. (2018). Patterns and processes in shorebird survival rates: a global review. *Ibis* 160: 723–741.

Monteith, K.M., Andrews, C., and Smiseth, P.T. (2012). Post-hatching parental care masks the effects of egg size on offspring fitness: a removal experiment on burying beetles. *Journal of Evolutionary Biology* 25: 1815–1822.

Moore, A.J., Brodie, E.D. III, and Wolf, J.B. (1997). Interacting phenotypes and the evolutionary process: I. Direct and indirect genetic effects of social interactions. *Evolution* 51: 1352–1362.

Moss, J.B., and Moore, A.J. (2021). Constrained flexibility of parental cooperation limits adaptive responses to harsh conditions. *Evolution* 75: 1835–1849.

Nussbaum, R.A., and Schultz, D.L. (1989). Coevolution of parental care and egg size. *The American Naturalist* 133: 591–603.

Parker, G., and Macnair, M. (1978). Models of parent-offspring conflict. I. Monogamy. *Animal Behaviour* 26: 97–110.

Pörtner, H.-O., Roberts, D.C., Tignor, M.M.B., Poloczanska, E., Mintenbeck, K., Alegría, A., Craig, M., Langsdorf, S., Löschke, S., Möller, V., Okem, A., and Rama, B. (2022). *Climate change 2022: impacts, adaptation and vulnerability – contribution of working group II to the sixth assessment report of the intergovernmental panel on climate change*. Cambridge University Press.

Potticary, A.L., Otto, H.W., McHugh, J.V., and Moore, A.J. (2023). Spatiotemporal variation in the competitive environment, with implications for how climate change may affect a species with parental care. *Ecology and Evolution* 13: e9972.

Remeš, V., and Martin, T.E. (2002). Environmental influences on the evolution of growth and developmental rates in passerines. *Evolution* 56: 2505–2518.

Reyes, E., Thrasher, P., Bonsall, M.B., and Klug, H. (2016). Population-level density dependence influences the origin and maintenance of parental care. *PLoS ONE* 11(4): e0153839.

Reynolds, J.D., Goodwin, N.B., and Freckleton, R.P. (2002). Evolutionary transitions in parental care and live bearing in vertebrates. *Philosophical Transactions of the Royal Society, B: Biological Sciences* 357: 269–281.

Reynolds, J.D., and Székely, T. (1997). The evolution of parental care in shorebirds: life histories, ecology, and sexual selection. *Behavioral Ecology* 8: 126–134.

Roper, M., Green, J.P., Salguero-Gómez, R., and Bonsall, M.B. (2023). Inclusive fitness forces of selection in an age-structured population. *Communications Biology* 6: 909.

Royle, N.J., Alonzo, S.H., and Moore, A.J. (2016). Co-evolution, conflict and complexity: what have we learned about the evolution of parental care behaviours? *Current Opinion in Behavioral Sciences* 12: 30–36.

Royle, N.J., Russell, A.F., and Wilson, A.J. (2014). The evolution of flexible parenting. *Science* 345: 776–781.

Royle, N.J., Smiseth, P.T., and Kölliker, M. (2012). *The Evolution of parental care*. Oxford University Press.

Rubenstein, D.R. (2011). Spatiotemporal environmental variation, risk aversion, and the evolution of cooperative breeding as a bet-hedging strategy. *Proceedings of the National Academy of Sciences of the United States of America* 108: 10816–10822.

Shine, R. (1978). Propagule size and parental care: the "safe harbor" hypothesis. *Journal of Theoretical Biology* 75: 417–424.

Shine, R. (1989). Alternative models for the evolution of offspring size. *The American Naturalist* 134: 311–317.

Smiseth, P.T., Darwell, C.T., and Moore, A.J. (2003). Partial begging: an empirical model for the early evolution of offspring signalling. *Proceedings of the Royal Society B: Biological Sciences* 270: 1773–1777.

Smiseth, P.T., Kölliker, M., and Royle, N.J. (2012). What is parental care? In: *The evolution of parental care* (eds. Royle, N.J., Smiseth, P.T., and Kölliker, M.), 40–61. Oxford University Press.

Smith, C.C., and Fretwell, S.D. (1974). The optimal balance between size and number of offspring. *The American Naturalist* 108: 499–506.

Stearns, S.C. (1976). Life-history tactics: a review of the ideas. *The Quarterly Review of Biology* 51: 3–47.

Stearns, S.C. (1982). The role of development in the evolution of life histories. In: Evolution and development: report of the Dahlem workshop on evolution and development Berlin 1981, May 10–15, pp. 237–258. Springer.

Stearns, S.C. (1992). *The evolution of life histories*. Oxford University Press.

Stiver, K.A., and Alonzo, S.H. (2009). Parental and mating effort: is there necessarily a trade-off? *Ethology* 115: 1101–1126.

Stockley, P., and Hobson, L. (2016). Paternal care and litter size coevolution in mammals. *Proceedings of the Royal Society B: Biological Sciences* 283: 20160140.

Summers, K., McKeon, C.S., and Heying, H. (2006). The evolution of parental care and egg size: a comparative analysis in frogs. *Proceedings of the Royal Society B: Biological Sciences* 273: 687–692.

Székely, T., and Reynolds, J.D. (1995). Evolutionary transitions in parental care in shorebirds. *Proceedings of the Royal Society B: Biological Sciences* 262: 57–64.

Tallamy, D.W., and Brown, W.P. (1999). Semelparity and the evolution of maternal care in insects. *Animal Behaviour* 57: 727–730.

Trivers, R.L. (1974). Parent-offspring conflict. *Integrative and Comparative Biology* 14: 249–264.

Vincze, O., Kosztolányi, A., Barta, Z., Küpper, C., Alrashidi, M., Amat, J.A., Argüelles Ticó, A., Burns, F., Cavitt, J., and Conway, W.C. (2017). Parental cooperation in a changing climate: fluctuating environments predict shifts in care division. *Global Ecology and Biogeography* 26: 347–358.

Visser, M.E., and Both, C. (2005). Shifts in phenology due to global climate change: the need for a yardstick. *Proceedings of the Royal Society B: Biological Sciences* 272: 2561–2569.

Warkentin, K.M. (2011). Environmentally cued hatching across taxa: embryos respond to risk and opportunity. *Integrative and Comparative Biology* 51: 14–25.

Webb, J., Szekely, T., Houston, A., and McNamara, J. (2002). A theoretical analysis of the energetic costs and consequences of parental care decisions. *Philosophical Transactions of the Royal Society, B: Biological Sciences* 357: 331–340.

West, H.E., and Capellini, I. (2016). Male care and life history traits in mammals. *Nature Communications* 7: 11854.

Westrick, S.E., Moss, J.B., and Fischer, E.K. (2023). Who cares? An integrative approach to understanding the evolution of behavioural plasticity in parental care. *Animal Behaviour* 200: 225–236.

Wilson, E.O. (2000). *Sociobiology: the new synthesis*. Harvard University Press.

Winemiller, K., and Rose, K. (1992). Patterns of life-history diversification in North American fishes- implications for population regulation. *Canadian Journal of Fisheries and Aquatic Sciences* 49: 2196–2218.

Wong, J.W., Meunier, J., and Kolliker, M. (2013). The evolution of parental care in insects: the roles of ecology, life history and the social environment. *Ecological Entomology* 38: 123–137.

7

Sex Allocation

Jun Abe[1] and Stuart A. West[2]

[1] *Faculty of Science, Kanagawa University, Yokohama, Japan*
[2] *Department of Biology, Oxford University, Oxford, UK*

7.1 Introduction

All sexually reproducing organisms are confronted with the decision of how to allocate resources to male *vs*. female production. This decision, termed sex allocation, is one of the most crucial life history traits that determine fitness (Charnov 1982, Frank 1998, Hardy 2002, West 2009). In species with separate sexes, individuals must decide what proportion of sons and daughters are produced. In species in which the same individual has the functions of both sexes, they are faced with the decision of reproductive efforts into each sex function in simultaneous hermaphroditic species and the order and timing of the sex change in sequential hermaphroditic species (sex changers).

Sex allocation is a social behaviour because it has fitness consequences for both the individuals that act on the behaviour and other individuals in the population (Hamilton 1967, 1971, Frank 1998, West et al. 2007, West 2009). Sex allocation thus provides a model trait for understanding how natural selection acts on social behaviours and represents one of the most productive areas in the field of social behaviour (Charnov 1982, Frank 1998, Hardy 2002, West 2009). Many theoretical studies predict variations in sex allocation dependent on different environmental factors, and many empirical studies support these theoretical predictions. More specific theoretical models have been developed and tested to match the biology of particular species. As well as explaining sex allocation, work in this area has also illuminated several general issues, such as the precision and the constraints of adaptation.

In this chapter, we first illustrate why equal investment in both sexes is often favoured and represents the 'null model'. Then we illustrate that when the assumptions for equal investment or outcomes are violated, natural selection can favour biased sex ratios.

7.2 Fisher's Theory

Why do most sexually reproducing organisms show unbiased (50% male) sex ratios (hereafter defined as proportion of males)? Düsing (1884) and Fisher (1930) provided an answer (Gardner 2023). Supposing that males and females are equally costly to produce, the most adaptive sex ratio depends on relative reproductive success obtained by a male and a female. Natural selection favours producing the sex that returns more reproductive success. While females produce offspring, males achieve their reproductive success by mating with females. Male reproductive success is influenced by the population sex ratio. If the population sex ratio is male-biased, individual males compete with a larger number of males for mating opportunities with a smaller number of females. Each male is expected to, on average, mate with less than one female, and so the average reproductive success of a male will be lower than that of a female. This means that any genes producing a higher proportion of females will spread in the population, and so the population sex ratio would become less male-biased. Conversely, in female-biased population, males will mate with an average of more than one female. Genes producing more males will be favoured by selection, and population sex ratio will become less female-biased. Therefore, the evolution of

Life History Evolution: Traits, Interactions, and Applications, First Edition. Edited by Michal Segoli and Eric Wajnberg.
© 2025 John Wiley & Sons Ltd. Published 2025 by John Wiley & Sons Ltd.

Table 7.1 The expected fitness consequence of a mother producing different offspring sex ratios under different population sex ratios.

Population sex ratio (% male)	25%	50%	75%
Fitness consequence of a mother producing 25% sons	$3 \times 4 + 1 \times \frac{0.75}{0.25} \times 4$ $= 24$	$3 \times 4 + 1 \times \frac{0.5}{0.5} \times 4$ $= 16$	$3 \times 4 + 1 \times \frac{0.25}{0.75} \times 4$ $= 13.33$
Fitness consequence of a mother producing 50% sons	$2 \times 4 + 2 \times \frac{0.75}{0.25} \times 4$ $= 32$	$2 \times 4 + 2 \times \frac{0.5}{0.5} \times 4$ $= 16$	$2 \times 4 + 2 \times \frac{0.25}{0.75} \times 4$ $= 10.67$
Fitness consequence of a mother producing 75% sons	$1 \times 4 + 3 \times \frac{0.75}{0.25} \times 4$ $= 40$	$1 \times 4 + 3 \times \frac{0.5}{0.5} \times 4$ $= 16$	$1 \times 4 + 3 \times \frac{0.25}{0.75} \times 4$ $= 8$

Each female produces four offspring. The fitness consequence of a mother is represented by the expected number of grand-offspring, which is the average reproductive success of daughters plus that of sons. The reproductive success of daughters equals the number of daughters multiplied by offspring number per female. The reproductive success of sons equals the number of sons multiplied by the expected mean mating frequency of a male, which is the proportion of females divided by the proportion of males in the population times offspring number per female. In a female-biased population (25% males), the reproductive success of a daughter is expected to be smaller than that of a son ($4 < 0.75/0.25 \times 4$), and a mother producing more sons (75% sons) will be favoured. In a male-biased population (75% males), the reproductive success of a daughter is expected to be greater than that of a son ($4 > 0.25/0.75 \times 4$), and a mother producing more daughters (25% sons) will be favoured.

sex ratio is frequency-dependent, favouring the rarer sex (Gardner 2023). Consequently, the population sex ratio will approach equality, at which point the average reproductive success of a male and a female is identical (Table 7.1).

The overall productivity in the population, defined as the total reproductive success of the population, is higher in a more female-biased population (Table 7.2). If each individual in the population produced a more female-biased offspring sex ratio, then the total reproductive success would be higher, and the population sex ratio could grow larger. However, natural selection acts on fitness of individuals, not the whole population. The production of a cooperative female-biased sex ratio would usually be outcompeted by selfish sex ratios, and so a female-biased sex ratio would not normally be evolutionarily stable. As shown above, in a female-biased population, individuals producing more sons gain increased reproductive success, and population sex ratios shift towards 50%, which is the optima for individuals. This context of sex ratio evolution is therefore analogous to other social dilemmas such as the 'tragedy of the commons' in which the selfish interests of individuals conflict with and win over the interests of the group (Hamilton 1967, 1971, Frank 1998, West et al. 2002, West 2009).

The adaptive sex allocation is also influenced by the relative costs of producing males and females. Natural selection favours producing the sex that achieves higher reproductive success relative to costs imposed to produce the sex. When the frequency-dependent evolution reaches the equilibrium, at which a male and a female have the same return on reproductive success per unit of cost, total investment of resources in males and females is predicted to be equal:

$$C_M N_M = C_F N_F \tag{7.1}$$

where C_M and C_F are average cost to produce a male and a female, respectively, and N_M and N_F are the number of males and females, respectively. This equation indicates that if males and females impose different costs to produce, sex ratio should be biased towards producing more individuals of the less costly sex.

Table 7.2 The whole reproductive success of a population under different population sex ratios, assuming that all individuals in the population produce the same offspring sex ratio.

Population sex ratio (% male)	25%	50%	75%
The whole reproductive success of a population	$0.75 N_P \times 4 + 0.25 N_P \times \frac{0.75}{0.25} \times 4$ $= 6 N_P$	$0.5 N_P \times 4 + 0.5 N_P \times \frac{0.5}{0.5} \times 4$ $= 4 N_P$	$0.25 N_P \times 4 + 0.75 N_P \times \frac{0.25}{0.75} \times 4$ $= 2 N_P$

The population contains N_P individuals, and each female produces four offspring. The whole reproductive success of a population is represented here by the expected number of grand-offspring, which is the average reproductive success of all daughters plus that of all sons in the population, as shown in Table 7.1.

The equal investment in the two sexes is not affected by aspects of breeding system, such as the number of times that a female typically mates. As long as the reproductive success of each individual is determined randomly (with respect to offspring sex ratios), reproductive skew among individuals does not influence the average reproductive success of the sexes. For example, consider a polygynous species with 10% males monopolizing mating with females in the population (high skew). While 90% males cannot obtain any reproductive success, 10% males are expected to have 10 times the reproductive success of females. The average reproductive success has not changed, and Equation (7.1) still holds.

Analogously, differential mortality between the sexes is also not influential, provided that the mortality takes place after the period of parental investment, and it occurs randomly among individuals (Leigh 1970). The loss of reproductive success by the reduction in the individuals of the sex with higher mortality is exactly balanced by the increased reproductive success of the surviving individuals of the sex. However, if differential mortality takes place within the period of parental investment, it affects investment in the sexes and hence the optimal sex allocation (Kahn et al. 2015). If the individuals of either sex die during parental investment, investment in the individuals terminates at the point of death, which diminishes average investment in an individual of the sex. Equation (7.1) predicts that reduction in investment in one sex should be compensated by the increased number of individuals of this sex. Therefore, a larger proportion of the sex with higher mortality is predicted to be produced, while sex ratio will be reversed after the period of differential mortality.

7.3 Interaction with Relatives

Fisher's (1930) theory implicitly assumes that individuals disperse for a long distance, or at least away from their natal area, and related individuals do not interact more commonly than with other members of the population. However, if interactions with relatives, like competition or cooperation, influence the reproductive success of relatives, then selection can favour biased sex allocation to decrease competition (local resource competition (LRC) or local mate competition (LMC)) or increase cooperation (local resource enhancement (LRE)).

7.3.1 Local Resource Competition

When individuals of one sex are more likely to compete, this can favour sex allocation biased towards the other sex (Clark 1978, Taylor 1981). This is termed LRC. Clark (1978) found that male-biased sex ratios in the prosimian primate galago, *Otolemur crassicaudatus*, could be explained by a social structure where females remain in their natal sites contrary to males, and compete with their mothers and sisters for food resources. A comparative study across primate species showed that while sex ratios were unbiased in species where both sexes dispersed, they were biased towards dispersive sex in species where either females or males dispersed (Silk and Brown 2008). LRC has also been observed in mammals, birds and insects (West 2009).

7.3.2 Local Mate Competition

If mating takes place locally among offspring produced by one or a small number of females, selection favours a female-biased sex ratio (Hamilton 1967). This process is a special case of LRC and is termed LMC. While Fisher's (1930) theory assumes that males compete for mating opportunities with unrelated males after dispersal, the limited dispersal of males leads to LMC. In this situation, a female-biased sex ratio is favoured because it reduces competition among related males and increases the number of mates for the males (Taylor 1981). As predicted by LMC theory, female-biased sex ratios were confirmed in a wide range of organisms that mate locally, including protozoa, nematodes, insects, mites, vertebrates and plants (West et al. 2005, West 2009).

One of the major factors that determines the degree of LMC is the number of mothers reproducing in their natal patches. Suppose a population consisting of discrete patches, in which a variable number of mothers produce offspring, and the developed offspring mate within the patch, before inseminated females disperse for new patches. In this situation, a more female-biased sex ratio is predicted with decreasing number of mothers in a patch because it avoids competition between related males. Conversely, a higher sex ratio is predicted with increasing number of mothers because it is favoured to compete with unrelated males. The evolutionarily stable sex ratio (proportion males) is predicted to be $(N-1)/2N$ for diploid species, where N is the number of mothers in a patch. When only one mother reproduces in a patch ($N = 1$), the sex ratio is predicted to be zero, which is interpreted as the minimum proportion of sons that can inseminate all the daughters within each patch. This can represent a very small number of males since each of them can generally inseminate a large number of females. As the number of mothers increases (larger N), the sex ratio approaches asymptotically towards 50%, which is the evolutionarily stable sex ratio in a panmictic population predicted by Fisher's (1930) theory (Gardner 2023). For haplodiploid species, although the sex ratio is predicted to be a little more female-biased, the qualitative pattern is the same as that for diploid species (Figure 7.1; Frank 1985, Herre 1985).

LMC theory has gained considerable empirical support. First, comparing among species or populations, the sex ratio is more female-biased in species or populations that experience higher LMC (West 2009). Second, females facultatively adjust offspring sex ratio in response to the number of mothers in many species, including parasitoid wasps, fig wasps, mites and protozoa (West et al. 2005, Reece et al. 2008, West 2009). For example, a laboratory experiment showed that the gregarious parasitoid wasp *Nasonia vitripennis* (Figure 7.1b), in which multiple offspring develop from a single host fly pupa and the offspring mate within their natal patches, increases sex ratio with the number of females laying eggs in the same host patches in accordance with the prediction of the LMC model (Figure 7.1a; Werren 1983). The same pattern of the sex ratio adjustment was confirmed in this species in the field, where the number of females laying eggs on a patch was determined using molecular markers (Burton-Chellew et al. 2008).

Studies of LMC have also allowed the precision of adaptation to be studied. First, the extent to which organisms behave perfectly can depend on the underlying mechanism to process and respond to information about the environment. Shuker and West (2004) examined how females assess the number of females laying eggs in the same patches in *N. vitripennis*. They disentangled direct information about the number of other females from indirect information about eggs laid by the females, using females who could not actually lay eggs, as their ovipositors were cut off. Their results showed that females primarily use the indirect cue from the presence of eggs, rather than the direct cue from the presence of other females. This suggests that there could be discrepancies between observed sex ratio and the predictions of most theoretical models, which assume that females directly assess the number of other females. More generally, this result emphasizes the importance of knowing the underlying mechanisms to fully understand the adaptation of the behaviour. The adaptation of sex ratio adjustment can also be influenced by underlying genetic or even genomic control (Wajnberg 1993, Cook et al. 2018, Pannebakker et al. 2020).

Second, another possible factor that can influence the precision of adaptive behaviour is selection pressure. Adaptive behaviour is more likely to evolve in situations that organisms experience more frequently in nature. Herre (1987) confirmed this by comparing the sex ratios of different pollinating fig wasp species, in which a variable number of females enter a fig fruit to lay eggs and develop offspring mates within the fig fruit. The average sex ratio observed in each species was close to the theoretical prediction for the number of ovipositing females that the species encountered most frequently in the field. Furthermore, greater sex ratio adjustment in response to female number was observed in species where the number of

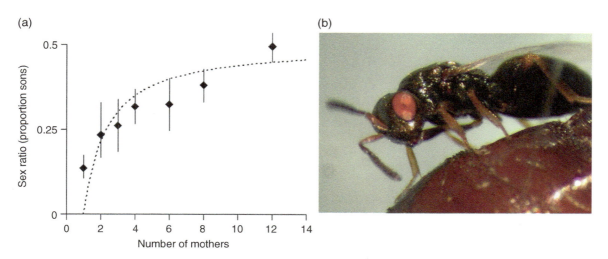

Figure 7.1 Offspring sex ratio (proportion of sons) depending on the number of mothers laying eggs in a patch in the parasitoid wasp *Nasonia vitripennis*. (a) The average sex ratio observed (diamond) increased with the number of mothers in accordance with the prediction of the LMC model for haplodiploid species (dotted line). Data are represented with means ± 1.96 SE. (*Source:* Werren (1983) / Oxford University Press). (b) Photo of a red-eyed mutant female by David Shuker and Stuart West.

females in a fig fruit was more variable. Finally, a smaller variance in sex ratio was observed in situations encountered more frequently, as a stronger selection causes more precise sex ratio adjustment (West and Herre 1998b).

Macke et al. (2011) provided the first experimental evolution evidence that sex ratio evolves in response to environmental conditions. They examined the spider mite, *Tetranychus urticae*. Individuals of this spider mite develop and mate within patches on leaves, corresponding to the assumption of the LMC model. By controlling the number of mothers in each generation, three types of replicated lines were created. In LMC+ and LMC− lines, one or 10 females were placed together, respectively, and allowed to develop and mate on a leaf patch, and mothers for the next generation were randomly selected after all the individuals developed on all the patches were mixed. In panmixia lines, 100 females were allowed to produce offspring on a large leaf patch, and mothers for the next generation were randomly selected from individuals developed in the patch.

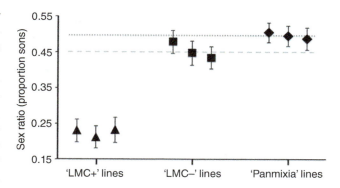

Figure 7.2 Offspring sex ratio after the experimental evolution in the mite *Tetranychus urticae*. Symbols represent mean sex ratio with 95% confidence intervals of each experimental line. The dotted and dashed lines represent the predicted sex ratio for panmixia and LMC− lines, respectively. (*Source:* Macke et al. (2011) With permission of American Association for the Advancement of Science - AAAS).

After up to 54 generations, observed sex ratios were both qualitatively and quantitatively consistent with sex ratios predicted by the LMC model for the number of mothers in each type of line (Figure 7.2). To investigate the ability of females to facultatively adjust their offspring sex ratio after the process of experimental evolution, the sex ratio of the three different selection regimes was also examined under both high and low LMC conditions (a single female or 40 females ovipositing in a patch, respectively). Females from LMC− and panmixia lines changed their sex ratio between low and high LMC conditions, whereas females from LMC+ lines produced a similar female-biased sex ratio under both conditions. This suggests that the ability of facultative adjustment was also influenced by the selection regime, with females who evolved under strong LMC conditions losing such ability.

The degree of LMC is also influenced by other factors, including variable clutch size among ovipositing females, partial dispersal of males and asymmetrical competition between males (Suzuki and Iwasa 1980, Werren 1980, Frank 1986, West and Herre 1998a, Shuker et al. 2005). The role of these different factors has been studied both theoretically and empirically (West 2009).

Another possible factor is limited dispersal by females. If dispersal by females is limited, relatedness between ovipositing females increases, and more female-biased sex ratio is likely to be favoured to decrease competition between related sons and increase the number of potential mates for the sons. However, limited dispersal by females also increases competition for resources between the females (LRC; Bulmer 1986, Frank 1986, Taylor 1988). Consequently, the extent to which limited dispersal favours a more female-biased sex ratio can depend upon life history details (Gardner et al. 2009, Rodrigues and Gardner 2015, Iritani et al. 2021, Chokechaipaisarn and Gardner 2022). This influence of limited dispersal on sex ratio evolution is analogous to that of limited dispersal on the evolution of cooperation (West et al. 2002, Lehmann and Rousset 2010).

Abe et al. (2021) found that females of the parasitoid wasp *Melittobia australica* adjusted offspring sex ratio depending on their dispersal status (Figure 7.3c). Females parasitize the larvae and pupae of solitary wasps and bees, and their life history matches the assumptions of LMC models with males not dispersing from host nest cells where they emerged. The sex ratio behaviour of *Melittobia* species was one of the outstanding puzzles in the field of sex allocation because many laboratory experiments showed that the female-biased sex ratio (about 1–5% males) changes little with increasing the number of ovipositing females, in sharp contrast to the sex ratios of other organisms and predictions by LMC models (Figure 7.3a; Matthews et al. 2009, Abe et al. 2014). However, Abe et al. (2021) found that there were two types of sex ratio patterns in the field. In *M. australica*, as their hosts clump spatially in patches, females inseminated in their emerged host nest cells can find new hosts either within the same host patches without dispersal, or in new patches after dispersal. An analysis with molecular markers indicated that genetically related females that walked to a host in the same patch (without dispersal) produced constantly female-biased sex ratios, as observed in the laboratory, while unrelated females that dispersed from different patches increased their offspring sex ratio by increasing the number of ovipositing females, as predicted by LMC theory (Figure 7.3b). As a laboratory experiment showed that females cannot directly recognize the kin of other females to adjust sex ratio, females are likely to use dispersal status as an indirect cue of relatedness.

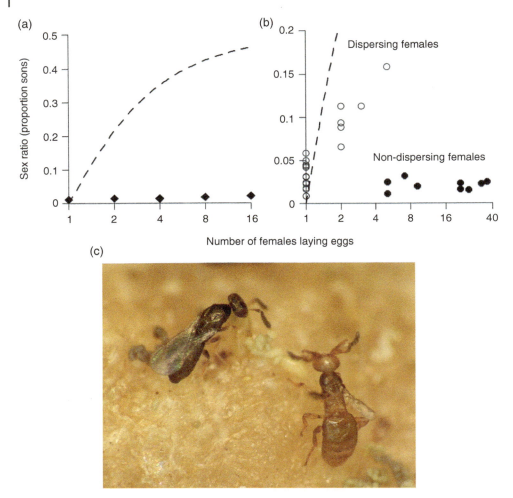

Figure 7.3 Offspring sex ratios in the parasitoid wasp *Melittobia australica*. (a) In laboratory experiments, females did not produce less biased sex ratios when there were more females laying eggs in a patch (solid diamonds). This contrasts with the prediction of local mate competition theory (dashed line). (b) In the field, dispersing females (open circles), but not non-dispersing females (solid circles), produced less biased sex ratios when there were more females laying eggs in a patch, which is qualitatively consistent with the prediction (dashed line). Note that the scales of y-axes are different between (a) and (b). (a) From (*Source:* Adapted from Abe et al. 2003.), and (b) from (*Source:* Adapted from Abe et al. 2021). (c) Photo of a black full-winged female (left) and an orange short-winged male (right) by Jun Abe.

7.3.3 Local Sperm Competition in Simultaneous Hermaphrodites

Simultaneous hermaphrodites are organisms whose individuals are capable of both male and female reproduction at the same time and occur in most plants, some invertebrates and some fish. Some species are able to self-fertilize (in a single individual), but others are not. The theory of LMC can also apply to these organisms (Charnov 1982). In this case, theory predicts investment ratio towards biased female function within the same individual. LMC occurs with simultaneous hermaphrodites when there is competition between related sperms, provided by the same sperm donor, for the fertilization of a given set of eggs in a sperm recipient (local sperm competition; Schärer 2009). This favours sex allocation towards female function in self-fertilizing species or in a small mating group, while a lower bias is favoured with larger mating group sizes (Charnov 1982, Schärer 2009, Schärer and Pen 2013).

The prediction of LMC theory for simultaneous hermaphrodites has been widely supported by empirical studies. Field studies found that male allocation increased with increasing population density in a barnacle and a fish species (Raimondi and Martin 1991, Hart et al. 2010). Self-fertilizing species or populations showed more female-biased sex allocation in plants and mussels (Brunet 1992, Johnston et al. 1998). Laboratory experiments in the flatworms showed a phenotypically plastic sex allocation in response to group size or the number of mating partners, in which individuals invested more towards male functions if they were kept in larger groups (Figure 7.4; Schärer 2009, Janicke et al. 2013; Ramm et al. 2019, Singh and Schärer 2022). The phenotypic plasticity of relative investments into male and female functions remained flexible even after maturation (Brauer et al. 2007). Such plasticity may be one of the selective advantages of simultaneous hermaphroditism over separate sexes when environmental factors in the habitats fluctuate frequently and unpredictably (Michiels 1998). Simultaneous hermaphroditism is also predicted to be favoured at low population densities, where individuals have difficulty finding mating partners, and this idea has been supported by a phylogenetic comparison (Charnov 1982, Puurtinen and Kaitala 2002, Eppley and Jesson 2008).

7.3.4 Local Resource Enhancement

Cooperative interactions between relatives by either sex can favour biased sex ratios towards the more cooperative sex (Taylor 1981). This is termed LRE. If offspring of one sex tend to remain in their natal group and help the reproduction of their parents, as occurs in many cooperative mammals and birds, selection can favour producing more of the sex that will help (Pen and Weissing 2000, Wild 2006).

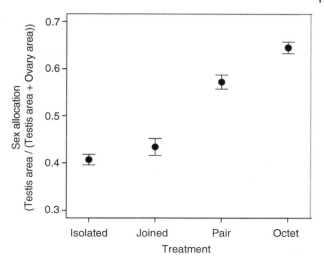

Figure 7.4 Sex allocation (relative investment in male function) increases with increasing group size in the simultaneous hermaphrodite flatworm *Macrostomum lignano*. Individuals were kept either alone (Isolated), except on the final day, when kept in pairs (Joined), in pairs (Pair) or in groups of eight (Octet). Means and SEs are represented. (*Source:* Adapted from Ramm et al. (2019)).

Komdeur et al. (1997) provided a striking example of the Seychelles warbler (*Acrocephalus sechellensis*) (Figure 7.5e). This species can experience both LRE and LRC, depending on environmental conditions. On high-quality territories, which contain more insects for food, producing daughters is advantageous for parents, because daughters are likely to remain to help their parents. In contrast, on low-quality territories, which have fewer insects, producing daughters is costly for parents because they will remain and increase competition for resources on the territories. As predicted, offspring sex ratios were female-biased on high-quality territories (LRE) and male-biased on low-quality territories (LRC; Figure 7.5a–c). In addition, the advantage of producing helpers also depends on the number of helpers that already exist in the territories – more than two helpers are costly even in high-quality territories. As predicted, parents on high-quality territories with no or one helper produced female-biased sex ratios, while parents with two helpers produced mostly males (Figure 7.5d). When one of two helpers was experimentally removed from high-quality territories, parents switched from all-male broods to female-biased broods. Finally, when the warblers were translocated to high-quality territories on a neighbouring island, where the species was not originally distributed, parents transferred from low-quality territories switched from producing male-biased to female-biased sex ratios, while parents transferred from high-quality territories kept producing female-biased sex ratios.

LRE can also explain other examples of biased sex ratios in birds, mammals and insects. One issue is that sex ratio adjustment appears to occur in some species but not in others. Griffin et al. (2005) examined this variation across different species and demonstrated that offspring sex ratios show larger adjustments in species where parents can obtain larger benefits from helpers. A meta-analysis across bird and mammal species found a highly significant correlation between the degree of sex ratio adjustment and the benefit of helpers (Griffin et al. 2005). However, more recent meta-analyses across bird species including subsequent studies found no correlation between sex ratio adjustment and the sex ratio of helpers, or between the production of helping sex and the number of current helpers (Khwaja et al. 2017), the latter of which contradicts the results of previous meta-analyses (West and Sheldon 2002, West et al. 2005). This suggests the possibility of confounding factors that could negate the benefit of helpers and the importance of evaluating both the benefit and cost of sex ratio adjustment.

LRE also occurs when the dispersal of either sex is limited and the individuals of that sex cooperate within the same generation (Iritani et al. 2021). This type of LRE is documented in a few bee species, in which the siblings of females cooperatively breed their offspring and produce female-biased offspring sex ratio (Schwarz 1988, Martins et al. 1999). Tang et al. (2014) found quasi-sociality in the parasitoid wasp *Sclerodermus harmandi*, in which multiple females facilitate the exploitation of hosts through cooperative host suppression and brood care. Contrary to the general pattern observed in many

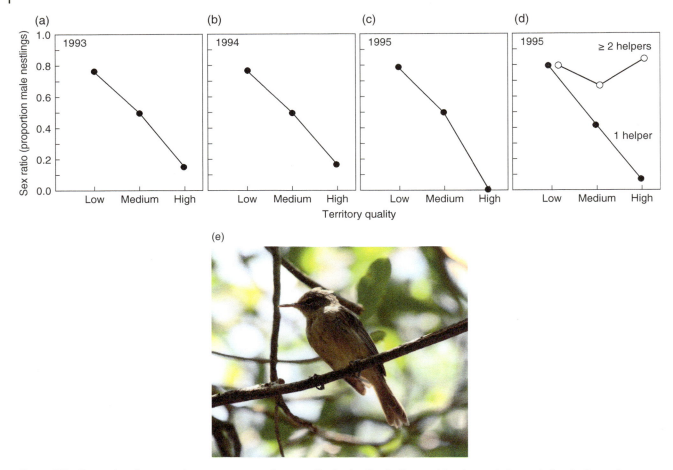

Figure 7.5 Sex ratio adjustment in response to territory quality in the Seychelles warbler *Acrocephalus sechellensis*. Sex ratios were male-biased in low-quality territories and female-biased in high-quality territories across different years (a–c). In high-quality territories, the sex ratio was male-biased with more than one helper (open circles), while it was female-biased with one helper (solid circles) in 1995 (d). (*Source:* Adapted from Komdeur et al. (1997)). (e) *Source:* Remi Jouan / Wikimedia / CC BY SA 3.0.

parasitoid wasps, a larger number of ovipositing females on a host produce a larger number of offspring per ovipositing female, specifically when they attack a relatively large host (Tang et al. 2014). While this parasitoid wasp is also subjected to LMC, the sex ratio was more female-biased than predicted solely by an LMC model. Instead, the observed pattern of sex ratio was fitted more to prediction by a model combining the effects of LMC and LRE (Iritani et al. 2021). However, as it was also insufficient to quantitatively explain the further female-biased sex ratio of *S. harmandi*, other effects such as infanticide and reproductive dominance among ovipositing females could play a role (Lehtonen et al. 2023).

7.4 Environmental Condition

Fisher's (1930) theory assumes that the average reproductive success of males and females does not vary with environmental conditions. However, if they differentially vary with environmental conditions, selection favours individuals to adjust sex allocation accordingly (Charnov 1982). If reproductive success increases with an environmental variable more rapidly for males than females, then individuals are predicted to switch from investing in females to males (Figure 7.6a). If female reproductive success increases more rapidly, the reverse is predicted (Figure 7.6b). This idea is termed the Trivers and Willard hypothesis or conditional sex allocation theory (Trivers and Willard 1973) and has been applied to variable situations including facultative sex ratio adjustment, environmental sex determination and sex change in sequential hermaphrodites.

Figure 7.6 Conceptual model for reproductive success between males and females. The reproductive success of males increases either more (a) or less (b) steeply than that of females. In species with separate sexes, selection favours the production of sons when the fitness of male is greater than that of female – this is above point τ in (a), but below point τ in (b). In sequential hermaphrodites (sex changers), individuals are favoured to be female below point τ and switch to male at the point τ in (a), and *vice versa* in (b).

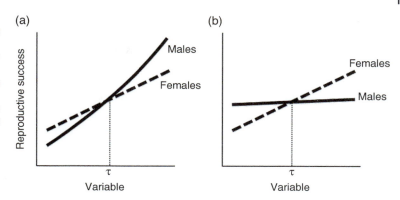

7.4.1 Facultative Sex Ratio Adjustment

When reproductive success obtained through male and female offspring differs and parents can predict the relative reproductive success based on environmental conditions, the parents are predicted to adjust the sex ratio of the offspring conditionally. The possible environmental conditions that affect reproductive success include maternal condition, mate attractiveness and resources for offspring growth (West 2009).

Conditional sex allocation theory was originally developed for mammals, specifically ungulates, where reproductive success of sons and daughters is influenced by maternal conditions. It envisages polygynous species, in which higher-conditioned males disproportionately acquire more mates and males have more variable reproductive success than females (Figure 7.6a, in which x-axis is maternal rank). Clutton-Brock et al. (1984) provided the first empirical support for this prediction with red deer *Cervus elaphus* (Figure 7.7b). In red deer herds, the rank of females, which correlates with their condition, had a variable effect on the quality of their offspring, with higher-ranking females producing heavier offspring. Clutton-Brock et al. (1984) showed that sons produced by higher-ranking mothers achieved a greater lifetime reproductive success, while the lifetime reproductive success of daughters was less dependent on maternal rank, as assumed by Trivers and Willard's hypothesis (Figure 7.7a). As predicted by the hypothesis, the authors also found that higher-ranking mothers produced higher proportion of sons, with 61%, 54% and 47% males produced by high-, medium- and low-ranking mothers, respectively (Clutton-Brock et al. 1984).

Since then, many empirical studies with ungulate species have examined the adjustment of offspring sex ratio in response to maternal condition. Several studies have provided positive relationships between maternal condition and offspring sex

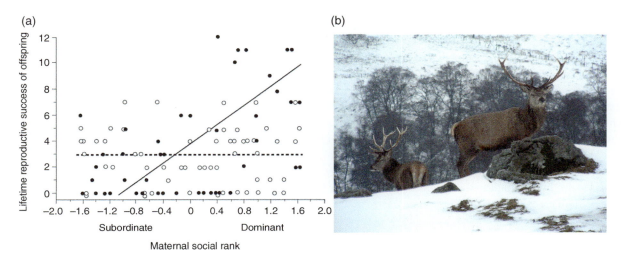

Figure 7.7 (a) The lifetime reproductive success of sons (solid circle) increases with maternal social rank, while that of daughters (open circle) does not in the red deer *Cervus elaphus*. The solid and dashed lines represent regression lines for sons and daughters, respectively. (*Source:* Adapted from Clutton-Brock et al. (1984)). (b) *Source:* Alan Stewart / Wikimedia / CC BY SA 2.0.

ratio, while others have reported no relationship or negative relationships. A meta-analysis across ungulate species found that there is a weak but significant effect of maternal condition on sex ratio (Sheldon and West 2004). The variable results and the overall weak relationship are likely to be explained by differences across species in the degree to which maternal condition affects the reproductive success of sons and daughters. The meta-analysis also showed that relationships between maternal condition and sex ratio were stronger in species with more sexual dimorphism, in which the strength of sexual selection is presumably stronger and male reproductive success is likely to be highly variable (Sheldon and West 2004). Alternatively, contrary to the assumption of the Trivers and Willard hypothesis that male reproductive success is influenced by maternal rank, in some species, maternal condition can also be transmitted to daughters. If this occurs, better-conditioned mothers would favour producing daughters, which shifts offspring sex ratio in the opposite direction to the original prediction (Figure 7.6). The overall sex ratio will depend on the balance between the two opposing effects (Wild and West 2007, West 2009).

Conditional sex allocation can also be favoured in response to mate attractiveness (Burley 1981). Females mating with more attractive males are predicted to produce more sons because the females are expected to gain more reproductive success from sons, which will inherent the attractiveness from their fathers, than daughters (Figure 7.6a, in which x-axis is mate attractiveness). This situation was formally investigated by theoretical models highlighting feedback between sex allocation and sexual selection (Fawcett et al. 2011, Booksmythe et al. 2013). In addition, several empirical studies have tested this prediction. For example, Sheldon et al. (1999) provided clear support for blue tits. The males of blue tits have ultraviolet plumage ornamentations on the top of their heads, which plays a role as a sexually selected trait to show their viability. Experimentally manipulating the ultraviolet reflectance with sunblock showed that females mated with males having brighter reflectance produced a higher proportion of sons (Sheldon et al. 1999). However, studies conducted later in the same species reported mixed evidence with inconsistent trends of sex ratio across years or across populations (Parker and Birkhead 2013).

In this area, multiple meta-analyses have been carried out, mainly focusing on birds. West and Sheldon (2002) included only the studies that had clear *a priori* predictions about the direction of effects and found a relatively strong relationship between mate attractiveness and sex ratio. However, subsequent meta-analyses including studies with broader criteria found that the mean effect size was very small but still significant in the predicted direction (Cassey et al. 2006, Booksmythe et al. 2017). While these analyses provide overall support for sex ratio adjustment on mate attractiveness, the average effect is subtle. One possible explanation is that selection pressure acting on sex ratio adjustment for mate attractiveness is weak, due to co-evolutionary feedback between sexual selection and sex allocation as predicted by theoretical studies (Fawcett et al. 2011, Booksmythe et al. 2013). In this case, large sample sizes are required to show significant results (West and Sheldon 2002). Alternatively, multiple factors can potentially influence the decision of sex ratio adjustment, which may diminish the detectability of the selection effect of mate attractiveness (West 2009). A more recent meta-analysis found that the mean effect size was considerably larger for traits that were shown to have a function for sexual selection than for traits in which such a function was not shown (Szász et al. 2019). This suggests that it is important to investigate not only the pattern of sex ratio adjustment but also the underlying assumptions, specifically the fitness consequence of sex ratio adjustment to make *a priori* predictions (West 2009).

Another factor in conditional sex allocation is resources for offspring growth (Charnov et al. 1981). This has been studied with solitary parasitoid wasps (Godfray 1994, West 2009). In many parasitoid wasps, the body size is expected to have a strong effect on female reproductive success with larger females producing a larger number of eggs, while it is expected to have a smaller effect on male reproductive success (Figure 7.6b, in which x-axis is host size or quality). In solitary parasitoids, where only one individual can develop per host, the size of emerging individuals correlates with the size of hosts, enabling ovipositing females to predict the reproductive success of sons and daughters based on the size of their hosts. In this situation, females are predicted to lay female eggs in relatively larger hosts and male eggs in relatively smaller hosts (Charnov et al. 1981). This prediction is supported by a huge number of empirical studies (West and Sheldon 2002). A meta-analysis also found that the mean effect size was larger for idiobiont wasps, in which the development of the host ceases at the point of parasitism, than koinobiont wasps, in which the host continues developing after parasitism, making it harder for ovipositing females to predict the final size of their offspring (West and Sheldon 2002). This result emphasizes the importance of environmental predictability for conditional sex allocation.

Comparing the overall weak tendencies with maternal condition in mammals and mate attractiveness in birds *vs.* stronger sex ratio shifts with host size in parasitoid wasps, this could reflect constraints imposed by the mechanism of sex determination. Mammals and birds have genetic (chromosomal) sex determination, in which offspring sex ratio may be determined by the random segregation of the heteromorphic sex chromosomes at meiosis. In contrast, wasps have haplodiploid sex determination that is likely to allow precise control of the offspring sex by determining whether an egg is fertilized (diploid eggs develop into females) or not (haploid eggs develop into males). However, there are some remarkable examples of sex ratio adjustment in mammals and birds, such as the Seychelles warbler (Figure 7.5). In addition, when controlling for selective pressure, meta-analyses did not find significant differences in the degree of sex ratio adjustment between species with genetic and haplodiploid sex determination (West and Sheldon 2002, West et al. 2005). This suggests that the proximate mechanism of sex determination might not fully constrain adaptive sex allocation. The diverse results in the studies of mammals and birds could be caused by the complex life histories and multiple factors that influence sex allocation (Wild and West 2007). Recently, there has been growing physiological evidence for sex ratio manipulation around conception in mammals and birds, which suggests that manipulation occurs through differential production or mortality of the heterogametes, or sex-specific embryonic mortality (Douhard 2017, Navara 2018).

Finally, conditional sex allocation can also be associated with sex ratio adjustment in species with overlapping generations. If generations overlap, and if the degree of overlap differs between the sexes and between the generations, the intensity of competition among each sex and then the relative reproductive success between males and females may vary between generations (Figure 7.6, in which x-axis is generation as a discrete rather than continuous variable). In this case, selection can favour females adjusting offspring sex ratio according to generation (Werren and Charnov 1978, Seger 1983). Females producing generations in which one sex is less likely to survive to future generations are predicted to overproduce the better-surviving sex because the individuals of that sex will achieve reproductive success in multiple generations. In contrast, females producing the other generations are predicted to overproduce the other sex because the individuals of the surviving sex are carried over from the previous generation. Sex ratio adjustment can also be favoured if population recruitment varies between generations (West and Godfray 1997).

Kahn et al. (2013) found that the observed sex ratio of the mosquitofish *Gambusia holbrooki*, in three populations, varies between generations and that observations are mostly consistent with theoretical predictions (Figure 7.8b). Field data showed that there are two breeding peaks in spring and autumn, suggesting two discrete generations per year, and that they have a suitable life history to test the above prediction. Spring-born males and females reproduced in autumn but could not survive the winter. Autumn-born males and females survived the winter and reproduced in spring, when all the males died, while some of the females survived to reproduce again in autumn. In this case, autumn-born females have a chance to reproduce in two seasons, while spring-born females compete with both spring- and autumn-born females. As predicted by theory, the sex ratio was male-biased in spring and female-biased in autumn. Furthermore, the biases were stronger in populations in which the degree of overlap between generations was greater (Figure 7.8a). The observed sex ratio correlated with sex ratio predicted by a theoretical model that was developed to match the biology of this species and parameterized using data from the populations.

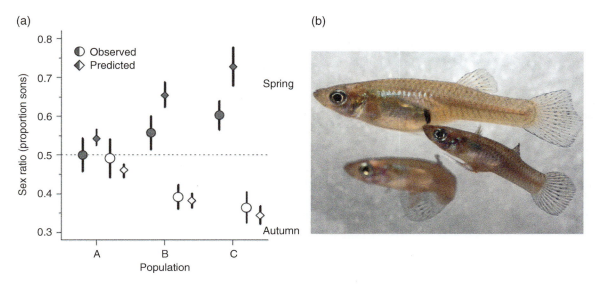

Figure 7.8 (a) Observed sex ratio (circles) consistent with predicted sex ratio (diamonds) with respect to generation overlap in the mosquitofish *Gambusia holbrooki*. Open and solid symbols represent autumn- and spring-born generations, respectively. Spring-born females reproduced with autumn-born females in the previous year with a relatively lower rate in population A, a moderate rate in population B and a relatively higher rate in population C, while male generations do not overlap. Mean ± 95% confidence intervals are represented for observed sex ratios (circles) and predicted mean and ranges are represented for theoretical predictions (diamonds). (*Source:* Adapted from Kahn et al. (2013)). (b) *Source:* Photo by Damien Esquerré.

7.4.2 Environmental Sex Determination

In some animals and plants, sex is determined by the environment that the individuals experience during embryonic development. Environmental sex determination is common in reptiles, with different patterns in different species. Males are produced at a relatively high temperature in some lizards and crocodiles, at a relatively low temperature in many turtles, and within a specific range of temperature in snapping turtles, while females are produced at the other temperature. Evolutionary transitions between environmental (temperature-dependent) sex determination and genetic sex determination appear to be quite flexible within reptiles and even between populations in a species (Pennell et al. 2018).

Conditional sex allocation theory can also be applied to explain sex allocation in species with environmental sex determination. If males and females are expected to gain different reproductive success depending on an environmental variable (Figure 7.6), embryos are predicted to develop into the sex that will have greater reproductive success, which is termed the differential fitness hypothesis (Charnov and Bull 1977). This differential fitness hypothesis has been clearly supported by empirical studies in the silverside fish and the brackish water shrimp, where both the relative reproductive success of the sexes and sex allocation patterns were demonstrated (Conover 1984, McCabe and Dunn 1997). Sex in the species is determined by temperature and photoperiod, which are indicators of the time of year. Individuals laid earlier have a longer growing period, resulting in a larger body size when mature. As predicted, the sex gaining a greater reproductive advantage from body size (females in the silverside fish and males in the brackish water shrimp) tended to be produced relatively earlier in the breeding season.

Pen et al. (2010) examined conditions causing divergence in sex determination mechanisms in the snow skink *Carinascincus ocellatus* (Figure 7.9e). The small viviparous reptile has two sex-determining systems depending on their habitats: temperature-dependent sex determination in lowland populations and genetic sex determination in highland populations. Field data showed that warmer conditions increased female reproductive success throughout their lifetime with an earlier date of birth and a longer growth period in a lowland population, but not in a highland population, which had a shorter season and a more synchronized birth. Furthermore, the highland population had a relatively high variation in temperature between years, which could make it difficult to predict the climate of a year before reproduction. A simulation model was developed considering the genetic mechanisms of sex determination for reptiles and parameterized with the field data taken in the lowland and highland population. The model predicted the evolution of sex-determining systems, in which temperature had a strong effect on sex ratio in the lowland population but not in the highland population (Figure 7.9a–d). The results indicated that both climatic effects on the relative reproductive success of male and female offspring and fluctuations in the climatic effect were responsible for the evolutionary divergence in sex determination.

The above study provided an adaptive explanation for the divergent sex-determining systems within a reptilian species. However, there is limited empirical evidence for the benefits of environmental sex determination across other reptiles (Warner and Shine 2008). Although different environments during developing period may cause differences in the lifetime reproductive success of individuals in short-lived species, this is not likely to apply to relatively long-lived species, as seen in some reptiles (Warner 2011, Pennell et al. 2018). The adaptive significance of temperature-dependent sex determination in reptiles actually remains an enigma.

7.4.3 Sex Change in Sequential Hermaphrodites

Sequential hermaphrodite organisms, including a variety of fish, invertebrates and plants, change their sex during a lifetime (Freeman et al. 1980, Charnov 1982, Policansky 1982). The direction of sex change is either female to male (protogyny) or male to female (protandry), although a minority of species show bidirectional sex change. A logic analogous to conditional sex allocation theory can apply to sex changes, which is termed the size-advantage model (Ghiselin 1969).

Sex change is an adaptive form of sex allocation in species where individuals continue to grow after they mature and reproduce. The size-advantage model suggests that sex change is favoured when the reproductive success increases with growth, and the rate of increase differs between the sexes (Ghiselin 1969, Warner 1975, Charnov 1982). If the reproductive success more rapidly increases with age or size in males than in females (Figure 7.6a, in which x-axis is age or size), individuals can maximize their lifetime reproductive success by starting reproduction as females and switching later to males (protogynous sex change). If the reverse holds (Figure 7.6b), protandrous (male first) sex change is favoured. While female reproductive success, which correlates with fecundity, generally increases with age or size, male reproductive success depends on mating system. Protogyny is predicted to occur in polygynous mating systems where old and large males can monopolize mating with females. In contrast, protogyny will be adaptive in monogamous mating systems where male age and size have little effect on their reproductive success.

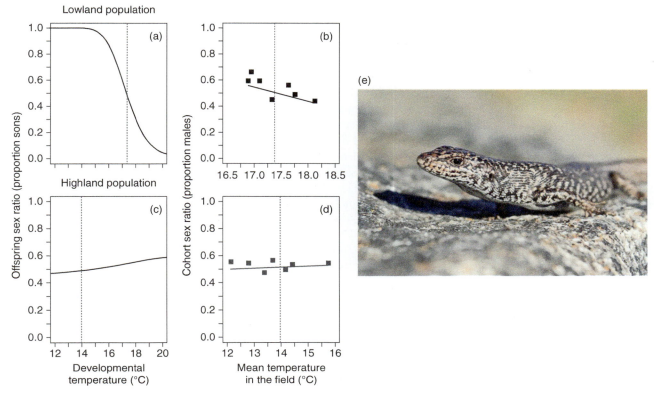

Figure 7.9 Predicted sex-determining systems (a, c) and consequent cohort sex ratio (b, d) for the lowland and highland populations in the snow skink *Carinascincus ocellatus*. An individual-based simulation model, which was parameterized with the field data, predicted reaction norms for offspring sex ratio in response to developmental temperature (a, c). The predicted reaction norm was strongly temperature-dependent in the lowland (a), but not in the highland (c). Cohort sex ratio predicted by the reaction norms (lines) resembled observed sex ratio in the field (squares) both in the lowland (b) and the highland (d). (*Source:* Adapted from Pen et al. (2010)). (e) *Source:* Photo by Mandy Caldwell.

The size-advantage hypothesis is largely supported by numerous empirical studies that compare the relationship between mating systems and the direction of sex changes (Allsop and West 2004, Munday et al. 2006, West 2009, Vega-Frutis et al. 2014). For example, in fish, many species in groupers, wrasses and parrotfish have polygynous mating systems, in which large males monopolize mating in a territorial or haremic social structure and show protogynous sex change (Erisman et al. 2009, Kazancıoğlu and Alonzo 2010). In contrast, the clownfish *Amphiprion percula*, in which a monogamous pair inhabits a sea anemone, sometimes with smaller non-breeders, shows protandrous sex change (Buston 2003). In these species, the removal of the largest individual from a breeding group leads to the sex change of the next largest individual.

Predictions of the size-advantage model were also supported by comparative studies. These studies focused mainly on fish, which exhibit various sexual systems, involving gonochorism (separate sexes) and sequential hermaphroditism. Comparative phylogenetic analysis confirmed that the evolution of sex change in labrid fish (wrasses and parrotfish) is correlated to the strength of male size advantage (Kazancıoğlu and Alonzo 2010). Further, a more recent study on the same group found that evolutionary transitions from gonochorism to protogynous sex change are likely to occur with increasing male size advantage, and protogynous sex change is less likely to be lost under strong size advantage (Hodge et al. 2020).

Male size advantage is also expected to be associated with spawning mode and the level of sperm competition. Polygynous species, where males use their size advantage to monopolize groups of females, spawn in pairs (without competing males) in territories or haremic groups, reducing the intensity of sperm competition. Contrarily, promiscuous species, where the reproductive advantage of large males is diminished and gonochorism is predicted, spawn in groups (with competing males), leading to intense sperm competition. As predicted, a comparative study on groupers found that evolutionary transitions in sexual patterns from protogyny to gonochorism were strongly associated with transitions in mating group structure from paired spawning to group spawning (Erisman et al. 2009). Likewise, relative gonad size in males, which is an indicator of the level of sperm competition, was smaller in protogynous species than in gonochoristic species across a wide range of fish taxa or within a specific taxon (Molloy et al. 2007, Erisman et al. 2009, Pla et al. 2020). However, contrary to the prediction, a study on seabreams showed that males in protandrous species had smaller relative gonad sizes than males in gonochoristic species (Pla et al. 2020). Although a protandrous species is also likely to exhibit low levels of sperm competition in monandry and paired spawning, this result may be explained if small-sized males need to invest greatly in the gonad to fertilize eggs released by large and fecund females in protandrous species (Pla et al. 2020).

Conditional sex allocation theory predicts adaptive sex allocation for individuals but not the consequences for the overall population sex ratio. Multiple factors influencing the sex ratio make it complicated to predict and explain variation in population sex ratios (Frank 1990, West 2009). However, an exception is sex-changing organisms, where a relatively clear prediction can be made that the population sex ratio should be biased towards the first sex (Allsop and West 2004). This is because overall reproductive success in the population is identical between males and females, and the average reproductive success of the first sex should be smaller than that of the second sex (Figure 7.6). In accordance with this prediction, a comparative study with data from fish and invertebrates found significantly biased sex ratios towards males or females in populations of protandrous and protogynous species, respectively (Allsop and West 2004). Similarly, the population sex ratios of protogynous species were more female-biased than those of gonochoristic species in a comparison across variable fish species (Erisman et al. 2009).

7.5 Future Directions

We have reviewed some of the many cases where sex allocation theory can explain variation in observed sex ratios both within and between species. A key future step is to explain variation in the extent to which species shift their sex allocation facultatively. In some cases, meta-analyses have shown consistent patterns with theoretical predictions but in others they have not (West and Sheldon 2002, Sheldon and West 2004, Griffin et al. 2005, West et al. 2005, Cassey et al. 2006, Booksmythe et al. 2017, Khwaja et al. 2017, Szász et al. 2019). It is unclear to what extent this represents variable selection, constraints on sex allocation, or the selfish interests of sex allocation distorters (Burt and Trivers 2006, Werren 2011, see also Chapter 14, this volume). An extreme case is temperature-dependent sex determination in reptiles, where there is considerable variation to be explained (Warner 2011, Pennell et al. 2018).

A key aspect of research on sex allocation is the strong mutual interplay between theoretical and empirical work, and it will be extremely useful to continue this. Empirical studies could investigate the underlying assumptions that could influence sex allocation, and theoretical studies could consider species-specific effects (Pen et al. 2010, Kahn et al. 2013, Abe et al. 2021). The rapid advancement of molecular techniques has allowed the development of molecular markers to investigate sex allocation in natural populations and studies of the genetics of sex allocation (Burton-Chellew et al. 2008, Douhard 2017, Cook et al. 2018, Navara 2018, Pannebakker et al. 2020, Abe et al. 2021). It is also useful to examine how sex allocation can interact with other traits, in which case coevolutionary dynamics can alter the predictions of sex allocation theory (Booksmythe et al. 2013, Macke et al. 2014, Abe and Kamimura 2015). Finally, because sex allocation can have a large influence on population growth, it can also be utilized for applied purposes, such as aiding biological control or conservation projects (Robertson et al. 2006, Ode and Hardy 2008, West 2009; see also Chapter 22, this volume).

Acknowledgements

We thank Michal Segoli and Eric Wajnberg for providing an opportunity to contribute to this book and for careful comments on the previous versions of the manuscript, and Ian C. W. Hardy for helpful comments. This work was partially supported by a Japan Society for the Promotion of Science grant-in-aid for scientific research (JSPS KAKENHI Grant 21K06353 and 21KK0267) and the ERC (834164).

References

Abe, J., Iritani, R., Tsuchida, K., Kamimura, Y., and West, S.A. (2021). A solution to a sex ratio puzzle in *Melittobia* wasps. *Proceedings of the National Academy of Sciences of the United States of America* 118: e2024656118.

Abe, J., and Kamimura, Y. (2015). Sperm economy between female mating frequency and male ejaculate allocation. *The American Naturalist* 185(3): 406–416.

Abe, J., Kamimura, Y., Kondo, N., and Shimada, M. (2003). Extremely female-biased sex ratio and lethal male–male combat in a parasitoid wasp, *Melittobia australica* (Eulophidae). *Behavioral Ecology* 14(1): 34–39.

Abe, J., Kamimura, Y., and West, S A. (2014). Inexplicably female-biased sex ratios in *Melittobia* wasps. *Evolution* 68: 2709–2717.

Allsop, D.J., and West, S.A. (2004). Sex-ratio evolution in sex changing animals. *Evolution* 58(5): 1019–1027.

Booksmythe, I., Mautz, B., Davis, J., Nakagawa, S., and Jennions, M.D. (2017). Facultative adjustment of the offspring sex ratio and male attractiveness: a systematic review and meta-analysis. *Biological Reviews* 92(1): 108–134.

Booksmythe, I., Schwanz, L.E., and Kokko, H. (2013). The complex interplay of sex allocation and sexual selection. *Evolution* 67(3): 673–678.

Brauer, V.S., Schärer, L., and Michiels, N.K. (2007). Phenotypically flexible sex allocation in a simultaneous hermaphrodite. *Evolution* 61(1): 216–222.

Bulmer, M.G. (1986). Sex ratio theory in geographically structured populations. *Heredity* 56(1): 69–73.

Burley, N. (1981). Sex ratio manipulation and selection for attractiveness. *Science* 211: 721–722.

Burt, A., and Trivers, R. (2006). *Genes in conflict*. Harvard University Press.

Burton-Chellew, M.N., Koevoets, T., Grillenberger, B.K., Sykes, E.M., Underwood, S.L., Bijlsma, K., Gadau, J., van De Zande, L., Beukeboom, L.W., West, S.A., and Shuker, D.M. (2008). Facultative sex ratio adjustment in natural populations of wasps: cues of local mate competition and the precision of adaptation. *The American Naturalist* 172(3): 393–404.

Buston, P. (2003). Size and growth modification in clownfish. *Nature* 424(6945): 145–146.

Brunet, J. (1992). Sex allocation in hermaphroditic plants. *Trends in Ecology & Evolution* 7(3): 79–84.

Cassey, P., Ewen, J.G., and Møller, A.P. (2006). Erratum: revised evidence for facultative sex ratio adjustment in birds. *Proceedings of the Royal Society B: Biological Sciences* 273: 3129–3130.

Charnov, E.L. (1982). *The theory of sex allocation*. Princeton University Press.

Charnov, E.L., and Bull, J. (1977). When is sex environmentally determined? *Nature* 266(5605): 828–830.

Charnov, E.L., Los-Den Hartogh, R.L., Jones, W.T., and van Den Assem, J. (1981). Sex ratio evolution in a variable environment. *Nature* 289: 27–33.

Chokechaipaisarn, C., and Gardner, A. (2022). Density-dependent dispersal promotes female-biased sex allocation in viscous populations. *Biology Letters* 18(8): 20220225.

Clark, A.B. (1978). Sex ratio and local resource competition in a prosimian primate. *Science* 201 (4351): 163–165.

Clutton-Brock, T.H., Albon, S.D., and Guiness, F.E. (1984). Maternal dominance, breeding success, and birth sex ratios in red deer. *Nature* 308: 358–360.

Conover, D.O. (1984). Adaptive significance of temperature-dependent sex determination in a fish. *The American Naturalist* 123(3): 287–313.

Cook, N., Boulton, R.A., Green, J., Trivedi, U., Tauber, E., Pannebakker, B.A., Ritchie, M. G., and Shuker, D.M. (2018). Differential gene expression is not required for facultative sex allocation: a transcriptome analysis of brain tissue in the parasitoid wasp *Nasonia vitripennis*. *Royal Society Open Science* 5(2): 171718.

Douhard, M. (2017). Offspring sex ratio in mammals and the Trivers–Willard hypothesis: in pursuit of unambiguous evidence. *BioEssays* 39(9): 1700043.

Düsing, C. (1884). *Die regierung des geschlechtsverhältnisses bei der vermehrung der menschen, tiere und pflanzen*. Fischer.

Eppley, S.M., and Jesson, L.K. (2008). Moving to mate: the evolution of separate and combined sexes in multicellular organisms. *Journal of Evolutionary Biology* 21(3): 727–736.

Erisman, B.E., Craig, M.T., and Hastings, P.A. (2009). A phylogenetic test of the size-advantage model: evolutionary changes in mating behavior influence the loss of sex change in a fish lineage. *The American Naturalist* 174(3): E83–E99.

Fawcett, T.W., Kuijper, B., Weissing, F.J., and Pen, I. (2011). Sex-ratio control erodes sexual selection, revealing evolutionary feedback from adaptive plasticity. *Proceedings of the National Academy of Sciences of the United States of America* 108(38): 15925–15930.

Fisher, R.A. (1930). *The genetical theory of natural selection*. Clarendon.

Frank, S.A. (1985). Hierarchical selection theory and sex ratios. II. On applying the theory, and a test with fig wasps. *Evolution* 39: 949–964.

Frank, S.A. (1986). Hierarchical selection theory and sex ratios. I. General solutions for structured populations. *Theoretical Population Biology* 29: 312–342.

Frank, S.A. (1990). Sex allocation theory for birds and mammals. *Annual Review of Ecology and Systematics* 21, 13–55.

Frank, S.A. (1998). *Foundations of social evolution*. Princeton University Press.

Freeman, D C., Harper, K.T., and Charnov, E. L. (1980). Sex change in plants: old and new observations and new hypotheses. *Oecologia* 47(2): 222–232.

Gardner, A. (2023). The rarer-sex effect. *Philosophical Transactions of the Royal Society of London. Series B: Biological Sciences* 378(1876): 20210500.

Gardner, A., Arce, A., and Alpedrinha, J. (2009). Budding dispersal and the sex ratio. *Journal of Evolutionary Biology* 22(5): 1036–1045.

Ghiselin, M.T. (1969). The evolution of hermaphroditism among animals. *The Quarterly Review of Biology* 44(2): 189–208.

Godfray, H.C.J. (1994). *Parasitoids: behavioral and evolutionary ecology*. Princeton University Press.

Griffin, A.S., Sheldon, B C., and West, S.A. (2005). Cooperative breeders adjust offspring sex ratios to produce helpful helpers. *The American Naturalist* 166(5): 628–632.

Hamilton, W.D. (1967). Extraordinary sex ratios. *Science* 156: 477–488.

Hamilton, W.D. (1971). Selection of selfish and altruistic behaviour in some extreme models. In: *Man and beast: comparative social behavior* (eds. Eisenberg, J.F., and Dillon, D.S.), 57–91. Smithsonian Press.

Hardy, I.C.W. (2002). *Sex ratios: concepts and research methods*. Cambridge University Press.

Hart, M.K., Kratter, A.W., Svoboda, A.-M., Lawrence, C.L., Craig Sargent, R., and Crowley, P.H. (2010). Sex allocation in a group-living simultaneous hermaphrodite: effects of density at two different spatial scales. *Evolutionary Ecology Research* 12: 189–202.

Herre, E.A. (1985). Sex ratio adjustment in fig wasps. *Science* 228(4701): 896–898.

Herre, E.A. (1987). Optimality, plasticity and selective regime in fig wasp sex ratios. *Nature* 329: 627–629.

Hodge, J.R., Santini, F., and Wainwright, P. C. (2020). Correlated evolution of sex allocation and mating system in wrasses and parrotfishes. *The American Naturalist* 196(1): 57–73.

Iritani, R., West, S.A., and Abe, J. (2021). Cooperative interactions among females can lead to even more extraordinary sex ratios. *Evolution Letters* 5(4): 370–384.

Janicke, T., Marie-Orleach, L., De Mulder, K., Berezikov, E., Ladurner, P., Vizoso, D.B., and Schärer, L. (2013). Sex allocation adjustment to mating group size in a simultaneous hermaphrodite. *Evolution* 67(11): 3233–3242.

Johnston, M.O., Das, B., and Hoeh, W.R. (1998) Negative correlation between male allocation and rate of self-fertilization n a hermaphroditic animal. *Proceedings of the National Academy of Sciences of the United States of America* 95: 617–620.

Kahn, A.T., Jennions, M.D., and Kokko, H. (2015). Sex allocation, juvenile mortality and the costs imposed by offspring on parents and siblings. *Journal of Evolutionary Biology* 28(2): 428–437.

Kahn, A.T., Kokko, H., and Jennions, M.D. (2013). Adaptive sex allocation in anticipation of changes in offspring mating opportunities. *Nature Communications* 4: 1603.

Kazancioğlu, E., and Alonzo, S.H. (2010). A comparative analysis of sex change in labridae supports the size advantage hypothesis. *Evolution* 64(8): 2254–2264.

Khwaja, N., Hatchwell, B.J., Freckleton, R.P., and Green, J.P. (2017). Sex allocation patterns across cooperatively breeding birds do not support predictions of the repayment hypothesis. *The American Naturalist* 190: 547–556.

Komdeur, J., Daan, S., Tinbergen, J., and Mateman, C. (1997). Extreme adaptive modification in sex ratio of the Seychelles warbler's eggs. *Nature* 385(6616): 522–525.

Lehmann, L., and Rousset, F. (2010). How life history and demography promote or inhibit the evolution of helping behaviours. *Philosophical Transactions of the Royal Society, B: Biological Sciences* 365(1553): 2599–2617.

Lehtonen, J., Malabusini, S., Guo, X., and Hardy, I.C.W. (2023). Individual- and group-level sex ratios under local mate competition: consequences of infanticide and reproductive dominance. *Evolution Letters* 7(1): 13–23.

Leigh, E.G. (1970). Sex ratio and differential mortality between the sexes. *The American Naturalist* 104: 205–210.

Macke, E., Magalhães, S., Bach, F., and Olivieri, I. (2011). Experimental evolution of reduced sex ratio adjustment under local mate competition. *Science* 334(6059): 1127–1129.

Macke, E., Olivieri, I., and Magalhães, S. (2014). Local mate competition mediates sexual conflict over sex ratio in a haplodiploid spider mite. *Current Biology* 24(23): 2850–2854.

Martins, R.P., Antonini, Y., Silveira, A., and West, S.A., (1999). Seasonal variation in the sex allocation of a neotropical solitary bee. *Behavioral Ecology* 10(4): 401–408.

Matthews, R.W., González, J.M., Matthews, J.R., and Deyrup, L D. (2009). Biology of the parasitoid *Melittobia* (Hymenoptera: Eulophidae). *Annual Review of Entomology* 54: 251–266.

McCabe, J., and Dunn, A.M. (1997). Adaptive significance of environmental sex determination in an amphipod. *Journal of Evolutionary Biology* 10(4): 515–527.

Michiels, N.K. (1998). Mating conflicts and sperm competition in simultaneous hermaphrodites. In: *Sperm competition and sexual selection* (eds. Birkhead, T.R., and Møller, A.P.), 219–254. Academic Press.

Molloy, P.P., Goodwin, N.B., Côté, I.M., Reynolds, J.D., and Gage, M.J.G. (2007). Sperm competition and sex change: a comparative analysis across fishes. *Evolution* 61(3): 640–652.

Munday, P.L., Buston, P.M., and Warner, R.R. (2006). Diversity and flexibility of sex-change strategies in animals. *Trends in Ecology & Evolution* 21(2): 89–95.

Navara, K.J. (2018). *Choosing sexes*. Springer.

Ode, P.J., and Hardy, I.C.W. (2008). Parasitoid sex ratios and biological control. In: *Behavioral ecology of insect parasitoids: from theoretical approaches to field applications* (eds. Wajnberg, E., Bernstein, C., and van Alphen, J.), 253–291. Blackwell Publishing.

Pannebakker, B.A., Cook, N., van Den Heuvel, J., van De Zande, L., and Shuker, D.M. (2020). Genomics of sex allocation in the parasitoid wasp *Nasonia vitripennis*. *BMC Genomics* 21(1): 499.

Parker, G.A., and Birkhead, T.R. (2013). Polyandry: the history of a revolution. *Philosophical Transactions of the Royal Society, B: Biological Sciences* 368(1613): 20120335.

Pen, I., Uller, T., Feldmeyer, B., Harts, A., While, G.M., and Wapstra, E. (2010). Climate-driven population divergence in sex-determining systems. *Nature*, 468(7322): 436–438.

Pen, I., and Weissing, F.J. (2000). Sex-ratio optimization with helpers at the nest. *Proceedings of the Royal Society B: Biological Sciences* 267: 539–544.

Pennell, M.W., Mank, J.E., and Peichel, C.L. (2018). Transitions in sex determination and sex chromosomes across vertebrate species. *Molecular Ecology* 27(19): 3950–3963.

Pla, S., Benvenuto, C., Capellini, I., and Piferrer, F. (2020). A phylogenetic comparative analysis on the evolution of sequential hermaphroditism in seabreams (Teleostei: Sparidae). *Scientific Reports* 10(1): 3606.

Policansky, D. (1982). Sex change in plants and animals. *Annual Review of Ecology and Systematics* 13: 471–495.

Puurtinen, M., and Kaitala, V. (2002). Mate-search efficiency can determine the evolution of separate sexes and the stability of hermaphroditism in animals. *The American Naturalist* 160(5): 645–660.

Raimondi, P.T., and Martin, J.E. (1991). Evidence that mating group size affects allocation of reproductive resources in a simultaneous hermaphrodite. *The American Naturalist* 138, 1206–1217.

Ramm, S.A., Lengerer, B., Arbore, R., Pjeta, R., Wunderer, J., Giannakara, A., Berezikov, E., Ladurner, P., and Schärer, L. (2019). Sex allocation plasticity on a transcriptome scale: socially sensitive gene expression in a simultaneous hermaphrodite. *Molecular Ecology* 28(9): 2321–2341.

Reece, S.E., Drew, D.R., and Gardner, A. (2008). Sex ratio adjustment and kin discrimination in malaria parasites. *Nature* 453: 609–614.

Robertson, B.C., Elliott, G.P., Eason, D.K., and Gemmell, N.J. (2006) Sex allocation theory aids species conservation. *Biology Letters* 2: 229–231.

Rodrigues, A.M.M., and Gardner, A. (2015). Simultaneous failure of two sex-allocation invariants: implications for sex-ratio variation within and between populations. *Proceedings of the Royal Society B: Biological Sciences* 282(1810): 20150570.

Schärer, L. (2009). Tests of sex allocation theory in simultaneously hermaphroditic animals. *Evolution* 63(6): 1377–1405.

Schärer, L., and Pen, I. (2013). Sex allocation and investment into pre- and post-copulatory traits in simultaneous hermaphrodites: the role of polyandry and local sperm competition. *Philosophical Transactions of the Royal Society, B: Biological Sciences* 368(1613): 20120052.

Schwarz, M.P. (1988). Local resource enhancement and sex ratios in a primitively social bee. *Nature* 331(6154): 346–348.

Seger, J. (1983). Partial bivoltinism may cause alternating sex-ratio biases that favour eusociality. *Nature* 301(5895): 59–62.

Sheldon, B.C., Andersson, S., Griffith, S.C., Örnborg, J., and Sendecka, J. (1999). Ultraviolet colour variation influences blue tit sex ratios. *Nature* 402(6764): 874–877.

Sheldon, B.C., and West, S.A. (2004). Maternal dominance, maternal condition, and offspring sex ratio in ungulate mammals. *The American Naturalist* 163(1): 40–54.

Shuker, D.M., Pen, I., Duncan, A.B., Reece, S.E., and West, S.A. (2005). Sex ratios under asymmetrical local mate competition: theory and a test with parasitoid wasps. *The American Naturalist* 166: 301–316.

Shuker, D.M., and West, S.A. (2004). Information constraints and the precision of adaptation: sex ratio manipulation in wasps. *Proceedings of the National Academy of Sciences of the United States of America* 101(28): 10363–10367.

Silk, J.B., and Brown, G.R. (2008). Local resource competition and local resource enhancement shape primate birth sex ratios. *Proceedings of the Royal Society B: Biological Sciences* 275(1644): 1761–1765.

Singh, P., and Schärer, L. (2022). Evolution of sex allocation plasticity in a hermaphroditic flatworm genus. *Journal of Evolutionary Biology* 35(6): 817–830.

Suzuki, Y., and Iwasa, Y. (1980). A sex ratio theory of gregarious parasitoids. *Researches on Population Ecology* 22(2): 366–382.

Szász, E., Garamszegi, L.Z., and Rosivall, B. (2019). What is behind the variation in mate quality dependent sex ratio adjustment? A meta-analysis. *Oikos* 128(1): 1–12.

Tang, X., Meng, L., Kapranas, A., Xu, F., Hardy, I.C.W., and Li, B. (2014). Mutually beneficial host exploitation and ultra-biased sex ratios in quasisocial parasitoids. *Nature Communications* 5: 4942.

Taylor, P.D. (1981). Intra-sex and inter-sex sibling interactions as sex determinants. *Nature* 291: 64–66.

Taylor, P.D. (1988). Inclusive fitness models with two sexes. *Theoretical Population Biology* 34(2): 145–168.

Trivers, R.L., and Willard, D.E. (1973). Natural selection of parental ability to vary the sex ratio of offspring. *Science* 179(4068): 90–92.

Vega-Frutis, R., Macías-Ordóñez, R., Guevara, R., and Fromhage, L. (2014). Sex change in plants and animals: a unified perspective. *Journal of Evolutionary Biology* 27(4): 667–675.

Wajnberg, E. (1993). Genetic variation in sex allocation in a parasitic wasp. Variation in sex pattern within sequences of oviposition. *Entomologia Experimentalis et Applicata* 69: 221–229.

Warner, R.R. (1975). The adaptive significance of sequential hermaphroditism in animals. *The American Naturalist* 109(965): 61–82.

Warner, D.A. (2011). Sex determination in reptiles. In: *Hormones and reproduction of vertebrates, reptiles* (eds. Norris, D.O., and Lopez, K.H.), 1–38. Academic Press.

Warner, D.A., and Shine, R. (2008). The adaptive significance of temperature-dependent sex determination in a reptile. *Nature* 451: 566–569.

Werren, J.H. (1980). Sex ratio adaptations to local mate competition in a parasitic wasp. *Science* 208(4448): 1157–1159.

Werren, J.H. (1983). Sex ratio evolution under local mate competition in a parasitic wasp. *Evolution* 37(1): 116–124.

Werren, J.H. (2011). Selfish genetic elements, genetic conflict, and evolutionary innovation. *Proceedings of the National Academy of Sciences of the United States of America* 108(2): 10863–10870.

Werren, J.H., and Charnov, E.L. (1978). Facultative sex ratios and population dynamics. *Nature* 272(5651): 349–350.

West, S.A. (2009). *Sex allocation*. Princeton University Press.

West, S.A., and Godfray, H.C J. (1997). Sex ratio strategies after perturbation of the stable age distribution. *Journal of Theoretical Biology* 186: 213–221.

West, S.A., Griffin, A.S., and Gardner, A. (2007) Social semantics: altruism, cooperation, mutualism, strong reciprocity and group selection. *Journal of Evolutionary Biology* 20: 415–432.

West, S.A., and Herre, E.A. (1998a). Partial local mate competition and the sex ratio: a study on non-pollinating fig wasps. *Journal of Evolutionary Biology* 11(5): 531–548.

West, S. A., and Herre, E.A. (1998b). Stabilizing selection and variance in fig wasp sex ratios. *Evolution* 52(2): 475–485.

West, S A., Pen, I., and Griffin, A. (2002). Cooperation and competition between relatives. *Nature* 296: 72–75.

West, S.A., and Sheldon, B.C. (2002). Constraints in the evolution of sex ratio adjustment. *Science* 295(5560): 1685–1688.

West, S.A., Shuker, D.M., and Sheldon, B.C. (2005). Sex-ratio adjustment when relatives interact: a test of constraints on adaptation. *Evolution* 59(6): 1211–1228.

Wild, G. (2006). Sex rations when helpers stay at the nest. *Evolution*, 60(10): 2012–2022.

Wild, G., and West, S.A. (2007). A sex allocation theory for vertebrates: combining local resource competition and condition-dependent allocation. *The American Naturalist* 170(5): E112–E128.

8

Life History Evolution

Complex Life Cycles Across Animal Diversity

Andreas Heyland[1], Konstantin Khalturin[2], and Vincent Laudet[2,3]

[1] *Integrative Biology, University of Guelph, Guelph, ON, Canada*
[2] *Institute of Cellular and Organismic Biology (ICOB), Academia Sinica, Taipei, Taiwan*
[3] *Marine Eco-Evo-Devo Unit, Okinawa Institute of Science and Technology (OIST), Okinawa, Japan*

8.1 Integration of Metamorphic Development Within the Life Cycle

Life histories are characterized by transitions, which are typically associated with major fitness consequences for the organisms. In animals, these may include, but are not limited to, fertilization, hatching, metamorphic development (MD) and settlement (the transition from the planktonic to the benthic environment in many marine invertebrate groups), reproduction and death. In contrast to most of these transitions, MD does not occur in all animals but is widespread among invertebrates (especially marine invertebrates with bentho-planktonic life cycles and insects) and some vertebrate species (most teleost fish and amphibians) (Figure 8.1). While there is no consensus among scientists on a definition of metamorphosis for animals, efforts to define the term typically include the changes in habitat, drastic morphological changes, changes in behaviour and feeding mode and many others (Bishop et al. 2006a). Comparing this life history transition between animals and focusing on the developmental, physiological and ecological processes involved allows us to draw inferences for the evolutionary history of this process in animals.

Indirect development is characterized by a larval stage within the life cycle and a metamorphic transition. It can be contrasted with direct development, where the embryos directly transition to the juvenile or adult form without an intermittent larval stage. A larva is a pre-reproductive dispersal or growth stage that has notably different morphology and function than the pre-reproductive juvenile or reproductive adult. While a discussion of larval diversity and evolution is beyond the scope of this chapter, we refer the reader to some comprehensive reviews of this fascinating topic (Orton 1953, Strathmann 1985, Smith 1997, Allen and Pernet 2007, Arenas-Mena 2010, Truman 2019, Strathmann 2020, De Robertis and Tejeda-Muñoz 2022). Still, as larvae typically occupy different ecological niches from the adults and do not reproduce, we need to consider how, for example, larval mortality, dispersal, growth and other performance metrics might ultimately impact fitness. This is particularly relevant when comparing organisms that produce large numbers of offspring with low survival rates in contrast to organisms that produce few offspring with a relatively high survival rate. Such reproductive strategies have vastly different implications for larval life, MD and post-metamorphic performance.

8.2 The Regulation of Metamorphic Development by Hormones

In all species studied to date, MD is tightly regulated by both internal and external factors. These factors contribute to the timing of this process within the life cycle and the environment they live in. The fitness consequences of the timing are critical, as initiating the metamorphic transition and transforming into a pre-reproductive juvenile is irreversible (Bishop et al. 2006a, Heyland and Moroz 2006, Hadfield et al. 2015). Environmental factors in the larval and adult habitat, such as food availability, predators, parasites and a range of stressors, can impact the timing of the process, and, depending on the species, the importance of these factors can vary. For example, while increased food levels typically lead to an

Life History Evolution: Traits, Interactions, and Applications, First Edition. Edited by Michal Segoli and Eric Wajnberg.
© 2025 John Wiley & Sons Ltd. Published 2025 by John Wiley & Sons Ltd.

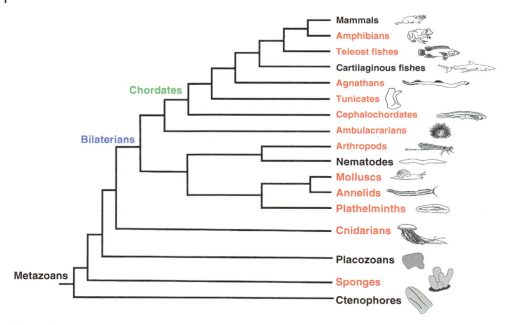

Figure 8.1 Metamorphic development with drastic morphological changes and a transition between habitats is widespread among animals. Simplified phylogeny showing major animal phyla with those that evolved metamorphic development, at least once, in red. Metamorphosis likely evolved independently several times in animal evolution.

acceleration of MD in frogs (Rose 2005), the same factor can lead to deceleration in some marine invertebrate species (Pechenik et al. 1996). Therefore, the direction and extent of the developmental plasticity are a consequence of the evolutionary history of the organisms, their physiological and developmental requirements, including specific constraints, and the ecological context in which the metamorphic transition unfolds. The timing is not just critical for the habitat change. It also impacts the morphological, physiological and developmental sequence of events. For example, in indirectly developing frog species, the limbs must appear before the shrinkage of the tail and the intestine, and visual system and other critical organ systems need to be transformed in time for the terrestrial or semi-terrestrial lifestyle of the adult (Laudet 2011, McMenamin and Parichy 2013). These processes on the cellular level involve differentiation, de-differentiation and apoptosis, among several other mechanisms. While we still lack many details on how these processes are regulated at the cellular level, convergent evidence from multiple lineages shows that the endocrine system and hormonal and peptide signalling play a critical role (Figure 8.2). Despite the fascinating diversity in the actual hormone types and peptides, the common principle of endocrine action and the nervous system is at the heart of MD and the metamorphic transition.

8.3 Review of Metamorphic Mechanisms Across Taxa with Ecological and Evolutionary Considerations

Knowledge and understanding of the endocrine, developmental and cellular mechanisms underlying MD are very much concentrated on specific species, including frogs, insects and some bilaterian and non-bilaterian marine invertebrates. While the development between the groups is divergent, some general principles can be identified and help define the metamorphic process in these species. Metamorphosis generally refers to the entire process, while MD is used to describe the various and often slow changes leading up to more dramatic changes that can unravel within short periods of time. The period of rapid change is often referred to as metamorphic climax (frogs; see Figure 8.2), settlement (marine invertebrates) and eclosion (insects) and can be associated with the change between habitats (Figure 8.3). Below, we will outline and discuss the endocrine and neuro-endocrine systems and how these mechanisms are relevant in both an ecological and evolutionary context.

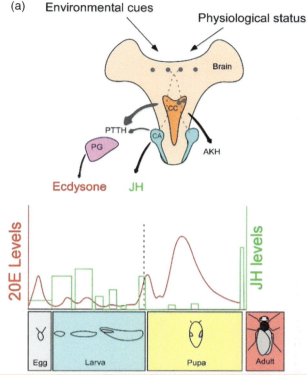
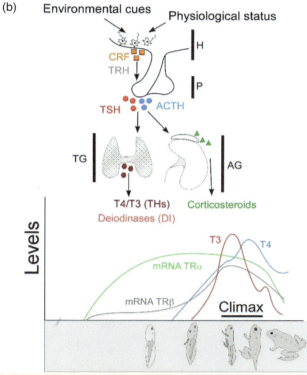

Primer – Endocrine regulation of metamorphic development in holometabolous insects

20E: 20 hydroxy-ecdysone is produced by the prothoracic gland (PG) in response to prothoracicotropic hormone (PTTH) from the corpora cardiaca (CC) or the corpora allata (CA). Photoperiod, temperature, food availability and humidity are among environmental factors impacting the production and release of 20E. Other signaling systems, such as insulin like growth factors (IGF), juvenile hormone (JH), allatotropins and ecdysteroids themselves influence ecdysteroid release.

JH: Juvenile hormone is released by the corpora allate (CA). It maintains larval characters and inhibits the initiation of pupa formation. JH levels are elevated during early larval instars, promoting growth and preventing premature metamorphosis. In late larval instars, JH levels gradually decline, allowing the initiation of metamorphic events and the maturation of imaginal discs. Prior to pupation, there is a significant decrease in JH levels, often referred to as the "JH titer drop". This drop in JH is a trigger for the onset of pupal development. Later in pupal development, prior to adult emergence, JH levels rise again. This increase in JH is involved in the initiation of reproductive maturation and the development of adult characteristics.

PG: The prothoracic gland produces ecdysteroids (ecdysone, then transformed into 20E in peripheral tissues) in response to PTTH from the CC or the CA.

PTTH: Prothoracicotropic hormone is produced from the CC and is a major regulator of ecdysone release from the CA.

CC: A pair of neurohemal organs located near the brain that store and release a range of neuropeptides, some are directly or indirectly important for metamorphic development, i.e. PTTH.

CA: A pair of retrocerebral endocrine glands in insect releasing JH.

AKH: Adipokinetic hormone is associated with metamorphosis via its main function on nutrient mobilization and energy utilization to support tissue remodeling, growth, and differentiation.

Met: Methoprene-tolerant is an intracellular receptor for JH, that maintains larval status (status quo function of JH) by promoting the expression of JH responsive genes, and repressing 20E response genes. (not indicated in a)

Downstream targets: The main downstream regulators of JH and ecdysteroids (20E) in insect metamorphic development include IGF, TOR (target of rapamycin), BR-C (Broad-complex), Kr-h1 (Krüppel homolog 1) and various hormone receptors and transcription factors as well as genes coding for proteins involved in hormone biosynthesis and metabolism. (not indicated in a)

Primer – Endocrine regulation of metamorphic development in frogs

THs: Thyroid hormones are tyrosine derivatives with a specific number of iodine atoms. In vertebrates THs are produced by the thyroid gland (TG), primarily thyroxine (T4), which is de-iodinated in target tissues to T3 (3,5,3'-triiodothyronine) by deiodinases (DI).

TRH: Thyrotropin-releasing hormone produced by the hypothalamus (H) and released in response to environmental and physiological signals.

TSH: Thyroid-stimulating hormone is produced by the pituitary gland (P) in response to TRH stimulation leading to the production of THs by the TG.

CRF: Corticotropin releasing factor plays a key role in activating TSH and TH production in frog tadpoles. CRF acts through receptors CRFR1 and CRFR2, expressed in the hypothalamus (H) and peripheral tissues. CRF is also involved in regulating the stress axis.

ACTH: Adrenocorticotropic hormone is produced by the pituitary gland (P) and activates the production of corticosteroids from the adrenal gland (AG) and is the main regulator of the stress axis.

AG: The adrenal gland produces corticosteroids in response to drenocorticotropic hormone (ACTH) stimulation.

TG: The thyroid gland produces THs, primarily thyroxine (T4).

TR: Thyroid hormone receptors are transcription factors with both a DNA-binding domain and a ligand binding domain. The DNA binding occurs via specific TH response elements (TRE) in the promoter region of target genes. In *Xenopus*, the two TH receptors, thyroid hormone receptor alpha (TRα) and thyroid hormone receptor beta (TRβ), regulate the various transformations that occur during metamorphosis. TRα is the predominant TH receptor during premetamorphosis, and maintains the larval state by repressing the expression of genes that are activated during metamorphic development. In contrast, TRβ promotes metamorphic development. The peak (climax) of TH levels coincides with TRβ levels and is required for the development and differentiation of many organs and tissues such as limbs, tail, and intestine. TRβ also promotes programmed cell death, in tissues that are no longer required in the juvenile/adult frog, such as the larval gills and the tail. The balance between TRα and TRβ expression is critical for regulating the timing and extent of the various developmental processes that occur during *Xenopus* metamorphosis.

DI: De-iodinases regulate the activation and degradation of THs in the target tissues.

GR: Glucocorticoid receptor have been increasingly considered to complement TH signaling in metamorphosis. The interplay between THs and corticoids occurs at multiple levels, including gene regulation, modulation of receptor expression, and control of hormone production.

Figure 8.2 Hormones and peptides are major players in the regulation of metamorphic development in insects (a) and frogs (b). The neuro-endocrine system integrates environmental input (light and temperature), with the developmental and physiological state of the organism. (a) Neuro-endocrine axis in insects with main endocrine signalling and synthesis components. The primer explains and outlines main functions of these components (see also Figure 8.3 for details on life cycle). (b) Hypothalamus-pituitary-endocrine gland axis in frogs and other vertebrates with main endocrine signalling and synthesis components. The primer explains and outlines main functions of these components (see also Figure 8.3 for details on life cycle). The systems in insects and frogs have evolved independently and use different types of hormones but show similarity in the hierarchical integration of sensory, neuro-endocrine and endocrine systems. Other animals discussed in this chapter use the same hormones but often lack centralized neuro-endocrine centres and glands.

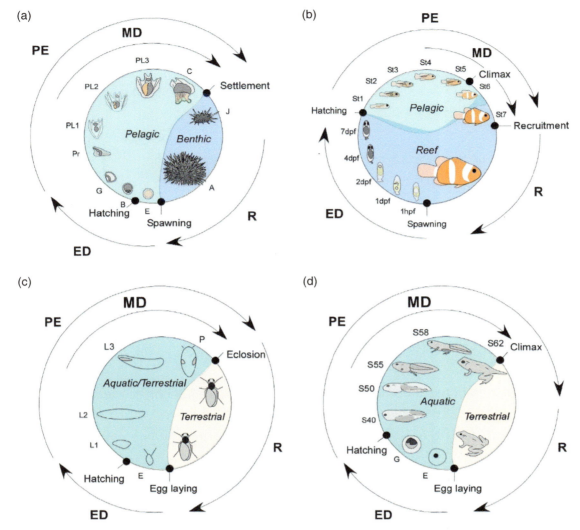

Figure 8.3 Metamorphic life cycles feature important similarities and differences. In indirectly developing sea urchins (a) larvae hatch before gastrulation (G), form a prism larval (Pr) and disperse in the plankton (pelagic environment) for up to several weeks. During that time, the pluteus larva (PL1-3) develops the juvenile rudiment that will form juvenile structures inside the larva. Rudiment formation marks the beginning of metamorphic development (MD) and ends with the competent larva (C) undergoing settlement, the transition into the benthic habitat in response to settlement cues. Several months to years after settlement, the juvenile becomes reproductive (R), completing the life cycle. In anemonefish (b) the parents take care of the eggs for seven days. After hatching, the larvae are taken by currents in the open ocean, where they grow in the plankton. After *ca*. two weeks, the larvae localize a reef, actively swim to pass the reef crest and then locate a specific habitat (here a giant sea anemone) in which they recruit. Holometabolous insects (c) hatch from the egg and develop via several larval stages (L1–L3) to the pupa (P). Within the larva, imaginal discs form from which adult structures are formed in the pupa. This process, including pupa formation and differentiation, is called the MD. After eclosion (hatching from the pupa), the insect is reproductive and completes the life cycle. Frogs (d) form tadpoles after hatching, which develop via a series of stages (S: Gosner stage) into a juvenile frog. Adult structures (legs, visual system, nervous system, etc.) develop in a specific sequence during MD, and during metamorphic climax, the tadpole tail undergoes cell death. The tadpole will undergo a habitat shift between the larval and juvenile stages, and frogs will become reproductive after an extended period of time in the new habitat. A: Adult; B: Blastula; dpf: Days post-fertilization; E: Egg; ED: Embryonic development; hpf: Hours post-fertilization; J: Juvenile; PE: Post-embryonic development; St: Stage.

8.3.1 Vertebrate Metamorphosis

Within the vertebrates, metamorphosis, in the way it is considered in this chapter, is restricted to amphibians, teleost fish and jawless vertebrates. In teleost fish, a range of life histories have been described, and the term larva is used across a range of taxonomic groups, even though the ecological transition between habitats as well as the morphological changes can be drastically different between groups (Miller and Kendall 2019). For example, flounders undergo a transition from the plankton to the benthic environment, completely rebuilding and transitioning their morphology and physiology, including extremely drastic changes such as eye migration (Schreiber 2013). Similarly, some fish species transition from freshwater to saltwater, or *vice versa* during their life cycle (Watanabe et al. 2014). Other fish species undergo much more subtle changes when they transition from a larval to an adult fish. In this section, we will be focusing on amphibian metamorphosis (specifically frogs; see Figure 8.3) to describe the mechanisms regulating the process. We will also discuss some migrating gobies and coral fish metamorphosis as examples of how MD is integrated with environmental cues and signals, as they experience critical habitat changes during MD. Thyroid hormone (TH) has been shown to play a pivotal role in frog metamorphosis, as exemplified by work on the African clawed frog, *Xenopus leavis* (Buchholz et al. 2006). We summarize the details of the endocrine signalling system underlying frog MD and provide a primer for relevant components in Figure 8.2.

8.3.1.1 Environmental Regulation of Vertebrate Metamorphosis

As discussed above, we have a lot of detailed knowledge about the regulation of metamorphosis in model organisms such as the frog *Xenopus laevis*. However, developmental processes and transitions are regulated by a range of environmental factors. Here we will review a few examples to illustrate this critical connection between the internal signalling systems and the external cues and factors.

8.3.1.1.1 Environmental Regulation of Metamorphosis in Frogs

Desiccation, or extreme drying out, can impact the survival of frogs, especially at the tadpole stage (Denver et al. 1998, Brady and Griffiths 2000). In response to drought conditions, predation risk and competition for limited resources, tadpoles undergo accelerated or synchronized metamorphosis. This adaptive response allows them to escape deteriorating aquatic habitats and accelerate the transition to terrestrial environments. However, early metamorphosis can result in smaller froglets that may be less well adapted to their future habitat, reflecting a trade-off between current survival and future reproductive success. Studies on spade-foot toads in arid environments have revealed unique adaptations for rapid metamorphosis, enabling them to take advantage of temporary breeding pools. This accelerated metamorphosis is dependent on increased production and action of TH and corticosterone, indicating molecular adaptations for swift transformation (Denver et al. 1998, Buchholz and Hayes 2005, Gomez-Mestre et al. 2013).

8.3.1.1.2 Environmental Regulation of Metamorphosis in Coral Reef Fish

Coral reef fish have a distinct life cycle that makes them an excellent model for studying metamorphosis in an ecological context (Figure 8.3). Many species of coral reef fish lay pelagic eggs directly in the water, often in large spawning aggregations. After hatching, the larvae grow in the plankton for several weeks (from two in damselfish to 64 in porcupine fish), benefiting from reduced predation and the opportunity for dispersal (Bonhomme and Planes 2000, Leis and McCormick 2002). When the larvae have grown enough to settle on a reef habitat, a process of metamorphosis is triggered by an increase in TH levels, involving changes in behaviour, morphology and physiology better suited to the reef life (McCormick et al. 2002, Laudet 2011, Holzer et al. 2017).

As the larvae swim onto the reef crest during their entry into the reef, one can easily catch them using crest nets (Lecchini et al. 2006, Barth et al. 2015). This allows the study of synchronized metamorphosing animals in natural populations. Studies on surgeonfish and anemonefish have shown that metamorphosis from pelagic larvae to reef-associated juveniles is regulated by TH, similar to other species such as amphibians and flatfish. The peak of TH is initiated early on while the larvae are still in the open ocean (Holzer et al. 2017, Besson et al. 2020, Roux et al. 2023). Understanding the triggers that initiate metamorphosis is challenging, but the duration of the pelagic larval phase (PLP) provides insights. The PLP can be influenced by various factors, such as larval behaviour, ocean currents, lunar cycle and local environmental conditions such as proximity to reef (signalled by sound and odour), and these may be key signals triggering metamorphosis (Leis 2006, Barth et al. 2015). Intraspecific variation in PLP suggests regulation at this level, indicating a link between ecological conditions (*e.g.*, distance to the reef, quality of this reef in terms of coral cover, etc.), metamorphosis and TH and corticoid levels.

The metamorphosis of coral reef fish is closely linked to their metabolism (Holzer et al. 2017, Roux et al. 2023). When fish enter the reef, they face challenges such as finding food resources, maturing sensory systems, transforming their diet, adapting their digestive system and avoiding predators. Metabolic shifts occur during metamorphosis, with larvae relying on glycolysis while juveniles use fatty acid β-oxidation and the Krebs cycle as their main energy source (Roux et al. 2023). The metabolic transition is associated with morphological and behavioural changes. Experimental alterations in lipid metabolism accelerate the development of metamorphosis endpoints. The metabolic pace of metamorphosis is regulated to align with available energy production (Roux et al. 2023). THs play a critical role in this process (Chocron et al. 2012, Sinha et al. 2018, Roux et al. 2023). Environmental conditions can influence TH signalling, potentially affecting the quality of juveniles. Indeed, studies on convict surgeonfish have shown that changes in TH levels, due to pollution or relocation, delay sensory organ maturation, impair ecological function and reduce predator avoidance (Holzer et al. 2017, Besson et al. 2020). Developmental plasticity and phenotypic disparity in response to environmental diversity highlight the acclimation of fish to their local environment during metamorphosis. The role of hormonal variation in these situations requires further investigation to understand the local acclimation of juvenile fish to their environment.

8.3.1.1.3 Environmental Regulation of Metamorphosis in Diadromous Fish

Many fish species migrate between freshwater and seawater and *vice versa*. All these are collectively called diadromous fish and these migrations, whatever the direction and age at which they occur, imply physiological, morphological and behavioural change, most often spectacular, hence considered as metamorphosis. We should distinguish the anadromous fish that migrate from the sea to reproduce in freshwater like the salmon from the catadromous fish like the eels that reproduce in the sea.

When the young salmons hatch in freshwater, they grow there for one or two years as parrs and then migrate towards the river mouth, change pigmentation, acquire osmoregulatory ability to live in seawater and enter the sea as smolt. The environmental factors that trigger downstream migration are usually level or rate of water flow and/or water temperature but also the growth of the individual fish (Jensen et al. 2012). These factors can stimulate migration differently in different populations and species, reflecting different adaptations to ensure high survival at sea entry (Jensen et al. 2012). This process, called smoltification, has been considered a metamorphic transition, and THs and glucocorticoids are important in its regulation (Laudet 2011, Björnsson et al. 2012, Lorgen et al. 2015, Watanabe et al. 2016). Similarly, when catadromous Atlantic eel larvae migrate from the sea to freshwater, this corresponds to a metamorphosis that is controlled by THs, but the environmental factors that control TH levels in the sea are rarely considered (Kawakami 2023). These hormones, together with corticoids, therefore appear to be a useful tool used by organisms to trigger, regulate and coordinate their life history transitions (Wada 2008).

Many gobies in the Indo-Pacific region have evolved a unique life cycle known as amphidromy. These fish typically spawn in freshwater rivers, with males guarding their eggs in nests below stones. After rapid hatching, the larvae are swept downstream into estuaries or the sea, where they live in the plankton as normal marine fish larvae and grow for approximately 30 days to more than half a year. They then migrate back to the river, completing their life cycle and reproducing. This migration back to the river, with the associated morphological, physiological and behavioural changes, is their metamorphosis and is marked by the transformation from a pelagic marine larva to a freshwater juvenile. THs play a significant role in this process. For example, in *Sicyopterus lagocephalus*, TH levels were found to be the highest when morphological changes (such as the change in the position of the mouth) were most important when the young fish entered the river. Treatment with either T3 (3,5,3'-triiodothyronine) or goitrogens that block TH synthesis will either accelerate of slow down this transformation in a manner very similar to what has been observed in other fish (Taillebois et al. 2011, Ellien et al. 2020).

Amphidromous gobies exhibit specific adaptations to support their life history strategy. They possess a remarkable sense of finding freshwater systems to recruit into and often travel long distances (up to 100 km). Some species have modified paired fins that act as suction cups, enabling them to swim in the upper reaches of rivers and even climb waterfalls.

Additionally, there are instances where some gobies have evolved landlocked populations and species that are unable to migrate between freshwater and marine environments due to geological or environmental barriers. In these landlocked situations, gobies have undergone evolutionary changes to adapt to exclusively freshwater habitats. They have larger eggs that hatch late, giving rise to large fry, but are fewer in number compared to amphidromous populations, suggesting a shift towards direct development like what is seen in tree frogs (see below). Examples of landlocked populations include *Rhinogobius* sp., which are landlocked above waterfalls in Okinawa, and *Gymnogobius isaza* in Lake Biwa, Japan. These populations demonstrate the adaptability of gobies to adapt to different environments and provide insights into the evolutionary processes involved in habitat isolation, leading to specialized behaviours and unique adaptations. The study of amphidromous gobies and their landlocked populations may therefore offer valuable information about the interconnectedness of freshwater and marine ecosystems, the role of THs in metamorphosis and the evolutionary responses of organisms to habitat changes. For example, it would be interesting to understand how changes in TH signalling explain the various developmental and morphological changes observed in landlocked *vs.* migratory species.

8.3.1.2 Evolutionary Patterns of Life History Diversity Within and Across Vertebrate Taxa

Given that our molecular knowledge of vertebrate metamorphosis has been gained mostly from studies done in amphibians and specifically in frogs, it is not surprising that this group of animals is also the one for which deviations from the classical metamorphosing pattern have been observed. There are two main types of variation in metamorphosis patterns that occur in amphibians: paedomorphosis, also often referred to as neoteny in salamanders, and direct development in frogs.

Neoteny is a type of heterochronic change in evolution (that is, a change in the rate of event in a descendant *vs.* the ancestor species) during which there is a slowing down of the development of the somatic tissues when compared to the germline. As a result, we find in these species animals that are morphologically like larvae but that can reproduce. This is referred to as paedomorphosis, whose etymology means 'having the form of a child'. This term is better than neoteny as it is more neutral: it depicts a trait (a phenotypic young stage with mature gonads) without discussing the mechanism (as we can reach such state by either neoteny, *i.e.*, slowing down of somatic development, or by progenesis, *i.e.*, acceleration of the germline development). What we observe in paedomorphic amphibians is the retention of larval character in an adult that can reproduce. This is observed in various types of salamanders that keep their external gills (typical of a larva, and hence the term perennibranchiate is often used to describe these animals). This can be facultative as in axolotl of the genus *Ambystoma*, in which metamorphosis can be triggered after TH treatment, or obligatory, as in the mudskipper genus *Necturus* and its cave-dwelling relative *Proteus*, which have been described as resistant to THs since exogenous application to THs does not produce any visible effect (Elinson and del Pino 2012, Johnson and Voss 2013). In both cases, it has been shown that TH responses appear essentially normal. The thyroid hormone receptors (TRs) have the same affinity for the ligands and are activated similarly to what we know from the TRs in *Xenopus* (Safi et al. 2004, 2006). In the obligatory paedomorphic salamanders, there is in fact no resistance to TH, as exogenous TH treatment induces a molecular response (Safi et al. 2006). In this species, the absence of metamorphosis is therefore likely a downstream effect, that is, the loss of TH-dependent control of key genes required for tissue transformation in the gills. Pushing the model forward, we can even predict that a post-embryonic period of higher TH levels and high TH sensitivity reminiscent of normal metamorphosis occurs in this species, but is not spectacular and has remained unnoticed, if present (Safi et al. 2006, Laudet 2011).

The case of Ambystomatid salamanders is more complicated because there is a wide diversity of species and contexts (Everson et al. 2021). The most widely studied species, *Ambystoma mexicanum*, is referred to as the axolotl, whose perenibranchiate larvae fail to reproduce due to insufficient TH production. Indeed, injection of TH causes a normal metamorphosis both molecularly and morphologically (reviewed in Johnson and Voss 2013). The frequency of metamorphosing salamanders varies inside the *Ambystoma* species complex, and this is at least in part driven by environmental factors such as the amount of iodine in the water (see, *e.g.*, Doyle and Whiteman 2008). However, genetic analysis and, in particular, interspecific crosses have shown that there is a genetic basis for the level of paedomorphosis (Voss et al. 2000). Therefore, it is likely that the absence of metamorphosis in *Ambystoma* is both environmental and linked to genetic difference in the control of TH production, which is upstream in the signalling cascade (Johnson and Voss 2013).

There are many interesting cases in salamanders and, in addition to *Necturus* and *Ambystoma*, other genera contain pedomorphic species, such as *Amphiuma*, *Pseudobranchus* and sirens, as well as the giant salamanders *Cryptobranchus* and *Andrias*, but these have not been molecularly studied and the basis of their lack of metamorphosis is still unknown (Hoverman 2008, Johnson and Voss 2013). In alpine lakes, the normally transforming *Triturus alpestris* has been shown to display facultative paedomorphosis linked to extreme environmental conditions, mostly differences in prey availability and temperature in altitude lakes *vs.* those in valleys (Denoël et al. 2005).

The larval stage can apparently be lost from the life cycle of amphibians with direct development, and this is particularly spectacular in anurans, in which the tadpoles and froglets exhibit very different morphologies (Elinson and del Pino 2012). The most common case is the Puerto Rican tree frog, *Eleutherodactylus coqui*. This frog lives in trees, far from ponds and, as an adaptation to this specific lifestyle, does not lay eggs in water. One consequence of this adaptation is a change in developmental mode, *i.e.*, direct development, in which the metamorphosis, with TH regulation, occurs within the egg (Callery and Elinson 2000). Specifically, TH is needed to complete the development of this species, and there is actually a change of order (displacement) of metamorphosis and hatching in this and other similar species (Heinicke et al. 2007, Hedges et al. 2008).

It is important to point out that the omission of the tadpole stages in *E. coqui* is associated with numerous other adaptations in embryos. Indeed, rather than laying thousands of small eggs (less than 1.8 mm in diameter), *E. coqui* laid a small number of large eggs (Elinson 2021). These large eggs are full of reserves that allow them to sustain the long incubation and the metamorphosis that occurs within the eggs. Here again, we see a trade-off between fecundity and specific life history traits because of a peculiar ecology. For example, during the long development of the egg, the tail is modified into a vascularized and membranous structure that allows gas exchanges (Elinson and del Pino 2012). Other spectacular adaptations concern the nutritional endoderm, a novel tissue derived from germ layer specification that provides nutrition to the embryo from the huge amount of yolk (Buchholz et al. 2007). These frogs are therefore fascinating models to study how the metamorphic process can change over evolutionary time.

The analysis of deviations from the classical biphasic life history separated by metamorphosis is much less advanced in teleost fish than in amphibians. To our knowledge, there is no clear example of paedomorphic fish with a clear reproductive larva. Still, there are cases indicative of paedomorphosis. In addition, there are several well-known examples of direct development and probably many more subtle variations that have not been characterized yet. Miniature cyprinids and gobies, such as species of the genus *Danionella*, *Paedocypris*, *Sundadanio*, *Aphia* and *Schindleria*, show obvious signs of retention of larvae characters in reproductive adults, but this can be linked to their evolution towards miniaturization rather than to an absent or delayed metamorphosis (Johnson and Brothers 1993, Mesa et al. 2005, Britz and Conway 2009, Conway et al. 2021). There have also been hypothesized cases of paedomorphosis in deep-sea fish, but this requires further investigation, and both molecular and endocrinological studies could illuminate the function of hormones in these life histories (Moore 1994).

The case of direct development is clearer and is well exemplified by the damselfish *Acanthochromis* (Kavanagh 2000) and *Altrichthys* (Bernardi et al. 2017), as well as by the apogonids *Pterapogon kauderni* and *Quinca mirifica* (Vagelli and Volpedo 2004, Vagelli 2019). In these fish, we find a very similar situation to the Puerto Rican tree frog: big eggs with a lot of yolk, allowing a long development, and, as a trade-off, a smaller number of eggs. What comes out of the eggs in these species is a small juvenile fish, not a larva. However, the ecological reason for this peculiar developmental strategy is still unknown. In apogons, it may be linked to buccal incubation, and in all known cases in fish, it is associated with parental care (see also Chapter 6, this volume). Therefore, the strategy could be to produce fewer eggs and to avoid the high predation associated with larval life. Hence, the parents invest, with long parental care, in the eggs and fry to allow a higher number of their progeny to reach adulthood. Up to now, in none of these species do we have a molecular or endocrinological analysis, and we, therefore, cannot tell if THs and/or corticoid signalling play a significant role. It would be very interesting to determine if, as in the Puerto Rican tree frog, THs are needed for the completion of embryonic development.

8.3.2 Insect Metamorphosis

MD in insects occurs within the holometabolous life cycle and is generally considered a derived life history mode, which evolved from ametaboly (direct development without final moult) and hemimetaboly (direct development with final moult) (Bellés 2020). Holometabolous insects form a pupa after larval development, which then transitions into the reproductive adult (Figures 8.2 and 8.3). Importantly, imaginal discs start forming during the embryonic phase in insects, preparing the primordial structures of adult appendages (Dubrovsky 2005). Many processes during the transition are regulated by juvenile hormones (JHs), ecdysteroids (specifically 20-hydroxyecdysone; 20E) and a range of other hormones (Figure 8.2). To review the detailed physiological functions of these hormones in a comparative context (within arthropods) is beyond the scope of this chapter, and we refer the reader to reviews on the topic (*e.g.*, Truman and Riddiford 1999, Hiruma and Kaneko 2013, Truman and Riddiford 2019) and the primer in Figure 8.2.

8.3.2.1 Environmental Control of Insect Metamorphosis

The endocrine, genetic and developmental mechanisms outlined above are tightly integrated with environmental signals within the insect's life history. These include, but are not limited to, food availability, photoperiod and temperature. Cues and signals linked to these environmental changes impact hormone and peptide production, release and function and therefore regulate developmental timing and growth during larval development and metamorphosis.

In insects, nutrient signalling is mediated through the insulin/insulin-like growth factor signalling (IIS) pathway, a highly conserved pathway that plays a pivotal role in regulating various physiological processes, including metabolism, growth, reproduction, ageing and the stress response (see also Chapters 2 and 9, this volume). Activation of the pathway involves insulin-like peptides (ILPs) and encompasses a complex regulatory network that coordinates nutrient-responsive growth signalling pathways, such as IIS and target of rapamycin (TOR), the Forkhead box O (FOXO) transcription factor and the phosphatidylinositol 3-kinase (PI3K) signalling cascade. Additionally, interactions with other hormonal pathways, such as stress-related hormones and ecdysone signalling during MD, further contribute to its regulation.

The interaction between insulin and JH in insect metamorphosis remains incompletely understood (Maestro et al. 2009). Recent evidence suggests that insulin-related peptides, along with ecdysteroids and JH, form a complex regulatory network influenced by the insect's nutritional status, highlighting their essential role in modulating developmental processes (Badisco et al. 2013). This network, along with TOR signalling, plays a significant role in growth regulation. In the tobacco hornworm *Manduca sexta*, JH controls the duration of growth by affecting the hormonal response to critical weight, which serves as a checkpoint for the transition from growth to the pupal stage (Mirth and Riddiford 2007). The decline in JH levels triggers hormonal changes, starting from the attainment of critical weight, leading to a terminal growth period and an increase in circulating ecdysone, which halts body growth (Nijhout and Williams 1974).

Starvation results in high levels of JH synthesis in the corpora allata (CA) and delays the critical weight transition. Removal of the CA results in earlier critical weight attainment at a smaller size, while the application of JH suppresses this transition and delays metamorphosis, leading to a larger size (Nijhout and Williams 1974). Interestingly, the effects of JH on growth rate seem to be dependent on the FOXO gene and are negatively regulated by insulin-like growth factor (IGF) signalling in *D. melanogaster* (Mirth et al. 2014).

Additionally, during metamorphic tissue remodelling in the cotton bollworm *Helicoverpa armigera*, 20E increases the expression of insulin-like growth factor 2 (IGF-2), which promotes the growth and proliferation of the adult midgut and fat body, preventing their death. IGF-2-like hormone activity marks progenitor cells of the imaginal fat body alongside larval fat body cells. Exploring whether the regulation of insulin ligands by steroid hormones is conserved across species would be an intriguing avenue for future research (Zhao et al. 2022).

The timing of ecdysis events in insects, including egg hatching, pupation and eclosion, can be influenced by the circadian clock and hormonal regulation (Yamamoto et al. 2017). The interplay between circadian clocks and hormone release, such as JH and ecdysteroids, determines the gate for progressing through metamorphosis. In *D. melanogaster*, specific neurons in the optic lobe and peripheral clock cells in the prothoracic gland (PG) drive the eclosion rhythm (Myers et al. 2003). Hormones like pre-ecdysis-triggering hormone, ecdysis-triggering hormone and eclosion hormone, along with ecdysteroids, play a vital role in triggering and coordinating the hormonal cascade necessary for ecdysis (Zitnan et al. 2007). The exact interaction between the circadian clock and the hormonal cascade is not fully understood, but the rhythmic expression of clock genes in pacemaker neurons regulates the timing of circadian activities (Short et al. 2016). Prothoracicotropic hormone (PTTH) neurons are associated with the brain's circadian timekeeping system, and the long-wavelength-sensitive opsin mediates light-dark cues for circadian clock entrainment (Vafopoulou et al. 2007, Wertman and Bleiker 2019).

Temperature has significant effects on insect physiology since they are ectotherms. High temperatures can accelerate development, leading to a premature metamorphosis. Such changes can impact other traits such as size, which has major implications for fitness (Nijhout et al. 2014; see also Chapter 1, this volume). Temperature stress during pre-metamorphic stages in the wood tiger moth *Arctia plantaginis* can also have carry-over effects to post-metamorphic stages and fitness consequences via changes in reproductive output (Galarza et al. 2019). Finally, desiccation tolerance, regulated in part by hormones, is another important trait. For example, severe desiccation conditions can modulate 20E levels in *Chironomus ramosus* and *D. melanogaster*, leading to desiccation-mediated delays in metamorphosis, which can also have fitness consequences (Thorat and Nath 2018).

8.3.2.2 Evolution of Insect Metamorphosis

Several reviews have been written over the years on the evolution of indirect development and metamorphosis in insects (Bellés and Santos 2014, Jindra 2019, Truman 2019, Truman and Riddiford 2019, Bellés 2020). The purpose of this section is to re-iterate the main hypotheses in the context of endocrine signals and discuss how our understanding of proximate mechanisms underlying MD can translate into evolutionary meaning.

As stated above, metamorphic (holometabolous) development in insects is evolutionarily derived from direct development (ametabolous and hemimetabolous). Selective forces in this process likely included niche differentiation between larvae and adults as well as enhanced dispersal potential driving the evolution of flight (Erezyilmaz 2006, Bellés 2019, Truman 2019). The analysis of mechanisms underlying this evolutionary transition has been broadly concerned with the similarities and differences between the holometabolous and hemimetabolous life cycles and the homology between larval stages (Erezyilmaz 2006). The Hinton hypothesis (Hinton, 1963) assumes homology between the larval stages of hemimetabolous and holometabolous insects and explains the evolution of the pupa by modifications of a hemimetabolous late larval stage. In contrast, the Berlese hypothesis (Berlese 1913) does not assume homology between larvae. Instead, it proposes that nymphal stages of ancestral hemimetabolous insects were reduced to one non-moulting pupal stage in holometabolous insects. By comparing JH and 20E levels and functions within the development of holometabolous and hemimetabolous species, both hypotheses and their modifications have incorporated extensive molecular and endocrine information over the years. This includes, for example, the role of JH and downstream genes in preventing precocious metamorphosis in both groups as well as shifts in timing of JH synthesis. Hypotheses also consider temporal and spatial changes of key downstream targets, which have been implicated in the interaction of JH and 20E signalling, such as Krüppel homolog1 (Kr-h1), broad complex (BR-C) and the E93 gene, in the so-called MEKRE93 pathway (Bellés and Santos 2014, Bellés 2019, Truman 2019). Which hypothesis is better at explaining the evolution of holometaboly is ultimately dependent on gaining better insight into how morphogenesis, cell death and remodelling are orchestrated in post-embryonic development. It also depends on gaining better insights into JH function during embryogenesis in holometabolous insects (Jindra 2019).

8.3.3 Regulation of the Bentho-Planktonic Life Cycles in Invertebrates

Many animal phyla have species with bentho-planktonic life cycles in aquatic but especially marine environments (Figure 8.1). These life histories are characterized by a larval pelagic phase in post-embryonic, pre-reproductive development. While larval forms are diverse, they often function as dispersal stages within the life cycle. This is in contrast to insects, where dispersal mainly occurs in the adult stage (see above). Before marine larvae transition from the pelagic to the benthic environment, MD begins inside the larva and forms juvenile structures. The timing of the habitat transition (settlement) is dependent on both environmental and developmental signals. The regulation of MD and settlement remains unknown for many species, but in some groups, such as the echinoderms, cnidarians, molluscs, polychaetes, tunicates and cephalochordates, some progress has been made in elucidating these processes. Previous reviews have discussed the commonalities in terms of endocrine and molecular mechanisms between these divergent groups (reviewed in Heyland and Moroz 2006). There are four general processes that occur during the metamorphic transition of animals: differentiation of juvenile/adult structures, degeneration of larval structures, metamorphic competence (the stage directly preceding the habitat transition) and the ecological/habitat transition itself. Importantly, the order of these events varies greatly between organisms and lineages (reviewed in Heyland and Moroz 2006). Among marine invertebrate larvae, metamorphic competence can last for extensive periods of time and depends on the availability of suitable settlement sites and their associated cues (Hodin 2006, Hadfield et al. 2015). Settlement, in contrast, can occur very quickly in marine environments, and major morphological changes can be associated with it, either before or after the attachment process. Across species, hormonal and neuro-endocrine regulation, apoptosis, the activation of specific signalling pathways and gene expression changes have been identified as critical components of MD and settlement among marine invertebrate species. Here we focus on a few mechanisms involving hormones and peptides, which have been identified across several taxonomic groups. We then discuss them both in an evolutionary and ecological context.

8.3.3.1 Settlement Cues and Signals

Many species select their future habitat based on the presence of conspecifics. While the fitness effects of these positive associations are somewhat obvious from a benthic perspective for organisms with bentho-planktonic life cycles, it is important to recognize that the timing and response to these cues depend on a range of environmental and ecological factors in the plankton, such as temperature, currents, light, predation and many others (Pawlik 1992, Rodriguez et al. 1993, McClintock and Baker 2001, Doll et al. 2022). Identifying specific chemical molecules responsible for inducing settlement has proven challenging, with most reports describing classes of chemical components or various bacteria whose presence on the surface serves as a trigger. Still, progress has been made to identify and characterize species-specific cues from biofilms in benthic communities that are suitable as future settlement sites (Hadfield 2011, Dobretsov and Rittschof 2020). While biofilms can provide reliable information for larvae in terms of potential future settlement sites, their composition can be extremely diverse, and finding signals common to biofilms in different habitats has been challenging. Lipopolysaccharides (LPS) from bacteria have been identified as an induction cue in serpulid polychaetes, and these compounds might have a broader significance in the settlement process across marine invertebrate species (Freckelton et al. 2022). Numerous studies have explored the stimuli that lead to larval-to-polyp transformation in various cnidarian groups (Leitz and Wagner 1993). Metabolites from brown algae have been shown to induce settlement in the hydroid *Coryne uchidae* (Kato et al. 1975). Furthermore, phospholipids and polysaccharides from *Pseudoalteromonas* and *Alcaligenes* have been identified as individual or combined inducers of *Hydractinia* planula settlement (Rischer et al. 2022). While the response to specific cues depends on the developmental stage of the larva, it is also important to recognize that a range of settlement strategies have evolved across species. For example, some species can be extremely opportunistic about when and where they settle, while others can delay the process for extensive periods of time until suitable habitats (cues) are found (Bishop et al. 2006b, Heyland et al. 2011). The mechanisms underlying both the strategies and the plasticity within those strategies require more research (Heyland et al. 2011).

8.3.3.2 Immune System-Related Signals

As organisms transition from one habitat to another, while undergoing major morphological changes in many species, it makes intuitive sense to consider the immune system and the role it might play in both the mechanisms and the evolution of MD and settlement. For example, in one ascidian species, genes related to innate immunity, including complement signalling and proteins with von Willebrand factor domains (a domain involved in cell adhesion, extracellular matrix proteins and integrin receptors; Whittaker and Hynes 2002), have been shown to contribute to the deconstruction and reconstruction processes during tail regression (Davidson and Swalla 2002). Furthermore, the LPS system discussed above has strong immunogenicity, and its ubiquitous expression by Gram-negative bacteria and structural variability make it a likely target of evolutionarily conserved innate immune receptors in larvae. In molluscs, immune-related genes have been observed to increase in expression during MD, suggesting the development of innate immunity in response to settlement cues (Balseiro et al. 2013, Liu et al. 2015, Dyachuk 2016). The pro-phenoloxidase (pro-PO) system is an important component of invertebrate innate immunity and is activated by binding to bacterial polysaccharides or fungal zymosan, leading to the production of phenoloxidase (PO). In the oyster *Crassostrea gigas*, PO expression and activity during development suggest that while embryos are initially protected via maternal provisioning, they eventually acquire their own pro-PO/PO defence system. This is particularly important during the settlement process, where exposure to external pathogens can lead to high mortality rates (Bado-Nilles et al. 2008, Thomas-Guyon et al. 2009). The pro-opiomelanocortin (POMC) system, like the pro-PO, is a precursor protein that is cleaved into peptides with diverse functions in neuro-modulation, neurotransmission, hormone regulation and growth factors (Day 2009). The POMC system is involved in various physiological processes such as pigmentation, steroidogenesis, energy homeostasis and immune responses. Hemocytes play a significant role in producing and transporting these peptides. Studies on the mussel *Mytilus edulis* have shown correlations between invertebrate and mammalian neuro-immune functions, particularly in POMC pathways involving tyrosine, dopamine (DA) and opioid processing (Malagoli et al. 2017). The evolutionary conservation of peptidergic signalling systems, including POMC, suggests the importance of inter- and intracellular communication efficiency in immune responses. These findings also highlight the potential for further research in understanding the mechanisms and pathways involved in metamorphosis and immune responses in bivalves, with implications for enhancing immunity and survival rates during this critical stage (Joyce and Vogeler 2018).

8.3.3.3 The Neuro-endocrine System

The neuro-endocrine system of marine invertebrate larvae with bentho-planktonic life cycles and metamorphosis plays a key role in detecting and transmitting environmental signals. As with many marine invertebrates, sea urchins require specific environmental cues from microbial biofilms to settle into the benthic environment (Figure 8.3). A range of neurotransmitter signalling systems have been found to be implicated in the settlement response in echinoids, including dopamine 2,3, L-DOPA (L-3,4-dihydroxyphenylalanine), glutamine, glutamic acid, nitric oxide (NO) and histamine (HA) (reviewed in Heyland and Moroz 2006, Heyland et al. 2011). This work is further supported by studies localizing neurotransmitters and peptides in components of the larval nervous system associated with the settlement of sea urchins and other echinoderms (Bisgrove and Burke 1987, Beer et al. 2001, Elphick and Thorndyke 2005, Burke et al. 2006, Elphick 2014, Katow et al. 2020). These same neurotransmitters have also been found to be involved in the settlement process in a range of other marine invertebrate groups.

In larvae of the sea urchin *Lytechinus pictus*, NO and guanosine $3',5'$-cyclic monophosphate (cGMP) negatively regulate metamorphosis, and nitric oxide synthase (NOS)-expressing neurons likely have a chemosensory role during settlement and metamorphosis (Bishop and Brandhorst 2001, 2007). HA, which can be both endogenously synthesized in animals and has exogenous origins from microorganisms (Sutherby et al. 2012), has been shown to induce the settlement response in sea urchins (Swanson et al. 2004). Research on the signalling system preceding settlement suggests that HA is a modulator of metamorphic competence in the sea urchin *Strongylocentrotus purpuratus* (Sutherby et al. 2012, Lutek et al. 2018) and may interact with both NO and TH within the MD signalling network in sea urchins (Heyland et al. 2011). Neuro-endocrine pathways play a crucial role in the process of metamorphosis in bivalve species. In particular, the dopaminergic, adrenergic and serotonergic pathways have been implicated. Studies on oyster larvae have shown that exposure to L-DOPA, a precursor of DA, induces settlement behaviour, whereas exposure to norepinephrine (NE) or epinephrine (EPI) triggers metamorphosis without attachment (reviewed in Joyce and Vogeler 2018). In the abalone *Haliotis asinina*, NO is required for the induction and progression of MD, and exogenous application of NO donors was found to significantly enhance MD (Ueda and Degnan 2014). Serotonin (5-HT), another neurotransmitter, is believed to be involved in the regulation of metamorphosis in bivalves. Its concentration increases before metamorphosis and putative serotonin receptors have been found in bivalve larvae (Cann-Moisan et al. 2002). NO signalling has also been shown to play a role in ascidian metamorphosis. In *Herdmania momus*, exogenous application of NO donors induces complete metamorphosis, suggesting that NO acts as a positive regulator of metamorphosis (Ueda and Degnan 2013). In *Ciona intestinalis*, NO signalling has been proposed as a potential modulator of apoptosis in the tail (Comes et al. 2007). In contrast to these findings, other research suggests that NO can function as an inhibitor of tail regression in *C. intestinalis* and *Boltenia villosa* (Bishop et al. 2001, Comes et al. 2007).

8.3.3.4 Hormonal Systems as Regulators of Metamorphic Development

As outlined above for vertebrates, hormonal systems play a key role in larval development of frogs and fish, and recent evidence suggests that some of these endocrine systems are not restricted to vertebrates. Extensive work on TH signalling in echinoids reveals a function of these hormones in larval and MD. Specifically: (1) sea urchins have the ability to synthesize THs from incorporated iodine (Heyland et al. 2006a, 2006b, Miller and Heyland 2010, 2013) and specific cells in embryos and larvae express TH synthesis enzymes, deiodinases and TH receptors (Cocurullo et al. 2023); (2) T4 (3,5,3′,5′-tetraiodothyronine), and to a lesser extent T3, accelerate larval development towards settlement (Heyland and Hodin 2004, Heyland et al. 2004), potentially via inhibiting NO signalling (a negative regulator of MD in sea urchins; Bishop et al. 2006b); (3) THs induce larval arm retraction via programmed cell death (Wynen and Heyland 2021, Wynen et al. 2022); and (4) THs induce and modulate skeletal growth and differentiation in both embryonic and larval development (Taylor and Heyland 2018, Taylor et al. 2023). Recent work, using single-cell transcriptomics (Cocurullo et al. 2023) and whole embryo/larval transcriptomics (Taylor et al. 2023), suggests that T4 binds to the sea urchin orthologue of αVβ3 integrin dimer (αPβG), which has been shown to function as a TH membrane receptor in mammals (Cayrol et al. 2015, Davis et al. 2021, Taylor et al. 2023). Sea urchin αPβG is expressed in skeletal mesenchyme cells and binds fluorescently labelled T4 (Cocurullo et al. 2023, Taylor et al. 2023). Interestingly, sea urchin nuclear hormone receptor beta (TRβ) orthologue expression is upregulated in larval stages of the sea urchin in response to both T4 and T3 treatment, and putative downstream targets with TH response elements (TREs) are differentially enriched (Taylor et al. 2023).

Similarly, the metamorphic transition in cephalochordates and urochordates is partially mediated by THs. Exposure of *Branchiostoma floridae* (Cephalochordata) larvae to Triiodothyroacetic acid (TRIAC) advances the timing of the larval to adult transition, which can be suppressed by coadministration of the TR inhibitor NH_3 (Paris et al. 2008b). Inhibitors of TH synthesis slow down the rate of MD in pre-transition amphioxus larvae, but coadministration of T3 restores the normal rate (Paris et al. 2010). The TR has been identified in the genome of the amphioxus species *B. floridae*, and its cDNA has been cloned (Paris et al. 2008a). In tunicate (Urochordata), metamorphosis TR signalling has been investigated in two species, *Styela clava* and *C. intestinalis*. In *S. clava*, the TR is highly expressed during larval tail regression, suggesting its role in metamorphosis (Carosa et al. 1998, Wei et al. 2020). This finding supports the hypothesis that the TR is required for MD. It has also been proposed that thyroid-like hormone signalling in tunicates may occur through the induction of the TR by TH, with the TR acting as a transcription factor in its unliganded state, as it has been proposed in other contexts (Fondell et al. 1993, Wen and Shi 2015, Morthorst et al. 2022).

Various components of thyroid-like hormone signalling have been identified in molluscs. A TR orthologue has been annotated in several mollusc species, including gastropods and bivalves, and gene activity has been shown to be involved in MD (Wang et al. 2009, Huang et al. 2015, Morthorst et al. 2022). Recent studies have also found evidence of an unexpected signalling system in molluscs. The analysis of several molluscan genomes identified ecdysone receptor (EcR) and related transcription factors in molluscs, and a preliminary work suggests involvement in metamorphosis induction and settlement (Vogeler et al. 2016). Furthermore, molluscs also have proteins belonging to the insulin superfamily, such as molluscan insulin-related peptides (MIPs), which regulate growth processes and may play a role in metamorphosis (Smit et al. 1988). Understanding these endocrine pathways in molluscs and their relationship with neurotransmitters can provide insights into larval development and metamorphic competence.

Retinoic acid (RA) signalling, another important hormone, has been proposed to be involved in the development and regulation of MD in sea stars (Asteroida) and feather stars (Crinoida) (Yamakawa and Wada 2022). Pharmacological and gene expression analysis combined with gain and loss of function studies suggest that RA signalling is required for the commencement of MD after larval settlement in sea stars (Yamakawa et al. 2018, 2019). The expression of genes related to the RA cascade, including the nuclear receptor RXR (Retinoid X Receptor) and retinol dehydrogenase, is also induced during polyp-to-jellyfish transition in *Aurelia aurita*. Laboratory experiments have shown that RA addition and incubation with a synthetic peptide derived from the CL390 protein can activate medusa formation (Fuchs et al. 2014). The combination of these signal cascades in *A. aurita* mirrors the mechanisms of metamorphosis observed in vertebrates, insects and some other marine invertebrate species (see above), where RXR acts as a co-receptor for species-specific hormone receptors.

8.3.3.5 Environmental Stressors Affecting Metamorphic Development and Settlement Among Marine Invertebrates

A range of environmental factors and stressors impact MD and settlement in echinoderms and other marine invertebrate species. Detailing the existing breadth of studies would be beyond the scope of this review. We therefore focus on some examples that are representative across groups.

A meta-analysis across marine invertebrate groups considering temperature, salinity and pH stress during embryo and larval stages suggests synergistic interactions between stressors on survival and other fitness-related traits (Przeslawski et al. 2015). In comparison to embryos, larvae were more vulnerable to pH and temperature stress, which is likely due to specific protection mechanisms that embryos have before hatching. Interaction types vary significantly across stressors, ontogenetic stages and biological responses but are rather consistent across phyla. Finally, calcifying larvae are more strongly impacted by ocean acidification than non-calcifying larvae (Przeslawski et al. 2015), as a pH decrease can weaken the skeleton and make it more energetically expensive to produce it. Ocean acidification due to increased CO_2 levels in the atmosphere is a factor that is broadly studied, but the impact of pH changes on MD and settlement among marine invertebrates has been less addressed.

While the PLP is distinct from the benthic juvenile and adult phases in marine invertebrate larvae, larval experience has been shown to impact post-settlement performance in a range of species (Pechenik et al. 1998). In metamorphic echinoids, newly settled larvae typically experience extremely high mortality rates because a functional juvenile mouth is not present for several days. Therefore, juveniles typically source larval reserves to bridge this metabolically challenging period. New data on nutrient markers (IGF/TOR/FOXO) has revealed nutrient stress post-settlement and that pre-settlement starvation directly impacts post-settlement expression of these stress markers as well as the juvenile morphology (Fadl et al. 2017, 2019). While reduced pH has been linked to increased larval mortality and stunted growth in indirectly developing sea urchins (Dorey et al. 2022), research also suggests latent effects, such as delayed growth and increased mortality in the juvenile and adult stages due to adverse larval experiences (Dubois 2014, Dorey et al. 2022). Still, important variation exists between species, which suggests that some species are more tolerant to acidification conditions than others (Espinel-Velasco et al. 2020).

In sea urchins, a broad-scale environmental cue for settlement and metamorphosis is turbulent shear in near-shore environments (Gaylord et al. 2013). By accelerating development, shear forces can make larvae responsive to specific settlement cues at an earlier time within their life cycle (Gaylord et al. 2013). The mechanism underlying this process appears to be linked to metamorphic competence and associated swimming behaviour in echinoids (Ferner et al. 2019).

8.3.3.6 Evolution of Metamorphosis Among Marine Invertebrate Groups

Metamorphosis has evolved several times independently among animal taxa (Hadfield 2000). How many times this transition happened is debatable, but based on current evidence, metamorphosis evolved independently in at least lophotrochozoans, ecdysozoans and deuterostomes (Hadfield 2000, Hodin 2006). Additionally, MD, and with that, the loss of a larval form within the life cycle, has occurred frequently as well (Strathmann 1978, 1993). These patterns raise the question about what drives gains and losses of MD and the associated presence or absence of larval forms. A common argument that can be found across theories about the evolution of metamorphosis emphasizes selection of different ontogenetic stages of life history, resulting in a conflict. Evolutionarily, this conflict is resolved by the emergence of divergent morphologies within the same organism, hence the evolution of larval forms and metamorphosis (Strathmann 1993, Bishop and Brandhorst 2003, Bishop et al. 2006a, Heyland and Moroz 2006, Hodin 2006, Hadfield et al. 2015, Truman 2019, Bellés 2020, Albecker et al. 2021).

In a range of marine invertebrate phyla, indirect development (including a larval stage after the embryo) has evolved into a mode without a larval stage (direct development) and there are very few examples of indirect development evolving from direct development among marine invertebrate phyla (Reitzel et al. 2006). Still, both life history strategies include a settlement process, in which planktonic stages undergo a transition from the planktonic to the benthic habitat (Reitzel et al. 2006, Heyland et al. 2011). Therefore, the term MD is associated with the larval form (indirect development) and with the formation of a juvenile rudiment inside the larva (Bishop et al. 2006a, Reitzel et al. 2006). The evolutionary ecological mechanisms underlying the loss of MD have been reviewed extensively (*e.g.*, Pechenik 1999, Hadfield et al. 2015, Ten Brink et al. 2020, Croll and Hadfield 2023) and involve mortality risk, food availability and dispersal potential in the plankton *vs.* the benthos. These factors also impact the timing of MD. For example, gastropod molluscs, ribbon worms, echinoderms and urochordates feature a rapid settlement process that is preceded (or followed in the case of ascidians) by slower morphological changes and preparation of the juvenile body plan inside the larval body (Hadfield 2000, Bishop and Brandhorst

2003, Heyland and Moroz 2006, Hodin 2006). This process provides a unique opportunity to study the interaction between two parallel developmental programmes in the same organism.

Molecular and cellular mechanisms underlying the loss of indirect development are less understood (but see Raff 1992). Still, as outlined earlier, the mechanisms regulating MD in animals also appear to play a role in the evolutionary changes between life histories (indirect development to direct development). Mechanisms regulating the various processes involved in metamorphosis, including the receptor–ligand interaction, have also been hypothesized to drive evolutionary changes in MD (gains and losses of metamorphic life histories in this case; Bishop and Brandhorst 2003, Heyland and Moroz 2006, Hodin 2006, Truman 2019). While some examples from frogs and insects provide evidence that changes in metamorphic hormone systems contributed to life history changes over evolutionary time (see also section on amphibians and insects above; Truman 2019, Westrick et al. 2022), less work has been conducted on marine invertebrate species. In echinoids, THs play a key role in MD, specifically the development and specialization of the larval and juvenile skeletons. When comparing directly developing (or some non-feeding echinoids) with indirectly developing species, the skeleton of the larva is strongly reduced and often absent in direct developers (Wray 1992, Hart 1996, Wray 1996). Similarly, THs regulate MD in Cephalochordates (Paris et al. 2008a, 2008b, 2010, Laudet 2011), which suggests that the regulation of MD by THs is a conserved feature for chordates (Laudet 2011). However, as outlined above, THs signalling has now been documented in a range of marine invertebrate groups, including lophotrochozoan species and the mechanisms for this process are not restricted to nuclear hormone receptors. This suggests that TH signalling may have been co-opted independently for the regulation of MD in animals. This argument is further supported by the fact that THs likely existed before the evolution of animals, and prokaryotes have been shown to incorporate iodine, iodinate tyrosine and to metabolize THs (review in Miller and Heyland 2010). Similarly, THs have been shown to be produced by phytoplankton and, therefore, provide a potential exogenous hormone source in addition or instead of endogenous synthesis. The up-regulation of endogenous TH synthesis could be a driver for the evolution of non-feeding and indirect development, as species will not have access to exogenous sources once they lose the ability to feed (Heyland and Hodin 2004, Heyland and Moroz 2005, Heyland et al. 2011).

8.4 Anthropogenic Environmental Impacts and Global Climate Change

Metamorphosis is a critical life history transition in animals, and understanding the integration of endocrine and genetic mechanisms with the environment is essential to understanding how it is triggered and regulated and how it can impact the future life of the juvenile and adult. Anthropogenic stressors, including those associated with global climate change, have significant impacts on metamorphosis and can affect larvae and juvenile stages differently (see also Chapter 18, this volume). Ocean acidification, temperature changes and altered food availability, which are all possible consequences of climate change, can all influence survival, performance and development during metamorphosis by acting on the hormonal systems controlling metamorphosis (Kurihara 2008, Whiteley 2011). However, the proximate causes and long-term consequences of these responses remain poorly studied, limiting our understanding of the underlying mechanisms and biodiversity risks (Albecker et al. 2021, Lowe et al. 2021, Suzuki and Toh 2021). Exploring how climate variability affects organisms with metamorphic life histories can provide insights into the complex interactions existing between metamorphosis and environmental factors (Lowe et al. 2021).

Insights into the scope and extent of phenotypic plasticity, which allows organisms to exhibit different traits in response to environmental changes, is crucial for understanding population responses to climate change (Suzuki and Toh 2021). Genetic assimilation, the process by which genetic variations underlying phenotypic responses in novel environments are selected, plays a significant role in this context (Kelly 2019). Hormones, as discussed in this chapter, regulate trait development and homeostasis, serving as a link between the environment and development (Levis and Pfennig 2019, Lema 2020). Changes in hormone levels can result in phenotypic differences, and the extent to which they influence phenotypic evolution depends on the level of genetic variation. Hormonally mediated traits can evolve through shifts in life history transitions or changes in adult phenotypes, as observed in amphibians and insects (Hirashima et al. 2000, Gomez-Mestre et al. 2013, Xu et al. 2020, Denver 2021). Understanding these evolutionary processes and their limitations is crucial for predicting the responses of organisms with metamorphic life histories to climate change (Suzuki and Toh 2021).

For coral reef fish, which are good ecological models, environmental stressors, such as pollutants and increased ocean temperatures, can have profound effects on the development, physiology and gene expression of organisms with metamorphic life histories (see also Chapters 18 and 19, this volume). Elevated temperatures have been found to affect growth, metabolism and gene expression in clownfish larvae, potentially accelerating their rate of development, dispersal capacity and settlement time (Moore et al. 2023). These studies very well emphasize the importance of investigating early life stages to understand the effects of ocean warming on coral reef fish and highlight the molecular mechanisms involved in the larval response to elevated temperatures (reviewed in Pouil et al. 2022, Schunter et al. 2022). The adaptive potential of species to climate change relies on phenotypic traits that increase tolerance to fluctuating conditions (Chelgren et al. 2006, Radchuk et al. 2013). Developmental or physiological constraints and trade-offs may limit the adaptive potential of metamorphosis-specific phenotypic traits, emphasizing the importance of studying the strength of selection during metamorphosis and the effects of environmental stressors (Benard 2004, Mueller et al. 2012, Martin et al. 2018, Gahm et al. 2021, Lowe et al. 2021, Roux et al. 2023).

In conclusion, investigating the impacts of climate variability and other human-induced stressors on survival, performance and development during metamorphosis is essential for understanding the responses and adaptive potential of organisms with metamorphic life histories. Phenotypic plasticity, genetic assimilation and trade-offs between functions at pre- and post-metamorphic stages play critical roles in shaping the evolutionary trajectories of these organisms. Furthermore, the effects of environmental stressors on sensory development, behaviour and ecological interactions have significant implications for population dynamics and community resilience. Future research should focus on unravelling the mechanisms underlying these responses, including the role of hormones and the interplay between genetic and environmental factors, to inform conservation and management strategies in the face of human-induced environmental changes.

References

Albecker, M.A., Wilkins, L.G.E., Krueger-Hadfield, S.A., Bashevkin, S.M., Hahn, M.W., Hare, M.P., Kindsvater, H.K., Sewell, M.A., Lotterhos, K.E., and Reitzel, A.M. (2021). Does a complex life cycle affect adaptation to environmental change? Genome-informed insights for characterizing selection across complex life cycle. *Proceedings of the Royal Society B: Biological Sciences* 288: 20212122.

Allen, J.D., and Pernet, B. (2007). Intermediate modes of larval development: bridging the gap between planktotrophy and lecithotrophy. *Evolution & Development* 9: 643–653.

Arenas-Mena, C. (2010). Indirect development, transdifferentiation and the macroregulatory evolution of metazoans. *Philosophical Transactions of the Royal Society, B: Biological Sciences* 365: 653–669.

Badisco, L., Wielendaele, P.V., and Broeck, J.V. (2013). Eat to reproduce: a key role for the insulin signaling pathway in adult insects. *Frontiers in Physiology* 4: 202.

Bado-Nilles, A., Gagnaire, B., Thomas-Guyon, H., Le Floch, S., and Renault, T. (2008). Effects of 16 pure hydrocarbons and two oils on haemocyte and haemolymphatic parameters in the Pacific oyster, *Crassostrea gigas* (Thunberg). *Toxicology in Vitro* 22: 1610–1617.

Balseiro, P., Moreira, R., Chamorro, R., Figueras, A., and Novoa, B. (2013). Immune responses during the larval stages of *Mytilus galloprovincialis*: metamorphosis alters immunocompetence, body shape and behavior. *Fish & Shellfish Immunology* 35: 438–447.

Barth, P., Berenshtein, I., Besson, M., Roux, N., Parmentier, E., Banaigs, B., and Lecchini, D. (2015). From the ocean to a reef habitat: how do the larvae of coral reef fishes find their way home? A state of art on the latest advances. *Vie et Milieu* 65(2): 91–100.

Beer, A.J., Moss, C., and Thorndyke, M. (2001). Development of serotonin-like and SALMFamide-like immunoreactivity in the nervous system of the sea urchin *Psammechinus miliaris*. *Biological Bulletin* 200: 268–280.

Bellés, X. (2019). The innovation of the final moult and the origin of insect metamorphosis. *Philosophical Transactions of the Royal Society, B: Biological Sciences* 374: 20180415.

Bellés, X. (2020). Insect metamorphosis: from natural history to regulation of development and evolution. *Journal of Experimental Zoology. Part B, Molecular and Developmental Evolution* 334: 381–382.

Bellés, X., and Santos, C.G. (2014). The MEKRE93 (Methoprene tolerant-Krüppel homolog 1-E93) pathway in the regulation of insect metamorphosis, and the homology of the pupal stage. *Insect Biochemistry and Molecular Biology* 52: 60–68.

Benard, M.F. (2004). Predator-induced phenotypic plasticity in organisms with complex life histories. *Annual Review of Ecology, Evolution, and Systematics* 35: 651–673.

Berlese, A. (1913). Intorno alle metamorfosi degli insetti. *Redia* 9: 121–137.

Bernardi, G., Crane, N.L., Longo, G.C., and Quiros, A.L. (2017). The ecology of *Altrichthys azurelineatus* and *A. curatus*, two damselfishes that lack a pelagic larval phase. *Environmental Biology of Fishes* 100: 111–120.

Besson, M., Feeney, W.E., Moniz, I., François, L., Brooker, R.M., Holzer, G., Metian, M., Roux, N., Laudet, V., and Lecchini, D. (2020). Anthropogenic stressors impact fish sensory development and survival via thyroid disruption. *Nature Communications* 11: 3614.

Bisgrove, B.W., and Burke, R.D. (1987). Development of the nervous system of the pluteus larva of *Strongylocentrotus droebachiensis*. *Cell and Tissue Research* 248: 335–343.

Bishop, C.D., Bates, W.R., and Brandhorst, B.P. (2001). Regulation of metamorphosis in ascidians involves NO/cGMP signaling and HSP90. *Journal of Experimental Zoology* 289: 374–384.

Bishop, C.D., and Brandhorst, B.P. (2001). NO/cGMP signaling and HSP90 activity represses metamorphosis in the sea urchin *Lytechinus pictus*. *Biological Bulletin* 201: 394–404.

Bishop, C.D., and Brandhorst, B.P. (2003). On nitric oxide signaling, metamorphosis, and the evolution of biphasic life cycles. *Evolution & Development* 5(5): 542–550.

Bishop, C.D., and Brandhorst, B.P. (2007). Development of nitric oxide synthase-defined neurons in the sea urchin larval ciliary band and evidence for a chemosensory function during metamorphosis. *Developmental Dynamics* 236: 1535–1546.

Bishop, C., Erezyilmaz, D., Flatt, T., Georgiou, C., Hadfield, M., Heyland, A., Hodin, J., Jacobs, M., Maslakova, S., and Pires, A. (2006a). What is metamorphosis? *Integrative and Comparative Biology* 46: 655–661.

Bishop, C.D., Huggett, M.J., Heyland, A., Hodin, J., and Brandhorst, B.P. (2006b). Interspecific variation in metamorphic competence in marine invertebrates: the significance for comparative investigations into the timing of metamorphosis. *Integrative and Comparative Biology* 46: 662–682.

Björnsson, B.T., Einarsdottir, I.E., and Power, D. (2012). Is salmon smoltification an example of vertebrate metamorphosis? Lessons learnt from work on flatfish larval development. *Aquaculture* 362: 264–272.

Bonhomme, F., and Planes, S. (2000). Some evolutionary arguments about what maintains the pelagic interval in reef fishes. *Environmental Biology of Fishes* 59: 365–383.

Brady, L.D., and Griffiths, R.A. (2000). Developmental responses to pond desiccation in tadpoles of the British anuran amphibians (*Bufo*, *B. calamita* and *Rana temporaria*). *Journal of Zoology* 252: 61–69.

Britz, R., and Conway, K.W. (2009). Osteology of *Paedocypris*, a miniature and highly developmentally truncated fish (Teleostei: Ostariophysi: Cyprinidae). *Journal of Morphology* 270: 389–412.

Buchholz, D.R., and Hayes, T.B. (2005). Variation in thyroid hormone action and tissue content underlies species differences in the timing of metamorphosis in desert frogs. *Evolution & Development* 7: 458–467.

Buchholz, D.R., Paul, B.D., Fu, L., and Shi, Y.-B. (2006). Molecular and developmental analyses of thyroid hormone receptor function in *Xenopus laevis*, the African clawed frog. *General and Comparative Endocrinology* 145: 1–19.

Buchholz, D.R., Singamsetty, S., Karadge, U., Williamson, S., Langer, C.E., and Elinson, R.P. (2007). Nutritional endoderm in a direct developing frog: a potential parallel to the evolution of the amniote egg. *Developmental Dynamics* 236: 1259–1272.

Burke, R.D., Angerer, L.M., Elphick, M.R., Humphrey, G.W., Yaguchi, S., Kiyama, T., Liang, S., Mu, X., Agca, C., Klein, W.H., Brandhorst, B.P., Rowe, M., Wilson, K., Churcher, A.M., Taylor, J.S., Chen, N., Murray, G., Wang, D., Mellott, D., Olinski, R., Hallböök, F., and Thorndyke, M.C. (2006). A genomic view of the sea urchin nervous system. *Developmental Biology* 300: 434–460.

Callery, E.M., and Elinson, R.P. (2000). Thyroid hormone-dependent metamorphosis in a direct developing frog. *Proceedings of the National Academy of Sciences of the United States of America* 97: 2615–2620.

Cann-Moisan, C., Nicolas, L., and Robert, R. (2002). Ontogenic changes in the contents of dopamine, norepinephrine and serotonin in larvae and postlarvae of the bivalve *Pecten maximus*. *Aquatic Living Resources* 15: 313–318.

Carosa, E., Fanelli, A., Ulisse, S., Di Lauro, R., Rall, J.E., and Jannini, E.A. (1998). *Ciona intestinalis* nuclear receptor 1: a member of steroid/thyroid hormone receptor family. *Proceedings of the National Academy of Sciences of the United States of America* 95: 11152–11157.

Cayrol, F., Díaz Flaqué, M.C., Fernando, T., Yang, S.N., Sterle, H.A., Bolontrade, M., Amorós, M., Isse, B., Farías, R.N., Ahn, H., Tian, Y.F., Tabbò, F., Singh, A., Inghirami, G., Cerchietti, L., and Cremaschi, G.A. (2015). Integrin αvβ3 acting as membrane receptor for thyroid hormones mediates angiogenesis in malignant T cells. *Blood* 125: 841–851.

Chelgren, N.D., Rosenberg, D.K., Heppell, S.S., and Gitelman, A.I. (2006). Carryover aquatic effects on survival of metamorphic frogs during pond emigration. *Ecological Applications* 16: 250–261.

Chocron, E.S., Sayre, N.L., Holstein, D., Saelim, N., Ibdah, J.A., Dong, L.Q., Zhu, X., Cheng, S.-Y., and Lechleiter, J.D. (2012). The trifunctional protein mediates thyroid hormone receptor-dependent stimulation of mitochondria metabolism. *Molecular Endocrinology* 26: 1117–1128.

Cocurullo, M., Paganos, P., Wood, N.J., Arnone, M.I., and Oliveri, P. (2023). Molecular and cellular characterization of the TH pathway in the sea urchin *Strongylocentrotus purpuratus*. *Cells* 12(2): 272.

Comes, S., Locascio, A., Silvestre, F., d'Ischia, M., Russo, G.L., Tosti, E., Branno, M., and Palumbo, A. (2007). Regulatory roles of nitric oxide during larval development and metamorphosis in *Ciona intestinalis*. *Developmental Biology* 306: 772–784.

Conway, K.W., Kubicek, K.M., and Britz, R., (2021). Extreme evolutionary shifts in developmental timing establish the miniature *Danionella* as a novel model in the neurosciences. *Developmental Dynamics* 250: 601–611.

Croll, R.P., and Hadfield, M.G. (2023). Development and metamorphic loss of the musculature in larvae of the nudibranch *Phestilla sibogae*: a functional ontogeny. *Acta Zoologica* 104: 231–254.

Davidson, B., and Swalla, B.J. (2002). A molecular analysis of ascidian metamorphosis reveals activation of an innate immune response. *Development* 129: 4739–4751.

Davis, P.J., Mousa, S.A., and Lin, H.-Y. (2021). Nongenomic actions of thyroid hormone: the integrin component. *Physiological Reviews* 101: 319–352.

Day, R. (2009). Proopiomelanocortin. In: *Encyclopedia of neuroscience* (ed. Squire, L.R.), 1139–1141. Academic Press.

De Robertis, E.M., and Tejeda-Muñoz, N. (2022). Evo-devo of *Urbilateria* and its larval forms. *Developmental Biology* 487: 10–20.

Denoël, M., Joly, P., and Whiteman, H.H. (2005). Evolutionary ecology of facultative paedomorphosis in newts and salamanders. *Biological Reviews of the Cambridge Philosophical Society* 80: 663–671.

Denver, R.J. (2021). Stress hormones mediate developmental plasticity in vertebrates with complex life cycles. *Neurobiology of Stress* 14: 100301.

Denver, R.J., Mirhadi, N., and Phillips, M. (1998). Adaptive plasticity in amphibian metamorphosis: response of *Scaphiopus hammondii tadpoles* to habitat desiccation. *Ecology* 79: 1859–1872.

Dobretsov, S., and Rittschof, D. (2020). Love at first taste: induction of larval settlement by marine microbes. *International Journal of Molecular Sciences* 21(3): 731.

Doll, P.C., Caballes, C.F., Hoey, A.S., Uthicke, S., Ling, S.D., and Pratchett, M.S. (2022). Larval settlement in echinoderms: a review of processes and patterns. *Oceanography and Marine Biology: An Annual Review* 60: 433–494.

Dorey, N., Butera, E., Espinel-Velasco, N., and Dupont, S. (2022). Direct and latent effects of ocean acidification on the transition of a sea urchin from planktonic larva to benthic juvenile. *Scientific Reports* 12, 5557.

Doyle, J.M., and Whiteman, H.H. (2008). Paedomorphosis in *Ambystoma talpoideum*: effects of initial body size variation and density. *Oecologia* 156, 87–94.

Dubois, P. (2014). The skeleton of postmetamorphic echinoderms in a changing world. *Biological Bulletin* 226, 223–236.

Dubrovsky, E.B., (2005). Hormonal cross talk in insect development. *Trends in Endocrinology and Metabolism* 16, 6–11.

Dyachuk, V.A. (2016). Hematopoiesis in Bivalvia larvae: cellular origin, differentiation of hemocytes, and neoplasia. *Developmental & Comparative Immunology* 65: 253–257.

Elinson, R.P. (2021). Development of a non-amphibious amphibian – an interview with a coquí. *The International Journal of Developmental Biology* 65: 171–176.

Elinson, R.P., and del Pino, E.M. (2012). Developmental diversity of amphibians. *WIREs Developmental Biology* 1: 345–369.

Ellien, C., Causse, R., Werner, U., Teichert, N., and Rousseau, K. (2020). Looking for environmental and endocrine factors inducing the transformation of *Sicyopterus lagocephalus* (Pallas 1770) (Teleostei: Gobiidae: Sicydiinae) freshwater prolarvae into marine larvae. *Aquatic Ecology* 54: 163–180.

Elphick, M.R. (2014). SALMFamide *salmagundi*: the biology of a neuropeptide family in echinoderms. *General and Comparative Endocrinology* 205: 23–35.

Elphick, M.R., and Thorndyke, M.C. (2005). Molecular characterisation of SALMFamide neuropeptides in sea urchins. *Journal of Experimental Biology* 208: 4273–4282.

Erezyilmaz, D. (2006). Imperfect eggs and oviform nymphs: a history of ideas about the origins of insect metamorphosis. *Integrative and Comparative Biology* 46(6): 795–807.

Espinel-Velasco, N., Agüera, A., and Lamare, M. (2020). Sea urchin larvae show resilience to ocean acidification at the time of settlement and metamorphosis. *Marine Environmental Research* 159: 104977.

Everson, K.M., Gray, L.N., Jones, A.G., Lawrence, N.M., Foley, M.E., Sovacool, K.L., Kratovil, J.D., Hotaling, S., Hime, P.M., Storfer, A., Parra-Olea, G., Percino-Daniel, R., Aguilar-Miguel, X., O'Neill, E.M., Zambrano, L., Shaffer, H.B., and Weisrock, D.W. (2021). Geography is more important than life history in the recent diversification of the tiger salamander complex. *Proceedings of the National Academy of Sciences of the United States of America* 18: e2014719118.

Fadl, A.E.A., Mahfouz, M.E., El-Gamal, M.M.T., and Heyland, A. (2017). New biomarkers of post-settlement growth in the sea urchin *Strongylocentrotus purpuratus*. *Heliyon* 3: e00412.

Fadl, A.E.A., Mahfouz, M.E., El-Gamal, M.M.T., and Heyland, A. (2019). Onset of feeding in juvenile sea urchins and its relation to nutrient signalling. *Invertebrate Reproduction & Development* 63: 11–22.

Ferner, M.C., Hodin, J., Ng, G., and Gaylord, B. (2019). Brief exposure to intense turbulence induces a sustained life-history shift in echinoids. *Journal of Experimental Biology* 222(4): 1–13.

Fondell, J.D., Roy, A.L., and Roeder, R.G. (1993). Unliganded thyroid-hormone receptor inhibits formation of a functional preinitiation complex – implications for active repression. *Genes & Development* 7: 1400–1410.

Freckelton, M.L., Nedved, B.T., Cai, Y.-S., Cao, S., Turano, H., Alegado, R.A., and Hadfield, M.G. (2022). Bacterial lipopolysaccharide induces settlement and metamorphosis in a marine larva. *Proceedings of the National Academy of Sciences of the United States of America* 119: e2200795119.

Fuchs, B., Wang, W., Graspeuntner, S., Li, Y., Insua, S., Herbst, E.M., Dirksen, P., Böhm, A.M., Hemmrich, G., Sommer, F., Domazet-Lošo, T., Klostermeier, U.C., Anton-Erxleben, F., Rosenstiel, P., Bosch, T.C., and Khalturin, K. (2014). Regulation of polyp-to-jellyfish transition in *Aurelia aurita*. *Current Biology* 24: 263–273.

Gahm, K., Arietta, A.A., and Skelly, D.K. (2021). Temperature-mediated trade-off between development and performance in larval wood frogs (*Rana sylvatica*). *Journal of Experimental Zoology Part A, Ecological and Integrative Physiology* 335: 146–157.

Galarza, J.A., Dhaygude, K., Ghaedi, B., Suisto, K., Valkonen, J., and Mappes, J. (2019). Evaluating responses to temperature during pre-metamorphosis and carry-over effects at post-metamorphosis in the wood tiger moth (*Arctia plantaginis*). *Philosophical Transactions of the Royal Society, B: Biological Sciences* 374: 20190295.

Gaylord, B., Hodin, J., and Ferner, M.C. (2013). Turbulent shear spurs settlement in larval sea urchins. *Proceedings of the National Academy of Sciences of the United Stated of America* 110(17): 6901–6906.

Gomez-Mestre, I., Kulkarni, S., and Buchholz, D.R. (2013). Mechanisms and consequences of developmental acceleration in tadpoles responding to pond drying. *PLoS ONE* 8(12): e84266.

Hadfield, M.G. (2000). Why and how marine-invertebrate larvae metamorphose so fast. *Seminars in Cell and Developmental Biology* 11: 437–443.

Hadfield, M.G. (2011). Biofilms and marine invertebrate larvae: what bacteria produce that larvae use to choose settlement sites. *Annual Review of Marine Science* 3: 453–470.

Hadfield, M.G., Carpizo-Ituarte, E.J., del Carmen, K., and Nedved, B.T. (2015). Metamorphic competence, a major adaptive convergence in marine invertebrate larvae. *American Zoologist* 41: 1123–1131.

Hart, M.W. (1996). Evolutionary loss of larval feeding: development, form and function in a facultatively feeding larva, *Brisaster latifrons*. *Evolution* 50: 174–187.

Hedges, S.B., Duellman, W.E., and Heinicke, M.P. (2008). New World direct-developing frogs (Anura: Terrarana): molecular phylogeny, classification, biogeography, and conservation. *Zootaxa* 1737: 1–182.

Heinicke, M., Duellman, W., and Hedges, S. (2007). Major Caribbean and Central American frog faunas originated by ancient oceanic dispersal. *Proceedings of the National Academy of Sciences of the United States of America* 104: 10092–10097.

Heyland, A., Degnan, S., and Reitzel, A.M. (2011). Emerging patterns in the regulation and evolution of marine invertebrate settlement and metamorphosis. In: *Mechanisms of life history evolution: the genetics and physiology of life history traits and trade-offs* (eds. Flatt, T., and Heyland, A.), 29–42. Oxford University Press.

Heyland, A., and Hodin, J. (2004). Heterochronic developmental shift caused by thyroid hormone in larval sand dollars and its implications for phenotypic plasticity and the evolution of nonfeeding development. *Evolution* 58: 524–538.

Heyland, A., and Moroz, L.L. (2005). Cross-kingdom hormonal signaling: an insight from thyroid hormone functions in marine larvae. *Journal of Experimental Biology* 208: 4355–4361.

Heyland, A., and Moroz, L.L. (2006). Signaling mechanisms underlying metamorphic transitions in animals. *Integrative and Comparative Biology* 46: 743–759.

Heyland, A., Price, D.A., Bodnarova-buganova, M., and Moroz, L.L. (2006a). Thyroid hormone metabolism and peroxidase function in two non-chordate animals. *Journal of Experimental Zoology. Part B, Molecular and Developmental Evolution* 306: 551–566.

Heyland, A., Reitzel, A.M., and Hodin, J. (2004). Thyroid hormones determine developmental mode in sand dollars (Echinodermata: Echinoidea). *Evolution & Development* 6: 382–392.

Heyland, A., Reitzel, A.M., Price, D.A., and Moroz, L.L. (2006b). Endogenous thyroid hormone synthesis in facultative planktotrophic larvae of the sand dollar *Clypeaster rosaceus*: implications for the evolutionary loss of larval feeding. *Evolution & Development* 8: 568–579.

Hinton, H.E., (1963). The origin and function of the pupal stage. *Proceedings of the Royal Entomological Society A: General Entomology* 38: 77–85.

Hirashima, A., Rauschenbach, I.Y., and Sukhanova, M.J. (2000). Ecdysteroids in stress responsive and nonresponsive *Drosophila virilis* lines under stress conditions. *Bioscience, Biotechnology, and Biochemistry* 64: 2657–2662.

Hiruma, K., and Kaneko, Y. (2013). Hormonal regulation of insect metamorphosis with special reference to juvenile hormone biosynthesis. *Current Topics in Developmental Biology* 103: 73–100.

Hodin, J. (2006). Expanding networks: signaling components in and a hypothesis for the evolution of metamorphosis. *Integrative and Comparative Biology* 46: 719–742.

Holzer, G., Besson, M., Lambert, A., François, L., Barth, P., Gillet, B., Hughes, S., Piganeau, G., Leulier, F., and Viriot, L. (2017). Fish larval recruitment to reefs is a thyroid hormone-mediated metamorphosis sensitive to the pesticide chlorpyrifos. *eLife* 6: e27595.

Hoverman, J. (2008). Digit reduction, body size, and paedomorphosis in salamanders. *Evolution & Development* 10: 449–463.

Huang, W., Xu, F., Qu, T., Zhang, R., Li, L., Que, H., and Zhang, G. (2015). Identification of thyroid hormones and functional characterization of thyroid hormone receptor in the Pacific oyster *Crassostrea gigas* provide insight into evolution of the thyroid hormone system. *PLoS ONE* 10(12): e0144991.

Jensen, A.J., Finstad, B., Fiske, P., Hvidsten, N.A., Rikardsen, A.H., and Saksgard, L. (2012). Timing of smolt migration in sympatric populations of Atlantic salmon (*Salmo salar*), brown trout (*Salmo trutta*), and Arctic char (*Salvelinus alpinus*). *Canadian Journal of Fisheries and Aquatic Sciences* 69: 711–723.

Jindra, M. (2019). Where did the pupa come from? The timing of juvenile hormone signalling supports homology between stages of hemimetabolous and holometabolous insects. *Philosophical Transactions of the Royal Society, B: Biological Sciences* 374: 20190064.

Johnson, G.D., and Brothers, E. (1993). *Schindleria*: a paedomorphic goby (Teleostei: Aobioidei). *Bulletin of Marine Science* 52: 441–471.

Johnson, C.K., and Voss, S.R. (2013). Salamander paedomorphosis: linking thyroid hormone to life history and life cycle evolution. *Current Topics in Developmental Biology* 103: 229–258.

Joyce, A., and Vogeler, S. (2018). Molluscan bivalve settlement and metamorphosis: neuroendocrine inducers and morphogenetic responses. *Aquaculture* 487: 64–82.

Kato, T., Kumanireng, A.S., Ichinose, I., Kitahara, Y., Kakinuma, Y., and Kato, Y. (1975). Structure and synthesis of active component from a marine alga, *Sargassum tortile*, which induces the settling of swimming larvae of *Coryne uchidai*. *Chemistry Letters* 4: 335–338.

Katow, H., Yoshida, H., and Kiyomoto, M. (2020). Initial report of γ-aminobutyric acidergic locomotion regulatory system and its 3-mercaptopropionic acid-sensitivity in metamorphic juvenile of sea urchin, *Hemicentrotus pulcherrimus*. *Scientific Reports* 10: 778.

Kavanagh, K.D. (2000). Larval brooding in the marine damselfish *Acanthochromis polyacanthus* (Pomacentridae) is correlated with highly divergent morphology, ontogeny and life-history traits. *Bulletin of Marine Science* 66: 321–337.

Kawakami, Y. (2023). Sensitivity of *Anguilliformes leptocephali* to metamorphosis stimulated by thyroid hormone depends on larval size and metamorphic stage. *Comparative Biochemistry and Physiology. Part A, Molecular & Integrative Physiology* 276: 111339276.

Kelly, M. (2019). Adaptation to climate change through genetic accommodation and assimilation of plastic phenotypes. *Philosophical Transactions of the Royal Society, B: Biological Sciences* 374: 20180176.

Kurihara, H. (2008). Effects of CO_2-driven ocean acidification on the early developmental stages of invertebrates. *Marine Ecology Progress Series* 373: 275–284.

Laudet, V. (2011). The origins and evolution of vertebrate metamorphosis. *Current Biology* 21: 726–737.

Lecchini, D., Polti, S., Nakamura, Y., Mosconi, P., Tsuchiya, M., Remoissenet, G., and Planes, S. (2006). New perspectives on aquarium fish trade. *Fisheries Science* 72: 40–47.

Leis, J.M. (2006). Are larvae of demersal fishes plankton or nekton? *Advances in Marine Biology* 51: 57–141.

Leis, J.M., and McCormick, M.I. (2002). The biology, behavior, and ecology of the pelagic, larval stage of coral reef fishes. In: *Coral reef fishes: dynamics and diversity in a complex ecosystem* (ed. Sale, P.F.), 171–199. Academic Press.

Leitz, T., and Wagner, T. (1993). The marine bacterium *Alteromonas espejiana* induces metamorphosis of the hydroid *Hydractinia echinata*. *Marine Biology* 115: 173–178.

Lema, S.C. (2020). Hormones, developmental plasticity, and adaptive evolution: endocrine flexibility as a catalyst for 'plasticity-first' phenotypic divergence. *Molecular and Cellular Endocrinology* 502: 110678.

Levis, N.A., and Pfennig, D.W. (2019). Phenotypic plasticity, canalization, and the origins of novelty: Evidence and mechanisms from amphibians. *Seminars in Cell & Developmental Biology* 88: 80–90.

Liu, Z., Zhou, Z., Wang, L., Jiang, S., Wang, W., Zhang, R., and Song, L. (2015). The immunomodulation mediated by a delta-opioid receptor for [Met5]-enkephalin in oyster *Crassostrea gigas*. *Developmental & Comparative Immunology* 49: 217–224.

Lorgen, M., Casadei, E., Król, E., Douglas, A., Birnie, M.J., Ebbesson, L.O., Nilsen, T.O., Jordan, W.C., Jørgensen, E.H., and Dardente, H. (2015). Functional divergence of type 2 deiodinase paralogs in the Atlantic salmon. *Current Biology* 25: 936–941.

Lowe, W.H., Martin, T.E., Skelly, D.K., and Woods, H.A. (2021). Metamorphosis in an era of increasing climate variability. *Trends in Ecology & Evolution* 36: 360–375.

Lutek, K., Dhaliwal, R.S., van Raay, T.J., and Heyland, A. (2018). Sea urchin histamine receptor 1 regulates programmed cell death in larval *Strongylocentrotus purpuratus*. *Scientific Reports* 8: 4002.

Maestro, J.L., Cobo, J., and Bellés, X. (2009). Target of rapamycin (TOR) mediates the transduction of nutritional signals into juvenile hormone production. *Journal of Biological Chemistry* 284: 5506–5513.

Malagoli, D., Mandrioli, M., Tascedda, F., and Ottaviani, E. (2017). Circulating phagocytes: the ancient and conserved interface between immune and neuroendocrine function. *Biological Reviews* 92: 369–377.

Martin, T.E., Tobalske, B., Riordan, M.M., Case, S.B., and Dial, K.P. (2018). Age and performance at fledging are a cause and consequence of juvenile mortality between life stages. *Science Advances* 4: eaar1988.

McClintock, J.B., and Baker, B.J. (2001). *Marine chemical ecology*. CRC Press.

McCormick, M., Makey, L., and Dufour, V. (2002). Comparative study of metamorphosis in tropical reef fishes. *Marine Biology* 141: 841–853.

McMenamin, S.K., and Parichy, D.M. (2013). Metamorphosis in teleosts. *Current Topics in Developmental Biology* 103: 127–165.

Mesa, M.L., Arneri, E., Caputo, V., and Iglesias, M. (2005). The transparent goby, *Aphia minuta*. *Reviews in Fish Biology and Fisheries* 15: 89–109.

Miller, A.E., and Heyland, A. (2010). Endocrine interactions between plants and animals: implications of exogenous hormone sources for the evolution of hormone signaling. *General and Comparative Endocrinology* 166: 455–461.

Miller, A.E., and Heyland, A. (2013). Iodine accumulation in sea urchin larvae is dependent on peroxide. *Journal of Experimental Biology* 216: 915–926.

Miller, B., and Kendall, A.W. (2019). *Early life history of marine fishes*. University of California Press.

Mirth, C.K., and Riddiford, L.M. (2007). Size assessment and growth control: how adult size is determined in insects. *BioEssays* 29: 344–355.

Mirth, C.K., Tang, H., Makohon-Moore, S.C., Salhadar, S., Gokhale, R.H., Warner, R.D., Koyama, T., Riddiford, L.M., and Shingleton, A.W. (2014). Juvenile hormone regulates body size and perturbs insulin sgnaling in *Drosophila*. *Proceedings of the National Academy of Sciences of the United States of America* 111(19): 7018–7023.

Moore, J. (1994). What is the role of paedomorphosis in deep-sea fish evolution? *Fourth Indo-Pacific Fish Conference* 1: 448–461.

Moore, B., Jolly, J., Izumiyama, M., Kawai, E., Ryu, T., and Ravasi, T. (2023). Clownfish larvae exhibit faster growth, higher metabolic rates and altered gene expression under future ocean warming. *Science of the Total Environment* 873: 162296.

Morthorst, J.E., Holbech, H., Croze, N.D., and LeBlanc, G.A. (2022). Thyroid-like hormone signaling in invertebrates and its potential role in initial screening of thyroid hormone system disrupting chemicals. *Integrated Environmental Assessment and Management*, 19(1): 63–82.

Mueller, C.A., Augustine, S., Kooijman, S.A., Kearney, M.R., and Seymour, R.S. (2012). The trade-off between maturation and growth during accelerated development in frogs. *Comparative Biochemistry and Physiology. Part A, Molecular & Integrative Physiology* 163: 95–102.

Myers, E.M., Yu, J., and Sehgal, A. (2003). Circadian control of eclosion: interaction between a central and peripheral clock in *Drosophila melanogaster*. *Current Biology* 13: 526–533.

Nijhout, H.F., Riddiford, L.M., Mirth, C., Shingleton, A.W., Suzuki, Y., and Callier, V. (2014). The developmental control of size in insects. *WIREs Developmental Biology* 3: 113–134.

Nijhout, H.F., and Williams, C.M. (1974). Control of moulting and metamorphosis in the tobacco hornworm, *Manduca sexta* (L.): cessation of juvenile hormone secretion as a trigger for pupation. *Journal of Experimental Biology* 61: 493–501.

Orton, G.L. (1953). The systematics of vertebrate larvae. *Systematic Zoology* 2: 63–75.

Paris, M., Brunet, F., Markov, G.V., Schubert, M., and Laudet, V. (2008a). The amphioxus genome enlightens the evolution of the thyroid hormone signaling pathway. *Development Genes and Evolution* 218: 667–680.

Paris, M., Escriva, H., Schubert, M., Brunet, F., Brtko, J., Ciesielski, F., Roecklin, D., Vivat-Hannah, V., Jamin, E.L., Cravedi, J.-P., Scanlan, T.S., Renaud, J.-P., Holland, N.D., and Laudet, V. (2008b). Amphioxus postembryonic development reveals the homology of chordate metamorphosis. *Current Biology* 18: 825–830.

Paris, M., Hillenweck, A., Bertrand, S., Delous, G., Escriva, H., Zalko, D., Cravedi, J.-P., and Laudet, V. (2010). Active metabolism of thyroid hormone during metamorphosis of amphioxus. *Integrative and Comparative Biology* 50: 63–74.

Pawlik, J.R. (1992). Chemical ecology of the settlement of benthic marine invertebrates. *Oceanography and Marine Biology* 97(2): 193–207.

Pechenik, J.A. (1999). On the advantages and disadvantages of larval stages in benthic marine invertebrate life cycles. *Marine Ecology Progress Series* 177: 269–297.

Pechenik, J.A., Estrella, M.S., and Hammer, K. (1996). Food limitation stimulates metamorphosis of competent larvae and alters postmetamorphic growth rate in the marine prosobranch gastropod *Crepidula fornicata*. *Marine Biology* 127: 267–275.

Pechenik, J.A., Wendt, D.E., and Jarrett, J.N. (1998). Metamorphosis is not a new beginning: larval experience influences juvenile performance. *BioScience* 48: 901–910.

Pouil, S., Besson, M. and Metian, M., (2022). Anemonefishes as models. In: *Ecotoxicology, evolution, development and ecology of Aaemonefishes* (eds. Laudet, V., and Ravasipp, T.), 275–284. CRC Press.

Przeslawski, R., Byrne, M., and Mellin, C. (2015). A review and meta-analysis of the effects of multiple abiotic stressors on marine embryos and larvae. *Global Change Biology* 21: 2122–2140.

Radchuk, V., Turlure, C., and Schtickzelle, N. (2013). Each life stage matters: the importance of assessing the response to climate change over the complete life cycle in butterflies. *Journal of Animal Ecology* 82: 275–285.

Raff, R.A. (1992). Direct-developing sea urchins and the evolutionary reorganization of early development. *BioEssays* 14: 211–218.

Reitzel, A.M., Sullivan, J.C., and Finnerty, J.R. (2006). Qualitative shift to indirect development in the parasitic sea anemone *Edwardsiella lineata*. *Integrative and Comparative Biology* 46: 827–837.

Rischer, M., Guo, H., and Beemelmanns, C. (2022). Signalling molecules inducing metamorphosis in marine organisms. *Natural Product Reports* 39: 1833–1855.

Rodriguez, S.R., Ojeda, F.P., and Inestrosa, N.C. (1993). Settlement of benthic marine invertebrates. *Marine Ecology Progress Series* 97: 193–207.

Rose, C.S. (2005). Integrating ecology and developmental biology to explain the timing of frog metamorphosis. *Trends in Ecology & Evolution* 20: 129–135.

Roux, N., Miura, S., Dussenne, M., Tara, Y., Lee, S.-H., de Bernard, S., Reynaud, M., Salis, P., Barua, A., and Boulahtouf, A. (2023). The multi-level regulation of clownfish metamorphosis by thyroid hormones. *Cell Reports* 42(7): 112661.

Safi, R., Bertrand, S.P., Marchand, O., Duffraisse, M., de Luze, A., Vanacker, J.-M., Maraninchi, M., Margotat, A., Demeneix, B., and Laudet, V. (2004). The Axolotl (*Ambystoma mexicanum*), a neotenic amphibian, expresses functional thyroid hormone receptors. *Endocrinology* 145: 760–772.

Safi, R., Vlaeminck-Guillem, V., Duffraisse, M., Seugnet, I., Plateroti, M., Margotat, A., Duterque-Coquillaud, M., Crespi, E.J., Denver, R.J., Demeneix, B., and Laudet, V. (2006). Pedomorphosis revisited: thyroid hormone receptors are functional in *Necturus maculosus*. *Evolution & Development* 8: 284–292.

Schreiber, A.M. (2013). Flatfish: an asymmetric perspective on metamorphosis. In: *Current topics in developmental biology* (ed. Shi, Y.-B.), 167–194. Academic Press.

Schunter, C., Donelson, J., Munday, P., and Ravasi, T. (2022). Resilience and adaptation to local and global environmental change. In: *Ecotoxicology, evolution, development and ecology of Aaemonefishes* (eds. Laudet, V., and Ravasipp, T.), 253–274. CRC Press.

Short, C.A., Meuti, M.E., Zhang, Q., and Denlinger, D.L. (2016). Entrainment of eclosion and preliminary ontogeny of circadian clock gene expression in the flesh fly, *Sarcophaga crassipalpis*. *Journal of Insect Phsiology* 93–94: 28–35.

Sinha, R.A., Singh, B.K., and Yen, P.M. (2018). Direct effects of thyroid hormones on hepatic lipid metabolism. *Nature Reviews Endocrinology* 14: 259–269.

Smit, A.B., Vreugdenhil, E., Ebberink, R.H.M., Geraerts, W.P.M., Klootwijk, J., and Joosse, J. (1988). Growth-controlling molluscan neurons produce the precursor of an insulin-related peptide. *Nature* 331: 535–538.

Smith, A.B. (1997). Echinoderm larvae and phylogeny. *Annual Review of Ecology and Systematics* 28: 219–241.

Strathmann, R.R. (1978). The evolution and loss of feeding larval stages of marine invertebrates. *Evolution* 32: 894–906.

Strathmann, R.R. (1985). Feeding and nonfeeding larval development and life-history evolution in marine invertebrates. *Annual Review of Ecology and Systematics* 16: 339–361.

Strathmann, R.R. (1993). Hypotheses on the origins of marine larvae. *Annual Review of Ecology and Systematics* 24: 89–117.

Strathmann, R.R. (2020). Multiple origins of feeding head larvae by the early Cambrian. *Canadian Journal of Zoology* 98: 761–776.

Sutherby, J., Giardini, J.-L., Nguyen, J., Wessel, G., Leguia, M., and Heyland, A. (2012). Histamine is a modulator of metamorphic competence in *Strongylocentrotus purpuratus* (Echinodermata: Echinoidea). *BMC Developmental Biology* 12: 14.

Suzuki, Y, and Toh, L. (2021). Constraints and opportunities for the evolution of metamorphic organisms in a changing climate. *Frontiers in Ecology and Evolution* 9: 2021.

Swanson, R.L., Williamson, J.E., de Nys, R., Kumar, N., Bucknall, M.P., and Steinberg, P.D. (2004). Induction of settlement of larvae of the sea urchin *Holopneustes purpurascens* by histamine from a host alga. *Biological Bulletin* 206: 161–172.

Taillebois, L., Keith, P., Valade, P., Torres, P., Baloche, S., Dufour, S., and Rousseau, K. (2011). Involvement of thyroid hormones in the control of larval metamorphosis in *Sicyopterus lagocephalus* (Teleostei: Gobioidei) at the time of river recruitment. *General and Comparative Endocrinology* 173: 281–288.

Taylor, E., and Heyland, A. (2018). Thyroid hormones accelerate initiation of skeletogenesis via MAPK (ERK1/2) in larval sea urchins (*Strongylocentrotus purpuratus*). *Frontiers in Endocrinology* 9: 439.

Taylor, E., Wynen, H., and Heyland, A. (2023). Thyroid hormone membrane receptor binding and transcriptional regulation in the sea urchin *Strongylocentrotus purpuratus*. *Frontiers in Endocrinology* 14: 1195733.

Ten Brink, H., Onstein, R.E., and de Roos, A.M. (2020). Habitat deterioration promotes the evolution of direct development in metamorphosing species. *Evolution* 74: 1826–1850.

Thomas-Guyon, H., Gagnaire, B., Bado-Nilles, A., Bouilly, K., Lapègue, S., and Renault, T. (2009). Detection of phenoloxidase activity in early stages of the Pacific oyster *Crassostrea gigas* (Thunberg). *Developmental & Comparative Immunology* 33: 653–659.

Thorat, L., and Nath, B.B. (2018). Insects with survival kits for desiccation tolerance under extreme water deficits. *Frontiers in Physiology* 9: 2018.

Truman, J.W. (2019). The evolution of insect metamorphosis. *Current Biology* 29: R1252–R1268.

Truman, J.W., and Riddiford, L.M. (1999). The origins of insect metamorphosis. *Nature* 401: 447–452.

Truman, J.W., and Riddiford, L.M. (2019). The evolution of insect metamorphosis: a developmental and endocrine view. *Philosophical Transactions of the Royal Society, B: Biological Sciences* 374: 20190070.

Ueda, N., and Degnan, S.M. (2013). Nitric oxide acts as a positive regulator to induce metamorphosis of the ascidian *Herdmania momus*. *PLoS ONE* 8(9): e72797.

Ueda, N., and Degnan, S.M. (2014). Nitric oxide is not a negative regulator of metamorphic induction in the abalone *Haliotis asinina*. *Frontiers in Marine Science* 1: 21.

Vafopoulou, X., Steel, C., and Terry, K.L. (2007). Neuroanatomical relations of prothoracicotropic hormone neurons with the circadian timekeeping system in the brain of larval and adult *Rhodnius prolixus* (Hemiptera). *The Journal of Comparative Neurology* 503(4): 511–524.

Vagelli, A.A., (2019). The reproductive biology and embryology of *Quinca mirifica*, an apogonid with direct development that poduces non-functional oocytes. *Copeia* 107: 36–60.

Vagelli, A.A., and Volpedo, A.V. (2004). Reproductive ecology of *Pterapogon kauderni*, an endemic apogonid from Indonesia with direct development. *Environmental Biology of Fishes* 70: 235–245.

Vogeler, S., Bean, T.P., Lyons, B.P., and Galloway, T.S. (2016). Dynamics of nuclear receptor gene expression during Pacific oyster development. *BMC Developmental Biology* 16: 33.

Voss, S.R., Shaffer, H.B., Taylor, J., Safi, R., and Laudet, V. (2000). Candidate gene analysis of thyroid hormone receptors in metamorphosing vs. nonmetamorphosing salamanders. *Heredity* 85: 107–114.

Wada, H., (2008). Glucocorticoids: mediators of vertebrate ontogenetic transitions. *General and Comparative Endocrinology* 156, 441–453.

Wang, S., Zhang, S., Zhao, B., and Lun, L. (2009). Up-regulation of C/EBP by thyroid hormones: a case demonstrating the vertebrate-like thyroid hormone signaling pathway in amphioxus. *Molecular and Cellular Endocrinology* 313: 57–63.

Watanabe, Y., Grommen, S.V., and De Groef, B. (2016). Corticotropin-releasing hormone: mediator of vertebrate life stage transitions? *General and Comparative Endocrinology* 228: 60–68.

Watanabe, S., Iida, M., Lord, C., Keith, P., and Tsukamoto, K. (2014). Tropical and temperate freshwater amphidromy: a comparison between life history characteristics of Sicydiinae, ayu, sculpins and galaxiids. *Reviews in Fish Biology and Fisheries* 24: 1–14.

Wei, J., Zhang, J., Lu, Q., Ren, P., Guo, X., Wang, J., Li, X., Chang, Y., Duan, S., Wang, S., Yu, H., Zhang, X., Yang, X., Gao, H., and Dong, B. (2020). Genomic basis of environmental adaptation in the leathery sea squirt (*Styela clava*). *Molecular Ecology Resources* 20: 1414–1431.

Wen, L., and Shi, Y.-B. (2015). Unliganded thyroid hormone receptor α controls developmental timing in *Xenopus tropicalis*. *Endocrinology* 156: 721–734.

Wertman, D.L., and Bleiker, K.P. (2019). Shedding new light upon circadian emergence rhythmicity in the mountain pine beetle (Coleoptera: Curculionidae: Scolytinae). *The Canadian Entomologist* 151(3): 273–277.

Westrick, S.E., Laslo, M., and Fischer, E.K. (2022). The big potential of the small frog *Eleutherodactylus coqui*. *eLife* 11: e73401.

Whiteley, N.M. (2011). Physiological and ecological responses of crustaceans to ocean acidification. *Marine Ecology Progress Series* 430: 257–271.

Whittaker, C.A., and Hynes, R.O. (2002). Distribution and evolution of von Willebrand/integrin A domains: widely dispersed domains with roles in cell adhesion and elsewhere. *Molecular Biology of the Cell* 13: 3369–3387.

Wray, G.A. (1992). The evolution of larval morphology during the post-Paleozoic radiation of echinoids. *Paleobiology* 18: 258–287.

Wray, G.A. (1996). Parallel evolution of nonfeeding larvae in echinoids. *Society of Systematic Biologists* 45: 308–322.

Wynen, H., and Heyland, A. (2021). Hormonal regulation of programmed cell death in sea urchin metamorphosis. *Frontiers in Ecology and Evolution* 9: 733787.

Wynen, H., Taylor, E., and Heyland, A. (2022). Thyroid hormone-induced cell death in sea urchin metamorphic development. *Journal of Experimental Biology* 225(23): jeb244560.

Xu, L.C., Nunes, C., Wang, V.R., Saito, A., Chen, T., Basak, P., Chang, J.J., Koyama, T., and Suzuki, Y. (2020). Distinct nutritional and endocrine regulation of prothoracic gland activities underlies divergent life history strategies in *Manduca sexta* and *Drosophila melanogaster*. *Insect Biochemistry and Molecular Biology* 119: 103335.

Yamakawa, S., Morino, Y., Honda, M., and Wada, H. (2018). The role of retinoic acid signaling in starfish metamorphosis. *EvoDevo* 9: 10.

Yamakawa, S., Morino, Y., Honda, M., and Wada, H. (2019). Regulation of metamorphosis by environmental cues and retinoic acid signaling in the lecithotrophic larvae of the starfish *Astropecten latespinosus*. *The Biological Bulletin* 237: 213–226.

Yamakawa, S., and Wada, H. (2022). Machinery and developmental role of retinoic acid signaling in echinoderms. *Cells* 11(3): 523.

Yamamoto, M., Nishimura, K., and Shiga, S. (2017). Clock and hormonal controls of an eclosion gate in the flesh fly *Sarcophaga crassipalpis*. *Zoological Science* 34(2): 151–160.

Zhao, Y.M., Wang, X.P., Jin, K.Y., Dong, D.J., Reiff, T., and Zhao, X.F. (2022). Insulin-like growth factor 2 promotes tissue-specific cell growth, proliferation and survival during development of *Helicoverpa armigera*. *Cells* 11(11): 1799.

Zitnan, D., Kim, Y.J., Zitnanová, I., Roller, L., and Adams, M.E. (2007). Complex steroid-peptide-receptor cascade controls insect ecdysis. *General and Comparative Endocrinology* 153: 88–96.

9

Social Living and Life History Evolution, with a Focus on Ageing and Longevity

Judith Korb[1,2] and Volker Nehring[1]

[1] *Evolutionary Biology & Ecology, University of Freiburg, Freiburg, Germany*
[2] *Research Institute for the Environment and Livelihoods, Charles Darwin University, Casuarina Campus, NT0909 Darwin, Australia*

9.1 Introduction

Sociality has produced some of the most striking life histories. Sociality can evolve when resources are available locally and the chances for reproduction elsewhere are slim. Under such circumstances, it can be more profitable to delay reproduction and stay with the parents than to disperse (*e.g.*, West et al. 2007, Korb 2008a). Instead of reproducing themselves, offspring can become helpers who contribute to raising their siblings (*e.g.*, Korb 2008a, Boomsma 2009, Queller and Strassmann 2009, Korb and Heinze 2016). Since there is a high probability that helpers and their siblings share the same alleles, helpers can indirectly increase their fitness (*e.g.*, Griffin and West 2003). Overall, the indirect fitness of individuals safely staying with the parents can then outweigh the direct fitness of individuals who often die during dispersal. Such cooperative breeding systems exist in many taxa including mammals, birds and insects (Korb and Heinze 2008, Boomsma 2009, Cornwallis et al. 2010, Lukas and Clutton-Brock 2012, Rubenstein and Abbot 2017). Cooperative breeders are characterized by helpers retaining their full physiological potential for reproduction. Depending on their own status, the season and other conditions, the helpers may at any time decide to disperse and reproduce themselves. Helpers can also potentially inherit the nest for direct reproduction once the parents die, which is typically a safer route than dispersal (Korb and Schneider 2007, Korb and Heinze 2008, Leadbeater et al. 2011, Lukas and Clutton-Brock 2012).

Some insect lineages have become eusocial (see Box 9.1: Superorganismality). In these lineages, offspring helping is obligatory, and most offspring are workers that are unlikely to directly reproduce (obligate eusociality, *sensu* Bourke 1999). A few lineages have even passed a major evolutionary transition and can be considered 'superorganisms', where workers are sterile; they have entirely lost the capability to reproduce (superorganism *sensu stricto*, Korb and Heinze 2016, Bernadou et al. 2021). Superorganisms are analogous to multicellular organisms, where the reproductives (queens and kings) represent the germ line and the workers the replaceable somatic tissue (Boomsma 2009, Korb et al. 2021). Sociality is thus a gradient from cooperative breeders to superorganisms (Figure 9.1).

The life cycle of an obligate eusocial insect colony begins with a founding phase when the queen (and in termites also a king) fends for herself until she has produced enough workers so she can delegate colony maintenance and focus on reproduction alone. The founding phase is characterized by high mortality and is followed by a so-called ergonomic phase when the colony grows in worker number and potentially stored resources (Oster and Wilson 1978). In this phase, the colony becomes stable and well-defended, and the mortality of the queen and the entire colony drops. Once the colony is large enough, the queen begins to produce sexual offspring that will later leave the colony to mate and establish new colonies (reproductive phase; Oster and Wilson 1978). In most species, the production of sexuals is synchronized across colonies of the same population and follows an annual pattern.

The life histories of individuals appear to be affected by their social lives. Lifespans of eusocial reproductives are often extraordinarily long, and breeders that have helpers available typically live longer than solitary breeders (Downing et al. 2021). Social insect queens can become decades old (Keller and Genoud 1997, Korb and Thorne 2017). In this chapter, we focus on the ultimate and proximate factors leading to such extraordinarily long lives, and we also investigate how lifespan interacts with other life history traits at both the individual and the colony level.

Life History Evolution: Traits, Interactions, and Applications, First Edition. Edited by Michal Segoli and Eric Wajnberg.
© 2025 John Wiley & Sons Ltd. Published 2025 by John Wiley & Sons Ltd.

> **Box 9.1 The superorganismality concept: the gradient of social evolution**
>
> Superorganismality is a concept that classifies the degrees of sociality according to the degree of cooperation and conflict between interactors (Queller and Strassmann 2009). High superorganismality is characterized by strong cooperation and low conflict. An evolutionary increase in superorganismality affects the life histories of individuals. We distinguish three levels of superorganismality, which are associated with the degree to which helpers/workers can reproduce:
>
> Cooperative breeders: Some adults do not breed but instead become helpers and provide alloparental care to the offspring of breeders, which they are typically related to. Helpers retain full reproductive totipotency and can at any time become breeders themselves, either by inheriting the natal breeding position or by dispersing and founding their own nest/territory.
>
> Eusocial animals: Animals that live in groups with permanent reproductive division of labour, overlapping generations and brood care. Reproductives receive help from workers who have a reduced reproductive potential.
>
> Superorganisms *sensu stricto*: A group of eusocial animals with sterile workers that have completely lost the ability to reproduce themselves; there is no conflict over reproduction any more. To distinguish the term from the colloquial metaphor that regards all eusocial insects as superorganisms due to their division of labour, the term 'superorganism *sensu stricto*' (Korb and Heinze 2016) has been suggested.

9.2 Ultimate Causes of Long Reproductive Lifespans

Is increased longevity the cause or consequence of social evolution? This common question is hard to answer. Reciprocal feedback processes make it difficult to disentangle what came first. However, the widespread occurrence of long lifespans of reproductives across social lineages (Keller and Genoud 1997, Sherman and Jarvis 2002, Blumstein and Møller 2008, Williams and Shattuck 2015) strongly suggests that both traits coevolve.

9.2.1 Theoretical Models

Theoretical models seem to provide support for both: that a social life fosters high survivorship and low mortality of reproductives, and *vice versa*, that low mortality facilitates the evolution of a social life. However, as we will show in the following, a differentiated view indicates specific patterns of reciprocal feedback loops across the sociality gradient. This view is based on the well-supported assumption that highly social systems with few fertile individuals evolved from cooperative breeder-like systems with offspring that temporarily delay breeding and act as helpers at the nest (*e.g.*, Alexander 1974, Korb and Heinze 2008).

Lucas and Keller (2020) argued that sociality evolved first and then lifespan expanded with increasing social complexity, namely, with the transition to eusociality. In line with this idea, Kreider et al. (2022), using individual-based simulations, showed that long lifespans can be an evolutionary consequence of cooperative breeding. Ageing (*i.e.*, the gradual decline of body functioning with time; we use the term interchangeable with senescence) generally starts after reaching reproductive maturity (Hamilton 1966; see also Chapter 2, this volume). Helping delays reproductive maturity and hence selects for increased lifespans, especially when helpers 'queue up' and older helpers are more likely to become breeders. In the models, lower genetic relatedness among helpers and reproductives resulted in longer lifespans. This is probably because less related helpers gain less indirect fitness from helping, so that nest inheritance (which allows for direct fitness) is more important for them. This does not fully line up with the empirical data, as cooperative breeders are generally related and there is no evidence of a connection between variation in relatedness and variation in lifespan (Downing et al. 2015, Griesser et al. 2017).

Kreider et al. (2022), however, did not address the origin of cooperative breeding. This was done by Wild and Korb (2017), who investigated the origin of cooperative breeding as a two-step process: the evolution of delayed dispersal followed by the evolution of helping, as supported by the empirical literature (*e.g.*, Brown 1974, Emlen 1982, Brown 1987, Hatchwell and Komdeur 2000, Griesser et al. 2017). The models show that high survival of individuals (*i.e.*, low extrinsic mortality of both breeders and individuals who delay dispersal) favours the evolution of both steps (Wild and Korb 2017). This implies that long lifespans due to low extrinsic mortality can be a pre-adaptation for social evolution and can foster the transition to helping and cooperative breeding. This is in line with predictions made by studies that focus specifically on mortality (Arnold and Owens 1998, Beauchamp 2014). Here, a low mortality especially of breeders limits opportunities for

Survival ↑ and fecundity ↑ (pre-adaptation):
evolution of cooperative breeding

Cooperative breeding

→ Increased survival of breeders due to helping
and
selection for increased longevity because of delayed breeding

Social insects of
low social complexity

Increasing group sizes
Increased division of labour

Specialization of castes with

(1) Reproductives:
- Extrinsic mortality: ↓
(After colony foundation)
→ Longevity: ↑
- Evolution of sexual/social maturity
→ Longevity: ↑

(2) Workers (and soldiers):
- Extrinsic mortality: ↑
(Especially while foraging)
→ Longevity: ↓

Social insects with
sterile workers

Superorganisms *sensu stricto*

Reproductives:
germline of a colony

Workers:
disposable soma

Life history trade-offs shifted from within
individuals to between castes

Figure 9.1 Social evolution and the evolution of longevity. High survival facilitates the evolution of delayed dispersal and helping (*i.e.*, cooperative breeding). The occurrence of helpers in cooperative breeders can further improve the survival of breeders. Helpers mature later, which is expected to lead to increased longevity. In social insects and mole-rats, reinforcing feedback mechanisms can lead from monomorphic species with low social complexity to the evolution of highly complex species with morphological castes and pronounced longevity differences (*e.g.*, most ants, most foraging termites and social mole-rats). These mechanisms include decreased extrinsic mortality of reproductives due to protection and improved nutrition and high random extrinsic mortality of foraging workers and soldiers during defence. Along with this, the reproductive potential of workers declines. In parallel, sexual/social maturity evolves, which strongly selects for an increased lifespan of reproductives. With the evolution of sterile workers, workers become comparable to disposable soma, and queens become the germline of a superorganismic colony. Life history trade-offs, for example, between fecundity and longevity, shift from within an individual to between castes (*i.e.*, queens with high fecundity and high longevity as opposed to workers with no fecundity and short lifespans). →: consequence, ↑: increase, ↓: decrease. Depicted in photos are yellow-bellied marmots (*Marmota flaviventris*), long-tailed tits (*Aegithalos caudatus*), wood-dwelling *Cryptotermes secundus* termites, *Belonogaster sp.* wasps, *Macrotermes bellicosus* termites with sterile workers and ants (*Crematogaster missouriensis*) (in order of appearance from top left to bottom right). *Source:* Photo credits: yellow-bellied marmot: Svenja Kroeger, long-tailed tits: Volker Salewski, ants: Jürgen Heinze, remaining: Judith Korb (Adapted from Korb and Heinze 2021).

independent reproduction due to low rates of breeder turnover. Limited opportunities for independent reproduction then facilitate the evolution of cooperative breeding. Furthermore, Wild and Korb (2017) also showed that high fecundity, in addition to high survival, can foster the evolution of cooperative breeding. A co-occurrence of both traits, high longevity and high fecundity, is most likely to favour the evolution of cooperative breeding. This might explain the absence of the fecundity–longevity trade-off in eusocial species in which queens can live very long and be extraordinarily fecund (Monroy Kuhn and Korb 2016) and suggests that the reversal of the trade-off is not a consequence but can be a prerequisite for social evolution. Such a positive association might occur when only individuals who are in the best condition become breeders, while less-fit individuals become helpers (Blacher et al. 2017). Yet, as a caveat, Wild and Korb (2017) investigated 'only' the origin but not the maintenance of cooperative breeding. Models that track the co-evolutionary feedback loops of lifespan, fecundity and sociality are still lacking.

The most extreme intraspecific divergence of lifespans is found in eusocial organisms, such as social insects and the eusocial mole-rats. Here, within the same colony, queens can live for decades, while workers often have lifespans of only a few months (see below). Originally, this lifespan divergence was explained by differences in caste-specific extrinsic mortality (Keller and Genoud 1997). Within the nest, queens are well protected from extrinsic mortality, while workers, especially at an older age, experience high extrinsic mortality while foraging. However, although plausible, these verbal arguments cannot fully explain the lifespan divergence between queens and workers (Kreider et al. 2021, Kramer et al. 2022). Instead, models show that monogamous social Hymenoptera with sterile workers evolve large lifespan divergences when antagonistic pleiotropy affects castes differently due to reproductive division of labour (*i.e.*, mutations that increase queen fecundity decrease worker lifespans more than queen lifespans) (Kreider et al. 2021). This was the case regardless of the degree of caste-specific extrinsic mortality. In addition, models by Kramer et al. (2022) have shown, for social Hymenoptera, that caste-specific extrinsic mortality can explain differences in queen-to-worker intrinsic mortality only to a minor extent. In addition, the delayed production of sexual offspring (sexual/social maturity *sensu* Korb and Heinze 2021; see below) was a major factor in selecting for increased queen longevity. Indeed, in social insect species with sterile workers (superorganisms *sensu stricto*), there are two types of maturity (see also below): (1) the 'normal' maturity, when the first eggs are laid, and (2) the sexual/social maturity, when the first sexual offspring are produced. In social insects with sterile workers, the first laid eggs develop 'only' into sterile workers that do not propagate genes directly to the next generation and hence are evolutionary less relevant. Only after several years, sexual/social maturity is reached when the queen produces sexual offspring that propagate genes. Accordingly, selection against ageing is strong until this second type of maturity, the sexual/social maturity, is reached (Kramer et al. 2022). In addition, polygyny or worker reproduction reduced the queen-to-worker lifespan divergence (Kramer et al. 2022) as was also empirically observed (Keller and Genoud 1997).

To summarize, the current theoretical results (*e.g.*, Arnold and Owens 1998, Kokko and Ekmann 2002, Wild and Korb 2017) suggest that long lifespans due to low extrinsic mortality facilitate the evolution of social life and cooperative breeding. Delayed breeding as observed in cooperative breeders and social insects further selects for extended lifespans of breeders and queens because it delays reproductive maturity (Korb and Heinze 2021, Kramer et al. 2022, Kreider et al. 2022). By contrast, caste-specific extrinsic mortality seems to play a minor role in explaining lifespan divergence between queens and workers. This pattern is in line with a testable scenario presented by Korb and Heinze (2021), who, in addition, propose specific factors leading to positive feedback between lifespan and fecundity in reproductives across the sociality spectrum (Figure 9.1). More work is needed to further explore these co-evolutionary relationships.

9.2.2 Empirical Evidence

9.2.2.1 Vertebrates

Sociality is associated positively with longevity in vertebrates. For birds, it was generally assumed that cooperative breeding is associated with increased longevity (*e.g.*, Arnold and Owens 1998). Yet, earlier studies suffered, for instance, from biased sampling or a lack of phylogenetic control (*e.g.*, Cockburn 2003, Cockburn et al. 2017). More recently, phylogenetically paired comparisons revealed that adults of cooperatively breeding bird species have a 10% increased annual survival compared to their pair-bonding counterparts, though maximum longevity does not differ (Beauchamp 2014). In addition, a study that reconstructed the ancestral states at transitions to cooperative breeding across the bird phylogeny showed that a longer lifespan increased the likelihood of evolving cooperative breeding (Downing et al. 2015), as has been suggested by Brown (1987) and Griesser et al. (2017), and which is in line with theory (*e.g.*, Kokko and Ekmann 2002, Wild and Korb 2017). Intraspecifically, a more recent study showed convincingly that individual Seychelles warblers who receive more help age more slowly (Hammers et al. 2019). This implies a causal explanation for increased annual survival of adult birds (see Beauchamp 2014) after the evolution of cooperative breeding.

In mammals, sociality seems to be positively associated with lifespan, but probably only in African mole-rats (Bathyergidae) (Rubenstein and Abbot 2017, Thorley 2020) and perhaps in humans (see Chapter 11, this volume). Among the Bathyergidae, the breeders of eusocial species – the naked mole-rat *Heterocephalus glaber* and several *Fukomys* species – live for 20 years and longer, which is extraordinarily long for a rodent (*e.g.*, Buffenstein and Jarvis 2002, Dammann and Burda 2006). A comprehensive analysis applying causal inference modelling of the full sociality spectrum from solitary to eusocial species revealed that breeders of social mole-rats live longer than solitary ones (Dammann et al. 2022). However, in the most recent study on mammals with the largest data set controlling for confounding factors such as body mass, no strong evidence was found across all mammals that cooperative breeders have longer lifespans than their non-cooperative counterparts (Thorley 2020).

9.2.2.2 Social Arthropods

As already said above, in the classical social insects, the eusocial Hymenoptera and termites, reproductives (queens, and in termites also kings) can have extraordinarily long lifespans of up to several decades (Keller and Genoud 1997, Keller 1998, Korb and Thorne 2017). Thus, they live much longer than their non-reproducing colony nestmates, the workers, and than adult solitary insects. This suggests that the evolution of eusociality has increased longevity in reproductives. However, there is considerable variation in longevity between species within lineages, and the degree of social complexity varies as well (*i.e.*, degree of 'superorganismality', after Queller and Strassmann 2009, Bernadou et al. 2021). Thus, the question arises whether queen longevity (or the queen-to-worker longevity ratio) is associated with social complexity among the social insects. Surprisingly few studies have formally tested for such an association. Even controlled comparisons of social insects *vs.* their non-social/subsocial sister taxa are rare. This is at least partly due to a lack and inconsistencies in records of longevity data. Here, we summarize what is known, separately for each lineage.

In bees, eusociality evolved at least five times independently (Hughes et al. 2008). Very generally, individuals of solitary species and workers are short-lived, while queens have extended lifespans (reviewed in Wcislo and Fewell 2017). In eusocial species from temperate regions, queens live a little more than one year, including overwintering in diapause, while queens of perennial colonies can have a lifespan of multiple years (Michener 1974, Winston 1991). Worker longevity is generally restricted to a few weeks. No formal analysis exists that relates longevity to social complexity.

Eusociality evolved twice in vespid wasps, and the lifespan of solitary wasps and workers is typically much shorter than that of queens, although considerable climate effects exist (reviewed in Hunt and Toth 2017). In primitively eusocial species of the seasonal tropics, queen and worker longevity differ little (data for two species suggest that queens live about twice as long as workers), and colonies generally last only a few months. As wasp sociality probably originated in these regions (Pickett and Carpenter 2010), longevity differences between castes might have been less pronounced at the origin of eusociality (Toth et al. 2016). In temperate regions, however, queens generally live close to a year, while workers may have a lifespan of a few weeks only (Toth et al. 2016), reflecting a pronounced queen-to-worker longevity ratio. A formal analysis of vespid wasps using the limited available data revealed that worker longevity tends to decrease with increasing social complexity (Toth et al. 2016), supporting a former suggestion by Strassmann (1985).

Ants originated from wasps colonizing the leaf-litter (Rabeling et al. 2008), and it has been suggested that the subterranean lifestyle reduced extrinsic mortality and led to longer lifetime durations (Carey 2001), a potential pre-adaptation for the evolution of sociality. All ants are social, so comparisons with solitary ants are impossible. However, ant workers of basal ant lineages such as the Ponerinae, which are assumed to resemble ancestral ants, still have the capability to reproduce sexually (Hughes et al. 2008). In a few ant genera, queens have been lost secondarily, and worker-like individuals have taken over reproduction (sexually mostly in the Ponerinae, clonally in other lineages; Peeters 1991). This means that comparisons between superorganismal, obligately eusocial and cooperatively breeding species are possible in principle. Lifespan data are scarce, though, in particular for the ponerine lineages. Ponerine ant reproductive lifespans anecdotally range around 2–3 years in species where workers regularly reproduce (*Harpegnathos saltator*: Liebig and Poethke 2004; *Platythyrea punctata*: Hartmann and Heinze 2003). This is below the average of ten years for ant queens listed by Keller and Genoud (1997) and also lower than the lifespan of queens of ponerine species in which workers do not typically reproduce (*Rhytidoponera purpurea ca.* 12.5 years; Hölldobler and Wilson 1990). However, reproductives do live longer than non-reproductives in all these examples, which could be a mechanistic effect of reproduction rather than a consequence of sociality. Indeed, physiological changes made when reproduction begins may coincidentally prolong lives (see section on proximate mechanisms below). While some of the ant species with sterile workers have the longest-lived queens known, there does not seem to be an overall effect of the transition to superorganisms on queen or worker lifespan (*e.g.*, among the leaf-cutting ants, Keller and Genoud 1997), and queens of some species with sterile workers live for less than a year (*Cardiocondyla obscurior*: Heinze and Schrempf 2012). The combination of high ecological variation, little available life history data and many confounding factors (*e.g.*, polygyny and founding mode, see below) may be the reason for a lack of a robust comparative analysis for the ants.

Termites evolved eusociality only once, independently from the social Hymenoptera, in the Blattodean lineage. While all termites are eusocial due to the presence of a sterile soldier caste, the degree of sociality (reflected in the degree of the workers' reproductive potential, which presents a gradient in superorganismality; see Box 9.1) varies considerably among workers of different species. In wood-dwelling species, for example, in which workers never leave their substrate to forage, workers generally resemble helpers of cooperative breeders (Korb 2007, 2008b). They are totipotent immatures that develop via moults into reproductives that either disperse to found new colonies or become replacement reproductives if the reproductives of the natal nest die. By contrast, in foraging species (*i.e.*, species in which workers sooner or later during colony development leave the nest to forage), workers have reduced reproductive options and can either only become replacement reproductives within the natal nest (pluripotent workers) or are sterile (Korb and Hartfelder 2008, Roisin and Korb 2011). A first comparison of longevities between these species and their sister taxon, the *Cyptocercus* woodroaches, suggests that totipotent workers of wood-dwelling termites have similar lifespans as woodroaches, while worker lifespan decreased in species in which workers lost totipotency, from pluripotent (*i.e.*, Mastotermitidae, Rhinotermitidae) to sterile (Termitidae) (Figure 9.2a) (Korb and Thorne 2017). At the same time, the queen's lifespan seems to increase (Figure 9.2a). Thus, overall, the queen-to-worker lifespan ratio increased, especially in species with sterile workers (*i.e.*, in the Termitidae),

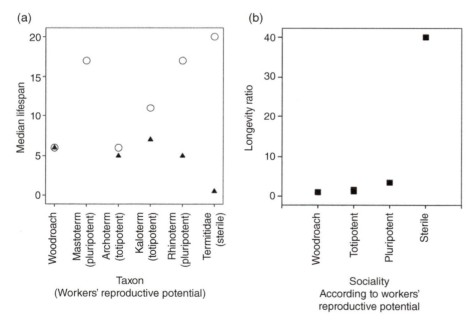

Figure 9.2 Longevity data for termites. Shown are (a) median lifespans across species for workers (black triangles) and queens (open circles) of different termite families (with their workers' reproductive potential) and the subsocial sister taxon, the woodroaches. In the woodroaches, reproducing females are equivalent to queens and workers, as all breeding females function as reproductives and, at the same time, also as workers because they also care for the brood. Data are based on results for *Cryptocercus punctulatus*. In (b), the queen-to-worker longevity ratio is given with data sorted by degree of sociality, defined by the workers' reproductive potential (from totipotent, where all reproductives develop from workers; to pluripotent, where workers can only become reproductives that inherit the natal breeding position but can never found their own colony; to sterile workers, which generally cannot reproduce). In (b), each data point presents one family of (a) to consider phylogenetic dependency, as the represented species within a family share the same degree of social complexity. Note that there are two overlapping data points for species with totipotent workers. Totipotent workers: Archoterm = Archotermopsidae *sensu stricto* (*i.e.*, excluding Hodotermitidae) and Kaloterm = Kalotermitidae; pluripotent workers: Rhinoterm: non-wood-dwelling Rhinotermitidae (*i.e.*, without wood-dwelling totipotent species for which no data are available); sterile workers: Termitidae. Mastoterm = Mastotermitidae, a family with a single extant representative, *Mastotermes darwiniensis*. This species has pluripotent workers. As worker longevity of this species is unknown, the queen-to-worker longevity ratio cannot be calculated. Based on data summarized in Shellman-Reeve (1997) and Korb and Thorne (2017).

due to a pronounced decrease in worker longevity (Figure 9.2). This coarse analysis is preliminary. However, given the scarcity of longevity data and the fact that worker sterility evolved probably only once in termites (*i.e.*, in the Termitidae), it will be challenging to carry out more detailed comparative analyses. Yet, support for a link between workers' reproductive potential and caste-specific ageing in termites also comes from experimental life history studies and molecular analyses (Monroy Kuhn et al. 2019, Rau and Korb 2021, Lin et al. 2023). For instance, totipotent workers of the Kalotermitidae *Cryptotermes secundus* invest strongly in anti-ageing mechanisms, especially under environmental stress conditions (Rau and Korb 2021), while this is less the case in sterile workers of the Termitidae *Macrotermes bellicosus* (Elsner et al. 2018).

Concerning other social arthropods, most social aphids and thrips have annual or biannual life cycles, with relatively short individual lifespans (reviewed in Abbot and Chapman 2017). No quantitative comparisons of longevity between social and non-social species exist. However, eusociality (here it is the occurrence of altruistic soldiers rather than altruistic workers) seems to be more common in species that have longer-lasting, slow-growing galls and longer generation times (*e.g.*, Chapman et al. 2008). For spiders, also no formal analyses exist but some hints suggest a positive correlation between sociality and longevity when considering total lifespan, including developmental time (reviewed in Avilés and Guevara 2017).

9.2.2.3 Conclusion: Ultimate Causes

To conclude, although formal analyses are rare, the data for social invertebrates, birds and mole-rats imply that there is a positive association between sociality and longevity. This was also concluded in a review that included fish and reptiles (Rubenstein and Abbot 2017). Causes for this association seem to be that offspring of longer-lived species have a higher likelihood of staying in the nest and becoming helpers (Kokko and Ekmann 2002, Downing et al. 2015, Wild and Korb 2017, da Silva 2022) (Figure 9.1). As long as helpers/workers are totipotent to become reproductives, delayed breeding leads to delayed maturity. This should select for delayed onset of ageing processes of all individuals (helpers/workers and breeders), as selection against ageing is generally maximal until reaching maturity and only starts to decline after the onset of maturity (Hamilton 1966, Kreider et al. 2022) (Figure 9.1). In parallel, helping can increase the extrinsic mortality of helpers/workers and reduce that of breeders/reproductives. Such caste-specific extrinsic mortality should then lead to caste-specific ageing patterns with shorter-lived helpers/workers and longer-lived breeders/reproductives (Figure 9.1). Positive feedback mechanisms may reinforce the reproductive-to-worker longevity ratio, depending on species-specific ecology and degree of worker sterility. In addition, in eusocial species, a second type of maturity emerges: sexual/social maturity, the point in the life cycle at which the reproductives or the queen start to produce sexual offspring (new queens and kings) (after Korb and Heinze 2021, Jaimes-Nino et al. 2022, Kramer et al. 2022, Kreider et al. 2022) (see also below). From an evolutionary perspective, only the sexual offspring are relevant, as they are the only individuals able to reproduce by founding new generations of colonies. As the production of sexual offspring often only starts several years after colony foundation, delayed sexual/social maturity strongly selects for an increased lifespan of reproductives. Within a superorganismality framework, workers become increasingly more soma-like and a germline/soma division emerges at the colony level. With the evolution of worker sterility, conflicts over reproduction between reproductives and workers disappear, and a complete germline/soma division is established with workers largely corresponding to disposable soma cells (superorganisms *sensu stricto*; Figure 9.1). During such evolutionary processes, life history trade-offs gradually shift from within an organism to between castes. Sterile workers take over the burden from queens, and hence workers become very short-lived and queens very long-lived, which results in an apparent absence of a fecundity–longevity trade-off in reproductives (Monroy Kuhn et al. 2019, Lin et al. 2023; see also Chapter 2, this volume).

9.2.3 Links Between Sociality, Longevity and Other Life History Traits

Social evolution affects not only lifespan but also other life history traits. With the transition from solitary life to cooperative breeding, reaching maturity is delayed because individuals have to wait longer for breeding opportunities. With the evolution of (sterile) worker castes, sexual/social maturity became increasingly more important for life history evolution, and it further delays the onset of ageing.

Most recently, it has been reported for ants, termites and eusocial mole-rats that queens do not age gradually but rather suddenly 'drop dead' without previously showing signs of senescence (Buffenstein and Jarvis 2002, Dammann and Burda 2006, Heinze and Schrempf 2012, Monroy Kuhn et al. 2021, Jaimes-Nino et al. 2022). In addition, fecundity does not seem to decline with age. How social insects achieve this is largely unclear, given that most solitary organisms gradually age. Typically, queen fecundity (egg-laying rate) increases during the first years until a plateau is reached (*e.g.*, Korb 2008a, Jaimes-Nino et al. 2022). In the ant *C. obscurior,* this plateau is maintained until the queen dies (Heinze and Schrempf

2012, Jaimes-Nino et al. 2022). With few exceptions in highly seasonal regions, queens continuously lay eggs during the course of their lifetime, with a fitness peak late in life (continuusparity, after Jaimes-Nino et al. 2022). This life history strategy is postulated to be distinct from the iteroparity (*i.e.*, individuals have several cycles of reproduction during the course of their lifetime) and semelparity (*i.e.*, individuals have a single episode of reproduction during the course of their lifetime) of non-eusocial taxa. It is expected to select for increased longevity because there is constant selection for reproduction (Jaimes-Nino et al. 2022).

Social insect reproductives have astonishing egg-laying rates despite their long lives, which results in total lifetime offspring numbers unparalleled in solitary organisms. They apparently overcame the otherwise common fecundity–longevity trade-off (Monroy Kuhn and Korb 2016). The most fecund caste, the queens (and in termites also the kings), has a much longer lifespan than the non-reproducing workers, and – in contrast to solitary organisms (Wolfner 1997) – mating and egg production lead to an increase in lifespan, at least in some species, as shown in Schrempf et al. (2005). The causes of this apparent re-shaping are still unclear. Reproductives might be individuals that are in better condition even before reproduction starts, and hence the costs of reproduction, apparently do not wear on them like they would on the average individual (Blacher et al. 2017; but see Schrempf et al. 2017). In addition, costs of queen reproduction might be paid by workers taking over the burden of queens, for instance, via increased feeding of the reproductive female. The trade-off would then have shifted from within individuals (reproductives) to between castes. The increase in the queen-to-worker longevity ratio with sociality in termites (Figure 9.2) and the shortened lifespan of workers with sociality in vespid wasps (Toth et al. 2016) might support this view.

An interesting life history trait in social evolution is sex ratio (see also Chapter 7, this volume), in particular in the social Hymenoptera. These animals are haplodiploid, which leads to relatedness asymmetries. When queens only mate once, workers are more closely related to their sisters than to their brothers. Because the altruism of social insects is largely driven by indirect fitness, workers should favour the production of new queens over that of new males. In contrast, queens favour a 1:1 sex ratio because they are equally related to their daughters and sons (Trivers and Hare 1976). Indeed, social hymenopteran sex ratios are typically skewed in the way the workers would favour, and there is some evidence of conflict between queens and workers over the sex ratio of the sexual offspring (Trivers and Hare 1976, Ratnieks et al. 2006, Meunier et al. 2008). When queens mate more than once, these conflicts become milder because, with each additional father, the average relatedness between sisters decreases.

In summary, not only lifespan but also other life history traits have been affected by social evolution, from fecundity to age at maturity and type of maturity to the mode of parity and the number and sex ratio of the offspring. We will also discuss effects on body size in the following Section 9.3.

9.3 Colony Life History in Obligatory Eusocial Insects

While the queen's life history traits are important for her fitness and thus under strong selection, they cannot be analysed in isolation. A queen's fitness crucially depends on her 'extended phenotype', her workers and their labour. In fact, the colony as a whole can be subject to selection and possesses life history traits that can be optimized (Bourke 2007). Most importantly, after the transition of eusocial colonies into superorganisms *sensu stricto*, the sexual maturity of queens is no longer reached when they produce the first egg. Queens typically produce workers long before they produce sexuals, and only the sexuals will found new colonies. Hence, the colony's maturity (and from an evolutionary point of view also that of the queen) is only reached once the first batch of sexuals is produced (sexual/social maturity; after Korb and Heinze 2021). Interestingly, in Hymenoptera, the decision about which of the queen's offspring develop into sexuals (and hence also when the production of sexuals begins and their sex ratio) is often not made by the queen but by the workers, who feed the larvae and, through their provisioning, can regulate the larvae's developmental trajectories (Ratnieks et al. 2006). Age at sexual/social maturity and offspring sex ratio are thus colony level and not individual traits. This has strong implications for the queen's lifespan because senescence only begins once an individual has reached maturity (see above).

Colony size at sexual/social maturity varies greatly between social insect species, from a few to millions of individuals. Across species, large colonies live longer, which parallels the body size to lifespan correlation observed, *e.g.*, among mammals and birds (Speakman 2005, Kramer and Schaible 2013). One reason might be that queens of larger colonies are better protected because they have more defenders, and that there is more redundancy in the workforce, which makes total colony failure less likely (Oster and Wilson 1978, Bourke 1999, Kramer and Schaible 2013). The larger the colonies are, the more individuals can specialize (Negroni et al. 2016, da Silva 2022). At one extreme, a solitary individual must carry out all tasks from foraging to reproduction; it must thus be a generalist. On the other extreme, in a colony with millions of individuals, each individual can specialize in just one task to the point that leaf-cutting ant workers specialize in different phases of leaf processing and can thus be more efficient than a generalist ever would be (Wilson 1980). Such specialization is not restricted to behaviour but also affects physiological and morphological traits. The more worker and queen phenotypes can diverge, the larger the queen-to-worker lifespan ratio can evolve (Oster and Wilson 1978, Bourke 1999, Kramer and Schaible 2013). Indeed, larger colonies also appear to be more energy-efficient in that smaller colonies use more energy per biomass. Across species, small colonies live faster and die younger (Shik et al. 2012).

Interestingly, the size and number of sexual offspring that a colony produces are traded off. Producing more sexuals means that each one can receive fewer resources (see also Chapter 3, this volume). In ants, colony size affects how colonies resolve this trade-off. Species with larger colonies produce fewer sexuals per worker, but each of these sexuals becomes larger than in species with smaller colonies (Shik 2008). This increases the fitness of each individual offspring because larger queens have better chances of surviving colony foundation.

Two additional traits greatly alter the life history of obligatory eusocial colonies: queen number and dispersal mode (see also Chapter 10, this volume). While the typical ancestral eusocial hymenopteran colony had only one queen and the sexuals dispersed by swarming, multi-queen colonies have evolved in different species (Hughes et al. 2008). These queens can be present simultaneously (mostly in ants; Boomsma et al. 2014) or be produced sequentially, for example, to allow dispersal through the splitting of colonies (*e.g.*, in the honeybee). In any case, additional queens can act as replacements for each other, which makes the colony less dependent on the survival of each single queen. Hence, individual and colonial lifespans become uncoupled. As a consequence, multi-queen colonies in social Hymenoptera typically live longer, but queens in multi-queen species die younger (Keller and Genoud 1997, Heinze 2017) (in agreement with corresponding theoretical models, see above; Kramer et al. 2022). The possibility of colonies containing more than one queen also allows for a new mode of dispersal: instead of swarming, sexuals can be re-adopted into their mother colony after mating, and queens then leave with workers to form new colonies. This type of colony budding relieves queens of the high mortality during the solitary founding phase, which is equivalent to high juvenile mortality of solitary species. Given that a high ratio of juvenile to adult mortality typically selects for longer reproductive lives, selection for longer lives of polygynous queens is thus somewhat relaxed (Stearns 1976, Kramer et al. 2022). The extreme of this lifestyle is reached in unicolonial ants, where the buds never really part from their mother colony but stay attached so that they form invasive supercolonies that can occupy square kilometres of land, like the Argentine ant and other invasive species do. Because these colonies contain many queens that are constantly replaced, they can grow without limit like some plants do and are virtually immortal (Helanterä et al. 2009).

9.4 Proximate Mechanisms

Research on mostly short-lived model organisms identified molecular networks underlying longevity and ageing that are highly conserved, from *Caenorhabditis elegans* and *Drosophila melanogaster* to mice, rats and humans (*e.g.*, Kenyon 2010, 2011, Partridge et al. 2011). They include the nutrient-related insulin/insulin-like growth factor signalling (IIS) and the target of rapamycin (TOR) pathways, which sense the availability of carbohydrates and amino acids/proteins, respectively (*e.g.*, Partridge et al. 2011, Gems and Partridge 2013). They interact with endocrine pathways – in insects, most importantly the life-shortening juvenile hormone (JH), in mammals, *e.g.*, glucocorticoids and thyroid hormones – to regulate fecundity and longevity (*e.g.*, Flatt et al. 2005). In addition, oxidative stress can be an important cause of ageing, when reactive oxygen species (ROS) produced during metabolism damage macromolecules like proteins or lipids (free radical theory of ageing) (*e.g.*, Sohal and Weindruch 1996, Finkel and Holbrook 2000). Furthermore, the shortening of telomeres (repetitive sequences at the end of chromosomes) during replication can lead to damage of genes and replicative senescence of cells (telomere attrition). Research focusing mainly on mammals currently recognizes 14 hallmarks of ageing, including loss of proteostasis and epigenetic alterations (Schmauck-Medina et al. 2022). For insects, the TI-J-LiFe (short for TOR/IIS-JH-Lifespan Fecundity) network was recently formulated as the molecular backbone of ageing and life history research (Korb et al. 2021) (Figure 9.3). This network pinpoints the major molecular mechanisms involved in regulating ageing and other life history traits (especially fecundity). It should help to guide research by partitioning mechanisms into upstream nutrient-related signalling components (TOR/IIS), endocrine components (JH; for vertebrates, corresponding hormones

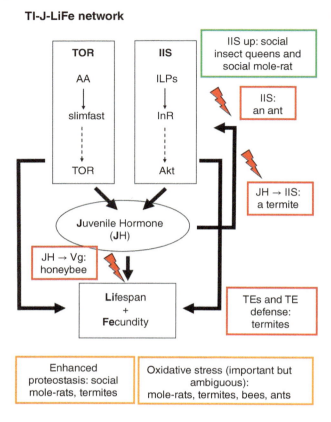

Figure 9.3 The TI-J-LiFe network with major results for eusocial animals. The TI-J-LiFe network is a molecular framework that represents the molecular underpinnings of life history traits, especially those related to longevity and fecundity (Adapted from Korb et al. 2021). Green: shared across several social insect lineages and social mole-rats. Red: major changes and re-wirings underlying an increased lifespan of reproductives in social insects. Orange: changes associated with an increased lifespan in social organisms. Bolts indicate rewiring of pathways. TOR: target of rapamycin, IIS: insulin/insulin-like growth factor signalling, JH: juvenile hormone, AA: amino acids, ILPs: insulin-like peptides, Vg: Vitellogenin, TE: transposable elements, InR: insulin receptor, Akt: a serine/threonine kinase, slimfast: an AA transporter.

like glucocorticoids) and downstream components that often directly affect lifespan and fecundity (LiFe), like immunity, oxidative stress, telomere attrition, epigenetic alternations, and transposable element (TE) activity.

In the following, we will summarize what is known about whether and how these molecular pathways change with an increase in lifespan in (social) organisms. For a general overview of mechanisms underlying ageing in model organisms, we refer, *e.g.*, to the following reviews: Kenyon (2010, 2011), Partridge et al. (2011), Gems and Partridge (2013) and Promislow et al. (2021).

9.4.1 Social Vertebrates

In birds and mammals, little research concentrated specifically on the mechanisms of ageing in social species compared to non-social species, except for mole-rats (see below) and a recent comparative study on mammals (Zhu et al. 2023). The latter compared brain transcriptomes of 94 mammals, distinguishing between solitary, pair-living and group-living species but not cooperatively breeding species. The authors identified especially hormone- and immunity-related pathways to be associated with social organization and longevity. As there are so few studies investigating mechanisms associated with lifespan in social birds and mammals, we summarize mechanisms that have been associated with a long lifespan in these taxa in general, with a focus on mole-rats, for which comparative studies exist.

In birds, a long lifespan has been associated with enhanced DNA damage repair, cell proliferation control, immunity and especially anti-oxidative mechanisms (*e.g.*, Wirthlin et al. 2018, Holtze et al. 2021). Helpers at the nest in cooperatively breeding birds have been found to reduce the oxidative stress of the breeding pairs, suggesting a potential mechanism for the longer lives of individuals who have helpers available (Cram et al. 2015). The strongest support for the important role of telomere attrition in ageing also comes from birds. Across bird species, telomere erosion correlates negatively with lifespan, and short-lived species seem to have a faster telomere shortening per year than long-lived birds (Sudyka et al. 2016, Tricola et al. 2018).

In mammals, research on bats associated their long lifespans relative to non-flying mammals with inflammation control and efficient antiviral responses (summarized in Holtze et al. 2021). Bats have relatively low ROS production for their high metabolic rate, and they seem to be rather resistant to oxidative stress. In addition, they have efficient DNA repair mechanisms. In contrast to birds, telomere length does not seem to be associated with longevity in bats (*e.g.*, Foley et al. 2020).

Long-lived social mole-rats share several of these traits with bats, including specific adaptations of the immune system, efficient DNA repair and a good mitonuclear balance (*i.e.*, coordination of nuclear and mitochondrial parts of the respiratory chain protein synthesis) (reviewed in Dammann 2017, Holtze et al. 2021). In contrast to bats, long-lived social mole-rats have enhanced proteasome activity, suggesting that damaged proteins can be recycled efficiently (Rodriguez et al. 2014, Sahm et al. 2018a). In addition, a transcriptomic study that compared 17 rodents with different longevities identified important genes of the TOR and IIS pathways to be positively selected with the evolution of enhanced longevity in rodents (Sahm et al. 2018b). Another transcriptome study compared 15 tissues of long-lived breeders and age-matched short-lived non-breeders of two social *Fukomys* mole-rat species (Sahm et al. 2021). They detected a strong endocrine signal. It implies that an inhibitor of a supposed anti-ageing hormone is downregulated in breeders and that steroid hormone synthesis is shifted from life-shortening glucocorticoids to sex steroid production. In addition, the IIS pathway seems to be upregulated in breeders. This seems unexpected, as high IIS signalling has been associated with short lifespan in short-lived model organisms (but see below, results for social insects). For the naked mole-rat, *H. glaber*, in addition, high hypoxia- and hypercapnia-resistance was found (Park et al. 2017). This species also challenges the free radical theory of ageing. It has relatively high ROS release and higher oxidation levels of lipids, proteins and DNA than mice, but the damage does not increase with age (Andziak and Buffenstein 2006).

9.4.2 Social Arthropods

For social arthropods, there is a rising focus on molecular studies, as social insects became emerging model organisms of ageing research during the last decades (Jemielity 2005, Heinze and Schrempf 2008, Rueppell 2009, Rascon et al. 2011, Lucas and Keller 2014, Korb 2016, Monroy Kuhn and Korb 2016, Rodrigues and Flatt 2016, Korb and Heinze 2021). Moving along the TI-J-LiFe network (Figure 9.3), we will first concentrate on the upper part (TI-J) of the network including links to fecundity. Then we summarize findings for the importance of the downstream part (LiFe) of the network. The latter part focuses on lifespan-related downstream processes, especially oxidative stress, for which many studies have been performed.

9.4.2.1 The TI-J-Part: Rewiring of the I-J-Fecundity Axis

The TOR pathway has so far not been directly associated with the long lifespan of social insect queens (*e.g.*, Weitekamp et al. 2017, Chandra et al. 2018, Monroy Kuhn et al. 2019, 2021, Rau and Korb 2021). Yet, current evidence suggests that molecular rewiring of the I-J-part of the TI-J-LiFe network and its link to fecundity plays an important role in explaining the long life of queens and their concomitant high fecundity (*i.e.*, the overcoming of the fecundity–longevity trade-off): Genes from the IIS pathway have been associated with caste-specific ageing in social Hymenoptera and termites (*e.g.*, Kapheim 2017, Weitekamp et al. 2017, Chandra et al. 2018, Rau and Korb 2021, Séité et al. 2022, Rau et al. 2023). Across lineages, social insect queens are generally characterized by an upregulation of the IIS pathway compared to workers (*e.g.*, Corona et al. 2007, Libbrecht et al. 2013, Lin et al. 2021). This is unexpected as high IIS activity is associated with a shortened lifespan in solitary model organisms (*e.g.*, Partridge et al. 2011). Very simplified, these life-shortening IIS effects typically stem from IIS activating JH- and fecundity-related processes (Flatt et al. 2005) that underlie the fecundity–longevity trade-off. This molecular trade-off between a life-extending downregulated IIS pathway and a concomitantly reduced fecundity that characterizes solitary species led to the hypothesis of molecular rewiring along the I-J-fecundity axis in social insects (Corona et al. 2007, Rodrigues and Flatt 2016). Indeed, in the honey bee, a rewiring occurs between the endocrine JH pathway and fecundity-associated vitellogenin (Vg) expression (vitellogenin is a yolk precursor necessary for egg production) (*e.g.*, Amdam et al. 2004, Corona et al. 2007) (Figure 9.3). In contrast to non-social insects, JH and Vg titres are negatively correlated, and constant vitellogenesis does not require concurrently high JH titres in honeybee queens (*e.g.*, Fluri et al. 1981, Hartfelder 2000, Amdam et al. 2004). However, the JH-Vg rewiring is not typical for other social insect queens, which generally require high JH titres for high Vg expression (*e.g.*, bees and wasps: Kapheim 2017; ants: Libbrecht et al. 2013, but see also Pamminger et al. 2016; termites: Maekawa et al. 2010, Lin et al. 2021). Yet, other rewirings along the I-J-Fecundity axis have been found. In a termite, an uncoupling of the IIS part from the downstream JH-Lifespan/Fecundity part seems to exist (Lin et al. 2021, Rau et al. 2023) (Figure 9.3). In an ant, rewiring occurs within the IIS pathway so that Vg production is uncoupled (Yan et al. 2022) (Figure 9.3).

9.4.2.2 The LiFe-Part, with a Focus on Lifespan

Important pathways that are directly associated with lifespan and longevity include those underlying oxidative stress, body maintenance (including immunity), transposable elements (TEs) and telomere attrition. The role of oxidative stress in explaining the long lifespan of social insect queens is unclear (de Verges and Nehring 2016, Kramer et al. 2021). Some results imply that reproductives are better protected against oxidative stress than non-reproducing workers (*e.g.*, honeybee: Seehuus et al. 2006, Kennedy et al. 2021; ants: Lucas et al. 2016; termites: Tasaki et al. 2017, 2018). On the other hand, no evidence has been found in other studies and species (*e.g.*, ants: Parker et al. 2004, Schneider et al. 2011; termite: Rau and Korb 2021). A recent comparative study using termites, bees and ants focused on oxidative stress by determining the degree of protein oxidation and the expression of anti-oxidant genes (Kramer et al. 2021). Here, the signal was also inconsistent across species, although suggesting that oxidative stress might be a significant factor in caste-specific ageing in social insects.

There is also some evidence that honey bee and ant queens invest more in body maintenance, including immunity, than their workers (*e.*g., Lucas et al. 2016, Kennedy et al. 2021). In addition, in termites, but probably not in social Hymenoptera, TE activity has been associated with caste-specific ageing, with queens and kings being better protected against these jumping genes than workers (Elsner et al. 2018, Monroy Kuhn et al. 2019, Korb et al. 2021, Post et al. 2023). Telomere attrition is unlikely to explain caste-specific ageing in termites (Koubová et al. 2021) and there are only few studies with mixed evidence from ants (Jemielity et al. 2007, Bonasio et al. 2010).

Importantly, in line with the idea that queens suddenly die without previous signs of senescence (see above), in a termite and an ant queen, gene expression is relatively stable throughout a queen's life and is only strongly altered just before the queens die (Monroy Kuhn et al. 2021, Jaimes-Nino et al. 2022). This ageing signal in termite queens is associated with physiological upheaval and strong signs of oxidative stress, TE activity and loss of proteostasis (Monroy Kuhn et al. 2021).

A first study that compared the transcriptomes of young and old workers as well as queens of six species across social insect lineages (two termites, two bees and two ants) implies many species- and lineage-specific mechanisms of caste-specific ageing, especially in the lower LiFe-part of the TI-J-LiFe network (Korb et al. 2021) (Figure 9.3). Yet, there were also some shared signals, for instance, of JH-related genes and vitellogenin being involved in reproduction and caste-specific ageing. These results stress the importance of the I-J-fecundity axis in understanding the long life of social insect reproductives and their apparent overcoming of the fecundity–longevity trade-off (Figure 9.3).

9.4.3 Conclusion: Proximate Mechanisms

Our current evidence suggests that there are some commonalities across social lineages as distant as vertebrates and insects that underlie the evolution of a long lifespan. Most obviously, this is an (unexpected) upregulation of the IIS pathway (Figure 9.3). In particular, in the social insects, rewiring and uncoupling of links along the TI-J-LiFe network seem to have led to overcoming the fecundity–longevity trade-off (Figure 9.3). Similarly, overcoming endocrine pleiotropic effects seems to be crucial in vertebrates (glucocorticoids) and insects (JH) for evolving a long lifespan. Yet, many changes seem to be species/lineage-specific (Figure 9.3). In birds, overcoming telomere attrition appears to be linked with an increased lifespan, while this does not seem to play a role in other long-lived social lineages. In termites, but not in social Hymenoptera, protection against TE activity seems associated with the long lifespan of their queens. The role of vanquishing oxidative stress in long-lived species remains ambiguous. The lineage- or species-specific idiosyncrasies might not be unexpected because sociality evolved multiple times independently. The idiosyncrasies illustrate that the evolution of a long lifespan (and even the overcoming of the longevity-fecundity trade-off) is not constrained by hard-wired molecular networks underlying life history traits. Thus, although the molecular pathways (as reflected in the TI-J-LiFe network) are highly conserved and shared among nematodes, insects and mammals, they can become uncoupled and re-wired if ultimate forces select for a long lifespan, for instance, in social organisms.

9.5 Conclusion

The present chapter has shown that reproductive individuals of social species often live longer than those of solitary species. Current evidence suggests that, already at the origin of sociality, longer lifetime duration in solitary organisms increases the chances for the evolution of sociality because nest inheritance and overlapping generations are more likely when lives are longer (Figure 9.1). Then, social life further promotes the evolution of longer lifespans. This is because social organisms reach maturity later than non-social organisms. First, delayed breeding in cooperative breeders delays maturity, and then the evolution of sterile workers results in a second type of maturity, social/sexual maturity, when the production of new sexuals begins. In addition, when helpers/workers take over the burden of wear and tear and extrinsic mortality from breeders/reproductives, lifespan differences across castes become more pronounced, in particular when the helpers lose reproductive totipotency. Disentangling the precise co-evolutionary feedback loops in empirical systems is difficult due to lineage-specific idiosyncrasies and a paucity of life history data and quantitative systematic analyses. As long as some helpers/workers can become breeders/reproductives later on, lifespan differences are less pronounced. With the separation into workers and queens in eusocial species, workers have become increasingly disposable. Worker and queen phenotypes then diverge, and the queen-to-worker lifespan ratio increases, culminating in species with sterile workers, in which reproductives basically function as germline and workers as disposable soma. This is a major evolutionary transition towards superorganisms *sensu stricto*, so life history trade-offs are now resolved at the individual level but occur at the colony level (e.g., short-lived workers versus long-lived queens).

The specific molecular mechanisms underlying changes in ageing vary across social lineages with some commonalities (Figure 9.3). This indicates that mechanisms causing ageing are less evolutionary hard-wired across animals than previously thought (*e.g.*, Gems and Partridge 2013) and than the fact that they are regulated in insects by the relatively conserved TI-J-LiFe network implies.

The co-evolutionary feedback between sociality and life history parameters is only poorly understood because precise data on both life history and social behaviour and its ultimate consequences are rare, in particular for species at the transitions between levels of superorganismality. More research is required to cover the broad phylogenetic gaps that still exist, which would allow us to separate phylogenetic effects from those of social evolution. Even more basic records of individual longevity would be very valuable. Such records should include maximum but also median longevities together with sample sizes. Optimally, they should allow estimation of age-specific survival based on life table analyses. Researchers working on social insects should consider that individual and social maturity are not identical and that they change with the degree of sociality. Experiments on proximate mechanisms in insects should be designed with the idea in mind that ageing in queens and kings is probably not gradual but starts only shortly before these individuals die. They also should avoid conflating behavioural transitions during colony founding with ageing. Using the TI-J-LiFe network as a guideline in mechanistic studies will simplify comparative analyses, as it covers all major genes related to life histories.

Acknowledgements

We thank Michal Segoli and Eric Wajnberg for inviting us to contribute a chapter on social organisms and for their efforts in bringing this book together. We also thank them and an anonymous reviewer for their valuable comments, which helped to improve our chapter. This chapter would not have been possible without the Research Unit FOR2281 'So-long' funded by the German Research Foundation (DFG).

References

Abbot, P., and Chapman, T. (2017). Sociality in aphids and thrips. In: *Comparative social evolution* (eds. Rubenstein, D.R., and Abbot, P.), 154–187. Cambridge University Press.

Alexander, R.D. (1974). The evolution of social behavior. *Annual Review of Ecology and Systematics* 5: 325–383.

Amdam, G., Norberg, K., Fondrk, M.K., and Page, R.E. (2004). Reproductive ground plan may mediate colony-level selection effects on individual foraging behavior in honey bees. *Proceedings of the National Academy of Sciences of the United States of America* 101: 11350–11355.

Andziak, B., and Buffenstein, R. (2006). Disparate patterns of age-related changes in lipid peroxidation in long-lived naked mole rats and shorter-lived mice. *Aging Cell* 5: 525–532.

Arnold, K.E., and Owens, I.P.F. (1998). Cooperative breeding in birds: a comparative test of the life history hypothesis. *Proceedings of the Royal Society B: Biological Sciences* 265: 739–745.

Avilés, L., and Guevara, J. (2017). Sociality in spiders. In: *Comparative social evolution* (eds. Rubenstein, D.R., and Abbot, P.), 188–224. Cambridge University Press.

Beauchamp, G. (2014). Do avian cooperative breeders live longer? *Proceedings of the Royal Society B: Biological Sciences* 281: 20140844.

Bernadou, A., Kramer, B.H., and Korb, J. (2021). Major evolutionary transitions in social insects, the importance of worker sterility and life history trade-offs. *Frontiers in Ecology and Evolution* 9: 732907.

Blacher, P., Huggins, T.J., and Bourke, A.F.G. (2017). Evolution of ageing, costs of reproduction and the fecundity – longevity trade-off in eusocial insects. *Proceedings of the Royal Society B: Biological Sciences* 284: 20170380.

Blumstein, D.T., and Møller, A.P. (2008). Is sociality associated with high longevity in North American birds? *Biology Letters* 4: 146–148.

Bonasio, R., Zhang, G., Ye, C., Mutti, N.S., Fang, X., Qin, N., Donahue, G., Yang, P., Li, Q., Li, C., Zhang, P., Huang, Z., Berger, S.L., Reinberg, D., Wang, J., and Liebig, J. (2010). Genomic comparison of the ants *Camponotus floridanus* and *Harpegnathos saltator*. *Science* 329: 1068–1071.

Boomsma, J.J. (2009). Lifetime monogamy and the evolution of eusociality. *Philosophical Transactions of the Royal Society, B: Biological Sciences* 364: 191–207.

Boomsma, J.J., Huszár, D.B., Pedersen, J.S. (2014). The evolution of multiqueen breeding in eusocial lineages with permanent physically differentiated castes. *Animal Behaviour* 92: 241–252.

Bourke, A.F.G. (1999). Colony size, social complexity, and reproductive conflict in social insects. *Journal of Evolutionary Biology* 12: 245–257.

Bourke, A.F.G. (2007). Kin selection and the evolutionary theory of aging. *Annual Review of Ecology, Evolution, and Systematics* 38: 103–128.

Brown, J.L. (1974). Alternative routes to sociality in jays – with a theory for the evolution of altruism and communal breeding. *American Zoologist* 14: 63–80.

Brown, J.L. (1987). *Helping and communal breeding in birds. Monographs in behavior and ecology.* Princeton University Press.

Buffenstein, R., and Jarvis, J.U.M. (2002). The naked mole rat – a new record for the oldest living rodent. *Science of Aging Knowledge Environment* 21: pe7.

Carey, J.R. (2001). Demographic mechanisms for the evolution of long life in social insects. *Experimental Gerontology* 36: 713–722.

Chandra, V., Fetter-Pruneda, I., Oxley, P.R., Ritger, A.L., McKenzie, S.K., Libbrecht, R., and Kronauer, D.J.C. (2018). Social regulation of insulin signaling and the evolution of eusociality in ants. *Science* 361: 398–402.

Chapman, T.W., Crespi, B.J., and Perry, S.P. (2008). The evolutionary ecology of eusociality in Australian gall thrips: a 'model clades' approach. In: *Ecology of social evolution* (eds. Korb, J., and Heinze, J.), 57–83. Springer.

Cockburn, A. (2003). Cooperative breeding in oscine passerines: does sociality inhibit speciation. *Proceedings of the Royal Society B: Biological Sciences* 270: 2207–2214.

Cockburn, A., Hatchwell, B.J., and Koenig, W.D. (2017). Sociality in birds. In *Comparative social evolution* (eds. Rubenstein, D.R., and Abbot, P.), 320–353. Cambridge University Press.

Cornwallis, C.K., West, S.A., Davis, K.E., and Griffin, A.S. (2010). Promiscuity and the evolutionary transition to complex societies. *Nature* 466: 969–972.

Corona, M., Velarde, R.A., Remolina, S., Moran-Lauter, A., Wang, Y., Hughes, K.A., and Robinson, G.E. (2007). Vitellogenin, juvenile hormone, insulin signaling, and queen honey bee longevity. *Proceedings of the National Academy of Sciences of the United States of America* 104: 7128–7133.

Cram, D.L., Blount, J.D., and Young, A.J. (2015). The oxidative costs of reproduction are group-size dependent in a wild cooperative breeder. *Proceedings of the Royal Society B: Biological Sciences* 282: 20152031.

da Silva, J. (2022). The extension of foundress life span and the evolution of eusociality in the Hymenoptera. *The American Naturalist* 199: E140–E155.

Dammann, P. (2017). Slow aging in mammals – lessons from African mole-rats and bats. *Seminars in Cell and Developmental Biology* 70: 154–163.

Dammann, P., and Burda, H. (2006). Sexual activity and reproduction delay ageing in a mammal. *Current Biology* 16: R117–R118.

Dammann, P., Šaffa, G., and Šumbera, R. (2022). Longevity of a solitary mole-rat species and its implications for the assumed link between sociality and longevity in African mole-rats (Bathyergidae). *Biology Letters* 18: 20220243.

de Verges, J., and Nehring, V. (2016). A critical look at proximate causes of social insect senescence: damage accumulation or hyperfunction? *Current Opinion in Insect Science* 16: 69–75.

Downing, P.A., Cornwallis, C.K., and Griffin, A.S. (2015). Sex, long life and the evolutionary transition to cooperative breeding in birds. *Proceedings of the Royal Society B: Biological Sciences* 282: 20151663.

Downing, P.A., Griffin, A.S., and Cornwallis, C.K. (2021). Hard-working helpers contribute to long breeder lifespans in cooperative birds. *Philosophical Transactions of the Royal Society, B: Biological Sciences* 376: 20190742.

Elsner, D., Meusemann, K., and Korb, J. (2018). Longevity and transposon defense, the case of termite reproductives. *Proceedings of the National Academy of Sciences of the United States of America* 115: 5504–5509.

Emlen, S.T. (1982). The evolution of helping. I. An ecological constraints model. *The American Naturalist* 119: 29–39.

Finkel, T., and Holbrook, N.J. (2000). Oxidants, oxidative stress and the biology of ageing. *Nature* 408: 239–247.

Flatt, T., Tu, M.P., and Tatar, M. (2005). Hormonal pleiotropy and the juvenile hormone regulation of *Drosophila* development and life history. *BioEssays* 27: 999–1010.

Fluri, P., Sabatini, A.G., Vecchi, M.A., and Wille, H. (1981). Blood juvenile hormone, protein and vitellogenin titres in laying and non-laying queen honeybees. *Journal of Apicultural Research* 20: 221–225.

Foley, N.M., Petit, E.J., Brazier, T., Finarelli, J.A., Hughes, G.M., Touzalin, F., Puechmaille, S.J., and Teeling, E.C. (2020). Drivers of longitudinal telomere dynamics in a long-lived bat species, *Myotis*. *Molecular Ecology* 29: 2963–2977.

Gems, D., and Partridge, L. (2013). Genetics of longevity in model organisms: debates and paradigm shifts. *Annual Review of Physiology* 75: 621–644.

Griesser, M., Drobniak, S.M., Nakagawa, S., Botero, C.A. (2017). Family living sets the stage for cooperative breeding and ecological resilience in birds. *PLoS Biology* 15(6): e2000483.

Griffin, A.S., and West, S.A. (2003). Kin discrimination and the benefit of helping in cooperatively breeding vertebrates. *Science* 302: 634–636.

Hamilton, W.D. (1966). The moulding of senescence by natural selection. *Journal of Theoretical Biology* 12: 12–45.

Hammers, M., Kingma, S.A., Spurgin, L.G., Bebbington, K., Dugdale, H.L., Burke, T., Komdeur, J., and Richardson, D.S. (2019). Breeders that receive help age more slowly in a cooperative breeding bird. *Nature Communications* 10: 1301.

Hartfelder, K. (2000). Insect juvenile hormone: from "status quo" to high society. *Brazilian Journal of Medical and Biological Research* 33: 157–177.

Hartmann, A., and Heinze, J. (2003). Lay eggs, live longer: division of labor and life span in a clonal ant species. *Evolution* 57:2 424–2429.

Hatchwell, B.J., and Komdeur, J. (2000). Ecological constraints, life-history traits and the evolution of cooperative breeding. *Animal Behaviour* 59: 1079–1086.

Heinze, J. (2017). Life-history evolution in ants: the case of *Cardiocondyla*. *Proceedings of the Royal Society B: Biological Sciences* 284: e20161406.

Heinze, J., and Schrempf, A. (2008). Aging and reproduction in social insects – a mini-review. *Gerontology* 54: 160–167.

Heinze, J., and Schrempf, A. (2012). Terminal investment: individual reproduction of ant queens increases with age. *PLoS ONE* 7(4): e35201.

Helanterä, H., Strassmann, J.E., Carrillo, J., and Queller, D.C. (2009). Unicolonial ants: where do they come from, what are they and where are they going? *Trends in Ecology & Evolution* 24: 341–349.

Hölldobler, B., and Wilson, E.O. (1990). *The ants*. Harvard University Press.

Holtze, S., Gorshkova, E., Brause, S., Cellerino, A., Dammann, P., Hildebrandt, T.B., Hoeflich, A., Hoffmann, S., Koch, P., Skulachev, M., Skulachev, V.P., Tozzini, E.T., and Sahm, A. (2021). Alternative animal models of aging research. *Frontiers in Molecular Biosciences* 8: 660959.

Hughes, W.O.H., Oldroyd, B.P., Beekman, M., and Ratnieks, F.L.W. (2008). Ancestral monogamy shows kin selection is key to the evolution of eusociality. *Science* 320: 1213–1216.

Hunt, J.H., and Toth, A.L.J. (2017). Sociality in wasps. In: *Comparative social evolution* (eds. Rubenstein, D.R., and Abbot, P.), 84–123. Cambridge University Press.

Jaimes-Nino, L.M., Heinze, J., and Oettler, J. (2022). Late-life fitness gains and reproductive death. *eLife* 11: 74695.

Jemielity, S. (2005). Long live the queen: studying aging in social insects. *Age* 27: 241–248.

Jemielity, S., Kimura, M., Parker, K.M., Parker, J.D., Cao, X., Aviv, A., and Keller, L. (2007). Short telomeres in short-lived males: what are the molecular and evolutionary causes? *Aging Cell* 6: 225–233.

Kapheim, K. (2017). Nutritional, endocrine, and social influences on reproductive physiology at the origins of social behavior. *Current Opinion in Insect Science* 22: 62–70.

Keller, L. (1998). Queen lifespan and colony characteristics in ants and termites. *Insectes Sociaux* 45: 235–246.

Keller, L., and Genoud, M. (1997). Extraordinary lifespans in ants: a test of evolutionary theories of ageing. *Nature* 389: 958–960.

Kennedy, A., Herman, J., and Rueppell, O. (2021). Reproductive activation in honeybee (*Apis mellifera*) workers protects against abiotic and biotic stress. *Philosophical Transactions of the Royal Society, B: Biological Sciences* 376: 20190737.

Kenyon, C.J. (2010). The genetics of ageing. *Nature* 464: 504–512.

Kenyon, C.J. (2011). The first long-lived mutants: discovery of the insulin/IGF-1 pathway for ageing. *Philosophical Transactions of the Royal Society, B: Biological Sciences* 366: 9–16.

Kokko, H., and Ekmann, J. (2002). Delayed dispersal as a route to breeding: territorial inheritance, safe havens, and ecological constraints. *The American Naturalist* 160: 468–484.

Korb, J. (2007). Workers of a drywood termite do not work. *Frontiers in Zoology* 4: e7.

Korb, J. (2008a). The ecology of social evolution in termites. In: *Ecology of social evolution* (eds. Korb, J., and Heinze, J.), 151–174. Springer.

Korb, J. (2008b). Termites, hemimetabolous diploid white ants? *Frontiers in Zoology* 5: 1–9.

Korb, J. (2016). Why do social insect queens live so long? Approaches to unravel the sociality-aging puzzle. *Current Opinion in Insect Science* 16: 104–107.

Korb, J., and Hartfelder, K. (2008). Life history and development – a framework for understanding developmental plasticity in lower termites. *Biological Reviews* 83: 295–313.

Korb, J., and Heinze, J. (2008). *Ecology of social evolution*. Springer.

Korb, J., and Heinze, J. (2016). Major hurdles for the evolution of sociality. *Annual Review of Entomology* 61: 297–316.

Korb, J., and Heinze, J. (2021). Ageing and sociality: why, when, and how does sociality change ageing patterns? *Philosophical Transactions of the Royal Society, B: Biological Sciences* 376: 20190727.

Korb, J., Meusemann, K., Aumer, D., Bernadou, A., Elsner, D., Feldmeyer, B., Foitzik, S., Heinze, J., Libbrecht, R., Lin, S., Majoe, M., Monroy Kuhn, J.M., Nehring, V., Negroni, M.A., Paxon, R.J., Séguret, A., Stoldt, M., and Flatt, T. (2021). Comparative transcriptomic analysis of the mechanisms underpinning ageing and fecundity in social insects. *Philosophical Transactions of the Royal Society, B: Biological Sciences* 376: 20190728.

Korb, J., and Schneider, K. (2007). Does kin structure explain the occurrence of workers in the lower termite *Cryptotermes secundus*? *Evolutionary Ecology* 21: 817–828.

Korb, J., and Thorne, J. (2017). Sociality in termites. In *Comparative social evolution* (eds. Rubenstein, D.R., and Abbot, P.), 124–153. Cambridge University Press.

Koubová, J., Pangrácová, M., Jankásek, M., Luksan, O., Jehlik, T., Brabcová, J., Jedlicka, P., Krivánek, J., Frydrychová, R.C., and Hanus, R. (2021). Long-lived termite kings and queens activate telomerase in somatic organs. *Proceedings of the Royal Society B: Biological Sciences* 288: 20210511.

Kramer, B.H., Nehring, V., Buttstedt, A., Heinze, J., Korb, J., Libbrecht, R., Meusemann, K., Paxton, R.J., Séguret, A., Schaub, F., and Bernadou, A. (2021). Oxidative stress and senescence in social insects: a significant but inconsistent link? *Philosophical Transactions of the Royal Society, B: Biological Sciences* 376: 20190732.

Kramer, B.H., and Schaible, R. (2013). Colony size explains the lifespan differences between queens and workers in eusocial Hymenoptera. *Biological Journal of the Linnean Society* 109: 710–724.

Kramer, B.H., van Doorn, G.S., Arani, B.M.S., and Pen, I. (2022). Eusociality and the evolution of aging in superorganisms. *The American Naturalist* 200: 63–80.

Kreider, J.J., Kramer, B.H., Komdeur, J., and Pen, I. (2022). The evolution of ageing in cooperative breeders. *Evolution Letters* 6: 450–459.

Kreider, J.J., Pen, I., and Kramer, B.H. (2021). Antagonistic pleiotropy and the evolution of extraordinary lifespans in eusocial organisms. *Evolution Letters* 5: 178–186.

Leadbeater, E., Carruthers, J.M., Green, J.P., Rosser, N.S., and Field, J. (2011). Nest inheritance is the missing source of direct fitness in a primitively eusocial insect. *Science* 333: 874–876.

Libbrecht, R., Corona, M., Wende, F., Azevedo, D.O., Serrao, J.E., and Keller, L. (2013). Interplay between insulin signaling, juvenile hormone, and vitellogenin regulates maternal effects on polyphenism in ants. *Proceedings of the National Academy of Sciences of the United States of America* 110: 1050–11055.

Liebig, J., and Poethke, H.-J. (2004). Queen lifespan and colony longevity in the ant *Harpegnathos saltator*. *Ecological Entomology* 29: 203–207.

Lin, S., Pen, I., and Korb, J. (2023). Effect of food restriction on survival and reproduction of a termite. *Journal of Evolutionary Biology* 36: 542–549.

Lin, S., Werle, J., and Korb, J. (2021). Transcriptomic analyses of the termite, *Cryptotermes secundus*, reveal a gene network underlying a long lifespan and high fecundity. *Communications Biology* 4: 384.

Lucas, E.R., and Keller, L. (2014). Ageing and somatic maintenance in social insects. *Current Opinion in Insect Science* 3: 1–6.

Lucas, E.R., and Keller, L. (2020). The co-evolution of longevity and social life. *Functional Ecology* 34: 76–87.

Lucas, E.R., Privman, E., and Keller, L. (2016). Higher expression of somatic repair genes in long-lived ant queens than workers. *Aging* 8: 1940–1949.

Lukas, D., and Clutton-Brock, T. (2012). Life histories and the evolution of cooperative breeding in mammals. *Proceedings of the Royal Society B: Biological Sciences* 279: 4065–4070.

Maekawa, K., Ishitani, K., Gotoh, H., Cornette, R., and Miura, T. (2010). Juvenile hormone titre and vitellogenin gene expression related to ovarian development in primary reproductives compared with nymphs and nymphoid reproductives of the termite *Reticulitermes speratus*. *Physiological Entomology* 35: 52–58.

Meunier, J., West, S.A., and Chapuisat, M. (2008) Split sex ratios in the social Hymenoptera: a meta-analysis. *Behavioral Ecology* 19: 382–390.

Michener, C.D. (1974). *The social behavior of the bees: a comparative study*. Harvard University Press.

Monroy Kuhn, J.M., and Korb, J. (2016). Editorial overview: social insects: aging and the re-shaping of the fecundity/longevity trade-off. *Current Opinion in Insect Science* 16: vii–x.

Monroy Kuhn, J.M., Meusemann, K., and Korb, J. (2019). Long live the queen, the king and the commoner? Transcript expression differences between old and young in the termite *Cryptotermes secundus*. *PLoS ONE* 14(2): e0210371.

Monroy Kuhn, J.M., Meusemann, K., and Korb, J. (2021). Disentangling the aging gene expression network of termite queens. *BMC Genomics* 22: 339.

Negroni, M.A., Jongepier, E., Feldmeyer, B., Kramer, B.H., and Foitzik, S. (2016). Life history evolution in social insects: a female perspective. *Current Opinion in Insect Science* 16: 51–57.

Oster, G.F., and Wilson, E.O. (1978). *Caste and ecology in the social insects*. Princeton University Press.

Pamminger, T., Treanor, D., Hughes, W.O.H. (2016). Pleiotropic effects of juvenile hormone in ant queens and the escape from the reproduction – immunocompetence trade-off. *Proceedings of the Royal Society B: Biological Sciences* 283: 20152409.

Park, T.J., Reznick, J., Peterson, B.I., Blass, G., Omerbasic, D., Bennett, N.C., Kuich, P.H.J.L., Zasada, C., Browe, B.M., Hamann, W., Applegate, D.T., Radke, M.H., Kosten, T., Lutermann, H., Gavaghan, V., Eigenbrod, O., Begay, V., Amoroso, V.F., Govind, V., Minshall, R.D., Smith, E.S.J., Larson, J., Gotthardt, M., Kempa, S., and Lewin, G.R. (2017). Fructose-driven glycolysis supports anoxia resistance in the naked mole-rat. *Science* 356: 307–311.

Parker, J.D., Parker, K.M., Sohal, B.H., Sohal, R.S., and Keller, L. (2004). Decreased expression of Cu–Zn superoxide dismutase 1 in ants with extreme lifespan. *Proceedings of the National Academy of Sciences of the United States of America* 101: 3486–3489.

Partridge, L., Alic, N., Bjedov, I., and Piper, M.D.W. (2011). Ageing in *Drosophila*: the role of the insulin/Igf and TOR signalling network. *Experimental Gerontology* 46: 376–381.

Peeters, C. (1991). The occurrence of sexual reproduction among ant workers. *Biological Journal of the Linnean Society* 44: 141–152.

Pickett, K.M., and Carpenter, J.M. (2010). Simultaneous analysis and the origin of eusociality in the Vespidae (Insecta: Hymenoptera). *Arthropod Systematics & Phylogeny* 68: 3–33.

Post, F., Bornberg-Bauer, E., Vasseur-Cognet, M., and Harrison, M.C. (2023). More effective transposon regulation in long-lived termite queens than in workers. *Molecular Ecology* 32: 369–380.

Promislow, D.E.L., Flatt, T., and Bonduriansky, R. (2021). The biology of aging in insects: from *Drosophila* to other insects and back. *Annual Review of Entomology* 67: 83–103.

Queller, D.C., and Strassmann, J.E. (2009). Beyond society: the evolution of organismality. *Philosophical Transactions of the Royal Society, B: Biological Sciences* 364: 3143–3155.

Rabeling, C., Brown, J.M., and Verhaagh, M. (2008). Newly discovered sister lineage sheds light on early ant evolution. *Proceedings of the National Academy of Sciences of the United States of America* 105: 14913–14917.

Rascon, B., Mutti, N.S., Tolfsen, C., and Amdam, G. (2011). Honey bee life history plasticity: development, behavior, and aging. In: *Mechanisms of life history evolution* (eds. Flatt, T., and Heyland, A.), 253–266. Oxford University Press.

Ratnieks, F.L.W., Foster, K.R., and Wenseleers, T. (2006). Conflict resolution in insect societies. *Annual Review of Entomology* 51: 581–608.

Rau, V., Flatt, T., and Korb, J. (2023). The remoulding of dietary effects on the fecundity/longevity trade-off in a social insect. *BMC Genomics* 24: 244.

Rau, V., and Korb, J. (2021). The effect of environmental stress on ageing in a termite species with low social complexity. *Philosophical Transactions of the Royal Society, B: Biological Sciences* 376: 20190739.

Rodrigues, M.A., and Flatt, T. (2016). Endocrine uncoupling of the trade-off between reproduction and somatic maintenance in eusocial insects. *Current Opinion in Insect Science* 16: 1–8.

Rodriguez, K.A., Osmulski, P.A., Pierce, A., Weintraub, S.T., Gaczynska, M., Buffenstein, R. (2014). A cytosolic protein factor from the naked mole-rat activates proteasomes of other species and protects these from inhibition. *Biochimica et Biophysica Acta* 1842: 2060–2072.

Roisin, Y., and Korb, J. (2011). Social organisation and the status of workers in termites. In: *Biology of termites: a modern synthesis* (eds. Bignell, E., Roisin, Y., and Lo, N.), 133–164. Springer.

Rubenstein, D.R., and Abbot, P. (2017). Social synthesis: opportunities for comparative social evolution. In: *Comparative social evolution* (eds. Rubenstein, D.R., and Abbot, P.), 427–452. Cambridge University Press.

Rueppell, O. (2009). Aging of social insects. In: *Organization of insect societies. From genome to sociocomplexity* (eds. Gadau, J., and Fewell, J.H.), 51–73. Harvard University Press.

Sahm, A., Bens, M., Henning, Y., Vole, C., Groth, M., Schwab, M., Hoffmann, S., Platzer, M., Szafranski, K., and Dammann, P. (2018a). Higher gene expression stability during aging in long-lived giant mole-rats than short-lived rats. *Aging* 10: 3938–3956.

Sahm, A., Bens, M., Szafranski, K., Holtze, S., Groth, M., Goerlach, M., Calkhoven, C., Mueller, C., Schwab, M., Kraus, J., Kestler, H.A., Cellerino, A., Burda, H., Hildebrandt, T., Dammann, P., and Platzer, M. (2018b). Long-lived rodents reveal signatures of positive selection in genes associated with lifespan. *PLoS Genetics* 14(3): e1007272.

Sahm, A., Platzer, M., Koch, P., Henning, Y., Bens, M., Groth, M., Burda, H., Begall, S., Ting, S., Goetz, M., van Daele, P., Staniszewska, M., Klose, J., Fragoso Costa, P., Hoffmann, S., Szafranski, K., and Dammann, P. (2021). Increased longevity due to sexual activity in mole-rats is associated with transcriptional changes in HPA stress axis. *eLife* 10: e57843.

Schmauck-Medina, T., Molière, A., Lautrup, S., Zhang, J., Chlopicki, S., Borland Madsen, H., Cao, S., Soendenbroe, C., Mansell, E., Vestergaards, M.B., Li, Z., Shiloh, Y., Opresko, P.L., Egly, J.M., Kirkwood, T., Verdin, E., Bohr, V.A., Cox, L.S., Stevnsner, T., Rasmussen, L.J., and Fang, E.F. (2022). New hallmarks of ageing: a 2022 Copenhagen ageing meeting summary. *Aging* 14: 6829–6839.

Schneider, S.A., Schrader, C., Wagner, A.E., Boesch-Saadatmandi, C., Liebig, J., Rimbach, G., and Roeder, T. (2011). Stress resistance and longevity are not directly linked to levels of enzymatic antioxidants in the ponerine ant *Harpegnathos saltator*. *PLoS ONE* 6(1): e14601.

Schrempf, A., Giehr, J., Röhrl, R., Steigleder, S., and Heinze, J. (2017). Royal Darwinian demons: enforced changes in reproductive efforts do not affect the life expectancy of ant queens. *The American Naturalist* 189: 436–442.

Schrempf, A., Heinze, J., and Cremer, S. (2005). Sexual cooperation: mating increases longevity in ant queens. *Current Biology* 15: 267–270.

Seehuus, S., Norberg, K., Gimsa, U., Krekling, T., and Amdam, G.V. (2006). Reproductive protein protects functionally sterile honey bee workers from oxidative stress. *Proceedings of the National Academy of Sciences of the United States of America* 103: 962–967.

Séité, S., Harrison, M.C., Sillam-Dusses, D., Lupoli, R., van Dooren, T.J.M., Robert, A., Poissonier, L.A., Lemainque, A., Renault, D., Acket, S., Andrieu, M., Viscarra, J., Sul, H.S., de Beer, Z.W., Bornberg-Bauer, E., and Vasseur-Cognet, M. (2022). Lifespan prolonging mechanisms and insulin upregulation without fat accumulation in long-lived reproductives of a higher termite. *Communications Biology* 5: 44.

Shellman-Reeve, J.S. (1997). The spectrum of eusociality in termites. In: *The evolution of social behavior in insects and arachnids* (eds. Choe, J.C., and Crespi, B.J.), 52–93. Cambridge University Press.

Sherman, P.W., and Jarvis, J.U.M. (2002). Extraordinary life spans of naked mole-rats (*Heterocephalus glaber*). *Journal of Zoology* 258: 307–311.

Shik, J.Z. (2008). Ant colony size and the scaling of reproductive effort. *Functional Ecology* 22: 674–681.

Shik, J.Z., Hou, C., Kay, A., Kaspari, M., and Gillooly, J.F. (2012). Towards a general life-history model of the superorganism: predicting the survival, growth and reproduction of ant societies. *Biology Letters* 8: 1059–1062.

Sohal, R.S., and Weindruch, R. (1996). Oxidative stress, caloric restriction, and aging. *Science* 273: 59–63.

Speakman, J.R. (2005). Body size, energy metabolism and lifespan. *Journal of Experimental Biology* 208: 1717–1730.

Stearns, S.C. (1976). Life-history tactics: a review of the ideas. *The Quarterly Review of Biology* 51: 3–47.

Strassmann, J.E. (1985). Worker mortality and the evolution of castes in the social wasp *Polistes exclamans*. *Insectes Sociaux* 32: 275–285.

Sudyka, J., Arct, A., Drobniak, S., Gustafsson, L., and Cichon, M. (2016). Longitudinal studies confirm faster telomere erosion in short-lived bird species. *Journal of Ornithology* 157: 373–375.

Tasaki, E., Kobayashi, K., Matsuura, K., and Iuchi, Y. (2017). An efficient antioxidant system in a long-lived termite queen. *PLoS ONE* 12(1): e0167412.

Tasaki, E., Kobayashi, K., Matsuura, K., and Iuchi, Y. (2018). Long-lived termite queens exhibit high Cu/Zn superoxide dismutase activity. *Oxidative Medicine and Cellular Longevity* 2018: 5127251.

Thorley, J. (2020). The case for extended lifespan in cooperatively breeding mammals: a re-appraisal. *PeerJ* 8: e9214.

Toth, A.L., Sumner, S., and Jeanne, R.L. (2016). Patterns of longevity across a sociality gradient in vespid wasps. *Current Opinion in Insect Science* 16: 28–35.

Tricola, G.M., Simons, M.J.P., Atema, E., Boughton, R.K., Brown, J.L., Dearborn, D.C., Divoky, G., Eimes, J.A., Huntington, C.E., Kitaysky, A.S., Juola, F.A., Lank, D.B., Litwa, H.P., Mulder, E.G.A., Nisbet, I.C.T., Okanoya, K., Safran, R.J., Schoech, S.J., Schreiber, E.A., Thompson, P.M., Verhulst, S., Wheelwright, N.T., Winkler, D.W., Young, R., Vleck, C.M., and Haussmann, M.F. (2018). The rate of telomere loss is related to maximum lifespan in birds. *Philosophical Transactions of the Royal Society, B: Biological Sciences* 373: 20160445.

Trivers, R.L., and Hare, H. (1976). Haplodiploidy and the evolution of the social insects. *Science* 191: 249–263.

Wcislo, W., and Fewell, J.H. (2017). Sociality in bees. In: *Comparative social evolution* (eds. Rubenstein, D.R., and Abbot, P.), 50–83. Cambridge University Press.

Weitekamp, C.A., Libbrecht, R., and Keller, L. (2017). Genetics and evolution of social behavior in insects. *Annual Reviews in Genetics* 51: 219–239.

West, S.A., Griffin, A.S., and Gardner, A. (2007). Evolutionary explanations for cooperation. *Current Biology* 17: 661–672.

Wild, G., and Korb, J. (2017). Evolution of delayed dispersal and subsequent emergence of helping, with implications for cooperative breeding. *Journal of Theoretical Biology* 427: 53–64.

Williams, S.A., and Shattuck, M.R. (2015). Ecology, longevity and naked mole-rats: confounding effects of sociality? *Proceedings of the Royal Society B: Biological Sciences* 282: 20141664.

Wilson, E.O. (1980). Caste and division of labor in leaf-cutter ants (Hymenoptera: Formicidae: Atta) I. The overall pattern in *A. sexdens*. *Behavioral Ecology and Sociobiology* 7: 143–156.

Winston, M.L. (1991). *The biology of the honey bee.* Harvard University Press.

Wirthlin, M., Lima, N.C., Guedes, R.I.M., Soares, A.E., Almeida, I.G.P., Cavaleiro, N.P., de Morais, G.L., Chaves, A.V., Howard, J.T., Teixeira, M., Schneider, P.N., Santos, F.R., Schatz, M.C., Felipe, M.S., Miyaki, C.Y., Aleixo, A., Schneider, M.P.C., Jarvis, E.D., Vasconcelos, A.T.R., Prosdocimi, F., and Mello, C.V. (2018). Parrot genomes and the evolution of heightened longevity and cognition. *Current Biology* 28: 4001–4008.

Wolfner, M.F. (1997). Tokens of love: functions and regulation of *Drosophila* male accessory gland products. *Insect Biochemisty and Molecular Biology* 27: 179–192.

Yan, H., Opachaloemphan, C., Carmona-Aldana, F., Mancini, G., Mlejnek, J., Descostes, N., Sieriebriennikov, B., Leibholz, A., Zhou, X., Ding, L., Traficante, M., Desplan, C., and Reinberg, D. (2022). Insulin signaling in the long-lived reproductive caste of ants. *Science* 377: 1092–1099.

Zhu, P., Liu, W., Zhang, X., Li, M., Liu, G., Yu, Y., Li, Z., Li, X., Du, J., Wang, X., Grueter, C.C., Li, M., and Zhou, X. (2023). Correlated evolution of social organization and lifespan in mammals. *Nature Communications* 14: 372.

10

Integrating Dispersal in Life History

Dries Bonte

Department of Biology, Ghent University, Gent, Belgium

10.1 Introduction

Temporal stability of the environment relative to an organism's lifespan is one of the major environmental determinants of life histories. In case of unstable environments, the *r/K* selection theory predicts that under minimal competition, organisms would benefit from life history strategies exploiting environments by having high reproductive efforts and short lifespans (*r*-strategies). On the other hand, in the case of stable environments, *K*-species with higher investments in competitive resistance are selected. The emphasis on reproduction, mortality and development rather than on the emerging population dynamics led to the translation of this theory to the akin slow-fast continuum (*e.g.*, Blackburn 1991, Salguero-Gómez et al. 2016; see also Chapter 5). Both the 'pace of life' and *r/K* selection theory should be regarded as a continuum with large potential to be integrated into more complex age- or stage-structured paradigms (Reznick et al. 2002).

The spatial components of fitness prospects are not explicitly part of these general paradigms but crucial to be considered for the evolution of life histories (Southwood 1977). In theory, organisms should display life history adaptations that maximize fitness not only within their local environment but also within the wider context of temporal and spatial changes. The relationship between the scale of spatio-temporal environmental variation, lifespan and dispersal capacity will then jointly impact fitness. Traits such as dormancy (Willis et al. 2014) or diapause (Dingle 1978, Joschinski and Bonte 2021) facilitate local persistence in unpredictable temporal environmental fluctuations. Alternatively, the dispersal of individuals, or their progeny, will maximize performance in spatio-temporally heterogeneous environments by either invoking risk-spreading mechanisms or the selection of locations where fitness can be maximized. Dispersal is therefore a central trait in life history (Bonte and Dahirel 2017).

Dispersal is mostly defined as the movement of organisms from the location of birth to the location of reproduction, with consequences for gene flow (Ronce 2007). This is a unique process during an individual's life, but its signature is pertinent across generations. Dispersal contrasts with frequent, routine movements related to, for instance, foraging and mate searching. As the latter type of routine movements, and their effects on fitness, take place at relatively small spatial and temporal scales within an individual's lifetime, they are usually considered as part of the local demography. Dispersal impacts the spread of genes, individuals and species at large scales and is therefore an additional, essential component to fitness and the organization of biodiversity on larger spatio-temporal scales (Weil et al. 2023).

Dispersal evolves as a risk-spreading mechanism, as a mechanism to reduce competition and mating with relatives, and as a mechanism to escape detrimental environmental conditions, induced by either external (*e.g.*, disturbances, diseases, and crowding) or internal (*e.g.*, body condition, and age) factors (Bowler and Benton 2005, Ronce 2007). The eventual dispersal rates and distances will, in general, evolve in relation to the expected fitness advantages of colonizing new habitats or joining other populations, but offset against the multiple costs associated with dispersal (Bonte et al. 2012). These costs are typically divided into energetic, time, risk and opportunity costs and may be levied directly or deferred during all phases of the dispersal process, including those prior to the actual dispersal event (see Table 10.1).

Dispersal maximizes fitness in spatially structured environments and is promoted when spatial variation of fitness benefits is high. Such conditions are prevalent when spatio-temporal environmental variation is high, but also when

Table 10.1 Definitions of the identified cost types of dispersal (from Bonte et al. 2012 / With permission of John Wiley & Sons).

Cost type	Definition
Energetic costs	Costs due to lost metabolic energy in movements. Energetic costs may also comprise costs related to the development of specific machinery associated with dispersal, *i.e.*, energetic expenditure for the construction of special dispersal organs and tissues (muscles, wings).
Time costs	Direct costs are due to the time invested in dispersal that cannot be invested in other activities.
Risk costs	Direct costs related to both mortality risks (*e.g.*, due to increased predation or settlement in unsuitable habitat) and deferred attrition costs by accumulated damage (*e.g.*, wing wear or wounding) or physiological changes.
Opportunity costs	Costs incurred by selecting the next-best choice available from several mutually exclusive options. Opportunity costs are typical in individuals giving up prior residence advantages and familiarity-related advantages. Opportunity costs also include the loss of any advantage derived from being locally adapted.

conditions leading to a loss of dispersal (*e.g.*, strong patch isolation or other conditions associated with strong dispersal costs) provoke strong kin-competition and/or inbreeding risks (Ronce 2007, Duputié and Massol 2013). Dispersal therefore evolves in a non-linear way with increasing patch isolation and may be further promoted when local relatedness is high. Population, and likely also community dynamics, are consequently both impacting and impacted by dispersal (Bonte and Dahirel 2017). Eco-evolutionary dynamics in meta-populations are thus driven by dispersal (Hanski 2012, Cheptou et al. 2017, Bonte et al. 2018). The co-variation between dispersal and any other (morphological, physiological, behavioural and life history) traits is referred to as a 'dispersal syndrome'. Its organization evolves from the tension between regional (meta-population) and local (population) levels of selection, but equally from evolutionary and ecological constraints. I here focus on the coupling between life history traits such as survival, reproduction, time until maturity and longevity and dispersal, so-called dispersal-life history syndromes.

It is important to note that, for animals and plants, let alone for microbial organisms (Chaudhary et al. 2022), different definitions of dispersal and research traditions have created separate research agendas. Plant dispersal has been traditionally approached from a population 'Eulerian' perspective on kernels of seed dispersal. This approach focusses on the spatial properties of the spread mechanisms and how the distribution of individuals changes over time. Animal biologists mostly follow a behavioural approach by taking the individual 'Lagrangian' perspective, in which single or multiple individuals are followed from the place of birth to the location of reproduction (Baguette et al. 2014). When studied from a biogeographic perspective, dispersal is defined as the success of lineages in crossing major geographic barriers, such as oceans, mountain ranges or deserts (Weil et al. 2023). While I will synthesize insights from all perspectives in this contribution, it is important to keep these different research agendas in mind to eventually recognize any bias in our understanding of dispersal evolution between protists, plants and animals. With some exceptions, the theory of dispersal evolution is based on the Lagrangian approach, as this individual perspective is most closely aligned to fitness. The simplifications of life histories by assuming annual life cycles, and thus non-overlapping generations, global or nearest neighbour dispersal distances, and most often non-informed immigration, enforce us therefore to unveil general principles, rather than generating predictions for specific species. In predictive ecology, however, Eulerian approaches are more commonly used to forecast spread processes, for instance, insect outbreaks, but mixed approaches seem to be promising as well to improve model performance (Baguette et al. 2014).

I start by outlining the importance of the full-lifecycle perspective for dispersal, subsequently review the available theory on dispersal-life history co-variation, and end with the available empirical evidence. I conclude this chapter by highlighting the needed steps to further advance our insights on life history evolution in spatially structured systems and why these insights are essential for our understanding of the organization of biodiversity across the world.

10.2 Dispersal as Part of the Life History

Dispersal is an individual-level process composed of traits related to departure (initiation of the eventual act of leaving natal habitat), transfer (movement) and settlement (finalization of the movement phase in novel habitat) (Clobert et al. 2009). This separation allows for a better understanding of the fitness costs associated with dispersal and their impacts on, or associations with reproduction, growth, maturation and survival traits (Bonte et al. 2012). Dispersal costs (Table 10.1) can be

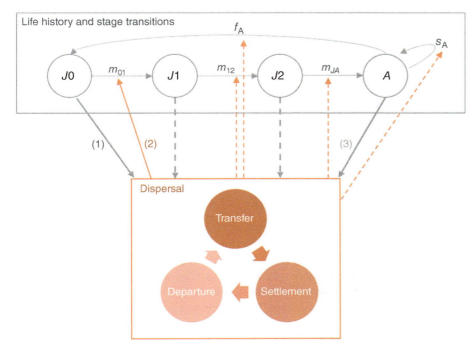

Figure 10.1 Dispersal costs and the integration life history (*Source:* adapted from Bonte et al. 2012). The life cycle of the organisms is stage-structured (J: juvenile, A: adult) with demographic parameters *m* representing transition probabilities from one age/size class to the next one (maturation rate being inversely related to the mortality at this stage), s_A the survival probability of the adults and f_A the fecundity or reproductive output of the adult stage. The relative abundance of the different stages and/or stage-specificity of dispersal impact the full dispersal strategy (grey arrows). Costs associated with dispersal feedback on the transition rates between the different life history stages and structure of the eventual demography (orange arrows). The thin dashed and full arrows indicate all possible feedback. The thin, full arrows represent an example of natal dispersal. Offspring densities (1) impact emigration, but costs associated with all stages eventually determine their survival (2). Dispersal rates are additionally impacted by adult densities through maternal effects (3). Wide arrows in the dispersal box show the sequence of the different dispersal phases.

incurred before, immediately, or later in life. Each of the three dispersal phases shows specific costs that feedback on life history (Figure 10.1). For example, trading off energy and reserves for reproduction to wing development in aphids (Roff and Bradford 1996, Dingle 2014) results in the well-documented, and evolvable, co-variation between reproduction, survival and dispersal. Any additional energy used for and during dispersal (*e.g.*, from injuries incurred during travelling) or time lost for foraging and mate location may impact other demographic traits. Depending on the timing of dispersal with respect to the life stages – immediately after birth, during maturation or at maturity – dispersal costs will be differentially incurred and impact life history decisions of the parents. For instance, when offspring disperse after emergence, dispersal costs will largely determine their full life history trajectory, with many individuals simply failing to survive the dispersal event. Often, these dispersal events are steered by the parent's reproductive strategies through, for instance, egg size – egg number trade-offs to maximize their genetic fitness (see also Chapter 3, this volume). Dispersal may thus be subject to a severe fitness conflict between parents and their offspring.

10.3 The Theory of Dispersal and Life Histories

The multi-causality of dispersal evolution is now widely acknowledged (Matthysen 2012) and earlier reviews have synthesized the evolutionary emergence of dispersal-life history syndromes (*e.g.*, Ronce and Clobert 2012). In brief, dispersal is caused by the interaction between environmental, demographic, genetic and developmental variation. Since these factors are at the basis of the evolution of life history traits, their co-variation will be shaped by complex cost–benefit structures. Because relevant selection pressures on dispersal are interacting with the evolution of life histories, concerted evolutionary strategies of dispersal, growth, ageing and reproduction are expected. The evolutionary origin of dispersal syndromes can be caused by mutually non-exclusive mechanisms ranging from genetic correlations among traits, responses of dispersal or life histories towards evolution of each one of these traits, and/or their joint dependent or independent evolution to the environment (Ronce and Clobert 2012, Cote et al. 2017).

The theory of dispersal-life history (co-)evolution is approached by: (1) models studying dispersal as the target trait evolving in response to a series of fixed life histories, (2) models having both dispersal and other traits jointly evolving in response to a common environmental pressure or (3) models with the evolution of specific dispersal – life history reaction norms in response to relevant environmental selection pressures. While for (1) and (2) dispersal syndromes are an outcome of evolutionary dynamics, they are in (3) the target of selection with the sign and strength of the co-variation representing the evolving 'trait'. I further elaborate on these three approaches below.

10.3.1 Dispersal Evolution in Response to Other Life History Attributes

Theory shows that dispersal prominently evolves in response to species' life histories like reproductive investments, density dependence and lifespan. Species characterized by a higher reproductive effort relative to the capacity of the system typically evolve at higher dispersal rates because increased dispersal alleviates local (kin) competition (Travis and Dytham 1999, Kisdi 2004, Bach et al. 2006). This finding is not emerging in simulations where dispersal is modelled as a density-dependent process based on the first principles of fitness maximization (Poethke and Hovestadt 2002). Under this assumption, high reproductive rates bring population sizes very quickly to the carrying capacity, thereby reducing spatial variation in densities and, hence, the probability of ending up in a low-density patch. Maternal investments that reduce fecundity and dispersal costs (e.g., by providing higher energy reserves to the dispersing larvae or seeds) increase the width and tail weight of the dispersal kernel, and thus the fraction of long-distance dispersers (Fronhofer et al. 2015). For species with adult dispersal, safer movement capacities, and thus any investment in reducing costs of transfer, increase evolutionary stable dispersal rates (Travis et al. 2012).

Dispersal and larger dispersal distances evolve theoretically under a delayed maturity and longer lifespan (see also Chapters 1 and 2, this volume) since prolonged longevity increases kin-competition, all else being equal. Conversely, reduced lifespans evolve under conditions that select against dispersal and raise local relatedness (Dytham and Travis 2006, Ronce and Promislow 2010). Species with multiple reproductive events (iteroparity) and overlapping generations are thus expected to evolve at increased level of dispersal (Taylor and Irwin 2000), because iteroparity increases (at least theoretically) local relatedness and therefore strengthens kin-competition and risks of inbreeding. The timing of dispersal within the life cycle is an important determinant of dispersal rates when it is modelled as a strategy to cope with competition in a temporary, varying environment (Johst and Brandl 1997). Since, under such assumptions, both population growth (the integration of mortality and reproduction during development) and dispersal are density dependent, their order within the life cycle will determine which process regulates densities, thereby rendering their impact on the alternative mechanism not relevant anymore.

Polygamous mating systems have a strong impact on fitness prospects for the different sexes (see also Chapters 5 and 6, this volume), leading to the evolution of higher dispersal rates in the sex competing most for mates. As explained earlier, spatio-temporal variance in the environment is a major driver of dispersal. Hence, sexes experiencing higher spatio-temporal variance in fitness expectations evolve higher dispersal rates (Gros et al. 2009). In polygamous systems, mean dispersal rates therefore evolve on average towards larger values compared to monogamous mating systems (Gandon 1999, Gandon and Rousset 1999, Perrin and Mazalov 1999). When local mate competition exceeds local resource competition, as mostly seen in males under polygyny or female promiscuity, male-biased dispersal is expected to evolve and be reinforced under inbreeding costs (Perrin and Mazalov 1999, see also Chapter 7, this volume). Other processes leading to imbalanced mating opportunities between sexes, like, for instance, male-killing endosymbiont manipulators (Bonte et al. 2009; see also Chapter 14, this volume), strongly impact spatio-temporal patterns in sex ratios. By means of an individual-based model, it was shown that the presence of such endosymbionts increases kin-competition in males, thereby promoting a male-biased dispersal. Such an increased male dispersal eventually induces an evolutionary rescue of the meta-population.

When mating opportunities are restricted, for instance, under low densities in ephemeral environments, high fecundity, high dispersal ability and the ability for selfing are theoretically expected to be jointly promoted (Baker and Stebbins 1965). Self-compatibility is therefore associated with an enhanced colonization capacity, a rule that is known as Baker's law. Because this general theory seemed to contradict some patterns in nature, Cheptou and Massol (2009) developed a meta-population model for the joint evolution of seed dispersal and self-fertilization when local pollen limitation varied stochastically over time. Under these conditions, dispersing outcrossers and non-dispersing (partial) selfers consistently evolved as alternative strategies.

10.3.2 Joint Evolution of Dispersal and Other Life History Attributes

Fixing one life history attribute while allowing dispersal to evolve provides strong insights into how organismal constraints in life histories affect dispersal rates or distances. However, dispersal, lifespan and reproduction may also evolve jointly in response to the same environment and cause the emergence of dispersal-life history trait co-variations. A slower evolutionary response of ageing to dispersal leads to the evolution of low dispersal-short lifespan syndromes under spatio-temporally stable environments (Dytham and Travis 2006). In environments characterized by high levels of disturbance (hence, spatio-temporally unstable environments), both high reproduction and dispersal are favoured because of on average low levels of local competition and high probabilities of patch extinction (Ronce et al. 2000). When dispersal evolution is jointly studied with the evolution of dormancy, a positive co-variation only evolves when the environment is unpredictable in space and time, hence in species living in ephemeral habitats (Buoro and Carlson 2014). Such arguments were already conceptualized by Southwood (1977). Traits can be genetically connected through pleiotropy (one gene coding for several traits) or phenotypically by developmental trade-offs. The importance of such trade-offs is well recognized in ecology. Dispersal-competition trade-offs are, for instance, an important stabilizing mechanism, leading to the coexistence of competing species in meta-populations (Levins and Culver 1971, Tilman 1994, Laroche et al. 2016).

In heterogeneous environments, informed emigration and immigration strategies allow different phenotypes to select their optimal environment (this is referred to as habitat matching choice). The balance between selection and dispersal determines the likelihood of any micro-geographic adaptation (Richardson et al. 2014). A non-random settlement process is, however, essential to allow both high dispersal and local adaptation to emerge (Holt 1987, Futuyma and Moreno 1988, Ravigné et al. 2009, Bolnick and Otto 2013). Habitat matching choice stabilizes meta-population dynamics, which, in turn, leads to the evolution of lower overall dispersal rates (Mortier et al. 2019). Dispersal rates also evolve in strong concert with cooperative behaviours and group living. The evolution of sociality is therefore promoted when relatedness is high (see also Chapter 9, this volume). Dispersal and cooperation will therefore coevolve in opposite directions until levels that balance the increased competition from limited dispersal (West et al. 2002). Since dispersal breaks down group relatedness and cohesion, sociality is thus more likely to evolve in spatio-temporally stable meta-populations.

10.3.3 The Evolution of Dispersal-Life History Reaction Norms

Life histories are largely shaped by the environment. Under food shortage or competition, for instance, lifespan and reproduction can be largely reduced, because of a reduced energy intake, but dispersal is promoted (Fronhofer et al. 2018). When the reaction norm has a genetic background, it may also be subject to selection. Conditional strategies, as expressed by the relationship between dispersal and the environmental conditions, might thus differently evolve according to the spatio-temporal dimensions of the meta-population and the meta-community (Fronhofer et al. 2018). Despite its likely omnipresence in nature, few theoretical studies have addressed how such dispersal-life history reaction norms or conditional dispersal evolve. Gyllenberg et al. (2008, 2011) and Kisdi et al. (2013) suggested that individual-level dispersal-body condition correlations can differ in direction depending on the spatio-temporal structure of the environment and dispersal costs. A positive co-variation between dispersal and fecundity evolves in meta-populations where costs of dispersal are low and advantages are high (Bonte and de la Peña 2009), so in connective meta-populations with substantial environmental stochasticity. In such meta-populations, individuals in the best condition profit maximally from the colonization of empty or low-density patches. Conversely, negative associations evolve in stable and more disconnected meta-populations (Figure 10.2a), where kin-competition prevails. From a lineage perspective, the emigration of individuals in the lowest condition is promoted as a mechanism to relieve local kin-competition.

Burton et al. (2010) show, by means of an individual-based model, that life histories determined by dispersal, reproduction and competitive ability, when jointly bound by an organism's total energy budget, are greatly affected by local selection pressures and spatial sorting processes during range expansions (Figure 10.2b; see also Chapter 20, this volume). Spatial sorting refers to the non-random accumulation of highly dispersive or highly fecund genotypes at the leading edge of expanding ranges (Chuang and Peterson 2016). This accumulation is potentially followed by assortative mating, which further increases spread rates and accelerates the eventual range expansion. When trade-offs are prevalent among reproduction, dispersal and competitive ability, enhanced dispersal and fecundity are selected at the cost of competitive ability. Within the core range, however, opposite strategies are promoted.

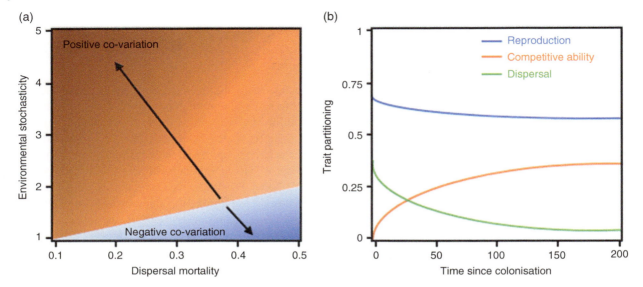

Figure 10.2 Evolution of dispersal syndromes in relation to spatio-temporal characteristics of the environment. (a) Sign and extent of the co-variation between dispersal and body condition changes. When patch turnover is large and dispersal costs are low in the meta-population (*i.e.*, low environmental stochasticity), individuals with high reproductive expectations disperse more. In stable meta-populations, and when dispersal costs increase, individuals in poor condition are the dispersing phenotypes (*Source:* adapted from Bonte and de la Peña 2009). (b) During range expansions, the allocation of resources to reproductive, competitive and dispersal changes. At the expanding margin (young populations, left at the *x*-axis), individuals evolve higher dispersal and reproductive rate and trade-off these traits with competitive ability, which is maximized in the oldest population at the range centre (*Source:* adapted from Burton et al. 2010).

10.4 Dispersal-Life History Co-variation in Nature

10.4.1 Empirical Evidence of Dispersal Syndromes Among Species

Correlations among life history and dispersal traits are mostly studied by meta-analyses at the interspecific level. In terrestrial plants, wind dispersal capacities are positively related to plant height and negatively to seed mass (Tamme et al. 2014, Augspurger et al. 2017). Large plant height and high seed mass are typically related to slow, competitive life histories (Thompson et al. 1993, 2002, Moles and Westoby 2004, Moles et al. 2005). Seed size is also associated with growth form, but the latter evolved independently from selection on dispersal (Thomson et al. 2018). Tall plants have a greater probability of developing special seed dispersal morphologies compared to small ones, suggesting that investments in dispersal effectively result from earlier specific life history adaptations to taller growth. For aquatic dispersal, seed size is positively correlated to dispersal distance, as smaller seeds become more easily entrapped in the riparian vegetation (de Jager et al. 2019). Small, round seeds were shown to better pass the gut system after ingestion, rendering plant species with these seed traits to be better dispersed by the vertebrates feeding on them (endozoochorous dispersal). Smaller seeds also seem to persist longer in the soil because of protective seed coating, lower maintenance energy and a deeper burial that decreases exposure to disturbances at the surface. Hence, a correlation emerges between modes of dispersal and longevity in seed banks (Thompson et al. 1993, 2002). Many plant species show strong variation in their association with mutualist frugivores that spread their seeds, but it remains to be studied whether traits that attract vectors co-vary with certain life history traits (Valenta and Nevo 2020).

Marine sessile organisms show positive correlations at the species level between dispersal, reproductive output and duration of the early larval stages. In these species, planktonic stages are responsible for dispersal, and offspring numbers, their size and the duration of the planktonic lifespan increase the potential distance they are moved in ocean currents (Burgess et al. 2016). Size and offspring number do often trade-off (see Chapter 3, this volume), so it can be expected that these spatial components contribute to the optimal life histories in this set of marine organisms. This offspring number/size-dispersal syndrome likely arises from the selection of traits with direct links to local fitness (*e.g.*, risk spreading by reproductive bet-hedging, resource acquisition through enhanced foraging, etc.). As such, it remains to be studied whether dispersal evolved as the main trait from the higher-outlined meta-population-level selection pressures or from local selection pressures on the fecundity and longevity life history traits. In many instances, like for reef fish, the evolution of positive associations between dispersal and fecundity is a likely outcome of joint adaptations to patchy and unpredictable environments.

In a set of 15 terrestrial and semi-terrestrial vertebrate and invertebrate orders, species life histories and ecology significantly influenced patterns of co-variation, with stronger evidence for the existence of dispersal-life history syndromes in aerial dispersers (Stevens et al. 2014). Overall, good dispersal abilities showed positive associations with fecundity and survival and are in line with expectations from extinction-colonization dynamics in meta-populations in which competition, and particularly kin-competition, promotes the joint evolution of dispersal, fast life histories and lower competitive abilities (*e.g.*, Olivieri et al. 1995, Clobert et al. 2004). Note that phylogenetic signals were not considered in this syndrome analysis. In a recent study, natal dispersal in birds was not found to be associated with some (coarsely defined) life history variables (Weeks et al. 2022). When dispersal is defined from a biogeographic angle (Weil et al. 2023), large body sizes and fast life histories jointly facilitate tetrapod (mammals, birds, reptiles and amphibians) dispersal success. In some clades, however, species with small bodies and/or slow life histories demonstrated a larger spread across biogeographic realms.

10.4.2 Phylogenetic Signals in Dispersal Syndromes

Phylogenetic signals in dispersal-related traits are usually substantial, implying that interspecific variation is embedded in evolutionary history. In birds, phylogenetic signals on dispersal distance are large and mediated by a positive co-variation with body size and associated properties like home or range size (Whitmee and Orme 2013). Over a wide range of terrestrial animals, on average 25% of the variation in dispersal could be explained by phylogenies, but the signal on dispersal and dispersal syndromes was stronger for orders with a higher proportion of aerially dispersing species such as birds, carabid beetles, dragonflies, butterflies and spiders (Stevens et al. 2014) and was independent of other life history traits. Such patterns can be explained by the higher energetic costs of flying compared to running animals, and the equal dependence of these costs on morphological traits that are much more phylogenetically conserved. A power law, for instance, typically scales travel speed, metabolic costs and animal body mass (*e.g.*, Cloyed et al. 2021), although maximal limits may arise from trade-offs with other vital functions like heat dissipation (Dyer et al. 2023). In plants, divergence in seed size was found to be phylogenetically associated with growth form, much more than divergences in dispersal mode or putative macro-ecological drivers (Moles et al. 2005). Evolutionary correlates can, however, be driven by other traits as well. In fig wasps, for instance, the phylogenetic signal is strong for dispersal capacity and mediated by interspecific variation in lipid content (Venkateswaran et al. 2017). Here, 'fast'-paced species displayed significantly higher flight durations, higher somatic lipid content and higher resting metabolic rates compared to the 'slow'-paced (longer-lived) species.

The phylogenetic signal of movement or dispersal-related traits like wing, body and seed size is substantial. However, when the realized dispersal propensity or distance is modelled as dependent variable, the phylogenetic signal is weak (Stevens et al. 2014, Weeks et al. 2022). Fandos et al. (2023) also showed in this respect that long-distance dispersal has weak phylogenetic conservatism, thereby representing the most labile property of the dispersal kernel. Dispersal-related traits can thus be strongly shaped by evolutionary history, but effective dispersal appears to be more labile among species and thus disconnected from other traits (Bonte and Dahirel 2017). Realized displacement distances are likely strongly determined by behavioural responses to settle or by external environmental factors like barriers and corridors. So, while dispersal has a strong genetic basis (Saastamoinen et al. 2018), it remains highly context- and condition-dependent. The rapid loss of dispersal in species colonizing islands (islands syndromes; Baeckens and van Damme 2020), the evolution of reduced dispersal in fragmented habitats (Bonte et al. 2003a, 2004, Cheptou et al. 2008), or increased dispersal at leading edges during range expansions (Shine et al. 2011, Chuang and Peterson 2016) represent solid examples of such a breakdown of the genetic signal in dispersal. Depending on the phylogenetic scale of interest and the research objective (*i.e.*, whether studied from an evolutionary or more applied ecological angle), trait co-variations may consequently strengthen, weaken, or even change in direction. At high levels of organization, like across multiple clades, large species are, for instance, associated with larger dispersal capacities because of their morphological and energetic capacities. However, within a clade, it is possible that smaller species are adapted to ephemeral habitats and thereby evolve larger dispersal capacities. Hence, statistically controlling for phylogeny or not may lead to opposite insights on the size–dispersal relationship (see Figure 10.3).

10.4.3 Micro-geographic Variation Among Populations

While evolutionary biology is built on individual life history variation, trait-based approaches in ecology have largely relied on the assumption of trait stability at the species level. This view only became recently surpassed with many studies across the tree of life demonstrating intraspecific variation to be putatively as large and ecologically as relevant compared to

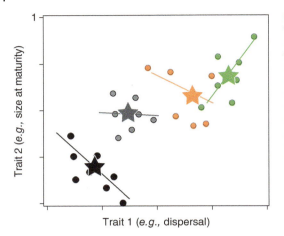

Figure 10.3 The strength and direction of dispersal syndromes can change when phylogenetic signals are considered or not. I here simplify phylogenetic signals to differences between four taxa (different colours) independent from their position (*i.e.*, their phylogenetic contrast according to the structure and branch length) along the phylogenetic tree. In this fictive example, larger species are overall associated with higher dispersal capacities. Star symbols represent the taxon's mean value. The phylogenetic signal is strong for size at maturity, and associations within each taxon change from negative (black, orange species), to positive (green species). This example thus shows us that phylogenetic constraints exist on dispersal from evolved differences in size, but different taxa evolved opposite syndromes, for instance, because of phylogenetically determined association with different environments.

interspecific variation (Bolnick 2001). Spatially differentiated selection pressures on dispersal and their integration into life histories can induce micro-geographic variation among populations (Richardson et al. 2014, Cheptou et al. 2017). Meta-populations characterized by on average strongly isolated patches represent population analogues to species-level island syndromes (with known evolution of dispersal loss and the evolution of either larger or smaller body sizes – Insular gigantism and dwarfism; Benítez-López et al. 2021). The joint dependent or independent selection in meta-populations of different levels of connectivity will likely generate syndromes with other life history traits like fecundity, size or longevity (*e.g.*, Saastamoinen et al. 2009, van Belleghem and Hendrickx 2014). In wind-dispersing plants (Soons and Heil 2002, Cheptou et al. 2008, Riba et al. 2009), arthropods (Bonte et al. 2003a, 2003b, 2007), but also more active dispersers like butterflies (Baguette et al. 2003, Schtickzelle et al. 2006), habitat fragmentation leads to the evolution of reduced dispersal. Some studies, however, show opposite patterns with fragmentation potentially leading to higher dispersal rates (Wang et al. 2011). Overall, it is not known whether these evolutionary changes in dispersal are associated with changes in other life history traits.

Dispersal heterogeneity (Soons and Heil 2002, Schtickzelle et al. 2006) can result from evolutionary and/or plastic responses. Breeding experiments and population genetics of the Glanville fritillary butterfly from the Åland Islands meta-population confirmed a strong genetic effect of dispersers being associated with specific *PGI-f* genotypes characterized by high flight metabolic rates and larger clutch sizes (Saastamoinen 2008). Common garden experiments over a single generation cannot separate genetic from maternal effects. These maternal effects can arise from intergenerational responses to environmental conditions that co-vary with levels of isolation. Smaller, more isolated populations may, for instance, experience poor nutritional conditions due to lower prey biomass or experience different climatic conditions that induce maternal effects towards the expression of an increased offspring dispersal (Mestre and Bonte 2012). Given that poor nutritional conditions may decrease fecundity and lifespan (Goossens et al. 2020), trait correlations differing in strength and even direction may emerge depending on the developmental conditions.

Experimental evolution in meta-populations provides strong evidence that changes in patch isolation affect dispersal-life history trait relationships (Lustenhouwer et al. 2023). In two-patch systems with manipulated connectivity (Tung et al. 2018a) or patch turnover (Friedenberg 2003), the evolution of dispersal was detected in fruit flies and nematodes, as well as various correlated responses of dispersal evolution and locomotor activity, exploratory tendency and aggression (Tung et al. 2018b). However, no correlations with body size, fecundity and longevity were, however, detected. In mites, similar experiments showed that meta-population typology, as determined by spatio-temporal variation in patch size but not the total meta-population capacity, leads to differences in the co-variation between reproduction, dispersal, longevity or sex ratio change that could not always be predicted from the available dispersal-life history theory (Fronhofer et al. 2014, De Roissart et al. 2016, Masier and Bonte 2020). These multivariate life history changes, however, have been shown to be adaptive (Bonte and Bafort 2019). In the same model species, fragmentation *per se*, so the increase of inter-patch distances, surprisingly did not affect the evolution of emigration rates but merely impacted starvation resistance as a cost-reducing trait (no food was provided outside the patches; Masier and Bonte 2020).

The spatial sorting of dispersive genotypes at expanding range fronts creates spatial divergence of dispersal along range and distribution gradients (Riba et al. 2009, Shine et al. 2011). The accumulation of distinct disperser genotypes at range fronts, thereby accelerating range expansion rates, becomes widely documented in nature and experimental settings across animals (Thomas et al. 2001, Phillips et al. 2006, Shine et al. 2011, Lombaert et al. 2014, van Petegem et al. 2016, Wolz et al. 2020, Clark et al. 2022, Narimanov et al. 2022), plants (Monty and Mahy 2010, Huang et al. 2015, Williams et al. 2016) and microbial species (Fronhofer and Altermatt 2015). Increased leg length and more straight movement patterns in cane toads at the edge of their invasion front are prominent textbook examples of such a spatial selection (Phillips et al. 2006; see also Chapter 20, this volume). While dispersal is considered as a principal trait under selection by spatial sorting (Narimanov et al. 2022), quite some experimental (*e.g.*, van Petegem et al. 2018) and common garden work (van Petegem et al. 2016, Wolz et al. 2020, Clark et al. 2022) showed fecundity to be equally, or even more strongly affected by this spatial selection pressure.

10.4.4 Condition-Dependence and Within-Population Variation

Dispersal is mediated by a strong genetic component across a wide variety of taxa (Saastamoinen et al. 2018). However, individuals from the same population do experience different developmental trajectories, or experience different levels of interactions with kin, their own or other species. In addition to genetic differences between individuals, plastic responses to this locally experienced environmental or demographic context (*e.g.*, competition and individual differences in resource acquisition) are therefore able to generate heterogeneity in dispersal and life history among individuals from the same population. Individual dispersal may therefore, like pace-of-life syndromes (van de Walle et al. 2023), be highly variable within species because of conditions experienced during life.

Individual heterogeneity in dispersal-life history associations may arise from experienced demographic and environmental variation prior to dispersal. It can also be the consequence of dispersal because costs have been incurred for either being philopatric (remaining in high densities; Spear et al. 1998) or dispersing (Bonte et al. 2012). Individual heterogeneity in dispersal conflicts with the view of dispersal as a random process, but it remains extremely difficult to understand whether such non-random dispersal is effectively an adaptive strategy, *i.e.*, whether it maximizes individual fitness or not. Studies linking life history traits to dispersal in vertebrates, depending on the proxies used, demonstrate fitness advantages (*e.g.*, Serrano and Tella 2012) or disadvantages (*e.g.*, Soulsbury et al. 2008) of dispersers relative to residents. Such individual heterogeneity can also be related to the existence of personality syndromes that entail series of morphological, physiological or behavioural specializations for dispersal (Cote et al. 2010), as well as transcriptomic divergence in genes related to metabolism, muscle development and immune responses (San-Jose et al. 2023). In spider mites, a physiological syndrome with dispersers characterized by lower concentrations of amino acids suggests the existence of dispersal-foraging trade-offs, but no direct correlation between fecundity and survival was found (Dahirel et al. 2019). In recent work across *Drosophila* species, expected trade-offs between dispersal and fecundity were also not found (De Araujo et al. 2023). Taken together, there is evidence that individuals who successfully dispersed can show a different phenotypic signature compared to individuals who remained philopatric, but this pattern may be highly taxon-specific and/or depending on the environmental context and the induced costs of dispersal in nature or in the experimental setup (see also Tigreros and Davidowitz 2019).

When dispersing individuals can only be identified after the effective dispersal event, it is not possible to understand whether the observed phenotypic divergence between dispersers and non-dispersers is a consequence or rather a cause of dispersal. Neither is it straightforward to understand exact life history-dispersal associations when dispersal is restricted to specific life stages that are also characterized by specific trait changes in, for instance, physiology (Goossens et al. 2020). Such studies, therefore, do not provide evidence of fitness advantages because the alternative strategy cannot be tested (*i.e.*, fitness of dispersive phenotypes that would have 'ended up' at home and thereby changed local demographic conditions). In a mite model system where mutants with a specific acaracide resistance were used as a genetic marker, reciprocal-transplant experiments unravelled the adaptiveness of non-random dispersal (Bonte et al. 2014). The experiment demonstrated that dispersal towards low-density environments improved fitness by reducing competition, but, more importantly, that this increased fitness was especially pronounced in dispersers compared to residents that were 'forced' to move. Alternatively, residents translocated to a typical colonizer environment where densities are low, experienced only limited fitness advantages compared to the effective dispersers from that population. This reflects patterns observed in fish and reptile systems, where dispersal was shown to be associated with bold, asocial personalities that perform better in new environments where competition is low (Cote and Clobert 2007, Cote et al. 2011).

10.5 Concluding Remarks and Outlook: Why Should We Care?

Dispersal and colonization processes contribute significantly to local demographic processes by respectively adding to or removing individuals (life stages) from the population. Dispersal is not a random process (Lowe and McPeek 2014) and is tightly integrated in life histories. It may be restricted to certain life stages, and, depending on the mechanisms and adaptations, associated with costs and benefits, which in turn impose feedback on reproductive and maturation traits within the life history. Understanding how dispersal is integrated into life histories is one of the major challenges to further develop a trait-based or mechanistic predictive ecology under, *e.g.*, climate change (Travis et al. 2013, Urban et al. 2016) and/or habitat fragmentation (Cote et al. 2017). Because any ecological outcome on population dynamics and spread will be the joint outcome of individual-level dispersal, maturation, reproduction and death, their association will be of major ecological importance. From a macro-ecological perspective, associations between dispersal and life history traits may also be useful to predict species range dynamics (Alzate and Onstein 2022).

Dispersal-life history associations showed, however, to be labile because of (1) strong evolutionary dynamics leading to substantial micro-geographic variation according to the spatio-temporal attributes of the environment and (2) the potential overruling environmental impact leading to plastic (so-called conditional) responses, including biotic and abiotic interactions (Bierbaum et al. 1989, Fronhofer et al. 2018). This lability should not be taken as an argument to abandon trait-based approaches to model demographics and distribution dynamics in species or some specific populations. Rather, combining a trait-based ecology (traits reflecting the capacity to disperse a certain distance) with a mechanistic framework based on first principles from theory (translated to the propensity to disperse) may provide a promising avenue for the development of hybrid (mechanistic and correlative) forecasting models for species biodiversity changes in a changing world. Considering individual heterogeneity in performance (Chardon et al. 2020) or environment-demography associations (Malchow et al. 2023) has already been shown to improve large-scale species distribution predictions. Forecasting events at smaller spatio-temporal scales will equally benefit from considering conditional responses of important demographic traits (Batsleer et al. 2022, Malchow et al. 2022).

Some outstanding questions remain: (1) many life history-dispersal associations emerge across the tree of life, but to what degree are these associations related to ecological and evolutionary parameters, including phylogeography? (2) Most available information on dispersal-life history syndromes comes from vascular plants, vertebrates and some well-studied groups of arthropods. How can we define, identify and understand the co-variation between dispersal and other life history traits in largely understudied but ecologically important taxa like fungi and other microbial organisms? (3) Does the observed variation in dispersal and dispersal syndromes represent a continuum, or can we, as for other ecological traits, identify discontinuities (Nash et al. 2014a, 2014b) and use such trait distributions as early warning signals for population collapse or as a tool to develop general biodiversity models? (4) We witness evolutionary patterns in dispersal syndromes, but to what degree are dispersal strategies optimized in nature? Since dispersal syndromes appear to be plastic, are they then mostly shaped by direct environmental factors through evolved reaction norms or just independent trajectories of the separate traits? Do associations with reproductive and developmental rates impose deviations from optimal dispersal rates (Brady et al. 2019), and if yes, do they represent maladaptations at the individual, deme or even meta-population level? (5) Not only dispersal rates or distances, but also dispersal costs may evolve. Under which conditions will selection act on the costs of dispersal rather than on emigration, transfer and settlement? How do these potential cost adaptations affect demographic rates? (6) Also, evidence is emerging that dispersing individuals are non-random members of a population and are associated with specific phenotypes. Is this divergence in signature important for the stability of meta-populations or even meta-communities? Conversely, do such meta-population-level selection pressures shape the evolution of such specific dispersal syndromes? Given that physiological and transcriptomic processes underlie changes in life history-dispersal variation (Goossens et al. 2020), can we identify and generalize such basic responses across taxa?

In conclusion, traditional life history traits like age at maturity, fecundity and longevity have been mostly studied from a mean-field perspective, thereby neglecting their spatial dimensions and dynamics. Here, I hope to have demonstrated that any understanding of life histories in natural populations needs to consider its spatial context. Dispersal therefore needs to be considered as an essential demographic trait, which is shaped by, and shaping life history strategies in spatially structured environments and populations.

References

Alzate, A., and Onstein, R.E. (2022). Understanding the relationship between dispersal and range size. *Ecology Letters* 25: 2303–2323.

Augspurger, C.K., Franson, S.E., and Cushman, K.C. (2017). Wind dispersal is predicted by tree, not diaspore, traits in comparisons of Neotropical species. *Functional Ecology* 31: 808–820.

Bach, L.A., Thomsen, R., Pertoldi, C., and Loeschcke, V. (2006). Kin competition and the evolution of dispersal in an individual-based model. *Ecological Modelling* 192: 658–666.

Baeckens, S., and van Damme, R. (2020). The island syndrome. *Current Biology* 30: R338–R339.

Baguette, M., Mennechez, G., Petit, S., and Schtickzelle, N. (2003). Effect of habitat fragmentation on dispersal in the butterfly *Proclossiana eunomia*. *Comptes Rendus Biologies* 326: 200–209.

Baguette, M., Stevens, V.M., and Clobert, J. (2014). The pros and cons of applying the movement ecology paradigm for studying animal dispersal. *Movement Ecology* 2: 13.

Baker, H.G., and Stebbins, G.L. (1965). *The genetics of colonizing species: proceedings of the first international union of biological sciences symposia on general biology*. 1st ed. Academic Press.

Batsleer, F., Maes, D., and Bonte, D. (2022). Behavioral strategies and the spatial pattern formation of nesting. *The American Naturalist* 199: E15–E27.

Benítez-López, A., Santini, L., Gallego-Zamorano, J. et al. The island rule explains consistent patterns of body size evolution in terrestrial vertebrates. *Nature Ecology and Evolution* 5, 768–786 (2021). https://doi.org/10.1038/s41559-021-01426-y

Bierbaum, T.J., Mueller, L.D., and Ayala, F.J. (1989). Density-dependent evolution of life-history traits in *Drosophila melanogaster*. *Evolution* 43: 382–392.

Blackburn, T.M. (1991). Evidence for a 'fast-slow' continuum of life-history traits among parasitoid hymenoptera. *Functional Ecology* 5: 65–74.

Bolnick, D.I. (2001). Intraspecific competition favours niche width expansion in *Drosophila melanogaster*. *Nature* 410: 463–466.

Bolnick, D.I., and Otto, S.P. (2013). The magnitude of local adaptation under genotype-dependent dispersal. *Ecology and Evolution* 3: 4722–4735.

Bonte, D., Baert, L., Lens, L., and Maelfait, J. (2004). Effects of aerial dispersal, habitat specialisation, and landscape structure on spider distribution across fragmented grey dunes. *Ecography* 27: 343–349.

Bonte, D., and Bafort, Q. (2019). The importance and adaptive value of life history evolution for metapopulation dynamics. *Journal of Animal Ecology* 88: 24–34.

Bonte, D., Bossuyt, B., and Lens, L. (2007). Aerial dispersal plasticity under different wind velocities in a salt marsh wolf spider. *Behavioral Ecology* 18: 438–443.

Bonte, D., and Dahirel, M. (2017). Dispersal: a central and independent trait in life history. *Oikos* 126: 472–479.

Bonte, D., and de la Pena, E. (2009). Evolution of body condition-dependent dispersal in metapopulations. *Journal of Evolutionary Biology* 22: 1242–1251.

Bonte, D., De Roissart, A., Wybouw, N., and van Leeuwen, T. (2014). Fitness maximization by dispersal: evidence from an invasion experiment. *Ecology* 95: 3104–3111.

Bonte, D., Hovestadt, T., and Poethke, H.-J. (2009). Sex-specific dispersal and evolutionary rescue in metapopulations infected by male killing endosymbionts. *BMC Ecology and Evolution* 9: 16.

Bonte, D., Lens, L., Maelfait, J., Hoffmann, M., and Kuijken, E. (2003a). Patch quality and connectivity influence spatial dynamics in a dune wolfspider. *Oecologia* 135: 227–233.

Bonte, D., Masier, S., and Mortier, F. (2018). Eco-evolutionary feedbacks following changes in spatial connectedness. *Current Opinion in Insect Science* 29: 64–70.

Bonte, D., van Dyck, H., Bullock, J.M., Coulon, A., Delgado, M., Gibbs, M., Lehouck, V., Matthysen, E., Mustin, K., Saastamoinen, M., Schtickzelle, N., Stevens, V.M., Vandewoestijne, S., Baguette, M., Barton, K., Benton, T.G., Chaput-Bardy, A., Clobert, J., Dytham, C., Hovestadt, T., Meier, C.M., Palmer, S.C.F., Turlure, C., and Travis, J.M.J. (2012). Costs of dispersal. *Biological Reviews* 87: 290–312.

Bonte, D., Vandenbroecke, N., Lens, L., and Maelfait, J.-P. (2003b). Low propensity for aerial dispersal in specialist spiders from fragmented landscapes. *Proceedings of the Royal Society B: Biological Sciences* 270: 1601–1607.

Bowler, D.E., and Benton, T.G. (2005). Causes and consequences of animal dispersal strategies: relating individual behaviour to spatial dynamics. *Biological Reviews* 80: 205–225.

Brady, S.P., Bolnick, D.I., Barrett, R.D.H., Chapman, L., Crispo, E., Derry, A.M., Eckert, C.G., Fraser, D.J., Fussmann, G.F., Gonzalez, A., Guichard, F., Lamy, T., Lane, J., McAdam, A.G., Newman, A.E.M., Paccard, A., Robertson, B., Rolshausen, G., Schulte, P.M., Simons, A.M., Vellend, M., and Hendry, A. (2019). Understanding maladaptation by uniting ecological and evolutionary perspectives. *The American Naturalist* 194: 495–515.

Buoro, M., and Carlson, S.M. (2014). Life-history syndromes: integrating dispersal through space and time. *Ecology Letters* 17: 756–767.

Burgess, S.C., Baskett, M.L., Grosberg, R.K., Morgan, S.G., and Strathmann, R.R. (2016). When is dispersal for dispersal? Unifying marine and terrestrial perspectives. *Biological Reviews* 9: 867–882.

Burton, O.J., Phillips, B.L., and Travis, J.M.J. (2010). Trade-offs and the evolution of life-histories during range expansion. *Ecology Letters* 13: 1210–20.

Chardon, N.I., Pironon, S., Peterson, M.L., and Doak, D.F. (2020). Incorporating intraspecific variation into species distribution models improves distribution predictions, but cannot predict species traits for a wide-spread plant species. *Ecography* 43: 60–74.

Chaudhary, V.B., Aguilar-Trigueros, C.A., Mansour, I., and Rillig, M.C. (2022). Fungal dispersal across spatial scales. *Annual Review of Ecology, Evolution, and Systematics* 53: 69–85.

Cheptou, P.-O., Carrue, O., Rouifed, S., and Cantarel, A. (2008). Rapid evolution of seed dispersal in an urban environment in the weed *Crepis sancta*. *Proceedings of the National Academy of Sciences of the United States of America* 105: 3796–3799.

Cheptou, P.-O., Hargreaves, A.L., Bonte, D., and Jacquemyn, H. (2017). Adaptation to fragmentation: evolutionary dynamics driven by human influences. *Philosophical Transactions of the Royal Society, B: Biological Sciences*, 372: 20160037.

Cheptou, P., and Massol, F. (2009). Pollination fluctuations drive evolutionary syndromes linking dispersal and mating system. *The American Naturalist* 174: 46–55.

Chuang, A., and Peterson, C.R. (2016). Expanding population edges: theories, traits, and trade-offs. *Global Change Biology* 22: 494–512.

Clark, E.I., Bitume, E.V., Bean, D.W., Stahlke, A.R., Hohenlohe, P.A., and Hufbauer, R.A. (2022). Evolution of reproductive life-history and dispersal traits during the range expansion of a biological control agent. *Evolutionary Applications* 15: 2089–2099.

Clobert, J., Ims, R.A., and Rousset, F. (2004). 13 – Causes, mechanisms and consequences of dispersal. In: *Ecology, genetics and evolution of metapopulations* (eds. Hanski, I., and Gaggiotti, O.E.), 307–335. Academic Press.

Clobert, J., Le Galliard, J.-F., Cote, J., Meylan, S., and Massot, M. (2009). Informed dispersal, heterogeneity in animal dispersal syndromes and the dynamics of spatially structured populations. *Ecology Letters* 12: 197–209.

Cloyed, C.S., Grady, J.M., Savage, V.M., Uyeda, J.C., and Dell, A.I. (2021). The allometry of locomotion. *Ecology* 102: e03369.

Cote, J., Bestion, E., Jacob, S., Travis, J., Legrand, D., and Baguette, M. (2017). Evolution of dispersal strategies and dispersal syndromes in fragmented landscapes. *Ecography* 40: 56–73.

Cote, J., and Clobert, J. (2007). Social personalities influence natal dispersal in a lizard. *Proceedings of the Royal Society B: Biological Sciences* 274: 383–390.

Cote, J., Clobert, J., Brodin, T., Fogarty, S., and Sih, A. (2010). Personality-dependent dispersal: characterization, ontogeny and consequences for spatially structured populations. *Philosophical Transactions of the Royal Society, B: Biological Sciences* 365: 4065–76.

Cote, J., Fogarty, S., Brodin, T., Weinersmith, K., and Sih, A. (2011). Personality-dependent dispersal in the invasive mosquitofish: group composition matters. *Proceedings of the Royal Society B: Biological Sciences* 278: 1670–1678.

Dahirel, M., Masier, S., Renault, D., and Bonte, D. (2019). The distinct phenotypic signatures of dispersal and stress in an arthropod model: from physiology to life history. *Journal of Experimental Biology* 222: jeb203596.

De Araujo, L.I., Karsten, M., and Terblanche, J.S. (2023). Flight-reproduction trade-offs are weak in a field cage experiment across multiple *Drosophila* species. *Current Research in Insect Science* 3: 100060.

de Jager, M., Kaphingst, B., Janse, E.L., Buisman, R., Rinzema, S.G.T., and Soons, M.B. (2019). Seed size regulates plant dispersal distances in flowing water. *Journal of Ecology* 107: 307–317.

De Roissart, A., Wybouw, N., Renault, D., van Leeuwen, T., Bonte, D., and Pfrender, M. (2016). Life history evolution in response to changes in metapopulation structure in an arthropod herbivore. *Functional Ecology* 30: 1408–1417.

Dingle, H. (1978). *Evolution of insect migration and diapause*. Series: Proceedings in life sciences. Springer.

Dingle, H. (2014). *Migration: the biology of life on the move*. Oxford University Press.

Duputié, A., and Massol, F. (2013). An empiricist's guide to theoretical predictions on the evolution of dispersal. *Interface Focus* 3: 20130028.

Dyer, A., Brose, U., Berti, E., Rosenbaum, B., and Hirt, M.R. (2023). The travel speeds of large animals are limited by their heat-dissipation capacities. *PLoS Biology* 21(4): e3001820.

Dytham, C., and Travis, J. (2006). Evolving dispersal and age at death. *Oikos* 113:530–538.

Fandos, G., Talluto, M., Fiedler, W., Robinson, R.A., Thorup, K., and Zurell, D. (2023). Standardised empirical dispersal kernels emphasise the pervasiveness of long-distance dispersal in European birds. *Journal of Animal Ecology* 92:158–170.

Friedenberg, N.A. (2003). Experimental evolution of dispersal in spatiotemporally variable microcosms. *Ecology Letters* 6: 953–959.

Fronhofer, E.A., and Altermatt, F. (2015). Eco-evolutionary feedbacks during experimental range expansions. *Nature Communications* 6: 6844.

Fronhofer, E.A., Joachim Poethke, H., and Dieckmann, U. (2015). Evolution of dispersal distance: maternal investment leads to bimodal dispersal kernels. *Journal of Theoretical Biology* 365: 270–279.

Fronhofer, E.A., Legrand, D., Altermatt, F., Ansart, A., Blanchet, S., Bonte, D., Chaine, A., Dahirel, M., De Laender, F., De Raedt, J., di Gesu, L., Jacob, S., Kaltz, O., Laurent, E., Little, C., Madec, L., Manzi, F., Masier, S., Pellerin, F., Pennekamp, F., Schtickzelle, N., Therry, L., Vong, A., Windandy, L., and Cote, J. (2018). Bottom-up and top-down control of dispersal across major organismal groups. *Nature Ecology & Evolution* 2: 1859–1863.

Fronhofer, E.A., Stelz, J.M., Lutz, E., Poethke, H.J., and Bonte, D. (2014). Spatially correlated extinctions select for less emigration but larger dispersal distances in the spider mite *Tetranychus urticae*. *Evolution* 68: 1838–1844.

Futuyma, D.J., and Moreno, G. (1988). The evolution of ecological specialization. *Annual Review of Ecology and Systematics* 19: 207–233.

Gandon, S. (1999). Kin competition, the cost of inbreeding and the evolution of dispersal. *Journal of Theoretical Biology* 200: 345–364.

Gandon, S., and Rousset, F. (1999). Evolution of stepping-stone dispersal rates. *Proceedings of the Royal Society B: Biological Sciences* 266: 2507–2513.

Goossens, S., Wybouw, N., van Leeuwen, T., and Bonte, D. (2020). The physiology of movement. *Movement Ecology* 8: 5.

Gros, A., Poethke, H.J., and Hovestadt, T. (2009). Sex-specific spatio-temporal variability in reproductive success promotes the evolution of sex-biased dispersal. *Theoretical Population Biology* 76: 13–18.

Gyllenberg, M., Kisdi, E., and Utz, M. (2008). Evolution of condition-dependent dispersal under kin competition. *Journal of Mathematical Biology* 57: 285–307.

Gyllenberg, M., Kisdi, É., and Utz, M. (2011). Body condition dependent dispersal in a heterogeneous environment. *Theoretical Population Biology* 79: 139–54.

Hanski, I. (2012). Eco-evolutionary dynamics in a changing world. *The Annals of the New York Academy of Sciences* 1249: 1–17.

Holt, R.D. (1987). Population dynamics and evolutionary processes: the manifold roles of habitat selection. *Evolutionary Ecology* 1: 331–347.

Huang, F., Peng, S., Chen, B., Liao, H., Huang, Q., Lin, Z., and Liu, G. (2015). Rapid evolution of dispersal-related traits during range expansion of an invasive vine *Mikania micrantha*. *Oikos* 124: 1023–1030.

Johst, K., and Brandl, R. (1997). Evolution of dispersal: the importance of the temporal order of reproduction and dispersal. *Proceedings of the Royal Society B: Biological Sciences* 264: 23–30.

Joschinski, J., and Bonte, D. (2021). Diapause and bet-hedging strategies in insects: a meta-analysis of reaction norm shapes. *Oikos* 130: 1240–1250.

Kisdi, É. (2004). Conditional dispersal under kin competition: extension of the Hamilton–May model to brood size-dependent dispersal. *Theoretical Population Biology* 66: 369–380.

Kisdi, E., Utz, M., and Gyllenberg, M. (2013). Evolution of condition-dependent dispersal. In: *Dispersal ecology and evolution* (eds. Clobert, J., Baguette, M., Benton, T., and Bullock, J.), 131–151. Oxford University Press.

Laroche, F., Jarne, P., Perrot, T., and Massol, F. (2016). The evolution of the competition–dispersal trade-off affects α- and β-diversity in a heterogeneous metacommunity. *Proceedings of the Royal Society B: Biological Sciences* 283: 20160548.

Levins, R., and Culver, D. (1971). Regional coexistence of species and competition between rare species. *Proceedings of the National Academy of Sciences of the United States of America* 68: 1246–1248.

Lombaert, E., Estoup, A., Facon, B., Joubard, B., Grégoire, J.-C., Jannin, A., Blin, A., and Guillemaud, T. (2014). Rapid increase in dispersal during range expansion in the invasive ladybird *Harmonia axyridis*. *Journal of Evolutionary Biology* 27: 508–517.

Lowe, W.H., and McPeek, M.A. (2014). Is dispersal neutral? *Trends in Ecology & Evolution* 29: 444–450.

Lustenhouwer, N., Moerman, F., Altermatt, F., Bassar, R.D., Bocedi, G., Bonte, D., Dey, S., Fronhofer, E.A., da Rocha, É.G., Giometto, A., Lancaster, L.T., Prather, R.B., Saastamoinen, M., Travis, J.M.J., Urquhart, C.A., Weiss-Lehman, C., Williams, J.L.,

Börger, L., and Berger, D. (2023). Experimental evolution of dispersal: unifying theory, experiments and natural systems. *Journal of Animal Ecology* 92: 1113–1123.

Malchow, A.K., Fandos, G., Kormann, U., Grüebler, M., Kéry, M., Hartig, F., and Zurell, D. (2022) Fitting individual-based models of species range dynamics to long-term monitoring data. bioRxiv. https://www.biorxiv.org/content/10.1101/2022.09.26.509574v2

Malchow, A.-K., Hartig, F., Reeg, J., Kéry, M., and Zurell, D. (2023). Demography-environment relationships improve mechanistic understanding of range dynamics under climate change. *Philosophical Transactions of the Royal Society, B: Biological Sciences* 378: 20220194.

Masier, S., and Bonte, D. (2020). Spatial connectedness imposes local- and metapopulation-level selection on life history through feedbacks on demography. *Ecology Letters* 23: 242–253.

Matthysen, E. (2012). Multicausality of dispersal: a review. In: *Dispersal ecology and evolution* (eds. Clobert, J., Baguette, M., Benton, T.G., and Bullock, J.M.), 1–17. Oxford University Press.

Mestre, L., and Bonte, D. (2012). Food stress during juvenile and maternal development shapes natal and breeding dispersal in a spider. *Behavioral Ecology* 23: 759–764.

Moles, A.T., Ackerly, D.D., Webb, C.O., Tweddle, J.C., Dickie, J.B., and Westoby, M. (2005). A brief history of seed size. *Science* 307: 576–580.

Moles, A.T., and Westoby, M. (2004). Seedling survival and seed size: a synthesis of the literature. *Journal of Ecology* 92: 372–383.

Monty, A., and Mahy, G. (2010). Evolution of dispersal traits along an invasion route in the wind-dispersed *Senecio inaequidens* (Asteraceae). *Oikos* 119: 1563–1570.

Mortier, F., Jacob, S., Vandegehuchte, M.L., and Bonte, D. (2019). Habitat choice stabilizes metapopulation dynamics by enabling ecological specialization. *Oikos* 128: 529–539.

Narimanov, N., Bauer, T., Bonte, D., Fahse, L., and Entling, M.H. (2022). Accelerated invasion through the evolution of dispersal behaviour. *Global Ecology and Biogeography* 31: 2423–2436.

Nash, K.L., Allen, C.R., Angeler, D.G., Barichievy, C., Eason, T., Garmestani, A.S., Graham, N.A.J., Granholm, D., Knutson, M., Nelson, R.J., Nyström, M., Stow, C.A., and Sundstrom, S.M. (2014a). Discontinuities, cross-scale patterns, and the organization of ecosystems. *Ecology* 95: 654–667.

Nash, K.L., Allen, C.R., Barichievy, C., Nyström, M., Sundstrom, S., and Graham, N.A.J. (2014b). Habitat structure and body size distributions: cross-ecosystem comparison for taxa with determinate and indeterminate growth. *Oikos* 123: 971–983.

Olivieri, I., Michalakis, Y., and Gouyon, P.-H. (1995). Metapopulation genetics and the evolution of dispersal. *The American Naturalist* 146: 202–228.

Perrin, N., and Mazalov, V. (1999). Dispersal and inbreeding avoidance. *The American Naturalist* 154: 282–292.

Phillips, B.L., Brown, G.P., Webb, J.K., and Shine, R. (2006). Invasion and the evolution of speed in toads. *Nature* 439: 803–803.

Poethke, H.J., and Hovestadt, T. (2002). Evolution of density–and patch-size-dependent dispersal rates. *Proceedings of the Royal Society B: Biological Sciences* 269: 637–645.

Ravigné, V., Dieckmann, U., and Olivieri, I. (2009). Live where you thrive: joint evolution of habitat choice and local adaptation facilitates specialization and promotes diversity. *The American Naturalist* 174: E141–E169.

Reznick, D., Bryant, M.J., and Bashey, F. (2002). r- and K-selection revisited: the role of population regulation in life-history evolution. *Ecology* 83: 1509–1520.

Riba, M., Mayol, M., Giles, B.E., Ronce, O., Imbert, E., van Der Velde, M., Chauvet, S., Ericson, L., Bijlsma, R., Vosman, B., Smulders, M.J.M., and Olivieri, I. (2009). Darwin's wind hypothesis: does it work for plant dispersal in fragmented habitats? *New Phytologist* 183: 667–677.

Richardson, J.L., Urban, M.C., Bolnick, D.I., and Skelly, D.K. (2014). Microgeographic adaptation and the spatial scale of evolution. *Trends in Ecology & Evolution* 29: 165–176.

Roff, D.A., and Bradford, M.J. (1996). Quantitative genetics of the trade-off between fecundity and wing dimorphism in the cricket *Allonemobius socius*. *Heredity* 76: 178–185.

Ronce, O. (2007). How does it feel to be like a rolling stone? Ten questions about dispersal evolution. *Annual Review of Ecology, Evolution, and Systematics* 38: 231–253.

Ronce, O., and Clobert, J. (2012). Dispersal syndromes. In: *Dispersal ecology and evolution* (eds. Clobert, J., Baguette, M., Benton, T., and Bullock, J.), 119–138. Oxford University Press.

Ronce, O., Perret, F., and Olivieri, I. (2000). Landscape dynamics and evolution of colonizer syndromes: interactions between reproductive effort and dispersal in a metapopulation. *Evolutionary Ecology* 14: 233–260.

Ronce, O., and Promislow, D.E.L. (2010). Kin competition, natal dispersal and the moulding of senescence by natural selection. *Proceedings of the Royal Society B: Biological Sciences* 277: 3659–3667.

Saastamoinen, M. (2008). Heritability of dispersal rate and other life history traits in the Glanville fritillary butterfly. *Heredity* 100: 39–46.

Saastamoinen, M., Bocedi, G., Cote, J., Legrand, D., Guillaume, F., Wheat, C.W., Fronhofer, E.A., Garcia, C., Henry, R., Husby, A., Baguette, M., Bonte, D., Coulon, A., Kokko, H., Matthysen, E., Niitepõld, K., Nonaka, E., Stevens, V.M., Travis, J.M.J., Donohue, K., Bullock, J.M., and del Mar Delgado, M. (2018). Genetics of dispersal. *Biological Reviews* 93: 574–599.

Saastamoinen, M., Ikonen, S., and Hanski, I. (2009). Significant effects of *Pgi* genotype and body reserves on lifespan in the Glanville fritillary butterfly. *Proceedings of the Royal Society B: Biological Sciences* 276: 1313–1322.

Salguero-Gómez, R., Jones, O.R., Jongejans, E., Blomberg, S.P., Hodgson, D.J., Mbeau-Ache, C., Zuidema, P.A., de Kroon, H., and Buckley, Y.M. (2016). Fast-slow continuum and reproductive strategies structure plant life-history variation worldwide. *Proceedings of the National Academy of Sciences of the United States of America* 113: 230–235.

San-Jose, L.M., Bestion, E., Pellerin, F., Richard, M., Di Gesu, L., Salmona, J., Winandy, L., Legrand, D., Bonneaud, C., Guillaume, O., Calvez, O., Elmer, K.R., Yurchenko, A.A., Recknagel, H., Clobert, J., and Cote, J. (2023). Investigating the genetic basis of vertebrate dispersal combining RNA-seq, RAD-seq and quantitative genetics. *Molecular Ecology* 32: 3060–3075.

Schtickzelle, N., Mennechez, G., and Baguette, M. (2006). Dispersal depression with habitat fragmentation in the bog fritillary butterfly. *Ecology* 87: 1057–1065.

Serrano, D., and Tella, J.L. (2012). Lifetime fitness correlates of natal dispersal distance in a colonial bird. *Journal of Animal Ecology* 81: 97–107.

Shine, R., Brown, G.P., and Phillips, B.L. (2011). An evolutionary process that assembles phenotypes through space rather than through time. *Proceedings of the National Academy of Sciences of the United States of America* 108: 5708–11.

Soons, M.B., and Heil, G.W. (2002). Reduced colonization capacity in fragmented populations of wind-dispersed grassland forbs. *Journal of Ecology* 90: 1033–1043.

Soulsbury, C.D., Baker, P.J., Iossa, G., and Harris, S. (2008). Fitness costs of dispersal in red foxes (*Vulpes vulpes*). *Behavioral Ecology and Sociobiology* 62: 1289–1298.

Southwood, T.R.E. (1977). Habitat, the templet for ecological strategies. *Journal of Animal Ecology* 46: 337–365.

Spear, L.B., Pyle, P., and Nur, N. (1998). Natal dispersal in the western gull: proximal factors and fitness consequences. *Journal of Animal Ecology* 67: 165–179.

Stevens, V.M., Whitmee, S., Le Galliard, J.F., Clobert, J., Bohning-Gaese, K., Bonte, D., Brändle, M., Matthias Dehling, D., Hof, C., Trochet, A., and Baguette, M. (2014). A comparative analysis of dispersal syndromes in terrestrial and semi-terrestrial animals. *Ecology Letters* 17: 1039–1052.

Tamme, R., Götzenberger, L., Zobel, M., Bullock, J.M., Hooftman, D.A.P., Kaasik, A., and Pärtel, A. (2014). Predicting species' maximum dispersal distances from simple plant traits. *Ecology* 95: 505–513.

Taylor, P.D., and Irwin, A.J. (2000). Overlapping generations can promote altruistic behavior. *Evolution* 54: 1135–1141.

Thomas, C.D., Bodsworth, E.J., Wilson, R.J., Simmons, A.D., Davies, Z.G., Musche, M., and Conradt, L. (2001). Ecological and evolutionary processes at expanding range margins. *Nature* 411: 577.

Thompson, K., Band, S.R., and Hodgson, J.G. (1993). Seed size and shape predict persistence in soil. *Functional Ecology* 7: 236–241.

Thompson, K., Rickard, L.C., Hodkinson, D.J., and Rees M. (2002) Seed dispersal, the search for trade-offs. In: *Dispersal ecology* (eds. Bullock, J.M., Kenward, R.E., and Hails, R.S.), 152–172. Blackwell Publishing.

Thomson, F.J., Letten, A.D., Tamme, R., Edwards, W., and Moles, A.T. (2018). Can dispersal investment explain why tall plant species achieve longer dispersal distances than short plant species? *New Phytologist* 217: 407–415.

Tigreros, N., and Davidowitz, G. (2019). Flight-fecundity tradeoffs in wing-monomorphic insects. In: *Advances in insect physiology* (ed. Jurenka, R.), 1–41. Academic Press.

Tilman, D. (1994). Competition and biodiversity in spatially structured habitats. *Ecology* 75: 2–16.

Travis, J.M.J., Delgado, M., Bocedi, G., Baguette, M., Bartoń, K., Bonte, D., Boulangeat, I., Hodgson, J.A., Kubisch, A., Penteriana, V., Saastamoinen, M., Stevens, V.M., and Bullock, J.M. (2013). Dispersal and species' responses to climate change. *Oikos* 122: 1532–1540.

Travis, J.M.J., and Dytham, C. (1999). Habitat persistence, habitat availability and the evolution of dispersal. *Proceedings of the Royal Society B: Biological Sciences* 266: 723.

Travis, J.M.J., Mustin, K., Bartoń, K.A., Benton, T.G., Clobert, J., Delgado, M.M., Dytham, C., Hovestadt, T., Palmer, S.C.F., van Dyck, H., and Bonte, D. (2012). Modelling dispersal: an eco-evolutionary framework incorporating emigration, movement, settlement behaviour and the multiple costs involved. *Methods in Ecology and Evolution* 3: 628–641.

Tung, S., Mishra, A., Gogna, N., Sadiq, M.A., Shreenidhi, P.M., Sruti, V.R.S., Dorai, K., and Dey, S. (2018a). Evolution of dispersal syndrome and its corresponding metabolomic changes. *Evolution* 72: 1890–1903.

Tung, S., Mishra, A., Shreenidhi, P.M., Sadiq, M.A., Joshi, S., Sruti, V.R.S., Dorai, K., and Dey, S. (2018b). Simultaneous evolution of multiple dispersal components and kernel. *Oikos* 127: 34–44.

Urban, M.C., Bocedi, G., Hendry, A.P., Mihoub, J.-B., Pe'er, G., Singer, A., Crozier, L.G., De Meester, L., Godsoe, W., Gonzalez, A., Hellmann, J.J., Holt, R.D., Huth, A., Johst, K., Krug, C.B., Leadley, P.W., Palmer, S.C., Pantel, J.H., Schmitz, A., Zollner, P.A., and Travis, J.M. (2016). Improving the forecast for biodiversity under climate change. *Science* 353: aad8466.

Valenta, K., and Nevo, O. (2020). The dispersal syndrome hypothesis: how animals shaped fruit traits, and how they did not. *Functional Ecology* 34: 1158–1169.

van Belleghem, S.M., and Hendrickx, F. (2014). A tight association in two genetically unlinked dispersal related traits in sympatric and allopatric salt marsh beetle populations. *Genetica* 142: 1–9.

van de Walle, J., Fay, R., Gaillard, J.-M., Pelletier, F., Hamel, S., Gamelon, M., Barbraud, C., Blanchet, F.F., Blumstein, D.T., Charmantier, A., Delord, K., Larue, B., Martin, J., Milot, E., Mayer, F., Rotella, F., Saether, B.-E., Teplitsky, C., van de Pol, M., van Vuren, D., Visser, M., Wells, C., Yarral, J.V., and Jenouvrier, S. (2023). Individual life histories: neither slow nor fast, just diverse. *Proceedings of the Royal Society B: Biological Sciences* 290: 20230511.

van Petegem, K.H.P., Boeye, J., Stoks, R., and Bonte, D. (2016). Spatial selection and local adaptation jointly shape life-history evolution during range expansion. *The American Naturalist* 188: 485–498.

van Petegem, K., Moerman, F., Dahirel, M., Fronhofer, E.A., Vandegehuchte, M.L., van Leeuwen, T., and Bonte, D. (2018). Kin competition accelerates experimental range expansion in an arthropod herbivore. *Ecology Letters* 21: 225–234.

Venkateswaran, V., Shrivastava, A., Kumble, A.L.K., and Borges, R.M. (2017). Life-history strategy, resource dispersion and phylogenetic associations shape dispersal of a fig wasp community. *Movement Ecology* 5: 25.

Wang, R., Ovaskainen, O., Cao, Y., Chen, H., Zhou, Y., Xu, C., and Hanski, I. (2011). Dispersal in the Glanville fritillary butterfly in fragmented versus continuous landscapes: comparison between three methods. *Ecological Entomology* 36: 251–260.

Weeks, B.C., O'Brien, B.K., Chu, J.J., Claramunt, S., Sheard, C., and Tobias, J.A. (2022). Morphological adaptations linked to flight efficiency and aerial lifestyle determine natal dispersal distance in birds. *Functional Ecology* 36: 1681–1689.

Weil, S.-S., Gallien, L., Nicolaï, M.P.J., Lavergne, S., Börger, L., and Allen, W.L. (2023). Body size and life history shape the historical biogeography of tetrapods. *Nature Ecology & Evolution* 7: 1467–1479.

West, S.A., Pen, I., and Griffin, A.S. (2002). Cooperation and competition between relatives. *Science* 296: 72–75.

Whitmee, S., and Orme, C.D.L. (2013). Predicting dispersal distance in mammals: a trait-based approach. *Journal of Animal Ecology* 82: 211–221.

Williams, J.L., Kendall, B.E., and Levine, J.M. (2016). Rapid evolution accelerates plant population spread in fragmented experimental landscapes. *Science* 353: 482–485.

Willis, C.G., Baskin, C.C., Baskin, J.M., Auld, J.R., Venable, D.L., Cavender-Bares, J., Donohue, K., de Casas, R.R., and The NESCent Germination Working Group (2014). The evolution of seed dormancy: environmental cues, evolutionary hubs, and diversification of the seed plants. *New Phytologist* 203: 300–309.

Wolz, M., Klockmann, M., Schmitz, T., Pekár, S., Bonte, D., and Uhl, G. (2020). Dispersal and life-history traits in a spider with rapid range expansion. *Movement Ecology* 8: 2.

11

The Evolution of Human Life Histories

Megan Arnot[1] and Ruth Mace[1,2]

[1] *Department of Anthropology, University College London, London, UK*
[2] *Toulouse School of Economics, Institute for Advanced Study at Toulouse, Toulouse, France*

11.1 Introduction

Humans display life history strategies broadly comparable to apes, but with a few important differences. We have several exaggerated ape traits, such as extremely slow growth, prolonged childhood and extremely high levels of investment in our offspring. Alongside these traits, humans also present many derived traits, such as having underdeveloped newborns, short inter-birth intervals, early weaning and high fertility (Jones 2011). Why humans differ from other apes is still a matter for discussion, and here we discuss the differences in the context of communal breeding. In addition to the differences between the life histories of humans and other apes, there is also a huge amount of variation in the timing of various life history traits both within and between human populations. However, the species-level predictions of variation in life history traits do not always match what we observe in human populations well. We make suggestions on how we could move forward within this field by formalizing the definitions of life history strategies and measurements of ecological harshness and constraints. Finally, we discuss the evolution of a life history trait that is unique to humans amongst primates: menopause and the prolonged post-reproductive lifespan (PRLS). Here, we summarize the existing evolutionary hypotheses for the presence of this trait and highlight the shortfalls in the current prevailing hypotheses and empirical research.

11.2 Life History Trade-Offs

Phenotypic variation is frequently interpreted and analysed by human behavioural ecologists from a life history framework, with the optimal allocation strategy being contingent upon environmental factors (Stearns and Koella 1986, Stearns 1992, Charnov, 1993). The first fundamental trade-off is between growth and reproduction (see also Chapter 1, this volume). Any energy invested into current growth means forgoing current reproduction, and *vice versa*. As a result, one must optimally allocate energy in response to the environment, as future reproduction is contingent upon an organism surviving long enough to do so. An example of this is some hunter-gather groups in central Africa of small stature. The term 'pygmy' is sometimes used to describe populations whose average male height is <155 cm (Cavalli-Sforza 1986) and often used to describe groups who live in the African tropics forests that have traditionally subsisted on a hunting and gathering lifestyle. It is thought that they may have evolved a small stature as a result of selection for early reproduction due to high mortality rates, with only 30–51% of children living till the age of 15 (Migliano et al. 2007, see also Chapter 1, this volume). While the reason for the high mortality is not clear (Migliano et al. 2007), a possible explanation is the high prevalence of diseases and infections such as conjunctivitis, chiggers, dental attrition, leprosy and yaws (Noireau et al. 1989, Walker and Hewlett 1990, Hewlett 1991, Louis et al. 1994, Lewis 1999). The second key trade-off is between offspring quantity and quality, known as the quantity–quality trade-off (Lack 1954; see also Chapter 3, this volume). This has been observed in the Agta, which is a small-scale, pre-industrial foraging society that resides in the Philippines. In this group, it was found that less nomadic groups had higher childhood mortality rates, possibly due to the increased levels of disease that have been observed

Life History Evolution: Traits, Interactions, and Applications, First Edition. Edited by Michal Segoli and Eric Wajnberg.
© 2025 John Wiley & Sons Ltd. Published 2025 by John Wiley & Sons Ltd.

to occur alongside sedentarization. Where there was high child mortality, there was increased fertility, which resulted in more children overall surviving to adulthood, suggesting a trade-off between offspring quality and quantity (Page et al. 2016).

How an organism allocates its energy is hypothesized to hinge on its risk of extrinsic mortality (which refers to death from an external force, such as predation; see de Vries et al. (2023) for a criticism of the idea) and resource availability (Stearns 1992). When the environment is harsher (*e.g.*, scarce resources, higher mortality), it is predicted that it would be optimal to preferentially allocate resources towards early reproduction and offspring quantity to ensure genetic propagation. On the other hand, in more 'forgiving' environments, organisms can age more slowly, allowing them to allocate more resources to growth-related activities and later life reproduction (see also Chapters 1 and 12, this volume). Events related to growth and reproduction (*e.g.*, weight at birth, age at sexual maturation, age at first birth, lifespan) are known as life history traits, and the age-specific patterning of life history traits in response to the local ecological conditions is indicative of an organism's life history strategy. Life history strategies have been modelled to lie on a fast–slow continuum (Promislow and Harvey 1990), where 'fast-strategists' preferentially allocate energy towards tasks related to current reproduction and offspring quantity, while 'slow-strategists' are more concerned with investing energy into early-life growth and late-life reproduction. It should be noted that life history strategies are relative. Hence, while a gorilla (*e.g.*) is a low strategist compared to a mouse, relative to a whale, it has a faster life history strategy.

11.3 The Life Histories of Great Apes

Broadly, primates are considered to have slow life histories (Jones 2011). They have a relatively late age at first birth, low total fertility and long lifespans (Table 11.1). Humans are somewhat outliers amongst primates in that we present a mixture of fast life history traits and exaggerated 'typical ape' life history traits (Table 11.1). Our babies are highly altricial and fat (Kuzawa 1998) and are then highly dependent on their mother following birth. We then have extremely prolonged childhoods followed by a period of adolescence, and, during both stages, we are still highly reliant on parents and other allocarers for food and other forms of provisioning (Kaplan 1994). Upon reaching adulthood, rapid reproduction is seen in many forager groups (Sadhir and Pontzer 2023), with parents often having multiple dependent offspring at once (Kramer 2014). Finally, we reach old age, at which point women experience reproductive senescence followed by an extended PRLS (Hawkes et al. 1998).

A summary of the timing of key life history traits in non-human great apes and contemporary hunter-gatherers can be found in Table 11.1. Parameters from hunter-gatherers are used because many aspects of the way they live – including their foraging patterns, living arrangements and population densities – are thought to broadly resemble those of ancestral humans before the invention of agriculture. For this reason, human behavioural ecologists assumed that contemporary foragers represent a lifestyle that all humans were engaged in until around 11,000 years ago (Marlowe 2005). Of course, this is not a completely accurate statement. Hunter-gatherers are as modern as you or I. Visits from anthropologists, charities, missionaries and general market integration have meant that the lifestyles of contemporary hunter-gatherers diverge from the 'ancestral condition'. Further, hunter-gatherer groups today do not present a uniform group, with variable hunting practices, religions, living arrangements, environments, technologies and diets. As such, the patterning of their life history traits is also highly variable. Despite this, hunter-gatherer groups have been highly influential in informing key models of human life history theory and human evolution (Fitzpatrick and Berbesque 2018). For this reason, available measures of life history traits from hunter-gather and foraging populations have been averaged in Table 11.1, with the global distribution of the mentioned groups being presented in Figure 11.1.

It is thought that the unique patterning of human life history traits is the product of a combination of our intelligence, our bipedal locomotion and our high levels of communal breeding in which resources are transferred across generations, as outlined in the following paragraphs (Kaplan et al. 2000). Like other apes, humans are altricial at birth. However, this altriciality is amplified in humans. At birth, the brain of human neonates is just 30% the size of adult brains, compared to chimpanzee neonates, who experience 40% of their brain growth *in utero* (DeSilva and Lesnik 2006). Humans are so altricial that it has been proposed by Portmann (1951) that humans would require an additional 9–12 months of gestation to be born in the neurological developmental state equivalent to a *Pan* (chimpanzee) neonate. As a result of being rather helpless at birth, human infants require a great deal of care. Human foragers do not become energetically self-sufficient until their late teens, meaning that – even after weaning – mothers and others have to provide children with calories, help and supervision

Table 11.1 Life history parameters of extant female great apes from wild populations compared with humans from hunter-gatherer populations.

Great ape species	Age at first birth (years)	Gestation length (days)	Age at weaning (years)	Inter-birth interval (years)	Age at last birth (years)	Maximum lifespan
Orangutan						
Pongo abelii	15.4[a]	260[h]	>8.8[k]	9.3[a]	>41[a]	53[a]
Pongo pygmaeus	15.7[b]	260[h]	6.5[l]	7.8[q]	—	—
Gorilla						
Gorilla gorilla	10.3[c]	255[h]	4.6[m]	4.2[r]	—	45[x]
Gorilla beringei	9.9[d]	255[i]	4.1[n]	3.9[d]	40.6[u]	43.7[y]
Bonobo						
Pan paniscus	14.2[e]	236[j]	4.5[e]	4.8[i]	—	>50[z]
Chimpanzee						
Pan troglodytes	11.5[f]	225[h]	4.7[o]	5.9[s]	42[v]	65[aa]
Human						
Homo sapiens	19.0[g]	270[h]	2.6[p]	3.3[t]	38.25[w]	>100[ab]

[a] Wich et al. (2004).
[b] Tilson et al. (1993).
[c] Czekala and Robbins (2001).
[d] Eckardt et al. (2016).
[e] Kuroda (1989).
[f] Walker et al. (2018).
[g] Average age of first birth from six foraging groups: Ju/'hoansi,19.2 years (Howell 2010); Hadza, 18.77 years (Blurton Jones 2016); Savannah Pumé, 16.09 years (Hackman and Kramer 2023); Hiwi, 20.5 years (Kaplan et al. 2000); Aché, 19.5 years (Hill and Hurtado 1996); Agta, 20.14 years (Goodman et al. 1985).
[h] Harvey et al. (1986).
[i] Harcourt et al. (1980).
[j] Hashimoto et al. (2022).
[k] No reliable data on the age of weaning in *P. abelii*, figure based on single observation of offspring suckling until their parents' death (Smith et al. 2017).
[l] van Noordwijk et al. (2013).
[m] Nowell and Fletcher (2007).
[n] Average of measures from Eckardt et al. (2016) and Robbins and Robbins (2021).
[o] Lonsdorf et al. (2020).
[p] Average weaning age from four foraging groups: Ju/'hoansi, 42 months (Alvarez, 2000); Hadza, 30 months (Blurton Jones 2016); Aché, 25 months (Hill and Hurtado 1996); Agta, 26 months (Konner 2005).
[q] Galdikas and Wood (1990).
[r] Average inter-birth interval of two *G. gorilla* populations: Bwindi, 56.4 months (Robbins et al. 2009); Karisoke, 47.8 months (Robbins et al. 2006).
[s] Thompson et al. (2007), composite measure based on data from Gombe, Mahale, Kibale, Budongo, Bossoi and Gambian communities of *P. troglodytes*.
[t] Average inter-birth interval from six foraging groups: Hadza, 38 months (Konner 2005); Efé, 38 months (Konner 2005); Aché, 37.6 months (Hill and Hurtado 1996); Agta, 36 months (Konner 2005); Ju/'hoansi, 41.3 months (Blurton Jones 1986, Howell, 2010); Hiwi, 45.1 (Kaplan et al. 2000).
[u] Robbins et al. (2006).
[v] Average age at last birth from four *P. troglodytes* populations: Bossou, 41 years (Sugiyama 2004); Gombe, 44 years (Goodall 1986); Mahale, 39 years (Nishida et al. 2003); Tai, 44 years (Boesch and Boesch-Achermann 2000).
[w] Average age at last birth from two foraging groups: Ju/'hoansi, 34.4 (Howell 2010); Aché, 42.1 (Hill and Hurtado 1996).
[x] Bronikowski et al. (2011).
[y] Robbins et al. (2006).
[z] Erwin et al. (2002).
[aa] Wood et al. (2017).
[ab] Dong et al. (2017).

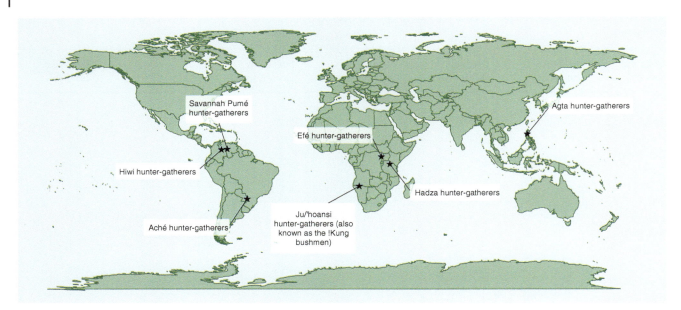

Figure 11.1 The global distribution of the hunter-gatherer groups that are referenced in this chapter.

(Kaplan 1994). Despite this, mothers frequently have more than one dependent offspring at a time – something that is not seen in other great apes. In other great ape species, mothers typically only have one dependent offspring at any given time, and provisioning responsibilities are negligible after weaning. Great ape mothers also spend a comparatively long time nursing their young (Emery Thompson and Sabbi 2019), not weaning them until between the ages of 4 and 8, depending on the species (Nowell and Fletcher 2007, van Noordwijk et al. 2013, Eckardt et al. 2016, Smith et al. 2017). Once the infant has been weaned, it becomes nutritionally self-sufficient, with post-weaning provisioning being seldom observed (Rapaport and Brown 2008). This contrasts with human groups, who wean their offspring at a comparatively young age. The Hadza, Aché and Agta – who are well-studied human forager populations – all spend approximately 2–2.5 years nursing their children (Hill and Hurtado 1996, Konner 2005, Blurton Jones 2016), while the Ju/'hoansi hunter-gatherers (also sometimes known as !Kung bushmen, but Ju/'hoansi is their own preferred name) spend slightly longer, not weaning their offspring until they are 3.5 years old on average (Alvarez 2000). In humans, the comparatively short time spent breastfeeding their children means that reproduction can happen more rapidly as they are not spending a long time amenorrhoeic. This is reflected in the average inter-birth intervals, which range from 3.75 years in the Hiwi (Kaplan et al. 2000) to three years in the Agta (Konner 2005). This has a narrower range compared to other great ape species, which span from 3.9 years in *G. gorilla* (Eckardt et al. 2016) to 9.3 years in *P. abelii* (Wich et al. 2004).

When nursing, the burden of care is very much the mother's own, as with other primates. However, once childhood is reached, caring responsibilities can be distributed to others in their social network. Childhood in humans is a period of cognitive and social development, in which valuable skills such as how to hunt, forage, work as part of a group, and socialize are learned (Bogin 1997, Pretelli et al. 2022). Childhood is not a life stage that is unique to humans, though it is much more pronounced in us compared to other great apes. This is likely because we require additional time to allow our brains to develop, as highlighted by the fact that physical growth is relatively slow between the ages of four and the onset of puberty (Rogol et al. 2000), but brain growth is rapid (Peterson et al. 2018).

Upon reaching puberty, human adolescents experience a growth spurt in which they reach their adult height (Bogin 2015). During this time, many secondary sexual characteristics are developed, such as increased body hair, breasts and wider hips. In humans, puberty is typically marked by the onset of menstruation, the development of breasts in girls, the deepening of the voice in males and the presence of pubic hair in both sexes. Puberty in males is typically around two years later than in females, which may be the result of sexual selection, with delayed physical maturation enhancing young males' competitive advantage in the 'mating market', and also increasing their success in hunting-related activities (Bogin 1994). Much like humans, other great apes experience a period of adolescence during which there are various physiological changes. Menstruation occurs in the females of all other species of great apes, though bleeding is far lighter than in humans and therefore difficult to measure in the wild, and may be a relatively rare event under conditions of natural fertility (the absence of artificial birth control) in both humans and other primates (Strassmann 1997). In the males of great apes, various distinct physiological changes occur: gorilla males develop a sagittal crest and silver hair between the ages of 10 and 16 years (Breuer et al. 2009); the testicles of male chimpanzees enlarge between the ages of approximately 8 and 10 years (Goodall 1986, Pusey 1990); and some male orangutans become 'flanged', exhibiting enlarged cheek 'flanges', an extremely large body size and long hair. Great apes also experience a growth spurt upon reaching adolescence (Berghänel et al. 2023), though this is less pronounced than it is in humans as other great apes grow faster during the earlier growth period (i.e., childhood) than humans do.

The reproductive behaviour of other great apes compared to humans also differs greatly. As previously stated, humans spend less time breastfeeding, hence amenorrhoeic, and so can resume reproduction more quickly. Looking at contemporary hunter-gatherer groups, their inter-birth intervals are approximately three years (Blurton Jones 1986, Hill and Hurtado 1996, Kaplan et al. 2000, Konner 2005), which contrasts with that of other great apes that range between four and nine years (Wich et al. 2004, Eckardt et al. 2016). The rapid reproduction of humans is thought to be facilitated by the fact that we are communal breeders. Alloparenting – defined as infant care provided by an individual other than the mother (Emmott and Page 2019) – is seldom observed in other great apes (though common in some other primates). Although there are instances in chimpanzees and gorillas where allocaring has been observed, it is not an obligate behaviour, and these species cannot be described as communal breeders (Bădescu et al. 2016, Grueter et al. 2019). This contrasts with humans, who live with, and breed in, large groups of both related and unrelated individuals (Hill et al. 2011). In all human societies, individuals other than the mother are involved in providing care for offspring. These alloparents may take the form of more formalized caregivers, such as nannies and teachers, or more informal carers like fathers, siblings, grandparents, neighbours, friends and other family members. These 'helpers at the nest' reduce the amount of work a mother needs to do, allowing her to reinvest energy into reproduction. This further contributes to shorter inter-birth intervals, allowing humans to have high fertility despite 'expensive' energetically demanding offspring. Help generally flows down the generations rather than up, so how important children are in helping their mothers reproduce is unclear. Stochastic dynamic programming models, which are a very natural way to model life history evolution, suggest children helping each other may not be especially important in reducing optimal birth intervals, whereas conditions that generate low infant mortality and competition between offspring for parental effort favour longer birth intervals (Mace 1998, Thomas et al. 2015).

Of course, investing energy in allocaring means diverting energy away from your own reproduction. Therefore, one could ask why one would choose to be an allocarer rather than reproducing directly. Many allcarers are those who are physiologically or behaviourally non-reproductive, such as children and grandparents. Allocaring is typically predicted by Hamilton's (1964) rule, which is formulated as $rB > c$, in which r represents the relatedness coefficient, B is the benefit to the recipient of the care and c is the cost to the provider of the care. In this sense, cooperation with kin is predicted to be beneficial so long as the relatedness to the recipient and the benefit derived from the act outweigh the costs incurred by the co-operator. For example, among the Hadza, time spent carrying children is positively predicted by relatedness (Crittenden and Marlowe 2008). In Samoa, a third gender – known as the fa'afafine – who do not reproduce as they are biologically male and only have sex with other biological males, show heightened avuncular tendencies towards their nieces and nephews, and therefore increase their indirect fitness by acting as highly investing alloparents (Vasey et al. 2007). Similarly, post-menopausal women have been shown to invest more in their grandchildren compared to women of the same age who are pre-menopausal (Hofer et al. 2019, Arnot and Mace 2021).

Communal breeding in humans is also thought to have selected for a life history trait that is unique to humans amongst primates: menopause. Unlike other great apes (and almost all other animals), human females have an exceptionally long period between their last birth and death. Looking at contemporary hunter-gatherer groups, we can see that a large proportion of individuals survive into their 70s and 80s (Hill and Hurtado 1996, Howell 2010, Blurton Jones 2016). Though there is a huge amount of variation (Towner et al. 2016), if we consider that the average age of last birth is approximately 38 in hunter-gatherers, then that means up to a third of a woman's life is lived post-reproductively. This extreme longevity beyond reproduction is not observed in other great apes, and the evolution of the extended PRLS will be discussed in detail later in this chapter.

11.4 Variation in Human Life History

Life history theory was developed to explain the variation in life history traits between species and to understand the physiological and behavioural differences also observed within species (Stearns and Koella 1986). Human behaviour is plastic, and therefore models predict that conditions causing faster and slower life history strategies to evolve at a species level should elicit a comparable behavioural and physiological response on an individual level. A great deal of empirical support has been found for this in humans, with harsher, more unpredictable environments being found to be predictive of an earlier menarche and accelerated reproduction (Wilson and Daly 1997, Nettle et al. 2010, Uggla and Mace 2016), indicating a response to ecological uncertainty.

As well as current ecological cues associated with individual-level life history traits, conditions present in early life are thought to act as a blueprint of what can be expected in adulthood and therefore somewhat determine one's life history strategy in adulthood (Csathó and Birkás 2018, Kavanagh and Kahl 2018). There is a large body of research showing that adverse childhood experiences, such as parent–offspring conflict and individual poverty, predict traits indicative of a faster life history strategy, such as accelerated puberty, earlier reproduction and smaller stature (Chisholm et al. 2005, Lawson and Mace 2008, Nettle et al. 2010).

Some evolutionary psychologists predicted that different life history strategies can be identified by examining behaviours relating to reproduction, cooperation, competition and future discounting. For example, promiscuity and a focus on short-term goals are behaviours associated with a fast life history strategy (Hill 1993, Sykorova and Flegr 2021). Psychometric tools have also been proposed as a means of measuring 'the speed' of one's life history strategy. The most popular recent one is the Arizona Life History Battery (Figueredo et al. 2004, 2007). According to this tool, a slow life history is characterized by high levels of conscientiousness, religiosity, prosociality, emotional stability, agreeableness and a secure attachment style. However, empirical evidence does not necessarily support its validity as an accurate measure of life history strategy (Kometani and Ohtsubo 2022), nor is it clear how the authors selected traits that they perceived to be characteristic of a slow life history strategy. In addition, the tool does not include any questions about traditional life history traits (*e.g.*, age at first birth) and efforts to validate the scale by correlating it with biological life history traits have not yet been fruitful (Copping et al. 2014, Mathes 2018, Mededovic 2020). As a result of this, it has been heavily criticized (Nettle and Frankenhuis 2019, 2020, Sear 2020).

There is little empirical evidence that life history traits do co-vary within humans in the way one would expect, should the fast–slow continuum be a good explanation for phenotypic plasticity at the individual level. In a sequence analysis of data from the United States, it was shown that life history traits did not cluster together in the expected way and that individuals did not necessarily follow a clear life history trajectory and could therefore not be binarized into 'fast' and 'slow' strategists (Sheppard and van Winkle 2020). Data from the United Kingdom also showed that age at menarche did not consistently predict the timing of other key life history traits, suggesting that puberty is not an accurate indicator of one's life history strategy (Lawn et al. 2020).

It may be that some contradictory findings within the literature are because the verbal models of life history strategies are not accurate enough, as variation in any one life history trait is linked to others and complex models of interactions are required. For example, negative childhood experiences are presumed to correlate with a faster life history strategy, but many events assumed to be stressful or adverse (*e.g.*, paternal absence and a high number of house moves) are not generalizable cross-culturally or temporally. For instance, a global study using data from 22 small-scale societies showed that populations with a high juvenile mortality rate had an earlier average age of puberty, menarche and first reproduction (Walker et al. 2006). However, there are many exceptions within the literature. A cross-cultural analysis found no evidence for father absence being predictive of accelerated puberty (Sear et al. 2019), and, in Brazil, it was found that early-life adversity predicted a delayed, rather than accelerated, menarche (Wells et al. 2019). Though the mechanism for this is not clear, it was hypothesized that this is due to a decreased calorie intake and high levels of infant mortality, which result in a greater proportion of energy being invested in immune function, resulting in slower maturation.

Further, it may be that the verbal and other models used to predict life history strategies in human studies are not specific enough. Life history differences were initially attributed to different risks of extrinsic mortality. However, few human studies include this as a measure in their models. Though there are a few exceptions (Wilson and Daly 1997, Uggla and Mace 2015, 2016), most research using human data uses vague proxies of extrinsic mortality like ecological harshness, negative childhood experiences and socio-economic status. It is likely that what is considered 'harsh' is both culturally and temporally specific, meaning that many predictions and results cannot be generalized.

Some have also criticized life history theory for relying heavily on intuitive logic (Moorad et al. 2019). A 'life fast, die young' paradigm is pervasive across both the evolutionary and non-evolutionary literature (*e.g.,* 'pace of life syndrome' in psychology; 'life course theory' in sociology). Despite this, many authors propose that the contribution of extrinsic mortality to the evolution and development of life history strategies is overstated and not supported by evolutionary theory (Moorad et al. 2019, André and Rousset 2020, de Vries et al. 2023). Instead, it may be that life history strategies are the result of density-dependent competition, in which it is not extrinsic mortality itself that affects life history evolution, but rather the intensity of competition that results from varying mortality rates (André and Rousset 2020). When extrinsic mortality is lower, the intensity of competition for resources increases because more individuals are maintained in the environment, meaning that traits that are optimal under intense competition (*e.g.,* higher levels of offspring investment, slow reproduction) will be favoured. Likewise, when there are high levels of extrinsic mortality, resource competition is relaxed, and so traits such as rapid reproduction can evolve in response to the lower population density. Therefore, a greater understanding of the effect of population density on competition in humans may be required to generate predictions pertaining to life history strategies.

11.5 Menopause and the Post-reproductive Lifespan

The termination of female fecundity is known as the 'menopause' and refers to the irreversible cessation of menstrual function. Following menopause, females tend to live at least ten years longer as sterile individuals. Life history theory has been used by researchers to model the timing of various life history traits (*e.g.,* age at first birth, menarche, etc.) and how they cluster with one another. However, despite menopause being a key life history trait, it is seldom modelled or analysed in the same way other traits that affect fitness are.

At a proximate level, menopause can be understood to occur because women are born with a finite number of eggs that deplete throughout their life due to ovulation and follicular atresia (the breakdown of ovarian follicles). However, all other mammals undergo a similar process of depletion, and yet the majority do not undergo menopause or live beyond the end of fertility outside of captive conditions. Other mammals' reproductive senescence occurs in tandem with the decline of the rest of their bodily systems (with the exception of some cetaceans, see below), meaning that the length of their reproductive lifespan and actual lifespan are correlated with one another (Walker and Herndon 2008, Alberts et al. 2013). In humans, females go through menopause around the age of 50 (Laisk et al. 2019), and then live upwards of 30 years beyond this. This period of post-reproductive life is confusing in an evolutionary sense, as why has selection not favoured life-long reproduction? As it is an evolutionary puzzle, many adaptive hypotheses have been proposed, with the most prevalent being explained and discussed here.

11.5.1 The Phylogenetic Patterning of Menopause and the Post-reproductive Lifespan

In other mammals, lifespan beyond menopause (if any) is typically extremely short. To quantify the PRLS across mammals, Levitis and Lackey (2011) developed a measure of post-reproductive representation (PrR), which describes the proportion of the animal's life spent post-reproductive. The PrR of 124 mammals is shown in Figure 11.2, where it can be seen that humans have a longer PrR (mean PrR = 0.524) than other mammals (mean PrR = 0.085). There is also a significant difference in PrR between captive (mean PrR = 0.139) and wild (mean PrR = 0.028) species. This is because, in captivity, the lifespan of animals is generally longer than it would be in the wild due to better provisioning, a lack of predation and veterinary care. Research has found that under such conditions, many non-human primates go through a process analogous to humans' menopausal transition, in that older female primates experience an irregularity in their menstrual cycles and a decline in oestrogen levels that eventually results in the termination of ovulation (Graham et al. 1979, Gould et al. 1981).

Outside of primates, there is evidence for a prolonged PRLS in five species of cetaceans: beluga whales, false killer whales, killer whales, narwhals and short-finned pilot whales (Olesiuk et al. 2005, Croft et al. 2017, Photopoulou et al. 2017, Ellis et al. 2018a). Collecting data on the reproductive status of cetaceans comes with many difficulties, as fertility status is typically inferred from post-mortem anatomical examination (Perrin et al. 1984). However, phylogenetic analysis suggests that menopause evolved multiple times in cetaceans (Ellis et al. 2018b), with most species stopping reproducing around the age of 40 but living decades beyond this (Croft et al. 2015). As cetaceans are the only other group of mammals to display a long PRLS in the wild, they are often compared when studying the evolution of menopause and the PRLS in humans, despite their phylogenetic distance from us and their difference in physical environment.

11.5.2 An Evolutionary Approach to Menopause

Given that menopause is phylogenetically rare and puzzling from an evolutionary standpoint, many have tried to situate it within evolutionary theory. Though some argue that it is a trait that has been selected for in and of itself, others argue that it is the by-product of other factors, such as our extreme longevity, cyclical fertility and sexual selection combined with constraints on fertility.

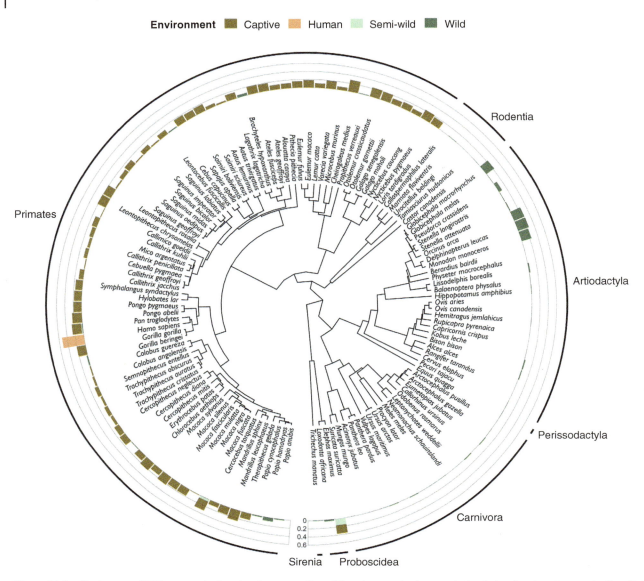

Figure 11.2 Phylogeny of 124 mammals showing the proportion of female years in the population being lived post-reproductive. Data on the duration of post-reproductive representation is taken from Alberts et al. (2013), Chapman et al. (2019), Croft et al. (2015), Ellis et al. (2018a), Levitis and Lackey (2011) and Photopoulou et al. (2017). Phylogeny generated from VertLife.org (https://vertlife.org/phylosubsets).

11.5.2.1 Menopause and the Post-reproductive Lifespan as a By-product
11.5.2.1.1 The Lifespan Artefact Hypothesis

Research looking at menopause in other primate species has shown that it is an event that sometimes occurs in non-natural settings, like zoos and reserves. In such environments, the lifespan of the primate can be artificially lengthened by virtue of better food provisioning, veterinary care and protection from predators. For example, chimpanzees living in Kibale National Park, Uganda, have been observed to live up to the age of 67 and live post-reproductively for about one-fifth of their adult lives, possibly because of favourable ecological circumstances resulting from living in a national park (*e.g.*, more stable food supplies) (Wood et al. 2023). As a result, some have theorized that human menopause is the product of a similar process and the result of our longer lifespans (Austad 1997, Peccei 2001). Here, menopause is simply an epiphenomenon, and the female reproductive system has just not yet 'caught up' with our longer lifespans. Though this is an appealing, intuitive hypothesis, it rests on the assumption that longevity is evolutionarily novel. Looking at contemporary hunter-gatherer populations, we can see that a large proportion of individuals live at least 20 years beyond the age of 50 (Hill and Hurtado 1996, Howell 2010, Blurton Jones 2016), which suggests humans may have been living well beyond the age of menopause for millennia. This hypothesis also implies that menopause is a recent trait. Menopause, of course, does not fossilize, and therefore it is hard to pick up in the archaeological record. However, Cerrito et al. (2020) found that menopause leaves a permanent change in dental cementum microstructure, offering a possible avenue for future research to understand when in human history it first emerged. At present, this method has not been used to study menopause in ancient hominins. Nonetheless, it has been estimated by Bogin (1999) that menopause and a PRLS could have emerged up to 1.8 million years ago, at which point hominin brain sizes were becoming larger, resulting in a prolonged period of dependency (childhood) and the necessity for cooperative breeding (*e.g.*, grandmothering). Therefore, at this time point, life history patterns are thought to have been becoming more like those of modern *Homo sapiens*, making it likely that menopause and the PRLS would have been present. This suggests that – if menopause was truly maladaptive – there would have been ample time for selection to work against it in some way. Whilst it is unappealing to evolutionary biologists to rely on an argument that hinges on constraints, it is of course possible that some physiological costs and benefits relating to producing fertile eggs, as yet not fully identified, may be limiting the number of eggs or duration of the fertile period in humans or other long-lived species. Hypotheses relying on constraints are hard to test.

11.5.2.1.2 Antagonistic Pleiotropy and Mutation Accumulation

Antagonistic pleiotropy is currently one of the most widely accepted explanations for why we age (Williams 1957), as highlighted in Chapter 2 of this volume. Pleiotropy refers to the phenomenon in which a single gene influences multiple traits in an organism. In the case of antagonistic pleiotropy, one of the traits influenced by the gene has a detrimental effect on the organism. Applied to menopause, it has been suggested that the rate of follicular atresia and the initial follicular stock provide a selective advantage in maintaining menstrual cycles at younger ages, with this advantage outweighing the costs associated with menopause occurring in midlife – a time at which life history factors are typically not expected to strongly impact fitness (Wood et al. 2001). In summary, if regular follicular is beneficial for pre-menopausal women, then it could be that menopause is simply a by-product of this process. However, this explanation does not identify why the selection for follicular depletion as a means of maintaining regular menstrual cycles led to the evolution of menopause long before the end of a woman's life (Wood et al. 2001, Laisk et al. 2019), nor does it explain why this process only happens in humans and no other mammals, like chimpanzees and elephants, whose life courses resemble those of humans but they do not have menopause (Lahdenpera et al. 2014).

Another possible antagonistic pleiotropic mechanism was recently proposed by Monaghan and Ivimey-Cook (2023), who suggest that the PRLS results from a rigorous screening process by oocyte mitochondria. The human brain requires embryos derived from high-quality oocytes for successful development. Consequently, a stringent quality control mechanism is in place for oocytes. Due to the natural degradation of oocytes, this quality control mechanism becomes more intense as the female ages, leading to the more rapid and frequent 'culling' of oocytes. Therefore, an antagonistic pleiotropic effect emerges: tough quality control prevents the fertilization of less viable oocytes, thus facilitating the production of high-performing offspring; but this quality control contributes to the rapid decline of the ovarian reserve in midlife, and so it carries the cost of time-limited fertility. While this hypothesis addresses why menopause only occurs in certain highly encephalized species (*i.e.*, humans and cetaceans), in and of itself it does not address why there is prolonged survival beyond menopause. Instead, the authors propose a synthesis between this 'metabolic hypothesis' and the Grandmother Hypothesis, which is addressed later in this chapter.

11.5.2.1.3 Male Mate Choice

The mate choice hypothesis of menopause suggests that menopause is the result of men preferring to mate with younger women (Morton et al. 2013), which is a preference observed across multiple cultures in humans (Buss 1989, Walter et al. 2020). This hypothesis is based on the assumption that, in our evolutionary history, males showed a preference for younger females, therefore making reproduction unlikely beyond a certain age. This created a selection shadow that led to the accumulation of genetic mutations, including one that was related to reproductive senescence. In this model, it is assumed that the male preference for younger women preceded the evolution of menopause, rather than it being present as a result of it. It is well documented that men report a preference for younger women (Buss 1989), but this is not universal in the animal kingdom. In chimpanzees, rhesus macaques, olive baboons and Bornean orangutans, males prefer to mate with older females (Conaway and Koford 1965, Lindburg 1971, Ransom 1981, Schürmann 1981, Muller et al. 2006), suggesting a male preference for female youth might have evolved in humans following menopause, not the other way around.

11.5.2.2 Menopause and the Post-reproductive Lifespan as an Adaptation

At present, there is limited empirical evidence in favour of the hypotheses that propose menopause is a constraint and the PRLS is an evolutionary by-product, related to the evolution of menopause. Importantly, these hypotheses often fail to distinguish between menopause and the extended PRLS, with most only attempting to explain the former, rather than the latter (or both). It should be noted that it is not always clear whether the authors are suggesting menopause is a trait that evolved as a result of selective pressures or whether it was a trait maintained by selective pressures (Thouzeau and Raymond 2017). This distinction is only important if we believe physiological or evolutionary constraints are important.

A fully evolved menopause and corresponding PRLS would mean that – at some point in our evolutionary history – the longevity of the female reproductive system would have been in line with other bodily systems, therefore allowing prolonged reproduction. A selective pressure (e.g., reproductive conflict) would have then caused menopause to occur earlier in the life course, thus causing menopause and the PRLS to evolve. On the other hand, a maintained menopause would mean that the female life expectancy got longer, and then some ecological factor (e.g., communal breeding) created an evolutionary pressure that meant early reproductive senescence was maintained as the women were able to offset the fitness costs of not being directly reproductive. These two pathways are shown graphically in Figure 11.3. Most literature refers to the evolution of menopause, thus alluding to the first pathway. However, it is not always explicit. It is currently not possible to know whether ancestral human females were reproducing into old age and then experienced a selective pressure against this (evolved menopause), or whether there has been a selective pressure against life-long reproduction (maintained menopause). Further discussion on adaptive theories will refer back to this differentiation.

11.5.2.2.1 The Mother Hypothesis

For all mammals, there are costs and benefits associated with reproduction, with the costs being particularly great for females who are responsible for gestating the offspring, giving birth to it and provisioning it postnatally. For humans,

(a) Hypothetical pathway to the evolution of menopause

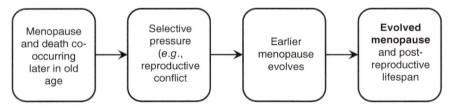

(b) Hypothetical pathway to the adaptive maintenance fo menopause

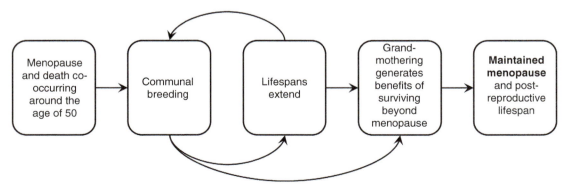

Figure 11.3 Hypothetical pathways to the (a) evolution of menopause and (b) the adaptive maintenance of menopause. Arrows indicate the direction of hypothesized pathways. *Source:* Adapted from Arnot (2021).

childbirth is particularly difficult due to the large size of the baby's head relative to the maternal birth canal (Wittman and Wall 2007). This mismatch between the foetal head and the mother's birth canal is the leading cause of obstructed labour, being responsible for 8% of all maternal deaths across the globe today (Dolea and AbouZhar 2003, World Health Organisation 2005, Say et al. 2014). Older women particularly experience a greater risk of both maternal and neonatal mortality, pregnancy-related illnesses (e.g., preeclampsia, hypertension) and poor foetal growth (Pavard et al. 2008, Cavazos-Rehg et al. 2015). The Lansing effect – the idea that offspring produced by older parents are of poorer 'quality', have a shorter lifespan and have lower reproductive value (Lansing 1947) – also may contribute to the costs of late-life childbearing. Data from pre-industrial Finland found that, between offspring born to mothers aged 16 and 50 years, lifetime reproductive success declined by 22% and individual fitness by 45% (Gillespie et al. 2013). Hence, the increased risk of complications during pregnancy and childbirth in older women and the possible threat to offspring fitness are thought to have acted as a selective pressure for early reproductive senescence, with menopause being selected to protect against the costs of later life pregnancies (Williams 1957, Alexander 1974). Under this model, reproduction in later life is not only assumed to affect the mother directly but would also indirectly negatively impact any existing offspring she has, as maternal mortality is a strong predictor of child mortality (Sear and Mace 2008). In summary, the Mother Hypothesis proposes that menopause is the product of selective pressure against life-long reproduction to protect against neonatal and maternal mortality.

However, for selection to have favoured menopause under these conditions, the risk of maternal mortality would have to be substantive enough to outweigh the fitness benefits of continued reproduction, and there is little empirical support for this. Firstly, data from contemporary hunter-gatherer groups suggests that maternal mortality is not as great a risk factor as previously thought. Even among the Aché hunter-gatherers, who traditionally have had little access to contemporary medicine, the odds of dying during childbirth were found to be 1 in 150 (Hill and Hurtado 1996). Some have even suggested that the female pelvis has gotten narrower in recent years due to medical interventions (such as caesareans) relaxing selective pressures (Pavlicev et al. 2020), therefore meaning that some of the current estimates of complications during birth are a product of aspects of modernity. Moreover, if the risk of maternal mortality was particularly strong in our evolutionary history, one might then question why there was not a strong selective pressure favouring phenotypes that were better able to carry and birth healthy infants late into life, given that it can be assumed the fitness benefits would be large. Finally, theoretical models do not predict that this hypothesis alone can explain post-reproductive longevity in human females (Thomas et al. 2015, Aimé et al. 2017). It may be that elements of the Mother Hypothesis contributed to the evolution or maintenance of menopause, but it fails to explain why women would then live so long post-reproductively. Therefore, as a standalone hypothesis, there is limited evidence that these conditions would neither maintain the menopause nor facilitate its evolution (Thouzeau and Raymond 2017).

11.5.2.2.2 The Grandmother Hypothesis

Perhaps the most famous hypothesis associated with menopause is the Grandmother Hypothesis. This builds on the Mother Hypothesis by acknowledging that the risk of maternal mortality alone might not be a sufficient adaptive explanation, but that the kin-selected benefits of the trait coupled with the risk associated with late-life birth might be (Hawkes et al. 1998, Shanley et al. 2007, Kim et al. 2012).

Allocarers assist in the provisioning and protection of human children, and their presence means that women can wean earlier and reduce their duration of lactational amenorrhea and therefore their inter-birth intervals, while also simultaneously caring for multiple generations of children, despite the great deal of investment children require (Lahdenpera et al. 2004). Genetic relatedness and possible benefits derived from helping primarily predict kin-directed care, and therefore a common allocarer is the grandmother. Models suggest that post-reproductive women can offset their lack of direct reproduction in later life by investing heavily in their grandchildren, which will result in increased indirect fitness (Shanley and Kirkwood 2001, Shanley et al. 2007, Aimé et al. 2017).

For post-reproductive women to offset the direct fitness losses that result from being post-menopausal, they would have to have a large positive effect on the survival and fertility of closely related kin, such as grandchildren. The effects of grandmother presence on grandchild outcomes have primarily been studied in natural fertility (both historical and contemporary) societies. Evidence from these populations has found that – in some circumstances – the presence of a grandmother can increase grandoffspring health and survival and also increase the fertility of their daughters, though this is contingent on social and cultural factors such as the grandoffspring sex and lineage (Sear et al. 2000, Gibson and Mace 2005, Engelhardt et al. 2019, Nenko et al. 2020, Chapman et al. 2021, Du et al. 2022). A review looking at the effects of kin-presence on child outcomes found that, in 69% of reviewed studies (9 out of 13), maternal grandmothers were associated with an increased probability of grandchild survival, with paternal grandmothers having the same effect in 53% of studies (9 out of 17) (Sear and Mace, 2008). For example, in Ethiopia, having a maternal grandmother who was able to assist with heavy domestic tasks increased the probability of a grandchild surviving to three years by 25% (Gibson and Mace 2005), and in The Gambia, maternal grandmothers improved the nutritional status of their grandchildren (Sear et al. 2000). Furthermore, more recent studies using data from pre-industrial Finland have shown that maternal grandmothers acted as a buffer against the detrimental effects of very short inter-birth intervals (Nenko et al. 2020) and that their presence had a positive effect on grandchild survival between the ages of two and five, which is when children would have typically been weaned (Chapman et al. 2021).

However, the benefit of having a maternal grandmother is not a universal phenomenon. In rural Malawi, the mortality rates of female grandchildren were highest in the presence of a maternal grandmother (Sear 2008). This was hypothesized to be due to conflict between female relatives over resources due to the population being matrilineal. Other studies have highlighted the benefits of paternal and male relatives, with paternal grandmothers investing more in their grandchildren than maternal grandmothers in rural Greece and Pakistan, possibly due to a combination of high paternity certainty and close proximity resulting from patrilocality (Pashos 2000, Chung et al. 2020). Additionally, data from western China has shown that both paternal and maternal grandfathers are associated with increased grandchild survivorship (Du et al. 2022).

The positive effects of grandmothering have also been documented in contemporary post-industrial societies, despite the possible confounding effects of contraception, available medical care, schooling and a neolocal residence pattern. Data from the United Kingdom has shown that maternal grandparents spend more time with their newborn grandchildren than paternal grandparents and are also more likely to provide their grandchildren with financial assistance and gifts (Pollet et al. 2009). In the United States, low birthweight infants born to teenage mothers who lived with their grandmother (in addition to their own mother) had better cognitive and health outcomes compared to babies who just lived with their mothers, even after controlling for maternal age (Pope et al. 1993).

Evidence has been found for a possible switch in time and energetic allocation relative to menopause status. Two studies using data from the United Kingdom (Arnot and Mace 2021) and Australia and the United States (Hofer et al. 2019) have shown that post-menopausal women dedicate more time to caring for their grandchildren compared to women of the same age who have not yet gone through menopause, even after controlling for relevant covariates. This is an indication of a behavioural adaptation towards caring for younger kin when physiologically post-reproductive, as a means of offsetting the costs of no longer being able to directly reproduce.

The large amount of data showing the tendency of grandmothers to invest heavily in their grandchildren has resulted in the Grandmother Hypothesis being one of the most popular adaptive explanations for the evolution of menopause. Another advantage of this hypothesis is that it also applies to cetaceans, who also experience menopause and a PRLS. Post-reproductive female killer whales (*Orcinus orca*) have been observed to be important hubs of ecological knowledge, with grandmothers passing on knowledge such as how to hunt and where the salmon reserves are (Brent et al. 2015). Calves who do not have a living grandmother have a 4.5 times greater mortality rate compared to calves with living grandmothers (Nattrass et al. 2019).

Despite the wealth of empirical evidence in support of the Grandmother Hypothesis, some have argued that the indirect fitness benefits of grandmothering are not sufficient to have selected for a PRLS over continued reproduction. Further, it is not necessarily clear why women would not be able to have grandchildren and also carry on having children themselves, particularly given that humans are communal breeders. In Asian elephants, older females have been observed to improve the survival outcomes of their grand-calves despite not experiencing a PRLS comparable to others, demonstrating that grandmothers can positively invest in their grandoffspring without complete reproductive cessation (Lahdenpera et al. 2016). Shanley et al. (2007) found that menopause was not costly when there was high maternal mortality and beneficial grandmothering and could have evolved in the context of parameters found in a natural fertility population of Gambian farmers. However, it is not clear if these costs and benefits would apply to hunter-gatherers, where there is less evidence for the significance of grandmaternal alloparenting. Nonetheless, it can be deduced from Shanley et al.'s (2007) model that, if small benefits arise in cases such as the grandmother stepping in in the rare cases of death of the mother, this can be a very substantial evolutionary advantage over time.

Taken together, it seems that if menopause is a maintained trait as opposed to one that was selected for in and of itself, then the Grandmother Hypothesis (possibly in combination with the Mother Hypothesis) may be sufficient as it acts as a selective pressure against the evolution of life-long reproduction in women. However, as we have shown, the benefits of grandmothers are not universal, and therefore, without knowing the ancestral residence patterns of humans, it is hard to formally calculate the fitness benefits and costs of being post-reproductive.

11.5.2.2.3 The Reproductive Conflict Hypothesis
Building on evidence in support of the Grandmother Hypothesis, Cant and Johnstone (2008) proposed the Reproductive Conflict Hypothesis which concentrates less on the kin-selected benefits of a PRLS and more on the fitness costs associated with inter-generational co-breeding under different residence patterns. This model is based on the assumption that menopause evolved under a patrilocal residence system, in which women leave their natal group and join that of their partners when they reproduce. A patrilocal residence system creates an age-dependent relatedness asymmetry that makes it adaptive for young females to invest in reproductive competition and for older females to stop reproducing. In patrilocal societies, when a young female leaves her natal group, she is related to no one in her new group. In contrast, older females in the younger females' new group will have high average relatedness by virtue of having been residing and reproducing there for a long time. The younger female has two options upon joining the new group: (1) she can help the older established females to reproduce, or (2) she can reproduce herself. As the young female is related to no one, she would not gain any inclusive fitness benefits from helping the older women to reproduce. Therefore, the adaptive strategy in this situation is for her to invest energy into her own reproduction. The older females also have two options: (1) they can forgo reproduction and help the younger females to reproduce, or (2) they can reproduce themselves. As the older females are established and therefore have higher relatedness to the group, they have less to lose from not reproducing. Therefore, in the case of older females, it may not be worth investing energy into reproductive competition with the younger females. Rather, it may be better for them to invest energy into increasing her inclusive fitness by investing in existing kin in the group. Hence, due to the asymmetry in relatedness between the women of the group and the associated differences in benefits of investing in reproductive competition, it is modelled that the younger females will 'win' the reproductive competition and therefore the right to directly reproduce, resulting in selection favouring the evolution of menopause in older women (Cant and Johnstone 2008).

This model was originally developed to explain menopause and an extended PRLS in humans. Since then, evidence for an analogous process of reproductive senescence in some species of cetaceans has been found, and therefore the model was amended slightly (Johnstone and Cant, 2010). The original model referred specifically to a patrilocal dispersal system, but cetaceans are not patrilocal. Due to difficulties in data collection, it is not completely clear what the social structures of cetaceans are. However, it is thought that they primarily display natal philopatry for both sexes, with non-local mating (Kasuya and Marsh 1984, Bigg et al. 1990, Heimlich-Boran 1993, O'Corry-Crowe et al. 1997, Palsbøll et al. 1997, Ellis et al. 2021). This social structure does not generate the reproductive conflict originally proposed to have caused menopause to evolve in humans, and therefore Johnstone and Cant (2010) amended the model accordingly. In humans, they modelled that patrilocality combined with local mating resulted in women becoming more related to their group with age, giving younger females (with low relatedness) the upper hand in reproductive competition. In philopatric cetaceans, individual female relatedness to other females in the group remains constant over time. However, relatedness to male group members increases with age, creating a relatedness structure where younger females should invest more in competition than older females. The reproductive conflict here is between co-breeding mothers and daughters. Empirical support for reproductive conflict in whales has been found, with a longitudinal data set from killer whales showing that the mortality hazard of progeny from the older generation of females is 1.7 times higher than that of progeny from younger females (Croft et al. 2017). This is consistent with younger females investing more in reproductive competition than older females, with the latter suffering a fitness cost as a result. It has also been observed that older females are more likely to offer help to male offspring, likely because they are not in direct reproductive competition with their sons (whose offspring reside outside the group). Therefore, they can increase their inclusive fitness via their sons by reducing the number of socially inflicted injuries they experience (Grimes et al. 2023) and passing on ecological knowledge (Brent et al. 2015) without incurring any costs via within-group competition (Foster et al. 2012).

Though data from cetaceans offers some support for the Reproductive Conflict Hypothesis, supporting data from human populations is lacking. Data from pre-industrial Finland and Gambia did show that simultaneous reproduction with in-laws is associated with a decline in offspring survivorship (Lahdenpera et al. 2012, Mace and Alvergne 2012). However, costs have been associated with co-breeding with many group members, not just in-laws. In a matrilineal population from western China, it was found that living with a sister was associated with decreased fertility (Ji et al. 2013); and, in Finland, the risk of offspring mortality as a juvenile increased by 23% if two co-residing women reproduced within two years of one another (Pettay et al. 2016). Furthermore, data from historical Norway found results converse to those predicted by the Reproductive Conflict Hypothesis, and showed that two women of different generations who co-bred with each other had greater reproductive success than those who did not, possibly because of learning and allocare opportunities that arise when co-breeding (Skjaervo and Roskaft 2013).

Researchers have also tested the Reproductive Conflict Hypothesis by looking at the age of menopause in contemporary societies, with the assumption that an earlier menopause will be experienced when there is inter-generational competition. This has been done in both western China (Yang et al. 2019) and Indonesia (Snopkowski et al. 2014), and it was predicted that patrilocal groups will have an earlier menopause than in other post-marital residence patterns. No evidence was found in support of this prediction in either country.

In addition to there being little empirical evidence for the Reproductive Conflict Hypothesis, it can also be brought into question because it is reliant upon the assumption that ancestral humans were patrilocal. Not only do we not know what the ancestral social structure of humans was, but we also do not know exactly at what point in our evolutionary history menopause evolved. Contemporary hunter-gatherer groups are often used as models of human evolution, but there is not one universal hunter-gatherer social structure, with cross-cultural research showing that both sexes are as likely to disperse (Hill et al. 2011). As a result, we cannot be sure whether the conditions required to create the age-specific levels of reproductive conflict were present in our evolutionary history. The cultural evolution of human kinship systems, post-marital residence and some other reproductive norms may be driven, at least in part, by the avoidance of reproductive and inter-generational conflict (Mace 2013, Ji et al. 2014). Ji et al. (2014) also note that, whereas in killer whales it is likely that reproductive conflict, if it occurs, occurs between females, in human societies it is possible that conflict between males might also be important, which could lead us to question why male–male inter-generational conflict does not result in male reproductive senescence.

Agent-based models have suggested that resource/reproductive conflict in combination with grandmothering might explain how menopause evolved (Thouzeau and Raymond 2017). However, the Grandmother Hypothesis and the Reproductive Conflict Hypothesis emphasize two different social systems. The latter hypothesis models that menopause evolved under a patrilocal residence pattern, while the former relies on the positive effect of grandmothers. As discussed earlier, much of the beneficial care comes from maternal grandmothers. If menopause evolved under a patrilocal residence pattern – as Cant and Johnstone (2008) suggest – then maternal grandmothers would not be there to help with alloparenting and get the inclusive fitness benefits required to offset the costs of being post-reproductive. Furthermore, patrilocality would mean that the paternal grandmother would be nearby, which under some conditions has been found to negatively affect grandchild health and survival. Any hypothesis that relies strongly on a particular form of sex-specific dispersal in our hunter-gatherer ancestors can be brought into question.

11.5.3 On Studying the Evolution of Menopause

A difficulty when researching menopause from an evolutionary perspective is that we are modelling the evolution (or maintenance) of a trait that has already evolved. Evolutionary anthropologists have previously used current variation in age at menopause to test evolutionary hypotheses relating to the emergence of menopause, with the assumption that an earlier menopause is a suitable proxy for the evolution of reproductive cessation. For instance, age at menopause has been used to empirically test the Reproductive Conflict Hypothesis (Cant and Johnstone 2008), in which an earlier age of menopause under a patrilocal residence pattern would be taken as evidence that reproductive conflict between generations may have resulted in selection for menopause in our evolutionary history (Snopkowski et al. 2014, Yang et al. 2019). There is some evidence that menopause timing varies optimally (Arnot and Mace 2020), meaning that menopause timing has been shaped by natural selection in a way that environmental variation triggers the production of the most optimal phenotype in that environment. However, the majority of research into variation in age of menopause is epidemiological and therefore not discussed from an evolutionary perspective. Even if age at menopause is plastic in a way that appears to be adaptive, it does not mean that this is indicative of how or why it evolved. There is a large body of evidence showing that a PRLS now has inclusive fitness benefits, and some evidence shows that individual-level variation in menopause timing can now be understood from an evolutionary framework. However, all the data we are using today is from populations that already have menopause and a PRLS. These studies show optimal variation of a trait within its usual range but do not show at all why the trait emerged in the first place. There are many cases in which the current function of a phenotype confers a fitness advantage, but the phenotype did not evolve as a result of natural selection for that function specifically. Fossil evidence suggests that the earliest feathers did not evolve for flying purposes but rather to assist in thermoregulation. Now many birds use their feathers for flight, and flight confers a fitness advantage, but feathers themselves are not an evolved adaptation for flight (Benton et al. 2019). As such, it could be that the prolonged period of infertility in females does appear to be adaptive today because it prevents reproductive conflict and increases the survival chances and health of younger kin. However, it does not mean that this is why the phenotype was selected for (or maintained) in the first place. At present, all we have is data from current (or historically recent) populations, and we should do with it the best we can. As there is evidence that menopause timing today varies in an optimal way, the age of menopause is still useful to use in evolutionary models if it is acknowledged that we are not necessarily modelling the evolutionary function of the trait, but whether the timing today optimally varies in a way that would be predicted by evolutionary theory.

11.6 Final Remarks

Life history evolution is hard to model, as changing one aspect of one life history trait shifts all the costs and benefits of others. Hence, for example, a costly childhood and menopause are all in the same evolutionary basket. Furthermore, models of human life history are hard to test because all extant humans have menopause, and costs and benefits today can hardly reflect the ancestral conditions in which these traits arose. On balance, we are minded to believe that human menopause evolved because some combination of grandmothering and mothering cannot be successful with too many costly human offspring, who need to learn how to survive and reproduce. Therefore, a non-fecund period of later life survival is important for these slow-growing, costly families and menopause assists with that (just as in killer whales, not dissimilar circumstances have also led to the evolution of a long PRLS). But that is not to say the case is closed. Significant challenges remain. There is not much evidence that grandmothers are heavily involved in childcare in hunter-gatherer populations, where menopause must have evolved (although they surely are in settled horticulturalists and farming populations). There is no definitive formal evolutionary model of the human menopause that all evolutionary anthropologists agree on. We are en route, but the evolution of human life history in general and menopause, in particular, is far from a closed book.

References

Aimé, C., Andre, J.B., and Raymond, M. (2017). Grandmothering and cognitive resources are required for the emergence of menopause and extensive post-reproductive lifespan. *PLoS Computational Biology* 13(7): e1005631.

Alberts, S.C., Altmann, J., Brockman, D.K., Cords, M., Fedigan, L.M., Pusey, A., Stoinski, T.S., Strier, K.B., Morris, W.F., and Bronikowski, A. M. (2013). Reproductive aging patterns in primates reveal that humans are distinct. *Proceedings of the National Academy of Sciences of the United States of America* 110(33): 13440–13445.

Alexander, R.D. (1974). The evolution of social behavior. *Annual Review of Ecology and Systematics* 5(1): 325–383.

Alvarez, H.P. (2000). Grandmother hypothesis and primate life histories. *American Journal of Physical Anthropology* 113(3): 435–450.

André, J.-B., and Rousset, F. (2020). Does extrinsic mortality accelerate the pace of life? A bare-bones approach. *Evolution and Human Behavior* 41(6): 486–492.

Arnot, M. (2021). The evolutionary ecology of menopause. PhD, University College London.

Arnot, M., and Mace, R. (2020). Sexual frequency is associated with age of natural menopause: results from the Study of Women's Health Across the Nation. *Royal Society Open Science* 7(1): 191020.

Arnot, M., and Mace, R. (2021). An evolutionary perspective on kin care directed up the generations. *Scientific Reports* 11(1): 14163.

Austad, S.N. (1997). Comparative aging and life histories in mammals. *Experimental Gerontology* 32(1–2): 23–38.

Bădescu, I., Watts, D.P., Katzenberg, M.A., and Sellen, D.W. (2016). Alloparenting is associated with reduced maternal lactation effort and faster weaning in wild chimpanzees. *Royal Society Open Science* 3(11): 160577.

Benton, M.J., Dhouailly, D., Jiang, B., and McNamara, M. (2019). The early origin of feathers. *Trends in Ecology & Evolution* 34(9): 856–869.

Berghänel, A., Stevens, J.M.G., Hohmann, G., Deschner, T., and Behringer, V. (2023). Adolescent length growth spurts in bonobos and other primates: mind the scale. *eLife* 12: RP86635.

Bigg, M.A., Olesiuk, P.F., Ellis, G.M., Ford, J.K.B., and Balcomb, K.C. (1990). Social organisation and geneaology of resident killer whales (*Orcinus orca*) in the coastal waters of British Columbia and Washington State. *Report of the International Whaling Commission* 12: 383–405.

Blurton Jones, N.G. (1986). Bushman birth Spacing – a test for optimal interbirth intervals. *Ethology and Sociobiology* 7(2): 91–105.

Blurton Jones, N.G. (2016). *Demography and evolutionary ecology of Hadza hunter-gatherers.* Cambridge University Press.

Boesch, C., and Boesch-Achermann, H. (2000). *The chimpanzees of the Tai forest: behavioural ecology and evolution.* Oxford University Press.

Bogin, B. (1994). Adolescence in evolutionary perspective. *Acta Paediatrica* 83: 29–35.
Bogin, B. (1997). Evolutionary hypotheses for human childhood. *Yearbook of Physical Anthropology* 40(25): 63–89.
Bogin, B. (1999). *Patterns of human growth*. Cambridge University Press.
Bogin, B. (2015). Human growth and development. In: *Basics in human evolution* (ed. Meuhlenbein, M.P.), 285–293. Academic Press.
Brent, L.J.N., Franks, D.W., Foster, E.A., Balcomb, K.C., Cant, M.A., and Croft, D.P. (2015). Ecological knowledge, leadership, and the evolution of menopause in killer whales. *Current Biology* 25(6): 746–750.
Breuer, T., Hockemba, M.B.N., Olejniczak, C., Parnell, R.J., and Stokes, E.J. (2009). Physical maturation, life-history classes and age estimates of free-ranging western gorillas – insights From Mbeli Bai, Republic of Congo. *American Journal of Primatology* 71(2): 106–119.
Bronikowski, A.M., Altmann, J., Brockman, D.K., Cords, M., Fedigan, L.M., Pusey, A., Stoinski, T., Morris, W.F., Strier, K.B., and Alberts, S.C. (2011). Aging in the natural world: comparative data reveal similar mortality patterns across primates. *Science* 331(6022): 1325–1328.
Buss, D.M. (1989). Sex differences in human mate preferences: evolutionary hypotheses tested in 37 cultures. *Behavioral and Brain Sciences* 12(1): 1–49.
Cant, M.A., and Johnstone, R.A. (2008). Reproductive conflict and the separation of reproductive generations in humans. *Proceedings of the National Academy of Sciences of the United States of America* 105(14): 5332–5336.
Cavalli-Sforza, L.L. (1986). *African pygmies*. Academic Press.
Cavazos-Rehg, P.A., Krauss, M.J., Spitznagel, E.L., Bommarito, K., Madden, T., Olsen, M.A., Subramaniam, H., Peipert, J.F., and Bierut, L.J. (2015). Maternal age and risk of labor and delivery complications. *Maternal and Child Health Journal* 19(6): 1202–1211.
Cerrito, P., Bailey, S.E., Hu, B., and Bromage, T.G. (2020). Parturitions, menopause and other physiological stressors are recorded in dental cementum microstructure. *Scientific Reports* 10(1): 5381.
Chapman, S.N., Jackson, J., Htut, W., Lummaa, V., and Lahdenpera, M. (2019). Asian elephants exhibit post-reproductive lifespans. *BMC Biology and Evolution* 19: 193.
Chapman, S.N., Lahdenperä, M., Pettay, J.E., and Lynch, R. (2021). Offspring fertility and grandchild survival enhanced by maternal grandmothers in a pre-industrial human society. *Scientific Reports* 11(1): 3652.
Charnov, E.L. (1993). *Life history invariants*. Oxford University Press.
Chisholm, J.S., Quinlivan, J.A., Petersen, R.W., and Coall, D.A. (2005). Early stress predicts age at menarche and first birth, adult attachment, and expected lifespan. *Human Nature* 16(3): 233–265.
Chung, E.O., Hagaman, A., LeMasters, K., Andrabi, N., Baranov, V., Bates, L.M., Gallis, J.A., O'Donnell, K., Rahman, A., Sikander, S., Turner, E.L., and Maselko, J. (2020). The contribution of grandmother involvement to child growth and development: an observational study in rural Pakistan. *BMJ Global Health* 5(8): e002181.
Conaway, C.H., and Koford, C.B. (1965). Estrous cycles and mating behavior in a free-ranging band of rhesus monkeys. *Journal of Mammalogy* 45(4): 577–588.
Copping, L.T., Campbell, A., and Muncer, S. (2014). Psychometrics and life history strategy: the structure and validity of the High K Strategy Scale. *Evolutionary Psychology* 12(1): 200–222.
Crittenden, A.N., and Marlowe, F.W. (2008). Allomaternal care among the Hadza of Tanzania. *Human Nature* 19(3): 249–262.
Croft, D.P., Brent, L.J., Franks, D.W., and Cant, M.A. (2015). The evolution of prolonged life after reproduction. *Trends in Ecology & Evolution* 30(7): 407–416.
Croft, D.P., Johnstone, R.A., Ellis, S., Nattrass, S., Franks, D.W., Brent, L.J., Mazzi, S., Balcomb, K.C., Ford, J.K., and Cant, M.A. (2017). Reproductive conflict and the evolution of menopause in killer whales. *Current Biology* 27(2): 298–304.
Csathó, A., and Birkás, B. (2018). Early-life stressors, personality development, and fast life strategies: an evolutionary perspective on malevolent personality features. *Frontiers in Psychology* 9: 305.
Czekala, N., and Robbins, M.M. (2001). Assessment of reproduction and stress through hormone analysis in gorillas. In: *Mountain gorillas: three decades of research at Karisoke* (eds. Robbins, M.M., Sicotte, P., and Stewart, K.J.), 317–340. Cambridge University Press.
de Vries, C., Galipaud, M., and Kokko, H. (2023). Extrinsic mortality and senescence: a guide for the perplexed. *Peer Community Journal* 3: e29.
DeSilva, J., and Lesnik, J. (2006). Chimpanzee neonatal brain size: implications for brain growth in *Homo erectus*. *Journal of Human Evolution* 51(2): 207–212.

Dolea, C., and AbouZhar, C. (2003). *Global burden of obstructed labour in the year 2000*. Evidence and Information for Policy, World Health Organization.

Dong, X., Milholland, B., and Vijg, J. (2017). Reply to: maximum human lifespan may increase to 125 years. *Nature* 546(7660): E21–E21.

Du, J., Page, A.E., and Mace, R. (2022). Grandpaternal care and child survival in a pastoralist society in western China. *Evolution and Human Behavior* 43(5): 358–366.

Eckardt, W., Fawcett, K., and Fletcher, A.W. (2016). Weaned age variation in the Virunga mountain gorillas (*Gorilla beringei beringei*): influential factors. *Behavioral Ecology and Sociobiology* 70(4): 493–507.

Ellis, S., Franks, D.W., Nattrass, S., Cant, M.A., Bradley, D.L., Giles, D., Balcomb, K.C., and Croft, D.P. (2018a). Postreproductive lifespans are rare in mammals. *Ecology and Evolution* 8(5): 2482–2494.

Ellis, S., Franks, D.W., Nattrass, S., Currie, T.E., Cant, M.A., Giles, D., Balcomb, K.C., and Croft, D.P. (2018b). Analyses of ovarian activity reveal repeated evolution of post-reproductive lifespans in toothed whales. *Scientific Reports* 8(1): 12833.

Ellis, S., Franks, D.W., Weiss, M.N., Cant, M.A., Domenici, P., Balcomb, K.C., Ellifrit, D.K., and Croft, D.P. (2021). Mixture models as a method for comparative sociality: social networks and demographic change in resident killer whales. *Behavioral Ecology and Sociobiology* 75: 75.

Emery Thompson, M., and Sabbi, K.H. (2019). Evolutionary demography of the great apes. In: *Human evolutionary demography* (eds. Burger, O., Lee, R., and Sear, R.), 1–70. Open Science Framework.

Emmott, E.H., and Page, A.E. (2019). Alloparenting. In: *Encyclopedia of evolutionary psychological science* (eds. Shackelford, T., and Weekes-Shackelford, V.), 210–223. Springer International Publishing.

Engelhardt, S.C., Bergeron, P., Gagnon, A., Dillon, L., and Pelletier, F. (2019). Using geographic distance as a potential proxy for help in the assessment of the grandmother hypothesis. *Current Biology* 29(4): 651–656.

Erwin, J.M., Hof, P.R., Ely, J.J., and Pearl, D.P. (2002). One gerontology: advancing understanding of aging through studies of great apes and other primates. In: *Aging in nonhuman primates* (eds. Erwin, J.M., and Hoff, P.R.), 1–21. Karger.

Figueredo, A.J., Vasquez, G., Brumbach, B.H., and Schneider, S.M. (2004). The heritability of life history strategy: the K-factor, covitality, and personality. *Social Biology* 51(3–4): 121–143.

Figueredo, A.J., Vasquez, G., Brumbach, B.H., and Schneider, S.M. (2007). The K-factor, covitality, and personality: a psychometric test of life history theory. *Human Nature* 18(1): 47–73.

Fitzpatrick, K., and Berbesque, J.C. (2018). Hunter-gatherer models in human evolution. In: *The international encyclopedia of anthropology* (ed. Callan, H.), 1–10. Wiley Blackwell.

Foster, E.A., Franks, D.W., Mazzi, S., Darden, S.K., Balcomb, K.C., Ford, J.K., and Croft, D.P. (2012). Adaptive prolonged postreproductive life span in killer whales. *Science* 337(6100): 1313.

Galdikas, B.M.F., and Wood, J.W. (1990). Birth spacing patterns in humans and apes. *American Journal of Physical Anthropology* 83(2): 185–191.

Gibson, M.A., and Mace, R. (2005). Helpful grandmothers in rural Ethiopia: a study of the effect of kin on child survival and growth. *Evolution and Human Behavior* 26(6): 469–482.

Gillespie, D.O.S., Russell, A.F., and Lummaa, V. (2013). The effect of maternal age and reproductive history on offspring survival and lifetime reproduction in preindustrial humans. *Evolution* 67(7): 1964–1974.

Goodall, J. (1986). *The chimpanzees of Gombe*. Harvard University Press.

Goodman, M.J., Estiokogriffin, A., Griffin, P.B., and Grove, J.S. (1985). Menarche, pregnancy, birth spacing and menopause among the Agta women foragers of Cagayan province, Luzon, the Philippines. *Annals of Human Biology* 12(2): 169–177.

Gould, K.G., Flint, M., and Graham, C.E. (1981). Chimpanzee reproductive senescence – a possible model for evolution of the menopause. *Maturitas* 3(2): 157–166.

Graham, C.E., Kling, O.R., and Steiner, R.A. (1979). Reproductive senescence in female nonhuman primates. In: *Ageing in nonhuman primates* (ed. Bowden, D.M.), 183–202. Van Nostrand Reinhold.

Grimes, C., Brent, L.J.N., Ellis, S., Weiss, M.N., Franks, D.W., Ellifrit, D.K., and Croft, D.P. (2023). Postreproductive female killer whales reduce socially inflicted injuries in their male offspring. *Current Biology* 33(15): 3250–3256.

Grueter, C.C., Hale, J., Jin, R.B., Judge, D., and Stoinski, T. (2019). Infant handling by female mountain gorillas: establishing its frequency, function, and (ir)relevance for life history evolution. *American Journal of Physical Anthropology* 168(4): 744–749.

Hackman, J., and Kramer, K.L. (2023). Kin networks and opportunities for reproductive cooperation and conflict among hunter-gatherers. *Philosophical Transactions of the Royal Society, B: Biological Sciences* 378(1868): 20210434.

Hamilton, W.D. (1964). The genetical evolution of social behaviour. *Journal of Theoretical Biology* 7(1): 1–16.

Harcourt, A.H., Fossey, D., Stewart, K.J., and Watts, D.H. (1980). Reproduction in wild gorillas and some comparison with chimpanzees. *Journal of Reproduction and Fertility. Supplement* 28(Supplement): 59–70.

Harvey, P.H., Martin, R.D., and Clutton-Brock, T.H. (1986). Life histories in comparative perspective. In: *Primate societies* (eds. Smuts, B.B., Cheney, D.L., Seyfarth, R.M., and Wrangham, R.W.), 181–196. Chicago University Press.

Hashimoto, C., Ryu, H., Mouri, K., Shimizu, K., Sakamaki, T., and Furuichi, T. (2022). Physical, behavioral, and hormonal changes in the resumption of sexual receptivity during postpartum infertility in female bonobos at Wamba. *Primates* 63(5): 109–121.

Hawkes, K., O'Connell, J.F., Jones, N.G., Alvarez, H., and Charnov, E.L. (1998). Grandmothering, menopause, and the evolution of human life histories. *Proceedings of the National Academy of Sciences of the United States of America* 95(3): 1336–1339.

Heimlich-Boran, J.R. (1993). Social organisation of the short-finned pilot whale, *Globicephala macrorhynchus*, with special reference to the comparative social ecology of delphanids. PhD Thesis, University of Cambridge.

Hewlett, B. (1991). *Intimate fathers: the nature and context of Aka pygmy paternal infant care.* The University of Michigan Press.

Hill, K. (1993). Life history theory and evolutionary anthropology. *Evolutionary Anthropology* 2(3): 78–88.

Hill, K., and Hurtado, A.M. (1996). *Ache life history: the ecology and demography of a foraging people.* Aldine de Gruyter.

Hill, K., Walker, R.S., Bozicevic, M., Eder, J., Headland, T., Hewlett, B., Hurtado, A.M., Marlowe, F.W., Wiessner, P., and Wood, B. (2011). Co-residence patterns in hunter-gatherer societies show unique human social structure. *Science* 331(6022): 1286–1289.

Hofer, M.K., Collins, H.K., Mishra, G.D., and Schaller, M. (2019). Do post-menopausal women provide more care to their kin? Evidence of grandparental caregiving from two large-scale national surveys. *Evolution and Human Behavior* 40(4): 355–364.

Howell, N. (2010). *Life histories of the Dobe !Kung: food, fatness, and well-being over the life-span.* University of California Press.

Ji, T., Wu, J.J., He, Q.Q., Xu, J.J., Mace, R., and Tao, Y. (2013). Reproductive competition between females in the matrilineal Mosuo of southwestern China. *Philosophical Transactions of the Royal Society, B: Biological Sciences* 368(1631): 20130081.

Ji, T., Xu, J.J., and Mace, R. (2014). Intergenerational and sibling conflict under patrilocality. A model of reproductive skew applied to human kinship. *Human Nature* 25(1): 66–79.

Johnstone, R.A., and Cant, M.A. (2010). The evolution of menopause in cetaceans and humans: the role of demography. *Proceedings of the Royal Society B: Biological Sciences* 277(1701): 3765–3771.

Jones, J.H. (2011). Primates and the evolution of long, slow life histories. *Current Biology* 21(18): R708–R717.

Kaplan, H. (1994). Evolutionary and wealth flows theories of fertility – empirical tests and new models. *Population and Development Review* 20(4): 753–791.

Kaplan, H., Hill, K., Lancaster, J., and Hurtado, A.M. (2000). A theory of human life history evolution: diet, intelligence, and longevity. *Evolutionary Anthropology* 9(4): 156–185.

Kasuya, T., and Marsh, H. (1984). Life history and reproductive biology of the short-finned pilot whale, *Globicephala macrorhynchus*, off the Pacific coast of Japan. *Report of the International Whaling Commission* 6: 259–310.

Kavanagh, P.S., and Kahl, B.L. (2018). Are expectations the missing link between life history strategies and psychopathology? *Frontiers in Psychology* 9: 89.

Kim, P.S., Coxworth, J.E., and Hawkes, K. (2012). Increased longevity evolves from grandmothering. *Proceedings of the Royal Society B: Biological Sciences* 279(1749): 4880–4884.

Kometani, A., and Ohtsubo, Y. (2022). Can impulsivity evolve in response to childhood environmental harshness? *Evolutionary Human Sciences* 4: e21.

Konner, M. (2005). Hunter-gatherer infancy and childhood. In: *Hunter-gatherer childhoods* (eds. Hewlett, B.S., and Lamb, M.E.), 19–64. Routledge.

Kramer, K.L. (2014). Why what juveniles do matters in the evolution of cooperative breeding. *Human Nature – An Interdisciplinary Biosocial Perspective* 25(1): 49–65.

Kuroda, S. (1989). Developmental retardation and behavioural characteristics of pygmy chimpanzees. In: *Understanding chimpanzees* (eds. Heltne, P.G., and Marquardt, L.A.), 184–193. Harvard University Press.

Kuzawa, C.W. (1998). Adipose tissue in human infancy and childhood: an evolutionary perspective. *Yearbook of Physical Anthropology* 41(S27): 177–209.

Lack, D. (1954). *The natural regulation of animal numbers.* Clarendon Press.

Lahdenpera, M., Gillespie, D.O.S., Lummaa, V., and Russell, A.F. (2012). Severe intergenerational reproductive conflict and the evolution of menopause. *Ecology Letters* 15(11): 1283–1290.

Lahdenpera, M., Lummaa, V., Helle, S., Tremblay, M., and Russell, A.F. (2004). Fitness benefits of prolonged post-reproductive lifespan in women. *Nature* 428(6979): 178–181.

Lahdenpera, M., Mar, K.U., and Lummaa, V. (2014). Reproductive cessation and post-reproductive lifespan in Asian elephants and pre-industrial humans. *Frontiers in Zoology* 11: 54.

Lahdenpera, M., Mar, K.U., and Lummaa, V. (2016). Nearby grandmother enchances calf survival and reproduction in Asian elephants. *Scientific Reports* 6: 27213.

Laisk, T., Tsuiko, O., Jatsenko, T., Horak, P., Otala, M., Lahdenpera, M., Lummaa, V., Tuuri, T., Salumets, A., and Tapanainen, J.S. (2019). Demographic and evolutionary trends in ovarian function and aging. *Human Reproduction Update* 25(1): 34–50.

Lansing, A.I. (1947). A transmissible, cumulative, and reversible factor in aging. *Journal of Gerontology* 2(3): 228–239.

Lawn, R.B., Sallis, H.M., Wootton, R.E., Taylor, A.E., Demange, P., Fraser, A., Penton-Voak, I.S., and Munafo, M.R. (2020). The effects of age at menarche and first sexual intercourse on reproductive and behavioural outcomes: a Mendelian randomization study. *PLoS ONE* 15(6): e0234488.

Lawson, D.W., and Mace, R. (2008). Sibling configuration and childhood growth in contemporary British families. *International Journal of Epidemiology* 37(6): 1408–1421.

Levitis, D.A., and Lackey, L.B. (2011). A measure for describing and comparing post-reproductive lifespan as a population trait. *Methods in Ecology and Evolution* 2(5): 446–453.

Lewis, I. (1999). Discrimination and access to health care: the case of nomadic forest hunter-gatherers in Africa. MSc, University of London.

Lindburg, D.G. (1971). The rhesus monkey in North India: an ecological and behavioural study. In: *Primate behaviour* (ed. Rosenblum, L.A.), 2–101. Academic Press.

Lonsdorf, E., Stanton, M., Pusey, A., and Murray, C. (2020). Sources of variation in weaned age among wild chimpanzees in Gombe National Park, Tanzania. *American Journal of Primatology* 171(3): 419–429.

Louis, F.J., Maubert, B., Lehesran, J.Y., Kemmegne, J., Delaporte, E., and Louis, J.P. (1994). High prevalence of anti-hepatitis-c virus-antibodies in a Cameroon rural forest area. *Transactions of the Royal Society of Tropical Medicine and Hygiene* 88(1): 53–54.

Mace, R. (1998). The co-evolution of human fertility and wealth inheritance strategies. *Philosophical Transactions of the Royal Society* 353: 389–397.

Mace, R. (2013). Cooperation and conflict between women in the family. *Evolutionary Anthropology: Issues, News, and Reviews* 22(5): 251–258.

Mace, R., and Alvergne, A. (2012). Female reproductive competition within families in rural Gambia. *Proceedings of the Royal Society B: Biological Sciences* 279(1736): 2219–2227.

Marlowe, F.W. (2005). Hunter-gatherers and human evolution. *Evolutionary Anthropology* 14(2): 54–67.

Mathes, E.W. (2018). Life history theory and tradeoffs: an obituary study. *Evolutionary Psychological Science* 4(4): 391–398.

Mededovic, J. (2020). On the incongruence between psychometric and psychosocial-biodemographic measures of life history. *Human Nature* 31(3): 341–360.

Migliano, A.B., Vinicius, L., and Lahr, M.M. (2007). Life history trade-offs explain the evolution of human pygmies. *Proceedings of the National Academy of Sciences of the United States of America* 104(51): 20216–20219.

Monaghan, P., and Ivimey-Cook, E.R. (2023). No time to die: evolution of a post-reproductive life stage. *Journal of Zoology* 321(1): 1–21.

Moorad, J., Promislow, D., and Silvertown, J. (2019). Evolutionary ecology of senescence and a reassessment of Williams' 'extrinsic mortality' hypothesis. *Trends in Ecology & Evolution* 34(6): 519–530.

Morton, R.A., Stone, J.R., and Singh, R.S. (2013). Mate choice and the origin of menopause. *PLoS Computational Biology*, 9(6): e1003092.

Muller, M.N., Thompson, M.E., and Wrangham, R.W. (2006). Male chimpanzees prefer mating with old females. *Current Biology* 16(22): 2234–2238.

Nattrass, S., Croft, D.P., Ellis, S., Cant, M.A., Weiss, M.N., Wright, B.M., Stredulinsky, E., Doniol-Valcroze, T., Ford, J.K.B., Balcomb, K.C., and Franks, D.W. (2019). Postreproductive killer whale grandmothers improve the survival of their grandoffspring. *Proceedings of the National Academy of Sciences of the United States of America* 116(52): 26669–26673.

Nenko, I., Chapman, S.N., Lahdenperä, M., Pettay, J.E., and Lummaa, V. (2020). Will granny save me? Birth status, survival, and the role of grandmothers in historical Finland. *Evolution and Human Behavior* 42(3): 239–246.

Nettle, D., Coall, D.A., and Dickins, T.E. (2010). Early-life conditions and age at first pregnancy in British women. *Proceedings of the Royal Society B: Biological Sciences* 278(1712): 1721–1727.

Nettle, D., and Frankenhuis, W.E. (2019). The evolution of life-history theory: a bibliometric analysis of an interdisciplinary research area. *Proceedings of the Royal Society B: Biological Sciences* 286(1899): 201900409.

Nettle, D., and Frankenhuis, W.E. (2020). Life-history theory in psychology and evolutionary biology: one research programme or two? *Philosophical Transactions of the Royal Society, B: Biological Sciences* 375(1803): 201904902.

Nishida, T., Corp, N., Hamai, M., Hasegawa, T., Hiraiwa-Hasegawa, M., Hosaka, K., Hunt, K.D., Itoh, N., Kawanaka, K., Matsumoto-Oda, A., Mitani, J.C., Nakamura, M., Norikoshi, K., Sakamaki, T., Turner, L., Uehara, S., and Zamma, K. (2003). Demography, female life history, and reproductive profiles among the chimpanzees of Mahale. *American Journal of Primatology* 59(3): 99–121.

Noireau, F., Carme, B., Apembet, J.D., and Gouteux, J.P. (1989). Loa loa and Mansonella perstans filariasis in the Chaillu mountains, Congo: parasitological prevalence. *Transactions of the Royal Society of Tropical Medicine and Hygiene* 83(4): 529–534.

Nowell, A.A., and Fletcher, A.W. (2007). Development of independence from the mother in Gorilla gorilla gorilla. *International Journal of Primatology* 28(2): 441–455.

O'Corry-Crowe, G.M., Suydam, R.S., Rosenberg, A., Frost, K.J., and Dizon, A.E. (1997). Phylogeography, population structure and dispersal patterns of the beluga whale *Delphinapterus leucas* in the western Nearctic revealed by mitochondrial DNA. *Molecular Ecology* 6(10): 970–995.

Olesiuk, P.F., Ellis, G.M., and Ford, J.K.B. (2005). Life history and population dynamics of northern resident killer whales (*Orcinus orca*) in British Columbia. Canadian Science Advisory Secretariat.

Page, A.E., Viguier, S., Dyble, M., Smith, D., Chaudhary, N., Salali, G.D., Thompson, J., Vinicius, L., Mace, R., and Migliano, A.B. (2016). Reproductive trade-offs in extant hunter-gatherers suggest adaptive mechanism for the Neolithic expansion. *Proceedings of the National Academy of Sciences of the United States of America* 113(17): 4694–4699.

Palsbøll, P.J., Heide-Jørgensen, M.P., and Dietz, R. (1997). Population structure and seasonal movements of narwhales, *Monodon monoceros*, determined from mtDNA analysis. *Heredity* 78(Pt 3): 284–292.

Pashos, A. (2000). Does paternal uncertainty explain discriminative grandparental solicitude? A cross-cultural study in Greece and Germany. *Evolution and Human Behavior* 21(2): 97–109.

Pavard, S., Metcalf, C.J.E., and Heyer, E. (2008). Senescence of reproduction may explain adaptive menopause in humans: a test of the "mother" hypothesis. *American Journal of Physical Anthropology* 136(2): 194–203.

Pavlicev, M., Romero, R., and Mitteroecker, P. (2020). Evolution of the human pelvis and obstructed labor: new explanations of an old obstetrical dilemma. *American Journal of Obstetrics and Gynecology* 222(1): 3–16.

Peccei, J.S. (2001). Menopause: adaptation or epiphenomenon? *Evolutionary Anthropology* 10(2): 43–57.

Perrin, W.F., Brownell, R.L.J., and DeMaster, D.P. (1984). Report on the workshop: reproduction in whales, dolphins and porpoises. Reports of the International Whaling Commission. Southwest Fisheries Science Center 1–28.

Peterson, M., Warf, B.C., and Schiff, S.J. (2018). Normative human brain volume growth. *Journal of Neurosurgery: Pediatrics* 21(5): 478–485.

Pettay, J.E., Lahdenpera, M., Rotkirch, A., and Lummaa, V. (2016). Costly reproductive competition between co-resident females in humans. *Behavioral Ecology* 27(6): 1601–1608.

Photopoulou, T., Ferreira, I.M., Best, P.B., Kasuya, T., and Marsh, H. (2017). Evidence for a postreproductive phase in female false killer whales Pseudorca crassidens. *Frontiers in Zoology* 14: 30.

Pollet, T.V., Nelissen, M., and Nettle, D. (2009). Lineage based differences in grandparental investment: evidence from a large British cohort study. *Journal of Biosocial Science* 41(3): 355–379.

Pope, S.K., Whiteside, L., Brooks-Gunn, J., Kelleher, K.J., Rickert, V.I., Bradley, R.H., and Casey, P.H. (1993). Low-birth-weight infants born to adolescent mothers. Effects of coresidency with grandmother on child development. *The Journal of the American Medical Association* 269(11): 1396–1400.

Portmann, A. (1951). *Biologische fragmente zu einer lehre vom menschen*. Benno Schwabe & Co.

Pretelli, I., Borgerhoff Mulder, M., and McElreath, R. (2022). Rates of ecological knowledge learning in Pemba, Tanzania: implications for childhood evolution. *Evolutionary Human Sciences* 4: e34.

Promislow, D.E.L., and Harvey, P.H. (1990). Living fast and dying young – a comparative-analysis of life-history variation among mammals. *Journal of Zoology* 220(3): 417–437.

Pusey, A.E. (1990). Behavioural changes at adolescence in chimpanzees. *Behaviour* 115(3/4): 203–246.

Ransom, T.W. (1981). *Beach troop of the Gombe*. Buckness University Press.

Rapaport, L.G., and Brown, G.R. (2008). Social influences on foraging behavior in young nonhuman primates: learning what, where, and how to eat. *Evolutionary Anthropology* 17(4): 189–201.

Robbins, M.M., Gray, M., Kagoda, E., and Robbins, A.M. (2009). Population dynamics of the Bwindi mountain gorillas. *Biological Conservation* 142(12): 2886–2895.

Robbins, M.M., and Robbins, A.M. (2021). Variability of weaning age in mountain gorillas (*Gorilla beringei beringei*). *American Journal of Physical Anthropology* 174(4): 776–784.

Robbins, A.M., Robbins, M.M., Gerald-Steklis, N., and Steklis, H.D. (2006). Age-related patterns of reproductive success among female mountain gorillas. *American Journal of Physical Anthropology* 131(4): 511–521.

Rogol, A.D., Clark, P.A., and Roemmich, J.N. (2000). Growth and pubertal development in children and adolescents: effects of diet and physical activity. *American Journal of Clinical Nutrition* 72(2): 521S–528S.

Sadhir, S., and Pontzer, H. (2023). Impact of energy availability and physical activity on variation in fertility across human populations. *Journal of Physiological Anthropology* 42: 1.

Say, L., Chou, D., Gemmill, A., Tuncalp, O., Moller, A.B., Daniels, J., Gulmezoglu, A.M., Temmerman, M., and Alkema, L. (2014). Global causes of maternal death: a WHO systematic analysis. *The Lancet Global Health* 2(6): E323–E333.

Schürmann, C. (1981). Courtship and mating behavior of wild orangutans in Sumatra. In: *Primate behavior and sociobiology* (eds. Chiarelli, A.B., and Corruccini, R.S.), 130–135. Springer.

Sear, R. (2008). Kin and child survival in rural Malawi – are matrilineal kin always beneficial in a matrilineal society? *Human Nature* 19(3): 277–293.

Sear, R. (2020). Do human 'life history strategies' exist? *Evolution and Human Behavior* 41(6): 513–526.

Sear, R., and Mace, R. (2008). Who keeps children alive? A review of the effects of kin on child survival. *Evolution and Human Behavior* 29(1): 1–18.

Sear, R., Mace, R., and McGregor, I.A. (2000). Maternal grandmothers improve nutritional status and survival of children in rural Gambia. *Proceedings of the Royal Society B: Biological Sciences*, 267(1453), 1641–1647.

Sear, R., Sheppard, P., and Coall, D.A. (2019). Cross-cultural evidence does not support universal acceleration of puberty in father-absent households. *Philosophical Transactions of the Royal Society, B: Biological Sciences* 374(1770): 20180124.

Shanley, D.P., and Kirkwood, T.B.L. (2001). Evolution of the human menopause. *BioEssays* 23(3): 282–287.

Shanley, D.P., Sear, R., Mace, R., and Kirkwood, T.B. (2007). Testing evolutionary theories of menopause. *Proceedings of the Royal Society B: Biological Sciences* 274(1628): 2943–2949.

Sheppard, P., and van Winkle, Z. (2020). Using sequence analysis to test if human life histories are coherent strategies. *Evolutionary Human Sciences* 2: e39.

Skjaervo, G.R., and Roskaft, E. (2013). Menopause: no support for an evolutionary explanation among historical Norwegians. *Experimental Gerontology* 48(4): 408–413.

Smith, T.M., Austin, C., Hinde, K., Vogel, E.R., and Arora, M. (2017). Cyclical nursing patterns in wild orangutans. *Science Advances* 3(5): e1601517.

Snopkowski, K., Moya, C., and Sear, R. (2014). A test of the intergenerational conflict model in Indonesia shows no evidence of earlier menopause in female-dispersing groups. *Proceedings of the Royal Society B: Biological Sciences* 281(1788): 20140580.

Stearns, S.C. (1992). *The evolution of life histories*. Oxford University Press.

Stearns, S.C., and Koella, J.C. (1986). The evolution of phenotypic plasticity in life-history traits – predictions of reaction norms for age and size at maturity. *Evolution* 40(5): 893–913.

Strassmann, B.I. (1997). The biology of menstruation in *Homo sapiens*: total lifetime menses, fecundity, and nonsynchrony in a natural-fertility population. *Current Anthropology* 38(1): 123–129.

Sugiyama, Y. (2004). Demographic parameters and life history of chimpanzees at Bossou, Guinea. *American Journal of Physical Anthropology* 124(2): 154–165.

Sykorova, K., and Flegr, J. (2021). Faster life history strategy manifests itself by lower age at menarche, higher sexual desire, and earlier reproduction in people with worse health. *Scientific Reports* 11(1): 11254.

Thomas, M.G., Shanley, D.P., Houston, A.I., McNamara, J.M., Mace, R., and Kirkwood, T.B. (2015). A dynamic framework for the study of optimal birth intervals reveals the importance of sibling competition and mortality risks. *Journal of Evolutionary Biology* 28(4): 885–895.

Thompson, M.E., Jones, J.H., Pusey, A.E., Brewer-Marsden, S., Goodall, J., Marsden, D., Matsuzawa, T., Nishida, T., Reynolds, V., Sugiyama, Y., and Wrangham, R.W. (2007). Aging and fertility patterns in wild chimpanzees provide insights into the evolution of menopause. *Current Biology* 17(24): 2150–2156.

Thouzeau, V., and Raymond, M. (2017). Emergence and maintenance of menopause in humans: a game theory model. *Journal of Theoretical Biology* 430: 229–236.

Tilson, R., Seal, U.S., Soemarna, K., Sumardja, E., Poniran, S., van Schaik, C.P., Leighton, M., Rijksen, H.D., and Eudey, A.A. (1993). Orangutan population and habitat viability analysis report. Orangutan population and habitat viability analysis worksop.

Towner, M.C., Nenko, I., and Walton, S.E. (2016). Why do women stop reproducing before menopause? A life-history approach to age at last birth. *Philosophical Transactions of the Royal Society, B: Biological Sciences* 371(1692): 20150147.

Uggla, C., and Mace, R. (2015). Effects of local extrinsic mortality rate, crime and sex ratio on preventable death in Northern Ireland. *Evolution, Medicine, and Public Health* 2015(1): 266–277.

Uggla, C., and Mace, R. (2016). Local ecology influences reproductive timing in Northern Ireland independently of individual wealth. *Behavioral Ecology* 27(1): 158–165.

van Noordwijk, M.A., Willems, E.P., Atmoko, S.S.U., Kuzawa, C.W., and van Schaik, C.P. (2013). Multi-year lactation and its consequences in Bornean orangutans (*Pongo pygmaeus wurmbii*). *Behavioral Ecology and Sociobiology* 67(5): 805–814.

Vasey, P.L., Pocock, D.S., and VanderLaan, D.P. (2007). Kin selection and male androphilia in Samoan *fa'afafine*. *Evolution and Human Behavior* 28(3): 159–167.

Walker, R., Gurven, M., Hill, K., Migliano, H., Chagnon, N., De Souza, R., Djurovic, G., Hames, R., Hurtado, A.M., Kaplan, H., Kramer, K., Oliver, W.J., Valeggia, C., and Yamauchi, T. (2006). Growth rates and life histories in twenty-two small-scale societies. *American Journal of Human Biology* 18(3): 295–311.

Walker, M.L., and Herndon, J.G. (2008). Menopause in nonhuman primates? *Biology of Reproduction* 79(3): 398–406.

Walker, P.L., and Hewlett, B.S. (1990). Dental-health diet and social-status among central African foragers and farmers. *American Anthropologist* 92(2): 383–398.

Walker, K.K., Walker, C.S., Goodall, J., and Pusey, A.E. (2018). Maturation is prolonged and variable in female chimpanzees. *Journal of Human Evolution* 114: 131–140.

Walter, K.V., Conroy-Beam, D., Buss, D.M., Asao, K., Sorokowska, A., Sorokowski, P., Aavik, T., Akello, G., Alhabahba, M.M., Alm, C., Amjad, N., Anjum, A., Atama, C.S., Duyar, D.A., Ayebare, R., Batres, C., Bendixen, M., Bensafia, A., Bizumic, B., Boussena, M., Butovskaya, M., Can, S., Cantarero, K., Carrier, A., Cetinkaya, H., Croy, I., Cueto, R.M., Czub, M., Dronova, D., Dural, S., Duyar, I., Ertugrul, B., Espinosa, A., Estevan, I., Esteves, C.S., Fang, L., Frackowiak, T., Garduno, J. C., Gonzalez, K.U., Guemaz, F., Gyuris, P., Halamova, M., Herak, I., Horvat, M., Hromatko, I., Hui, C.M., Jaafar, J.L., Jiang, F., Kafetsios, K., Kavcic, T., Kennair, L.E.O., Kervyn, N., Ha, T.T.K., Khilji, I.A., Kobis, N.C., Lan, H.M., Lang, A., Lennard, G.R., Leon, E., Lindholm, T., Linh, T.T., Lopez, G., Luot, N.V., Mailhos, A., Manesi, Z., Martinez, R., McKerchar, S.L., Mesko, N., Misra, G., Monaghan, C., Mora, E.C., Moya-Garofano, A., Musil, B., Natividade, J.C., Niemczyk, A., Nizharadze, G., Oberzaucher, E., Oleszkiewicz, A., Omar-Fauzee, M.S., Onyishi, I.E., Ozener, B., Pagani, A.F., Pakalniskiene, V., Parise, M., Pazhoohi, F., Pisanski, A., Pisanski, K., Ponciano, E., Popa, C., Prokop, P., Rizwan, M., Sainz, M., Salkicevic, S., Sargautyte, R., Sarmany-Schuller, I., Schmehl, S., Sharad, S., Siddiqui, R.S., Simonetti, F., Stoyanova, S.Y., Tadinac, M., Correa Varella, M A., Vauclair, CM., Vega, L.D., Widarini, D. A., Yoo, G., Zat'ková, M., and Zupančič, M. (2020). Sex differences in mate preferences across 45 countries: a large-scale replication. *Psychological Science* 31(4): 408–423.

Wells, J.C.K., Cole, T.J., Cortina-Borja, M., Sear, R., Leon, D.A., Marphatia, A.A., Murray, J., Wehrmeister, F.C., Oliveira, P.D., Goncalves, H., Oliveira, I.O., and Menezes, A.M.B. (2019). Low maternal capital predicts life history trade-offs in daughters: why adverse outcomes cluster in individuals. *Frontiers in Public Health* 7: 206.

Wich, S.A., Utami-Atmoko, S.S., Setia, T.M., Rijksen, H.D., Schurmann, C., and van Schaik, C. (2004). Life history of wild Sumatran orangutans (*Pongo abelii*). *Journal of Human Evolution* 47(6): 385–398.

Williams, G.C. (1957). Pleiotropy, natural-selection, and the evolution of senescence. *Evolution* 11(4): 398–411.

Wilson, M., and Daly, M. (1997). Life expectancy, economic inequality, homicide, and reproductive timing in Chicago neighbourhoods. *BMJ* 314(7089): 1271–1274.

Wittman, A.B., and Wall, L.L. (2007). The evolutionary origins of obstructed labor: bipedalism, encephalization, and the human obstetric dilemma. *Obstetrical & Gynecological Survey* 62(11): 739–748.

Wood, J.W., Holman, D.J., and O'Connor, K. A. (2001). The evolution of menopause by antagonistic pleiotropy. Center for Studies in Demography and Ecology Working Paper 01-04.

Wood, B.M., Negrey, J.D., Brown, J.L., Deschner, T., Thompson, M.E., Gunter, S., Mitani, J.C., Watts, D.P., and Langergraber, K.E. (2023). Demographic and hormonal evidence for menopause in wild chimpanzees. *Science* 382: eadd5473.

Wood, B.M., Watts, D.P., Mitani, J.C., and Langergraber, K.E. (2017). Favorable ecological circumstances promote life expectancy in chimpanzees similar to that of human hunter-gatherers. *Journal of Human Evolution* 105: 41–56.

World Health Organisation. (2005). Make every mother and child count. In: The world health report 2005. World Health Organisation.

Yang, Y., Arnot, M., and Mace, R. (2019). Current ecology, not ancestral dispersal patterns, influences menopause symptom severity. *Ecology and Evolution* 9(22): 12503–12514.

Part II

Interactions

12

Life History Traits in the Context of Predator–Prey Interactions

Joseph Travis

Department of Biological Science, Florida State University, Tallahassee, FL, USA

12.1 Introduction

No interaction unites the study of life history evolution with its ecological underpinnings like that of a predator with its prey. This coupling of evolutionary biology and ecology has deep roots in both disciplines. Adaptations to deter predation were among the first subjects explored by biologists captivated by Darwin's theory of natural selection (Endler 1986, Kimler and Ruse 2013). Ecologists have long probed the nature of the interaction, specifically whether predators are selective with respect to the developmental stage, age, body size or sex of potential prey (Slobodkin 1961). The subject of this chapter, the effort to understand the consequences of predation for the optimal life history traits of prey, can be traced at least to the insights of Schmalhausen (1949), who pointed out that predation pressure on adults should generate natural selection for earlier maturation and increased fertility.

The development of mathematical theory for the evolution of life histories catalysed the integration of predator–prey interactions with the study of life histories. That theory examined how different patterns of age-specific mortality could promote the evolution of different ages and body sizes at maturity, along with different patterns of age-specific reproductive effort (Gadgil and Bossert 1970, Law 1979, Michod 1979). To the ecologist familiar with the habits of different predators, the contrast between individuals experiencing different regimes of predation in different populations was an obvious opportunity for testing these theories.

The enormity of the literature in this area is testimony to the success of using predator–prey interactions to study life histories. The early empirical work designed to test life history theory described the covariation, across species or populations, between predation regimes and life histories, which appeared to match nicely with straightforward theoretical predictions (*e.g.,* Tinkle and Ballinger 1972). However, time would prove that theory would not always offer straightforward predictions and that, empirically, it would prove difficult to isolate the direct selective action of predators on the life history of prey.

In this chapter, I examine how the action of predators can drive the evolution of an optimal life history in the prey. I restrict my treatment to three parameters of a life history, juvenile growth rate, age and the size at maturity. Development patterns connect growth rate, age and size at maturity (see also Chapter 1, this volume) so these traits lend themselves to a unified treatment. They are also critical variables in the life history of harvested and managed species (see Chapter 23, this volume) so a close examination of how this suite of traits evolves in response to predation is of considerable applied importance.

I begin with a brief review of the types of predations important for the study of life history evolution. In the subsequent sections, I review theory and present several case studies of how predation affects the evolution of life histories, including adaptive phenotypic plasticity in life histories in response to perceived predation risk. In reviewing empirical work, I have emphasized studies from the large literature on marine and aquatic systems, which offer examples of every variety of predator–prey interaction. In the concluding section, I propose five directions for future research that can clarify questions that remain unresolved.

Life History Evolution: Traits, Interactions, and Applications, First Edition. Edited by Michal Segoli and Eric Wajnberg.
© 2025 John Wiley & Sons Ltd. Published 2025 by John Wiley & Sons Ltd.

12.2 Types of Predation

Predatory behaviour and attack patterns vary widely but, in weighing how predation might influence the evolution of life history traits, biologists have focused on three particular patterns.

First, some predators attack their prey without respect for developmental stage, age, size or sex. This can happen when an individual predator is much larger than the species on which it preys. For example, most of the large predatory fish in lower-elevation sections of Trinidadian streams are much larger than the guppies (*Poecilia reticulata*) and killifish (*Rivulus hartii*) on which they feed (Figure 12.1). Mark-recapture data indicate that those predators feed indiscriminately on their prey (Reznick et al. 1996, Reznick and Bryant 2007).

Second, a predator can act in size-limited fashion. A size-limited predator cannot subdue and consume an individual prey that is larger than some limiting size. For example, insect predators will readily prey upon small tadpoles but cannot physically handle larger ones (Travis et al. 1985, Cronin and Travis 1986). Crabs (*Carcinus* sp.) are adept at preying upon smaller mussels (*Mytilus* sp.) but are unable to break through the shells of larger ones (Kitching et al. 1959). This pattern is widespread. In fact, it is the dominant pattern of predation on juvenile fish (Persson et al. 1996, Sogard 1997). A size-limited predator preying upon immature, growing prey is especially interesting in the context of life history evolution. This is because the existence of a size refuge before or at metamorphosis or maturity sets the stage for the predator to be an agent of selection on growth rate, age and size at those life history transitions (Gleanson and Bengtson 1996, Allen 2008, Urban 2008, Furness and Reznick 2014).

Figure 12.1 Trinidadian guppies and some of their downstream predators in the Northern Range mountains of Trinidad. Upper left: Trinidadian guppies, *Poecilia reticulata*. Upper right: pike cichlid *Crenicichla alta*. Lower left: wolf fish, *Hoplias malabaricus*. Lower right: two-spot sardine, *Astyanax bimaculatus*. Source: Courtesy of Paul Bentzen.

Third, a predator can forage, attack and consume individuals selectively by prey body size, prey age or prey developmental stage. A predator might prey selectively on smaller or younger individuals because they are easier to capture. This is distinct from a predator's being unable to capture a prey individual larger than a particular size. Conversely, a predator might prey on larger individuals to maximize the return in energetic value for a given level of foraging effort and handling time. Natural size-selective predation of either smaller or larger individual prey is extremely common (Power et al. 1989, Wellborn 1994, Reichard et al. 2018, Santoyo-Brito et al. 2021).

Age-, stage- and size-specific predation can resemble one another but need not be the same phenomenon. The resemblance occurs because the three variables are usually associated with one another. Juveniles are usually smaller than adults and, in continuously growing individuals, smaller individuals are usually younger than larger individuals. Of course, in some species, large juveniles may overlap in body size with small adults and, if growth rates vary among individuals, age and size will not be perfectly correlated. While it can be difficult, empirically, to distinguish these three modes of predation, the distinction among them is important because, as I describe in the next section, theory for age-, stage- and size-specific predation need not lead to the same predictions.

The associations among age, stage and size can make it difficult to interpret observational data on predation unambiguously. In most cases, experiments are necessary to discern what, precisely, a selective predator is selecting. For example, an experiment manipulating the stage and size distributions of fish attacked by common egrets (*Ardea alba*) and snowy egrets (*Egretta thula*) demonstrated that the birds preyed selectively on adult fish but also on the larger of those adults (Trexler et al. 1994). Despite the difficulties, there are many other clear examples of age- or stage-specific predation (Richardson et al. 2006, Tatman et al. 2018, Pekkarinen et al. 2020).

12.3 Theory for Predator-Driven Life History Evolution

12.3.1 Overview

Most models for predator-driven life history evolution have a common protocol. First, build a model for an age-, stage- or size-structured population. The model includes, implicitly or explicitly, a cost of reproducing at one or more ages or stages that appear at a subsequent age or stage. This cost may be in either reduced survival or fecundity. In a model of a size-structured population, the cost of reproducing at a particular size is a subsequent reduction in growth rate. Most models also include a direct effect of body size on reproductive output, which sets up a potential trade-off between the advantages of early reproduction at a small size and the benefits of delayed reproduction that accrue through growing to a larger size before reproducing (see Chapter 1, this volume). Second, define a measure of fitness that will be maximized by the optimal life history strategy. Most models use either the innate growth rate of the model population, r_m, the net reproductive rate, R_0, or a quantity proportional to one or other of these measures. Third, examine how different patterns of age-, stage- or size-specific mortality, which are assumed to be imposed by different predation regimes, change the optimal values of life history components like growth rate, age and size at maturity. In many cases, the optimal life history can be found by examining the fitness landscape as a function of mortality rates (DeAngelis et al. 1985). In some cases, the optimal life history can be found only by identifying an evolutionarily stable strategy, meaning a life history strategy that cannot be invaded by any other strategy (Day et al. 2002, Gardmark et al. 2003).

Different models focus on different questions and include different additional details. First, some models examine age-specific mortality (Law 1979, Michod 1979), some, stage-specific mortality (Abrams and Rowe 1996) and others, size-specific mortality (Day et al. 2002, Taborsky et al. 2018). Second, some models allow continuous growth after maturity (Law 1979, Gardmark et al. 2003) while others specify growth to cease at maturity (Abrams and Rowe 1996, Taborsky et al. 2018). Taylor and Gabriel (1992) examined when selection should favour post-maturation growth.

Another distinction is between models that treat predation as a density-independent source of extrinsic mortality on the prey (Schaffer 1974, Law 1979, Michod 1979) and those that do not. The density-independent models treat the rate of extrinsic mortality as independent of the population density of the prey. The biological assumption implicit in these models is that there is negligible coupling between the rate of prey consumption and predator dynamics. This could apply either when a long-lived predator species feeds on many short-lived prey species or when a short-lived prey species is subject to predation from several omnivorous predator species.

Three types of models do not make this assumption. One class of such models retains density-independent predation rates but couples the population dynamics of the predator to its consumption of the prey (DeAngelis et al. 1985, Day et al. 2002, Gardmark et al. 2003). This coupling introduces feedback between predator and prey densities. A second class of models retain density-independent predation but couples the dynamics of the resources upon which the prey feed to the population dynamics of the prey (Abrams and Rowe 1996, Day et al. 2002). These models introduce feedback among the prey's resources, the population dynamics of the prey and the predator's consumption rates. A third class of model introduces density-dependent predation, in which predation rates decrease as prey density increases, using a type II functional response (Gardmark et al. 2003). This approach introduces feedback between consumption rates and prey densities, which could occur via either population dynamic responses of the predator or a behavioural response of predators. Gardmark et al. (2003) contrasted the effects of density-independent predation with those of density-dependent predation by incorporating different functional responses of predators to prey density but found no qualitative difference between the predictions from models with different functional responses.

These details are crucial for empirical work designed to test the predictions of theory. Predictions from models of age-specific mortality differ from those derived from models of stage-specific mortality. Models that link the dynamics of prey resources with the population dynamics of the prey or that couple the consumption rate of the predator with its own population dynamics can generate different predictions for the same pattern of age- or stage-specific mortality than those of simpler models. Moreover, as I describe below in the section on empirical evidence, the failure to disentangle the direct effects of predation from the indirect effects of resource levels can make it impossible to interpret empirical observations unambiguously.

12.3.2 Predictions from Theory

The situation most commonly invoked in discussions of predation and life history is when the predator acts as an agent of age-specific, density-independent mortality (Gadgil and Bossert 1970, Schaffer 1974, Law 1979, Michod 1979, Taylor and Gabriel 1992). Here, as extrinsic mortality in older age classes increases relative to mortality in younger age classes, selection favours a decrease in the age at maturity. In effect, it is advantageous for an individual to mature and reproduce before achieving an age at which it will encounter a higher risk of mortality. Older individuals could be more susceptible to predators if their behaviour or habitat use makes them more easily detected or more easily captured. By contrast, higher mortality in younger age classes favours an increase in the age at maturity, assuming that the gain in body size and fecundity is large enough to overcome the increased cumulative risk of mortality before maturing. Similar predictions emerge from simple models of size-dependent mortality (Roff 2001, 2002).

Theory predicts the opposite pattern when mortality is stage-specific (Abrams and Rowe 1996, Roff 2002): an increase in mortality in the adult stage, relative to the juvenile stage, will favour delayed maturation. In this case, the individual must mature in order to reproduce but, once it does, it encounters an increased risk of mortality. The best strategy is to delay maturity and grow much larger to obtain the benefit of increased fecundity once it reaches the adult stage. In the converse situation, an increase in mortality of the immature stage, relative to the adult stage, will favour an acceleration of maturation. Here, the best strategy is to escape the juvenile stage as soon as possible in order to capture the benefit of growing larger while in a stage with a lower risk of mortality.

When juveniles experience size-limited predation, meaning that predation is both stage- and size-specific (*i.e.*, predators are more likely to attack smaller juveniles), the intuitive prediction is that selection will favour a more rapid growth rate and/or accelerated maturation (Williams 1966). This is borne out in mathematical models of density-independent predation (DeAngelis et al. 1985, Day et al. 2002).

Theory for size-limited predation on adults not only can predict the evolution of earlier or later maturity, depending upon ecological details, but can also predict the emergence of a stable polymorphism for maturation age and size (Taborsky et al. 2018). The polymorphism occurs when the adult mortality rate decreases as body size increases, but there is a sharp drop in that rate at a particular body size, as could happen if predators are unable to prey upon an adult larger than some threshold body size. One strategy will be for earlier maturation at a size below the threshold and the other will be for later maturation at a size larger than the threshold. In the absence of a sharp drop in mortality rate, earlier or later maturation is favoured, depending upon how rapidly the mortality rate changes with body size. Two stable strategies can also emerge if adult mortality increases as body size increases and there is a sharp increase in the mortality rate at some body size, but the conditions promoting this result are very restrictive.

Some of these predictions are unchanged when predation is density-dependent or when the dynamics of the predator and/or prey resources are included in the mathematical models. For example, Coulson et al. (2022) showed that density-dependent juvenile mortality favoured a larger size at maturity, produced via delayed maturation, when the gain in adult body size and productivity overcame the cost of juvenile mortality, just as in models of density-independent predation. When the rate of juvenile mortality was very high and the gain in body size was insufficient to overcome the cost, early maturation at a smaller size was favoured.

The prediction that size-limited predation on juveniles will select for more rapid growth also holds in more complicated models when resources for the prey are abundant (Chase 1999) and there is no alternative escape from predation except to reach the size refuge (DeAngelis et al. 1985). On the other hand, when productivity is low, which extends appreciably the time needed to grow large enough to attain refuge from predator of larger size, the optimum life history is to mature early (Chase 1999). When there is an opportunity to invest in a defensive structure to deter predation, DeAngelis et al. (1985) show that rapid growth to attain the size of refuge is favoured only under low levels of predation. Higher levels of predation favour investing in the defensive structure rather than growing through the vulnerable stage.

Models that couple the population dynamics of predators with those of their prey can generate different predictions from those generated by models focused only on the demography of the prey. This happens because models of coupled dynamics include feedback among predator, prey and resources, which, in turn, create important indirect effects of predation (Abrams and Rowe 1996, Day et al. 2002, Gardmark et al. 2003).

The simplest indirect effect is that increased predation on juveniles will decrease their density, which can, in turn, increase the *per capita* level of resources fuelling juvenile growth. Abrams and Rowe (1996) examined the joint dynamics of prey and their resources and showed that when there is only predator-induced mortality, the direct effect of an increase in predation rate via increased predator density is to favour a decreased size at maturity. However, the resultant indirect effect of predators, which is an increase in *per capita* resources, favours maturing at a larger size. Which of these effects determines the net direction of selection depends heavily on the sensitivity of predator-induced mortality to the allocation of energy to growth. For example, if increased growth requires increased foraging and more exposure to predators, then the net effect will be to decrease the allocation of effort to growth to reduce risk, thereby making the direct effect of predation, which favours a smaller size at maturity, the predominant one. Alternatively, if increased growth can be achieved without increasing risk, then the indirect effect of increased resources will predominate, favouring a larger size at maturity. In the general context of this model, an increase in predation can produce almost any outcome, depending on the ecological details.

Day et al. (2002) expanded this theme with a model that included the dynamics of prey, resources and predator. They contrasted the direct and indirect effects between a system in which predator and prey densities were strongly coupled and one in which they were not (*i.e.*, consumption of prey did not affect predator dynamics). This contrast produced several divergent predictions. For example, when there is size-limited predation on juveniles, an increase in the density-independent mortality rate enhanced selection for rapid growth when there was weak coupling but not when there was strong coupling. This counter-intuitive result emerges because, with strong coupling, increased density-independent mortality reduced prey density and, ultimately, predator density as well. The reduced density of predators then relaxes selection for more rapid growth.

Embedding the prey within a larger trophic web can alter predictions from theory for the effects of age-specific harvesting in fisheries. Many fisheries biologists have attributed the long-term decrease in the age and size at maturity in hundreds of fishery stocks to an evolutionary response to harvesting older, larger individuals, as predicted by traditional theory for age-selective, density-independent predation, with harvesting being considered equivalent to size-selective predation (Heino et al. 2015, see also Chapter 23, this volume). When individuals in a harvested population are exposed to an additional source of mortality from density-dependent predation, the age at maturity can either decrease or increase, depending upon the age-specificity of the predator and the intensity of predation relative to the intensity of harvest (Gardmark et al. 2003). In addition, as discussed previously, the indirect effect of increased *per capita* resources may act to increase age and size at maturity, which could weaken the direct effect of harvesting.

The variety of possible predictions from theories for the evolution of growth rate, age and size at maturity reflects the rich possibilities of seemingly simple age-, stage- or size-specific effects of predators. This richness creates a substantial challenge for empirical studies. Assessing whether a given pattern in nature validates theoretical predictions requires the empiricist to decide which theory applies, which conditional factors are in play, and, perhaps most importantly, whether the direct effect of predation is actually the agent responsible. In the next section, I use selected case studies to evaluate how well we understand the connections between predation and life history evolution.

12.4 Empirical Evidence

The most straightforward way to test theoretical predictions is via a careful experiment that can incorporate the assumptions of a particular theory, test a specific prediction precisely and eliminate confounding factors. Whereas theory tells us what is possible, experiments reveal what is plausible.

There is a large body of laboratory work testing life history theory (Travis et al. 2023a). Here I focus on one artificial selection experiment and two laboratory natural selection experiments, following the terms defined in Fuller et al. (2005), who examined the direct role of size- or stage-specific mortality in driving rapid evolution in growth rate, age and size at maturity.

Edley and Law (1988) subjected experimental populations of the cladoceran, *Daphnia magna*, to either of two culling regimes, removal of small individuals or removal of large individuals. In this artificial selection experiment, they culled the animals by size without respect to age or stage (juvenile or reproductive). After approximately two dozen generations, populations subjected to the culling of smaller individuals evolved more rapid growth rates through the smallest size classes than those in the populations subjected to the culling of larger individuals, as predicted by theory for the response to size-limited predation. The evolution of rapid growth was limited to the size classes that were vulnerable to culling: individuals from the small cull populations in the larger size classes grew at comparable rates to individuals in the larger size classes in the large cull treatment. Individuals from the large cull treatment evolved to reproduce earlier and at a smaller size than those from the opposite treatment, again, in accord with predictions from age-specific or size-specific predation. In the experiment, there was a confounding of *per capita* resource levels with culling treatment but, for reasons explained in Edley and Law (1988), it likely was not sufficient to explain their results.

Spitze (1991) examined how experimental populations of *Daphnia pulex* evolved when exposed to 8–12 generations of stage-limited predation from the larvae of the midge *Chaoborus americanus*. The midge preys upon individuals in the earliest instars, with individuals in instars six and above invulnerable. Reproduction usually begins at or after the sixth instar and an individual can grow through 10–20 instars in their lifetime. In this experiment, individuals from populations in the *Chaoborus* treatment evolved more rapid growth rates and earlier maturation, when compared to controls without predation, matching the predictions from theory for stage-specific predation. This evolutionary change was a direct effect of predation. Spitze (1991) indeed took great pains to avoid confounding resource levels with predation treatment, adding algal cells on a regular, frequent basis to keep resource levels comparable between treatments and preclude an indirect effect of predation through resources.

Gasser et al. (2000) subjected experimental populations of *Drosophila melanogaster* to either high or low adult mortality, reporting results after 39 (low adult mortality) or 63 (high adult mortality) generations. Individuals from populations experiencing low adult mortality evolved longer development times and were larger at eclosion than individuals from populations experiencing high adult mortality. This result matches predictions from the theory for age-specific mortality. Just as in Spitze's (1991) work, there was no confounding of resource levels with the 'predation' treatment: adult densities were adjusted twice weekly and larval densities were maintained at the same levels in both treatments.

Whereas laboratory experiments tell us what is plausible, studies of natural populations demonstrate what has actually happened. There is a large literature on the associations, among species or populations, between variation in life histories and variation in ecological factors that might be selected for those life histories, such as population density or risk of predation (Roff 2002). There is also a history of scepticism about whether those associations reflect cause and effect. Stearns (1977) reviewed a substantial number of studies on many taxa, from invertebrates to mammals. The author concluded that very few studies have provided convincing evidence that the ecological factor associated with life history variation was actually promoting it (Tables 3–9 in Stearns 1977). Grainger and Levine (2022) performed a meta-analysis on 31 studies of associations between predation and life history. They found only 11 studies to have an effect size significantly different from zero. The only consistent result that they found was that increased predation on adults was associated with smaller body sizes at sexual maturity. While this result conforms to predictions from theory on age-specific predation, its usefulness as a test of theory must be approached with caution. The meta-analysis did not examine age at maturity, nor did it distinguish age-specific from stage-specific predation. Few of the studies used in the meta-analysis were able to separate direct from indirect effects of predators.

These critical assessments illustrate that testing theoretical predictions in natural populations is not an easy task. To be sure, it has not proven difficult to document associations among populations between the presence or absence of predators and life history trait variation in a prey (Crowl 1990, Wellborn 1994, Johnson and Belk 2001, Jennions and Telford 2002, Brandley et al. 2014). Others have documented associations between gradations of predation intensity and life history trait variation (Reznick and Endler 1982, Leibold and Tessier 1991, Urban 2008). In many cases (Reznick 1982, Leibold and Tessier 1991, Johnson 2001, Urban 2008), there are genetic bases to the phenotypic variation in life histories, so there is clearly an evolutionary phenomenon at hand. The question, however, is, what is the mechanism producing these associations? The answer requires disentangling the direct effects of predators from the indirect effects, to establish how much each contributes to the selection behind those population distinctions.

Two case studies illustrate how difficult it is to disentangle those effects. Urban (2008) examined larval growth in the salamander, *Ambystoma maculatum*, across a latitudinal gradient of predation intensity in southern New England, USA. Populations further south along the gradient live with a higher diversity of predators at higher predator densities than populations further north. More pointedly, populations further south have higher densities of two size-limited predators, the salamanders *A. opacum* and *Notophthalmus viridescens*. In nature, survival of *A. maculatum* larvae decreases as the average size of the two salamander predators increases (Urban 2007), implicating predation from these species as a major source of mortality. Because *A. maculatum* can grow into a size refuge from predation, Urban (2008), following the theory for the effects of size-limited predators, predicted that individuals from populations further south, which experience higher densities of size-limited predators, would grow faster than populations further north.

A common garden experiment verified that larval growth rates were highest in populations that experienced the higher densities of size-limited predators (Urban 2008). The question then became whether the different growth rates, which were genetically based, were the result of direct selection by predators or indirect effects. Field surveys indicated that the thermal environments of the natural ponds did not differ across the gradient so different growth rates were not a result of selection for thermal performance (Keen et al. 1984). It is unknown whether predation has a sufficiently large effect on resource availability to create an indirect effect that would be the stronger selection pressure for rapid growth. However, if this were the case, productivity would have to decrease from south to north to match the pattern of variation in growth rate. The lack of a thermal difference among the aquatic environments inhabited by the salamanders suggested no such gradient in productivity was likely. As a result, Urban (2008) concluded that differential predation was likely the real selective agent beneath the variation in growth patterns in these salamanders.

Detailed investigations pointed to the opposite conclusion for the elevational gradient in life histories documented in Trinidadian guppies, *P. reticulata* (Reznick and Travis 2019). Initial surveys (Reznick and Endler 1982) found that individuals from guppy populations living downstream with large, piscivorous predators (Figure 12.1) matured earlier and smaller than individuals from populations living upstream, where those predators are absent. This is the pattern expected from theory for age-specific predation. Common garden experiments indicated that these differences were genetically based (Reznick 1982). With these results in hand, guppies quickly became an emblematic example of predator-driven life history variation.

However, as Reznick and Endler (1982) pointed out, differential predation was not the only hypothesis for these differences. The later age and larger size at maturity in upstream populations could be the result of selection imposed by intraspecific competition for limited resources. The reasoning behind this hypothesis was that the release from heavy predation pressure creates an indirect effect of higher guppy density. Higher guppy density would, in turn, drive down the biomass of resources and create intense intraspecific competition. These higher-elevation populations would be density-regulated through competition, not limited by predation as in the downstream populations. Density-dependent selection, operating primarily through the effects of higher density on either juvenile survival or adult fecundity, would favour later age and larger size at maturity (Travis et al. 2023b).

Four lines of evidence indicate that the indirect effects of predators on guppy density and resources, not their direct effects on mortality rates, drive the differences between upstream and downstream populations in age and size at maturity. First, extensive field surveys demonstrated that upstream locations of guppies, which lack the large predators that inhabit downstream populations, have higher densities of guppies and lower standing crops of invertebrates and algae (Reznick et al. 2001). Guppies in the upstream populations have nutrient-poor diets of algae and detritus, in contrast with the richer diets of invertebrates observed in downstream populations (Zandona et al. 2011). These differences are consistent with the conditions that would generate density-dependent selection through resource limitation. Second, experimental manipulations of population densities in natural streams verified that low-predation populations experienced density-dependent dynamics (Reznick et al. 2012, Bassar et al. 2013). Third, laboratory and mesocosm experiments showed that low-predation guppies had greater competitive abilities than high-predation guppies (Potter et al. 2019) and were less sensitive to the depressant effects of density on vital rates (Bassar et al. 2013).

The fourth line of evidence emerged from recreating the historical pattern of guppy colonisation of upstream habitats by placing guppies from a downstream location with predators into four streams without guppies or predators (Travis et al. 2014).

Monthly censuses of these experimental populations over five years and annual common garden experiments demonstrated that delayed maturity in the male descendants of the transplanted fish did not begin evolving immediately after the transplanted populations were released from predation but only after the populations had grown to high densities (Reznick et al. 2019). In addition, in each of the four experimental populations, increased density was associated with decreased recruitment into the adult stage, not decreased adult survival (Travis et al. 2023b). This demographic effect of density is precisely what density-dependent selection theory requires for the evolution of later age and larger size at maturity.

The guppy story illustrates that the indirect effects of predation are more than a theoretical possibility. It is, however, not the only such illustration. Life history differences observed between populations of the killifish, *R. hartii*, that do or do not experience size-limited predation by guppies on their neonates, are maintained by the indirect effects of predation on resource availability (Walsh and Reznick 2008, 2011). Hulthen et al. (2021) demonstrated a remarkable result about the associations between predation regime and life history trait variation in the Bahamas mosquitofish, *Gambusia hubbsi*. Through a series of field observations and experiments, they inferred that divergent predation regimes were primarily responsible for variation in female reproductive rate, fecundity and offspring size, but variation in resource availability was responsible for variation in juvenile growth rate, male age and size at maturity.

In light of these results on indirect effects, it is fair to reconsider the well-documented associations between different predation regimes and variations in life history. This is not to argue that direct effects of predation are not important for those associations but to argue that, because indirect effects can mimic direct effects, the challenge for new work ought to be distinguishing those effects. One area in which the direct effects of predators are clearly important is the phenotypic plasticity of life history traits. There is a growing body of evidence that predation risk has driven the evolution of adaptive plasticity. In the next section, I review some of that evidence.

12.5 Adaptive Plasticity in Life Histories

Phenotypic plasticity is the ability of a genotype to express different phenotypes in different environments (Travis 2023). Adaptive plasticity, the ability of a genotype to express a different phenotype with high fitness in different environments, evolves when four conditions prevail. First, natural selection must favour different optimum values of traits in different environments. Second, those environments must vary with sufficient frequency either across generations or across spatial patches. Third, there must be a reliable cue for each environment that an individual can perceive in time to develop the appropriate phenotype. Fourth, there must be genetic variation for phenotypic plasticity such that the raw material to respond to fluctuating selection is present.

Each of those conditions is derived from mathematical theory, which specifies 'how different' optimum values must be, what 'sufficient frequency' of variation means, how 'reliable' the cue must be, and what constitutes an appropriate level of genetic variation for phenotypic plasticity. The theory also includes a cost to plasticity, meaning that a plastically produced phenotype will not be at the exact optimum value in every environment, even if it is 'close' to the optimum in most environments. These theories are reviewed elsewhere (*e.g.,* Murren et al. 2015, Scheiner 2019). The important point is that when the action of predators makes the optimum life history different from what it would be without predators, when the risk of predation is variable in time or space and when there is a reliable cue about the risk of predation, adaptive plasticity in the expression of the life history can be favoured.

Of course, demonstrating adaptive plasticity in life history in response to predation risk is a harder task than postulating it. A compelling demonstration must address all four of the required conditions listed above. As one might expect, few studies have investigated adaptive plasticity for any type of trait at this level of detail. In fact, some reviews have suggested that the prevalence of adaptive plasticity in general has been overestimated (Palacio-Lopez et al. 2015). Nonetheless, there are enough case studies to justify concluding that plasticity in life history expression can be an important adaptation to predation risk when risk varies spatially or temporally.

Studies of *Daphnia* provide excellent examples. Several studies on different species have demonstrated plasticity for both accelerated and delayed maturity, depending upon the action of the particular predator. While, as discussed earlier, larvae of the midge *C. americanus* are size-limited, which would favour delayed maturation, planktivorous fish feed selectively on larger *Daphnia* individuals, which would favour earlier maturation. When individual *D. galeata* were exposed to water that had housed one such planktivorous fish, the roach *Rutilus rutilus*, the individual *D. galeata* grew more slowly and matured in the fourth instar (Machacek 1991). Individuals in the control group matured in the fifth instar. A similar experiment with individual *D. hyalina* showed that individuals exposed to water that had housed the planktivorous fish, the ide *Leuciscus idus*, matured earlier and smaller than their compatriots in the control treatment (Stibor 1992). Pijanowska et al. (2022) showed that individual *D. magna* exposed to fish kairomones grew more slowly and matured earlier than individuals in a control treatment. By contrast, Spitze (1992) showed that individual *D. pulex* exposed to water inhabited by *C. americanus* larvae developed thicker carapaces and prominent neck teeth, both deterrents to predation (Krueger and Dodson 1981, Havel and Dodson 1984), along with delayed maturity, compared with individuals in the control treatment. Tollrian (1995) found that individual *D. pulex* exposed to concentrations of *Chaoborus flavicans* kairomones showed the same patterns. With clever experimentation, the author was also able to show that the delay in maturity was independent of the development of the carapace and neck teeth.

Similar plastic responses, in which the altered expression of life history traits is in the direction predicted by theory, are well-documented in many other species (Crowl and Covich 1990, Johnson 2001, Urban 2008, Bell et al. 2011, Gale et al. 2013, Raczynski et al. 2022). As in *Daphnia*, in which age at maturity is either accelerated or delayed, depending upon the specific predator to which the animals are exposed, so, too, have Trinidadian guppies been shown to respond differently to different predators. Individuals exposed to water containing kairomones from the large-bodied, non-selective predator *Crenicichla alta*, and alarm pheromones from other guppies, matured at smaller sizes than individuals in control treatments (Torres-Dowdall et al. 2012). Guppy growth patterns are such that smaller maturity is likely associated with earlier maturity. By contrast, male guppies exposed to visual and chemical cues from the size-limited predator *R. hartii*, delayed maturation and matured at a larger size than males in the control treatment (Gosline and Rodd 2008).

Some of these striking results raise an uncomfortable question: why do we see these patterns of plasticity? To be sure, in these cases, the directions of plasticity are pleasing validations of theoretical predictions. The problem is at a deeper level. Adaptive plasticity is an evolutionary response to selection in a variable environment in which different trait values are favoured in different environments. There is no doubt that an individual *Daphnia* should mature earlier when at risk of predation from fish that will consume older, larger individuals and later, at a larger size, in the absence of that predation risk. The question is whether that predation risk is sufficiently variable to promote the evolution of adaptive plasticity. If a high risk of fish predation is a constant fact of *Daphnia* life, we might expect selection to favour constitutive expression of earlier maturity, presuming, as theory does, that there is a cost to plasticity such that a constitutively expressed phenotype has a higher fitness in the presence of fish than one expressed via plasticity. That assumption of a cost also implies that if the developmental and physiological machinery necessary to express later maturation is expensive but never needed, selection should act to remove it. Yet, these empirical studies indicate that populations retain that machinery, even when predation risk appears to vary little from one generation to the next. Either there is more variation in the risk of fish predation than is often assumed or, as Murren et al. (2015) suggest, the cost of plasticity is much less than is often assumed.

A negligible cost to plasticity would explain problematic features of other case studies. For example, Crowl and Covich (1990) described remarkable life history plasticity in response to the risk of size-limited crayfish predation in the snail *Physella virgata*. However, the slower growth and delayed reproduction observed when they exposed snails to cues for the risk of predation did not differ between individuals from populations living with crayfish and those from populations that do not encounter crayfish. If plasticity is costly, then the machinery for it should be lost in populations that never encounter crayfish.

A similar phenomenon was observed in guppies in response to the risk of *C. alta* predation. The plastic response in growth rate and age at maturity did not differ between individuals from populations seemingly always exposed to those predators and individuals from populations never exposed to them (Torres-Dowdall et al. 2012). This case is more puzzling. It is certainly possible that the populations exposed to the larger predators experience variable intensities of predation pressure from one generation to another, which would favour the retention of the plastic response. However, the predator-free populations occur upstream above barrier waterfalls that stop the upstream movement of the larger predators, so they are certainly never exposed to those predators. Either the cost of retaining the plastic response in these populations is negligible or there is enough upstream migration of guppies to steadily introduce the genes for the plastic response into a population in which they confer no benefit.

Of course, there may be a significant cost to plasticity and predation risk may be much more variable than it appears in the systems described in these studies. Most studies quantify fitness of alternative phenotypes in different environments and verify the reliability of the cue. They fail to quantify the degree to which the supposedly variable environment really is variable (Burgess and Marshall 2014). As a result, we cannot know if the retention of plasticity where it appears unnecessary is a contradiction of theory or a reflection of insufficient ecological information. This issue deserves closer attention.

12.6 Future Directions

Five questions loom large as directions for future research. First, what is the relative contribution of direct and indirect effects of predators to the associations between life history variation and the intensity of predation? On the theoretical side of this question, there is a need for developing more theories for the effects of predators when those effects are embedded in realistic food webs, especially when predation is density-dependent (Stearns, 2000). On the empirical side of the question, there is a need for more work that will distinguish the direct effects of predators from their indirect effects. The laboratory studies of Spitze (1992) and Gasser et al. (2000) examined direct effects of predation while controlling resource levels. The horizon now ought to be examining how direct and indirect effects combine to influence trait evolution (Hulthen et al. 2021, Pijanowska et al. 2022).

Second, how do life histories evolve when the interaction between predator and prey depends upon the body sizes of individuals in both populations? Our models for life history evolution in the context of predator–prey interactions are, for the most part, models of density-independent mortality in a structured prey population or models of unstructured predator populations preying upon structured prey populations. In many predator–prey systems, attack rates and functional responses depend upon the relative sizes of predator and prey (Travis et al. 1985, Semlitsch and Gibbons 1988, Jara 2008, Fonseca et al. 2018). The numerical dynamics when two structured populations interact can be quite complicated (Persson et al. 1998, Bassar et al. 2017, de Roos 2021). Whether these complications introduce novel results for life history evolution, as they do for questions of species coexistence, remains to be determined. On the empirical side, new methods for quantifying size-structured interactions that are easily integrated into theory (Bassar et al. 2017, Anaya-Rojas et al. 2021) make it possible to test predictions from new theory in a wider range of natural populations than is covered by the existing literature.

Third, why does life history plasticity persist where theory suggests it should not? This is a more general problem about plasticity, beyond that appearing in predator–prey interactions. Nonetheless, some of the most salient examples of the phenomenon are examples of seemingly adaptive plastic responses by prey to the risk of predation. These are ideal empirical models with which to attack this neglected problem.

Fourth, does adaptive life history plasticity in response to the risk of predation extend across generations? There is a rapidly growing literature on transgenerational plasticity, which occurs when the environmental experience of a parent influences the phenotype of its offspring (reviewed in Levell et al. 2023). This literature includes many cases in which the stress caused in a parent by its perception of predation risk affects the phenotype of its offspring (e.g., McGhee et al. 2021). However, it is unclear how much of this effect is adaptive. The meta-analysis of MacLeod et al. (2022) revealed no consistency in how transgenerational plasticity in response to predation risk influenced specific phenotypic traits in offspring. More research can clarify if this inconsistency reflects differences in how the plasticity of different types of traits is controlled via epigenetic mechanisms (e.g., differences between life history traits and behavioural traits) or whether there is genuine idiosyncrasy among populations and species in the evolution of transgenerational plasticity.

Fifth, do different types of traits that exhibit predator-induced plasticity share a common epigenetic control? Many cases of predator-induced plasticity involve plastic changes in traits in addition to life history traits, like the concomitant delay in maturity and thicker carapace expressed by *D. pulex* in response to *C. flavicans* kairomones (Tollrian 1995). An obvious question is whether there is independent epigenetic control of the plasticity in each trait, which would suggest independent evolution of adaptive plasticity. Tollrian's (1995) experiments indicated that this was the case. It is unknown whether this result is a general one.

One avenue for pursuing the problem is through transcriptomics, or the analysis of mRNA concentrations in different conditions. A transcriptomic analysis of predator-induced plasticity in *D. pulex* revealed that exposure to predation risk increased the expression of genes involved in cuticle formation as well as proteins required for resource reallocation (Rozenberg et al. 2015). The study of the epigenetic control of adaptive plasticity is growing rapidly (Fraser et al. 2014, Ghalambor et al. 2015, Tills et al. 2018, Szyf 2021) and this approach can illuminate whether traits that respond jointly to a stimulus like exposure to predation risk share overlapping networks of gene expression and control. This knowledge would enable new tests of hypotheses about plasticity. In particular, it would enable a test of whether the epigenetic control of traits involved in alternative routes to escape predation forces a trade-off between them, as predicted by theory (DeAngelis et al. 1985).

Acknowledgements

I am grateful to Drs. Michael Cortez, Anja Felmy, Alirio Rosales and the editors for comments on a previous draft that led to a much-improved manuscript. My studies of predator–prey interactions and life histories have been made possible by funds provided by the United States National Science Foundation and the Florida State University. I acknowledge current support from the National Science Foundation award DEB 2100163.

References

Abrams, P.A. and Rowe, L. (1996). The effects of predation on the age and size of maturity of prey. *Evolution* 50: 1052–1061.

Allen, J.D. (2008). Size-specific predation on marine invertebrate larvae. *Biological Bulletin* 214: 42–49.

Anaya-Rojas, J.M., Bassar, R.D., Potter, T. et al. (2021). The evolution of size-dependent competitive interactions promotes species coexistence. *Journal of Animal Ecology* 90: 2704–2717.

Bassar, R.D., Lopez-Sepulcre, A., Reznick, D.N., and Travis, J. (2013). Experimental evidence for density-dependent regulation and selection on Trinidadian guppy life histories. *The American Naturalist* 181: 25–38.

Bassar, R.D., Travis, J., and Coulson, T. (2017). Predicting coexistence in species with continuous ontogenetic niche shifts and competitive asymmetry. *Ecology* 98: 2823–2836.

Bell, A.M., Dingemanse, N.J., Hankison, S.J. et al. (2011). Early exposure to nonlethal predation risk by size-selective predators increases somatic growth and decreases size at adulthood in threespined sticklebacks. *Journal of Evolutionary Biology* 24: 943–953.

Brandley, M.C., Kuriyama, T., and Hasegawa, M. (2014). Snake and bird predation drive the repeated convergent evolution of correlated life-history traits and phenotype in the Izu Island Scincid lizard (*Plestiodon latiscutatus*). *PLoS ONE* 9 (3): e92233.

Burgess, S.C. and Marshall, D.J. (2014). Adaptive parental effects: the importance of estimating environmental predictability and offspring fitness appropriately. *Oikos* 123: 769–776.

Chase, J.M. (1999). To grow or to reproduce? The role of life-history plasticity in food web dynamics. *The American Naturalist* 154: 571–586.

Coulson, T., Felmy, A., Potter, T. et al. (2022). Density-dependent environments can select for extremes of body size. *Peer Community in Evolutionary Biology* 2: e49.

Cronin, J.T. and Travis, J. (1986). Size-limited predation on larva *Rana areolata* (Anura, Ranidae) by two species of backswimmer (Insecta, Hemiptera, Notonectidae). *Herpetologica* 42: 171–174.

Crowl, T.A. (1990). Life-history strategies of a fresh-water snail in response to stream permanence and predation – balancing conflicting demands. *Oecologia* 84: 238–243.

Crowl, T.A. and Covich, A.P. (1990). Predator-induced life-history shifts in a fresh-water snail. *Science* 247: 949–951.

Day, T., Abrams, P.A., and Chase, J.M. (2002). The role of size-specific predation in the evolution and diversification of prey life histories. *Evolution* 56: 877–887.

de Roos, A.M. (2021). Dynamic population stage structure due to juvenile-adult asymmetry stabilizes complex ecological communities. *Proceedings of the National Academy of Sciences of the United States of America* 118: e2023709118.

DeAngelis, D.L., Kitchell, J.A., and Post, W.M. (1985). The influence of naticid predation on evolutionary strategies of bivalve prey – conclusions from a model. *The American Naturalist* 126: 817–842.

Edley, M.T. and Law, R. (1988). Evolution of life histories and yields in experimental populations of *Daphnia magna*. *Biological Journal of the Linnean Society* 34: 309–326.

Endler, J.A. (1986). Defense against predators. In: *Predator-prey relationships: perspectives and approaches from the study of lower vertebrates* (ed. M.E. Feder and G.V. Lauder), 109–134. University of Chicago Press.

Fonseca, M.M., Pallini, A., Lima, E., and Janssen, A. (2018). Ontogenetic stage-specific reciprocal intraguild predation. *Oecologia* 188: 743–751.

Fraser, B.A., Janowitz, I., Thairu, M. et al. (2014). Phenotypic and genomic plasticity of alternative male reproductive tactics in sailfin mollies. *Proceedings of the Royal Society B: Biological Sciences* 281: 20132310.

Fuller, R.C., Baer, C.F., and Travis, J. (2005). How and when selection experiments might actually be useful. *Integrative and Comparative Biology* 45: 391–404.

Furness, A.I. and Reznick, D.N. (2014). The comparative ecology of a killifish (*Rivulus hartii*) across aquatic communities differing in predation intensity. *Evolutionary Ecology Research* 16: 249–265.

Gadgil, M. and Bossert, W.H. (1970). Life historical consequences of natural selection. *The American Naturalist* 104: 1–24.

Gale, B.H., Johnson, J.B., Schaalje, G.B., and Belk, M.C. (2013). Effects of predation environment and food availability on somatic growth in the livebearing fish *Brachyrhaphis rhabdophora* (Pisces: Poeciliidae). *Ecology and Evolution* 3: 326–333.

Gardmark, A., Dieckmann, U., and Lundberg, P. (2003). Life-history evolution in harvested populations: the role of natural predation. *Evolutionary Ecology Research* 5: 239–257.

Gasser, M., Kaiser, M., Berrigan, D., and Stearns, S.C. (2000). Life-history correlates of evolution under high and low adult mortality. *Evolution* 54: 1260–1272.

Ghalambor, C.K., Hoke, K.L., Ruell, E.W. et al. (2015). Non-adaptive plasticity potentiates rapid adaptive evolution of gene expression in nature. *Nature* 525: 372–375.

Gleason, T.R. and Bengston, D.A. (1996). Growth, survival and size-selective predation mortality of larval and juvenile inland silversides, *Menidia beryllina* (Pisces; Atherinidae). *Journal of Experimental Marine Biology and Ecology* 199: 165–177.

Gosline, A.K. and Rodd, F.H. (2008). Predator-induced plasticity in guppy (*Poecilia reticulata*) life history traits. *Aquatic Ecology* 42: 693–699.

Grainger, T.N. and Levine, J.M. (2022). Rapid evolution of life-history traits in response to warming, predation and competition: a meta-analysis. *Ecology Letters* 25: 541–554.

Havel, J.E. and Dodson, S.I. (1984). *Chaoborus* predation on typical and spined morphs of *Daphnia pulex* – behavioral observations. *Limnology and Oceanography* 29: 487–494.

Heino, M., Diaz Pauli, B., and Dieckmann, U. (2015). Fisheries-induced evolution. *Annual Review of Ecology, Evolution, and Systematics* 46: 461–480.

Hulthen, K., Hill, J.S., Jenkins, M.R., and Langerhans, R.B. (2021). Predation and resource availability interact to drive life-history evolution in an adaptive radiation of livebearing fish. *Frontiers in Ecology and Evolution* 9: 619277.

Jara, F.G. (2008). Tadpole-odonate larvae interactions: influence of body size and diel rhythm. *Aquatic Ecology* 42: 503–509.

Jennions, M.D. and Telford, S.R. (2002). Life-history phenotypes in populations of *Brachyrhaphis episcopi* (Poeciliidae) with different predator communities. *Oecologia* 132: 44–50.

Johnson, J.B. (2001). Adaptive life-history evolution in the livebearing fish *Brachyrhaphis rhabdophora*: genetic basis for parallel divergence in age and size at maturity and a test of predator-induced plasticity. *Evolution* 55: 1486–1491.

Johnson, J.B. and Belk, M.C. (2001). Predation environment predicts divergent life-history phenotypes among populations of the livebearing fish *Brachyrhaphis rhabdophora*. *Oecologia* 126: 142–149.

Keen, W.H., Travis, J., and Juilianna, J. (1984). Larval growth in three sympatric *Ambystoma* species: species differences and the effects of temperature. *Canadian Journal of Zoology* 62: 1043–1047.

Kimler, W. and Ruse, M. (2013). Mimicry and camouflage. In: *The Cambridge encyclopedia of Darwin and evolutionary thought* (ed. M. Ruse), 139–145. Cambridge University Press.

Kitching, J.A., Sloane, J.F., and Ebling, F.J. (1959). The ecology of Louch Ine. 8. Mussels and their predators. *Journal of Animal Ecology* 28: 331–341.

Krueger, D.A. and Dodson, S.I. (1981). Embryological induction and predation ecology in *Daphnia pulex*. *Limnology and Oceanography* 26: 219–223.

Law, R. (1979). Optimal life histories under age-specific predation. *The American Naturalist* 114: 399–417.

Leibold, M. and Tessier, A.J. (1991). Contrasting patterns of body size for *Daphnia* species that segregate by habitat. *Oecologia* 86: 342–348.

Levell, S.T., Bedgood, S.A., and Travis, J. (2023). Plastic maternal effects of social density on reproduction and fitness in the least killifish, *Heterandria formosa*. *Ecology and Evolution* 13: e10074.

Machacek, J. (1991). Indirect effect of planktivorous fish on the growth and reproduction of *Daphnia galeata*. *Hydrobiologia* 225: 193–197.

Macleod, K.J., Monestier, C., Ferrari, M.C.O. et al. (2022). Predator-induced transgenerational plasticity in animals: a meta-analysis. *Oecologia* 200: 371–383.

McGhee, K.E., Barbosa, A.J., Bissell, K. et al. (2021). Maternal stress during pregnancy affects activity, exploration and potential dispersal of daughters in an invasive fish. *Animal Behaviour* 171: 41–50.

Michod, R.E. (1979). Evolution of life histories in response to age-specific mortality factors. *The American Naturalist* 113: 531–550.

Murren, C.J., Auld, J.R., Callahan, H. et al. (2015). Constraints on the evolution of phenotypic plasticity: limits and costs of phenotype and plasticity. *Heredity* 115: 293–301.

Palacio-Lopez, K., Beckage, B., Scheiner, S., and Molofsky, J. (2015). The ubiquity of phenotypic plasticity in plants: a synthesis. *Ecology and Evolution* 5: 3389–3400.

Pekkarinen, A.J., Kumpula, J., and Tahvonen, O. (2020). Predation costs and compensations in reindeer husbandry. *Wildlife Biology* 2020: 00684.

Persson, L., Andersson, J., Wahlstrom, E., and Eklov, P. (1996). Size-specific interactions in lake systems: predator gape limitation and prey growth rate and mortality. *Ecology* 77: 900–911.

Persson, L., Leonardsson, K., De Roos, A.M. et al. (1998). Ontogenetic scaling of foraging rates and the dynamics of a size-structured consumer-resource model. *Theoretical Population Biology* 54: 270–293.

Pijanowska, J., Bednarska, A., Dawidowicz, P., and Stibor, H. (2022). Food level modifies the life-history response of *Daphnia* under chemically-induced predation stress. *Fundamental and Applied Limnology* 196: 325–339.

Potter, T., King, L., Travis, J., and Bassar, R.D. (2019). Competitive asymmetry and local adaptation in Trinidadian guppies. *Journal of Animal Ecology* 88: 330–342.

Power, M.E., Dudley, T.L., and Cooper, S.D. (1989). Grazing catfish, fishing birds, and attached algae in a Panamanian stream. *Environmental Biology of Fishes* 26: 285–294.

Raczynski, M., Stoks, R., and Sniegula, S. (2022). Warming and predation risk only weakly shape size-mediated priority effects in a cannibalistic damselfly. *Scientific Reports* 12: 17324.

Reichard, M., Lanes, L.E.K., Polacik, M. et al. (2018). Avian predation mediates size-specific survival in a Neotropical annual fish: a field experiment. *Biological Journal of the Linnean Society* 124: 56–66.

Reznick, D. (1982). The impact of predation on life-history evolution in Trinidadian guppies – genetic basis of observed life-history patterns. *Evolution* 36: 1236–1250.

Reznick, D.N., Bassar, R.D., Handelsman, C.A. et al. (2019). Eco-evolutionary feedbacks predict the time course of rapid life-history evolution. *The American Naturalist* 194: 671–692.

Reznick, D.N., Bassar, R.D., Travis, J., and Rodd, F.H. (2012). Life-history evolution in guppies VIII: the demographics of density regulation in guppies (*Poecilia reticulata*). *Evolution* 66: 2903–2915.

Reznick, D. and Bryant, M. (2007). Comparative long-term mark-recapture studies of guppies (*Poecilia reticulata*): differences among high and low predation localities in growth and survival. *Annales Zoologici Fennici* 44: 152–160.

Reznick, D., Butler, M.J., and Rodd, H. (2001). Life-history evolution in guppies. VII. The comparative ecology of high- and low-predation environments. *The American Naturalist* 157: 126–140.

Reznick, D.N., Butler, M.J., Rodd, F.H., and Ross, P. (1996). Life-history evolution in guppies (*Poecilia reticulata*). 6. Differential mortality as a mechanism for natural selection. *Evolution* 50: 1651–1660.

Reznick, D. and Endler, J.A. (1982). The impact of predation on life-history evolution in Trinidadian guppies (*Poecilia reticulata*). *Evolution* 36: 160–177.

Reznick, D.N. and Travis, J. (2019). Experimental studies of evolution and eco-evo dynamics in guppies (*Poecilia reticulata*). *Annual Review of Ecology, Evolution, and Systematics* 50: 335–354.

Richardson, J.M.L., Gunzburger, M.S., and Travis, J. (2006). Variation in predation pressure as a mechanism underlying differences in numerical abundance between populations of the poeciliid fish *Heterandria formosa*. *Oecologia* 147: 596–605.

Roff, D.A. (2001). Age and size at maturity. In: *Evolutionary ecology: concepts and case studies* (ed. C.W. Fox, D.A. Roff, and D.J. Fairbairn), 99–112. Oxford University Press.

Roff, D.A. (2002). *Life history evolution*. Sinauer Associates.

Rozenberg, A., Parida, M., Leese, F. et al. (2015). Transcriptional profiling of predator-induced phenotypic plasticity in *Daphnia pulex*. *Frontiers in Zoology* 12: 18.

Santoyo-Brito, E., Perea-Fox, S., Nunez, H., and Fox, S.F. (2021). Maternal care and secretive behaviour of neonates in the highly social lizard *Liolaemus leopardinus* (Squamata: Liolaemidae) from the central Chilean Andes may relate to size-specific bird predation. *Behaviour* 158: 195–223.

Schaffer, W.M. (1974). Selection for optimal life histories: effects of age structure. *Ecology* 55: 291–303.

Scheiner, S.M. (2019). The theory of the evolution of plasticity. In: *The theory of evolution* (ed. S.M. Scheiner and D.P. Mindell), 254–272. University of Chicago Press.

Schmalhausen, I.I. (1949). *Factors of evolution*. Blakiston Press.

Semlitsch, R.D. and Gibbons, J.W. (1988). Fish predation in size-structured populations of treefrog tadpoles. *Oecologia* 75: 321–326.

Slobodkin, L.B. (1961). *Growth and regulation of animal populations*. Holt, Rinehart, and Winston.

Sogard, S.M. (1997). Size-selective mortality in the juvenile stage of teleost fishes: a review. *Bulletin of Marine Science* 60: 1129–1157.

Spitze, K. (1991). *Chaoborus* predation and life-history evolution in *Daphnia pulex*: temporal pattern of population diversity, fitness, and mean life history. *Evolution* 45: 82–92.

Spitze, K. (1992). Predator-mediated plasticity of prey life history and morphology: *Chaoborus americanus* predation on *Daphnia pulex*. *The American Naturalist* 139: 229–247.

Stearns, S.C. (1977). Evolution of life-history traits: critique of theory and a review of data. *Annual Review of Ecology and Systematics* 8: 145–171.

Stearns, S.C. (2000). Life history evolution: successes, limitations, and prospects. *Naturwissenschaften* 87: 476–486.

Stibor, H. (1992). Predator-induced life-history shifts in a fresh-water cladoceran. *Oecologia* 92: 162–165.

Szyf, M. (2021). Experience and the genome: the role of epigenetics. In: *New horizons in evolution* (ed. S.P. Wasser and M. Frenkel-Morgenstern), 45–76. Elsevier.

Taborsky, B., Heino, M., and Dieckmann, U. (2018). Life-history multistability caused by size-dependent mortality. *The American Naturalist* 192: 62–71.

Tatman, N.M., Liley, S.G., Cain, J.W., and Pitman, J.W. (2018). Effects of calf predation and nutrition on elk vital rates. *Journal of Wildlife Management* 82: 1417–1428.

Taylor, B.E. and Gabriel, W. (1992). To grow or not to grow: optimal resource allocation for *Daphnia*. *The American Naturalist* 139: 248–266.

Tills, O., Truebano, M., Feldmeyer, B. et al. (2018). Transcriptomic responses to predator kairomones in embryos of the aquatic snail *Radix balthica*. *Ecology and Evolution* 8: 11071–11082.

Tinkle, D.W. and Ballinger, R.E. (1972). *Sceloporus undulatus*: study of intraspecific comparative demography of a lizard. *Ecology* 53: 570–584.

Tollrian, R. (1995). Predator-induced morphological defenses: costs, life-history shifts, and maternal effects in *Daphnia pulex*. *Ecology* 76: 1691–1705.

Torres-Dowdall, J., Handelsman, C.A., Reznick, D.N., and Ghalambor, C.K. (2012). Local adaptation and the evolution of phenotypic plasticity in Trinidadian guppies (*Poecilia reticulata*). *Evolution* 66: 3432–3443.

Travis, J. (2023). Phenotypic plasticity. In: *Oxford bibliography of ecology* (ed. D. Gibson). Oxford University Press. https://doi.org/10.1093/obo/9780199830060-0242.

Travis, J., Bassar, R.D., Coulson, T. et al. (2023b). Population regulation and density-dependent demography in the Trinidadian guppy, *Poecilia reticulata*. *The American Naturalist* 202: 413–432.

Travis, J., Bassar, R.D., Coulson, T. et al. (2023a). Density-dependent selection. *Annual Review of Ecology, Evolution, and Systematics* 54: 85–105.

Travis, J., Keen, W.H., and Juilianna, J. (1985). The role of relative body size in a predator-prey relationship between dragonfly naiads and larval anurans. *Oikos* 45: 59–65.

Travis, J., Reznick, D., Bassar, R.D. et al. (2014). Do eco-evo feedbacks help us understand nature? Answers from studies of the Trinidadian guppy. *Advances in Ecological Research* 50: 1–40.

Trexler, J.C., Tempe, R.C., and Travis, J. (1994). Size-selective predation of sailfin mollies by two species of heron. *Oikos* 69: 250–258.

Urban, M.C. (2007). Predator size and phenology shape prey survival in temporary ponds. *Oecologia* 154: 571–580.

Urban, M.C. (2008). Salamander evolution across a latitudinal cline in gape-limited predation risk. *Oikos* 117: 1037–1049.

Walsh, M.R. and Reznick, D.N. (2008). Interactions between the direct and indirect effects of predators determine life history evolution in a killifish. *Proceedings of the National Academy of Sciences of the United States of America* 105: 594–599.

Walsh, M.R. and Reznick, D.N. (2011). Experimentally induced life-history evolution in a killifish in response to the introduction of guppies. *Evolution* 65: 1021–1036.

Wellborn, G.A. (1994). Size-biased predation and prey life histories: a comparative study of freshwater amphiphod populations. *Ecology* 75: 2104–2117.

Williams, G.C. (1966). *Adaptation and natural selection*. Princeton University Press.

Zandona, E., Auer, S.K., Kilham, S.S. et al. (2011). Diet quality and prey selectivity correlate with life histories and predation regime in Trinidadian guppies. *Functional Ecology* 25: 964–973.

13

Life History Trait Evolution in the Context of Host–Parasite Interactions

Alison B. Duncan[1], Giacomo Zilio[2], and Oliver Kaltz[1]

[1] ISEM, University of Montpellier, CNRS, IRD, EPHE, Montpellier, France
[2] CEFE, University of Montpellier, CNRS, EPHE, IRD, Montpellier, France

13.1 Introduction

Host and parasite life histories are intrinsically linked and very much shaped by one another. Hosts are constantly evolving to counter the negative effects of their parasites. Parasites in turn evolve mechanisms to counter different host defences. This antagonistic relationship leads to reciprocal (co)evolutionary changes in both players (Woolhouse et al. 2002). Models of host–parasite coevolution, in their most simple form, consider a process of two interacting loci, with variable numbers of alleles conferring either resistance of the host to the parasite or allowing the parasite to infect the host (Bell and Smith 1987). Sometimes an arbitrary 'growth cost' is added to resistance or infectivity alleles, preventing their fixation in the population, and allowing continued coevolutionary cycling.

Over the past decades, models of coevolution have become more sophisticated (Buckingham and Ashby 2022). However, it is probably no understatement that organisms and coevolutionary processes are far more complex than these models assume. Namely, hosts have a life history, with various developmental milestones between birth and death and investment decisions to be made regarding growth, reproduction and dispersal, all shaped by evolution (Table 13.1). Parasites interfere with these investments and steal energy from the host necessary to carry out their own developmental programme, which ultimately manifests itself as virulence (parasite-induced fitness reductions to the host). Because parasites reduce host fitness, any measures restoring it should be selectively favoured. This can be the allocation of resources to immune defence, which can come at a cost to other life history traits. This may be the first thing that springs to mind when considering how hosts overcome parasites. However, hosts can also defend themselves by modifying their life history itself either to reduce the risk of or compensate for fitness losses from infection. In this chapter, we focus on these latter possibilities, reviewing theoretical work as well as empirical studies. A number of very good reviews already highlight plasticity in life history shifts to buffer parasites (Hochberg et al. 1992, Koella et al. 1998, Agnew et al. 2000, Duffield et al. 2017). This plasticity may well be an evolved, adaptive response to parasitism, though in this chapter we aim to highlight cases where fixed host life history changes have evolved in response to parasites. We also focus on certain recent developments, namely the role of protective symbionts as modifiers of resistance *vs.* tolerance evolution (see also Chapter 14, this volume) and the impact of parasites on host dispersal evolution (Zilio et al. 2024).

While it may be true that antagonistic species like predators are running only for dinner (and not for life, like their prey), this certainly does not hold for (obligate) parasites. Parasites have a life history, with exactly the same milestones as any other organism (Table 13.1), but with the added twist that it needs to match that of their host. Challenges are numerous, from finding the right entry point to start an infection, replicating in particular host compartments, to deflecting immune responses or hijacking host behaviours in order to promote transmission. Here, we concentrate on a very simple and basic challenge: that of keeping the host alive long enough, while extracting its resources, and transmitting to new hosts before the current one dies. This is the crux of virulence evolution, a central topic in evolutionary ecology (see also Chapter 24, this volume). We will revisit some basic theories and show how virulence depends on external factors (background host

---

This is ISEM publication number **ISEM 2024-207**.

Life History Evolution: Traits, Interactions, and Applications, First Edition. Edited by Michal Segoli and Eric Wajnberg.
© 2025 John Wiley & Sons Ltd. Published 2025 by John Wiley & Sons Ltd.

Table 13.1 Parallels between life history traits in hosts and parasites during juvenile and adult stages. Unique parasite traits (in bold) are those for which direct parallels are not directly obvious for the host.

		Host	Parasite
Juvenile		Birth	Infection
		Growth rate	Within-host replication
Adult		Age/size at maturity	Time of onset for transmission
		Reproduction	Transmission
		Asexual or sexual	Asexual or sexual
		Number and size of offspring	Number and size of transmission stages
		Sex allocation	Sex allocation
		Provisioning	Provisioning
		Production of resting stages	Production of resting stages
			Vertical (from parent to offspring) and/or horizontal (among all individuals in a population)
		Dispersal	Dispersal
		Rate and/or timing	Rate/timing
			Active, passive (*e.g.*, wind or **with host/vector**)

mortality) and other factors in the within-host environment, related to host immune responses, but also co-infection. We emphasize that behind 'virulence' lies a whole suite of potential life history traits that can evolve and that, considering their joint action and trade-offs between them, may be important for predicting virulence evolution. This will be worked out in more detail for scenarios where parasites travel with infected hosts in spatially structured populations.

We provide an overview of both theoretical and empirical results showing how host and parasite life histories evolve in response to one another. The empirical examples mentioned include observational studies of natural populations, work that combines field and laboratory investigation and experimental evolution studies in the laboratory. A common theme throughout this chapter is the importance of scaling up from individual to population or even meta-population level. This allows the integration of eco-evolutionary feedback. We highlight when epidemiological feedbacks are important for driving evolutionary shifts in life history and how this may affect coevolution.

13.2 Host Life History Evolution in Response to Parasites

Hosts can modify different life history traits in response to parasitism to increase their reproductive fitness (Table 13.2). Minchella and Loverde (1981) were the first to document this, showing that *Biomphalaria glabrata* snails exposed to trematode parasites (*Schistosoma mansoni*) increased egg laying. This is in line with classic life history theory based on trade-offs: when there is a threat to future reproduction, it is better to invest in current reproduction (Williams 1966, Stearns 1989). Any modification in life history traits intensifying current reproductive effort in response to a risk of reproduction ending prematurely is termed 'terminal investment' (or fecundity compensation, when referring to increased offspring production only) (Duffield et al. 2017). Thus, parasites are just one of many environmental factors that can change mortality patterns, which can in turn impact life history shifts.

As stated above, there are a number of very good reviews about changes in host life history as a plastic response triggered by actual infection or cues in the environment indicating a risk of parasitism across diverse taxa (Michalakis and Hochberg 1994, Koella et al. 1998, Agnew et al. 2000, Valenzuela-Sanchez et al. 2021), and how responses may depend on both abiotic and biotic conditions (Duffield et al. 2017). What is less clear is how host life history traits evolve in response to parasitism. In this chapter we focus on the latter, trying to better understand when parasites cause evolution of life history trait shifts in host populations.

13.2.1 Evolutionary Shifts in Reproductive Effort and Phenology: Some Theory and Empirical Evidence

A number of theoretical studies have investigated how host life history traits can evolve in response to parasitism independent of physiological resistance (and any associated costs). Hochberg et al. (1992) showed that hosts should evolve earlier age

Table 13.2 Host life history trait modifications in response to parasitism.

Life history trait	Host species	Parasite species	Laboratory study	Field study	Evolved or plastic response	Reference
Age at maturity	*Arabidopsis thaliana* (plant)	*Pseudomonas syringae* (bacterium) *Xanthomonas campestris* (bacterium) *Peronospora parasitica* (oomycete)	Shift to earlier flowering in infected plants		Plastic	Korves and Bergelson (2003)
	Daphnia magna (crustacean)	*Glugoides intestinalis* (microsporidian)	Shift to earlier reproduction in presence of parasite		Genetic variation	Chadwick and Little (2005)
	Male and female *Litoria verreauxii alpina* (frog)	*Batrachochytrium dendrobatidis* (fungus)		Comparison infected *vs.* uninfected populations. Earlier age at maturity in infected populations	Unknown	Scheele et al. (2017)
	Male and female *Sarcophilus harrisii* (Tasmanian devil)	Devil facial tumour disease		Compared before and after disease introduced. Earlier age at maturity after disease arrival in population	Unknown	Jones et al. (2008)
Smaller size at maturity	*Cerithidea californica* (snail)	Trematodes		Negative correlation between parasite prevalence and size at maturity	Possibly evolved as differences remained following reciprocal transfer into opposing environments, but do not control for parental effects	Lafferty (1993)
	Male *Litoria verreauxii alpina* (frog)	*Batrachochytrium dendrobatidis* (fungus)		Comparison infected *vs.* uninfected populations. Smaller size in infected populations	Unknown	Scheele et al. (2017)
	Male and female *Perca fluviatilis* Perch (fish)	Unknown		Multiple time-points before and after epidemic/parasite introduction. Ongoing reduction in size following parasite outbreak	Indicative of evolution	Ohlberger et al. (2011)
Reproductive effort	*Biomphalaria glabrata* (snail)	*Schistosoma mansoni* (trematode)	Increased egg laying following exposure to the parasite (regardless if became infected) compared to unexposed snails		Plastic	Minchella and Loverde (1981)
	Male *Litoria verreauxii alpina* (frog)	*Batrachochytrium dendrobatidis* (fungus)	Individuals collected as eggs from four field populations and raised in laboratory. Higher baseline reproductive effort in males from infected *vs.* uninfected populations when unexposed to the parasite		Indicative of evolution, but do not control for parental effects	Brannelly et al. (2021)

(*Continued*)

Table 13.2 (Continued)

Life history trait	Host species	Parasite species	Laboratory study	Field study	Evolved or plastic response	Reference
	Silene latifolia (plant)	*Microbotryum violaceum* (fungus)	Increased reproductive effort in males from both parasite-infected and uninfected populations following parasite exposure	Increased number of flowers in infected plants	Plastic	Shykoff and Kaltz (1997)
	Male *Tenebrio molitor* (insect)	Immune challenge	Invest more in cuticular hydrocarbons and glandular pheromones making them more attractive to females		Plastic	Nielsen and Holman (2011)
Production of resting stages	*Daphnia magna*	*Pasteuria ramosa* (bacterium)	Individuals collected from field population. Susceptibility measured in the laboratory: More susceptible genotypes engage in sexual reproduction and the production of resting stages to escape parasite epidemic		Genetic variation for trait, indicative of adaptive evolution	Mitchell et al. (2004), Duncan et al. (2006)
Semelparity (single reproductive event)	*Sarcophilus harrisii* (Tasmanian devil)	Devil facial tumour disease		Compared before and after disease. Single reproductive event following disease outbreak, compared to multiple reproductive events prior	Unknown	Jones et al. (2008)

The table shows the life history trait modification in the host (first column) followed by the host species in response to the parasite species. The table also denotes whether the result was obtained from a field or laboratory study or both. It also states whether the response is plastic, *i.e.*, context-dependent, which may itself be adaptive, or an evolved fixed response. Note that this is not an exhaustive literature review. The aim is to provide examples of host life history shifts in response to parasitism across different taxa. In some cases, it is not entirely clear whether shifts in life history are an adaptive response in the host or parasite manipulation (Duffield et al. 2017). Some studies control for this by implementing an immune challenge (parasite exposure or heat-killed pathogens) instead of infection (see Reaney and Knell 2010).

at maturity in response to more virulent parasites (Figure 13.1a). This finding is in line with classic life history theory that earlier age at maturity and investment in reproduction is expected to evolve under increased extrinsic mortality, here being caused by parasitism (Williams 1966, Stearns 1989). Koella and Restif (2001) corroborated the findings of Hochberg et al. (1992) but also showed that earlier maturity would in turn select for increased virulence, which would reduce the force of infection (Koella and Restif 2001). The increase in host mortality reduces the frequency of infected hosts and thus the rate susceptible individuals become infected. This, in turn, relaxes selection on earlier maturity because there is reduced parasite prevalence (Koella and Restif 2001). Similarly, Gandon et al. (2002a) found that selection on reproductive effort in response to parasitism depended on epidemiological feedbacks: virulence initially selects for increased reproductive effort, but reductions in the force of infection (parasite prevalence or the intensity of parasite-mediated selection) in turn relaxes selection on reproductive effort (Gandon et al. 2002a).

However, depending on the underlying disease ecology, very different predictions can be made. For example, host phenology may evolve to minimize infection risk, by reducing the overlap with the reproduction time of the parasite. In line with this, a theoretical study showed that parasitism with a castrating parasite can select for later, rather than earlier, host emergence in a season, balancing infection risk against the time available for reproduction (MacDonald and Brisson 2023). This may be relevant for interactions between species of the plant genus *Silene* with a sterilising fungal pathogen *Microbotryum* spp. (Bernasconi et al. 2009). This pathogen is transmitted by flower-visiting insect vectors and leads to an increased infection risk, especially in female plants (Shykoff et al. 1996b). Because the fungus needs time to grow from contaminated flowers into the plant and from there into the roots for permanent infection, it has been suggested that this pathogen selects for later flowering in the year (Biere and Antonovics 1996), decreased floral display (Shykoff et al. 1996a) and shorter flower lifespan (Shykoff et al. 1996b, Kaltz and Shykoff 2001). This could lead to even more profound changes, such as the evolution of dioecy (separation of male and female function) or the shift from perennial life cycles to an annual cycle (Thrall et al. 1993).

A number of studies have compared host life history traits among parasite-infected *vs.* uninfected natural populations across different taxa (Table 13.2). The majority of these studies are correlational, thus although consistent with shifts in life history due to parasitism, observed differences may be due to alternative unmeasured variables. However, some studies have attempted to demonstrate that life history changes have occurred in response to parasite-mediated selection. One documented an ongoing decline in size at maturation in male and female perch in Lake Windemere, UK, over more than two decades following the outbreak of a parasite in 1976 (Ohlberger et al. 2011). Although the authors cannot be sure, they argue that the

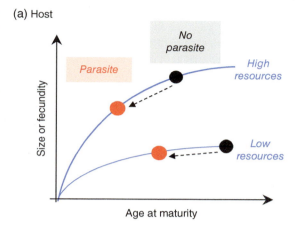

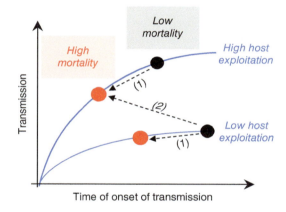

Figure 13.1 Graphical illustrations of host and parasite life history change in response to mortality threats. (a) Relationship between host age at maturity and size (≈fecundity), for low and high resource levels. The black points show the optimal age at maturity. Under low resources, hosts grow more slowly, reproduce later and have lower fecundity than under high resources. Parasite-induced host mortality selects for a shift along each curve (dashed arrow), towards earlier reproduction accompanied by reduced size (and) fecundity (red points). (b) Analogous to (a), relationship between parasite time of onset of transmission (latency) and transmission (*e.g.*, transmission stage production), for low vs. high host resource exploitation strategies (*i.e.*, low vs. high virulence). The black points show the optimal latency under low host extrinsic mortality. Low-exploitation parasites grow more slowly within the host, start transmitting later and have less transmission than high-exploitation parasites. High extrinsic host mortality leads to an evolutionary shift towards an earlier onset of transmission in two possible ways (dashed arrows): (1) Shifts occur along the curve, leading to earlier onset and similar transmission. (2) Shift occurs by switching curves, *i.e.*, by increasing host exploitation, leading to earlier onset and increased transmission.

fact this change was observed in two genetically distinct populations and that their methods account for other explanatory variables, this decline in size is consistent with adaptive genetic changes. A similar response has been observed in the Tasmanian devil, with a shift to semelparity (single episode of reproduction) and earlier age at maturation in less than ten years following the arrival of devil facial tumour disease (Jones et al. 2008). Other studies have shown that changes in host life history remain following reciprocal transplant into environments without parasites (Lafferty 1993) or in a common garden setting bringing individuals from parasitized and parasite-free populations into the laboratory and following development (Brannelly et al. 2021; Table 13.2).

Koella and Boëte (2002) investigated how selection on a life history trait in the absence of parasites correlated with immunity. They selected for earlier or later pupation in the mosquito *Aedes aegypti* for 10 generations. Following selection, they found that earlier age at pupation was associated with a reduced ability to encapsulate and melanise a Sephadex bead mimicking an infectious agent. This result indicates an evolutionary trade-off between an important life history trait (developmental time) and parasite resistance. Though most studies have investigated the inverse, how parasite-mediated selection on resistance is correlated with changes in life history traits (see the following section).

13.2.2 Costs of Resistance and a Brief Perspective on Tolerance

Parasite-mediated selection can indirectly affect host life history traits when they are genetically correlated with resistance, that is, the capacity to prevent infection or reduce parasite load (qualitative or quantitative resistance, respectively). Genetic correlations between resistance and host life history traits form the basis of a large body of theory (Gillespie 1975, Bowers et al. 1994, Antonovics and Thrall 1997, Agrawal and Lively 2002). In such a scenario, being resistant comes at the cost of lower reproduction or longevity in the absence of parasites. Costs of resistance are thought to be why genetic variation for resistance remains and why resistance does not go to fixation in host populations (*e.g.*, Boots and Bowers 1999, Boots and Haraguchi 1999, Gandon et al. 2002b). Indeed, classic experimental evolution work showed that the emergence of resistance is associated with reduced competitive ability (Lenski 1988). This cost can effectively stall coevolution between bacteria and phages (Lenski and Levin 1985). Subsequently, various experiments have indicated that accumulating costs of resistance might constrain escalating coevolutionary arms-race type cycles of selection for ever-increasing resistance and infectivity, and promote the transition to fluctuating selection dynamics (Hall et al. 2011, Frickel et al. 2016). Other studies using experimental evolution in the laboratory have also demonstrated the evolution of costs: selection for increased resistance in the presence of parasites negatively impacted other life history traits (*e.g.*, reproduction, longevity and growth rate) in the absence of parasites (Luong and Polak 2007, Boots 2011, Duncan et al. 2011, Meaden et al. 2015). Another approach is to use inbred, isogenic lines representing standing genetic variation in a population (or species), as done with the pyralid moth *Plodia interpunctella*. This revealed a negative correlation between susceptibility to a baculovirus and development time (Bartlett et al. 2018). These studies highlight that parasites can affect host life history evolution when these traits are genetically linked to resistance.

Costs of resistance have also been found in natural host populations (Auld et al. 2013, Gibson et al. 2013, Susi et al. 2015). For example, in plants, quantitative and qualitative resistance was shown to be negatively correlated with reduced seed set or flower production (Simms and Rausher 1987, Biere and Antonovics 1996). However, how these costs evolve and shape resistance polymorphism evolution is not always clear (Auld et al. 2013).

Mechanistically, costs of resistance may arise due to the allocation of limited resources (Bajgar et al. 2015), immunopathology (Graham et al. 2010, McGonigle et al. 2017) or pleiotropy among different gene functions (Martin and Tate 2023). For example, in Soay sheep, higher levels of heritable immunity are associated with increased immunopathology and reduced reproduction (but higher over-winter survival for females) (Graham et al. 2010). Another study used experimental evolution of three *Drosophila* species to study the proximate physiological mechanisms linking resistance and its correlated life history costs. Experimental selection for resistance to the parasitoid *Leptopilina boulardi* led to higher levels of resistance in the three species and was associated with reduced competitive ability, at least when measured at lower resource levels (McGonigle et al. 2017). Resistance in each species was explained by higher levels of circulating haemocytes, although not via the same mechanisms (McGonigle et al. 2017): in *D. mauritiana* and *D. simulans* haemocyte production increased, whereas in *D. melanogaster* sessile haemocytes were activated and released into the bloodstream (McGonigle et al. 2017). As there was no increase in total haemocytes in the latter case, the authors argue that the observed life history cost may be due to immunopathology rather than the result of a trade-off over energy allocation. Note that, in a separate study, evolved resistance in *D. melanogaster* to *L. boulardi* was, however, associated with increased haemocyte production, possibly creating a conflict of resource allocation to sugars for growth or to immunity (Leitao et al. 2020).

Theoretical models of resistance evolution often implement very simple mechanisms of associated fitness costs. These are well matched in (and often inspired by) bacteria–phage systems. One way bacteria can evolve resistance to phage is by modifying cell surface receptors (Mayo-Munoz et al. 2023, 2024), which prevents attachment of phage and subsequent infection, but at the same time disrupts morphological and functional aspects of the bacterial cell membrane, leading to reduced growth, motility or biofilm formation (Egido et al. 2022). Negative side effects sometimes also include increased susceptibility to antibiotics or loss of 'virulence factors' (e.g., reduced endotoxin production and thus lowered pathogenicity in humans), which makes these evolutionary life history changes interesting from a therapeutic perspective (Gurney et al. 2020).

Of course, if fitness costs emerge from allocation constraints, resource availability in the environment becomes important. In line with this, the moth P. interpunctella evolved resistance more easily in high-resource environments (Boots 2011). In the bacterium Pseudomonas fluorescens, resource-rich conditions lower the costs of resistance to phage, thus speeding up coevolutionary rates and allowing higher levels of resistance in experimental cultures (Lopez-Pascua et al. 2011). Higher resource environments can also induce a shift of host–parasite coevolution from fluctuating selection to arms-race dynamics (Lopez-Pascua et al. 2014). Similarly, Pseudomonas aeruginosa is more likely to evolve costly constitutive resistance to phage at high resource supply, whereas low-resource conditions favour the evolution of a less costly alternative, via acquired immunity through the clustered regularly interspaced short palindromic repeats system (CRISPR) (Westra et al. 2015). The fact that (experimental) evolution of resistance evolved to higher levels in permissive conditions (Boots 2011) indicates that resource availability might modify trajectories, for example towards tolerance rather than resistance (see e.g., Zeller and Koella 2017).

Indeed, tolerating an infection rather than fighting it may be an alternative evolutionary option (Kutzer and Armitage 2016), particularly if resistance bears heavy costs and/or if it is easily counteracted by parasite evolution (Little et al. 2010, Best et al. 2014). Tolerance is when hosts limit the fitness consequences of parasite infection. It can manifest itself as increased survival or fecundity, without impacting parasite load (Restif and Koella 2003, Kutzer and Armitage 2016). In this sense, tolerance is often taken as the slope of fitness against parasite load (Raberg et al. 2007; but see Little et al. 2010). Mortality tolerance is beneficial for parasite fitness, increasing the infectious period (Best et al. 2008). In contrast, fecundity tolerance can negatively impact parasite fitness via a trade-off with mortality tolerance, thus increasing host mortality and reducing the infectious period (Budischak and Cressler 2018). Vale and Little (2012) demonstrate genetic variation for fecundity tolerance in a natural population of Daphnia magna infected with the bacteria Pasteuria ramosa, showing that this is indeed a trait on which selection can act. Furthermore, they highlight that this type of tolerance is equivalent to fecundity compensation (increased reproductive effort when infected), suggesting that many life history shifts are in fact a manifestation of tolerance (Vale and Little 2012).

It is not entirely clear when hosts should evolve resistance vs. mortality or fecundity tolerance. However, whichever evolves will strongly affect transmission dynamics and possible eco-evolutionary feedbacks. For instance, mortality tolerance has the potential to facilitate transmission and disease outbreaks. This, in turn, may remove constraints on parasite evolution and favour more virulent and more transmissible variants. Such feedbacks are particularly relevant in a medical context, when certain infection damage-limiting drugs reduce the host's illness (i.e., make them more tolerant), without directly targeting infection development (Vale et al. 2014).

13.2.3 Scaling Up: Parasite Effects on Host Dispersal Evolution

Dispersal is a central life history trait and a major driver of eco-evolutionary spatial dynamics, transporting genes and individuals across landscapes (Kubisch et al. 2014, Bonte and Dahirel 2017; see also Chapter 10, this volume). Some theoretical work has suggested that host geographical ranges might be constrained by parasites due to associated costs (e.g., virulence and/or mortality; Hochberg and Ives 1999, Holt and Keitt 2000, Price and Kirkpatrick 2009). However, this research did not consider host dispersal as a trait that can evolve, as has been the case in more recent years (Phillips et al. 2006, Lombaert et al. 2014, Fronhofer and Altermatt 2015, Saastamoinen et al. 2018), or that might interact with parasites. Indeed, recent theoretical studies show that parasites can promote dispersal evolution (reviewed in Zilio et al. 2024), by causing fluctuations in host population density. The fluctuations reduce environmental predictability and thereby reduce overall fitness expectations across the meta-population. These conditions favour increased dispersal, as a means to escape the detrimental conditions imposed by parasites (Chaianunporn and Hovestadt 2011, Chaianunporn and Hovestadt 2019). The underlying mechanism is similar to that in single-species scenarios, where dispersal evolves in response to increased local extinction risk, caused by reductions in habitat quality (Duputié and Massol 2013).

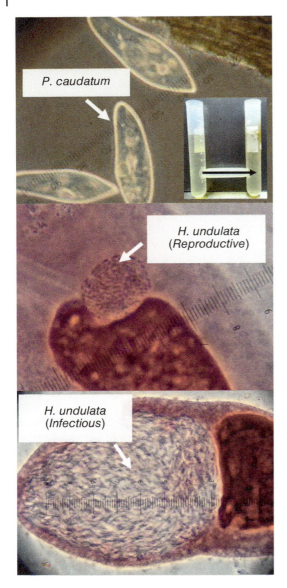

Figure 13.2 Aquatic model system, used for experimental evolution in the context of dispersal, virulence and vertical vs. horizontal transmission. The freshwater protozoan *Paramecium caudatum* naturally disperses in interconnected microcosms (top). The propensity to disperse has a genetic basis and readily responds to selection. The *Paramecium* become infected with the bacterial parasite *Holospora undulata* while feeding. These parasites have a mixed transmission mode, with reproductive stages (middle, *H. undulata* at early reproductive stage of infection of the host micronucleus) being vertically transmitted during host division, and infectious forms (bottom, *H. undulata* at fully developed infectious stage) being released when the host divides or dies.

Empirical studies have demonstrated proximate effects of parasites on host dispersal in a number of systems (Zilio et al. (2021) and references therein), but only rarely investigate dispersal evolution. Taylor and Buckling (2013) studied the experimental evolution of motility (as a proxy of dispersal) of the bacterium *P. aeruginosa* facing attack by a phage. Regular exposure to phage on soft agar plates, potentially allowing the bacteria to swim away from phage, indeed favoured the evolution of increased motility. However, motility evolved even if phage were placed on the agar in such a way that swimming increased rather than decreased contact risk for the bacteria. This speaks against dispersal as an escape mechanism and suggests that a change in motility might have been a by-product of other aspects of phage-mediated selection. Yet, Koskella et al. (2011) found no evidence of a relationship between motility and resistance or growth rate for a large collection of natural bacterial isolates.

In a recent study, Zilio et al. (2023b) investigated evolutionary processes in experimental range expansions of the ciliate host *Paramecium caudatum*, in the presence and absence of its bacterial parasite *Holospora undulata*. These experiments allowed natural dispersal of the *Paramecium* between interconnected microcosms (Figure 13.2). One main result was that *Paramecium* evolving in infected populations were more resistant but also dispersed less than *Paramecium* from parasite-free populations. This shows that the presence of parasites does not necessarily select for increased dispersal, and it suggests that selection for parasite resistance might trade off with dispersal. Nonetheless, parasite-mediated selection did not totally obstruct spatial selection. As expected, *Paramecium* from infected 'range fronts' (a treatment involving frequent dispersal into new patches) evolved higher dispersal rates than their 'range core' counterparts (a treatment involving populations remaining in place in the same patch). This result mirrors data from the core and front populations of a biological invasion of the cane toad *Rhinella marina* (see also Chapter 20, this volume) and its nematode parasite *Rhabdias pseudosphaerocephala*, showing rapid evolutionary changes in several traits for both players, such as resistance, infectivity and dispersal (Mayer et al. 2021, Shine et al. 2021, Brown et al. 2024, Schlippe-Justicia et al. 2022), all of which suggest coevolutionary dynamics and patterns of local adaptation with dispersal evolution as a main driver.

13.2.4 Scaling Down: Symbionts Mediating Host Life History Evolution

Recent years have seen an increasing interest in the role of symbionts, or the microbiota, as an extension of the host phenotype. One of the biggest challenges is to understand the evolutionary implications for the different players involved (Decaestecker et al. 2024). Here we briefly highlight certain aspects in the context of host–parasite interactions, notably how symbionts can modify life history evolution.

Protective symbionts can change host evolution in the presence of parasites (see also Chapter 14, this volume). In particular, symbionts permit the evolution of alternative modes of parasite defence, which may have consequences for life history evolution, or *vice versa* (Martinez et al. 2016, Vorburger and Perlman 2018, Rafaluk-Mohr et al. 2022). This was demonstrated through the experimental evolution of *D. melanogaster* resistance against a virus. When *Drosophila* populations harboured

Wolbachia symbionts, there was no selection for a host gene conferring resistance to the virus, while this resistance gene readily increased in frequency in *Wolbachia*-free host populations (Martinez et al. 2016). Thus, symbionts (here *Wolbachia*) may modify selection for host resistance genes in populations, and eventually lead to a kind of 'host addiction' to their symbionts (Martinez et al. 2016). Along the same lines, in *Caenorhabditis elegans*, alternative tolerance strategies evolved in response to the pathogenic bacterium *Staphylococcus aureus* in host populations that harboured the protective symbiont *Enterococcus faecalis* (Rafaluk-Mohr et al. 2022). In the presence of the symbiont, hosts evolved mortality tolerance (improved survival when infected), whereas unprotected hosts evolved fecundity tolerance (improved fecundity when infected) (Rafaluk-Mohr et al. 2022). The authors discuss how this difference in tolerance strategy may be due to the fact that *E. faecalis* reduces the virulence of *S. aureus*: lower virulence is predicted to select for mortality tolerance (Kutzer and Armitage 2016).

The mechanisms of symbiont-mediated protection are not always clear. They may result from direct microbe-to-microbe interactions within the host (Hamilton et al. 2016), competition for available space or resources (Paredes et al. 2016), but also indirectly through the upregulation of the host's general stress response (Sørensen et al. 2003) or of its immune system (Dong et al. 2009), in such a way that prevents further infection with other parasites. For example, the mosquito gut microbiota upregulates immune genes that reduce infection with malaria (Dong et al. 2009).

Protective symbionts certainly play a role in determining host–parasite associations in nature. A clear example of this is the selection for and spread of the bacterial endosymbiont *Spiroplasma* in natural populations of *Drosophila neotestacea* in North America over a period of two decades due to the protection it provides against a sterilising nematode (Jaenike et al. 2010). Note, however, that the beneficial effects of protective symbionts can be context-dependent such that, under certain conditions (*e.g.*, the absence of parasites against which they protect), their presence becomes costly (Drew et al. 2021). This means that their role may shift along the mutualism–parasitism continuum, depending on prevailing conditions, favouring or preventing their fixation (Drew et al. 2021).

Even so, it is exciting that symbionts or microbiota can contribute to the host's 'extended genotype', and potentially its extended phenotype as well (Fellous and Salvaudon 2009; see also Chapter 14, this volume). A mosaic of host and foreign genes (symbionts, microbiota) may interact and respond to selection in potentially complex ways and open novel (co)evolutionary trajectories (Fellous and Salvaudon 2009). Coevolution between a parasite and a protective symbiont may be the driver behind host–parasite coevolutionary dynamics. In this context, future work is needed to uncover the role of protective symbionts driving life history evolution in response to parasites. Are there parallels between costs associated with harbouring a protective symbiont and host resistance genes? Do they lead to similar patterns of host–parasite coevolution? How much is the expression of standard life history traits (*e.g.*, age/size at maturity, reproductive effort) in the face of parasitism due to protective symbionts?

13.2.5 Complexity: The Importance of Eco-evolutionary Feedbacks and Multispecies Interactions

The above works highlight that parasites can drive changes in host life history evolution, which is often linked to trade-offs with other traits such as resistance, but also other life history traits such as dispersal. These studies also show that eco-evolutionary feedbacks are important for driving selection on host life history. These feedbacks can arise due to the resource environment affecting host density, and in turn, parasite density and thus the force of infection. Feedbacks may produce counter-intuitive outcomes. For example, the 'resistance is futile' effect shows selection against resistance when parasite prevalence is high and resistance is costly in terms of reduced fecundity (Walsman et al. 2023). This is because, if all genotypes become infected, the one with the lowest resistance will have the highest fecundity (Walsman et al. 2023). There is support for this scenario in a number of populations where there is a negative correlation between resistance and prevalence and predictions were qualitatively matched in a laboratory experiment, whereby resistance declined when parasite prevalence was high (Walsman et al. 2023). This illustrates that epidemiological feedbacks can drive host life history evolution in populations and that outcomes are not necessarily intuitive.

Furthermore, we often make the simplifying assumption that a given host species is interacting with only one parasite species against which specific resistance evolves. In reality, hosts will encounter many parasites and symbionts simultaneously, imposing different fitness-associated costs and making it difficult to evolve resistance to all of these parasites. In such a scenario, one possibility is that life history shifts alone, without resistance evolution, enable hosts to maximize fitness when faced with multiple parasites. Alternatively, life history shifts may occur in response to the most commonly encountered or more virulent parasites (Gandon et al. 2002a). A better understanding of the parasite community to which hosts are exposed should reveal strategies employed to fight parasites and whether life history shifts, evolved or plastic, provide a general response.

13.3 Parasite Life History Evolution in Response to Hosts: The Case of Virulence

Parasite virulence (broadly defined as parasite-induced harm to the host) evolution is one central theme in studies on the evolutionary ecology of host–parasite interactions (Frank 1996, Alizon et al. 2009, Cressler et al. 2016; see also Chapter 24, this volume). Classic theoretical models treat virulence as a trait in its own right (Frank 1996, Alizon et al. 2009), or even as a 'shared trait', determined by parasite and host properties (Restif and Koella 2003). In the context of this chapter, we prefer to treat virulence as an emergent trait. In this sense, virulence is a by-product of diverting host resources and allocating them to parasite development. In other words, it is the consequence of the infection life cycle, or parasite life history.

Standard textbook formalisations of virulence evolution typically define virulence as parasite-induced host mortality (Figure 13.3). In their simplest form, using susceptible-infected-recovered (SIR) models and assuming a virulence-transmission trade-off, theoretical models predict the evolution of an optimal level of intermediate virulence that maximizes parasite fitness (R_0; see Figure 13.3). The virulence-transmission trade-off is essential to these models because it links an increase in transmission (numerator in equation in Figure 13.3) with an unavoidable increase in host mortality (denominator in equation in Figure 13.3) through host exploitation. This is radically different from older views, which assumed that in the long run, parasites evolve to become less and less harmful (Ewald 1983).

Even though being widely accepted as 'conventional wisdom', there is surprisingly little unequivocal empirical evidence for this hypothesis (Acevedo et al. 2019) and references therein (Doumayrou et al. 2013, Godinho et al. 2023). In part, this has to do with the fact that R_0 is an over-simplification. Indeed, not all parasites act on host survival, *e.g.*, if virulence causes reductions in fecundity, it does not necessarily evolve towards an optimum (O'Keefe and Antonovics 2002). Furthermore, parasites may use multiple transmission routes, which differ in their impact on host fitness (*e.g.*, vertical and horizontal transmission; Ebert 2013). Moreover, basic models do not take into account the dynamical nature of parasite fitness via epidemiological feedbacks. Importantly, the number of available susceptible hosts (S) changes dynamically with the progression of an epidemic (*i.e.*, the changes in infection prevalence) and with evolutionary change (Lion and Metz 2018). Erudite overviews of this topic and its subtleties can be found in, *e.g.*, Frank (1996), Alizon et al. (2009), Cressler et al. (2016) (see also, Chapter 24, this volume).

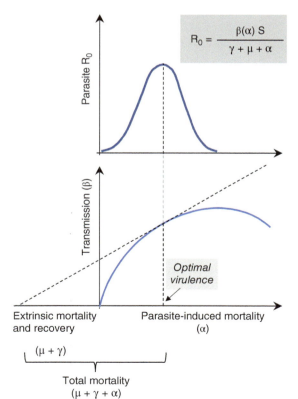

Figure 13.3 Classic model of virulence evolution. Parasite fitness is defined as the number of new infections produced over the lifetime of a single infected host (R_0) and virulence by the rate of parasite-induced host mortality (α). R_0 is calculated as the ratio of the number of newly infected hosts, divided by the sum of three sources of mortality for the parasite (β = transmission rate, S = number of susceptible hosts, γ = host recovery rate, μ = extrinsic host mortality rate). For a saturating relationship between transmission rate (β) and virulence (α) (lower panel), the virulence optimum maximises R_0 (upper panel). Graphically, the optimum is the tangent of the line that passes through the x-axis at ($\gamma + \mu + \alpha$).

13.3.1 Why Consider Virulence Evolution from a Life History Perspective?

In parasite life history terms, virulence evolution is about how to make best use of the infectious period between the onset of transmission and parasite-induced host death (Day 2003). This relates to standard life history dilemmas of investment in reproduction *vs.* survival, or current *vs.* future reproduction (Stearns 1989). The longer host death is postponed, the more host resources are needed to maintain transmission. Similarly, diverting host resources to current transmission reduces resources available for future transmission and host survival. Virulence thus represents the mortality cost resulting from reproductive effort (exploitation of host resources; Day 2003). It is the consequence of life history decisions regulating developmental time (within-host growth), age at maturity (latency, onset of transmission) and fecundity (number and quality of transmission propagules) or dispersal (transmission into distant patches).

The above baseline model (Figure 13.3) depicts a rather simple scenario. It can be adapted to various real-life life history strategies: parasites may continuously release transmission stages (be iteroparous) or release them in a single bout (be semelparous). Accordingly, they may become obligate killers, reduce host fecundity to some degree, or castrate the host, thereby killing the host genetically, but keeping it alive for transmission. Life history evolution may also involve more complex infection life cycles, with different transmission pathways (vertical/horizontal), the use of vectors or alternating hosts.

If virulence is indeed a life history epiphenomenon, we need to decompose it into the relevant underlying traits to better understand its evolution (de Roode et al. 2008, Hall and Ebert 2012, Alizon and Michalakis 2015). With multiple traits involved, it is also important to study the (change in the) genetic correlations and trade-offs between them (Mackinnon and Read 1999, Nørgaard et al. 2021). It is also important to understand how trait change scales up from disease development at the individual level to the epidemiological dynamics within or across populations (Mideo et al. 2011). With virulence evolution being the leitmotif, these different aspects are the guidelines throughout the discussion developed below.

13.3.2 Sources of Shortened Infection Lifespan: Host Mortality and Infection Clearance

Basic theory predicts that increased parasite-independent host background mortality leads to higher optimum levels of parasite virulence (van Baalen and Sabelis 1995). In terms of life history, the increased probability of host death shortens parasite infection lifespan. Thus, if future reproduction is compromised, the expected evolutionary response is increased investment in early reproduction, through faster development and/or earlier age at maturity (Stearns 1989), meaning faster parasite within-host growth and/or shorter latency time, and hence higher virulence (Day 2003) (Figure 13.1b).

Cooper et al. (2002) tested the evolutionary consequences of modifying infection lifespan in a serial passage experiment (*i.e.*, the artificial transfer of parasites between hosts over multiple host generations), using a virus of the gypsy moth *Lymantria dispar*. They found that viral selection lines passaged early after infection (transferred to new hosts on day 5) evolved higher levels of host mortality than lines passaged four days later (transferred to new hosts on day 9). Consistent with a shift towards earlier transmission, early-transferred viruses produced larger numbers of transmission propagules (inclusion bodies) over the first days after infection, compared to late-transfer viruses. Interestingly, however, early-transferred viruses produced smaller inclusion bodies (and thus presumably containing fewer virions), while their overall production of inclusion bodies was not different from the late-transferred lines. This indicates that the shift towards early transmission traded off with qualitative and quantitative aspects of virus fecundity (transmission). Similar constraints between early and late reproduction were reported for a selection experiment on a bacterial parasite of *Daphnia* (Auld et al. 2014a) and are also well-known in bacteriophages. In phages, the time to host cell burst (lysis time, or latency) is negatively correlated with burst size (the number of viral particles per cell), such that shifts towards earlier lysis time come at the expense of phage productivity, and thus the capacity to spread in a host population (Bull et al. 2004, Kannoly et al. 2022). Costs of shortened latency may also arise through trade-offs with offspring quality, such as the survival of phages in the environment (Elena 2001, Kannoly et al. 2022), potentially adding other extrinsic factors to the equation.

Using the bacterial parasite *H. undulata* of the protist *P. caudatum*, Nidelet et al. (2009) performed a serial passage experiment, similar to that done by Cooper et al. (2002). They also found that parasites from an early host-killing treatment produced higher levels of host mortality, compared to parasites from a late-killing treatment. The killing treatments consisted of crushing infected hosts and recovering the infectious stages of the parasite to start a new passage cycle. Accordingly, parasites from the early-killing treatment evolved an earlier switch from reproductive to infectious stages produced inside the host, which is a shorter latency time. Furthermore, reproductive parasite load was associated with this switch, suggesting that faster within-host replication was responsible for the faster switch towards transmission in the early parasites (reproductive stages are converted into infectious stages, similar to *Plasmodium*, see below and Figure 13.2).

These examples provide proof of principle of the theory in constrained serial passage experiments, bypassing population level dynamical feedbacks between the life and death of hosts, transmission and selection (Ebert 1998a). In a more realistic experimental setup, Silva and Koella (2024) carried out a selection experiment, where, instead of artificially shortening infection lifespan, they collected spores for the next generation of infection (in a serial passage experiment) from the first 50 naturally killed infected adult females of the mosquito *Anopheles gambiae*, or from the last 50 killed. After multiple passages, they found that early killers imposed less mortality in larvae and adults, and also had a later onset of the production of transmission spores, along with lower spore loads. These findings are exactly the opposite of the predicted outcomes and of the findings reported in the above studies. The authors speculate that a longer infectious period might intensify within-host competition and intra-host selection for fast-growing and virulent variants. They conclude that it may be important to allow natural transmission dynamics, which might uncover additional trade-offs, such as those between the timing of spore production, the quantity and quality of the transmission stages, or their survival in the environment (as seen above for viral pathogens).

Ebert and Mangin (1997) undertook a more integrative approach by testing the effect of host background mortality in freely evolving long-term populations of *D. magna* infected with the microsprodian parasite *Glugoides intestinalis*. They imposed extrinsic host mortality by repeatedly removing 80% of the hosts and replacing them with uninfected naive individuals, in order to keep population size constant. Ironically, just like in Silva and Koella (2024), the experiment produced the 'wrong' results: parasites from the replacement lines took more time to kill their hosts and produced fewer spores at the time of host death, compared to the non-replacement control. There has been some debate regarding the experimental design of this study (Ebert 1998b, Hochberg 1998, Day and Gandon 2007). Possibly, the fact that susceptible hosts were not added to the control treatment inadvertently increased levels of multiple infections and thereby selected for increased virulence (Ebert and Mangin 1997, Day and Gandon 2007). Note, however, the generally overlooked result that replacement lines, although producing fewer spores, had a higher per-spore transmission probability (Ebert and Mangin 1997). This could mean that the parasite responded to background mortality with a life history change (make better, rather than more spores) not considered by virulence theory, or alternatively, that stronger within-host competition in the control treatment (without removing hosts) led to a strategy of producing more offspring, but of lower quality. Ultimately, then, it would have been interesting to let parasites from the two long-term treatments compete under a 'replacement' regime to test which evolved strategy is more successful under this condition (see Elena 2001).

Ebert and Mangin (1997)'s study nicely illustrates the strength of experimental evolution. It can replicate true eco-evolutionary dynamics and give sufficient degrees of freedom to the study organism to produce surprising results. Importantly, however, since not everything is controlled, one needs to think carefully about what kind of ecological scenario is envisaged: Ebert and Mangin (1997)'s replacement treatment mimics an open system, with immigration from an uninfected source population, whereas the control treatment is a closed system without immigration. As shown by Day and Gandon (2007), these experimental choices are crucial because, from an evolutionary epidemiology point of view, selection by background mortality works indirectly, through modifying the availability of susceptible hosts in the population (see also Lion and Metz 2018). Thus, in an alternative mortality treatment without immigration, one could let populations deal with the loss of individuals naturally, without experimental interference. This could then be seen as a comparison between disturbed, boom-and-bust populations and undisturbed populations with more constant population sizes (see also Choo et al. 2003; and below).

Finally, infection lifespan may not only be terminated by host death, but also through the action of the host's immune system, or the action of drug treatment. This may favour the evolution of plastic responses, with parasites adjusting their development once the immune response kicks in. For example, after experimentally enhancing the immune response of infected hosts (rodents) through the injection of a cytokine, a parasitic nematode accelerated its maturation at the larval stage and boosted overall production of transmission stages (Babayan et al. 2010). This mirrors some of the above constitutive evolutionary changes, but, unlike in Cooper et al. (2002)'s study, a shift towards earlier age at maturity did not trade off with fecundity. Thus, the observed response can be likened to terminal investment (see above). Virulence was not measured in this study (Babayan et al. 2010).

In contrast, Reece et al. (2010) found that sublethal anti-malaria drug doses decreased, rather than increased, investment of *Plasmodium* parasites into the production of transmission stages. This was due to a decreased conversion rate of asexual reproductive stages into sexual transmission stages, possibly because drugs keep densities below a conversion signal (through quorum sensing). Reduced conversion might be an adaptive response, as diverting resources into current survival can ensure future transmission (Mideo et al. 2008). Interestingly, drug-resistant *Plasmodium* showed no conversion modification under drug treatment (Reece et al. 2010).

The contrasting results found in these two studies have interesting implications. The *Plasmodium* case shows that a response to a shortened infection lifespan does not necessarily involve adaptive life history plasticity, the alternative being drug resistance (Reece et al. 2010). The nematode case implies that immune system-enhancing treatments might have the unwanted effect of increasing rates of transmission (Babayan et al. 2010), a point generally to consider in non-transmission blocking vaccines (Gandon et al. 2001).

13.3.3 Russian Doll Effect: Breaking Down the Virulence-Transmission Trade-Off into its Components

The previous section showed that a classic prediction from virulence evolution (higher extrinsic host mortality increases virulence) translates into a life history question of parasite infection lifespan (Day 2003, Perlman 2009). If future reproduction is compromised, a pathogen should evolve to invest into more and earlier reproduction. At least in well-controlled experiments, this evolutionary shift is also associated with higher virulence (Cooper et al. 2002, Nidelet et al. 2009). As such, this is broadly consistent with the assumption of a virulence-transmission trade-off, underlying general theory.

Explicit investigations on the existence of the virulence-transmission trade-off have been accumulating over recent years, sometimes with mixed results, especially regarding the shape of the relationship (Doumayrou et al. 2013, Acevedo et al. 2019). It has also been shown that the trade-off can be differently expressed under contrasting ecological conditions, for example, when opportunities for transmission change the intensity of within-host density regulation (Godinho et al. 2023). Moreover, when breaking down virulence into actual parasite life history traits, we see additional trade-offs emerging, for example, between latency (time of onset of transmission) and the quality of transmission stages. Latency-transmission trade-offs are rarely included in virulence evolution models, and this might change predictions for virulence evolution (Saad-Roy et al. 2020). Additional trade-offs may render things even more complex. For plant pathogens, it is well known that 'infectivity' genes allowing host resistance to be overcome can impair subsequent within-host growth (Cruz et al. 2000, Bahri et al. 2009), suggesting a trade-off between infectivity and virulence, analogous to the costs of resistance (see above). Whether this virulence-infectivity trade-off holds as a general rule is unclear (Adiba et al. 2010, Leggett et al. 2012, Auld et al. 2014b, Gowler et al. 2023).

13.3.4 Where to Put the Reproductive Effort: Vertical *vs.* Horizontal Transmission

Many parasites have a mixed mode of transmission (Ebert 2013), that is the life history option to invest in either horizontal transmission (infectious spread between any individuals in a population), or vertical transmission (from parent to offspring). Total fitness is then determined by the sum of contributions through both pathways (Lipsitch et al. 1995). Depending on epidemiological conditions, one or the other pathway may be more rewarding (Lipsitch et al. 1995), so parasites need to make a 'decision' about where to put their reproductive effort. In particular, there is an intrinsic trade-off between the two pathways: while horizontal transmission should increase with increasing host exploitation (and thus damage to host survival or fecundity), vertical transmission aligns host and parasite fitness (parasite transmission is proportional to host fecundity), and thus host exploitation should be minimal. Moreover, in addition to the differential costs of reproductive effort in the two pathways, vertical and horizontal transmission can also represent two ecologically different life history choices. For example, vertical transmission does not rely on the presence of susceptible hosts and might therefore facilitate the establishment of infection in newly colonized patches. Vertical transmission might also facilitate the evolution of mutualistic host–symbiont relationships, even though phylogenetic evidence for this is weak (Sachs et al. 2011, Drew et al. 2021).

If horizontal transmission trades off with vertical transmission, we might expect a virulence continuum, whereby species with more horizontal transmission should become increasingly virulent (Ewald 1987). Such a pattern was found for tropical wasp species parasitising fig trees (Herre 1993), but otherwise, many counter-examples exist (parasites with high horizontal transmission, but not being virulent), and in certain systems, a trade-off between the two transmission modes does not exist in the first place (Ebert 2013). Moreover, even if a trade-off exists, Lipsitch et al. (1996) show that predictions are more complex, due to epidemiological feedbacks modulating the availability of susceptible hosts (see also Day and Gandon 2007, Weitz et al. 2019). Simply speaking, with increasing horizontal transmission, populations go from an epidemic into an endemic phase, with fewer and fewer susceptible hosts available, thereby diminishing the returns from additional investment in horizontal transmission (Lipsitch et al. 1996, Day and Gandon 2007).

Experimental studies on the evolution of mixed-mode parasites are mostly based on the simple qualitative prediction that parasites evolve higher or lower virulence, depending on which pathway provides more opportunities for transmission. Accordingly, certain studies artificially blocked either vertical or horizontal transmission in serial passage-like experiments (Drew et al. 2021). Thus, during multiple passages with exclusive horizontal transmission, Stewart et al. (2005) showed that a virus of barley (*Hordeum vulgare*) evolved with increased horizontal transmission success and became more virulent (causing reductions in seed production), even though within-host multiplication did not appear to be the cause of host fitness reduction. Conversely, passages of exclusive vertical seed transmission led to an increase in vertical transmission fidelity, reduced virulence and also reduced horizontal transmissibility (Stewart et al. 2005). Similarly, in a classic experiment using a temperate bacteriophage of *Escherichia coli*, Bull et al. (1991) blocked access to horizontal transmission, thereby forcing the phage into exclusive vertical transmission (during host cell division). In the complementary treatment, only phage originating from lysis (=horizontal transmission) were passaged. After 150 generations of exclusive vertical transmission, the vertical-trained phage had lost the lytic pathway and outcompeted the horizontal-trained phage in the absence of susceptible hosts (*i.e.*, horizontal transmission was not possible). However, when susceptible bacteria were added, the horizontal-trained phage held the upper hand. Along the same lines, when this biological system was allowed to alternate between periods of vertical and horizontal transmission, phage that had experienced more rounds of horizontal transmission was found to produce more transmission stages (virus titer) and cause stronger reductions in bacterial cell density, compared to phage that had only rarely gone through rounds of horizontal transmission (Bull et al. 1991).

What kinds of real-life ecological scenarios would favour investment in one or the other transmission pathway? As already seen above, theoretical models have identified the availability of susceptible hosts for horizontal transmission as one key factor driving evolution. Thus, Turner et al. (1998) carried out a modified version of Bull et al. (1991)'s experiment, using a bacterial plasmid as a 'symbiont'. In a vertical transmission long-term treatment, bacterial conjugation and plasmid transfer were not possible and thus all transmission was vertical, while, in two other treatments, vertical and horizontal transmission was unconstrained and different numbers of susceptible, plasmid-free bacteria were added. However, this manipulation of available hosts (none, few, many) for horizontal transmission had no clear effect on the evolution of the cost of carrying the plasmid for the host (*i.e.*, virulence), nor on the evolution of the conjugation rate (*i.e.*, the rate of horizontal transmission) of the plasmids (Turner et al. 1998). The authors speculate that horizontal transmission events may have been too infrequent (relative to vertical transmission) to produce consistent and similar evolutionary responses across the different replicates of the same long-term treatment.

Berngruber et al. (2013) placed the question of host availability in the context of emerging epidemics. In a theoretical model, the authors show that during the epidemic phase of an infection outbreak, more virulent variants of a mixed-mode parasite (*i.e.*, investing relatively more into horizontal transmission) have a selective advantage over the less virulent variants, because susceptible hosts represent a non-limiting resource and thus extra-investment in horizontal transmission is beneficial. However, when dynamics reach the endemic phase, fewer hosts are available, producing diminishing returns for investment in horizontal transmission (Lipsitch et al. 1996) and therefore favouring less virulent variants. These predictions were confirmed in competition experiments using latent bacteriophage of *E. coli*, which exists as a wildtype variant with a balanced ratio of lysis *vs.* lysogenesis (*i.e.*, horizontal *vs.* vertical transmission), or as a mutant variant killing more than 90% of infected hosts through lysis. Indeed, the virulent mutant rose to very high frequency when the experiment was started with a large number of susceptible hosts, but decreased in frequency when all hosts were already infected at the beginning of the experiment (Berngruber et al. 2013).

Based on similar reasoning, but focusing on variation in the opportunities for vertical transmission, Magalon et al. (2010) conducted an experiment with the above-mentioned *Paramecium–Holospora* system (Figure 13.2). This bacterial parasite produces specific morphs for vertical transmission (reproductive forms, transmitted during mitotic host division) and horizontal transmission (infectious forms). There is a plastic switch in the production of the two cell types (Kaltz and Koella 2003). In rapidly dividing populations, infected hosts carry primarily reproductive forms, whereas in populations at carrying capacity, reproductive forms are massively converted into infectious forms, causing more damage to the host (Restif and Kaltz 2006). In a long-term experiment, a serial dilution treatment produced frequent episodes of host population growth and therefore opportunity for vertical transmission, while baseline populations were kept near carrying capacity. After more than 150 host generations, parasites from the dilution treatment had a higher fidelity of vertical transmission and had less impact on host division and survival, compared to parasites from the baseline treatment. While these evolutionary changes did not initially appear to trade off with horizontal transmission capacity, parasites from dilution lines were found to have nearly completely lost their horizontal infectivity after two additional years of the experiment (Dusi et al. 2015). These results are consistent with the simple prediction that increased opportunities for vertical transmission leading to concomitant evolutionary adaptations, improve the efficacy of this pathway. In the long run, these changes can cause an evolutionary trade-off with horizontal transmission, as predicted by theory (Lipsitch et al. 1996). More broadly speaking, the study also illustrates an explicit ecological scenario, where parasite evolution is mediated through changes in host life history and host population dynamics.

Looking at environmental conditions from a somewhat different angle, Zilio et al. (2023a) studied the consequences of resource availability on the evolution of parasite transmission mode. Resource availability impacts not only host fecundity but also the transitions between different developmental states. This is important in the case of the microsporidian parasite *Edhazardia aedis*, which has tuned vertical and horizontal transmission to the different life cycle states of its host, the mosquito *A. aegypti* (Agnew and Koella 1999). Horizontal transmission is associated with the killing of infected larvae, while vertical transmission occurs through the eggs of infected females. Providing hosts with high or low resource availability, corresponding to permissive vs. restrictive ecological conditions, Zilio et al. (2023a) altered host development time (longer development under low resources), and thereby the time window for the evolving parasite to switch its development to maximize transmission from the host larval or adult state. After ten host generations, parasites evolving under permissive conditions tended to have a better vertical transmission potential, with emerging infected females more likely to harbour the right spore type for vertical transmission, have greater fecundity and live longer, compared to when infected with parasites from the low-food treatment. This is broadly consistent with the prediction that conditions favouring host survival and reproduction select for more benign parasite strategies, ensuring chances of vertical transmission.

13.3.5 Evolution of Virulence and Dispersal: Small and Big Worlds

Movement, dispersal and gene flow drive the spread of epidemics in networks and at global scales (Bahl et al. 2011, Pettersson et al. 2018) and determine the distribution of genetic variation and patterns of local adaptation (Ostfeld et al. 2005, Parratt and Laine 2016). In recent years, theoretical work has explored the interplay between spatial structure, dispersal and virulence evolution (Lion and Gandon 2016). A series of models, based on the virulence-transmission trade-off and adding spatially explicit settings, have shown that optimal virulence depends on levels of dispersal (Lion and Boots 2010). Namely, less virulent parasites evolve under low rather than high dispersal (Boots et al. 2004, Kamo et al. 2007). Under low dispersal, parasite transmission is limited to the availability of local hosts only (Kamo and Boots 2006, Lion et al. 2006). This increases the risk of 'self-shading' (*i.e.*, increased parasite kin competition) and parasite extinction, thereby favouring more prudent host exploitation strategies and lower virulence. In contrast, high dispersal can create 'small worlds', where the spatial structure disappears and more transmissible and virulent parasites are favoured (Boots and Sasaki 1999, Wild et al. 2009). By increasing dispersal, parasites have higher chances of encountering a new host population, achieving transmission not only in the local neighbourhood but also in more distant patches (Kamo and Boots 2006, Lion et al. 2006). Empirical tests manipulating population structure and connectedness in bacterial–virus (*E. coli*-phages) and insect–virus (*P. interpunctella*-PiGV) species supported such theoretical predictions (Kerr et al. 2006, Boots and Mealor 2007, Berngruber et al. 2015; but see Noël et al. 2023).

Typically, the above theory makes no specific assumptions about the way parasites disperse. However, predictions might change when the biology of the specific host–parasite systems are taken into account. In fact, in many cases, parasites rely on their host for dispersal. Actively moving hosts impose additional selective pressures on the parasites, and virulence can evolve towards higher (Griette et al. 2015) or lower levels (Osnas et al. 2015), depending on the relationships between virulence and dispersal. High virulence is favoured at the front of an epidemic when host availability is high and there is little interference with dispersal, the reason being that host exploitation and transmission are not limited by the availability of susceptible hosts (Griette et al. 2015, Lion and Gandon 2016). However, in various systems infection reduces host activity and movement, potentially imposing a virulence-dispersal trade-off. In this scenario, low-virulence parasite strains are predicted to travel with the host at the epidemic front, escaping more competitive (and more virulent) strains through faster dispersal (Osnas et al. 2015). This is consistent with observations in natural migratory populations of the monarch butterfly *Danaus plexippus* infected with the protozoan *Ophryocystis elektroscirrha* (Bradley and Altizer 2005), and in the bacterial parasite *Mycoplasma gallisepticum* travelling with geographically expanding populations of wild house finches (*Carpodacus mexicanus*) (Hawley et al. 2007). Similarly, in laboratory experiments with *D. magna* (Nørgaard et al. 2019), less virulent and more dispersive strains of the bacterial parasite *P. ramosa* were more successful at establishing in recently colonized host populations. Testing the predictions of Osnas et al. (2015), Nørgaard et al. (2021) carried out a long-term selection experiment in the aquatic *Paramecium*/*Holospora* system (see above), mimicking the ecological conditions at a range core and range edge of expanding host (and parasite) populations. Consistent with predictions, they found the emergence of an evolutionary virulence-dispersal trade-off. Compared to parasites from the range core treatment, range front parasites were less virulent and facilitated the dispersal of infected hosts. These changes were further associated with a longer development time to produce infectious stages and a shift towards vertical transmission.

The above examples highlight the importance of dispersal as a life history trait affecting parasite evolution. Clearly, parasites have various means of dispersal, either passively (wind, water currents), together with their host or actively by their own movement (*e.g.*, parasitic insects). Our main point here, in the context of virulence evolution, is that these life history details can create trait relationships that play out at the meta-population scale and add additional complexity (modifying basic predictions) to the classic evolutionary scenarios built on the more locally acting virulence-transmission trade-off. This further suggests that parasites, just like hosts, evolve multi-trait 'dispersal syndromes' (Cote et al. 2022), combining dispersal capacity with traits determining parasite development and transmission. Understanding the joint evolution of these traits requires appropriate empirical approaches and corresponding multivariate statistical tools (Zilio et al. 2023b). Similar challenges apply to theoretical treatments. So far, only very few models have studied the simultaneous evolution of interaction traits (resistance/infectivity, virulence) and dispersal in host and/or parasites (Drown et al. 2013, Chaianunporn and Hovestadt 2019, Deshpande et al. 2021). This task is far from trivial because the more parameters are allowed to (co)evolve, the more complex eco-evolutionary feedbacks and their spatio-temporal dynamics may become (Zilio et al. 2024).

13.3.6 Very Small Worlds: Parasite Life History Evolution in Response to Multiple Infections

The real world can in fact be really small, such that multiple parasites (strains or species) end up in the same host individual or co-circulate in the same population. This is the norm rather than the exception, and it may lead to interactions (positive or negative) between parasites and thus have profound effects on their life history evolution (Alizon et al. 2013, Zele et al. 2018). The majority of theoretical models investigating multiple infections consider competitive interactions between parasites and focus on consequences for virulence evolution (Alizon et al. 2013). So-called super-infection scenarios assume that one parasite rapidly replaces another in a host. These models typically predict that super-infection selects for more virulent parasites because these are more competitive, and hence more likely to take over the host. The interesting twist here is that good 'super-infectors' only need to be efficient within-host competitors, but not necessarily good at transmitting the infection (Alizon et al. 2013). This can lead to a scenario of evolutionary branching and co-existence with moderately virulent parasites that evolve to infect susceptible hosts (Alizon 2013).

In contrast to super-infection, co-infection scenarios assume that different parasites co-exist in the within-host environment (Alizon 2013; see Sofonea et al. 2017, for additional multiple infection possibilities). Co-infection models predict the evolution of more virulent parasites if they compete indirectly via the host immune system or for host resources (Frank 1996, Choisy and de Roode 2010, Alizon 2013, Alizon et al. 2013). Yet, there are no direct empirical tests of this prediction. In contrast, selection for lower virulence is expected if interactions among parasites are direct, via toxins or the production of public goods (*i.e.*, resources produced by a parasite that can be exploited by the whole parasite community; West and Buckling 2003, Gardner et al. 2004, Alizon et al. 2013). The production of costly toxins that kill competitors reduces their capacity for within-host growth and thus explains the evolution of lower virulence (compared to levels evolving under single-parasite conditions; Gardner et al. 2004). Conversely, the presence of multiple parasites reduces the benefit of producing public goods, as they may be exploited by genetically unrelated cheaters not paying the energetic costs of their production. Thus, selection is expected to favour lower public good production, which in turn leads to lower within-host growth, and thus lower virulence (West and Buckling 2003). Accordingly, experimental evolution of co-infecting parasites with direct interactions found selection for lower virulence. This was observed in mixed-strain treatments of the bacterial insect pathogen *Bacillus thuringiensis* (Garbutt et al. 2011), and in the pathogenic bacterium *S. aureus*, when evolving in *C. elegans* hosts carrying the protective microbe *E. faecalis* (Ford et al. 2016). In the latter case, *S. aureus* produces siderophores, public good molecules that scavenge iron and are used by the entire bacterial community. In co-infections with the protective microbe, *S. aureus* evolved to produce fewer siderophores and was less virulent, causing lower levels of host mortality (Ford et al. 2016), as predicted by the theory.

Other experimental studies show that the presence of competitor parasites in co-infection affects not only virulence but can induce life history shifts as well (Alizon et al. 2013, Zele et al. 2018). For instance, Duncan et al. (2015) found that the microsporidian *Vavraia culicis* increased the rate of spore production in co-infections with another microsporidian *E. aedis* in the mosquito host *A. aegyptii*. Similarly, the relationship between parasite load and speed to sporulation became positive in co-infected hosts (mixed-strain infections) but was negative in single (strain) infections of the fungus *Podosphaera plantaginis* in the plant host *Plantago lanceolata* (Laine and Makinen 2018). Both studies indicate that parasites can shift towards earlier timing and investment in transmission in response to co-infections. Consistent with this being an adaptive response, an experimental evolution study found that evolving the bacteriophage ϕ2 (of *P. fluorescens*) in populations composed of both single and co-infections *vs.* single infections only, selected for plasticity in lysis time. When these phages were in co-infection, they lysed their host sooner (Leggett et al. 2013).

In real-word scenarios, the occurrence of co-infected hosts will most likely be the result of sequential rather than simultaneous infections. When infections occur sequentially it is often the first parasite to arrive that has an advantage (Karvonen et al. 2019). However, an existing parasite can make a host more susceptible to secondary infections (Zele et al. 2018), which can in turn lead to within-host competition (Kamiya et al. 2018). Zilio and Koella (2020) corroborated the finding that *V. culicis* increases its spore production rate in co-infection with *E. aedis*, but only when it was the first parasite to infect the mosquito host (Zilio and Koella 2020). The sequence of parasite arrival in a host can also modify infection patterns in natural populations (Halliday et al. 2017, Clay et al. 2019). This work highlights the potential consequences of the sequence of parasite arrival for long-term epidemiological and evolutionary trajectories (Karvonen et al. 2019), but how this actually plays out remains to be elucidated.

While the above examples focus on the competitive within-host interactions in the classic 'multiple infection' context, the advent of studies on microbiota has made it clear that not all these interactions are antagonistic. Indeed, various studies have demonstrated that host microbiota can protect from, or facilitate, parasite infections (Azambuja et al. 2005, Koch and Schmid-Hempel 2011, Stevens et al. 2021). A recent study using *C. elegans* highlighted that an incomplete immunity induced by host microbiota can lead to the evolution of high virulence in the parasite (Hoang et al. 2024), similar to what is predicted for imperfect vaccination (Read et al. 2015). In contrast, parasites evolved on immune-compromised hosts exhibited an accumulation of mutations and lower virulence (Hoang et al. 2024). More studies need to explore the ecological processes, underlying mechanisms and evolutionary implications. Changes in these community interactions might have long-term consequences for host and parasite life history evolution. On the one hand, the use of probiotic microbes/protective symbionts could help against the spread of infectious disease (Ippolito et al. 2018), or be applied to human health and food production (McKenzie et al. 2018). On the other, by removing or adding a focal species in the microbial interaction network, we may face the risks of increased parasite virulence.

13.4 Concluding Remarks

This chapter shows how mortality threats and fecundity reductions drive host and parasite life history shifts, consistent with classic life history theory (Stearns 1989). One peculiarity is that these threats are mediated by the reciprocal player. For the host, parasites impose mortality or fitness reductions via virulence. For the parasite, mortality can parallel that of the host, and be imposed by the host's immune system or by competitors in the (host) environment.

Shifts in life history traits are not isolated but instead are embedded amongst other traits. This includes classic life history trade-offs (fecundity *vs.* survival, current *vs.* future reproduction), but also relationships with typical interaction traits, such as resistance, infectivity or virulence. This highlights the need to take a multi-trait approach to gain a better understanding of trait covariances and when trade-offs may constrain (co)evolution (Zilio et al. 2023b).

Of particular importance is the understanding that host–parasite epidemiological and (co)evolutionary interactions are dynamic, driven by concomitant changes in demography and evolution, giving rise to eco-(co)evolutionary feedbacks. Many of the above studies highlight this point, but still few have explored coevolution of life history traits. Just as hosts and parasites engage in coevolutionary cycles of resistance and infectivity, there may also be ongoing coevolution between, say, host age at maturity and parasite latency time. In this respect, Hood et al. (2015) show that radiation of *Rhagoletis pomonella* fruit fly races via adaptation to different host fruits led to a shift in the timing of eclosion associated with a parallel shift for their sympatric parasitoid wasp populations.

Finally, by considering dispersal as an integral part of host and parasite life histories, a meta-population perspective is added. The importance of spatial structure has long been recognized in the fields of epidemiology and (co)evolution (Ostfeld et al. 2005, Lion and Gandon 2016), but the notion that dispersal can evolve, jointly with other life history traits, and produce feedbacks across spatial scales is still relatively new (Zilio et al. 2024). An understanding of these dynamics also requires the integration of spatial modifiers, such as spatial network structure (Höckerstedt et al. 2022, Deshpande et al. 2024) or environmental heterogeneity (Fabre et al. 2022), suggesting that different optima of exploitation (or defence) strategies and life histories can evolve in different environments, depending on the available resources (Hochberg and van Baalen 1998).

A better understanding of these processes may improve our capacity to predict the spatial spread of (emergent) diseases and their evolving features, but also the fate of biological invasions (Poulin 2017). Chapter 24 in this volume places some of the questions addressed here in an explicit interventional context, enabling the better implementation of public health strategies.

In summary, future work needs to integrate multiple life history traits, including dispersal for both host and parasite, to garner a better understanding of the eco-evolutionary dynamics of host–parasite interactions, across multiple scales. It remains to be seen whether coevolution between life history traits can drive reciprocal changes in both players, in the same way in which classic theory views the coevolution between resistance and infectivity. Finally, (co)evolution likely involves additional players. Specifically, the role of symbiont–symbiont coevolution driving host and parasite life history traits may be greatly underestimated. Hosts often harbour multiple symbionts (*e.g.*, parasites, protective symbionts, microbiome) that can coevolve amongst themselves (symbiont–symbiont (co)evolution) and drive (evolutionary) change in the host phenotype.

References

Acevedo, M.A., Dillemuth, F.P., Flick, A.J., Faldyn, M.J., and Elderd, B.D. (2019). Virulence-driven trade-offs in disease transmission: a meta-analysis. *Evolution* 73: 636–647.

Adiba, S., Huet, M., and Kaltz, O. (2010). Experimental evolution of local parasite maladaptation. *Journal of Evolutionary Biology* 23: 1195–1205.

Agnew, P., and Koella, J.C. (1999). Life history interactions with environmental conditions in a host-parasite relationship and the parasite's mode of transmission. *Evolutionary Ecology* 13: 67–89.

Agnew, P., Koella, J.C., and Michalakis, Y. (2000). Host life-history responses to parasitism. *Microbes and Infection* 2: 891–896.

Agrawal, A.A., and Lively, C.M. (2002). Infection genetics: gene-for-gene versus matching-alleles models and all points in between. *Evolutionary Ecology Research* 4: 79–90.

Alizon, S. (2013). Co-infection and super-infection models in evolutionary epidemiology. *Interface Focus* 3: 20130031.

Alizon, S., de Roode, J.C., and Michalakis, Y. (2013). Multiple infections and the evolution of virulence. *Ecology Letters* 16: 556–567.

Alizon, S., Hurford, A., Mideo, N., and van Baalen, M. (2009). Virulence evolution and the trade-off hypothesis: history, current state of affairs and the future. *Journal of Evolutionary Biology* 22: 2452–2459.

Alizon, S., and Michalakis, Y. (2015). Adaptive virulence evolution: the good old fitness-based approach. *Trends in Ecology & Evolution* 30: 248–254.

Antonovics, J., and Thrall, P.H. (1997). The cost of resistance and the maintenance of genetic polymorphism in host-pathogen systems. *Proceedings of the Royal Society B: Biological Sciences* 257: 105–110.

Auld, S.K., Hall, S.R., Housley Ochs, J., Sebastian, M., and Duffy, M.A. (2014a). Predators and patterns of within-host growth can mediate both among-host competition and evolution of transmission potential of parasites. *The American Naturalist* 184(Suppl 1): S77–S90.

Auld, S.K., Penczykowski, R.M., Housley Ochs, J., Grippi, D.C., Hall, S.R., and Duffy, M.A. (2013). Variation in costs of parasite resistance among natural host populations. *Journal of Evolutionary Biology* 26: 2479–2486.

Auld, S.K.J.R., Wilson, P.J., and Little, T.J. (2014b). Rapid change in parasite infection traits over the course of an epidemic in a wild host-parasite population. *Oikos* 123: 232–238.

Azambuja, P., Garcia, E.S., and Ratcliffe, N.A. (2005). Gut microbiota and parasite transmission by insect vectors. *Trends in Parasitology* 21: 568–572.

Babayan, S.A., Read, A.F., Lawrence, R.A., Bain, O., and Allen, J.E. (2010). Filarial parasites develop faster and reproduce earlier in response to host immune effectors that determine filarial life expectancy. *PLoS Biology* 8(10): e1000525.

Bahl, J., Nelson, M.I., Chan, K.H., Chen, R.B., Vijaykrishna, D., Halpin, R.A., Stockwell, T.B., Lin, X.D., Wentworth, D.E., Ghedin, E., Guan, Y., Peiris, J.S.M., Riley, S., Rambaut, A., Holmes, E.C., and Smith, G.J.D. (2011). Temporally structured metapopulation dynamics and persistence of influenza A H3N2 virus in humans. *Proceedings of the National Academy of Sciences of the United States of America* 108: 19359–19364.

Bahri, B., Kaltz, O., Leconte, M., de Vallavieille-Pope, C., and Enjalbert, J. (2009). Tracking costs of virulence in natural populations of the wheat pathogen, *Puccinia striiformis* f.sp. *tritici*. *BMC Evolutionary Biology* 9: 26.

Bajgar, A., Kucerova, K., Jonatova, L., Tomcala, A., Schneedorferova, I., Okrouhlik, J., and Dolezal, T. (2015). Extracellular adenosine mediates a systemic metabolic switch during immune response. *PLoS Biology* 13(4): e1002135.

Bartlett, L.J., Wilfert, L., and Boots, M. (2018). A genotypic trade-off between constitutive resistance to viral infection and host growth rate. *Evolution* 72: 2749–2757.

Bell, G. and Smith, J. M. (1987) Short-term selection for recombination among mutually antagonistic species. *Nature* 328: 66–68.

Bernasconi, G., Antonovics, J., Biere, A., Charlesworth, D., Delph, L.F., Filatov, D., Giraud, T., Hood, M.E., Marais, G.A.B., McCauley, D., Pannell, J.R., Shykoff, J.A., Vyskot, B., Wolfe, L.M., and Widmer, A. (2009). Silene as a model system in ecology and evolution. *Heredity* 103: 5–14.

Berngruber, T.W., Froissart, R., Choisy, M., and Gandon, S. (2013). Evolution of virulence in emerging epidemics. *PLoS Pathogens* 9(3): e1003209.

Berngruber, T.W., Lion, S., and Gandon, S. (2015). Spatial structure, transmission modes and the evolution of viral exploitation strategies. *PLoS Pathogens* 11(4): e1004810.

Best, A., White, A., and Boots, M. (2008). Maintenance of host variation in tolerance to pathogens and parasites. *Proceedings of the National Academy of Sciences of the United States of America* 105: 20786–20791.

Best, A., White, A., and Boots, M. (2014). The coevolutionary implications of host tolerance. *Evolution* 68: 1426–1435.

Biere, A., and Antonovics, J. (1996). Sex-specific costs of resistance to the fungal pathogen *Ustilago violacea* (*Microbotryum violaceum*) in *Silene alba*. *Evolution* 50: 1098–1110.

Bonte, D., and Dahirel, M. (2017). Dispersal: a central and independent trait in life history. *Oikos* 126: 472–479.

Boots, M. (2011). The evolution of resistance to a parasite is determined by resources. *The American Naturalist* 178: 214–220.

Boots, M., and Bowers, R.G. (1999). Three mechanisms of host resistance to microparasites – avoidance, recovery and tolerance- show different evolutionary dynamics. *Journal of Theoretical Biology* 201: 13–23.

Boots, M., and Haraguchi, Y. (1999). The evolution of costly resistance in host-parasite systems. *The American Naturalist* 153: 359–370.

Boots, M., Hudson, P.J., and Sasaki, A. (2004). Large shifts in pathogen virulence relate to host population structure. *Science* 303: 842–844.

Boots, M., and Mealor, M. (2007). Local interactions select for lower pathogen infectivity. *Science* 315: 1284–1286.

Boots, M., and Sasaki, A. (1999). 'Small worlds' and the evolution of virulence: infection occurs locally and at a distance. *Proceedings of the Royal Society B: Biological Sciences* 266: 1933–1938.

Bowers, R.G., Boots, M., and Begon, M. (1994). Life-history trade-offs and the evolution of pathogen resistance: competition between host strains. *Proceedings of the Royal Society B: Biological Sciences* 257: 247–253.

Bradley, C.A., and Altizer, S. (2005). Parasites hinder monarch butterfly flight: implications for disease spread in migratory hosts. *Ecology Letters* 8: 290–300.

Brannelly, L.A., Webb, R.J., Jiang, Z., Berger, L., Skerratt, L.F., and Grogan, L.F. (2021). Declining amphibians might be evolving increased reproductive effort in the face of devastating disease. *Evolution* 75: 2555–2567.

Brown, G.P., Shine, R., and Rollins, L.A. (2024). A biological invasion modifies the dynamics of a host-parasite arms race. *Proceedings of the Royal Society B: Biological Sciences* 291: 20232403.

Buckingham, L.J., and Ashby, B. (2022). Coevolutionary theory of hosts and parasites. *Journal of Evolutionary Biology* 35: 205–224.

Budischak, S.A., and Cressler, C.E. (2018). Fueling defense: effects of resources on the ecology and evolution of tolerance to parasite infection. *Frontiers in Immunology* 9: 2453.

Bull, J.J., Molineux, I.J., and Rice, W.R. (1991). Selection of benevolence in a host-parasite system. *Evolution* 45: 875–882.

Bull, J.J., Pfennig, D.W., and Wang, I.N. (2004) Genetic details, optimization and phage life histories. *Trends in Ecology & Evolution* 19: 76–82.

Chadwick, W., and Little, T.J. (2005). A parasite-mediated life-history shift in *Daphnia magna*. *Proceedings of the Royal Society B: Biological Sciences* 272: 505–509.

Chaianunporn, T., and Hovestadt, T. (2011). The role of mobility for the emergence of diversity in victim-exploiter systems. *Journal of Evolutionary Biology* 24: 2473–2484.

Chaianunporn, T., and Hovestadt, T. (2019). Dispersal evolution in metacommunities of tri-trophic systems. *Ecological Modelling* 395: 28–38.

Choisy, M., and de Roode, J.C. (2010). Mixed infections and the evolution of virulence: effects of resource competition, parasite plasticity, and impaired host immunity. *The American Naturalist* 175: E105–E118.

Choo, K., Williams, P.D., and Day, T. (2003). Host mortality, predation and the evolution of parasite virulence. *Ecology Letters* 6: 310–315.

Clay, P.A., Dhir, K., Rudolf, V.H.W., and Duffy, M.A. (2019). Within-host priority effects systematically alter pathogen coexistence. *The American Naturalist* 193: 187–199.

Cooper, V.S., Reiskind, M.H., Miller, J.A., Shelton, K.A., Walther, B.A., Elkinton, J.S., and Ewald, P.W. (2002). Timing of transmission and the evolution of virulence of an insect virus. *Proceedings of the Royal Society B: Biological Sciences* 269: 1161–1165.

Cote, J., Dahirel, M., Schtickzelle, N., Altermatt, F., Ansart, A., Blanchet, S., Chaine, A.S., De Laender, F., De Raedt, J., Haegeman, B., Jacob, S., Kaltz, O., Laurent, E., Little, C. J., Madec, L., Manzi, F., Masier, S., Pellerin, F., Pennekamp, F., Therry, L., Vong, A., Winandy, L., Bonte, D., Fronhofer, E.A., and Legrand, D. (2022). Dispersal syndromes in challenging environments: a cross-species experiment. *Ecology Letters* 25: 2675–2687.

Cressler, C.E., Mc, L.D., Rozins, C., Van Den Hoogen, J., and Day, T. (2016). The adaptive evolution of virulence: a review of theoretical predictions and empirical tests. *Parasitology* 143: 915–930.

Cruz, C.M.V., Bai, J.F., Oña, I., Leung, H., Nelson, R.J., Mew, T.W., and Leach, J.E. (2000). Predicting durability of a disease resistance gene based on an assessment of the fitness loss and epidemiological consequences of avirulence gene mutation. *Proceedings of the National Academy of Sciences of the United States of America* 97: 13500–13505.

Day, T. (2003). Virulence evolution and the timing of disease life-history events. *Trends in Ecology & Evolution* 18: 113–118.

Day, T., and Gandon, S. (2007). Applying population-genetic models in theoretical evolutionary epidemiology. *Ecology Letters* 10: 876–888.

de Roode, J.C., Yates, A.J., and Altizer, S. (2008). Virulence-transmission trade-offs and population divergence in virulence in a naturally occurring butterfly parasite. *Proceedings of the National Academy of Sciences of the United States of America* 105: 7489–7494.

Decaestecker, E., van de Moortel, B., Mukherjee, S., Gurung, A., Stoks, R., and De Meester, L. (2024). Hierarchical eco-evo dynamics mediated by the gut microbiome. *Trends in Ecology & Evolution* 39: 165–174.

Deshpande, J.N., Dakos, V., Kaltz, O., and Fronhofer, E.A. (2024). Landscape structure drives eco-evolution in host-parasite systems. bioRxiv 2023.2010.2024.563775.

Deshpande, J.N., Kaltz, O., and Fronhofer, E.A. (2021). Host-parasite dynamics set the ecological theatre for the evolution of state- and context-dependent dispersal in hosts. *Oikos* 130: 121–132.

Dong, Y., Manfredini, F., and Dimopoulos, G. (2009). Implication of the mosquito midgut microbiota in the defense against malaria parasites. *PLoS Pathogens* 5(5): e1000423.

Doumayrou, J., Avellan, A., Froissart, R., and Michalakis, Y. (2013). An experimental test of the transmission-virulence trade-off hypothesis in a plant virus. *Evolution* 67: 477–486.

Drew, G.C., Stevens, E.J., and King, K.C. (2021). Microbial evolution and transitions along the parasite-mutualist continuum. *Nature Reviews Microbiology* 19: 623–638.

Drown, D.M., Dybdahl, M.F., and Gomulkiewicz, R. (2013). Consumer-resource interactions and the evolution of migration. *Evolution* 67: 3290–3304.

Duffield, K.R., Bowers, E.K., Sakaluk, S.K., and Sadd, B.M. (2017). A dynamic threshold model for terminal investment. *Behavioural Ecology and Sociobiology* 71: 185.

Duncan, A.B., Agnew, P., Noel, V., and Michalakis, Y. (2015). The consequences of co-infections for parasite transmission in the mosquito *Aedes aegypti*. *Journal of Animal Ecology* 84: 498–508.

Duncan, A.B., Fellous, S., and Kaltz, O. (2011). Reverse evolution: selection against costly resistance in disease-free microcosm populations of *Paramecium caudatum*. *Evolution* 65: 3462–3474.

Duncan, A.B., Mitchell, S.E., and Little, T.J. (2006). Parasite-mediated selection and the role of sex and diapause in Daphnia. *Journal of Evolutionary Biology* 19: 1183–1189.

Duputié, A., and Massol, F. (2013). An empiricist's guide to theoretical predictions on the evolution of dispersal. *Interface Focus* 3(6): 20130028.

Dusi, E., Gougat-Barbera, C., Berendonk, T.U., and Kaltz, O. (2015). Long-term selection experiment produces breakdown of horizontal transmissibility in parasite with mixed transmission mode. *Evolution* 69: 1069–1076.

Ebert, D. (1998a). Experimental evolution of parasites. *Science* 282: 1432–1435.

Ebert, D. (1998b). Infectivity, multiple infections, and the genetic correlation between within-host growth and parasite virulence: a reply to Hochberg. *Evolution* 52: 1869–1871.

Ebert, D. (2013). The epidemiology and evolution of symbionts with mixed-mode transmission. *Annual Review of Ecology, Evolution, and Systematics* 44: 623–643.

Ebert, D., and Mangin, K.L. (1997). The influence of host demography on the evolution of virulence of a microsporidian gut parasite. *Evolution* 51: 1828–1837.

Egido, J.E., Costa, A.R., Aparicio-Maldonado, C., Haas, P.J., and Brouns, S.J.J. (2022). Mechanisms and clinical importance of bacteriophage resistance. *FEMS Microbiology Reviews* 46(1): fuab048.

Elena, S.F. (2001). Evolutionary history conditions the timing of transmission in vesicular stomatitis virus. *Infection, Genetics and Evolution* 1: 151–159.

Ewald, P.W. (1983). Host-parasite relations, vectors, and the evolution of disease severity. *Annual Review of Ecology and Systematics* 14: 465–485.

Ewald, P.W. (1987). Transmission modes and evolution of the parasitism-mutualism continuum. *Annals of the New York Academy of Sciences* 503: 295–306.

Fabre, F., Burie, J.B., Ducrot, A., Lion, S., Richard, Q., and Djidjou-Demasse, R. (2022). An epi-evolutionary model for predicting the adaptation of spore-producing pathogens to quantitative resistance in heterogeneous environments. *Evolutionary Applications* 15: 95–110.

Fellous, S., and Salvaudon, L. (2009). How can your parasites become your allies? *Trends in Parasitology* 25: 62–66.

Ford, S.A., Kao, D., Williams, D., and King, K.C. (2016). Microbe-mediated host defence drives the evolution of reduced pathogen virulence. *Nature Communications* 7: 13430.

Frank, S.A. (1996). Models of parasite virulence. *The Quarterly Review of Biology* 71: 37–78.

Frickel, J., Sieber, M., and Becks, L. (2016). Eco-evolutionary dynamics in a coevolving host-virus system. *Ecology Letters* 19: 450–459.

Fronhofer, E.A., and Altermatt, F. (2015). Eco-evolutionary feedbacks during experimental range expansions. *Nature Communications* 6: 844.

Gandon, S., Agnew, P., and Michalakis, Y. (2002a). Coevolution between parasite virulence and host life-history traits. *The American Naturalist* 160: 374–388.

Gandon, S., Mackinnon, M.J., Nee, S., and Read, A.F. (2001). Imperfect vaccines and the evolution of pathogen virulence. *Nature* 414: 751–756.

Gandon, S., van Baalen, M., and Jansen, A.A. (2002b). The evolution of parasite virulence, superinfection and host resistance. *The American Naturalist* 159: 658–669.

Garbutt, J., Bonsall, M.B., Wright, D.J., and Raymond, B. (2011). Antagonistic competition moderates virulence in *Bacillus thuringiensis*. *Ecology Letters* 14: 765–772.

Gardner, A., West, S.A., and Buckling, A. (2004). Bacteriocins, spite and virulence. *Proceedings of the Royal Society B: Biological Sciences* 271: 1529–1535.

Gibson, A.K., Petit, E., Mena-Ali, J., Oxelman, B., and Hood, M.E. (2013). Life-history strategy defends against disease and may select against physiological resistance. *Ecology and Evolution* 3: 1741–1750.

Gillespie, J.H. (1975). Counting the cost of disease resistance. *Trends in Ecology & Evolution* 13: 8–9.

Godinho, D.P., Rodrigues, L.R., Lefevre, S., Delteil, L., Mira, A.F., Fragata, I.R., Magalhaes, S., and Duncan, A.B. (2023). Limited host availability disrupts the genetic correlation between virulence and transmission. *Evolution Letters* 7: 58–66.

Gowler, C.D., Essington, H., O'Brien, B., Shaw, C.L., Bilich, R.W., Clay, P.A., and Duffy, M.A. (2023). Virulence evolution during a naturally occurring parasite outbreak. *Evolutionary Ecology* 37: 113–129.

Graham, A.L., Hayward, A.D., Watt, K.A., Pilkington, J.G., Pemberton, J.M., and Nussey, D.H. (2010). Fitness correlates of heritable variation in antibody responsiveness in a wild mammal. *Science* 330: 662–665.

Griette, Q., Raoul, G,. and Gandon, S. (2015). Virulence evolution at the front line of spreading epidemics. *Evolution* 69: 2810–2819.

Gurney, J., Brown, S.P., Kaltz, O., and Hochberg, M.E. (2020). Steering phages to combat bacterial pathogens. *Trends in Microbiology* 28: 85–94.

Hall, M.D., and Ebert, D. (2012). Disentangling the influence of parasite genotype, host genotype and maternal environment on different stages of bacterial infection in *Daphnia magna*. *Proceedings of the Royal Society B: Biological Sciences* 279: 3176–3183.

Hall, A.R., Scanlan, P.D., Morgan, A.D., and Buckling, A. (2011). Host-parasite coevolutionary arms races give way to fluctuating selection. *Ecology Letters* 14: 635–642.

Halliday, F.W., Umbanhowar, J., and Mitchell, C.E. (2017). Interactions among symbionts operate across scales to influence parasite epidemics. *Ecology Letters* 20: 1285–1294.

Hamilton, P.T., Peng, F., Boulanger, M.J., and Perlman, S.J. (2016). A ribosome-inactivating protein in a *Drosophila* defensive symbiont. *Proceedings of the National Academy of Sciences of the United States of America* 113: 350–355.

Hawley, D.M., Davis, A.K., and Dhondt, A.A. (2007). Transmission-relevant behaviours shift with pathogen infection in wild house finches (*Carpodacus mexicanus*). *Canadian Journal of Zoology* 85: 752–757.

Herre, E. A. (1993). Population-structure and the evolution of virulence in nematode parasites of fig wasps. *Science* 259: 1442–1445.

Hoang, K.L., Read, T.D., and King, K.C. (2024). Incomplete immunity in a natural animal-microbiota interaction selects for higher pathogen virulence. *Current Biology* 34: 1357–1363

Hochberg, M.E. (1998). Establishing genetic correlations involving parasite virulence. *Evolution* 52: 1865–1868.

Hochberg, M.E., and Ives, A.R. (1999). Can natural enemies enforce geographical range limits? *Ecography* 22: 268–276.

Hochberg, M.E., Michalakis, Y., and de Meeus, T. (1992). Parasitism as a constraint on the rate of life-history evolution. *Journal of Evolutionary Biology* 5: 491–504.

Hochberg, M.E., and van Baalen, M. (1998). Antagonistic coevolution over productivity gradients. *The American Naturalist* 152: 620–634.

Höckerstedt, L., Numminen, E., Ashby, B., Boots, M., Norberg, A., and Laine, A.L. (2022). Spatially structured eco-evolutionary dynamics in a host-pathogen interaction render isolated populations vulnerable to disease. *Nature Communications* 13: 6018.

Holt, R.D., and Keitt, T.H. (2000). Alternative causes for range limits: a metapopulation perspective. *Ecology Letters* 3: 41–47.

Hood, G.R., Forbes, A.A., Powell, T.H., Egan, S.P., Hamerlinck, G., Smith, J.J., and Feder, J.L. (2015). Sequential divergence and the multiplicative origin of community diversity. *Proceedings of the National Academy of Sciences of the United States of America* 112: E5980–E5989.

Ippolito, M.M., Denny, J.E., Langelier, C., Sears, C.L., and Schmidt, N.W. (2018). Malaria and the microbiome: a systematic review. *Clinical Infectious Diseases* 67: 1831–1839.

Jaenike, J., Unckless, R., Cockburn, S.N., Boelio, L.M., and Perlman, S.J. (2010). Adaptation via symbiosis: recent spread of a *Drosophila* defensive symbiont. *Science* 329: 212–215.

Jones, M.E., Cockburn, A., Hamede, R., Hawkins, C., Hesterman, H., Lachish, S., Mann, D., McCallum, H., and Pemberton, D. (2008). Life-history change in disease-ravaged Tasmanian devil populations. *Proceedings of the National Academy of Sciences of the United States of America* 105: 10023–10027.

Kaltz, O., and Koella, J.C. (2003) Host growth conditions regulate the plasticity of horizontal and vertical transmission in *Holospora undulata*, a bacterial parasite of the protozoan *Paramecium caudatum*. *Evolution* 57: 1535–1542.

Kaltz, O., and Shykoff, J.A. (2001). Male and female *Silene latifolia* plants differ in per-contact risk of infection by a sexually transmitted disease. *Journal of Ecology* 89: 99–109.

Kamiya, T., Mideo, N., and Alizon, S. (2018). Coevolution of virulence and immunosuppression in multiple infections. *Journal of Evolutionary Biology* 31: 995–1005.

Kamo, M., and Boots, M. (2006). The evolution of parasite dispersal, transmission, and virulence in spatial host populations. *Evolutionary Ecology Research* 8: 1333–1347.

Kamo, M., Sasaki, A., and Boots, M. (2007). The role of trade-off shapes in the evolution of parasites in spatial host populations: an approximate analytical approach. *Journal of Theoretical Biology* 244: 588–596.

Kannoly, S., Singh, A., and Dennehy, J.J. (2022). An optimal lysis time maximizes bacteriophage fitness in quasi-continuous culture. *MBio* 13(3): e0359321.

Karvonen, A., Jokela, J., and Laine, A.L. (2019). Importance of sequence and timing in parsite coinfections. *Trends in Parasitology* 35: 109–118.

Kerr, B., Neuhauser, C., Bohannan, B.J., and Dean, A.M. (2006). Local migration promotes competitive restraint in a host-pathogen 'tragedy of the commons'. *Nature* 442: 75–78.

Koch, H., and Schmid-Hempel, P. (2011). Socially transmitted gut microbiota protect bumble bees against an intestinal parasite. *Proceedings of the National Academy of Sciences of the United States of America* 108: 19288–19292.

Koella, J.C., Agnew, P., and Michalakis, Y. (1998). Coevolutionary interactions between host life histories and parasite life cycles. *Parasitology* 116: S47–S55.

Koella, J.C., and Boëte, C. (2002). A genetic correlation between age at pupation and melanization immune response of the yellow fever mosquito *Aedes aegypti*. *Evolution* 56: 1074–1079.

Koella, J.C., and Restif, O. (2001). Coevolution of parasite virulence and host life history. *Ecology Letters* 4: 207–214.

Korves, T.M., and Bergelson, J. (2003). A developmental response to pathogen infection in *Arabidopsis*. *Plant Physiology* 133: 339–347.

Koskella, B., Taylor, T.B., Bates, J., and Buckling, A. (2011). Using experimental evolution to explore natural patterns between bacterial motility and resistance to bacteriophages. *The ISME Journal* 5: 1809–1817.

Kubisch, A., Holt, R.D., Poethke, H.-J., and Fronhofer, E.A. (2014). Where am I and why? Synthesizing range biology and the eco-evolutionary dynamics of dispersal. *Oikos* 123: 5–22.

Kutzer, M.A., and Armitage, S.A. (2016). Maximising fitness in the face of parasites: a review of host tolerance. *Zoology* 119: 281–289.

Lafferty, K.D. (1993). The marine snail, *Cerithidea californica*, matures at smaller sizes where parasitism is high. *Oikos* 68: 3–11.

Laine, A.L., and Makinen, H. (2018). Life-history correlations change under coinfection leading to higher pathogen load. *Evolution Letters* 2: 126–133.

Leggett, H.C., Benmayor, R., Hodgson, D.J., and Buckling, A. (2013). Experimental evolution of adaptive phenotypic plasticty in a parasite. *Current Biology* 23: 139–142.

Leggett, H.C., Cornwallis, C.K., and West, S.A. (2012). Mechanisms of pathogenesis, infective dose and virulence in human parasites. *PLoS Pathogens* 8(2): e1002512.

Leitao, A.B., Arunkumar, R., Day, J.P., Geldman, E.M., Morin-Poulard, I., Crozatier, M., and Jiggins, F.M. (2020). Constitutive activation of cellular immunity underlies the evolution of resistance to infection in *Drosophila*. *eLife* 9: e59095.

Lenski, R.E. (1988). Experimental studies of pleiotropy and epistasis in *Escherichia coli*. I. Variation in competitive fitness among mutants resistant to virus T4. *Evolution* 42: 425–432.

Lenski, R.E., and Levin, B.R. (1985). Constraints on the coevolution of bacteria and virulent phage: a model, some experiments, and predictions for natural communities. *The American Naturalist* 125: 585–602.

Lion, S., and Boots, M. (2010). Are parasites "prudent" in space? *Ecology Letters* 13: 1245–1255.

Lion, S., and Gandon, S. (2016). Spatial evolutionary epidemiology of spreading epidemics. *Proceedings of the Royal Society B: Biological Sciences* 283: 20161170.

Lion, S., and Metz, J.A.J. (2018). Beyond R_0 maximisation: on pathogen evolution and environmental dimensions. *Trends in Ecology & Evolution* 33: 458–473.

Lion, S., van Baalen, M., and Wilson, W.G. (2006). The evolution of parasite manipulation of host dispersal. *Proceedings of the Royal Society B: Biological Sciences* 273: 1063–1071.

Lipsitch, M., Nowak, M.A., Ebert, D., and May, R.M. (1995). The population dynamics of vertically and horizontally transmitted parasites. *Proceedings of the Royal Society B: Biological Sciences* 260: 321–327.

Lipsitch, M., Siller, S., and Nowak, M.A. (1996). The evolution of virulence in pathogens with vertical and horizontal transmission. *Evolution* 50: 1729–1741.

Little, T.J., Shuker, D.M., Colegrave, N., Day, T., and Graham, A.L. (2010). The coevolution of virulence: tolerance in perspective. *PLoS Pathogens* 6(9): e1001006.

Lombaert, E., Estoup, A., Facon, B., Joubard, B., Grégoire, J.C., Jannin, A., Blin, A., and Guillemaud, T. (2014) Rapid increase in dispersal during range expansion in the invasive ladybird *Harmonia axyridis*. *Journal of Evolutionary Biology* 27: 508–517.

Lopez-Pascua, L., Gandon, S., and Buckling, A. (2011). Abiotic heterogeneity drives parasite local adaptation in coevolving bacteria and phages. *Journal of Evolutionary Biology* 25: 187–195.

Lopez-Pascua, L., Hall, A.R., Best, A., Morgan, A.D., Boots, M., and Buckling, A. (2014). Higher resources decrease fluctuating selection during host-parasite coevolution. *Ecology Letters* 17: 1380–1388.

Luong, L.T., and Polak, M. (2007). Costs of resistance in the *Drosophila – Macrocheles* system: a negative genetic correlation between ectoparasite resistance and reproduction. *Evolution* 61: 1391–1402.

MacDonald, H., and Brisson, D. (2023). Parasite-mediated selection on host phenology. *Ecology and Evolution* 13: e10107.

Mackinnon, M.J., and Read, A.F. (1999). Genetic relationships between parasite virulence and transmission in the rodent malaria *Plasmodium chabaudi*. *Evolution* 53: 689–703.

Magalon, H., Nidelet, T., Martin, G., and Kaltz, O. (2010). Host growth conditions influence experimental evolution of life-history and virulence of a parasite with vertical and horizontal transmission. *Evolution* 64: 2126–2138.

Martin, R.A., and Tate, A.T. (2023). Pleiotropy promotes the evolution of inducible immune responses in a model of host-pathogen coevolution. *PLoS Computational Biology* 19(4): e1010445.

Martinez, J., Cogni, R., Cao, C., Smith, S., Illingworth, C.J., and Jiggins, F.M. (2016). Addicted? Reduced host resistance in populations with defensive symbionts. *Proceedings of the Royal Society B: Biological Sciences* 283: 20160778.

Mayer, M., Shine, R., and Brown, G.P. (2021). Rapid divergence of parasite infectivity and host resistance during a biological invasion. *Biological Journal of the Linnean Society* 132: 861–871.

Mayo-Munoz, D., Pinilla-Redondo, R., Birkholz, N., and Fineran, P.C. (2023). A host of armor: prokaryotic immune strategies against mobile genetic elements. *Cell Reports* 42: 112672.

Mayo-Munoz, D., Pinilla-Redondo, R., Camara-Wilpert, S., Birkholz, N., and Fineran, P.C. (2024). Inhibitors of bacterial immune systems: discovery, mechanisms and applications. *Nature Reviews Genetics* 25: 237–254.

McGonigle, J.E., Leitao, A.B., Ommeslag, S., Smith, S., Day, J.P., and Jiggins, F.M. (2017). Parallel and costly changes to cellular immunity underlie the evolution of parasitoid resistance in three *Drosophila* species. *PLoS Pathogens* 13(10): e1006683.

McKenzie, V.J., Kueneman, J.G., and Harris, R.N. (2018). Probiotics as a tool for disease mitigation in wildlife: insights from food production and medicine. *Annals of the New York Academy of Sciences* 1429: 18–30.

Meaden, S., Paszkiewicz, K., and Koskella, B. (2015). The cost of phage resistance in a plant pathogenic bacterium is context-dependent. *Evolution* 69: 1321–1328.

Michalakis, Y., and Hochberg, M.E. (1994). Parasitic effects on host life-history traits: a review of recent studies. *Parasite* 1: 291–294.

Mideo, N., Day, T., and Read, A.F. (2008). Modelling malaria pathogenesis. *Cellular Microbiology* 10: 1947–1955.

Mideo, N., Nelson, W.A., Reece, S.E., Bell, A.S., Read, A.F., and Day, T. (2011). Bridging scales in the evolution of infectious disease life-histories: Application. *Evolution* 65: 3298–3310.

Minchella, D.J., and Loverde, T.L. (1981). A cost of increased early reproductive effort in the snail *Biomphalaria glabrata*. *The American Naturalist* 118: 876–881.

Mitchell, S.E., Read, A.F., and Little, T.J. (2004). The effect of a pathogen epidemic on the genetic structure and reproductive strategy of the crustacean *Daphnia magna*. *Ecology Letters* 7: 848–858.

Nidelet, T., Koella, J.C., and Kaltz, O. (2009). Effects of shortened host life span on the evolution of parasite life history and virulence in a microbial host-parasite system. *BMC Evolutionary Biology* 9: 65.

Nielsen, M.L., and Holman, L. (2011). Terminal investment in multiple sexual signals: immune-challenged males produce more attractive pheromones. *Functional Ecology* 26: 20–28.

Noël, E., Lefèvre, S., Varoqui, M., and Duncan, A.B. (2023). The scale of competition impacts parasite virulence evolution. *Evolutionary Ecology* 37: 153–163.

Nørgaard, L.S., Phillips, B.L., and Hall, M.D. (2019). Infection in patchy populations: contrasting pathogen invasion success and dispersal at varying times since host colonization. *Evolution Letters* 3: 555–566.

Nørgaard, L.S., Zilio, G., Saade, C., Gougat-Barbera, C., Hall, M.D., Fronhofer, E.A., and Kaltz, O. (2021). An evolutionary trade-off between parasite virulence and dispersal at experimental invasion fronts. *Ecology Letters* 24: 739–750.

O'Keefe, K.J., and Antonovics, J. (2002). Playing by different rules: the evolution of virulence in sterilizing pathogens. *The American Naturalist* 159: 597–605.

Ohlberger, J., Langangen, O., Edeline, E., Olsen, E.M., Winfield, I.J., Fletcher, J.M., James, J.B., Stenseth, N.C., and Vollestad, L.A. (2011). Pathogen-induced rapid evolution in a vertebrate life-history trait. *Proceedings of the Royal Society B: Biological Sciences* 278: 35–41.

Osnas, E.E., Hurtado, P.J., and Dobson, A.P. (2015). Evolution of pathogen virulence across space during an epidemic. *The American Naturalist* 185: 332–342.

Ostfeld, R.S., Glass, G.E., and Keesing, F. (2005). Spatial epidemiology: an emerging (or re-emerging) discipline. *Trends in Ecology & Evolution* 20: 328–336.

Paredes, J.C., Herren, J.K., Schupfer, F., and Lemaitre, B. (2016). The role of lipid competition for endosymbiont-mediated protection against parasitoid wasps in *Drosophila*. *MBio* 7(4): 10.1128/mbio.01006-16.

Parratt, S.R., and Laine, A.L. (2016). The role of hyperparasitism in microbial pathogen ecology and evolution. *The ISME Journal* 10: 1815–1822.

Perlman, R.L. (2009). Life histories of pathogen populations. *International Journal of Infectious Diseases* 13: 121–124.

Pettersson, J.H.O., Bohlin, J., Dupont-Rouzeyrol, M., Brynildsrud, O.B., Alfsnes, K., Cao-Lormeau, V.M., Gaunt, M.W., Falconar, A.K., De Lamballerie, X., Eldholm, V., Musso, D., and Gould, E.A. (2018). Re-visiting the evolution, dispersal and epidemiology of Zika virus in Asia. *Emerging Microbes & Infections* 7: 79.

Phillips, B.L., Brown, G.P., Webb, J.K., and Shine, R. (2006). Invasion and the evolution of speed in toads. *Nature* 439: 803–803.

Poulin, R. (2017). Invasion ecology meets parasitology: advances and challenges. *International Journal for Parasitology: Parasites and Wildlife* 6: 361–363.

Price, T.D., and Kirkpatrick, M. (2009). Evolutionarily stable range limits set by interspecific competition. *Proceedings of the Royal Society B: Biological Sciences* 276: 1429–1434.

Raberg, L., Sim, D., and Read, A.F. (2007). Disentangling genetic variation for resistance and tolerance to infectious diseases in animals. *Science* 318: 812–814.

Rafaluk-Mohr, C., Gerth, M., Sealey, J.E., Ekroth, A.K.E., Aboobaker, A.A., Kloock, A., and King, K.C. (2022). Microbial protection favors parasite tolerance and alters host-parasite coevolutionary dynamics. *Current Biology* 32: 1593–1598 e3.

Read, A.F., Baigent, S.J., Powers, C., Kgosana, L.B., Blackwell, L., Smith, L.P., Kennedy, D.A., Walkden-Brown, S.W., and Nair, V.K. (2015). Imperfect vaccination can enhance the transmission of highly virulent pathogens. *PLoS Biology* 13(7): e1002198.

Reaney, L.T., and Knell, R.J. (2010). Immune activation but not male quality affects female current reproductive investment in a dung beetle. *Behavioral Ecology* 21: 1367–1372.

Reece, S.E., Ali, E., Schneider, P., and Babiker, H.A. (2010). Stress, drugs and the evolution of reproductive restraint in malaria parasites. *Proceedings of the Royal Society B: Biological Sciences* 277: 3123–3129.

Restif, O., and Kaltz, O. (2006). Condition-dependent virulence in a horizontally and vertically transmitted bacterial parasite. *Oikos* 114: 148–158.

Restif, O., and Koella, J.C. (2003). Shared control of epidemiological traits in a coevolutionary model of host-parasite interactions. *The American Naturalist* 161: 827–836.

Saad-Roy, C.M., Wingreen, N.S., Levin, S.A., and Grenfell, B.T. (2020). Dynamics in a simple evolutionary-epidemiological model for the evolution of an initial asymptomatic infection stage. *Proceedings of the National Academy of Sciences of the United States of America* 117: 11541–11550.

Saastamoinen, M., Bocedi, G., Cote, J., Legrand, D., Guillaume, F., Wheat, C.W., Fronhofer, E.A., Garcia, C., Henry, R., Husby, A., Baguette, M., Bonte, D., Coulon, A., Kokko, H., Matthysen, E., Niitepold, K., Nonaka, E., Stevens, V.M., Travis, J.M.J., Donohue, K., Bullock, J.M., and Delgado, M.D. (2018). Genetics of dispersal. *Biological Reviews* 93: 574–599.

Sachs, J.L., Skophammer, R.G., and Regus, J.U. (2011). Evolutionary transitions in bacterial symbiosis. *Proceedings of the National Academy of Sciences of the United States of America* 108: 10800–10807.

Scheele, B.C., Skerratt, L.F., Hunter, D.A., Banks, S.C., Pierson, J.C., Driscoll, D.A., Byrne, P.G., and Berger, L. (2017). Disease-associated change in an amphibian life-history trait. *Oecologia* 184: 825–833.

Schlippe Justicia, L., Mayer, M., Shine, R., Shilton, C., and Brown, G. P. (2022). Divergence in host–parasite interactions during the cane toad's invasion of Australia. *Ecology and Evolution* 12: e9220

Shine, R., Alford, R.A., Blennerhasset, R., Brown, G.P., DeVore, J L., Ducatez, S., Finnerty, P., Greenlees, M., Kaiser, S.W., McCann, S., Pettit, L., Pizzatto, L., Schwarzkopf, L., Ward-Fear, G., and Phillips, B.L. (2021). Increased rates of dispersal of free-ranging cane toads (*Rhinella marina*) during their global invasion. *Scientific Reports* 11: 23574.

Shykoff, J.A., Bucheli, E., and Kaltz, O. (1996a). Anther smut disease in *Dianthus silvester* (Caryophyllaceae): natural selection on floral traits. *Evolution* 51: 383–392.

Shykoff, J.A., Bucheli, E., and Kaltz, O. (1996b). Flower lifespan and disease risk. *Nature* 379: 779–779.

Shykoff, J.A., and Kaltz, O. (1997). Effects of the anther smut fungus *Microbotryum violaceum* on host life-history patterns in *Silene latifolia* (Caryophyllaceae). *International Journal of Plant Sciences* 158: 164–171.

Silva, L.M., and Koella, J.C. (2024). Complex interactions in the life cycle of a simple parasite shape the evolution of virulence. bioRxiv. https://doi.org/10.1101/2024.01.28.577571

Simms, E.L., and Rausher, M.D. (1987). Costs and benefits of plant resistance to herbivory. *The American Naturalist* 130: 570–581.

Sofonea, M.T., Alizon, S., and Michalakis, Y. (2017). Exposing the diversity of multiple infection patterns. *Journal of Theoretical Biology* 419: 278–289.

Sørensen, J.G., Kristensen, T.N., and Loeschcke, V. (2003). The evolutionary and ecological role of heat shock proteins. *Ecology Letters* 6: 1025–1037.

Stearns, S.C. (1989). Trade-offs in life-history evolution. *Functional Ecology* 3: 259–268.

Stevens, E.J., Bates, K.A., and King, K.C. 2021). Host microbiota can facilitate pathogen infection. *PLoS Pathogens* 17(5): e1009514.

Stewart, A.D., Logsdon, J.M., and Kelley, S.E. (2005). An empirical study of the evolution of virulence under both horizontal and vertical transmission. *Evolution* 59: 730–739.

Susi, H., Laine, A.-L., and Power, A. (2015). The effectiveness and costs of pathogen resistance strategies in a perennial plant. *Journal of Ecology* 103: 303–315.

Taylor, T.B., and Buckling, A. (2013). Bacterial motility confers fitness advantage in the presence of phages. *Journal of Evolutionary Biology* 26: 2154–2160.

Thrall, P.H., Biere, A., and Antonovics, J. (1993). Plant life-history and disease susceptibility – the occurrence of *Ustilago violacea* on different species within the Caryophyllaceae. *Journal of Ecology* 81: 489–498.

Turner, P.E., Cooper, V.S., and Lenski, R.E. (1998). Tradeoff between horizontal and vertical modes of transmission in bacterial plasmids. *Evolution* 52: 315–329.

Vale, P.F., Fenton, A., and Brown, S.P. (2014). Limiting damage during infection: lessons from infection tolerance for novel therapeutics. *PLoS Biology* 12(1): e1001769.

Vale, P.F., and Little, T.J. (2012). Fecundity compensation and tolerance to a sterilizing pathogen in *Daphnia*. *Journal of Evolutionary Biology* 25: 1888–1896.

Valenzuela-Sanchez, A., Wilber, M.Q., Canessa, S., Bacigalupe, L.D., Muths, E., Schmidt, B.R., Cunningham, A.A., Ozgul, A., Johnson, P.T.J., and Cayuela, H. (2021). Why disease ecology needs life-history theory: a host perspective. *Ecology Letters* 24: 876–890.

van Baalen, M., and Sabelis, M.W. (1995). The dynamics of multiple infection and the evolution of virulence. *The American Naturalist* 146: 881–910.

Vorburger, C., and Perlman, S.J. (2018). The role of defensive symbionts in host-parasite coevolution. *Biological Reviews* 93: 1747–1764.

Walsman, J.C., Duffy, M.A., Caceres, C.E., and Hall, S.R. (2023). "Resistance is futile": weaker selection for resistance by abundant parasites increases prevalence and depresses host density. *The American Naturalist* 201: 864–879.

Weitz, J.S., Li, G. L., Gulbudak, H., Cortez, M.H., and Whitaker, R.J. (2019). Viral invasion fitness across a continuum from lysis to latency. *Virus Evolution* 5(1): vez006.

West, S.A., and Buckling, A. (2003). Cooperation, virulence and siderophore production in bacterial parasites. *Proceedings of the Royal Society B: Biological Sciences* 270: 37–44.

Westra, E.R., van Houte, S., Oyesiku-Blakemore, S., Makin, B., Broniewski, J.M., Best, A., Bondy-Denomy, J., Davidson, A., Boots, M., and Buckling, A. (2015). Parasite exposure drives selective evolution of constitutive versus inducible defense. *Current Biology* 25: 1043–1049.

Wild, G., Gardner, A., and West, S.A. (2009). Adaptation and the evolution of parasite virulence in a connected world. *Nature* 459: 983–986.

Williams, G.G. (1966). Natural selection, the cost of reproduction, and a refinement of lack's principle. *The American Naturalist* 100: 687–690.

Woolhouse, M.E.J., Webster, J.P., Domingo, E., Charlesworth, B., and Levin, B.R. (2002). Biological and biomedical implications of the co-evolution of pathogens and their hosts. *Nature Genetics* 32: 569–577.

Zele, F., Magalhaes, S., Kefi, S., and Duncan, A.B. (2018). Ecology and evolution of facilitation among symbionts. *Nature Communications* 9: 4869.

Zeller, M., and Koella, J.C. (2017). The role of the environment in the evolution of tolerance and resistance to a pathogen. *The American Naturalist* 190: 389–397.

Zilio, G., Deshpande, J.N., Duncan, A.B., Fronhofer, E.A., and Kaltz, O. (2024). Dispersal evolution and eco-evolutionary dynamics in antagonistic species interactions. *Trends in Ecology & Evolution* 39: 666–676. https://doi.org/10.1016/j.tree.2024.03.006

Zilio, G., Kaltz, O., and Koella, J.C. (2023a). Resource availability for the mosquito *Aedes aegypti* affects the transmission mode evolution of a microsporidian parasite. *Evolutionary Ecology* 37: 31–51.

Zilio, G., and Koella, J.C. (2020). Sequential co-infections drive parasite competition and the outcome of infection. *Journal of Animal Ecology* 89: 2367–2377.

Zilio, G., Nørgaard, L.S., Gougat-Barbera, C., Hall, M.D., Fronhofer, E.A., and Kaltz, O. (2023b). Travelling with a parasite: the evolution of resistance and dispersal syndromes during experimental range expansion. *Proceedings of the Royal Society B: Biological Sciences* 290: 20221966.

Zilio, G., Nørgaard, L.S., Petrucci, G., Zeballos, N., Gougat-Barbera, C., Fronhofer, E.A., and Kaltz, O. (2021). Parasitism and host dispersal plasticity in an aquatic model system. *Journal of Evolutionary Biology* 34: 1316–1325.

14

How Do Microbial Symbionts Shape the Life Histories of Multicellular Organisms?

Elad Chiel[1] and Yuval Gottlieb[2]

[1] *Department of Biology and Environment, University of Haifa-Oranim, Tivon, Israel*
[2] *Koret School of Veterinary Medicine, The Hebrew University of Jerusalem, Rehovot, Israel*

14.1 Introduction

'It is a microbial world'. Until about 30 years ago, such a statement may have evoked responses of doubt or even aversion, but today it is common knowledge that every living creature hosts and even depends on microbial partners to survive, develop and reproduce. Over a hundred years of research revealed the indispensable ties between microbes and multicellular organisms, and the fascinating mechanisms behind these ties are increasingly understood with the recent huge leaps in molecular biology and sequencing technologies.

Pioneering evolutionary theories on the origin of eukaryotic cells from merged prokaryotic organisms – Symbiogenesis by C. Mereschkowski in 1910 (Kowallik and Martin 2021), Symbionticism by I. Wallin (Wallin 1927) and the Serial Endosymbiotic Theory by L. Margulis (Sagan 1967) – were the first to consider microbial interactions as a promoter for gaining novel traits. Since then, numerous new findings led to the formation of the hologenome theory of evolution (Zilber-Rosenberg and Rosenberg 2008), which argues that since all animals and plants establish symbiotic relationships with microorganisms, selection is acting on both host and microbe genomes as a consortium to shape the holobiont fitness and adaptation under certain environments, and therefore the holobiont can be considered as a unit for natural selection. Nevertheless, this theory is criticized for being too simplistic and only relevant in cases of obligate interactions (see definition in the next section) with strict vertical transmission, such as mitochondria in eukaryotic cells or nutritional symbionts of sap-feeding insects (Table 14.1). Indeed, as will be detailed below, there are ample examples of symbioses between multicellular hosts and macro- and microorganisms in which each partner has an independent life stage(s) that can evolve independently. Douglas and Werren (2016) argued that hosts and microbes should be considered instead as ecological communities with a broad range of interactions, transmission pathways and levels of fidelity. A similar view was suggested by Stencel and Wloch-Salamon (2018), writing that 'the idea of the holobiont (a host and its associated microbes) represents, in fact, a unit of cooperation: a system built of cooperating units, some of which may be co-inherited along with the host (and thus act along with the host as a single Darwinian individual), while others are linked only via functional integration'. On the theoretical level, criticism was raised, for example by Koskella and Bergelson (2020), who claimed that the term 'holobiont' is not specific enough to conceptually advance the field, and that '... we would do well to retain an emphasis on the behaviour of individuals species that comprise communities while appreciating the significance of these interactions to both host and symbiont ecology and evolution'. Lastly, Morris et al. (2018) add, 'The definitions and significance of the terms 'holobiont' and 'hologenome' are understood differently by different authors. [...] researchers in disparate fields [...] may be misled by the hologenome concept into believing in a unified "superorganism" that does not exist'.

Alongside this debate, there seems to be a consensus that microbial symbionts have profound effects on many aspects of their hosts' biology. The study of microbiome is a flourishing research field, with numerous evidence of the importance of the microbiome to human and animal health (Ross et al. 2019, Peixoto et al. 2021, VanEvery et al. 2022, Rock and Turnbaugh 2023), which is also gaining public awareness. For example, Yong (2016) wrote: 'Every one of us is a zoo in our own right – a colony enclosed within a single body. A multispecies collective. An entire world'. In this chapter, we review selected examples that highlight the centrality of microbial symbionts in the biology of their hosts, with an emphasis on life history traits.

Life History Evolution: Traits, Interactions, and Applications, First Edition. Edited by Michal Segoli and Eric Wajnberg.
© 2025 John Wiley & Sons Ltd. Published 2025 by John Wiley & Sons Ltd.

Table 14.1 Selected examples of nutritional symbioses between eukaryote hosts and microbes.

Host	Obligate nutritional symbiont(s)	Nutrients provided by the symbionts to the host	Nutrients provided by the host to the symbionts	References
Hemipteran insects that feed on plant sap	Specific bacteria (each host superfamily harbours a unique lineage)	Essential amino acids, vitamins, carotenoids	Non-essential amino acids, sugars	Douglas (2006)
Blood-feeding arthropods, leeches and vampire bats	Specific bacteria	Mostly B vitamins	Not specifically known	Rio et al. (2016), Zepeda Mendoza et al. (2018)
Lower termites	Flagellates, bacteria, Archaea	Lignocellulose digestion, nitrogen fixation	Uric acid	Scharf and Peterson (2021)
Fungi-farming insects (higher termites, wood-feeding beetles, ants from the tribe Attini)	Fungi and bacteria	Lignocellulose digestion, nitrogen fixation	Not specifically known	Li et al. (2021)
Cnidarians	Dinoflagellate algae (genus *Symbiodinium*)	photosynthetic products, oxygen	Waste products	Kirk and Weis (2016)
Legumes	Rhizobia (nitrogen-fixing bacteria)	Fixed nitrogen	Photosynthetic products	Udvardi and Poole (2013)
Ruminants	Multiple bacteria, fungi and protozoa species	Lignocellulose digestion	Not specifically known	Mizrahi et al. (2021)

14.2 Categories of Microbial Symbiosis

Symbiotic interactions can be classified by several, somewhat overlapping, criteria:

- By the outcome of the interaction. (1) Mutualism, *i.e.*, when both organisms benefit from the interaction; (2) parasitism, *i.e.*, when one organism benefits and the other is harmed; (3) commensalism, *i.e.*, when one organism benefits and the other is neither benefited nor harmed; or (4) amensalism, *i.e.*, when one organism causes harm to the other without any cost or benefits to itself. The interpretation of neutral effect on one side in commensalism and amensalism may merely be because the actual nature of the interaction (beneficial or detrimental) is not fully understood. Moreover, the nature of the interaction between species may change, *e.g.*, from mutualism to parasitism and *vice versa*, according to environmental conditions and genetic factors.
- By the degree of dependence. (1) Obligate symbiosis, *i.e.*, when one or both organisms cannot survive, develop and reproduce without the other; or (2) facultative symbiosis, *i.e.*, when the interaction is not crucial for survival, development or reproduction. Interactions can be obligate for one organism and facultative for the other.
- By the symbiont population genetic structure, which is derived from the symbionts' transmission mode: (1) Closed symbiosis, *i.e.*, symbionts are clonal within hosts due to strict vertical transmission; (2) open symbiosis, *i.e.*, symbionts are acquired from the environment and can be horizontally transmitted, therefore they undergo frequent recombination; or (3) mixed symbiosis, *i.e.*, symbionts' transmission is mostly vertical but horizontal acquisitions also occur, leading to some degree of recombination (Perreau and Moran 2021).

14.3 How Microbial Symbionts Are Involved in Essential Biological Functions of Their Hosts?

Microbial symbionts can influence their hosts in multiple ways, such as by supplying them with essential nutrients, conferring them protection against pathogens and natural enemies and mediating or altering their reproduction. These effects are ultimately interconnected in determining the life history traits of host populations.

Causality of symbiont effect on host life history traits is tested by comparison of symbiotic and aposymbiotic individuals, or by describing changes in the microbial community under various conditions. However, the mechanisms (or paths) of these effects are not easily resolved, as the specific molecular and metabolic interactions which lead to the phenotypic outcome, and the selection directions and states, are currently mostly unknown. In the following sections, we summarize evidence for symbionts' effects on life history phenotypes related to key biological functions.

14.4 Nutritional Microbial Symbionts

With few exceptions (*e.g.*, Hammer et al. 2019), most animals house microbial symbionts in the digestive tract which play a central part in food digestion. In fact, specialized diets of unbalanced food sources (*e.g.*, plant sap, blood, and wood; see Table 14.1) evolved via the interactions with microbes that synthesize the missing nutrients or break down food molecules that the host cannot utilize by itself. A recent meta-analysis demonstrated that B vitamins are the only limiting nutrients consistently associated with the evolution of obligate symbiosis in insects. Additionally, on an evolutionary scale, the cooperation with nutritional symbionts has led to an immense species proliferation of herbivore insects, whereas in strict blood-feeders species diversification has been very limited (Cornwallis et al. 2023). Generally, the association with symbionts that are relevant for nutrition is obligate for the host, such that if the symbiosis is disrupted, the host's development is severely impaired, usually leading to its demise. From the microbial symbiont's side, the association in some cases is also obligate (*e.g.*, nutritional symbionts of blood and plant sap-feeding arthropods), such that the symbionts have lost their ability to survive apart from their host, but in other cases, the microbial symbiont can be free-living. A well-known example of the latter is that of cnidarians and their unicellular algal symbionts, *Symbiodinium* (also known as Zooxanthellae, Figure 14.1). *Symbiodinium* are housed in gastrodermal cells of cnidarians and provide them with most of their photosynthetic products. In return, *Symbiodinium* receives inorganic nutrients from the host (Fransolet et al. 2012). The cnidarian hosts cannot survive without *Symbiodinium*, whereas *Symbiodinium* can survive freely in the seawater. Increased seawater temperatures, irradiation and pollution result in *Symbiodinium* loss by the cnidarians, a phenomenon known as coral bleaching, leading to a collapse of the entire reefs' ecosystems (van Woesik et al. 2022). One of the approaches to save cnidarians from this fate is by supplying them with probiotic bacteria (Maire and van Oppen 2022).

Many plants also rely on symbionts to supply them nutrients, most notably plants that house nitrogen-fixing bacteria and/or mycorrhizal fungi (see also Chapter 17, this volume). About 85% of land plants depend on mycorrhizal fungi for growth and survival (most of them with arbuscular mycorrhiza) which function as an extension of the roots: the fungi transport water and minerals from the soil to the plant, and the plant transfers sugars produced by photosynthesis to the fungi (Brundrett and Tedersoo 2018). There are two groups of nitrogen-fixing soil bacteria that establish tight symbiotic interactions with plants: *Rhizobia* (Proteobacteria) associated with legumes, and *Frankia* (Actinobacteria) associated with Fagales, Rosales and Cucurbitales. Both groups of bacteria induce the formation of root nodules by the plant, where they are housed and fix atmospheric nitrogen (Figure 14.2). Mycorrhizal fungi and nitrogen-fixing bacteria have substantial positive synergistic effects on the fitness and life history of the host plants, although the effects vary considerably (Larimer et al. 2010, 2014, van der Heijden et al. 2015, Pahua et al. 2018, Afkhami et al. 2020, 2021).

Ultimately, affecting the nutritional state of a host can result in further effects on certain life history traits such as longevity, size, maturity rate, and hence reproductive parameters. For example, the basic biological requirement of nutritional

Figure 14.1 A sea anemone, *Exaiptasia pallida*, inoculated with algal cells of *Breviolum minutum* (previously known as *Symbiodinium minutum*). Scale bar: 1 mm. *Source:* Jinkerson et al., (2022) / ELSEVIER / CC BY-NC 4.0.

Figure 14.2 A *Trifolium* sp. plant (left) and its root nodules (the whitish bulbs) harbouring nitrogen-fixing bacteria of the genus *Rhizobium* (right). *Source:* Photos: E. Chiel.

supplementation of B vitamins directly affects fertilisation and consequent sex ratio in whiteflies (Wang et al. 2020a) and fertility in ticks (Duron et al. 2018, Ben-Yosef et al. 2020). On the other hand, negative effect of the adult's reproduction time was found in the grain beetle *Oryzaephilus surinamensis* harbouring a nutritional symbiont (Engl et al. 2020). Indirect effects on nutrition can arise from symbionts inducing longer ingestion time to their hosts, such as in the beetle *Sitobion miscanthi* where their symbionts are essential to harden their cuticle during adult development, and thus also affect their longevity and survival rates (Li et al. 2023).

14.5 Reproductive Microbial Symbionts

Microbial symbionts can have major effects on the reproduction mode and fitness of their hosts, from the pre-mating stage to the post-natal offspring development. Accumulating evidence from mammals (including humans), fish and birds highlights an interaction between the host endocrine system, particularly corticosteroid hormones, and its microbiome, both in the gut and in the reproductive organs, but the direction of these interactions is not always clear (Comizzoli et al. 2021, Sisk-Hackworth et al. 2023). During the pre-birth stages, the hormones–microbiome connection can influence mating signals (*e.g.*, odours, colours, and behaviours), sexual receptivity, sperm quality and a healthy course of pregnancy, which are attributed both to gut, vaginal and uterine microbiome (Comizzoli et al. 2021, Adnane and Chapwanya 2022, Gudnadottir et al. 2022, Holyoak et al. 2022, Kwon and Lee 2022, Hugon and Golos 2023). During birth, the neonate is inoculated with cervix, and vagina-inhabiting microbiota which gradually establishes in the offspring and may have long-term influence on its health and development (Sarkar et al. 2021, Enav et al. 2022, Woodruff et al. 2022), although this is not always supported (Dos Santos et al. 2023). During infancy, the maternal skin, oral and milk microbiome are constantly transmitted to the offspring and have long-lasting effects on its well-being (Enav et al. 2022, Porro et al. 2023). An important aspect of this topic is the transitions in microbiome community structure that occur in captive animals and the correlation of these transitions to reduction in the animals' reproductive success and health. Therefore, this aspect is gaining increasing attention for animal conservation (Dallas and Warne 2022, Crates et al. 2023). For all the above, it is often difficult to demonstrate causality, and actual life history parameter measurements are mostly lacking. Moreover, focusing on certain microbes can simplify our understanding of the consequences of symbionts and reproduction-related traits. A few examples are presented here: The extensively studied model organism, the pea aphid, is parthenogenetic under favourable conditions. However, in conditions that induce sexual reproduction, it was demonstrated that facultative symbionts decrease the pea aphids' longevity and fecundity, and one symbiont (*Spiroplasma*) can induce male killing (see below) (Simon et al. 2011). Another example comes from the compost earthworm, *Eisenia andrei*, harbouring a mixed nephridial community of symbionts. In various combinations, the symbionts promote the production of more cocoons and offspring, and faster sexual maturity

(Viana et al. 2018). An example of increased mating success is found in the bug *Riptortus pedestris* harbouring the symbiont *Caballeronia*. Although it does not affect survival, the symbiont positively affects male body weight and size. Moreover, symbiotic males have larger hind legs, which is beneficial in male–male competition. This is probably because symbiotic males have enhanced feeding efficiency. Moreover, symbiotic males display enhanced movement and flight capacity (Jung and Lee 2023).

A completely different set of microbial effects on reproduction is 'reproductive manipulations', which are common in arthropods. The term refers to several symbiotic bacterial species that manipulate the reproduction of their hosts to produce more female offspring (at the expense of male offspring) probably because these symbionts are transmitted to the offspring only by the mother, whereas their presence in the males is a dead end (from the point of view of the symbiont). To date, several bacterial lineages have been found to be reproductive manipulators: *Wolbachia pipientis* (the most studied and famous one), *Cardinium*, *Rickettsia* spp., *Spiroplasma* spp., *Arsenophonus nasoniae* and *Rickettsiella* (Drew et al. 2019, Doremus and Hunter 2020, Perlmutter and Bordenstein 2020, Pollmann et al. 2022). There are four known types of reproductive manipulations: (1) Parthenogenesis: symbiont-carrying individuals are females that produce female offspring only. This manipulation was documented only in haplodiploid hosts (mainly in parasitoid Hymenoptera) infected with either *Wolbachia*, *Cardinium* and *Rickettsia*. (2) Feminization: symbiont-carrying males are converted to phenotypic functional females. This manipulation was documented in hosts infected with either *Wolbachia* or *Cardinium* in isopods, spiders, lepidopteran and hemipteran insects. (3) Male killing: symbiont-carrying males die during embryonic development, freeing more resources for the female offspring. This manipulation was documented in hosts infected with either *Wolbachia, Rickettsia, Arsenophonus* or *Spiroplasma* in lepidopteran and dipteran insects. (4) Cytoplasmic incompatibility (CI): a cross between a symbiont-carrying male and a symbiont-free female is incompatible, resulting in a gradual decrease of symbiont-free individuals from the population. This manipulation was demonstrated in multiple insects and acari taxa.

These bacteria are also termed 'reproductive parasites', because, apparently, their presence primarily affects their own transmission and dispersal, and only secondarily affects their host's life history in various ways. Thus, to explain the selection state and stability of these widespread phenomena, numerous studies measured the effects of reproductive manipulators on different life history traits. For example, double infection of the mite *Tetranychus truncates* with *Wolbachia* and *Spiroplasma* induced CI, but, in addition, females produced more eggs relative to other strains. This double infection did not affect host survival, but reduced female's and male's developmental time (Zhang et al. 2018). Other examples of reproductive manipulator effects are related to nutrition and defence, as described in this chapter for other symbiotic interaction types. Sometimes, these so-called 'reproductive manipulators' symbionts do not manipulate reproduction at all but affect other life history traits, such as in the cabbage root fly *Delia radicum*, which is infected with *Wolbachia* which does not induce any reproductive manipulation but reduces egg hatch rate, improves larval viability and prolongs development time (Lopez et al. 2018).

In the past *ca*. 20 years, *Wolbachia*-induced CI has been utilized for controlling mosquitos (Incompatible Insect Technique, IIT). This is done by rearing mosquitos that have been transfected with CI-inducing *Wolbachia* and then releasing large numbers of *Wolbachia*-infected males in areas infested with *Wolbachia*-free wild mosquito populations. Mosquito females usually mate only once; therefore, the more *Wolbachia*-infected males are released the higher the chances of copulation between *Wolbachia*-infected males and wild *Wolbachia*-free females, resulting in failure of egg fertilisation. Another approach is to release both males and females that are *Wolbachia*-infected. The rationale here is that not only does *Wolbachia* spread in the population via CI, but *Wolbachia* also blocks the multiplication of mosquito-transmitted pathogens (*Plasmodium* [the causative agent of malaria], dengue, chikungunya, yellow fever, zika, and West Nile virus diseases), and shortens the life of the adult mosquitos. Therefore, the outcome of this approach is a replacement of the disease-transmitting mosquitos with non-transmitting mosquitos (Kaur et al. 2021, Gong et al. 2023). IIT is currently implemented in 14 countries in Asia, Latin America and Oceania (https://www.worldmosquitoprogram.org/).

14.6 Defensive Microbial Endosymbionts

Another central way in which microbial symbionts affect the life history of their hosts is by improving their tolerance to pathogens, natural enemies and abiotic mortality factors. In return, the symbionts receive housing and nutrients, and in some cases also guaranteed vertical transmission to the host's offspring. These mutualistic interactions are usually facultative and vary according to site-specific selection pressures. One of the best-studied defensive symbiosis systems is that of aphids and a variety of facultative bacterial symbionts that confer tolerance against pathogens (Scarborough et al. 2005), heat stress (Montllor et al. 2002) and parasitoids (Oliver et al. 2010; see Figure 14.3). Protection from parasitoids is conferred

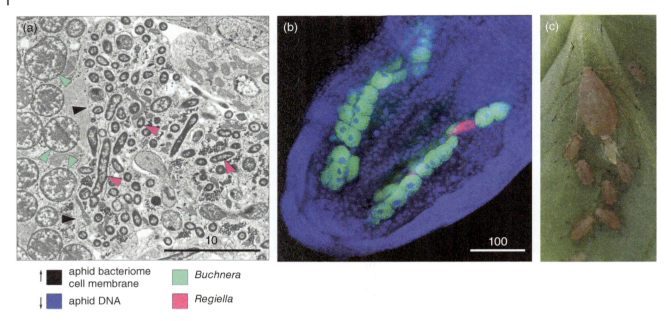

Figure 14.3 (a) Transmission electron micrograph showing elongate cells of the facultative aphid symbiont *Regiella insecticola* within a bacteriocyte (pink arrows), and nearby bacteriocytes containing the obligate nutritional symbiont *Buchnera aphidicola* (green arrows). Black arrows indicate the bacteriome cell membrane (*Source:* Photo by J. White and N. Moran / CC BY 4.0). (b) Position of symbiont-containing bacteriocytes within the abdomen as revealed by fluorescent *in situ* hybridisation using diagnostic probes. Blue is a general DNA stain, highlighting aphid nuclei, red indicates *R. insecticola* and green indicates *B. aphidicola* (*Source:* Photo by R. Koga / CC BY 4.0. (c) An adult and nymphs of the pea aphid, *Acyrthsiphon pisum*, the host of these symbionts. Scales are in µm in (a) and (b) and in mm in (c). *Source:* The International Aphid Genomics Consortium (2010) / PUBLIC LIBRARY OF SCIENCE (PLOS) / Public domain.

by a bacteriophage residing in the bacterial symbiont *Hamiltonella defensa*. The bacteriophage encodes for a toxin that eliminates the parasitoid's egg (oviposited into the body of the aphid). However, the aphid invests energy and nutrients to maintain and propagate the bacterial symbionts, and that comes at the expense of the aphid's fitness (Oliver et al. 2006, Łukasik et al. 2013). For example, *H. defensa* reduces the lifespan and lifetime reproduction of aphids (Vorburger and Gouskov 2011), and the symbiont *Candidatus* Regiella insecticola reduces the number of winged offspring and alters the timing of sexual reproduction, although these effects are host genotype-dependent (Leonardo and Mondor 2006). Thus, the interaction between the three partners, defensive symbionts–aphids–aphid parasitoids, operates as a feedback loop: in habitats where parasitoids are abundant, aphids that do not harbour the defensive symbionts are strongly selected against, whereas in habitats with few or no parasitoids, unprotected aphids have an advantage over the protected ones. Upper trophic levels, such as hyperparasitoids (parasitoids that attack parasitoids), are indirectly affected by this loop as well (Pekas et al. 2023). Such a feedback loop is likely relevant to many other host–symbiont–antagonist systems. Other parasitoid species of the families Braconidae and Ichneumonidae harbour their own defensive symbionts: Polydnaviruses which are injected into the host's body along with the eggs and inactivate the immune system of the parasitoid's host (a caterpillar), thus protecting the parasitoid eggs from elimination (Strand and Burke 2020). Polydnaviruses are integrated into their hosts' genome, therefore it is not possible to measure the effects of these viruses on life history traits, such as lifespan and fecundity of the hosts.

Defensive symbiosis is ubiquitous also between plants and microbes. Many plants, for example, harbour endophytic fungi that produce bioactive molecules (such as alkaloids) which deter or harm herbivores and pathogens and aid the plant to tolerate heat, salinity and other abiotic stresses. An interesting example is the grass *Dichanthelium lanuginosum* var. *thermale* which grows in hot and acidic soils in geothermal areas (like in Yellowstone Park, USA). The grass harbours the endophytic fungus *Curvularia protuberata*, which harbours a mycovirus (*Curvularia* thermal tolerance virus, CThTV). Both are essential for the plant's heat tolerance (Márquez et al. 2007, Morsy et al. 2010). Due to these important beneficial effects, endophytic fungi are a prolific research ground for the improvement of agricultural crops as well as in search for medicinal applications (Gupta et al. 2020). For example, endophytic fungi can increase the growth rate, reproductive yield and biomass of rice (Redman et al. 2011). Additional prominent examples of microbial defensive symbiosis are listed in Table 14.2.

Table 14.2 Selected examples of defensive symbioses between eukaryote hosts and microbes.

Host	Defensive symbiont(s)	Defensive effect	Examples for effects on life history traits
Various plants	Endophytic fungi	Protection from / tolerance to herbivores, pathogens and abiotic factors (Yan et al. 2019)	Increasing the growth rate, reproductive yield, and biomass of rice (Redman et al. 2011)
Bugula neritina (a Bryozoan)	*Candidatus* Endobugula sertula (a bacterium)	Chemical protection from predators (Lopanik 2014, Morita and Schmidt 2018)	Increased fecundity (Mathew et al. 2016)
Santia spp. (marine isopods)	Ectosymbiotic cyanobacteria	Chemical protection from predators, food (Wahl et al. 2012, Flórez et al. 2015)	No information found
Aphids	*Hamiltonella defensa* (a bacterium) carrying a bacteriophage	Protection from parasitic wasps (Oliver and Higashi 2019)	**Reduces the lifespan and lifetime reproduction of aphids in the absence of parasitoids** (Vorburger and Gouskov 2011)
	Candidatus Regiella insecticola (a bacterium)	Protection from pathogenic fungi (Scarborough et al. 2005)	**Reduces the number of winged offspring and alters the timing of sexual reproduction** (Leonardo and Mondor 2006)
	Serratia symbiotica (a bacterium)	Heat tolerance (Montllor et al. 2002)	**Reduced fecundity, reproduction time and body mass** (Skaljac et al. 2018, Pons et al. 2019)
	Candidatus Fukatsuia symbiotica (a bacterium)	Protection from parasitic wasps, pathogenic fungi and heat tolerance (Heyworth and Ferrari 2015)	**Reduced fecundity** (Heyworth et al. 2020)
Drosophila hydei	*Spiroplasms* (a bacterium)	Protection from parasitic wasps	Increasing survival and fecundity of flies that survived the parasitoid attack (Xie et al. 2014)
Various beetles	*Sodalis* spp. and other bacteria	Desiccation tolerance (producing tyrosine which is essential for cuticle sclerotisation) (Vigneron et al. 2014, Engl et al. 2018, Duplais et al. 2021, Dell'Aglio et al. 2023, Kiefer et al. 2023)	**Prolong larval development time** (Su et al. 2022)
Nasonia vitripennis (a parasitic wasp)	Gut bacteria	Pesticide degradation (Wang et al. 2020b)	No information found
Bumble bees	Gut microbiota	Protection from an intestinal parasite (Koch and Schmid-Hempel 2011)	No information found
Green frogs (*Lithobates clamitans*)	Gut microbiota	Thermal tolerance (Fontaine et al. 2022)	**Decreased body mass** (Fontaine et al. 2022)

Text in bold indicates a negative effect of the symbiont(s) on the host.

14.7 Diapause and Microbial Symbionts

Diapause is an integral and critical part of many animals' life cycles, enabling them to survive periods of harsh environmental conditions. The involvement of microbial symbionts in the diapause of their hosts was reviewed extensively by Mushegian and Tougeron (2019) and Carey and Assadi-Porter (2017). Here we highlight the main points and review recent publications. The lack of food substrates during diapause starvation leads to drastic changes in the gut lumen and microbiota, *e.g.*, in a coral (Brown et al. 2022), insects (Liu et al. 2016, Bosmans et al. 2018, Didion et al. 2021), a copepod (Datta et al. 2018), bears (Sommer et al. 2016), lemurs (Greene et al. 2022), an alligator (Tang et al. 2019), a frog (Weng et al. 2016) and a chipmunk (Zhou et al. 2022). The contribution of the symbionts to the diapausing host is related to the effects described above on nutrition and/or defence. During diapause, the host relies on fat reserves, which need to be accumulated in advance and evidently microbial symbionts play a role in this process. For example, survival rates of diapausing females of the mosquito *Culex pipiens* that were grown under sterile conditions were significantly lower compared to the control group. Further, these females had high carbohydrate levels but did not accumulate lipid reserves, suggesting an inability to process ingested sugars necessary for diapause-associated lipid accumulation (Didion et al. 2021). Axenic diapausing larvae of the parasitoid *Nasonia vitripennis* had lower levels of glucose and glycerol, and consequently lower body weight than conventional diapausing larvae (Dittmer and Brucker 2021). The gut microbiome also functions in recycling nitrogen waste molecules in diapausing animals, such as in a hemipteran bug (Kashima et al. 2006), a frog (Wiebler et al. 2018) and a squirrel (Regan et al. 2022), thereby redirecting the nitrogen back to the host. Hibernating eutherian mammals accumulate brown adipose tissue for thermal insulation, and gut microbiota is involved in this as well (Chevalier et al. 2015, Sommer et al. 2016, Moreno-Navarrete and Fernandez-Real 2019). Hibernating digger wasps harbour a bacterial symbiont that protects them from pathogens during hibernation by producing a mixture of antibiotics (Koehler et al. 2013).

14.8 Concluding Remarks

In this chapter, we reviewed the diverse influences of microbial symbionts on the life history and fitness of their hosts. Observational and experimental evidence tightly links microbial symbionts and variations in their host life history traits. In many cases, the mechanisms which determine the nature of the interactions are not known, or not fully understood, due to confounding factors that are identified as the trait driver. These may include behavioural change, induction of genetic repertoire and metabolic activity that is induced by overlooked microbes.

Microbe–host interactions are dynamic, changing on both ecological and evolutionary timescales. Horizontal gene transfers from symbionts to hosts, for example, occur frequently and provide the host with new traits (Soucy et al. 2015, Husnik and McCutcheon 2017). With evolutionary time, if microbial symbionts lose too many genes, they lose their ability to live independently, which brings us back to the holobiont/hologenome controversy mentioned in the introduction of the chapter. The best-known examples are probably mitochondria and chloroplasts, whose ancient ancestors were free-living bacteria, and with time became organelles. Thus, the vast indirect microbial effect on life history traits remains to be further discovered. Nevertheless, the plethora of evidence on the centrality of microbial symbionts in the biology of higher organisms has changed the way we understand nature, and holds promise for conservation biology, as well as for animal, plant and human health.

References

Adnane, M., and Chapwanya, A. (2022). A review of the diversity of the genital tract microbiome and implications for fertility of cattle. *Animals* 12(4): 460.

Afkhami, M.E., Almeida, B.K., Hernandez, D.J., Kiesewetter, K N., and Revillini, D.P. (2020). Tripartite mutualisms as models for understanding plant-microbial interactions. *Current Opinion in Plant Biology* 56: 28–36.

Afkhami, M.E., Friesen, M.L., and Stinchcombe, J.R. (2021). Multiple mutualism effects generate synergistic selection and strengthen fitness alignment in the interaction between legumes, rhizobia and mycorrhizal fungi. *Ecology Letters* 24(9): 1824–1834.

Ben-Yosef, M., Rot, A., Mahagna, M., Kapri, E., Behar, A., and Gottlieb, Y. (2020). Coxiella-like endosymbiont of *Rhipicephalus sanguineus* is required for physiological processes during ontogeny. *Frontiers in Microbiology* 11: 516585.

Bosmans, L., Pozo, M.I., Verreth, C., Crauwels, S., Wäckers, F., Jacquemyn, H., and Lievens, B. (2018). Hibernation leads to altered gut communities in bumblebee queens (*Bombus terrestris*). *Insects* 9(4): 188.

Brown, A.L., Sharp, K., and Apprill, A. (2022). Reshuffling of the coral microbiome during dormancy. *Applied and Environmental Microbiology* 88(23): 23.

Brundrett, M.C., and Tedersoo, L. (2018). Evolutionary history of mycorrhizal symbioses and global host plant diversity. *New Phytologist* 220(4): 1108–1115.

Carey, H.V., and Assadi-Porter, F.M. (2017). The hibernator microbiome: host-bacterial interactions in an extreme nutritional symbiosis. *Annual Review of Nutrition* 37: 477–500.

Chevalier, C., Stojanović, O., Colin, D.J., Suarez-Zamorano, N., Tarallo, V., Veyrat-Durebex, C., Rigo, D., Fabbiano, S., Stevanović, A., Hagemann, S., Montet, X., Seimbille, Y., Zamboni, N., Hapfelmeier, S., and Trajkovski, M. (2015). Gut microbiota orchestrates energy homeostasis during cold. *Cell* 163(6): 1360–1374.

Comizzoli, P., Power, M.L., Bornbusch, S.L., and Muletz-Wolz, C.R. (2021). Interactions between reproductive biology and microbiomes in wild animal species. *Animal Microbiome* 3(1): 87.

Cornwallis, C.K., van't Padje, A., Ellers, J., Klein, M., Jackson, R., Kiers, E.T., West, S.A., and Henry, L.M. (2023). Symbioses shape feeding niches and diversification across insects. *Nature Ecology & Evolution* 7(7): 1022–1044.

Crates, R., Stojanovic, D., and Heinsohn, R. (2023). The phenotypic costs of captivity. *Biological Reviews* 98(2): 434–449.

Dallas, J.W., and Warne, R.W. (2022). Captivity and animal microbiomes: potential roles of microbiota for influencing animal conservation. *Microbial Ecology* 85(3): 820–838.

Datta, M.S., Almada, A.A., Baumgartner, M.F., Mincer, T.J., Tarrant, A.M., and Polz, M.F. (2018). Inter-individual variability in copepod microbiomes reveals bacterial networks linked to host physiology. *The ISME Journal* 12(9): 2103–2113.

Dell'Aglio, E., Lacotte, V., Peignier, S., Rahioui, I., Benzaoui, F., Vallier, A., Da Silva, P., Desouhant, E., Heddi, A., and Rebollo, R. (2023). Weevil carbohydrate intake triggers endosymbiont proliferation: a trade-off between host benefit and endosymbiont burden. *MBio* 14(2): e03333–22.

Didion, E.M., Sabree, Z.L., Kenyon, L., Nine, G., Hagan, R.W., Osman, S., and Benoit, J.B. (2021). Microbiome reduction prevents lipid accumulation during early diapause in the northern house mosquito, *Culex pipiens*. *Journal of Insect Physiology* 134: 104295.

Dittmer, J., and Brucker, R.M. (2021). When your host shuts down: larval diapause impacts host-microbiome interactions in *Nasonia vitripennis*. *Microbiome* 9(1): 85.

Doremus, M.R., and Hunter, M.S. (2020). The saboteur's tools: common mechanistic themes across manipulative symbioses. *Advances in Insect Physiology* 58: 317–353.

Dos Santos, S.J., Pakzad, Z., Albert, A.Y.K., Elwood, C.N., Grabowska, K., Links, M.G., Hutcheon, J.A., Maan, E.J., Manges, A.R., Dumonceaux, T.J., Hodgson, Z.G., Lyons, J., Mitchell-Foster, S.M., Gantt, S., Joseph, K.S., van Schalkwyk, J.E., Hill, J.E., and Money, D.M. (2023). Maternal vaginal microbiome composition does not affect development of the infant gut microbiome in early life. *Frontiers in Cellular and Infection Microbiology* 13: 1144254.

Douglas, A.E. (2006). Phloem-sap feeding by animals: problems and solutions. *Journal of Experimental Botany* 57(4): 747–754.

Douglas, A.E., and Werren, J.H. (2016). Holes in the hologenome: why host-microbe symbioses are not holobionts. *MBio* 7(2): 02099–15.

Drew, G.C., Frost, C.L., and Hurst, G.D. (2019). Reproductive parasitism and positive fitness effects of heritable microbes. In *Encyclopedia of life sciences*, 1–8. John Wiley & Sons, Ltd.

Duplais, C., Sarou-Kanian, V., Massiot, D., Hassan, A., Perrone, B., Estevez, Y., Wertz, J.T., Martineau, E., Farjon, J., Giraudeau, P., and Moreau, C.S. (2021). Gut bacteria are essential for normal cuticle development in herbivorous turtle ants. *Nature Communications* 12(1): 676.

Duron, O., Morel, O., Noël, V., Buysse, M., Binetruy, F., Lancelot, R., Loire, E., Ménard, C., Bouchez, O., Vavre, F., and Vial, L. (2018). Tick-bacteria mutualism depends on B vitamin synthesis pathways. *Current Biology* 28(12): 1896–1902.e5.

Enav, H., Bäckhed, F., and Ley, R.E. (2022). The developing infant gut microbiome: a strain-level view. *Cell Host & Microbe* 30(5): 627–638.

Engl, T., Eberl, N., Gorse, C., Krüger, T., Schmidt, T.H.P., Plarre, R., Adler, C., and Kaltenpoth, M. (2018). Ancient symbiosis confers desiccation resistance to stored grain pest beetles. *Molecular Ecology* 27(8): 2095–2108.

Engl, T., Schmidt, T.H.P., Kanyile, S.N., and Klebsch, D. (2020). Metabolic cost of a nutritional symbiont manifests in delayed reproduction in a grain pest beetle. *Insects* 11(10): 717.

Flórez, L.V., Biedermann, P.H.W., Engl, T., and Kaltenpoth, M. (2015). Defensive symbioses of animals with prokaryotic and eukaryotic microorganisms. *Natural Product Reports* 32(7): 904–936.

Fontaine, S.S., Mineo, P.M., and Kohl, K.D. (2022). Experimental manipulation of microbiota reduces host thermal tolerance and fitness under heat stress in a vertebrate ectotherm. *Nature Ecology & Evolution* 6(4): 405–417.

Fransolet, D., Roberty, S., and Plumier, J.C. (2012). Establishment of endosymbiosis: the case of cnidarians and *Symbiodinium*. *Journal of Experimental Marine Biology and Ecology* 420–421: 1–7.

Gong, J.T., Li, T.P., Wang, M.K., and Hong, X.Y. (2023). *Wolbachia*-based strategies for control of agricultural pests. *Current Opinion in Insect Science* 57: 101039.

Greene, L.K., Andriambeloson, J.B., Rasoanaivo, H.A., Yoder, A.D., and Blanco, M.B. (2022). Variation in gut microbiome structure across the annual hibernation cycle in a wild primate. *FEMS Microbiology Ecology* 98(7): fiac07.

Gudnadottir, U., Debelius, J.W., Du, J., Hugerth, L.W., Danielsson, H., Schuppe-Koistinen, I., Fransson, E., and Brusselaers, N. (2022). The vaginal microbiome and the risk of preterm birth: a systematic review and network meta-analysis. *Scientific Reports* 12(1): 7626.

Gupta, S., Chaturvedi, P., Kulkarni, M.G., and van Staden, J. (2020). A critical review on exploiting the pharmaceutical potential of plant endophytic fungi. *Biotechnology Advances* 39: 107462.

Hammer, T.J., Sanders, J.G., and Fierer, N. (2019). Not all animals need a microbiome. *FEMS Microbiology Letters* 366(10): 117.

Heyworth, E.R., and Ferrari, J. (2015). A facultative endosymbiont in aphids can provide diverse ecological benefits. *Journal of Evolutionary Biology* 28(10): 1753–1760.

Heyworth, E.R., Smee, M.R., and Ferrari, J. (2020). Aphid facultative symbionts aid recovery of their obligate symbiont and their host after heat stress. *Frontiers in Ecology and Evolution* 8: 56.

Holyoak, G.R., Premathilake, H.U., Lyman, C.C., Sones, J.L., Gunn, A., Wieneke, X., and DeSilva, U. (2022). The healthy equine uterus harbors a distinct core microbiome plus a rich and diverse microbiome that varies with geographical location. *Scientific Reports* 12: 14790.

Hugon, A.M., and Golos, T.G. (2023). Non-human primate models for understanding the impact of the microbiome on pregnancy and the female reproductive tract. *Biology of Reproduction* 109(1): 14790.

Husnik, F., and McCutcheon, J.P. (2017). Functional horizontal gene transfer from bacteria to eukaryotes. *Nature Reviews Microbiology* 16(2): 67–79.

Jinkerson, R.E., Russo, J.A., Newkirk, C.R., Kirk, A.L., Chi, R.J., Martindale, M.Q., Grossman, A.R., Hatta, M., and Xiang, T. (2022). Cnidarian-Symbiodiniaceae symbiosis establishment is independent of photosynthesis. *Current Biology* 32(11): 2402–2415.

Jung, M., and Lee, D.H. (2023). Effect of gut symbiont *Caballeronia insecticola* on life history and behavioral traits of male host *Riptortus pedestris* (Hemiptera: Alydidae). *Journal of Asia-Pacific Entomology* 26(2): 102085.

Kashima, T., Nakamura, T., and Tojo, S. (2006). Uric acid recycling in the shield bug, *Parastrachia japonensis* (Hemiptera: Parastrachiidae), during diapause. *Journal of Insect Physiology* 52(8): 816–825.

Kaur, R., Shropshire, J.D., Cross, K.L., Leigh, B., Mansueto, A.J., Stewart, V., Bordenstein, S.R., and Bordenstein, S.R. (2021). Living in the endosymbiotic world of *Wolbachia*: a centennial review. *Cell Host & Microbe* 29(6): 879–893.

Kiefer, J.S.T., Bauer, E., Okude, G., Fukatsu, T., Kaltenpoth, M., and Engl, T. (2023). Cuticle supplementation and nitrogen recycling by a dual bacterial symbiosis in a family of xylophagous beetles. *The ISME Journal* 17(7): 1029–1039.

Kirk, N.L., and Weis, V.M. (2016). Animal–*Symbiodinium* symbioses: foundations of coral reef ecosystems. In: *The mechanistic benefits of microbial symbionts. Advances in environmental microbiology*, Vol 2 (ed. Hurst, C.), 269–294. Springer.

Koch, H., and Schmid-Hempel, P. (2011). Socially transmitted gut microbiota protect bumble bees against an intestinal parasite. *Proceedings of the National Academy of Sciences of the United States of America* 108(48): 19288–19292.

Koehler, S., Doubský, J., and Kaltenpoth, M. (2013). Dynamics of symbiont-mediated antibiotic production reveal efficient long-term protection for beewolf offspring. *Frontiers in Zoology* 10(1): 3.

Koskella, B., and Bergelson, J. (2020). The study of host–microbiome (co)evolution across levels of selection. *Philosophical Transactions of the Royal Society, B: Biological Sciences* 375(1808): 20190604.

Kowallik, K.V., and Martin, W.F. (2021). The origin of symbiogenesis: an annotated English translation of Mereschkowsky's 1910 paper on the theory of two plasma lineages. *Biosystems* 199: 104281.

Kwon, M.S., and Lee, H.K. (2022). Host and microbiome interplay shapes the vaginal microenvironment. *Frontiers in Immunology* 13: 3226.

Larimer, A.L., Bever, J.D., and Clay, K. (2010). The interactive effects of plant microbial symbionts: a review and meta-analysis. *Symbiosis* 51(2): 139–148.

Larimer, A.L., Clay, K., and Bever, J.D. (2014). Synergism and context dependency of interactions between arbuscular mycorrhizal fungi and rhizobia with a prairie legume. *Ecology* 95(4): 1045–1054.

Leonardo, T.E., and Mondor, E.B. (2006). Symbiont modifies host life-history traits that affect gene flow. *Proceedings of the Royal Society B: Biological Sciences* 273(1590): 1079–1084.

Li, X., Sun, Y., Tian, X., Wang, C., Li, Q., Li, Q., Zhu, S., Lan, C., Zhang, Y., Li, X., Ding, R., and Zhu, X. (2023). *Sitobion miscanthi* L type symbiont enhances the fitness and feeding behavior of the host grain aphid. *Pest Management Science* 79(4): 1362–1371.

Li, H., Young, S.E., Poulsen, M., and Currie, C.R. (2021). Symbiont-mediated digestion of plant biomass in fungus-farming insects. *Annual Review of Entomology*, 66: 297–316.

Liu, W., Li, Y., Guo, S., Yin, H., Lei, C.L., and Wang, X.P. (2016). Association between gut microbiota and diapause preparation in the cabbage beetle: a new perspective for studying insect diapause. *Scientific Reports* 6(1): 38900.

Lopanik, N.B. (2014). Chemical defensive symbioses in the marine environment. *Functional Ecology* 28(2): 328–340.

Lopez, V., Cortesero, A.M., and Poinsot, D. (2018). Influence of the symbiont *Wolbachia* on life history traits of the cabbage root fly (*Delia radicum*). *Journal of Invertebrate Pathology* 158: 24–31.

Łukasik, P., van Asch, M., Guo, H., Ferrari, J., and Godfray, H.C.J. (2013). Unrelated facultative endosymbionts protect aphids against a fungal pathogen. *Ecology Letters* 16(2): 214–218.

Maire, J., and van Oppen, M.J.H. (2022). A role for bacterial experimental evolution in coral bleaching mitigation? *Trends in Microbiology* 30(3): 217–228.

Márquez, L.M., Redman, R.S., Rodriguez, R.J., and Roossinck, M.J. (2007). A virus in a fungus in a plant: three-way symbiosis required for thermal tolerance. *Science* 315(5811): 513–515.

Mathew, M., Bean, K.I., Temate-Tiagueu, Y., Caciula, A., Mandoiu, I.I., Zelikovsky, A., and Lopanik, N.B. (2016). Influence of symbiont-produced bioactive natural products on holobiont fitness in the marine bryozoan, *Bugula neritina* via protein kinase C (PKC). *Marine Biology* 163(2): 440.

Mizrahi, I., Wallace, R.J., and Moraïs, S. (2021). The rumen microbiome: balancing food security and environmental impacts. *Nature Reviews Microbiology* 19(9): 553–566.

Montllor, C.B., Maxmen, A., and Purcell, A.H. (2002). Facultative bacterial endosymbionts benefit pea aphids *Acyrthosiphon pisum* under heat stress. *Ecological Entomology* 27(2): 189–195.

Moreno-Navarrete, J.M., and Fernandez-Real, J.M. (2019). The gut microbiota modulates both browning of white adipose tissue and the activity of brown adipose tissue. *Reviews in Endocrine and Metabolic Disorders* 20(4): 387–397.

Morita, M., and Schmidt, E.W. (2018). Parallel lives of symbionts and hosts: chemical mutualism in marine animals. *Natural Product Reports* 35(4): 357–378.

Morris, J.J., Bordenstein, S., and Theis, K. (2018). What is the hologenome concept of evolution? *F1000Research* 2018: 7, 1664.

Morsy, M.R., Oswald, J., He, J., Tang, Y., and Roossinck, M.J. (2010). Teasing apart a three-way symbiosis: transcriptome analyses of *Curvularia protuberata* in response to viral infection and heat stress. *Biochemical and Biophysical Research Communications* 401(2): 225–230.

Mushegian, A.A., and Tougeron, K. (2019). Animal-microbe interactions in the context of diapause. *Biological Bulletin* 237(2): 180–191.

Oliver, K.M., Degnan, P.H., Burke, G.R., and Moran, N.A. (2010). Facultative symbionts in aphids and the horizontal transfer of ecologically important traits. *Annual Review of Entomology* 55(1): 247–266.

Oliver, K.M., and Higashi, C. (2019). Variations on a protective theme: *Hamiltonella defensa* infections in aphids variably impact parasitoid success. *Current Opinion in Insect Science* 32: 1–7.

Oliver, K.M., Moran, N.A., and Hunter, M.S. (2006). Costs and benefits of a superinfection of facultative symbionts in aphids. *Proceedings of the Royal Society B: Biological Sciences* 273(1591): 1273–1280.

Pahua, V.J., Stokes, P.J.N., Hollowell, A.C., Regus, J.U., Gano-Cohen, K.A., Wendlandt, C.E., Quides, K.W., Lyu, J.Y., and Sachs, J.L. (2018). Fitness variation among host species and the paradox of ineffective rhizobia. *Journal of Evolutionary Biology* 31(4): 599–610.

Peixoto, R.S., Harkins, D.M., and Nelson, K.E. (2021). Advances in microbiome research for animal health. *Annual Review of Animal Biosciences* 9: 289–311.

Pekas, A., Tena, A., Peri, E., Colazza, S., and Cusumano, A. (2023). Competitive interactions in insect parasitoids: effects of microbial symbionts across tritrophic levels. *Current Opinion in Insect Science* 55: 101001.

Perlmutter, J.I., and Bordenstein, S.R. (2020). Microorganisms in the reproductive tissues of arthropods. *Nature Reviews Microbiology* 18(2): 97–111.

Perreau, J., and Moran, N.A. (2021). Genetic innovations in animal-microbe symbioses. *Nature Reviews Genetics* 23(1): 23–39.

Pollmann, M., Moore, L.D., Krimmer, E., D'Alvise, P., Hasselmann, M., Perlman, S.J., Ballinger, M.J., Steidle, J.L.M., and Gottlieb, Y. (2022). Highly transmissible cytoplasmic incompatibility by the extracellular insect symbiont *Spiroplasma*. *iScience* 25(5): 104335.

Pons, I., Renoz, F., Noël, C., and Hance, T. (2019). New insights into the nature of symbiotic associations in aphids: infection process, biological effects, and transmission mode of cultivable *Serratia symbiotica* bacteria. *Applied and Environmental Microbiology* 85(10): e02445-18.

Porro, M., Kundrotaite, E., Mellor, D.D., and Munialo, C.D. (2023). A narrative review of the functional components of human breast milk and their potential to modulate the gut microbiome, the consideration of maternal and child characteristics, and confounders of breastfeeding, and their impact on risk of obesity later in life. *Nutrition Reviews* 81(5): 597–609.

Redman, R.S., Kim, Y.O., Woodward, C.J.D.A., Greer, C., Espino, L., Doty, S.L., and Rodriguez, R.J. (2011). Increased fitness of rice plants to abiotic stress via habitat adapted symbiosis: a strategy for mitigating impacts of climate change. *PLoS ONE* 6(7): e14823.

Regan, M.D., Chiang, E., Liu, Y., Tonelli, M., Verdoorn, K.M., Gugel, S.R., Suen, G., Carey, H.V., and Assadi-Porter, F.M. (2022). Nitrogen recycling via gut symbionts increases in ground squirrels over the hibernation season. *Science* 375(6579): 460–463.

Rio, R.V.M., Attardo, G.M., and Weiss, B.L. (2016). Grandeur alliances: symbiont metabolic integration and obligate arthropod hematophagy. *Trends in Parasitology* 32(9): 739–749.

Rock, R.R., and Turnbaugh, P. J. (2023). Forging the microbiome to help us live long and prosper. *PLoS Biology* 21(4): e3002087.

Ross, A.A., Rodrigues Hoffmann, A., and Neufeld, J.D. (2019). The skin microbiome of vertebrates. *Microbiome* 7(1): 79.

Sagan, L. (1967). On the origin of mitosing cells. *Journal of Theoretical Biology* 14(3): 225–274.

Sarkar, A., Yoo, J.Y., Dutra, S.V.O., Morgan, K.H., and Groer, M. (2021). The association between early-life gut microbiota and long-term health and diseases. *Journal of Clinical Medicine* 10(3): 459.

Scarborough, C.L., Ferrari, J., and Godfray, H.C.J. (2005). Aphid protected from pathogen by endosymbiont. *Science* 310(5755): 1781.

Scharf, M.E., and Peterson, B.F. (2021). A century of synergy in termite symbiosis research: linking the past with new genomic insights. *Annual Review of Entomology* 66: 23–43.

Simon, J.C., Boutin, S., Tsuchida, T., Koga, R., Gallic, J.F., Frantz, A., Outreman, Y., and Fukatsu, T. (2011). Facultative symbiont infections affect aphid reproduction. *PLoS ONE* 6(7): e21831.

Sisk-Hackworth, L., Kelley, S.T., and Thackray, V.G. (2023). Sex, puberty, and the gut microbiome. *Reproduction* 165(2): R61–R74.

Skaljac, M., Kirfel, P., Grotmann, J., and Vilcinskas, A. (2018). Fitness costs of infection with *Serratia symbiotica* are associated with greater susceptibility to insecticides in the pea aphid *Acyrthosiphon pisum*. *Pest Management Science* 74(8): 1829–1836.

Sommer, F., Ståhlman, M., Ilkayeva, O., Arnemo, J. M., Kindberg, J., Josefsson, J., Newgard, C.B., Fröbert, O., and Bäckhed, F. (2016). The gut microbiota modulates energy metabolism in the hibernating brown bear *Ursus arctos*. *Cell Reports* 14(7): 1655–1661.

Soucy, S.M., Huang, J., and Gogarten, J.P. (2015). Horizontal gene transfer: building the web of life. *Nature Reviews Genetics* 16(8): 472–482.

Stencel, A., and Wloch-Salamon, D.M. (2018). Some theoretical insights into the hologenome theory of evolution and the role of microbes in speciation. *Theory in Biosciences* 137(2): 197–206.

Strand, M.R., and Burke, G.R. (2020). Polydnaviruses: evolution and function. *Current Issues in Molecular Biology* 34: 163–182.

Su, Y., Lin, H.C., Teh, L.S., Chevance, F., James, I., Mayfield, C., Golic, K.G., Gagnon, J.A., Rog, O., and Dale, C. (2022). Rational engineering of a synthetic insect-bacterial mutualism. *Current Biology* 32(18): 3925–3938.e6.

Tang, K.Y., Wang, Z.W., Wan, Q.H., and Fang, S.G. (2019). Metagenomics reveals seasonal functional adaptation of the gut microbiome to host feeding and fasting in the Chinese alligator. *Frontiers in Microbiology* 10: 2409.

The International Aphid Genomics Consortium (2010). Genome sequence of the pea aphid *Acyrthosiphon pisum*. *PLoS Biology* 8(2): e1000313.

Udvardi, M., and Poole, P.S. (2013). Transport and metabolism in legume-rhizobia symbioses. *Annual Review of Plant Biology* 64: 781–805.

van der Heijden, M.G.A., de Bruin, S., Luckerhoff, L., van Logtestijn, R.S.P., and Schlaeppi, K. (2015). A widespread plant-fungal-bacterial symbiosis promotes plant biodiversity, plant nutrition and seedling recruitment. *The ISME Journal* 10(2): 389–399.

van Woesik, R., Shlesinger, T., Grottoli, A.G., Toonen, R.J., Vega Thurber, R., Warner, M.E., Marie Hulver, A., Chapron, L., McLachlan, R.H., Albright, R., Crandall, E., DeCarlo, T.M., Donovan, M.K., Eirin-Lopez, J., Harrison, H.B., Heron, S.F., Huang, D., Humanes, A., Krueger, T., Madin, J.S., Manzello, D., McManus, L.C., Matz, M., Muller, E.M., Rodriguez-Lanetty, M., Vega-Rodriguez, M., Voolstra, C.R., and Zaneveld, J. (2022). Coral-bleaching responses to climate change across biological scales. *Global Change Biology* 28(14): 4229–4250.

VanEvery, H., Franzosa, E.A., Nguyen, L.H., and Huttenhower, C. (2022). Microbiome epidemiology and association studies in human health. *Nature Reviews Genetics* 24(2): 109–124.

Viana, F., Paz, L.C., Methling, K., Damgaard, C.F., Lalk, M., Schramm, A., and Lund, M.B. (2018). Distinct effects of the nephridial symbionts *Verminephrobacter* and *Candidatus* Nephrothrix on reproduction and maturation of its earthworm host *Eisenia andrei*. *FEMS Microbiology Ecology* 94(2): fix178.

Vigneron, A., Masson, F., Vallier, A., Balmand, S., Rey, M., Vincent-Monégat, C., Aksoy, E., Aubailly-Giraud, E., Zaidman-Rémy, A., and Heddi, A. (2014). Insects recycle endosymbionts when the benefit is over. *Current Biology* 24(19): 2267–2273.

Vorburger, C., and Gouskov, A. (2011). Only helpful when required: a longevity cost of harbouring defensive symbionts. *Journal of Evolutionary Biology* 24(7): 1611–1617.

Wahl, M., Goecke, F., Labes, A., Dobretsov, S., and Weinberger, F. (2012). The second skin: ecological role of epibiotic biofilms on marine organisms. *Frontiers in Microbiology* 3: 292.

Wallin, I.E. (1927). *Symbionticism and the origin of species*. Waverly Press.

Wang, Y. Bin, Ren, F.R., Yao, Y. L., Sun, X., Walling, L.L., Li, N.N., Bai, B., Bao, X.Y., Xu, X.R., and Luan, J.B. (2020a). Intracellular symbionts drive sex ratio in the whitefly by facilitating fertilization and provisioning of B vitamins. *The ISME Journal* 14(12): 2923–2935.

Wang, G.H., Berdy, B.M., Velasquez, O., Jovanovic, N., Alkhalifa, S., Minbiole, K. P.C., and Brucker, R. M. (2020b). Changes in microbiome confer multigenerational host resistance after sub-toxic pesticide exposure. *Cell Host & Microbe* 27(2): 213–224.e7.

Weng, F C.H., Yang, Y.J., and Wang, D. (2016). Functional analysis for gut microbes of the brown tree frog (*Polypedates megacephalus*) in artificial hibernation. *BMC Genomics* 17(13): 31–42.

Wiebler, J.M., Kohl, K.D., Lee, R.E., and Costanzo, J.P. (2018). Urea hydrolysis by gut bacteria in a hibernating frog: evidence for urea-nitrogen recycling in Amphibia. *Proceedings of the Royal Society B: Biological Sciences* 285: 20180241.

Woodruff, K.L., Hummel, G.L., Austin, K.J., Lake, S.L., and Cunningham-Hollinger, H.C. (2022). Calf rumen microbiome from birth to weaning and shared microbial properties to the maternal rumen microbiome. *Journal of Animal Science* 100(10): skac264.

Xie, J., Butler, S., Sanchez, G., and Mateos, M. (2014). Male killing *Spiroplasma* protects *Drosophila melanogaster* against two parasitoid wasps. *Heredity* 112(4): 399–408.

Yan, L., Zhu, J., Zhao, X., Shi, J., Jiang, C., and Shao, D. (2019). Beneficial effects of endophytic fungi colonization on plants. *Applied Microbiology and Biotechnology* 103(8): 3327–3340.

Yong, E. (2016). *I contain multitudes: the microbes within us and a grander view of life* (1st ed.). Harper Collins Publishers.

Zepeda Mendoza, M.L., Xiong, Z., Escalera-Zamudio, M., Runge, A.K., Thézé, J., Streicker, D., Frank, H.K., Loza-Rubio, E., Liu, S., Ryder, O.A., Samaniego Castruita, J.A., Katzourakis, A., Pacheco, G., Taboada, B., Löber, U., Pybus, O.G., Li, Y., Rojas-Anaya, E., Bohmann, K., Carmona Baez, A., Arias, C.F., Liu, S., Greenwood, A.D., Bertelsen, M.F., White, N.E., Bunce, M., Zhang, G., Sicheritz-Pontén, T., and Gilbert, M.P.T. (2018). Hologenomic adaptations underlying the evolution of sanguivory in the common vampire bat. *Nature Ecology & Evolution* 2(4): 659–668.

Zhang, Y.K., Yang, K., Zhu, Y.X., and Hong, X.Y. (2018). Symbiont-conferred reproduction and fitness benefits can favour their host occurrence. *Ecology and Evolution* 8(3): 1626–1633.

Zhou, J., Wang, M., and Yi, X. (2022). Alteration of gut microbiota of a food-storing hibernator, Siberian chipmunk *Tamias sibiricus*. *Microbial Ecology* 84(2): 603–612.

Zilber-Rosenberg, I., and Rosenberg, E. (2008). Role of microorganisms in the evolution of animals and plants: the hologenome theory of evolution. *FEMS Microbiology Reviews* 32(5): 723–735.

15

Ecological and Evolutionary Links Between Defences and Life History Traits in Plants

Xoaquín Moreira[1] and Luis Abdala-Roberts[2]

[1] *Misión Biológica de Galicia (CSIC), "Ecology and Evolution of Plant-Herbivore Interactions" Group, Pontevedra, Galicia, Spain*
[2] *Departamento de Ecología Tropical, Campus de Ciencias Biológicas y Agropecuarias, Universidad Autónoma de Yucatán, Mérida, México*

15.1 Evolutionary Ecology of Plant Defences Against Herbivores

Interactions between plants and insect herbivores date back approximately 350 million years (Labandeira 2007). Throughout their shared history, species on each side have engaged in coevolutionary arms races leading to defensive escalation (Ehrlich and Raven 1964, Futuyma and Agrawal 2009, Coley and Kursar 2014) and specialisation (Futuyma and Moreno 1988, Hardy and Otto 2014, Forrister et al. 2015). From the plant's perspective, these evolutionary dynamics have given rise to and shaped a wide array of potent anti-herbivore defences (Schoonhoven et al. 2005, Agrawal 2007), driving escape from herbivores in evolutionary time and promoting species diversification (Becerra et al. 2009, Cacho et al. 2015, Endara et al. 2017).

Plant defences can be broadly classified into tolerance and resistance (Núñez-Farfán et al. 2007). Tolerance is defined as the ability of plants to mitigate the negative fitness effects of herbivory after an attack has taken place (Strauss and Agrawal 1999, Stowe et al. 2000, Fornoni 2011), and is achieved by plant physiological and biochemical mechanisms such as increased photosynthetic rates and resource reallocation between tissues after damage which mediate regrowth capacity in vegetative and reproductive tissues (Agrawal et al. 1999, Fornoni 2011, Moreira et al. 2012, Robert et al. 2014). Resistance, on the other hand, is the capacity of plants to reduce or avoid herbivore damage (Rausher 2001, Agrawal 2011, Carmona et al. 2011) and includes both direct and indirect defensive traits (Heil 2008, Dicke and Baldwin 2010, Kessler and Heil 2011). Direct defence is conferred by chemical (*e.g.*, phenolics, terpenoids, and alkaloids) and physical (*e.g.*, thorns, spines, trichomes, and leaf toughness) traits that deter herbivores, reduce their consumption, or decrease their survival (Agrawal 2007, Carmona et al. 2011, Mithöfer and Boland 2012). Indirect defence involves anatomical or physical structures (*e.g.*, domatia and extrafloral nectaries) and chemical traits (*e.g.*, some types of volatile organic compounds) that provide shelter, food, or information on herbivore presence to predators and parasitoids (Dicke and Baldwin 2010, Kessler and Heil 2011, Turlings and Erb 2018) which in turn suppress herbivory and, in doing so, increase plant growth or reproduction (Hairston et al. 1960, Schmitz et al. 2000, Romero and Koricheva 2011). Finally, both types of resistance traits can be expressed constitutively, *i.e.*, fixed levels present in tissues in the absence of damage, or induced (*i.e.*, *de novo* synthesized) after herbivore damage which leads to an increase in defences relative to constitutive levels (Karban and Baldwin 1997, Dicke and van Loon 2000, Karban 2011).

Much of the research on plant defences has focused on understanding the ecological and evolutionary drivers of variability in plant allocation to different types of defensive traits, *e.g.*, direct *vs.* indirect (Ballhorn et al. 2008, Mooney et al. 2010, Rasmann et al. 2011), and the expression of constitutive *vs.* induced levels (Zhang et al. 2008, Moreira et al. 2014, Rasmann et al. 2015). In addition, and beyond defensive traits, a great deal of attention has been given to the reciprocal influence or links between defence allocation and traits related to other plant functions, namely growth and reproduction, from both ecological and evolutionary perspectives (reviewed by Johnson et al. 2015 and Züst and Agrawal 2017). In addressing these topics, plant defence theory has spanned different levels of biological organisation, ranging from within-plant (*e.g.*, among plant parts) to among-population or across species (McCall and Fordyce 2010, Kempel et al. 2011, Moreira et al. 2014, Eichenberg et al. 2015, Agrawal and Hastings 2019).

This prolific body of research has shed three key realizations. First, plant defences in many cases have not evolved independently of other plant functions (*e.g.*, growth and reproduction), and these interdependencies often arise due to plant

Life History Evolution: Traits, Interactions, and Applications, First Edition. Edited by Michal Segoli and Eric Wajnberg.
© 2025 John Wiley & Sons Ltd. Published 2025 by John Wiley & Sons Ltd.

endogenous resource allocation constraints (Hahn and Maron 2016, Züst and Agrawal 2017, Defossez et al. 2018, Pellissier et al. 2018, Cope et al. 2021). Second, herbivory and other interactions (*e.g.*, seed dispersal and pollination) many times jointly shape plant traits, leading to correlated evolution of plant defensive traits and reproductive or growth-related traits which are linked via pleiotropic effects or allocation constraints (Campbell 2015, Johnson et al. 2015). Together, these findings highlight that the study of plant defence evolution must be placed within a life history context that recognizes its interdependencies with growth- and reproduction-related traits. By the same token, insight into the evolution of the latter two functions can be informed by addressing evolutionary correlations with plant defences. Third, plant defence expression is highly abiotic context-dependent (*e.g.*, contingent on conditions of abiotic stress) with plants making allocation choices based on the costs (*vs.* benefits) of expressing different traits depending on the environmental setting in which they are found (Simms and Fritz 1990, Zangerl and Bazzaz 1992, Obeso 2002, Züst and Agrawal 2017). For example, abiotic conditions often determine plant resource use and allocation choices because environmental stress affects (*e.g.*, exacerbates) allocation constraints, and, in so doing, shapes the relative costs of defences and other plant functions (Stamp 2003, Züst and Agrawal 2017). These abiotic effects on plant phenotypes can also, in turn, affect interactions with plant-associated antagonists or mutualists and, in so doing, also alter their joint selection on plant defence- and growth- or reproduction-related traits (Abdala-Roberts and Mooney 2015).

In this chapter, we synthesize research showing linkages between defences and growth- and reproduction-related traits shaping life histories in land plants. We focus to a large extent on evidence of plant endogenous allocation constraints (*e.g.*, trade-offs), as this is one of the most studied mechanisms explaining (or assumed to explain) evolutionary links between plant defences and life history traits, particularly for those related to growth (for other plant-based endogenous mechanisms see Kempel et al. 2011, Whitehead et al. 2017). In the case of links to plant reproduction, much of the work (and, accordingly, our review) focuses on patterns of joint selection by herbivores and plant mutualists leading to correlated evolution of plant defences and reproductive life history features, including floral traits, mating and pollination system traits, as well as seed and fruit-related (*e.g.*, dispersal) traits. Finally, we briefly review empirical and theoretical work offering integrative views for understanding the correlated evolution of all three interrelated functions. Throughout this chapter, when available, we also address research on abiotic context-dependencies in these associations, as it is through this abiotic lens that we have gained much insight into how plant defences and life history traits have jointly evolved.

15.2 Correlated Evolution of Plant Defences and Life History Traits

15.2.1 Defences and Growth-Related Traits

15.2.1.1 Plant-Based Allocation Constraints

Associations between plant growth and defences can arise due to allocation constraints (see Table 15.1), as both functions draw from shared resource pools (Stamp 2003). For example, trade-offs have been shown between the production of secondary metabolites and the demand for photosynthates during plant development (Loomis 1932, Herms and Mattson 1992,

Table 15.1 A general summary of reviewed mechanisms underlying associations between plant defences and growth- or reproduction-related traits affecting plant life histories.

Plant endogenous-based processes	Joint selection related to interactions (*e.g.*, seed predation, pollination).
1) Physiological or allocation constraints leading to negative correlations (trade-offs). 2) Genetic linkages: Pleiotropy leading to positive or negative correlations.	1) Linked traits: Two or more linked traits (see left column) leading to trait combinations and ecological trade-offs between conflicting selection pressures. 2) Single traits jointly selected: Traits playing a dual role, also leading to ecological trade-offs.
Growth-defence (Herms and Mattson 1992, Mole 1994, Stamp 2003, Huot et al. 2014, Züst and Agrawal 2017, He et al. 2022, Monson et al. 2022). Defence-reproduction (Bazzaz et al. 1987, Herms and Mattson 1991, Ågren et al. 1999).	Growth-defence (Viola et al. 2010). Defence-reproduction (Bronstein et al. 2007, Carr and Eubanks 2014, Campbell 2015, Johnson et al. 2015, Whitehead et al. 2022).

We include a representative (yet not exhaustive) compilation of more in-depth reviews specialising in each pairwise link involving defences.

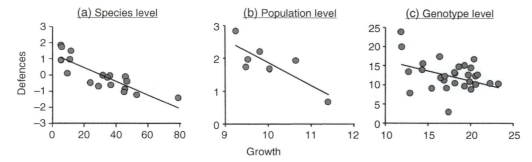

Figure 15.1 Trade-offs between different measures of plant growth and defences in pines (*Pinus* sp.). Panel (a) shows a negative correlation ($R^2 = 0.62$, $P < 0.001$) between height (in cm) and the concentration of resin produced in different plant parts for saplings belonging to 18 *Pinaceae* species (modified from Moreira et al. 2014). Panel (b) shows a negative correlation ($R^2 = 0.65$, $P = 0.028$) between plant height (in m) and resin yield (in g) of mature *Pinus pinaster* trees belonging to seven populations (values are population means; modified from Vázquez-González et al. 2022). Finally, panel (c) depicts a negative correlation ($R^2 = 0.15$, $P = 0.024$) between plant biomass (in g) and the concentration of resin produced in the stem (in mg g^{-1} d.w.) of *P. pinaster* saplings from 33 open-pollinated genetic families (Adapted from Sampedro et al. 2011).

Lerdau et al. 1994, Koricheva et al. 1998, Stamp 2004) (see Figure 15.1). Much of this work has shown that the strength or occurrence of growth-defence trade-offs is highly contingent on environmental conditions (reviewed by Züst and Agrawal 2017 and He et al. 2022), with abiotic manipulations or comparisons across abiotically contrasting sites shedding insight into their ecological and evolutionary basis (Fine et al. 2004, van Zandt 2007, Des Marais et al. 2013). Studies have shown that allocation constraints and resulting trade-offs involving several types of secondary metabolites (*e.g.*, phenolic compounds, terpenoids) often arise or become stronger under abiotic stress (*e.g.*, low water or resource availability and salinisation) or under high competition implying some degree of resource limitation (reviewed by Züst and Agrawal 2017 and He et al. 2022). These responses have been observed across diverse plant taxa, including, for example, willow (*Salix*) species (Salicaceae) (Glynn et al. 2007), the black poplar *Populus nigra* (Salicaceae) (Hale et al. 2005) and the loblolly pine *Pinus taeda* (Pinaceae) (Lombardero et al. 2000) in the case of trees, as well as in herbs such as the thale cress *Arabidopsis thaliana* (Brassicaceae) (Barto and Cipollini 2005) and the wild cabbage *Brassica oleracea* (Brassicaceae) (Moreira et al. 2018) to name only a few (reviewed by Züst and Agrawal 2017, He et al. 2022). By the same token, studies have also shown that resource addition (*e.g.*, soil nutrients) often (but not always, see ahead) ameliorates the expression of growth-defence trade-offs (*e.g.*, Sampedro et al. 2011). The strengthening of allocation constraints when plants are faced with abiotic stress has been attributed to impaired resource use and/or acquisition (Coley et al. 1985, Endara and Coley 2011) or higher physiological or biochemical demands associated with maintaining function under stress (Mithöfer and Boland 2012, Kant et al. 2015). Interestingly, recent mechanistic work at the biochemical and molecular level has reported that plants can, to some extent, actively modulate these constraints when faced with environmental challenges, *e.g.*, by regulating the joint allocation to growth and defences (and resulting trade-offs) via biochemical mechanisms involving interactions between growth-related primary (*e.g.*, auxin) and defence-related secondary (*e.g.*, salicylic acid) metabolisms (He et al. 2022).

For many systems, we now have a good hold on the types of environmental contexts promoting growth-defence trade-offs, but there has been substantial discussion on the underlying mechanisms. Among the multiple views, one holds that at moderate to high resource availability, resources allocated to growth in rapidly dividing tissues take place at the expense of secondary metabolism, as growth demands are high and prioritized over defences (the growth-differentiation-balance hypothesis; GDBH, Loomis 1932, Herms and Mattson 1992, Stamp 2004). This condition is expected to occur more frequently for herbaceous, particularly short-lived (*e.g.*, annual) plants, as these cannot draw on stored reserves to overcome allocation constraints (Herms and Mattson 1992, Stamp 2003). Long-lived plants, in contrast, can divert resources from reserves to attenuate allocation constraints and jointly allocate to growth and defences in rapidly growing tissues (*e.g.*, immature leaves; Stamp 2003). That said, there has been supportive evidence for growth-defence trade-offs from empirical studies in both herbaceous plants (*e.g.*, *A. thaliana*; Barto and Cipollini 2005) and trees (*e.g.*, *Salix* spp.; Glynn et al. 2007). However, comparative (inter-specific) studies involving resource manipulations remain scarce, thus preventing a robust test of these expectations and reaching generalisations. In line with this literature, a separate body of research has shown that environments with high competition (*e.g.*, for soil nutrients or light; in *Populus*: Donaldson et al. 2006) promote rapid plant growth, indicating prioritisation of growth over defences (reviewed by He et al. 2022 and Monson et al. 2022). This provides

Figure 15.2 Secretion of resin (viscous fluid exuded from ducts) from a young pine tree after the attack from the phloem feeder weevil *Hylobius abietis*. *Source:* Photo by Luis Sampedro.

similar evidence in favour of expected trade-offs between growth and defences, as competition interacts with resource availability and is shaped by the same growth-related mechanisms.

Another influential view on the mechanisms driving growth-defence associations holds that plants in resource-poor environments grow more slowly and in turn allocate more energy to producing defences (the Resource Availability Hypothesis; Coley et al. 1985, Endara and Coley 2011). The reason for this is that tissues lost to herbivory are costlier to replace under low resource availability and therefore plants prioritize defence at the expense of growth when resources are limited (Coley et al. 1985). Empirical studies have been largely supportive of this hypothesis across numerous plant taxa (reviewed by Endara and Coley 2011). For example, in a phylogenetically controlled study with 20 tropical tree species, Fine et al. (2004) found that seedlings of slow-growing species adapted to resource-poor soils outperformed in these soils seedlings of fast-growing species adapted to high-fertility soils and that this advantage was mediated by reduced herbivory, presumably because they are more highly defended than fast-growing species. Likewise, work on temperate trees reported that slow-growing pine species (*Pinus* spp.) had higher levels of constitutive chemical defences (*e.g.*, resin; Figure 15.2) than fast-growing ones (Moreira et al. 2014). Similar patterns have also been reported for mountainous herbaceous plants in temperate areas (Pellissier et al. 2016, Defossez et al. 2018).

Some studies, including recent work, have nonetheless pointed to substantial variability in the strength and sign of plant defence-growth associations, particularly within species (*i.e.*, among genotypes) (Hahn and Maron 2016, López-Goldar et al. 2020). Indeed, while a number of studies have found predicted negative genotypic correlations suggestive of trade-offs (*e.g.*, Donaldson et al. 2006, Osier and Lindroth 2006, Orians et al. 2010, Sampedro et al. 2011, Züst et al. 2015, Vázquez-González et al. 2020), studies have often failed to detect these negative correlations, even under abiotically stressful conditions (Ohnmeiss and Baldwin 1994, Messina et al. 2002, Tuller et al. 2018). For example, in an early study, Ohnmeiss and Baldwin (1994) found no trade-offs between biomass and nicotine production in the coyote tobacco *Nicotiana attenuata* (Solanaceae), and this finding remained consistent under nitrogen-limited soils. Similarly, subsequent work by Orians et al. (2010) found that intra-specific trade-offs were ephemeral and only present early in seedling development when root establishment is a priority in hybrids of two willow species (*Salix* spp.). Furthermore, counter to predictions of growth prioritisation over defences under high resource availability (GDBH), recent studies have found that plants often allocate simultaneously to growth and defence, in some cases even showing positive growth-defence correlations (Hahn and Maron 2016, López-Goldar et al. 2020, Hahn et al. 2021, Kichas et al. 2023). For example, López-Goldar et al. (2020) found a positive genotypic correlation between growth and constitutive chemical defences in maritime pine (*Pinus pinaster,* Pinaceae). In addition, Hahn et al. (2021) found that growth-defence associations ranged from negative to positive depending on the scale (individual to population) in a perennial herb (*Monarda fistulosa,* Lamiaceae). Finally, while abiotic stress can in some of these cases modulate intra-specific growth-defence relationships (consistent, broadly speaking, with previously revised studies; Züst and Agrawal 2017, He et al. 2022), this does not always lead to negative (or stronger negative) associations as previously held (Hahn and Maron 2016). Collectively, these studies question the widely held predominance of allocation constraints as drivers of growth-defence linkages, at least at the intra-specific level, and call for new predictions leveraging mechanisms across levels of biological organisation (intra- to inter-specific levels).

15.2.1.2 The Role of Herbivore Pressure

The above section reviewed cases for which resources or abiotic factors influence plant allocation costs or constraints and this in turn affects correlations between defences and growth-related traits. However, these links may also arise due to variation in top-down forcing, *i.e.*, herbivore attack. In this case, allocation constraints (or some other mechanistic link) would be a pre-requisite, but herbivory would jointly shape defences and growth-related traits to the extent that the latter mediate the risk of plants being attacked by herbivores (Zangerl and Rutledge 1996, Agrawal et al. 2006). These effects can take place, for example, if herbivore pressure (*e.g.*, measured as the likelihood of being attacked or amount of damage) is determined by how common (*i.e.*, abundant or frequent), temporally available (*i.e.*, ephemeral or long-lived) or easily detectable (*e.g.*, based on size) plants are. Longer-lived or larger plants, as well as less cryptic species (relative to biotic or abiotic backgrounds), are visually easier to detect by insect herbivores (*i.e.*, more apparent; Feeny 1976, Castagneyrol et al. 2013, Smilanich et al. 2016) and expected to evolve defences that are effective against most consumers, *i.e.*, quantitative defences which typically reduce plant digestibility (the Plant Apparency Theory, Feeny 1976; reviewed by Agrawal et al. 2006). In contrast, less apparent plants are more likely to exhibit qualitative defences (*i.e.*, toxic secondary compounds that are effective in small doses and relatively inexpensive for the plant to produce) which limit feeding by a subset of specialised herbivores (Feeny 1976, Strauss et al. 2015). This, therefore, sets up contrasting modes of defence that are evolutionarily linked to growth-related life histories interacting with herbivory risk.

Growth-related traits predictive of detectability or risk of attack by herbivores include features such as growth form, plant ontogenetic stage and leaf habit (*i.e.,* evergreen *vs.* deciduous). Among these, probably the best evidence comes from work assessing the effects of growth form: woody species are long-lived, larger in size, and therefore offer resources that are more available through time and easier to detect by herbivores (reviewed by Turcotte et al. 2014). In line with apparency theory, empirical studies show that woody species are more likely to be attacked by herbivores and, accordingly, invest more in quantitative defences relative to smaller and ephemeral non-woody (*e.g.*, herbaceous) species (reviewed by Turcotte et al. 2014 and Strauss et al. 2015).

Ontogeny, is another potentially important (albeit less studied from an apparency perspective) predictor of herbivory as older plants are larger in size, particularly for woody species, and therefore become easier to find by herbivores (Moreira et al. 2017, Cole et al. 2021). In the case of long-lived species, theory and empirical work pose that defence investment increases from the seedling to the sapling stage, once a greater pool of resources is stored and the plant can increase allocation to defences (Boege and Marquis 2005, Barton and Koricheva 2010, Smilanich et al. 2016, Barton and Boege 2017). Here, ontogenetic changes in allocation to growth presumably shape these ontogenetic changes in defence investment (*e.g.*, Orians et al. 2010, Cole et al. 2021), but ontogenetic variation in the risk of herbivory could also play a key role during plant development by affecting defences. Experimental designs aimed at teasing apart these mechanisms are, however, rare. Among the few studies, Moreira et al. (2017) found evidence suggesting both mechanisms could be at work: English oak (*Quercus robur,* Fagaceae) saplings had lower amounts of leaf herbivory than adults, consistent with apparency theory, but were also more defended (had higher concentrations of phenolic compounds) than adults which could have driven herbivory (i.e., consistent with a bottom-up allocation constraints mechanism). Similar findings were reported for quaking aspen (*Populus tremuloides,* Salicaceae) (Cole et al. 2021). By contrast, in a study with eight oak species, Galmán et al. (2019) found that herbivory did not significantly differ between ontogenetic stages whereas chemical defences (phenolic compounds) were higher for saplings than adults, and were unrelated to herbivory, rejecting both mechanisms and suggesting that secondary metabolites are involved in other functions (*e.g.,* coping with abiotic stress). In addition, a study by Ochoa-López et al. (2018) found that oviposition by the specialist butterfly *Euptoieta hegesia* was higher for seedlings than adult plants of a tropical coastal shrub *Turnera velutina*, presumably because they are less chemically defended, i.e., a bottom-up process. The latter two studies therefore provide more support for ontogenetic variation in growth-defence allocation constraints rather than apparency as a driver of plant growth (size)-defence relationships.

Finally, leaf habit (*i.e.,* evergreen *vs.* deciduous) could also be an important predictor of herbivory, reflecting joint evolutionary dynamics between growth and defence. Evergreen species are well-known to exhibit a conservative resource-use strategy with lower photosynthetic and respiration rates leading to reduced resource allocation to growth (Lohbeck et al. 2015), whereas deciduous species exhibit an exploitative resource-use strategy with higher growth rates (Reich et al. 1998, Poorter and Garnier 2007). At the same time, from the herbivore's perspective, evergreen species offer leaf tissue year-round and therefore provide more predictable and abundant resources for herbivores (Lim et al. 2015). Therefore, if growth trades off with defences, then the evolution of leaf habit has likely evolved in association with that of defences. For example, it was found that contrary to expectations based on apparency, leaf habit was generally not predictive of levels of leaf chemical defences (*e.g.*, phenolic compounds) in 56 oak (*Quercus*) species (Moreira and Pearse 2017). However, evergreen oak species had greater leaf toughness and specific leaf mass than deciduous species, traits predictive of resource use which can also affect herbivory, suggesting correlated evolution between leaf habit and allocation to leaf traits associated with physical or structural defence. Likewise, results from a latitudinal gradient study by Lim et al. (2015) including more than a thousand plant species also suggest that leaf habit affects herbivory via apparency effects. They found that leaf herbivory on evergreen species increased towards the equator whereas no association was found for deciduous species, arguing that this is because

Figure 15.3 A honeybee collecting nectar and a caterpillar feeding on flowers of the violet wild petunia *Ruellia nudiflora* (Acanthaceae). *Source:* Photo by Luis Salinas-Peba.

the latter are only attacked during the growing season and not (or very little) during the winter months, leading to a weaker year-round difference in herbivory across latitudes relative to evergreen species. However, associations with leaf defences were not tested, precluding an assessment of links between growth and defences as mediated by herbivory pressure. These results call for more work along these lines to better understand how leaf habit relates to apparency and plant defences.

15.2.2 Defences and Reproductive Traits

Evolutionary links between defences and reproduction are widespread (Bronstein et al. 2007, Jones and Agrawal 2017), and shape plant reproductive life histories (Eckhart and Seger 1999). However, studies considering plant allocation constraints between defences and reproduction (or other plant endogenous mechanisms) are comparatively less common (Herms and Mattson 1991, Obeso 2002, Cole et al. 2021) (see Table 15.1). Rather, most work (mainly empirical studies) has focused on the concurrent effects of herbivory and plant mutualisms (*e.g.*, pollination, frugivory/seed dispersal), often implying plant-mediated indirect interactions driving the joint evolution of plant defensive and reproductive traits (*e.g.*, early work by Strauss 1997 and Herrera et al. 2002, reviewed by Johnson et al. 2015 and Whitehead et al. 2022) (see Table 15.1). This involves, for example, cases of correlated selection when one plant trait affects the evolution of another trait (*e.g.*, through pleiotropic effects) leading to the evolution of trait combinations, as well as diffuse selection on a single trait when it plays a dual role defending against herbivory or attracting pollinators or seed dispersers (Figure 15.3). While not the focus of this section, there are also cases that do not involve plant defences but rather joint selection on reproductive traits by pollinators and pre-dispersal seed predators (*e.g.*, floral volatiles, morphology or display size; Miller et al. 2008, Brody and Irwin 2012, Abdala-Roberts et al. 2014, Campbell et al. 2022), including instances when a plant associate is an herbivore in early life stages and then a pollinator as an adult (*e.g.,* nursery mutualisms; Pellmyr 2003).

We next review work addressing patterns of joint selection on plant defences and reproduction; first those involving pollinators, and then those relating to frugivores/seed dispersers, in each case providing examples of plant traits selected upon by each group and the processes by which defence-reproduction interdependencies arise. When available, we point to known or suspected causality behind joint selective pressures, as well as their contingency on abiotic factors. These joint selective pressures have likely played a key role in the evolution of plant reproductive life histories by shaping, for example, mating and pollination systems.

15.2.2.1 Plant Defences and Pollination-Related Traits

There is ample evidence that herbivores and, by the same token, the defensive traits they select upon, drive (and are affected by) plant reproductive features, including floral traits and mating systems (reviewed by Campbell 2015 and Johnson et al. 2015). Studies at the intra-specific level have shown that herbivory can increase self-crossing rates (though in some cases, it can also promote outcrossing; Johnson et al. 2015) and in so doing affect the evolution of plant reproductive traits and mating systems (reviewed by Campbell 2015). Such effects can take place through various mechanisms, including inbreeding depression and purging of deleterious alleles associated with reproductive traits (Husband and Schemske 1996, Campbell and Kessler 2013), as well as by reducing pollinator visitation via changes in floral traits (*e.g.*, reduced flower display, changes in volatile emissions and presence of secondary metabolites in floral rewards; reviewed in Kessler and Halitschke 2009 and Campbell 2015). Conversely, there are also cases where pollinator-selected reproductive features affect plant defence evolution. Again, at the intra-specific level, studies have shown that plant selfing rates impact defences via inbreeding depression and the accumulation of deleterious mutations negatively affecting plant defences and resistance to herbivory (Carr and Eubanks 2014, Johnson et al. 2015). Overall, these studies thus support the idea that plant defences and reproduction are reciprocally linked ecologically and evolutionarily.

Macroevolutionary studies have proven highly useful for pinpointing plant traits and uncovering evolutionary links between herbivory and pollination across species. Studies with species of the genus *Dalechampia* have shown that flower resins, used as defences against florivores, were co-opted and selected upon as rewards by resin-collecting bees, whereas floral morphological traits (*e.g.*, bracts) for pollinator attraction evolved into physical defensive structures that protect floral sexual organs and developing fruits (Armbruster 1997, Armbruster et al. 2009). In some cases, evolutionary links arise because floral traits associated with pollinator attraction and defence are correlated due to allocation constraints or pleiotropy, as shown for flower colour which depends on the expression of secondary metabolites such as anthocyanins, as well as resin-secreting tissues in the inflorescences (reviewed by Johnson et al. 2015). For example, in species of *Hakea* (Proteaceae) red flower colouration is correlated with cyanide concentrations, such that the evolution of these chemical defences owing to herbivory has in turn shaped floral use by birds and the evolution of floral traits associated with bird pollination (Hanley et al. 2009). Similar dynamics are probably in place in other well-studied systems showing evidence of dual evolutionary effects of herbivores and pollinators, though explicit analyses of plant defences were not conducted (*e.g.*, *Yucca*, Pellmyr 2003; *Ficus*, Cook and Rasplus 2003, see also Chapter 17, this volume). Finally, a study by Campbell and Kessler (2013) on nightshades (Solanaceae) reported that evolutionary transitions from outcrossing (self-incompatibility) to self-crossing are associated with increased inducibility of plant defences. Here, the expectation is that herbivory acts as a driver of self-crossing in plants (see above), again highlighting the reciprocal nature of these evolutionary dynamics (Campbell 2015).

Evolutionary links between herbivory and plant mating systems also emphasize implications for the evolution of separate sexes (dioecy) from different variants of hermaphroditism such as gynodioecy (presence of female and hermaphrodite plants) or monoecy (both sexes in the same plant) (Ashman 2002, 2006; reviewed by Johnson et al. 2015). Empirical work has shown that herbivory is higher in pollen-producing hermaphrodite plants in gynodioecious species and in male plants of dioecious species (reviewed by Cornelissen and Stiling 2005 and Johnson et al. 2015), though recent work points to no difference in chemical defences between sex types and herbivory results are inconclusive (Sargent and McKeough 2022). In some systems, however, evidence for plant sex-biased herbivory and defences has been found and suggest defence-reproduction life-history linkages. For example, transitions from monoecy to dioecy in strawberries (*Fragaria* spp.) have been presumably favoured by sex-biased weevil herbivory on hermaphrodites (as compared to plants with separate sexes) and by increased self-crossing (through lower floral display and reduced pollinator attraction; Penet et al. 2009; but see Ashman et al. 2004). Similar patterns have been reported for other species, though other herbivory-related mechanisms are thought to be at work. For example, it has been argued that herbivory could promote the evolution of dioecy by directly affecting resource allocation between the sexes causing a reduction in resource acquisition and increased allocation to male rather than female function (*e.g.*, Krupnick and Weis 1999), possibly due to lower costs of the former. These sex-based differences in herbivore pressure are expected to lead to sexual dimorphism in defences, therefore pointing to an evolutionary link between defence and plant sex. Still, the general applicability of this prediction has been questioned and alternative causalities where herbivory is not the driver have been proposed (Avila-Sakar and Romanow 2012). For example, sexual dimorphism in plant defences could respond to other factors besides herbivore pressure (*e.g.*, abiotic stresses; Ashman 2006) which shape differences in plant defence allocation and in turn herbivory (third-party factors may similarly shape growth-defence correlations, see above). Female plants are expected to invest more resources into reproduction than males, such that allocation trade-offs are expected to favour slower growth and in turn higher investment in defensive traits relative to males (Eckhart and Seger 1999, Barrett and Hough 2013). We further discuss the three-way intersection between plant growth, defences and reproduction (sex) later in this chapter.

15.2.2.2 Plant Defences and Seed Dispersal-Related Traits

Evolutionary research on animal seed-dispersal systems can be traced back to several seminal reviews (*e.g.*, McKey 1975, Howe and Smallwood 1982, Janzen 1983) and also includes more recent syntheses (*e.g.*, Herrera 1982, Traveset et al. 2007) as well as special journal issues (see Carlo et al. 2022). Seed dispersal includes mechanisms related to attachment to animals (in the case of vertebrates such as mammals and birds), storage and eventual consumption of seeds whereby some seeds accidentally escape predation (*i.e.*, granivory) as well as the attraction of dispersers (usually vertebrates) with fleshy tissues (edible seed appendages or coverings) which provide nutrients (*i.e.*, frugivory) (Herrera 2002). Of these, frugivory will be the focus of this section (*sensu* Herrera 2002), as it is probably the most representative example of active dispersal thought to involve coevolutionary dynamics that enhance seed dispersal. Here too, as for pollinators, plants face joint effects of herbivores (and pathogens) attacking fruits and seed-dispersing frugivores, setting the stage for correlated evolution involving both defence and reproduction-related traits which shape plant life histories in various ways.

Evidence for frugivore-mediated evolution of fruit and seed traits has mounted in recent decades (reviewed by Valenta and Nevo 2020), including to a large extent traits associated with dispersal by vertebrates (*e.g.*, birds, primates) such as seed size, fruit crop size and fruit colour (*e.g.*, Jordano 1995, Schaefer et al. 2008, Galetti et al. 2013, Nevo et al. 2018). However, much of this work has not studied these traits from the perspective of plant antagonists even though organisms such as herbivores probably affect them. In this sense, fruit and seed secondary chemistry have received increased attention in recent years (reviewed by Nelson and Whitehead 2021 and Whitehead et al. 2022). Some authors have argued that dispersers or antagonists select upon secondary metabolites independently (*i.e.*, selection by one does not affect the other), whereas others argue for non-independent (e.g., diffusse) evolutionary dynamics, *i.e.*, compounds that have evolved in response to one type of interaction affect the other, often leading to ecological trade-offs (see Whitehead et al. 2022). For example, studies of understory tropical herbs of the genus *Piper* show that the concentration and composition of amides in the fruit pulp conferring resistance against enemies (mainly pathogens but also insect herbivores) also negatively affect fruit preference and gut retention time in seed-dispersing bats (Baldwin and Whitehead 2015, Whitehead et al. 2016). Similar findings have been reported for alkaloids and phenolic compounds in other plant taxa such as *Solanum* sp. and bird-dispersed tropical shrubs (Cipollini and Levey 1997, Cazetta et al. 2008). Likewise, traits such as fruit scents or fruit colour which have presumably evolved to attract mutualists can also attract plant enemies (Valenta and Nevo 2020).

Though less studied, there also appear to be evolutionary links between physical traits of fruits and joint selection by antagonists and frugivores (reviewed by Whitehead et al. 2022). In a recent study, Valenta et al. (2022) reported that fruit (skin) hardness cannot be entirely explained by mechanical constraints and likely reflects joint selection by frugivores and herbivores (for softer *vs.* harder skins, respectively). In addition, in an early study with 115 plant species, Rodgerson (1998) found that mechanical defence in seeds (measured as seed hardness) was greater in species adapted to ant dispersal and such species experienced much lower levels of seed predation.

Fruit or seed chemical and physical traits responding to joint effects of herbivores and seed dispersers are likely to affect crucial post-zygotic reproductive features such as ripening speed and its resulting effects on fruiting phenology (Cipollini and Levey 1997) and dispersal patterns determining the types of habitats seeds arrive in and the biotic and abiotic conditions therein (Nelson and Whitehead 2021). In addition, such fruit or seed chemical and physical traits also affect different aspects of seed physiology which influence dormancy (and thus propensity to form seed banks) and reproductive success (*e.g.*, Dalling et al. 2020). The resulting evolution of these jointly selected traits can lead to far-reaching effects on plant life histories.

15.3 Tripartite Views Shed Insight into the Evolution of Plant Life History Traits

We have shown that plant growth- and reproduction-related life history traits do not evolve independently from traits associated with plant defence, and that many times a single trait can play ecological roles associated with more than one function. Thus far, we have reviewed pairwise patterns involving joint evolution of defences and one other function. However, given that growth and reproduction are also evolutionarily linked (*e.g.*, via allocation costs or pleiotropy; reviewed by Obeso 2002 and Monson et al. 2022), a tripartite view becomes essential to better understand the implications for plant life history evolution (Obeso 2002). Yet, to date, relatively few studies have taken this integrative approach (but see Cole et al. 2021). We next briefly review areas of research that can serve as frameworks for achieving this three-way integration and motivate future mechanistic and evolutionary work, including the role of abiotic forcing.

Empirical and theoretical work on dioecious plants addressing the linkages between plant growth, reproduction and defences represents one such framework that could be further developed. Studies have shown that plant sexes differ in ecologically important traits, including physical and chemical traits associated with resistance to herbivores (reviewed by Barrett and Hough 2013). There is some evidence that female plants (relative to males) invest more resources into reproduction and less in growth (Delph 1990, Obeso 2002), with this in turn leading to higher defences if they trade off with growth (Cornelissen and Stiling 2005). Nonetheless, these patterns are highly variable across plant taxa (Sargent and McKeough 2022) and this is compounded by the fact that allocation costs are often difficult to detect (*e.g.*, because selection minimizes costs or due to methodological constraints in separating costs of different functions; see Obeso 2002). Still, further mechanistic research is needed, and abiotic manipulations could prove highly useful to this end (*e.g.*, in testing for allocation constraints) for understanding how environmental stress shapes growth-defence linkages as a function of plant mating systems. For example, the importance of ecological context in favouring dioecy has been pointed out (Ashman 2006), as well as the importance of teasing apart the relative roles of abiotic factors (*e.g.*, resource availability) and herbivory pressure (Johnson et al. 2015) to understand how growth-defence associations shape dioecy and other reproductive life histories in plants. Another potentially insightful approach could be to compare costs of reproduction and defence across species with contrasting growth strategies. For example, Obeso (2002) found highly variable patterns across more than 100 species, but in cases where growth-reproduction trade-offs were detected (allocation to reproduction reduced that to growth), these were contingent on growth form. Namely, reproductive costs were higher for females than males (as expected) for woody species whereas the inverse occurred for herbaceous species. We are not aware of analyses of this type including plant defences. In this sense, macroevolutionary studies would be highly useful to achieve a better understanding of the evolutionary drivers of dioecy and plant sex-based differences in growth and defence allocation.

Other potentially useful, yet largely undeveloped approaches lie in comparative tests of fruit or seed dispersal strategies across relevant axes of variation in plant growth strategies, including, for example, growth form or leaf habit. Previous work has shown that the proportion of vertebrate-dispersed seeds is higher for trees and shrubs compared to herbs, mediated by traits such as seed size, a pattern consistent across different types of ecological communities such as temperate and tropical dry forests (Herrera 2002). Accordingly, relationships between plant growth form and herbivore pressure (*e.g.*, via apparency, see section on the role of herbivore pressure) combined with reported links between dispersal traits and defences (see section on plant defences and seed dispersal-related traits) could set the stage for associations between dispersal strategies, fruit and seed defences, and growth-related traits. Likewise, recent work has put forward views on the use of life form (perennial *vs.* annual life cycles) as a syndrome and how correlations between growth, defences and reproduction have evolved along this axis and at the same time influenced associated life history traits (see Lundgren and Des Marais 2020). Insights gained can also increase our understanding of crop domestication effects on different plant functions and inform breeding strategies to minimize allocation costs or genetic linkages to optimize vegetative or reproductive traits. Recent work across species of *Physalis* (Solanaceae) expands on these views reporting on associations among perenniality, mating system, and plant defensive strategies (constitutive *vs.* induced defences) (Jacobsen 2022).

Finally, a plant ontogenetic perspective can provide another useful framework to understand better these three-way associations. By and large, studies on plant function correlations have been based on a single life stage (Barton and Boege 2017), despite the fact that endogenous mechanisms (*e.g.*, allocation constraints; Orians et al. 2010) and species interactions potentially underlying them change across plant (and insect) life stages. Even for theories such as the GDBH, where plant development is predicted to play a central role in structuring function correlations over time, studies are often short-term and neglect longer-term measurements to test for ontogenetic variation, particularly those including reproductive life stages (*e.g.*, Barto and Cipollini 2005). One noteworthy example is a recent study by Cole et al. (2021), who found evidence for trade-offs between growth and defence as well as between growth and reproduction in quaking aspen (*P. tremuloides*), but not between defence and reproduction. These patterns were contingent on plant ontogeny, indicating that trade-offs change in strength or are even precluded based on changes in allocation to different functions as plants grow older. For example, younger plants invest more in defence but gradually decrease their investment to reach the lowest levels before the onset of reproduction (*i.e.*, limited overlap in investment across ontogeny), in sharp contrast to growth which is fast initially and strongly competes for allocation to defences (Cole et al. 2021). These authors also found that plant sex explained observed patterns, because, for example, growth-reproduction trade-offs were stronger for female than male plants, allowing the latter to grow faster and reach a similar size to females (Cole et al. 2021). Overall, these findings point to heritable plant ontogenetic trajectories in patterns of allocation to all three functions, whereby fast initial growth appears to set the stage for subsequent allocation trade-offs which in some cases are sex specific.

15.4 Challenges for Future Research

It is important to always keep in mind that the expression and role in resistance of many plant traits putatively associated with defence often change as a function of the biotic and abiotic context (Karban 2011). For example, some traits will confer resistance against some herbivores but not others (*e.g.,* specialists *vs.* generalists), act synergistically or antagonistically with other traits or evolutionarily respond to abiotic factors initially and are later co-opted as anti-herbivore traits (Ali and Agrawal 2012). Research grounded on a solid natural history that recognizes (and helps describe) these context-dependencies is needed to build more robust and unbiased views on plant defence evolution and in turn understand how putative defensive traits relate to plant life histories. We next list what we believe are key challenges for paving a stronger integration between plant defence and plant life history research, in each case recognising and addressing the labile ecological and evolutionary nature of many plant traits putatively linked to defence.

An important challenge, which emerges from our review (also pointed out by Hahn and Maron 2016 and Agrawal and Hastings 2019), will be to leverage intra- and inter-specific variation in relationships between plant defences and other functions. Studies conducted at multiple levels of organisation will increase our understanding of how biotic and abiotic factors acting at different scales shape defence-life history correlated evolution.

The influence of abiotic factors forcing on defence-life history associations has been given more attention in some research contexts (*e.g.*, growth-defence associations) than in others (*e.g.*, defence-reproduction) (Obeso 2002, Züst and Agrawal 2017), and is strongly biased towards some factors (*e.g.*, nutrients, drought) over other equally important ones (*e.g.,* warming and soil salinity). A more expansive view of unexplored abiotic factors, combined with joint assessments of two or more drivers simultaneously, is needed to achieve a more nuanced and robust understanding of abiotic effects on plant defence-life history evolution.

The reviewed literature also shows the contrasting manner (and resulting biases) in how growth-defence *vs.* reproduction-defence linkages have been approached. In the former case, there has been a historical push towards addressing plant endogenous mechanisms (*e.g.*, allocation constraints) whereas the influence of correlated or diffuse selection from the perspective of species interactions (*e.g.*, competition *vs.* herbivory; see Hambäck and Beckerman 2003) has received considerably less attention (see Table 15.1). In contrast, the latter has focused more on joint selection from herbivory and plant mutualisms, but relatively little work has addressed plant-based endogenous mechanisms linking defensive and reproductive traits (Cole et al. 2021). Accordingly, an important challenge will be to reduce these biases and integrate these mechanisms for a better understanding of the evolution of these relationships, particularly as we move towards three-way perspectives (*i.e.,* simultaneous assessments of defence-growth-reproduction).

Another important consideration, also relevant for achieving integration, will be to expand defence measurements to incorporate multiple defensive traits and strategies, how these are linked, and how such evolutionary relationships between defences in turn feed into associations with growth and reproduction-related traits (Moreira et al. 2020). This can help, for example, to identify defensive traits that are more likely to be linked to life history evolution or drive plant defensive syndromes (*i.e.*, correlated expression of different groups of defensive traits) and how these relate to life history traits.

Finally, continued work at the biochemical or molecular level is essential to uncover plant-based mechanisms behind defence-life history links (Figueroa-Macías et al. 2021, He et al. 2022). This will aid in measuring or teasing apart the costs of different functions (or lack thereof), understanding when and how links are more likely to occur (*e.g.*, interactions between primary metabolism and secondary metabolism via different plant signalling pathways), as well as a mechanistic understanding of how abiotic factors shape such processes.

References

Abdala-Roberts, L., and Mooney, K.A. (2015). Plant and herbivore evolution within the trophic sandwich. In: *Trophic interactions: bottom-up and top-down interactions in aquatic and terrestrial ecosystems* (eds. Hanley, T.C.N., and La Pierre, K.J.), 339–363. Cambridge University Press.

Abdala-Roberts, L., Parra-Tabla, V., Campbell, D.R, and Mooney, K.A. (2014). Soil fertility and parasitoids shape herbivore selection on plants. *Journal of Ecology* 102: 1120–1128.

Agrawal, A.A. (2007). Macroevolution of plant defense strategies. *Trends in Ecology & Evolution* 22: 103–109.

Agrawal, A.A. (2011). Current trends in the evolutionary ecology of plant defense. *Functional Ecology* 25: 420–432.

Agrawal, A.A., and Hastings, A.P. (2019). Trade-offs constrain the evolution of an inducible defense within but not between plant species. *Ecology* 100: e02857.

Agrawal, A.A., Lau, J., and Hämback, P.A. (2006). Community heterogeneity and the evolution of interactions between plants and insect herbivores. *Quarterly Review of Biology* 81: 349–376.

Agrawal, A.A., Strauss, S.Y., and Stout, M.J. (1999). Costs of induced responses and tolerance to herbivory in male and female fitness components of wild radish. *Evolution* 53: 1093–1104.

Ågren, J., Danell, K., and Elmqvist, T. (1999). Sexual dimorphism and biotic interactions. In: *Gender and sexual dimorphism in flowering plants* (eds. Geber, M.A., Dawson, T.E., and Delph, L.F.), 217–246. Springer.

Ali, J.G., and Agrawal, A.A. (2012). Specialist versus generalist insect herbivores and plant defense. *Trends in Plant Science* 17: 293–302.

Armbruster, W.S. (1997). Exaptations link evolution of plant-herbivore and plant-pollinator interactions: a phylogenetic inquiry. *Ecology* 78: 1661–1672.

Armbruster, W.S., Lee, J., and Baldwin, B.G. (2009). Macroevolutionary patterns of defense and pollination in *Dalechampia* vines: adaptation, exaptation, and evolutionary novelty. *Proceedings of the National Academy of Sciences of the United States of America* 106: 18085–18090.

Ashman, T.L. (2002). The role of herbivores in the evolution of separate sexes from hermaphroditism. *Ecology* 83: 1175–1184.

Ashman, T.L. (2006). The evolution of separate sexes: a focus on the ecological context. In: *Ecology and evolution of flowers* (eds. Harder, L.D., and Barrett, S.C.H.), 204–222. Oxford University Press.

Ashman, T.L., Cole, D.H., and Bradburn, M. (2004). Sex-differential resistance and tolerance to herbivory in a gynodioecious wild strawberry. *Ecology* 85: 2550–2559.

Avila-Sakar, G., and Romanow, C.A. (2012). Divergence in defence against herbivores between males and females of dioecious plant species. *International Journal of Evolutionary Biology* 2012: e897157.

Baldwin, J.W., and Whitehead, S.R. (2015). Fruit secondary compounds mediate the retention time of seeds in the guts of Neotropical fruit bats. *Oecologia* 177: 453–466.

Ballhorn, D.J., Kautz, S., Lion, U., and Heil, M. (2008). Trade-offs between direct and indirect defences of lima bean (*Phaseolus lunatus*). *Journal of Ecology* 96: 971–980.

Barrett, S.C.H., and Hough, J. (2013). Sexual dimorphism in flowering plants. *Journal of Experimental Botany* 64: 67–82.

Barto, E.K., and Cipollini, D. (2005). Testing the optimal defense theory and the growth-differentiation balance hypothesis in *Arabidopsis thaliana*. *Oecologia* 46: 169–178.

Barton, K.E., and Boege, K. (2017). Future directions in the ontogeny of plant defence: understanding the evolutionary causes and consequences. *Ecology Letters* 20: 403–411.

Barton, K.E., and Koricheva, J. (2010). The ontogeny of plant defense and herbivory: characterizing general patterns using meta-analysis. *The American Naturalist* 175: 481–493.

Bazzaz, F.A., Chiariello, N.R., Coley, P.D., and Pitelka, L.F. (1987). Allocating resources to reproduction and defense. *BioScience* 37: 58–67.

Becerra, J.X., Noge, K., and Venable, D.L. (2009). Macroevolutionary chemical escalation in an ancient plant–herbivore arms race. *Proceedings of the National Academy of Sciences of the United States of America* 106: 18062–18066.

Boege, K., and Marquis, R.J. (2005). Facing herbivory as you grow up: the ontogeny of resistance in plants. *Trends in Ecology & Evolution* 20: 441–448.

Brody, A.K., and Irwin, R.E. (2012). When resources don't rescue: flowering phenology and species interactions affect compensation to herbivory in *Ipomopsis aggregata*. *Oikos* 121: 1424–1434.

Bronstein, J.L., Huxman, T.E., and Davidowitz, G. (2007). Plant-mediated effects linking herbivory and pollination. In: *Ecological communities: plant mediation in indirect interaction webs* (eds. Ohgushi, T., Craig, T.P., and Price, P.W.), 75–103. Cambridge University Press.

Cacho, N.I., Kliebenstein, D.J., and Strauss, S.Y. (2015). Macroevolutionary patterns of glucosinolate defense and tests of defense-escalation and resource availability hypotheses. *New Phytologist* 208: 915–927.

Campbell, S.A. (2015). Ecological mechanisms for the coevolution of mating systems and defence. *New Phytologist* 205: 1047–1053.

Campbell, D.R., Bischoff, M., Raguso, R.A., Briggs, H.M., and Sosenski, P. (2022). Selection of floral traits by pollinators and seed predators during sequential life history stages. *The American Naturalist* 199: 808–823.

Campbell, S.A., and Kessler, A. (2013). Plant mating system transitions drive the macroevolution of defense strategies. *Proceedings of the National Academy of Sciences of the United States of America* 110: 3973–3978.

Carlo, T.A., Cazetta, E., Traveset, A., Guimarães, P.R., and McConkey, K.R. (2022). Special issue: fruits, animals and seed dispersal: timely advances on a key mutualism. *Oikos* 2022: e09220.

Carmona, D., Lajeunesse, M.J., and Johnson, M.T.J. (2011). Plant traits that predict resistance to herbivores. *Functional Ecology* 25: 358–367.

Carr, D.E., and Eubanks, M.D. (2014). Interactions between insect herbivores and plant mating systems. *Annual Review of Entomology* 59: 185–203.

Castagneyrol, B., Giffard, B., Péré, C., and Jactel, H. (2013). Plant apparency, and overlooked driver of associational resistance to insect herbivory. *Journal of Ecology* 101: 418–429.

Cazetta, E., Schaefer, H.M., and Galetti, M. (2008). Does attraction to frugivores or defense against pathogens shape fruit pulp composition? *Oecologia* 155: 277–286.

Cipollini, M.L., and Levey, D.J. (1997). Secondary metabolites of fleshy vertebrate-dispersed fruits: adaptive hypotheses and implications for seed dispersal. *The American Naturalist* 150: 346–372.

Cole, C.T., Morrow, C.J., Barker, H.L., Rubert-Nason, K.F., Riehl, J.F.L., Köllner, T.G., Lackus, N.D., and Lindroth, R.L. (2021). Growing up aspen: ontogeny and trade-offs shape growth, defence and reproduction in a foundation species. *Annals of Botany* 127: 505–517.

Coley, P.D., Bryant, J.P., and Chapin. F.S. (1985). Resource availability and plant antiherbivore defense. *Science* 230: 895–899.

Coley, P.D., and Kursar, T.A. (2014). Is the high diversity in tropical forests driven by the interactions between plants and their pests? *Science* 343: 35–36.

Cook, J.M., and Rasplus, J.Y. (2003). Mutualists with attitude: coevolving fig wasps and figs. *Trends in Ecology & Evolution* 18: 241–248.

Cope, O.L., Keefover-Ring, K., Kruger, E.L., and Lindroth, R.L. (2021). Growth–defense trade-offs shape population genetic composition in an iconic forest tree species. *Proceedings of the National Academy of Sciences of the United States of America* 118: e2103162118.

Cornelissen, T., and Stiling, P. (2005). Sex-biased herbivory: a meta-analysis of the effects of gender on plant-herbivore interactions. *Oikos* 111: 488–500.

Dalling, J.W., Davis, A.S., Arnold, A.E., Sarmiento, C., and Zalamea, P.-C. (2020). Extending plant defense theory to seeds. *Annual Review of Ecology, Evolution, and Systematics* 51: 123–141.

Defossez, E., Pellissier, L., and Rasmann, S. (2018). The unfolding of plant growth form-defence syndromes along elevation gradients. *Ecology Letters* 21: 609–618.

Delph, L.F. (1990). Sex-ratio variation in the gynodioecious shrub *Hebe strictissima* (Scrophulariaceae). *Evolution* 44: 134–142.

Des Marais, D.L., Hernandez, K.M., and Juenger, T.E. (2013). Genotype-by-environment interaction and plasticity: exploring genomic responses of plants to the abiotic environment. *Annual Review of Ecology, Evolution, and Systematics* 44: 5–29.

Dicke, M., and Baldwin, I.T. (2010). The evolutionary context for herbivore-induced plant volatiles: beyond the "cry-for-help". *Trends in Plant Science* 15: 167–175.

Dicke, M., and van Loon, J. (2000). Multitrophic effects of herbivore-induced plant volatiles in an evolutionary context *Entomologia Experimentalis et Applicata* 97: 237–249.

Donaldson, J.R., Kruger, E.L., and Lindroth, R.L. (2006). Competition- and resource-mediated tradeoffs between growth and defensive chemistry in trembling aspen (*Populus tremuloides*). *New Phytologist* 169: 561–570.

Eckhart, V.M., and Seger, J. (1999). Phenological and developmental costs of male function in hermaphroditic plants. In: *Life history evolution in plants* (eds. Vuorisalo, T.O., and Mutikainen, P.K.), 195–213. Kluwer.

Ehrlich, P.R., and Raven, P.H. (1964). Butterflies and plants: a study in plant coevolution. *Evolution* 18: 586–608.

Eichenberg, D., Purschke, O., Ristok, C., Wessjohann, L., and Bruelheide, H. (2015). Trade-offs between physical and chemical carbon-based leaf defence: of intraspecific variation and trait evolution. *Journal of Ecology* 103: 1667–1679.

Endara, M.J., and Coley, P.D. (2011). The resource availability hypothesis revisited: a meta-analysis. *Functional Ecology* 25: 389–398.

Endara, M.-J., Coley, P.D., Ghabash, G., Nicholls, J.A., Dexter, K.G., Donoso, D.A., Stone, G.N., Pennington, R.T., and Kursar, T.A. (2017). Coevolutionary arms race versus host defense chase in a tropical herbivore–plant system. *Proceedings of the National Academy of Sciences of the United States of America* 114: E7499–E7505.

Feeny, P. (1976). Plant apparency and chemical defense. *Recent Advances in Phytochemistry* 10: 1–40.

Figueroa-Macías, J.P., García, Y.C., Núñez, M., Díaz, K., Olea, A.F., and Espinoza, L.J. (2021). Plant growth-defense trade-offs: molecular processes leading to physiological changes. *International Journal of Molecular Sciences* 22: 693.

Fine, P.V.A., Mesones, I., and Coley, P.D. (2004). Herbivores promote habitat specialization by trees in Amazonian forests. *Science* 305: 663–665.

Fornoni, J. (2011). Ecological and evolutionary implications of plant tolerance to herbivory. *Functional Ecology* 25: 399–407.

Forrister, M.L., Novotny, V., Panorska, A.K., Baje, L., Basset, Y., Butterill, P.T., Cizek, L., Coley, P.D., Dem, F., Diniz, I.R., Drozd, P., Fox, M., Glassmire, A.E., Hazen, R., Hrcek, J., Jahner, J.P., Kaman, O., Kozubowski, T.J., Kursar, T.A., Lewis, O.T., Lill, J., Marquis, R.J., Miller, S.E., Morais, H.C., Murakami, M., Nickel, H., Pardikes, N.A., Ricklefs, R.E., Singer, M.S., Smilanich, A.M., Stireman, J.O., Villamarín-Cortez, S., Vodka, S., Volf, M., Wagner, D.L., Walla, T., Weiblen, G.D., and Dyer, L.A. (2015). The global distribution of diet breadth in insect herbivores. *Proceedings of the National Academy of Sciences of the United States of America* 112: 442–447.

Futuyma, D.J., and Agrawal, A.A. (2009). Macroevolution and the biological diversity of plants and herbivores. *Proceedings of the National Academy of Sciences of the United States of America* 106: 18054–18061.

Futuyma, D.J., and Moreno, G. (1988). The evolution of ecological specialization. *Annual Review of Ecology and Systematics* 19: 207–233.

Galetti, M., Guevara, R., Cortes, M.C., Fadini, R., Von Matter, S., Leite, A.B., Labecca, F., Ribeiro, C.S. Carvalho, R.G. Collevatti, M.M. Pires, P.R. Guimaraes, P.H. Brancalion, T., Ribeiro, M.C., and Jordano, P. (2013). Functional extinction of birds drives rapid evolutionary changes in seed size. *Science* 340: 1086–1090.

Galmán, A., Abdala-Roberts, L., Covelo, F., Rasmann, S., and Moreira, X. (2019). Parallel increases in insect herbivory and defenses with increasing elevation for both saplings and adult trees of oak (*Quercus*) species. *American Journal of Botany* 106: 1558–1565.

Glynn, C., Herms, D.A., Orians, C.M., Hansen, R.C., and Larsson, S. (2007). Testing the growth-differentiation balance hypothesis: dynamic responses of willows to nutrient availability. *New Phytologist* 176: 623–634.

Hahn, P.G., Keefover-Ring, K., Nguyen, L.M.N., and Maron, J.L. (2021). Intraspecific correlations between growth and defence vary with resource availability and differ within and among populations. *Functional Ecology* 35: 2387–2396.

Hahn, P.G., and Maron, J.L. (2016). A framework for predicting intraspecific variation in plant defense. *Trends in Ecology & Evolution* 31: 646–656.

Hairston, N.G., Smith, F.E., and Slobodkin, L.B. (1960). Community structure, population control, and competition. *The American Naturalist* 44: 421–-425.

Hale, B.K., Herms, D.A., Hansen, R.C., Clausen, T P., and Arnold, D. (2005). Effects of drought stress and nutrient availability on dry matter allocation, phenolic glycosides, and rapid induced resistance of poplar to two lymantriid defoliators. *Journal of Chemical Ecology* 31: 2601–2620.

Hambäck, P.A., and Beckerman, A.P. (2003). Herbivory and plant resource competition: a review of two interacting interactions. *Oikos* 101: 26–37.

Hanley, M.E., Lamont, B.B., and Armbruster, W.S. (2009). Pollination and plant defence traits co-vary in Western Australian Hakeas. *New Phytologist* 182: 251–260.

Hardy, N.B., and Otto, S.P. (2014). Specialization and generalization in the diversification of phytophagous insects: tests of the musical chairs and oscillation hypotheses. *Proceedings of the Royal Society B: Biological Sciences* 281: 20132960.

He, Z., Webster, S., and Yang He, S. (2022). Growth–defense trade-offs in plants. *Current Biology* 32: R634–R639.

Heil, M. (2008). Indirect defence via tritrophic interactions. *New Phytologist* 178: 41–61.

Herms, D.A., and Mattson, W.J. (1991). Does reproduction compromise defense in woody plants? In: *Forest insect guilds: patterns of interaction with host trees* (eds. Baranchikov, Y.N., Mattson, W.J., Hain, F.P., and Payne, T.L.), 35–46. U.S. Department of Agriculture, Forest Service, Northeastern Forest Experiment Station.

Herms, D.A., and Mattson, W.J. (1992). The dilemma of plants: to grow or defend. *The Quarterly Review of Biology* 67: 283–335.

Herrera, C.M. (1982). Defense of ripe fruit from pests – its significance in relation to plant–disperser interactions. *The American Naturalist* 120: 218–241.

Herrera, C.M. (2002). Seed dispersal by vertebrates. In: *Plant-animal interactions: an evolutionary approach* (eds. Herrera, C.M., and Pellmyr, O.), 185–208. Blackwell Scientific.

Herrera, C.M., Medrano, M, Rey, P.J., Sanchez-Lafuente, A.M., Garcia, M.B., Guitian, J., and Manzaneda, A.J. (2002). Interaction of pollinators and herbivores on plant fitness suggests a pathway for correlated evolution of mutualism- and antagonism-related traits. *Proceedings of the National Academy of Sciences of the United States of America* 99: 16823–16828.

Howe, H F., and Smallwood, J. (1982). Ecology of seed dispersal. *Annual Review of Ecology and Systematics* 13: 201–228.

Huot, B., Yao, J., Montgomery, B.L., and He, S.Y. (2014). Growth–defense tradeoffs in plants: a balancing act to optimize fitness. *Molecular Plant* 7: 1267–-1287.

Husband, B.C., and Schemske, D.W. (1996). Evolution of the magnitude and timing of inbreeding depression in plants. *Evolution* 50: 54–70.

Jacobsen, D.J. (2022). Growth rate and life history shape plant resistance to herbivores. *American Journal of Botany* 109: 1074–1084.

Janzen, D.H. (1983). Dispersal of seed by vertebrate guts. In: *Coevolution* (eds. Futuyma, D.J., and Slatkin, M.), 232–262. Sinauer Associates Inc.

Johnson, M.T.J., Campbell, S.A., and Barrett, S.C.H. (2015). Evolutionary interactions between plant reproduction and defense against herbivores. *Annual Review of Ecology, Evolution, and Systematics* 46: 191–213.

Jones, P.L., and Agrawal, A.A. (2017). Learning in insect pollinators and herbivores. *Annual Review of Entomology* 62: 53–71.

Jordano, P. (1995). Angiosperm fleshy fruits and seed dispersers: a comparative analysis of adaptation and constraints in plant-animal interactions. *The American Naturalist* 145: 163–191.

Kant, M.R., Jonckheere, W., Knegt, B., Lemos, F., Liu, J., Schimmel, B.C.J., Villarroel, C.A., Ataide, L.M.S., Dermauw, W., Glas, J.J., Egas, M., Janssen, A., van Leeuwen, T., Schuurink, R.C., Sabelis, M.W., and Alba, J.M. (2015). Mechanisms and ecological consequences of plant defence induction and suppression in herbivore communities. *Annals of Botany* 115: 1015–1051.

Karban, R. (2011). The ecology and evolution of induced resistance against herbivores. *Functional Ecology* 25: 339–347.

Karban, R., and Baldwin, I.T. (1997). *Induced responses to herbivory*. The University of Chicago Press.

Kempel, A., Schadler, M., Chrobock, T., Fischer, M., and van Kleunen, M. (2011). Tradeoffs associated with constitutive and induced plant resistance against herbivory. *Proceedings of the National Academy of Sciences of the United States of America* 108: 5685–5689.

Kessler, A., and Halitschke, R. (2009). Testing the potential for conflicting selection on floral chemical traits by pollinators and herbivores: predictions and case study. *Functional Ecology* 23: 901–912.

Kessler, A., and Heil., M. (2011). The multiple faces of indirect defences and their agents of natural selection. *Functional Ecology* 25: 348–357.

Kichas, N.E., Pederson, G.T., Hood, S.M., Everett, R.G., and McWethy, D.B. (2023). Increased whitebark pine (*Pinus albicaulis*) growth and defense under a warmer and regionally drier climate. *Frontiers in Forests and Global Change* 6: 1089138.

Koricheva, J., Larsson, S., Haukioja, E., and Keinanen, M. (1998). Regulation of woody plant secondary metabolism by resource availability: hypothesis testing by means of meta-analysis. *Oikos* 83: 212–226.

Krupnick, G.A., and Weis, A.E. (1999). The effect of floral herbivory on male and female reproductive success in *Isomeris arborea*. *Ecology* 80: 135–149

Labandeira, C. (2007). The origin of herbivory on land: initial patterns of plant tissue consumption by arthropods. *Insect Science* 14: 259–275.

Lerdau, M., Litvak, M., and Monson, R. (1994). Plant chemical defense: monoterpenes and the growth-differentiation balance hypothesis. *Trends in Ecology & Evolution* 9: 58–61.

Lim, J.Y., Fine, P.V.A., and Mittelbach, G.G. (2015). Assessing the latitudinal gradient in herbivory. *Global Ecology and Biogeography* 24: 1106–1112.

Lohbeck, M., Lebrija-Trejos, E., Martínez-Ramos, M., Meave, J.A., Poorter, L., and Bongers F. (2015). Functional trait strategies of trees in dry and wet tropical forests are similar but differ in their consequences for succession. *PLoS ONE* 10(4): e0123741.

Lombardero, M.J., Ayres, M.P., Lorio, P.L., and Ruel, J.J. (2000). Environmental effects on constitutive and inducible resin defences of *Pinus taeda Ecology Letters* 3: 329–339.

Loomis, W.E. (1932). Growth-differentiation balance vs. carbohydrate-nitrogen ratio. *Proceedings of the American Society for Horticultural Science* 29: 240–245.

López-Goldar, X., Zas, R., and Sampedro, L. (2020). Resource availability drives microevolutionary patterns of plant defences. *Functional Ecology* 34: 1640–1652.

Lundgren, M.R., and Des Marais, D.L. (2020). Life history variation as a model for understanding trade-offs in plant-environment interactions. *Current Biology* 30: R180–R189.

McCall, A.C., and Fordyce, J.A. (2010). Can optimal defence theory be used to predict the distribution of plant chemical defences? *Journal of Ecology* 98: 985–992.

McKey, D. (1975). The ecology of coevolved seed dispersal systems. In: *Coevolution of animals and plants* (eds. Gilbert, L.E, and Raven, P.H.), 155–191. University of Texas Press.

Messina, F.J., Durham, S.L., Richards, J.H., and McArthur, E.D. (2002). Trade-off between plant growth and defense? A comparison of sagebrush populations. *Oecologia* 131: 43–51.

Miller, T.E.X., Tenhumberg, B., and Louda, S.M. (2008). Herbivore-mediated ecological costs of reproduction shape the life history of an iteroparous plant. *The American Naturalist* 171: 141–149.

Mithöfer, A., and Boland, W. (2012). Plant defence against herbivores: chemical aspects. *Annual Review of Plant Biology* 63: 431–450.

Mole, S. (1994). Trade-offs and constraints in plant-herbivore defense theory: a life-history perspective. *Oikos* 71: 3–12.

Monson, R.K., Trowbridge, A.M., Lindroth, R.L., and Lerdau, M.T. (2022). Coordinated resource allocation to plant growth–defense tradeoffs. *New Phytologist* 233: 1051–1066.

Mooney, K.A., Halitschke, R., Kessler, A., and Agrawal, A.A. (2010). Evolutionary trade-offs in plants mediate the strength of trophic cascades. *Science* 237: 1642–1644.

Moreira, X., Abdala-Roberts, L., Galmán, A., Bartlow, A.W., Berny-Mier y Teran, J.C., Carrari, E., Covelo, F., de la Fuente, M., Ferrenberg, S., Fyllas, N.M., Hoshika, Y., Lee, S.R., Marquis, R.J., Nakamura, M., Nell, C.S., Pesendorfer, M.B., Steele, M.A., Vázquez-González, C., Zhang, S., and Rasmann, S. (2020). Ontogenetic consistency in oak defence syndromes. *Journal of Ecology* 108: 1822–1834.

Moreira, X., Abdala-Roberts, L., Gols, R, and Francisco, M. (2018). Plant domestication decreases both constitutive and induced chemical defences by direct selection against defensive traits. *Scientific Reports* 8: 12678.

Moreira, X., Glauser, G., and Abdala-Roberts, L. (2017). Interactive effects of plant neighbourhood and ontogeny on insect herbivory and plant defensive traits. *Scientific Reports* 7: 4047.

Moreira, X., Mooney, K.A., Rasmann, S., Petry, W.K., Carrillo-Gavilán, A., Zas, R., and Sampedro, L. (2014). Trade-offs between constitutive and induced defences drive geographical and climatic clines in pine chemical defences. *Ecology Letters* 17: 537–546.

Moreira, X., and Pearse, I S. (2017). Leaf habit does not determine the investment in both physical and chemical defences and pair-wise correlations between these defensive traits. *Plant Biology* 19: 354–359.

Moreira, X., Zas, R., and Sampedro, L. (2012). Genetic variation and phenotypic plasticity of nutrient re-allocation and increased fine root production as putative tolerance mechanisms inducible by methyl-jasmonate in pine trees. *Journal of Ecology* 100: 810–820.

Nelson, A.S., and Whitehead, S.R. (2021). Fruit secondary metabolites shape seed dispersal effectiveness. *Trends in Ecology & Evolution* 36: 1113–1123.

Nevo, O., Valenta, K., Razafimandimby, D., Melin, A.D., Ayasse, M., and Chapman, C.A. (2018). Frugivores and the evolution of fruit color. *Biology Letters* 19: 20180377.

Núñez-Farfán, J., Fornoni, J., and Valverde, P.L. (2007). The evolution of resistance and tolerance to herbivores. *Annual Review of Ecology, Evolution, and Systematics* 38: 541–566.

Obeso, J.R. (2002). The costs of reproduction in plants. *New Phytologist* 155: 321–348.

Ochoa-López, S., Rebollo, R., Barton, K.E., Fornoni, J., and Boege, K. (2018). Risk of herbivore attack and heritability of ontogenetic trajectories in plant defense. *Oecologia* 187: 413–426.

Ohnmeiss, T.E., and Baldwin, I.T. (1994). The allometry of nitrogen to growth and an inducible defense under nitrogen-limited growth. *Ecology* 75: 995–1002.

Orians, C.M., Hochwender, C.G., Fritz, R.S., and Snäll, T. (2010). Growth and chemical defense in willow seedlings: trade-offs are transient. *Oecologia* 163: 283–290.

Osier, T.L., and Lindroth, R.L. (2006). Genotype and environment determine allocation to and costs of resistance in quaking aspen. *Oecologia* 148: 293–303.

Pellissier, L., Descombes, P., Hagen, O., Chalmandrier, L., Glauser, G., Kergunteuil, A,. Defossez, E., and Rasmann, S. (2018). Growth-competition-herbivore resistance trade-offs and the responses of alpine plant communities to climate change. *Functional Ecology* 32: 1693–1703.

Pellissier, L., Moreira, X., Danner, H., Serrano, M., Salamin, N., van Dam, N.M., and Rasmann, S. (2016). The simultaneous inducibility of phytochemicals related to plant direct and indirect defences against herbivores is stronger at low elevation. *Journal of Ecology* 104: 1116–1125.

Pellmyr, O. (2003). Yuccas, yucca moths, and coevolution: a review. *Annals of the Missouri Botanical Garden* 90: 35–55.

Penet, L., Collin, C.L., and Ashman, T.L. (2009). Florivory increases selfing: an experimental study in the wild strawberry, *Fragaria virginiana*. *Plant Biology* 11: 38–45.

Poorter, H., and Garnier, E. (2007). Ecological significance of inherent variation in relative growth rate and its components. In: *Functional plant ecology* (eds. Pugnaire, F.I., and Valladares, F.), 67–100. CRC Press.

Rasmann, S., Chassin, E., Bilat, J., Glauser, G., and Reymond, P. (2015). Trade-off between constitutive and inducible resistance against herbivores is only partially explained by gene expression and glucosinolate production. *Journal of Experimental Botany* 66: 2527–2534.

Rasmann, S., Erwin, A.C., Halitschke, R., and Agrawal, A.A. (2011). Direct and indirect root defences of milkweed (*Asclepias syriaca*): trophic cascades, trade-offs and novel methods for studying subterranean herbivory. *Journal of Ecology* 99: 16–25.

Rausher, M.D. (2001). Co-evolution and plant resistance to natural enemies. *Nature* 411: 857–864

Reich, P.B., Ellsworth, D.S., and Walters, M.B. (1998). Leaf structure (specific leaf area) modulates photosynthesis–nitrogen relations: evidence from within and across species and functional groups. *Functional Ecology* 12: 948–958.

Robert, C., Ferrieri, R.A., Schirmer, S., Babst, B.A., Schueller, M.J., Machado, R.A.R., Arce, C.C.M., Hibbard, B.E., Gershenzon, J., Turlings, T.C.J., and Erb, M. (2014). Induced carbon reallocation and compensatory growth as root herbivore tolerance mechanisms. *Plant, Cell & Environment* 37: 2613–2622.

Rodgerson, L. (1998). Mechanical defense in seeds adapted for ant dispersal. *Ecology* 79: 1669–1677.

Romero, G., and Koricheva, J. (2011). Contrasting cascade effects of carnivores on plant fitness: a meta-analysis. *Journal of Animal Ecology* 80: 696–704.

Sampedro, L., Moreira, X., and Zas, R. (2011). Costs of constitutive and herbivore-induced chemical defenses in pine trees emerge only under low resources availability. *Journal of Ecology* 99:818–827.

Sargent, R.D., and McKeough, A.D. (2022). New evidence suggests no sex bias in herbivory or plant defense. *The American Naturalist* 200: 435–447.

Schaefer, H.M., McGraw, K., and Catoni, C. (2008). Birds use fruit colour as honest signal of dietary antioxidant rewards. *Functional Ecology* 22: 303–310.

Schmitz, O.J., Hamback, P.A., and Beckerman, A.P. (2000). Trophic cascades in terrestrial systems: a review of the effects of carnivore removal on plants. *The American Naturalist* 155: 141–153.

Schoonhoven, L.M., van Loon, B., van Loon, J.J., and Dicke, M. (2005). *Insect-plant biology*. Oxford University Press on Demand.

Simms, E.L., and Fritz, R.S. (1990). The ecology and evolution of host-plant resistance to insects. *Trends in Ecology & Evolution* 5: 356–360.

Smilanich, A.M., Fincher, R.M., and Dyer, L.A. (2016). Does plant apparency matter? Thirty years of data provide limited support but reveal clear patterns of the effects of plant chemistry on herbivores. *New Phytologist* 210: 1044–1057.

Stamp, N. (2003). Out of the quagmire of plant defense hypotheses. *The Quarterly Review of Biology* 78: 23–55.

Stamp, N. (2004). Can the growth-differentiation balance hypothesis be tested rigorously? *Oikos* 107: 439–448.

Stowe, K.A., Marquis, R.J., Hochwender, C.G., and Simms, E.L. (2000). The evolutionary ecology of tolerance to consumer damage. *Annual Review of Ecology and Systematics* 31: 565–595.

Strauss, S.Y. (1997). Floral characters link herbivores, pollinators, and plant fitness. *Ecology* 78: 1640–1645.

Strauss, S.Y., and Agrawal, A.A. (1999). The ecology and evolution of plant tolerance to herbivory. *Trends in Ecology & Evolution* 14: 179–185.

Strauss, S.Y., Cacho, N I., Schwartz, M.W., Schwartz, A.C., and Burns, K.C. (2015). Apparency revisited. *Entomologia Experimentalis et Applicata* 157: 74–85.

Traveset, A., Robertson, A.W., Rodríguez-Pérez, J., Dennis, A.J., Schupp, E.W., Green, R.J., and Westcott, D.A. (2007). Seed dispersal: theory and its application in a changing world. In: *A review on the role of endozoochory on seed germination* (eds. Dennis, A.J., Schupp, E.W., Green, R.J., and Westcott, D.A), 78–103. CABI Publishing.

Tuller, J., Marquis, R.J., Andrade, S.M.M., Monteiro, A.B., and Faria, L.D.B. (2018). Trade-offs between growth, reproduction and defense in response to resource availability manipulations. *PLoS ONE* 13(8): e0201873.

Turcotte, M.M., Davies, T.J., Thomsen, C.J.M., and Johnson, M.T.J. (2014). Macroecological and macroevolutionary patterns of leaf herbivory across vascular plants. *Proceedings of the Royal Society B: Biological Sciences* 281: 20140555.

Turlings, T.C.J., and Erb, M. (2018). Tritrophic interactions mediated by herbivore-induced plant volatiles: mechanisms, ecological relevance, and application potential. *Annual Review of Entomology* 63: 433–452.

Valenta, K., Bhramdat, H.D., Calhoun, G.V., Daegling, D.J., and Nevo, O. (2022). Variation in ripe fruit hardness: a mechanical constraint? *Oikos* 2022: e08074.

Valenta, K., and Nevo, O. (2020). The dispersal syndrome hypothesis: how animals shaped fruit traits, and how they did not. *Functional Ecology* 34: 1158–1169.

van Zandt, P.A. (2007). Plant defense, growth, and habitat: a comparative assessment of constitutive and induced resistance. *Ecology* 88: 1984–1993.

Vázquez-González, C., Sampedro, L., López-Goldar, X., Solla, A., Vivas, M., Rozas, V., Lombardero, M.J., and Zas, R. (2022). Inducibility of chemical defences by exogenous application of methyl jasmonate is long-lasting and conserved among populations in mature *Pinus pinaster* trees. *Forest Ecology and Management* 518: 120280.

Vázquez-González, C., Sampedro, L., Rozas, V., and Zas, R. (2020). Climate drives intraspecific differentiation in the expression of growth-defence trade-offs in a long-lived pine species. *Scientific Reports* 10: 10584.

Whitehead, S.R., Obando Quesada, M.F., and Bowers, M.D. (2016). Chemical tradeoffs in seed dispersal: defensive metabolites in fruits deter consumption by mutualist bats. *Oikos* 125: 927–937.

Whitehead, S.R., Schneider, G.F., Dybzinski, R., Nelson, A.S., Gelambi, M., Jos, E., and Beckman, N.G. (2022). Fruits, frugivores, and the evolution of phytochemical diversity. *Oikos* 2022: e08332.

Whitehead, S R., Turcotte, M.M., and Poveda, K. (2017). Domestication impacts on plant–herbivore interactions: a meta-analysis. *Philosophical Transactions of the Royal Society B* 372: 20160034.

Viola, D.V., Mordecai, E.A., Jaramillo, A.G., Sistla, S.A., Albertson, L.K., Gosnell, J.S., Cardinale, B.J., and Levine, J.M. (2010). Competition–defense tradeoffs and the maintenance of plant diversity. *Proceedings of the National Academy of Sciences of the United States of America* 107: 17217–17222.

Zangerl, A.R., and Bazzaz, F.A. (1992). Theory and pattern in plant defense allocation. In: *Plant resistance to herbivores and pathogens, ecology, evolution and genetics* (eds. Fritz, R., and Simms, E.), 363–391. University of Chicago Press.

Zangerl, A.R., and Rutledge, C.E. (1996). Probability of attack and patterns of constitutive and induced defense: a test of optimal defense theory. *The American Naturalist* 147: 599–608.

Zhang, P.-J., Shu, J.-P.; Fu, C.-X., Zhou, Y., Hu, Y., Zalucki, M.P., and Liu, S.-S. (2008). Trade-offs between constitutive and induced resistance in wild crucifers shown by a natural, but not an artificial, elicitor. *Oecologia* 157: 83–92.

Züst, T., and Agrawal, A.A. (2017). Trade-offs between plant growth and defense against insect herbivory: an emerging mechanistic synthesis. *Annual Review of Plant Biology* 68: 513–534.

Züst, T., Rasmann, S., and Agrawal A.A. (2015). Growth–defense tradeoffs for two major anti-herbivore traits of the common milkweed *Asclepias syriaca*. *Oikos* 124: 1404–1415.

16

Are you in Synch?

How the Timing of Plant and Insect Life History Events Affects Pollination Interactions

Tamar Keasar[1] and Tzlil Labin[2]

[1] *Department of Biology and Environment, University of Haifa-Oranim, Tivon, Israel*
[2] *Department of Evolutionary and Environmental Biology, University of Haifa, Haifa, Israel*

16.1 Generalisation in Pollination Networks

The mutualistic interactions between plants and their pollinators have long been recognized to be highly generalized. Flowers are typically visited by numerous pollinator species, and individual pollinators feed on multiple flower species (Waser et al. 1996, Johnson and Steiner 2000). The composition of plant and pollinator assemblages varies between sites and years, and the links between interacting species are rewired between flowering seasons as well (Dupont et al. 2009). Highly specific long-term plant–pollinator associations, such as between figs and fig wasps (see Chapter 17, this volume), are the exception rather than the rule. Pollination networks are bipartite graphs that visualize the species (nodes) and interactions (links) recorded in plant–pollinator communities. Standard indices have been developed to calculate and compare specialisation across networks (Blüthgen et al. 2006). These indices consider the number of species that a plant or a pollinator interacts with, and the evenness of the interaction frequencies. Such comparisons reveal consistent macro-ecological patterns. For example, specialisation in pollination networks increases with habitat productivity and precipitation (Luna et al. 2022) and decreases with altitude (from low to high elevations; Trøjelsgaard and Olesen 2013, Hoiss et al. 2015, Classen et al. 2020).

The mechanisms that promote overall generalisation in pollination networks, and that account for the observed macro-ecological trends, are not sufficiently understood. For instance, the positive relationship between productivity and specialisation was attributed to a greater selectivity of foragers when food resources are abundant (Luna et al. 2022). A number of other interpretations have been proposed for the well-documented decline in network specialisation with elevation. Trøjelsgaard and Olesen (2013) considered the hypothesis that species-poor pollination networks are more generalized than species-rich ones because of shortage of interaction partners. This hypothesis was not supported in their study, as species richness in the networks that they analysed did not vary strongly with elevation, whereas network specialisation did. Classen et al. (2020) found that network specialisation declined with elevation on Mt. Kilimanjaro, and that temperature was the best predictor of specialisation level. They proposed that the low temperatures at high altitudes increase energetic demands and reduce foraging time windows for insects, favouring generalized interactions with plants. Similarly, pollination networks became more generalized with altitude in the Canary Islands (Lara-Romero et al. 2019). In this case, generalisation was driven by filtering of the pollinator community: at high elevations, large, generalized pollinators were dominant, probably due to their better thermal tolerance, while smaller specialist species declined. A large dataset of bumblebee foraging choices at three altitude zones in the Colorado Rocky Mountains, USA, also indicated higher generalisation at high altitudes (Miller-Struttmann and Galen, 2014). This was attributed to the short time window suitable for flowering and bee nesting at high elevations, which increases the overlap in flowering periods among plant and bee species in the community. Consequently, the number of plant taxa available to individual foragers increases, as does the number of potential pollinators per plant, promoting generalized foraging at the level of the plant–pollinator community. This interpretation implicates the phenologies of flower–insect communities as drivers of increasing generalisation on high mountains.

Building on these ideas, we discuss here a more general potential constraint on specialized pollination interactions: the difficulty of synchronising the life history stages of specific pairs of plant and insect species. We review the literature and describe a case study from our own research, to explore why perfect plant–pollinator synchrony is uncommon. We argue that weak plant–pollinator phenological synchrony generates opportunities for generalized interactions for specific species

Life History Evolution: Traits, Interactions, and Applications, First Edition. Edited by Michal Segoli and Eric Wajnberg.
© 2025 John Wiley & Sons Ltd. Published 2025 by John Wiley & Sons Ltd.

across years and sites. We then highlight studies that test whether network specialisation correlates positively with phenological synchrony, a pattern that would be expected if synchrony is a sufficient condition for specialized interactions at the community level.

16.2 What Drives Flowering Phenology?

The seasonal timing of flowering in plants responds to environmental signals, primarily temperature and photoperiod (Tooke and Battey 2010), but also to human manipulations such as fertilisation and mowing (Liu et al. 2017). Additional cues include snowmelt, in alpine environments (Jerome et al. 2021), and rainfall in arid ones (Peñuelas et al. 2004, Donoso et al. 2016). The effects of temperature can be rather complex, as many species require an accumulation of cold days (vernalisation) in winter, followed by warm spring temperatures, to start flowering. For instance, Tansey et al. (2017) found inter-annual variation in the timing of first flowering in 22 plant species from the temperate UK flora. The date of flowering advanced with increasing spring temperatures in all species and was also influenced by a chilling period in winter in the early-flowering species. Day length, on the other hand, predicted the onset of flowering in the late-season species.

Each of the environmental cues discussed above can shift the timing of first flowering, peak flowering and end flowering to different extents. Such shifts, along with species-specific differences in responsiveness, alter the composition of co-flowering plant communities under climate change (CaraDonna et al. 2014). They also affect the temporal synchrony of flowering between species within a plant community, and within species across sites. This was illustrated in an analysis of a 35-year dataset of flowering phenology from five communities along a 1267-metre elevation gradient in Arizona, USA (Fisogni et al. 2022). During this period (1984–2019), the phenological synchrony between different plant species within each of the communities declined. The decline was particularly steep in the lowest-elevation community. On the other hand, the flowering dates of species that occurred in more than one community became more synchronized between pairs of communities over time.

16.3 What Drives Pollinator Phenology?

Developing insect pollinators pass through egg, larval, pupal (if holometabolous) and adult stages. Many species are not active year-round, and instead spend weeks or months in diapause, dormancy or larval/pupal development. Developmental rates are mostly regulated by temperature, while diapause termination can be regulated by temperature or day length. A chilling period affects the timing of emergence from diapause in many temperate species (Marshall et al. 2020), while increased soil moisture is needed for emergence in a desert bee (Danforth 1999).

The effects of local temperatures on pollinators' onset of spring activity were confirmed in manipulative field studies. In some experiments, radiation-reflecting or radiation-absorbing surfaces were placed next to nests of overwintering bees, while other studies applied direct heating to the nests. Heating and radiation-absorbing materials advanced the bees' emergence, while the presence of reflective surfaces delayed emergence (Vinchesi et al. 2013, CaraDonna et al. 2018). Translocating nests with overwintering bees between high-elevation and low-elevation sites also resulted in changes in the timing of spring emergence. The bees emerged earlier at low elevations than at high elevations, regardless of the origin of the nests (Forrest and Thomson 2011).

16.4 Do Interacting Plant–Insect Species Share Similar Reaction Norms to Temperature?

While temperature regulates life history events in both plants and pollinators, the functions that describe each partner's phenotypic responses to temperature (reaction norms) may differ. This possibility has attracted much research attention in the context of climate change because changing temperatures could uncouple the phenology of plants and their pollinators. Such mismatches might disrupt pollination interactions (Kudo and Ida 2013).

Observational studies suggest that between-year temperature fluctuations indeed affect insect–plant phenological synchronisation in some communities. For example, the spring emergence dates of syrphid flies and the flowering onset of their forage plants could be predicted from the date of snowmelt in a 20-year dataset from Colorado, USA. However, the flies' activity periods were shorter than the blooming period, and the response of the flies to snow melt was slower than that of the flowers. As a result, the period of overlap between the flies and the flowers diminished in cold years (Iler et al. 2013). Similarly, the onset of flowering of *Corydalis ambigua*, an early-spring Japanese ephemeral plant, depended more strongly on snowmelt date than the emergence of its bumblebee pollinators (Kudo and Cooper 2019). Long-term phenological records of the specialist bee *Habropoda laboriosa* (Apidae) and its host plants (genus *Vaccinium* (Ericaceae)) also suggest that the phenology of all partners advances with increasing spring temperatures, but that the strength of the response varies between species and sites. Bees respond more strongly than plants, and more strongly in northern populations than in southern ones. This creates a potential for phenological mismatch and pollinator shortage in northern sites (Weaver and Mallinger 2022). A similar pattern was detected in a long-term observational dataset from the United States, which contains flowering records of *Claytonia virginica* (Montiaceae) and observations of its specialized pollinator *Andrena erigeniae* (Andrenidae). Flower phenology responded more strongly to temperature in warmer areas than in colder regions, while the timing of bee activity was more strongly influenced by temperature in the colder areas (Xie et al. 2022). The timing of flowering of *ca.* 1500 species in Germany advanced over the years 1980–2020, as did the activity period of *ca.* 300 insect pollinator species. Flowering phenology, which historically lagged behind that of the pollinators, advanced at a faster rate than insect phenology during the study period. This actually resulted in improved plant–insect synchronisation with time (Freimuth et al. 2022). The opposite trend, namely higher sensitivity to warming in pollinators than in plants, has been described as well (Robbirt et al. 2014). Yet other observational studies found similar advances in the activity of both plants and pollinators in response to warming (Phillimore et al. 2012, Sevenello et al. 2020, Cane 2021). In other words, both interaction partners showed similar reaction norms to changes in temperature in these communities.

Manipulative experiments provide further examples of the different sensitivity of some plants and pollinators to temperature changes. In a recent greenhouse study, a community of three annual plants and one solitary bee was grown at ambient or under warmer temperatures. The warmed plants started blooming earlier than the ambient temperature controls, whereas the timing of bee emergence was unaffected by the warming treatment (de Manincor et al. 2023). In a second study, cocoons of solitary bees were placed in natural grassland sites that differ in mean temperature and were monitored until emergence. The between-site variation in bee emergence dates was lower than the variation in blooming dates of local early-season flowers, suggesting lower sensitivity of the bees to temperature cues compared to the plants (Kehrberger and Holzschuh 2019). Forrest and Thomson (2011) reached a similar conclusion based on their experiment of nest transplantations between high- and low-elevation sites. In this experiment, bees that were moved to warmer sites advanced their spring emergence, but to a lesser extent than the surrounding plants.

In summary, variable and complex temperature-dependent reaction norms have been described for specific species of insects and flowers, both in observational and in experimental studies. However, the integrated responses of whole plant communities to environmental change (such as global warming) generally resemble the community-level response of their pollinators. Consequently, Bartomeus et al. (2011) and Ovaskainen et al. (2013) predicted high resilience of community-level pollination interactions to phenological mismatches, even if interactions between individual species are disrupted.

16.5 Species-Level Phenological Asynchrony and Generalized Pollination: A Case Study

Consider an insect-pollinated plant species with a broad distribution and a long flowering season. Given the effects of climate cues on flowering reviewed above, the timing of its flowering is expected to vary across space and time. For example, flowering may advance in warmer or wetter locations and years. As described above, potential pollinators may respond differently to these same climate cue variations, causing the pollinator communities that encounter the plant to also vary spatially and temporally. Under this simple scenario, the plant's pollination system can become highly generalized merely due to overlap with different pollinators in different years/locations. We draw on our ongoing work on the pollination ecology of *Anemone coronaria* to illustrate this notion.

16.5.1 *A. coronaria*'s Pollination Biology

The Mediterranean geophyte *A. coronaria* is one of about 150 species in the genus *Anemone* (Ranunculaceae). Like many of its congeners, *A. coronaria* has a broad geographical distribution. In Israel, where we study the species' pollination ecology, its range covers much of the country's steep climatic gradient (Figure 16.1), particularly in grazed habitats (Perevolotsky

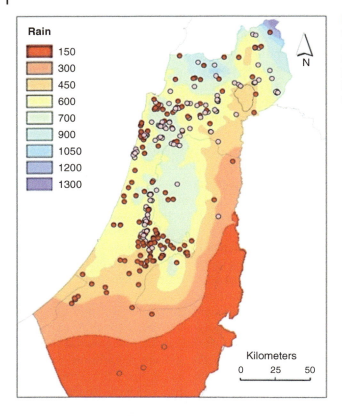

Figure 16.1 *Anemone coronaria*'s distribution in Israel, plotted on the background of annual rainfall isohyets. Data are based on a citizen science project in 2018–2019 (Y. Sapir and T. Keasar, unpublished data). Red symbols denote red-blooming populations, pink symbols indicate colour-polymorphic populations. Box 16.1 provides more information on the species' flower colour polymorphism.

et al. 2011). There is little genetic differentiation between natural populations across the plant's distribution range (Yonash et al. 2004, Dafni et al. 2020). A. coronaria has a long flowering season of several months throughout its distribution, and its flowers are colour-polymorphic (Box 16.1). Flowering in northern populations starts earlier and lasts longer than in southern populations. These features provide opportunities for interactions with multiple pollination partners.

A. coronaria blooms during winter and early spring. Self-fertilisation is limited by dichogamy (separation in time between a flower's male and female phases). This is achieved by protogyny: young flowers function as females, while older flowers function as males. Anemones rely on both insects and wind as vectors to transfer their pollen between individuals. Insects

---

Box 16.1 The Biology and Reproduction of *Anemone coronaria*

A. coronaria, the poppy anemone, is a perennial herb growing from a corm. The species has a Mediterranean distribution, and its blooming season extends from November to March. Blooming of an individual flower lasts *ca.* 2–3 weeks. A female stage lasting 2–3 days is followed by a much longer male phase. The whole plant as well as the corolla petals grow in size during blooming. The flowers attract honeybees, solitary bees, flies, beetles and other insects.

Flower colour is determined by the anthocyanin and flavonoid content of the sepals. The production of these pigments is genetically controlled. The main colour morphs are red, purple, and white. Many intermediate morphs exist as well (Figure 16.2). Red-only monomorphic populations dominate arid areas, while more mesic areas contain monomorphic and colour-polymorphic populations (Figure 16.1). The red morph also flowers later than the non-red morphs.

Because of the temporal separation of male and female stages, the species largely outcrosses. However, fertilisation between flowers from different sexual stages on the same plant (geitonogamy) is possible. The different colour morphs interbreed freely and produce viable seeds. Each flower produces *ca.* 200–300 woolly-coated achenes, which ripen *ca.* 3–7 weeks after fertilisation.

Figure 16.2 *A. coronaria*'s main colour morphs. *Source:* Photo by Na'ama Tessler.

commonly use the flowers as pollen sources (Horovitz 1976, Motten 1982, Horovitz 1991, Murphy and Vasseur 1995, Keasar et al. 2010), as shelters, or as mating sites (Horovitz 1976, Keasar et al. 2010, Dafni et al. 2020). Additional key features of *A. coronaria*'s reproductive biology are summarized in Box 16.1.

A. coronaria has been under cultivation as an ornamental plant for hundreds of years (Horovitz et al. 1975, Laura et al. 2006, Laura and Allavena 2007, Dhooghe et al. 2012). The environmental cues that affect the phenology of commercial cultivars under greenhouse conditions have been identified (Box 16.2). Here, we provide complementary information on the phenology of wild populations and its complex interplay with pollinator phenology.

Box 16.2 Phenology Regulation in Domesticated *A. coronaria*

In Mediterranean climates, plants grow mainly during the cool and rainy season. During the hot and dry season, perennials commonly enter summer dormancy, induced in response to long days and high temperatures (Ben-Hod et al. 1988). *A. coronaria*'s growth cycle is likewise regulated by temperature and photoperiod. *A. coronaria* plants from commercial cultivars that were grown under controlled temperatures and photoperiodic schedules entered dormancy regardless of illumination at high temperatures but required more than 11 hours of daylight to initiate dormancy at lower temperatures. Increasing day lengths and rising temperatures are therefore suggested to induce anemone dormancy under field conditions in the spring (Ben-Hod et al. 1988). Dormancy onset was also found to be related to corm size – the larger the corms, the later the onset of dormancy, allowing a longer growth period for the larger corms (Ben-Hod et al. 1989). Interestingly, experimental manipulations of temperature and photoperiod did not affect the time between planting of the corms and the onset of flowering. Some 50% of the corms in this experiment attained flowering within 65–75 days after corm planting, regardless of day length and temperature (Ben-Hod et al. 1989). Furthermore, only plants with sufficiently large corms and at least 5–6 leaves developed flowers. Thus, photo- and thermo-periodic effects on *A. coronaria*'s flowering seem to be indirect, being mediated through their influence on the induction and termination of dormancy (Ben-Hod et al. 1988). Wetting of the soil by early-winter rains is likely an additional cue for the termination of dormancy in natural populations. Controlled experiments to test irrigation effects on the timing of flowering are still lacking.

16.5.2 *A. coronaria*'s Flowering Phenology

To characterize between-year variation in *A. coronaria*'s flowering phenology, we analysed a historical dataset of citizen science flowering records (A. Shmida, unpublished data). Most observations ($n = 5869$) of flowering anemones were collected between 1980 and 2001, in different parts of Israel. At least 100 records are available for each year within this period. We considered six one-month periods between November and April (covering the winter and early spring seasons) and plotted the proportion of all flowering records per period, for each year separately. The seasonal distribution of flowering records varied considerably between years (Figure 16.3), as did the timing of flowering onset, peak, and end.

To estimate the spatial variability in the anemones' flowering phenology, we monitored permanent plots in 18 sites along the species' distribution range throughout 2021–2022 and 2022–2023, counting flowers every 2–3 weeks. Blooming started earlier in high-rainfall sites than in arid sites. For example, flowering onset in the most arid site (mean annual rainfall = 302 mm) occurred 35 days later than in the wettest site (annual rainfall = 636 mm) in 2022, and 47 days later in 2023. Accordingly, rainfall was a statistically significant predictor of the starting date of flowering (GLM with Gamma family and log link function, $P = 0.037$). Rainfall did not predict the end date of flowering ($P = 0.202$), nor did the duration of flowering (days from start to end) increase with rainfall ($P = 0.266$). Physiological traits associated with flower colour also contributed to the phenological variation: in the ten colour-polymorphic populations that we surveyed in 2022 and 2023, red flowers started flowering about 11 days later than the non-red flowers (paired *t*-test: $t_{17} = 3.581$, $P = 0.002$).

16.5.3 The Phenology of *A. coronaria*'s Insect Visitors

To assess the composition of the anemones' insect visitors, we observed visits in the field, using arrays of plucked anemones. The arrays contained equal numbers of red, white and purple flowers. The number of insect visits to each flower morph was recorded, and visitors that are potential pollinators were identified to order (bees, flies and beetles). We ran the experiments in three sites in Israel, situated in the northern (mesic), centre (intermediate) and southern (arid) parts of *A. coronaria*'s distribution range, three times along each of the 2022 and 2023 flowering seasons.

Bees and beetles were the dominant visitors to the flowers, and their proportions varied with date and site. Bees and flies dominated the early season, while beetles became the main visitors in mid- and late-season (Figure 16.4). Overall, in the three observation rounds combined, beetles were the most frequent visitors in the arid site, whereas bees were the dominant visitors in the intermediate and mesic sites (Figure 16.5). Beetle activity increased with ambient temperature (linear model: $P = 0.002$), whereas bee activity did not ($P = 0.570$).

Taken together, our observations indicate high variation between years (Figure 16.3) and sites in the timing of *A. coronaria*'s flowering. The assemblage of potential pollinators encountered by a flower varies with its geographical location and flowering date. Inter-annual variation in visitor composition, which is still undocumented for *A. coronaria*, may further increase the variability in flower–insect encounters across years. As a result, potential pollination partners from

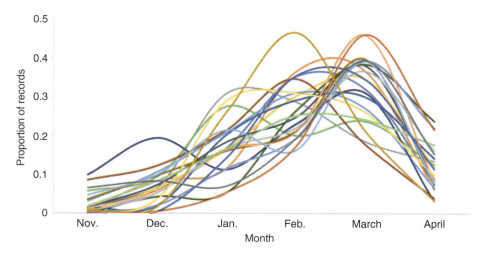

Figure 16.3 Proportions of *A. coronaria* blooming reports over the flowering season (November–April), collected over 22 years, during 1980–2001 in Israel. The lines are smoothed for clarity.

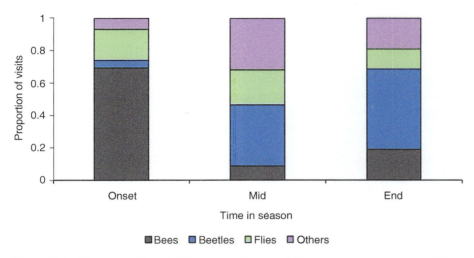

Figure 16.4 The composition of visitor taxa to *A. coronaria* flower arrays at the onset, middle and end of the 2022–2023 seasons. Data are pooled over the three observation sites.

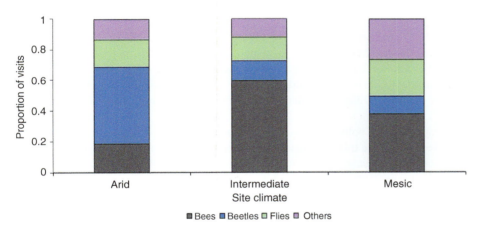

Figure 16.5 The composition of visitor taxa to *A. coronaria* flower arrays at the arid, intermediate and mesic observation sites in 2022–2023. Data are pooled over all observation rounds.

different insect orders coincide with our study plant over space and time. Interacting with many of these partners manifests as a generalized pollination system, which may reduce the plant's risk of pollen limitation.

16.6 Community-Level Phenology and Pollination Specialisation

After reviewing phenological variation in plants and their pollinators and illustrating it at the level of single species, we now extend the discussion to consider how phenology may affect specialisation in entire plant–pollinator communities. To maintain high specialisation in a pollination network, interacting flower and insect species must overlap in time. Thus, some level of phenological synchrony is a necessary condition for specialisation. However, who interacts with whom in flower–insect communities also depends on the abundance of the species and their morphology (Stang et al. 2006, Ornai and Keasar 2020). In particular, short-tongued insects such as flies often specialize in short-tubed flower species, whereas long-tongued insects (*e.g.*, butterflies) tend to be more generalized (Klumpers et al. 2019). The question is thus whether phenological synchrony is a good predictor of specialisation in pollination networks – in other words, whether well-synchronized plant–pollinator assemblages are more specialized than poorly synchronized pollination networks. Three studies suggest that this is not the case.

The first line of evidence is based on the observations of pollination networks along altitude gradients. As reviewed above, pollination interactions become less specialized with increasing elevation. This could result from a variety of mechanisms, one of them being low synchrony between plants and insects at high elevations. To address this hypothesis, Benadi et al. (2014) recorded flower–insect interactions and phenologies at different elevations in the Alps. Increasing altitude delayed the seasonal timing of insects and flowers to similar extents, suggesting that phenological asynchrony does not increase with elevation. Furthermore, there was no relationship between the specialisation level of specific insects and the extent of phenological overlap with their forage plants. A second study analysed flower–butterfly interactions and phenology over 17 years at a single site in Spain (Donoso et al. 2016). This study focused on network variation in time, rather than in space. Similar to Benadi et al.'s (2014) findings, the phenological mismatch between the pollinators and their food plants did not differ between specialists and generalists. The third study documented the phenological synchrony and the specialisation of interactions between flowers and their pollinators (insects and hummingbirds) in three tropical communities in Costa Rica. Specialized pollinators were less synchronized with their plant partners than generalized pollinators. The specialized pollinators also visited plant species with shorter flowering periods compared to generalized pollinators (Maglianesi et al. 2020). Beyond illustrating the disconnect between specialisation and phenological synchrony, this study suggests that specialized pollinators are more prone to phenological mismatch with flowers than generalists. This might expose mismatched specialists to higher fitness costs than generalists. Consistent with the idea of greater fitness costs to mismatched specialists than to generalists, higher phenological mismatches with flowers were correlated with lower nesting success and population growth rates in five species of solitary bees in Argentina (Vázquez et al. 2023). In this case, the adverse effects of the mismatches were more severe for foraging specialists than for generalists.

These three observational studies, focusing on the pollinators' viewpoint, suggest that specialist pollinator species are similarly or even less synchronized with their forage plants than generalist pollinators. Possible explanations for this finding include (1) sampling biases, e.g., failure to record some highly specialized pollinators (Benadi et al. 2014); (2) rigid foraging preferences of specialized pollinators, which reduce their potential to exploit food plants with matched phenologies (Maglianesi et al. 2020); and (3) differences in diet between the pollinators' larval and adult stages, which complicate scoring of their diet breadth. This explanation was raised with respect to butterflies, whose herbivorous larvae have often narrower diets than the nectarivore adults. The insects' phenology may be selected to maximize food availability to the larvae in these species (Donoso et al. 2016).

Although the existing evidence does not support a link between phenological synchrony and specialisation of pollination networks, much additional research is needed to validate this conclusion. In this respect, pollination networks from additional biomes should be analysed. The correlations between community-level specialisation and phenological synchrony ought to be explored from the plant perspective as well. This would involve comparing plant–pollinator synchrony levels between flower communities that vary in their range of pollinators.

Importantly, experiments that manipulate the network's synchrony and measure the resulting effects on specialisation are needed. In such experiments, the timing of plant flowering can be accelerated or delayed, using temperature and irrigation manipulations, in field cages that exclude insects. At the onset of flowering, the exclosure cages can be removed and the plants' interactions with the local pollinator communities recorded. We propose these ideas as exciting possibilities for future studies.

16.7 Concluding Remarks

The timing of life history events plays a key role in insect-mediated pollination mutualisms. To successfully interact, pollinators must mature in synchrony with the blooming of their food plants. These life history transitions are triggered by environmental cues, primarily temperature, in both insects and plants. However, reaction norms to these cues are species- and location-specific, hence the phenological events of insects and plants are rarely perfectly synchronized. Phenological asynchrony seems to enable generalized pollination interactions in many plant and insect species because they coincide with multiple interaction partners across time and space. From an evolutionary perspective, such generalism may reduce selection pressures for plant–pollinator synchrony. These observations raise the question of whether the extent of phenological synchrony between pollination partners predicts the specialisation of their interaction, another important life history trait. In other words, are specialized pollination networks better synchronized than generalized networks? Although initial observational studies do not support this hypothesis, additional tests are needed to evaluate it conclusively.

References

Bartomeus, I., Ascher, J.S., Wagner, D., Danforth, B.N., Colla, S., Kornbluth, S., and Winfree, R. (2011). Climate-associated phenological advances in bee pollinators and bee-pollinated plants. *Proceedings of the National Academy of Sciences of the United States of America* 108(51): 20645–20649.

Benadi, G., Hovestadt, T., Poethke, H.J., and Blüthgen, N. (2014). Specialization and phenological synchrony of plant–pollinator interactions along an altitudinal gradient. *Journal of Animal Ecology* 83(3): 639–650.

Ben-Hod, G., Kigel, J., and Steinitz, B. (1988). Dormancy and flowering in *Anemone coronaria* L. as affected by photoperiod and temperature. *Annals of Botany* 61(5): 623–633.

Ben-Hod, G., Kigel, J., and Steinitz, B. (1989). Photothermal effects on corm and flower development in *Anemone coronaria* L. *Scientia Horticulturae* 40(3): 247–258.

Blüthgen, N., Menzel, F., and Blüthgen, N. (2006). Measuring specialization in species interaction networks. *BMC Ecology* 6: 9.

Cane, J. (2021). Global warming, advancing bloom and evidence for pollinator plasticity from long-term bee emergence monitoring. *Insects* 12(5): 457.

CaraDonna, P.J., Cunningham, J.L., and Iler, A. (2018). Experimental warming in the field delays phenology and reduces body mass, fat content and survival: implications for the persistence of a pollinator under climate change. *Functional Ecology* 32(10): 2345–2356.

CaraDonna, P.J., Iler, A.M., and Inouye, D.W. (2014). Shifts in flowering phenology reshape a subalpine plant community. *Proceedings of the National Academy of Sciences of the United States of America* 111(13): 4916–4921.

Classen, A., Eardley, C.D., Hemp, A., Peters, M.K., Peters, R.S., Ssymank, A., and Steffan-Dewenter, I. (2020). Specialization of plant–pollinator interactions increases with temperature at Mt Kilimanjaro. *Ecology and Evolution* 10(4): 2182–2195.

Dafni, A., Tzohari, H., Ben-Shlomo, R., Vereecken, N.J., and Ne'eman, G. (2020). Flower colour polymorphism, pollination modes, breeding system and gene flow in *Anemone coronaria*. *Plants* 9: 397.

Danforth, B.N. (1999). Emergence dynamics and bet hedging in a desert bee, *Perdita portalis*. *Proceedings of the Royal Society B: Biological Sciences* 266(1432): 1985–1994.

de Manincor, N., Fisogni, A., and Rafferty, N.E. (2023). Warming of experimental plant–pollinator communities advances phenologies, alters traits, reduces interactions and depresses reproduction. *Ecology Letters* 26(2): 323–334.

Dhooghe, E., Grunewald, W., Reheul, D., Goetghebeur, P., and van Labeke, M.C. (2012). Floral characteristics and gametophyte development of *Anemone coronaria* L. and *Ranunculus asiaticus* L. (Ranunculaceae). *Scientia Horticulturae* 138: 73–80.

Donoso, I., Stefanescu, C., Martínez-Abraín, A., and Traveset, A. (2016). Phenological asynchrony in plant–butterfly interactions associated with climate: a community-wide perspective. *Oikos* 125(10): 1434–1444.

Dupont, Y.L., Padrón, B., Olesen, J.M., and Petanidou, T. (2009). Spatio-temporal variation in the structure of pollination networks. *Oikos* 118(8): 1261–1269.

Fisogni, A., de Manincor, N., Bertelsen, C.D., and Rafferty, N.E. (2022). Long-term changes in flowering synchrony reflect climatic changes across an elevational gradient. *Ecography* 2022: e06050.

Forrest, J.R., and Thomson, J.D. (2011). An examination of synchrony between insect emergence and flowering in Rocky Mountain meadows. *Ecological Monographs* 81(3): 469–491.

Freimuth, J., Bossdorf, O., Scheepens, J.F., and Willems, F.M. (2022). Climate warming changes synchrony of plants and pollinators. *Proceedings of the Royal Society B: Biological Sciences* 289(1971): 20212142.

Hoiss, B., Krauss, J., and Steffan-Dewenter, I. (2015). Interactive effects of elevation, species richness and extreme climatic events on plant–pollinator networks. *Global Change Biology* 21(11): 4086–4097.

Horovitz, A. (1976). Edaphic factors and flower colour distribution in the Anemoneae (Ranunculaceae). *Plant Systematics and Evolution* 126(3): 239–242.

Horovitz, A. (1991). The pollination syndrome of *Anemone coronaria* L.; an insect-biased mutualism. *Acta Horticulturae* 288: 283–287.

Horovitz, A., Bullowa, S., and Negbi, M. (1975). Germination characters in wild and cultivated Anemone coronaria L. *Euphytica* 24(1): 213–220.

Iler, A.M., Inouye, D.W., Høye, T.T., Miller-Rushing, A.J., Burkle, L.A., and Johnston, E.B. (2013). Maintenance of temporal synchrony between syrphid flies and floral resources despite differential phenological responses to climate. *Global Change Biology* 19(8): 2348–2359.

Jerome, D.K., Petry, W.K., Mooney, K.A., and Iler, A.M. (2021). Snow melt timing acts independently and in conjunction with temperature accumulation to drive subalpine plant phenology. *Global Change Biology* 27(20): 5054–5069.

Johnson, S.D., and Steiner, K.E. (2000). Generalization versus specialization in plant pollination systems. *Trends in Ecology & Evolution* 15(4): 140–143.

Keasar, T., Harari, A.R., Sabatinelli, G., Keith, D., Dafni, A., Shavit, O., Zylbertal, A., and Shmida, A. (2010). Red anemone guild flowers as focal places for mating and feeding by Levant glaphyrid beetles. *Biological Journal of the Linnean Society* 99(4): 808–817.

Kehrberger, S., and Holzschuh, A. (2019). How does timing of flowering affect competition for pollinators, flower visitation and seed set in an early spring grassland plant? *Scientific Reports* 9(1): 15593.

Klumpers, S.G., Stang, M., and Klinkhamer, P.G. (2019). Foraging efficiency and size matching in a plant–pollinator community: the importance of sugar content and tongue length. *Ecology Letters* 22(3): 469–479.

Kudo, G., and Cooper, E.J. (2019). When spring ephemerals fail to meet pollinators: mechanism of phenological mismatch and its impact on plant reproduction. *Proceedings of the Royal Society B: Biological Sciences* 286(1904): 20190573.

Kudo, G., and Ida, T.Y. (2013). Early onset of spring increases the phenological mismatch between plants and pollinators. *Ecology* 94(10): 2311–2320.

Lara-Romero, C., Seguí, J., Pérez-Delgado, A., Nogales, M., and Traveset, A. (2019). Beta diversity and specialization in plant–pollinator networks along an elevational gradient. *Journal of Biogeography* 46(7): 1598–1610.

Laura, M., and Allavena, A. (2007). *Anemone coronaria* breeding: current status and perspectives. *European Journal of Horticultural Science* 72(6): 241–247.

Laura, M., Allavena, A., Magurno, F., Lanteri, S., and Portis, E. (2006). Genetic variation of commercial *Anemone coronaria* cultivars assessed by AFLP. *Journal of Horticultural Science and Biotechnology* 81(4): 621–626.

Liu, Y., Miao, R., Chen, A., Miao, Y., Liu, Y., and Wu, X. (2017). Effects of nitrogen addition and mowing on reproductive phenology of three early-flowering forb species in a Tibetan alpine meadow. *Ecological Engineering* 99: 119–125.

Luna, P., Villalobos, F., Escobar, F., Neves, F.S., and Dáttilo, W. (2022). Global trends in the trophic specialisation of flower-visitor networks are explained by current and historical climate. *Ecology Letters* 25(1): 113–124.

Maglianesi, M.A., Hanson, P., Brenes, E., Benadi, G., Schleuning, M., and Dalsgaard, B. (2020). High levels of phenological asynchrony between specialized pollinators and plants with short flowering phases. *Ecology* 101(11): e03162.

Marshall, K.E., Gotthard, K., and Williams, C.M. (2020). Evolutionary impacts of winter climate change on insects. *Current Opinion in Insect Science* 41: 54–62.

Miller-Struttmann, N.E., and Galen, C. (2014). High-altitude multi-taskers: bumble bee food plant use broadens along an altitudinal productivity gradient. *Oecologia* 176: 1033–1045.

Motten, A.F. (1982) Autogamy and competition for pollinators in *Hepatica americana* (Ranunculaceae). *American Journal of Botany* 69(8): 1296–1305.

Murphy, S.D., and Vasseur, L. (1995). Pollen limitation in a northern population of *Hepatica acutiloba*. *Canadian Journal of Botany* 73(8): 1234–1241.

Ornai, A., and Keasar, T. (2020). Floral complexity traits as predictors of plant-bee interactions in a Mediterranean pollination web. *Plants* 9(11): 1432.

Ovaskainen, O., Skorokhodova, S., Yakovleva, M., Sukhov, A., Kutenkov, A., Kutenkova, N., Shcherbakov, A., Meyke, E., and Delgado, M.D.M. (2013). Community-level phenological response to climate change. *Proceedings of the National Academy of Sciences of the United States of America* 110(33): 13434–13439.

Peñuelas, J., Filella, I., Zhang, X., Llorens, L., Ogaya, R., Lloret, F., Comas, P., Estiarte, M., and Terradas, J. (2004). Complex spatiotemporal phenological shifts as a response to rainfall changes. *New Phytologist* 161(3): 837–846.

Perevolotsky, A., Schwartz-Tzachor, R., Yonathan, R., and Ne'eman, G. (2011). Geophytes-herbivore interactions: reproduction and population dynamics of *Anemone coronaria* L. *Plant Ecology* 212(4): 563–571.

Phillimore, A.B., Stålhandske, S., Smithers, R.J., and Bernard, R. (2012). Dissecting the contributions of plasticity and local adaptation to the phenology of a butterfly and its host plants. *The American Naturalist* 180(5): 655–670.

Robbirt, K.M., Roberts, D.L., Hutchings, M.J., and Davy, A.J. (2014). Potential disruption of pollination in a sexually deceptive orchid by climatic change. *Current Biology* 24(23): 2845–2849.

Sevenello, M., Sargent, R.D., and Forrest, J.R.K. (2020). Spring wildflower phenology and pollinator activity respond similarly to climatic variation in an eastern hardwood forest. *Oecologia* 193: 475–488.

Stang, M., Klinkhamer, P.G., and van Der Meijden, E. (2006). Size constraints and flower abundance determine the number of interactions in a plant–flower visitor web. *Oikos* 112(1): 111–121.

Tansey, C.J., Hadfield, J.D., and Phillimore, A.B. (2017). Estimating the ability of plants to plastically track temperature-mediated shifts in the spring phenological optimum. *Global Change Biology* 23(8): 3321–3334.

Tooke, F., and Battey, N.H. (2010). Temperate flowering phenology. *Journal of Experimental Botany* 61(11): 2853–2862.

Trøjelsgaard, K., and Olesen, J.M. (2013). Macroecology of pollination networks. *Global Ecology and Biogeography* 22(2): 149–162.

Vázquez, D.P., Vitale, N., Dorado, J., Amico, G., and Stevani, E.L. (2023). Phenological mismatches and the demography of solitary bees. *Proceedings of the Royal Society B: Biological Sciences* 290(1990): 20221847.

Vinchesi, A., Cobos, D., Lavine, L., and Walsh, D. (2013). Manipulation of soil temperatures to influence brood emergence in the alkali bee (*Nomia melanderi*). *Apidologie* 44: 286–294.

Waser, N.M., Chittka, L., Price, M.V., Williams, N.M., and Ollerton, J. (1996). Generalization in pollination systems, and why it matters. *Ecology* 77(4): 1043–1060.

Weaver, S.A., and Mallinger, R.E. (2022). A specialist bee and its host plants experience phenological shifts at different rates in response to climate change. *Ecology* 103(5): e3658.

Xie, Y., Thammavong, H.T., and Park, D.S. (2022). The ecological implications of intra- and inter-species variation in phenological sensitivity. *New Phytologist* 236(2): 760–773.

Yonash, Y., Jinggui, F., Shamay, A., Paz, N., Lavi, U., and Cohen, A. (2004). Phenotypic and genotypic analysis of a commercial cultivar and wild populations of *Anemone coronaria*. *Euphytica* 136: 51–62.

17

Life Histories in the Context of Mutualism

Renee M. Borges

Centre for Ecological Sciences, Indian Institute of Science, Bangalore, India

If you want to go far, go together.

Source: Possibly a Luo proverb from Africa.

17.1 Introduction

In mutualism, individuals of two or more species exchange benefits. Returns on benefits may not be immediate, and providing benefits could be costly. Hence, mutualism prevails when the net benefits to both partners are greater or equal to the costs (Bronstein 2001, Sachs et al. 2004). There are many advantages to mutualism. For example, the gametes and propagules of mutualistic partners go further in space as occurs in pollination or seed dispersal mutualisms (Ghazoul 2005) and allow new genetic combinations or the discovery of new germination sites. Although mutualisms are believed to have arisen by the co-option of antagonists (Johnson et al. 2021), they promote lineage longevity. Indeed, mutualisms tend to have a longer evolutionary history compared to antagonisms (Zeng and Wiens 2021) suggesting that mutualistic lineages persist further in time than those involved with antagonisms. In the absence of the vertical transmission of a mutualist during which the mutualist directly passes from parent to offspring, which is only possible for small-sized symbiotic organisms, partnerships need to be reassembled every generation. In long-lived plants or animals, several generations and lineages of mutualists may form transient associations during the lifespan of an individual mutualist. Therefore, a key requirement for the maintenance of mutualism is the regularity with which partners encounter each other in space and time.

A life history strategy of a species or a population is 'a set of coadapted traits designed by natural selection to solve particular ecological problems' (Stearns 1976). Consider hippopotami and mosquitoes, the former are large and live slow, the latter are small, live fast and die soon in absolute time. Yet, by the equal fitness paradigm (Brown et al. 2018), both hippopotami and mosquitoes are equally fit. They merely occupy different portions of life history strategy space. Inter-population variation in life history strategies may also occur in response to ecological variables such as seasonality and climate (Morrison and Hero 2003, Hedderson and Longton 2008, Behrman et al. 2015). A life history strategy is a multidimensional ensemble of traits that interact with each other and with physiological, behavioural and developmental traits in many ways that may constrain the ensemble (Stearns 1980). Considering life history's multidimensionality, when individuals of two species interact in a mutualistic interaction, the life histories of these species must intersect in ways that enable their beneficial engagement. Just as mutualism must overcome constraints to become established, several pathways could lead to mutualism breakdown among which, surprisingly, the reversal to antagonism is rarely found (Sachs and Simms 2006). This could be due to the loss of those traits during the evolution of the mutualistic partnership that are difficult to regain. For example, the parasitoid wasp *Asobara tabida* fails to produce oocytes in the absence of the once-parasitic *Wolbachia* bacteria (Dedeine et al. 2001). Regaining independent oogenesis is likely to be difficult. Furthermore, it is tempting to assume that life histories of interacting partners have coevolved such that the relevant ontogenetic stages that are pertinent to the interaction are available at the same time. However, it is equally plausible that only species with suitable exaptations and matching traits were able to pair up and subsequently coevolve. Whenever mutualism also involves symbiosis, which is when one partner lives on or within another partner, the likelihood of matching, facilitatory or coupled life history traits is greater.

17.1.1 Life History and Other Traits that Predispose Organisms to Enter into Mutualism

Some organisms are more likely to require interspecific service providers than others. Some non-mutually exclusive traits that make organisms more likely to benefit from the delivery of services are immobility, limited access to nutrients and an outcrossing breeding system. The benefits provided by mutualists are varied and include protection, transport, nutrition, shelters or rearing sites, reproduction and the control of competitors by interference or exploitative competition, *i.e.*, indirect mutualism *sensu* Wootton (1994) and Zhang et al. (2021). In indirect mutualism, a third partner influences the interaction between the other two, *e.g.*, a parasitoid wasp may control the local populations of pollinator wasps in the mutualism between figs and their pollinating wasps. Here the parasitoid serves as an indirect mutualist of the plant since it prevents over-exploitation of the flowers by pollinators that deposit eggs in the flowers where the larvae develop (Krishnan et al. 2015). The requirement for benefits may be lesser or greater at particular ontogenetic stages. Likewise, beneficial services may only be provided at certain stages in the life cycles of the mutualistic partners (Leichty and Poethig 2019). For example, young plant leaves may be most vulnerable to herbivores and require the most protection from mutualistic aggressive ants, while the ants themselves must have a colony size of a sufficient number of workers that could be recruited to the site of herbivore attack. Here plant ontogeny and ant colony development should match for an anti-herbivore benefit received from the ants. The plant must also be able to provide an equivalent benefit to the ants in terms of nutrients to feed the colony or nesting space. Plant and ant colony growth should therefore intersect at the appropriate ontogenetic stages. It is likely that all mutualisms are stage-structured such that reciprocal benefits are only possible at particular developmental stages (Nakazawa 2020).

17.1.2 Asymmetry in Partner Interaction and Consequences for Mutualism

Mutualisms have been examined theoretically and empirically from the perspective of asymmetry in the rewards that partners can offer or in the dependence of one partner on the other (Noë et al. 1991, Bshary and Bronstein 2004). However, the causes and traits responsible for such asymmetries are often not fully understood (Chomicki et al. 2020). There are different types of asymmetries between partner traits with consequences for the development of mutualism: (1) Size asymmetry. Smaller-sized partners may be more likely to enter mutualisms, the most successful being the irreversible evolution of mitochondria and chloroplasts from free-living bacteria. (2) Generation time asymmetry. Following from size asymmetry, there is usually generation time asymmetry with one partner having a much longer absolute generation time than the other. This may have consequences for coevolutionary rates. (3) Population size asymmetry. Size and generation time asymmetry can lead to differences in population size between partners. Hence, size, generation time and population size are covariates that could influence mutualism. (4) Specialisation asymmetry. One or more partners in a mutualism may be specialized while others are generalized in their interaction. In this sense, mutualistic networks are often nested and asymmetric (Vázquez et al. 2009), features which may contribute to their overall stability up to a certain level (Bascompte and Scheffer 2023), *e.g.*, a plant that is serviced by many species of seed dispersers is likely to be less prone to local extinction than one serviced by a single disperser species.

Since body size, generation time and population size co-vary, any of these variables could be the main driver that affects interaction asymmetry or degree of specialisation in a mutualism. It is likely that the degree of specialisation is greater when one of the partners is small or very large but that intermediate-sized partners have diffuse interactions. For example, small- and large-sized flowers are more likely to have specialized pollinators compared to those of intermediate size (Figure 17.1a). This assumes that the rewards in the larger flowers are only available to larger-sized visitors, and are relatively concealed, *e.g.*, within nectar spurs or tubes, compared to other large, open flowers such as those of *Rafflesia* which could attract several carrion fly species of a range of different sizes (Hidayati and Walck 2016, Zhao et al. 2022).

17.2 Mutualism Benefits and Life History Traits

There are many benefits from mutualism. In this chapter, examples from each type of benefit are provided along with the required life history compatibility between partners.

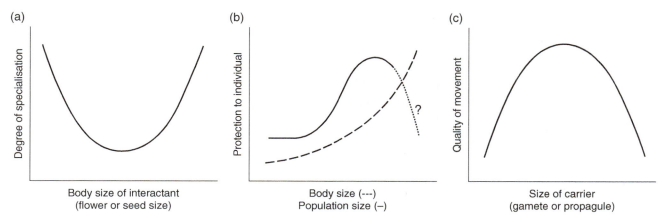

Figure 17.1 Hypothetical curves of responses between life history traits and features of mutualism. (a) Relationship between body size and the degree of specialisation of an interaction. (b) Relationship between body size or population size and the protection afforded to an interactant. Dotted line indicates ambiguity in the prediction. (c) Relationship between the size of the carrier of a gamete or a propagule such as a seed and the quality of the movement in terms of distance carried to an appropriate site for germination and propagule survival.

17.2.1 Protection

Physical protection afforded by a mutualist is likely to be positively related to body size of the defending partner and beyond a certain threshold could be negatively related to the population size of the same mutualistic species (Figure 17.1b). Body size and population size are negatively correlated and have different impacts on mutualism. For example, a single individual of a protective ant species, which is patrolling an individual plant in exchange for extrafloral nectar or nesting benefits on the same plant, is relatively ineffective in providing defence against herbivores. However, the entire ant colony that is either resident within the plant or includes that individual plant within its foraging territory will be effective in plant defence. It also follows that as ant size increases, an entire colony may not find residential space within a single plant, and a single plant may not be able to nourish such a colony. However, if plant size increases, then even larger-sized ants and/or a large colony may find residence and provide protection to a single individual plant. Larger colonies of protective ants are found in larger-sized individual plants in the association between *Cecropia* plants and *Azteca* ants in Brazil (Reis et al. 2022). Similar general principles would apply to patrolling ants that protect hemipteran aggregations in exchange for nutrition.

17.2.1.1 Bacterial Co-constructed Countershading Mechanism in Squids

Bioluminescent *Vibrio* bacteria are engaged in protection symbioses with marine fauna, *e.g.*, the squid light organ co-constructed by *Vibrio fischeri* and its squid host that provides countershading protection against visual predators (Nyholm and McFall-Ngai 2021). In this mutualism, the bacterial population within the light organ increases at night when the bioluminescence occurs. The generated light prevents a silhouette of the squid from forming under nocturnal moonlight and thereby the nocturnally foraging squid avoids predators. Here, bacteria gain nutrition while the squid host gains protection from predators. Despite this being a specialized interaction, the bacteria can live outside the squid. Dominant 'D' strains that outcompete other strains form hyper-aggregates at the entrance pores of the nascent light organ of juvenile squid to gain access to the organ and to help in its development (Visick et al. 2021). Each morning, in the adult squid, the light organ is voided of between 70% and 95% of the bacteria and the resident population grows to repopulate the light organ to reach the high density required for bioluminescence via quorum sensing at night. This daily process, which continues throughout the lifespan of the squid, may also provide the opportunity for the host to purge itself of uncooperative bacteria (Nyholm and McFall-Ngai 2021). Hence, the ability to colonize squid is as important a life history trait in Vibrionaceae as is persistence outside the squid host. The persistence trait is conferred by the accumulation of nutrients, *e.g.*, polyhydroxybutyrate (PHB) granules (Baker et al. 2019), derived from the nutrient-rich symbiotic and co-constructed environment of the light organ.

17.2.1.2 Ant–Plant Protective Interactions

While many plant species interact with opportunistic ants that visit their extrafloral nectaries, *i.e.*, myrmecophilic species, this section will only focus on protection mutualisms in myrmecophytic plants, *i.e.*, those that also harbour ants within them, since their interactions with ants are likely more specialized. Here plants gain protection from herbivory, and ants gain nest sites and nutrition. Plant structures that harbour ants are termed domatia. Although hundreds of plants over diverse families interact with ants, only a few ant species in about 30 genera interact mutualistically with myrmecophytic plants (Chomicki and Renner 2015, Nelsen et al. 2018). There is clear asymmetry: fewer ant species form mutualistic interactions with many more plants. This is a likely consequence of life history traits. Usually, a single colony of one species of mutualistic ants occupies a single individual plant, although parasitic ant species, *i.e.*, those that are non-protective, could also be co-residents (Kautz et al. 2012). The ant–plant mutualism is horizontally transmitted, and partners need to reassemble every ant and plant generation. Horizontal transmission is unlike vertical transmission. In the former, compatible partners need to encounter each other every generation, while, in the latter, a mutualistic symbiont is transferred from one generation to the next. It is unlikely that ant colony lifespan matches plant lifespan, especially in long-lived perennial plants, and therefore several successive ant lineages probably inhabit individual plants. The longevity of a single colony in an ant–plant mutualism has not been investigated. Myrmecophytic plants usually have lower chemical defences and have opted instead for protection by ants (Fonseca-Romero et al. 2019). Therefore, from an early life history stage, they must be colonized by protective ants. There is a well-defined ontogeny in the development of plant defence by ants (Quintero et al. 2013). For those ant-plants that house protective ants within hollow domatia (*e.g.*, stipular thorns or stems), the structures have to be robust and spacious at an early stage. This necessitates their precocious development, especially an increase in the primary diameter of twigs that support these structures (Brouat and McKey 2000). Stem domatia are the most common form of domatium and occur in a diversity of plant families (Chomicki and Renner 2015) suggesting either that the precociousness of stem thickening is an easy trait to develop or that lineages with thickened stems were co-opted for the development of this mutualism. Leaf–stem allometry coupled with enlargement of special organs such as the embryonic hypocotyl to form domatia are signatures of symbioses with ants (Chomicki and Renner 2019).

There should also be a close correspondence between plant traits and ant developmental traits for the protection mutualism to be maintained (Leal et al. 2023). An important life history trait is ant colony size. For some plant-ants, colony size is determined by the available nesting space within the individual host tree (Fonseca 1993, Marting et al. 2018). Ant worker size (measured by head width) may also increase with host plant size and the possibility of greater resources (Marting et al. 2018). The size of founding queens is also important since their head widths must be smaller than the opening of the domatia, especially in hollow stem domatia, or else they would be unable to enter the nesting space to initiate a colony. Domatium entrance size can serve as a filter against parasitic ants and appears to be responsive to local coevolutionary pressures (Brouat et al. 2001). However, domatium size also exhibits phenotypic plasticity and if domatium size is reduced, with corresponding reduction in entrance size, coevolved ants may be disadvantaged (Kokolo et al. 2020).

Selection for aggressive ants with colony sizes that can be accommodated within an individual plant is likely to occur. Since aggression correlates with anti-herbivore behaviour (Gaume et al. 2006, Heil 2013), even if two different founding queens establish colonies on the same individual plant, and if they differ in their aggression levels, then those modules of the plant occupied by the more aggressive colony will suffer less herbivory, and thereby show more photosynthesis and growth than the other plant modules. Thus, in the absence of active partner choice, the most efficient colonies are selected and rewarded (Hernández-Zepeda et al. 2018). Ant-plant saplings colonized by ants grew faster than those experimentally depleted of ants (Oliveira et al. 2015). Furthermore, defence by ants was greater in older plants, which are often larger and have more rewards (Trager and Bruna 2006), suggesting that partners are responsive to mutual benefits. Consequently, when myrmecophilic plants only provide nutrient rewards without providing housing to ants, larger plant individuals tend to be visited by a greater number of species and individuals from which they derive more anti-herbivore protection (Koch et al. 2016). The growth benefits that plants derive from ants are equivalent to those they derive from rhizobia and mycorrhizae (Gibert et al. 2019), making ant–plant mutualisms important to plant fitness. However, plant life history also plays a role with perennial plants deriving greater reproductive benefits than annual plants from mutualism with ants (Trager et al. 2010). This is because perennial plants are likely to be larger and therefore would have more resources to provision mutualistic resident ants.

The social organisation of ant species can also influence the likelihood of mutualism establishment. Polydomous ant colonies, *i.e.*, those that have multiple nests possibly distributed over several plants, could either be successful competitors, taking control over several individual plants (Federle et al. 1998) or be successful colonizers (Debout et al. 2009). Some myrmecophytic plants are clonal, *e.g.*, the leguminous tree *Humboldtia brunonis* (Dev et al. 2010), which is defended by a nomadic, polydomous and aggressive dolichoderine (Gaume et al. 2006). Clonality results in closely spaced plants that could be occupied by a single ant colony and might facilitate colonisation by polydomous ant species. Colonisation–competition trade-offs can also determine which ant species have net positive effects on their host plants as shown in myrmecophytic plants in which the mutualist is the better colonizer but may lose out to parasitic species that are better competitors and are more fecund (Yu et al. 2001, Debout et al. 2009). For example, in the interactions between the mutualistic *Azteca* ant, the parasitic *Allomerus* ant and the myrmecophytic plant *Cordia nodosa*, *Azteca* has the upper hand in low-density *Cordia* populations since its winged queens are better fliers and thus more efficient colonizers, while the parasite *Allomerus* is better represented in high-density plant populations (Yu et al. 2001). This three-way interaction can also be described in terms of a dispersal–fecundity trade-off between mutualists and parasites. Therefore, the life history traits of plants and ants are inextricably linked and whether a mutualism develops is greatly dependent on context. Overall, plants and ants have comparable ecological and life history strategies (Gibb et al. 2023) and species with compatible traits may enter into mutualistic interactions.

17.2.2 Nutrition

Nutritional mutualisms are very common in nature and are often associated with symbionts. The development of nutritional symbioses with microbes, for example, has enabled insects to diversify their feeding niches and consume plants that are deficient in nitrogen. This is because the microbes synthesize several essential amino acids for their insect hosts (Sudakaran et al. 2017, Reis et al. 2020; see also Chapter 14, this volume). Nutrition is the basis for many mutualisms with solid or liquid food as a reward as also occurs in ant–plant and ant–hemipteran interactions. The nutritional aspects of these ant-specific associations will not be discussed in this chapter.

17.2.2.1 Zooxanthellae and Corals

Zooxanthellae confer autotrophy on sessile coral holobionts. These dinoflagellates can regulate cell-cycle processes in their host corals, preventing apoptosis or death of host cells, and thereby ensure their own survival (Gorman et al. 2022). The greatest specialisation occurs in brooding corals in which the normally planktonic larvae of zooxanthellae are brooded within the maternal coral body and the daughter polyps acquire the symbiotic zooxanthellae during the brooding process (Baird et al. 2009, Goodbody-Gringley et al. 2018). Most corals acquire zooxanthellae from the marine environment while only a few (*ca.* 15%) acquire them via maternal inheritance (Schwarz et al. 2002). Brooding corals tend to be weedy with fast growth and rapid population turnover and are capable of long-distance dispersal. The brooding life history trait could even be characteristic of isolated coral populations (Darling et al. 2012) that would have difficulty in acquiring zooxanthellae in its absence.

The life history constraints that arise from sessility are applicable to corals and plants. Grime's C-S-R (competition-stress tolerance-ruderal) life history strategy axes developed for plants have been applied to corals (Grime 1977, Darling et al. 2012) with some success. In Grime's formulation, stress constitutes abiotic factors that limit plant growth such as light limitation and poor mineral content while disturbance constitutes biotic components that affect plant growth and reproduction such as herbivory as well as abiotic disturbances such as wind, fire and frosts (Grime 1977). In Grime's life history scheme, combinations of high and low stress coupled with high and low disturbance result in three possible viable strategies: competitive (C) plants that occupy areas with low stress and low disturbance, stress-tolerant (S) plants that occupy habitats with high stress but low disturbance, and ruderal (R) plants that are present in areas with low stress and high disturbance. The high stress and high disturbance combination is considered one in which no species can survive. C, S and R are extreme points on the life history axes and plant species could also occupy the continuum between these extremes. C plants invest in attaining large size to acquire control over resources and thus delay reproduction. S plants invest in repair and tolerance mechanisms and may be small or grow slowly to larger sizes. R plants invest in reproduction and put out large numbers of propagules (Pierce et al. 2017). It appears that many species of corals fit into the combination of traits predicted by Grime's (1977) life history classification for plants. Competitive corals have the fastest growth rate and stress-tolerant corals have the highest fecundity while brooding seems to predominate in weedy opportunistic species (Darling et al. 2012). The examination of coral traits and their zooxanthellae associations from a life history perspective requires much more investigation (Brandl et al. 2019).

17.2.2.2 Plants, Rhizobia and Mycorrhizae

Plants, with their sedentary lifestyles, have entered into nutritional mutualisms with mycorrhizal fungi and nodulating rhizobia, whose life histories have been shaped by their host interactions and host partner choice (Denison and Kiers 2011, Montoya et al. 2023). In the rare absence of these nutrition-garnering partners, plants have adopted strategies such as carnivory, development of cluster roots and even parasitism (Werner et al. 2018). Mycorrhizae and rhizobia can also have synergistic mutualistic effects on their host plants and on each other's productivity. However, such multiple mutualism effects have rarely been quantified (Afkhami et al. 2021).

Rhizobia (mostly α-proteobacteria and some β- and γ-proteobacteria; Lemaire et al. 2016) are nitrogen-fixing, non-spore-forming bacteria that can exist independently in the soil. In root nodules, they depend on plant carbon while providing nitrogen and develop into specialized forms termed bacterioids (Denison and Kiers 2011). Rhizobia can exhibit both fast and slow life histories while present within the same individual plant (Young 1996). Rhizobia within root nodules accumulate carbon in the form of PHB granules (as in the Vibrionaceae discussed earlier). However, PHB accumulation is likely to conflict with the carbon requirements of the host and the host may sanction those nodules where accumulation of PHB occurs at the cost of nitrogen fixation (Denison and Kiers 2011). An important life history strategy is for rhizobia to persist in the soil when hosts are unavailable or when they incur host sanctions and are expelled from senescent nodules. This persister stage is possibly linked to PHB stores, which may allow rhizobia to remain dormant for up to a century (Muller and Denison 2018). Rhizobia can increase their fitness by a factor of several millions by engaging in mutualism with plants (Denison and Kiers 2011). However, the chance that a single rhizobium cell will engage in mutualism with a plant host is 10^{-9} (Denison and Kiers 2004), thus providing strong selection on rhizobia to competitively locate suitable hosts.

Unlike rhizobia, arbuscular mycorrhizal fungi (AMF) are obligately associated with plants and are unable to survive, except in spore form, outside the plant host. Between 4% and 20% of an individual plant's carbon is allocated to AMF. In turn, AMF forage for phosphorus and also provide protection against biotic and abiotic stresses (Diagne et al. 2020). A single AMF hyphal cell is a multinucleate heterokaryon since the nuclei originate from different 'individuals' through anastomosis or merging of cells (Noë and Kiers 2018). This multinucleated state makes the interactions between plants and the communally produced mycorrhizal arbuscles a multi-partite interaction, similar to the many rhizobia genotypes within root nodules. It is possible that multinucleated and genetically heterogeneous spores arising from the more metabolically active arbuscles have greater fitness. Therefore, the interactions between plants and these multinucleated entities may be better characterized as a cooperative rather than an inter-individual mutualism (Noë and Kiers 2018).

The hyphae of AMF can grow up to 100 m in the soil. Therefore, they can explore a much greater area than plant roots. Moreover, in this process of soil exploration, they can anastomose with other hyphae in a complex network and also interact with multiple plant hosts (Denison and Kiers 2011). Consequently, the units and levels of selection that are applicable to AMF are complex since individuals are hard to identify, which adds to the difficulty of understanding partner selection in mycorrhizae (Werner and Kiers 2015). The life histories of mycorrhizae have also been examined from Grime's (1977) C-S-R framework (Chagnon et al. 2013). Accordingly, competitive mycorrhizae are those that acquire carbon rapidly from their hosts growing in carbon-rich areas and with a high requirement for phosphorus. In this way, the interests of C-selected mycorrhizae and plants growing in a carbon-rich area can be aligned. Similarly, stress-tolerant mycorrhizae (S-selected) are those that can persist even under carbon limitation or other stresses and would be most suited to interaction with stress-tolerant plants. Ruderal or R-selected mycorrhizae would be expected to have a high population growth rate and short life cycle and would be compatible with ruderal plants (Chagnon et al. 2013). Most research on mycorrhizae has involved ruderal fungi (Bennett and Groten 2022), and, therefore, the trait space of other plant and mycorrhizal life history combinations needs to be examined. Similarly, the persistence of AMF spores and their dispersal have received limited attention (Paz et al. 2021). Germination synchrony of spores with the phenology of host plants in seasonal environments is very likely given the obligate nature of the mutualism (Gemma and Koske 1988). Legumes with perennial rather than annual life histories benefit more from association with either rhizobia or arbuscular mycorrhiza (Primieri et al. 2022), suggesting that a longer time scale favours the development of more rewarding interactions.

17.2.3 Farming

This is a type of mutualism in which nutrients are acquired from a partner that dwells outside the cultivator. Farming appears widespread in non-human animals and fungi. Ants, termites and beetles farm fungi, mostly basidiomyctes (Biedermann and Vega 2020). Ants also farm epiphytes (Chomicki 2022), damselfish cultivate algae (Hata and Kato 2006) and fungi possibly cultivate bacteria (Pion et al. 2013). For this interaction to be successful, the farmer should be able to exercise control over the life history and genetic diversity of the crop. However, it is equally possible that the cultivar has imposed selection on farmer traits (Mueller 2002).

17.2.3.1 Fungus-Farming Ants

In all fungus-farming ants, mushroom formation is suppressed, and the ants propagate only a single asexual cultivar within the nest (Mueller 2002). Instead of mushrooms, the fungal hyphae of the highly evolved leaf-cutting ants produce asexually derived tubercular structures, termed gongylidia, which serve as food for the colony and, since they occur in clusters, can be readily harvested by the ants (Mueller 2002). In the lower attine ants, the fungus mycelium is a dikaryon containing unfused nuclei from two cells, while the higher attines farm fungi with a mycelium that exhibits facultative polyploidy (containing multiple nuclei), and the more highly evolved leaf-cutting ants farm fungi that are obligately polyploid (polykaryotic) (Kooij et al. 2015). Fully functional polyploidy of the crop fungi has apparently allowed leaf-cutting ants to harvest a variety of living plant parts, compared to the dead wood or leaf litter foraging of the lower attines, since the complex fungal crop genotypes appear to be more efficient at degrading and combating the chemical defences of live plant tissue. It therefore seems that the ploidy transition of the fungal crop has also been accompanied by a foraging transition in leaf-cutting ants from dead to living tissue (Kooij et al. 2015).

Polyploidy of the fungal crop is related to the colony size of the fungus-farming ants which ranges from a few hundred workers in the lower attines through a few thousand workers in the higher attines to a colony size greater by one to three orders of magnitude in leaf-cutting ants (Riveros et al. 2012, Kooij et al. 2015). Interestingly, brain size in attine ants decreases with greater colony size and is possibly indicative of greater task and caste specialisation as well as increased worker polymorphism (Riveros et al. 2012).

Colony size in leaf-cutting ants is also related to the mode of colony establishment. Queens of the lower attines undertake foraging themselves to nourish the first brood of workers while the large queens of the highly evolved *Atta* leaf-cutting ants establish their first set of workers through a claustral process in which trophic unfertilized eggs are laid and these are used to feed the first brood (Schultz 2022). As colony size increases, the male:female ratio decreases such that fewer males are produced. This is a trait that aligns with fungal interests since only females are responsible for the vertical transmission of fungal hyphae during claustral colony founding (Mueller 2002). There is also a gradient of specialisation with the fungal partner(s) from the more loosely interacting lower attines to the highly specialized leaf-cutting ants that associate obligately with *Leucoagaricus gongylophorus* (Mehdiabadi et al. 2012, Khadempour et al. 2016).

17.2.3.2 Fungus-Farming Termites

Like in fungus-farming ants, basidiocarp or mushroom formation is also suppressed in the fungus-farming termites (Macrotermitinae). However, when colonies decline, mushrooms could be produced (Nobre et al. 2011), which generate sexual spores that are likely wind-dispersed or dispersed by other agents such as earthworms in the soil. The dominant fungal mycelium is heterokaryotic and arises from the fusion of several homokaryotic mycelia. Within the colony, there is active selection of the fastest growing heterokaryotic strain, which is also facilitated by the fusion of clonally related mycelia until one strain predominates. This strain usually has the highest level of asexual spore production (Aanen 2006, Aanen et al. 2009). That an entire nest contains just one fungal strain after rigorous selection by the termites helps to stabilize the mutualism (Aanen et al. 2009) since the most favourable strain gains advantage without losing productivity by continual competition with other strains. As in fungus-farming ants, the fungal mycelium generates bite-sized asexually derived nodules termed mycotetes, which are consumed by the termites (Katariya et al. 2017; Figure 17.2). The longevity of termite nest mounds spans several decades and outlives the lifespan of the founder kings and queens which is a couple of decades. Nest stability and its constant remodelling and repair are key traits in the fungus–termite mutualism (Zachariah et al. 2017, Wisselink et al. 2020, Borges and Murthy 2023). Fungal transmission is possibly horizontal and occurs when workers pick up spores during foraging and introduce them into the colony. Vertical transmission via female or male alates occurs only in two unrelated genera (Korb and Aanen 2003) and is the exception to the ancestral state. This is unlike the strictly vertical transmission of the symbiont in fungus-farming ants.

17.2.4 Movement

Mutualisms that ensure propagule or gamete movement are important in otherwise sessile organisms or those with low mobility. Here there is often size asymmetry between partners with the partner involved in moving the propagules or gametes being much smaller in size than the other partner. Most examples involve plants, while others involve phoretic transport wherein an individual of one species is carried by another. Examples of mutualisms in phoresy are few (Borges 2022).

Figure 17.2 Examples of mutualisms. (a) Tiny fig wasps carrying pollen swarm outside the entrance of the urn-shaped fig inflorescence. (b) The mutualism between figs and pollinating fig wasps is influenced by third parties such as predatory ants and parasitic gall wasps that oviposit into the fig inflorescence from the outside using their long ovipositors. (c) Fungus-farming termites cultivate fungus gardens in cathedral-shaped earthen greenhouses. (d) Callow termite workers near the fungal crop, which consists of asexual white nodules termed mycotetes. (e) A queen of a fungus-farming termite. The size variation between the queen and workers is spectacular. *Source:* Picture credits: Ashok Mallik, Abdul Hakkim.

17.2.4.1 Movement in Plants

The quality and quantity of movement of plant gametes or propagules, *e.g.,* seed or pollen, are likely to be lower with small and large vectors but higher with intermediate-sized carriers (Figure 17.1c). Quality is measured in terms of the distance the propagule or gamete travels from the maternal plant, or the type of gametes received, especially in obligately outcrossing plants. The longest distance may not be the optimal distance for gene flow since gametes or seeds may be carried into unsuitable habitats. Therefore, philomatry (preference for the maternal site) and selfing may even be desirable when suitable growth conditions are patchily available and when genetic load has been purged. Some plants resort to selfing to avoid outcrossing while others resort to a type of bet-hedging producing both aerial (outcrossed) and subterranean cleistogamous (selfed) flowers, a strategy termed amphicarpy (Sadeh et al. 2009, Zhang et al. 2020, Sánchez-Martín et al. 2021). Seeds produced by the aerial flowers are lighter and easily dispersed while those produced by the subterranean flowers are fewer, heavier and have low dispersal, thus maintaining some amount of philomatry (Cheplick 1994, 2022). Amphicarpic plants are antitelechoric (Cheplick 2022) since their subterranean structures discourage telechory (longer-range seed dispersal) owing to the benefits of local adaptation. Therefore, some organisms may actively discourage mutualism, which leads to movement away from the maternal site, as part of their life history strategy. Seed dispersal mutualisms will not be discussed further in this chapter.

Brood site or nursery pollination mutualisms are usually highly specialized interactions between plants and insects (Sakai 2002, Kawakita and Kato 2017; but see De Medeiros et al. 2019, Haran et al. 2022) owing to the particular biology that is necessary to breed within plant tissues. This type of pollination mutalism occurs with insects that are either ovule parasites, pollen parasites or those that develop within senescent flowers. Nursery pollination mutualism occurs in diverse plant families and involves several insect groups such as wasps, moths, flies, beetles and thrips (Sakai 2002). Here, the plant obtains the benefits of pollen transfer and seed production, and the insect acquires sites for brood nutrition and brood development. Brood site pollination mutualisms are often beneficial parasitisms since the plant offers rewards that involve giving up its own fitness in terms of flowers or seeds in exchange for gamete movement. The biological intimacy in this mutualism should result in close evolutionary interaction (Hembry et al. 2018), and coevolutionary patterns (Cruaud et al. 2012; but see Satler et al. 2019). To facilitate biological intimacy, tightly orchestrated or flexible life histories are needed. For example, flexible diapause occurs in *Epicephala* seed-consuming moths, which enables them to track the phenology of their obligate *Glochidion* plant hosts (Finch et al. 2021).

Considering that there are already reproductive and life history trade-offs that occur within each plant and animal species independently of mutualism (Primack 1987, Zera and Harshman 2001, Borges 2011, Salguero-Gómez et al. 2016), there are several critical aspects of these mutualisms that are relevant to the beneficial coupling of partner biology (Borges 2021). Most relate to pollinator body size, which is an important life history trait, relative to the architecture of the inflorescence and the arrangement of potential brood sites. Since development time is often related to size (Roff 2002; see also Chapter 1, this volume), brood development time must also match brood site longevity. For example, if pollinators breed within flower ovules, then their development time must match the time taken for maturation of the fruit. This duration is equivalent to brood site longevity since the fruit will either be consumed or be dispersed after that time. If pollinator mating also occurs within brood sites, then factors related to local mate competition (LMC; see Chapter 7, this volume) are also important and relevant to pollinator reproductive strategy (Greeff and Kjellberg 2022). Sometimes, complete adult eclosion cannot take place within the brood site, possibly due to mismatch between brood development time and brood site longevity, and pupation may occur outside the brood site, *e.g.*, in the soil (Kawakita and Kato 2017). Many brood site pollinators are semelparous, *i.e.*, breed only once in their lifetime. Consequently, the choice of brood site is of utmost importance (Borges 2021). Once offspring eclose within the brood sites, the emerging adults must find suitable brood sites within their dispersal range in order to continue their life cycle.

In the fig–fig wasp brood site pollination mutualism, female fig wasps pollinate flowers and also lay eggs in pollinated flowers within the urn-shaped enclosed inflorescences (Figure 17.2). The wasps are semelparous and oviposit within only one inflorescence since they lose their wings during entry into the urn. The fig wasp pollinators die within the fig inflorescence and the next generation of wasps that develop within the urn leaves the inflorescence with pollen to continue the mutualistic interaction. Failure to pollinate an inflorescence or excessive egg-laying will result in the host fig tree imposing sanctions on cheating wasps by aborting the inflorescence (Borges 2021). The fig wasp pollinators are capital breeders that depend on nutritional capital that is stored during the larval stage (Stephens et al. 2009). Pollinator fig wasps are on the fast end of a slow–fast life history continuum with an adult lifespan of 24–48 hours (Ghara and Borges 2010). Since individual fig trees are usually synchronous in their reproductive phenology, the 'fast' wasps must locate trees other than their natal trees within a short time, as their natal trees would be devoid of suitable brood sites at the time of wasp dispersal from their nurseries. This implies that such pollinators must have enough resource capital to find and utilize new brood sites within their short lifespan. Some brood site pollinators can enter into a metabolically inactive state, *i.e.,* diapause, to wait for suitable brood sites. For example, yucca moth larvae pupate in the soil and could diapause for up to 30 years waiting for the appropriate conditions for plant and mutualist to coincide (Powell 2001). Potential brood sites that exist outside the dispersal ability of pollinators are lost opportunities for brood site colonisation. Fig wasp pollinators have been selected for fuel stores, flight ability and metabolic rates that match the spatial distribution of brood sites such that those wasps that service fig species with clumped distributions have lower fuel stores (capital) compared to those that service widely spaced species (Venkateswaran et al. 2017, 2018).

17.2.4.2 Phoretic Animal Movement

Phoresy is the transport of one species by another. Immobile or relatively sedentary species, *e.g.*, nematodes, mites and leeches, may engage in phoretic interactions. Phoresy enables movement out of deteriorating resource patches or dispersal to mating arenas and breeding sites. There are few examples of phoretic mutualisms since phoretic interactions are usually parasitic (Borges 2022). Many of the mutualistic interactions involve nematodes whose dauer juvenile stage is amenable to phoresy (Crook 2014). In the dauer stage, which could be triggered by declining resources, feeding ceases and the responsiveness of juvenile nematodes to chemical cues from potential dispersal agents increases. For example, mutualistic nematodes are carried by beetles to their nests where they help to manage a suitable microbiome of bacteria and fungi and are later transferred between beetle generations or may move between hosts during mating (Ledón-Rettig et al. 2018). Gall-inducing nematodes engage in mutualistic interactions with gall-inhabiting flies. Here, juvenile nematodes are carried onto suitable plant hosts by flies. The nematodes develop first into parthenogenetic females, which are the gall inducers. These parthenogenetic females lay eggs within the galls. The first batch consists of male eggs, which is then followed by a clutch of females. Later, mating occurs within the galls and the next crop of nematode juveniles enter adult female flies where they feed on oviducal material and are transferred to the meristematic and gall-inducible tissues of another plant to complete the life cycle. In this example, the phoretic or travelling nematode stage is parasitic while the gall-inducing stage is mutualistic. How this mutualism is stabilized is unknown (Borges 2022), but it clearly requires a high degree of synchrony between the gall inducers and the gall users.

17.3 Future Directions

Mutualisms have contributed to the origin and maintenance of biological diversity through coupled beneficial interactions (Bronstein 2021) and are therefore important to an understanding of the history of life. This chapter focuses on interactions between life history traits and associated factors that help to constitute mutualism. Species can engage in mutualisms when their life cycles and life histories overlap at the appropriate ontogenetic stages. However, there are often asymmetries in mutualism wherein one partner may be more dependent on the interaction than the other partner.

Mutualisms may also arise amidst conflicting interests such as predation or herbivory among partners. A pollinator that is a mutualist as an adult and an herbivore as a larva is in potential conflict with its host plant, especially if it is a specialist pollinator and/or herbivore. Such a pollinator may experience a trade-off between allocating resources to flight (which would make it a more effective pollinator as an adult) or to fecundity (which would make it a more efficient herbivore as a larva) (Davidowitz et al. 2022). Such conflicts are only beginning to be investigated.

Mutualisms are often investigated selectively based on the interest or expertise of the investigator. However, the consequences of all mutualisms occurring simultaneously in an organism must also be examined, *e.g.*, above- and below-ground processes have rarely been considered together in plants (De Deyn 2017). For example, ants are less attracted to the extrafloral nectar of plants with rhizobia owing to a decrease in nectar production (Godschalx et al. 2015), suggesting trade-offs between different mutualistic effects within the plants. Similarly, the presence of mycorrhizae reduced extrafloral nectary production (Laird and Addicott 2007), but they also increased pollinator visitation (Wolfe et al. 2005) as did rhizospheric bacteria (Magalhães et al. 2023). These examples suggest that a holistic view of interactions is required to understand the net benefits of multiple interactions on mutualistic partners as well as their direct and indirect effects. In addition, what may appear to be mutualistic (*i.e.*, food for defence exchange) may actually turn out to be manipulative and parasitic on closer investigation, as is the case for ant associations with nectar-organ caterpillars (Pierce and Dankowicz 2022). Some of these lycaenid caterpillars secrete neurogenic amines from their nectar organs that alter ant behaviour by suppressing dopamine levels resulting in 'drugged' ants that faithfully tend these caterpillars (Hojo et al. 2015). It is important, therefore, to examine mutualisms within the overall contexts and environments in which they occur. Examination of trait variation between individuals is also lacking and such investigation will allow a better understanding of the shifts between cooperation and antagonism (Moran et al. 2022).

The plethora of flower visitation and pollination networks that are available (*e.g.*, Sazatornil et al. 2016, Vizentin-Bugoni et al. 2018, Lázaro et al. 2020), while stressing the importance of key species, fall short of incorporating flowering phenology and adult pollinator lifespans into their analysis (E-Vojtkó et al. 2020, Guimarães 2020). However, some network studies are beginning to examine plant and pollinator phenology, pollinator sex differences in foraging as well as direct and indirect interactions including neutral and facilitatory effects (Sonne et al. 2020, Duchenne et al. 2021, Guzman et al. 2021, Simmons et al. 2021, Smith et al. 2021, Morán-López et al. 2022; see also Chapter 16, this volume). A glaring lacuna is the lack of incorporation of larval food availability into lepidopteran and dipteran networks. This deficiency makes the evolutionary longevity of these networks unpredictable, especially within the Anthropocene since the loss or untimely availability of larval food plants would undermine the long-term stability of the mutualisms. Furthermore, pollen and nectar availability data capture largely hymenopteran colony requirements and neglect the requirements of other important pollinator groups. Investigation of plant breeding systems, including clonality, and their incorporation into plant–pollinator networks would be valuable. Clonal plants exhibit either a guerilla (spatially dispersed) or a phalanx (aggregated) growth strategy. While the floral display of aggregated clonal plants is likely to be more attractive to pollinators than that of spatially dispersed plants, clonality *per se* and its interactions with pollinators have been scarcely investigated (Bittebiere et al. 2020).

The climatic tolerance limits of one mutualist may also restrict the evolution of other mutualistic partners. For example, in the aphid–*Buchnera* nutritional mutualism, *Buchnera* is an intracellular bacterial symbiont that synthesizes essential amino acids for its aphid host. Since the Cenozoic, the distribution of this highly successful mutualism has been restricted due to *Buchnera*'s low heat tolerance (Perkovsky and Wegierek 2016). Such constraints, which result in environmental mismatches between mutualists, are extremely relevant within the Anthropocene (Keeler et al. 2021). Mutualisms can also be disrupted by species with invasive life histories (Hays et al. 2022).

It is important to consider the mutualistic unit as a composite entity and to consider the combined fitness effects of joint phenotypes (Andersson 2000, O'Brien et al. 2021), a concept that is relevant for a holobiont perspective of those life forms that have been traditionally viewed as plants or animals but may not exist without symbiotic microbes (Borges 2017). Some fungi and algae have even entered into a mutualistic relationship to form composites that are viewed as separate entities, *i.e.*, lichens: a single lichen 'individual' is now considered to be composed of many fungi, algae and (cyano) bacteria that defy conventional delineation (Allen and Lendemer 2022, Spribille et al. 2022). Here too, multiple mutualism effects should be considered, but this approach is lacking. Lichens are believed to have long generation times (Ametrano et al. 2022), but little else is known about their life history. There is still much to be learned about lichenized fungi and their algae, especially since lichenisation has evolved independently at least ten times from ascomycete and basidiomycete fungi (Spribille et al. 2022).

It is valuable to consider the relative rates of evolution between mutualistic partners since this may determine the longevity of the interaction in evolutionary time. A faster rate of evolution in both partners may imply that mutualists can coevolve in tandem as may occur in a Red Queen-type of evolutionary scenario. In this scenario, both partners evolve at similar evolutionary rates so that they can stay in the same relative position with regard to each other. However, when partners are asymmetrical in their rates of evolution, it is possible that the partner with the slower rate of evolution will dictate the trajectory of the interaction in what has been termed the Red King effect (Bergstrom and Lachmann 2003). In this context, it is useful to realize that trees and shrubs that are long-lived have slower rates of molecular evolution compared to short-lived herbs (Smith and Donoghue 2008) and therefore the Red King effect is more likely in mutualistic interactions involving trees and shrubs compared to herbs. The impact of these broad-scale differences on mutualistic partnerships and their evolutionary trajectories needs more attention.

In summary, many aspects of life histories must align for mutualisms between individual plants, animals, microbes, fungi and composite entities to originate and be maintained over either ecological or evolutionary time. However, the perspective of life history has not always been included in empirical or theoretical considerations of mutualisms. This is an exciting and important area for research agendas in the future.

References

Aanen, D.K. (2006). As you reap, so shall you sow: coupling of harvesting and inoculating stabilizes the mutualism between termites and fungi. *Biology Letters* 2: 209–212.

Aanen, D.K., de Fine Licht, H.H., Debets, A.J., Kerstes, N.A., Hoekstra, R.F., and Boomsma, J.J. (2009). High symbiont relatedness stabilizes mutualistic cooperation in fungus-growing termites. *Science* 326: 1103–1106.

Afkhami, M.E., Friesen, M.L., and Stinchcombe, J.R. (2021). Multiple Mutualism Effects generate synergistic selection and strengthen fitness alignment in the interaction between legumes, rhizobia and mycorrhizal fungi. *Ecology Letters* 24: 1824–1834.

Allen, J.L., and Lendemer, J.C. (2022). A call to reconceptualize lichen symbioses. *Trends in Ecology & Evolution* 37: 582–589.

Ametrano, C.G., Lumbsch, H.T., Di Stefano, I., Sangvichien, E., Muggia, L., and Grewe, F. (2022). Should we hail the Red King? Evolutionary consequences of a mutualistic lifestyle in genomes of lichenized ascomycetes. *Ecology and Evolution* 12: e8471.

Andersson, J.O. (2000). Evolutionary genomics: is *Buchnera* a bacterium or an organelle? *Current Biology* 10: R866–R868.

Baird, A.H., Guest, J.R., and Willis, B.L. (2009). Systematic and biogeographical patterns in the reproductive biology of scleractinian corals. *Annual Review of Ecology, Evolution, and Systematics* 40: 551–571.

Baker, L.J., Freed, L.L., Easson, C.G., Lopez, J.V., Fenolio, D., Sutton, T.T., Nyholm, S.V., and Hendry, T.A. (2019). Diverse deep-sea anglerfishes share a genetically reduced luminous symbiont that is acquired from the environment. *eLife* 8: e47606.

Bascompte, J., and Scheffer, M. (2023). The resilience of plant–pollinator networks. *Annual Review of Entomology* 68: 363–380.

Behrman, E.L., Watson, S.S., O'Brien, K.R., Heschel, M.S., and Schmidt, P.S. (2015). Seasonal variation in life history traits in two *Drosophila* species. *Journal of Evolutionary Biology* 28: 1691–1704.

Bennett, A.E., and Groten, K. (2022). The costs and benefits of plant–arbuscular mycorrhizal fungal interactions. *Annual Review of Plant Biology* 73: 649–672.

Bergstrom, C.T., and Lachmann, M. (2003). The Red King effect: when the slowest runner wins the coevolutionary race. *Proceedings of the National Academy of Sciences of the United States of America* 100: 593–598.

Biedermann, P.H., and Vega, F.E. (2020). Ecology and evolution of insect-fungus mutualisms. *Annual Review of Entomology* 65: 431–455.

Bittebiere, A.K., Benot, M.L., and Mony, C. (2020). Clonality as a key but overlooked driver of biotic interactions in plants. *Perspectives in Plant Ecology, Evolution and Systematics*, 43: 125510.

Borges, R.M. (2011). Living long or dying young in plants and animals: Ecological patterns and evolutionary processes. In: *The field of biological aging: past, present and future* (ed. Olgun, A.), 61–82. Transworld Research Network.

Borges, R.M. (2017). Co-niche construction between hosts and symbionts: ideas and evidence. *Journal of Genetics* 96: 483–489.

Borges, R.M. (2021). Interactions between figs and gall-inducing fig wasps: adaptations, constraints, and unanswered questions. *Frontiers in Ecology and Evolution* 9: 685542.

Borges, R.M. (2022). Phoresy involving insects as riders or rides: life history, embarkation, and disembarkation. *Annals of the Entomological Society of America* 115: 219–231.

Borges, R.M., and Murthy T.G. (2023). Building castles on the ground: conversations between ecologists and engineers. *Journal of the Indian Institute of Science* 103: 1093–1104.

Brandl, S.J., Rasher, D.B., Côté, I.M., Casey, J.M., Darling, E.S., Lefcheck, J.S., and Duffy, J.E. (2019). Coral reef ecosystem functioning: eight core processes and the role of biodiversity. *Frontiers in Ecology and the Environment* 17: 445–454.

Bronstein, J.L. (2001). The costs of mutualism. *American Zoologist* 41: 825–839.

Bronstein, J.L. (2021). The gift that keeps on giving: why does biological diversity accumulate around mutualisms? In: *Plant-animal interactions: source of biodiversity* (eds. Del-Klaro, K., and Torezan-Silingardi, H.M.), 283–306. Springer.

Brouat, C., Garcia, N., Andary, C., and McKey, D. (2001). Plant lock and ant key: pairwise coevolution of an exclusion filter in an ant–plant mutualism. *Proceedings of the Royal Society B: Biological Sciences* 268: 2131–2141.

Brouat, C., and McKey, D. (2000). Origin of caulinary ant domatia and timing of their onset in plant ontogeny: evolution of a key trait in horizontally transmitted ant-plant symbioses. *Biological Journal of the Linnean Society* 71: 801–819.

Brown, J.H., Hall, C.A., and Sibly, R.M. (2018). Equal fitness paradigm explained by a trade-off between generation time and energy production rate. *Nature Ecology & Evolution* 2: 262–268.

Bshary, R., and Bronstein, J.L. (2004). Game structures in mutualistic interactions: what can the evidence tell us about the kind of models we need? *Advances in the Study of Behavior* 34: 59–102.

Chagnon, P.L., Bradley, R.L., Maherali, H., and Klironomos, J.N. (2013). A trait-based framework to understand life history of mycorrhizal fungi. *Trends in Plant Science* 18: 484–491.

Cheplick, G.P. (1994). Life history evolution in amphicarpic plants. *Plant Species Biology* 9: 119–131.

Cheplick, G.P. (2022). Philomatry in plants: why do so many species have limited seed dispersal? *American Journal of Botany* 109: 29–45.

Chomicki, G. (2022). Plant farming by ants: convergence and divergence in the evolution of agriculture. In: *The convergent evolution of agriculture in humans and insects* (eds. Schultz, T.R., Gawne, R., and Peregrine, P.N.), 161–174. Vienna Series in Theoretical Biology MIT Press.

Chomicki, G., Kiers, E.T., and Renner, S.S. (2020). The evolution of mutualistic dependence. *Annual Review of Ecology, Evolution, and Systematics* 51: 409–432.

Chomicki, G., and Renner, S.S. (2015). Phylogenetics and molecular clocks reveal the repeated evolution of ant-plants after the late Miocene in Africa and the early Miocene in Australasia and the Neotropics. *New Phytologist* 207: 411–424.

Chomicki, G., and Renner, S.S. (2019). Farming by ants remodels nutrient uptake in epiphytes. *New Phytologist* 223: 2011–2023.

Crook, M. (2014). The dauer hypothesis and the evolution of parasitism: 20 years on and still going strong. *International Journal for Parasitology* 44: 1–8.

Cruaud, A., Rønsted, N., Chantarasuwan, B., Chou, L.S., Clement, W.L., Couloux, A., Cousins, B., Genson, G., Harrison, R.D., Hanson, P.E. and Hossaert-Mckey, M., Jabbour-Zahab, R., Jousselin, E., Kerdelhué, C., Kjellberg, F., Lopez-Vaamonde, C.,

Peebles, J., Peng, Y-Q, Pereira, R.A.S., Schramm, T., Ubaidillah, R., van Noort, S., Weiblen, G.D., Yang, D-R., Yodpinyanee, A., Libeskind-Hadas, R., Cook, J.M., and Rasplus, J-Y. (2012). An extreme case of plant–insect codiversification: figs and fig-pollinating wasps. *Systematic Biology* 61: 1029–1047.

Darling, E.S., Alvarez-Filip, L., Oliver, T.A., McClanahan, T.R., and Côté, I.M. (2012). Evaluating life-history strategies of reef corals from species traits. *Ecology Letters* 15: 1378–1386.

Davidowitz, G., Bronstein, J.L., and Tigreros, N. (2022). Flight-fecundity trade-offs: a possible mechanistic link in plant–herbivore–pollinator systems. *Frontiers in Plant Science* 13: 979.

De Deyn, G.B. (2017). Plant life history and above–belowground interactions: missing links. *Oikos* 126: 497–507.

De Medeiros, B.A., Núñez-Avellaneda, L.A., Hernandez, A.M., and Farrell, B.D. (2019). Flower visitors of the licuri palm (*Syagrus coronata*): brood pollinators coexist with a diverse community of antagonists and mutualists. *Biological Journal of the Linnean Society* 126: 666–687.

Debout, G.D., Dalecky, A., Ngomi, A.N., and McKey, D.B. (2009). Dynamics of species coexistence: maintenance of a plant-ant competitive metacommunity. *Oikos* 118: 873–884.

Dedeine, F., Vavre, F., Fleury, F., Loppin, B., Hochberg, M.E., and Boulétreau, M. (2001). Removing symbiotic *Wolbachia* bacteria specifically inhibits oogenesis in a parasitic wasp. *Proceedings of the National Academy of Sciences of the United States of America* 98: 6247–6252.

Denison, R.F., and Kiers, E.T. (2004). Lifestyle alternatives for rhizobia: mutualism, parasitism, and forgoing symbiosis. *FEMS Microbiology Letters* 237: 187–193.

Denison, R.F., and Kiers, E.T. (2011). Life histories of symbiotic rhizobia and mycorrhizal fungi. *Current Biology* 21: R775–R785.

Dev, S.A., Shenoy, M., and Borges, R.M. (2010). Genetic and clonal diversity of the endemic ant-plant *Humboldtia brunonis* (Fabaceae) in the Western Ghats of India. *Journal of Biosciences* 35: 267–279.

Diagne, N., Ngom, M., Djighaly, P.I., Fall, D., Hocher, V., and Svistoonoff, S. (2020). Roles of arbuscular mycorrhizal fungi on plant growth and performance: importance in biotic and abiotic stressed regulation. *Diversity* 12: 370.

Duchenne, F., Fontaine, C., Teulière, E., and Thébault, E. (2021). Phenological traits foster persistence of mutualistic networks by promoting facilitation. *Ecology Letters* 24: 2088–2099.

E-Vojtkó, A., de Bello, F., Durka, W., Kuehn, I., and Goetzenberger, L. (2020). The neglected importance of floral traits in trait-based plant community assembly. *Journal of Vegetation Science* 31: 529–539.

Federle, W., Maschwitz, U., and Fiala, B. (1998). The two-partner ant-plant system of *Camponotus* (*Colobopsis*) sp. 1 and *Macaranga puncticulata* (Euphorbiaceae): natural history of the exceptional ant partner. *Insectes Sociaux* 45: 1–16.

Finch, J.T., Power, S.A., Welbergen, J.A., and Cook, J.M. (2021). Staying in touch: how highly specialised moth pollinators track host plant phenology in unpredictable climates. *BMC Ecology and Evolution* 21: 161.

Fonseca, C.R. (1993). Nesting space limits colony size of the plant-ant *Pseudomyrmex concolor*. *Oikos* 67: 473–482.

Fonseca-Romero, M.A., Fornoni, J., Del-Val, E., and Boege, K. (2019). Ontogenetic trajectories of direct and indirect defenses of myrmecophytic plants colonized either by mutualistic or opportunistic ant species. *Oecologia* 190: 857–865.

Gaume, L., Shenoy, M., Zacharias, M., and Borges, R.M. (2006). Co-existence of ants and an arboreal earthworm in a myrmecophyte of the Indian Western Ghats: anti-predation effect of the earthworm mucus. *Journal of Tropical Ecology* 22: 341–344.

Gemma, J.N., and Koske, R.E. (1988). Seasonal variation in spore abundance and dormancy of *Gigaspora gigantea* and in mycorrhizal inoculum potential of a dune soil. *Mycologia* 80: 211–216.

Ghara, M., and Borges, R.M. (2010). Comparative life-history traits in a fig wasp community: implications for community structure. *Ecological Entomology* 35: 139–148.

Ghazoul, J. (2005). Pollen and seed dispersal among dispersed plants. *Biological Reviews* 80: 413–443.

Gibb, H., Bishop, T.R., Leahy, L., Parr, C.L., Lessard, J.P., Sanders, N.J., Shik, J.Z., Ibarra-Isassi, J., Narendra, A., Dunn, R.R., and Wright, I.J. (2023). Ecological strategies of (pl)ants: towards a world-wide worker economic spectrum for ants. *Functional Ecology* 37: 13–25.

Gibert, A., Tozer, W., and Westoby, M. (2019). Plant performance response to eight different types of symbiosis. *New Phytologist* 222: 526–542.

Godschalx, A.L., Schädler, M., Trisel, J.A., Balkan, M.A., and Ballhorn, D.J. (2015). Ants are less attracted to the extrafloral nectar of plants with symbiotic, nitrogen-fixing rhizobia. *Ecology* 96: 348–354.

Goodbody-Gringley, G., Wong, K.H., Becker, D.M., Glennon, K., and de Putron, S.J. (2018). Reproductive ecology and early life history traits of the brooding coral, *Porites astreoides*, from shallow to mesophotic zones. *Coral Reefs* 37: 483–494.

Gorman, L.M., Konciute, M.K., Cui, G., Oakley, C.A., Grossman, A.R., Weis, V.M., Aranda, M., and Davy, S.K. (2022). Symbiosis with dinoflagellates alters cnidarian cell-cycle gene expression. *Cellular Microbiology* 2022: 3330160.

Greeff, J.M., and Kjellberg, F. (2022). Pollinating fig wasps' simple solutions to complex sex ratio problems: a review. *Frontiers in Zoology* 19: 3.

Grime, J.P. (1977). Evidence for the existence of three primary strategies in plants and its relevance to ecological and evolutionary theory. *The American Naturalist* 111: 1169–1194.

Guimarães Jr, P.R. (2020). The structure of ecological networks across levels of organization. *Annual Review of Ecology, Evolution, and Systematics* 51: 433–460.

Guzman, L.M., Chamberlain, S.A., and Elle, E. (2021). Network robustness and structure depend on the phenological characteristics of plants and pollinators. *Ecology and Evolution* 11: 13321–13334.

Haran, J., Procheş, Ş., Benoit, L., and Kergoat, G.J. (2022). From monocots to dicots: host shifts in Afrotropical derelomine weevils shed light on the evolution of non-obligatory brood pollination mutualism. *Biological Journal of the Linnean Society* 137: 15–29.

Hata, H., and Kato, M. (2006). A novel obligate cultivation mutualism between damselfish and *Polysiphonia* algae. *Biology Letters* 2: 593–596.

Hays, B.R., Riginos, C., Palmer, T.M., Doak, D.F., Gituku, B.C., Maiyo, N.J., Mutisya, S., Musila, S., and Goheen, J.R. (2022). Demographic consequences of mutualism disruption: browsing and big-headed ant invasion drive acacia population declines. *Ecology* 103: e3655.

Hedderson, T.A., and Longton, R.E. (2008). Local adaptation in moss life histories: population-level variation and a reciprocal transplant experiment. *Journal of Bryology* 30: 1–11.

Heil, M. (2013). Let the best one stay: screening of ant defenders by *Acacia* host plants functions independently of partner choice or host sanctions. *Journal of Ecology* 101: 684–688.

Hembry, D.H., Raimundo, R.L., Newman, E.A., Atkinson, L., Guo, C., Guimaraes Jr, P.R., and Gillespie, R.G. (2018). Does biological intimacy shape ecological network structure? A test using a brood pollination mutualism on continental and oceanic islands. *Journal of Animal Ecology* 87: 1160–1171.

Hernández-Zepeda, O.F., Razo-Belman, R., and Heil, M. (2018). Reduced responsiveness to volatile signals creates a modular reward provisioning in an obligate food-for-protection mutualism. *Frontiers in Plant Science* 9: 1076.

Hidayati, S.N., and Walck, J.L. (2016). A review of the biology of *Rafflesia*: what do we know and what's next? *Buletin Kebun Raya* 19: 67–78.

Hojo, M.K., Pierce, N.E., and Tsuji, K. (2015). Lycaenid caterpillar secretions manipulate attendant ant behavior. *Current Biology* 25: 2260–2264.

Johnson, C.A., Smith, G.P., Yule, K., Davidowitz, G., Bronstein, J.L., and Ferrière, R. (2021). Coevolutionary transitions from antagonism to mutualism explained by the co-opted antagonist hypothesis. *Nature Communications* 12: 2867.

Katariya, L., Ramesh, P.B., Gopalappa, T., and Borges, R.M. (2017). Sex and diversity: the mutualistic and parasitic fungi of a fungus-growing termite differ in genetic diversity and reproductive strategy. *Fungal Ecology* 26: 20–27.

Kautz, S., Ballhorn, D.J., Kroiss, J., Pauls, S.U., Moreau, C.S., Eilmus, S., Strohm, E., and Heil, M. (2012). Host plant use by competing acacia-ants: mutualists monopolize while parasites share hosts. *PLoS ONE* 7(5): e37691.

Kawakita, A., and Kato, M. (2017). Evolution and diversity of obligate pollination mutualisms. In: *Obligate pollination mutualism* (eds. Kato, M., and Kawakita, A.), 249–270. Springer.

Keeler, A.M., Rose-Person, A., and Rafferty, N.E. (2021). From the ground up: building predictions for how climate change will affect belowground mutualisms, floral traits, and bee behavior. *Climate Change Ecology* 1: 100013.

Khadempour, L., Burnum-Johnson, K.E., Baker, E.S., Nicora, C.D., Webb-Robertson, B.J.M., White III, R.A., Monroe, M.E., Huang, E.L., Smith, R.D., and Currie, C.R. (2016). The fungal cultivar of leaf-cutter ants produces specific enzymes in response to different plant substrates. *Molecular Ecology* 25: 5795–5805.

Koch, E.B., Camarota, F., and Vasconcelos, H.L. (2016). Plant ontogeny as a conditionality factor in the protective effect of ants on a Neotropical tree. *Biotropica* 48: 198–205.

Kooij, P.W., Aanen, D.K., Schiøtt, M., and Boomsma, J.J. (2015). Evolutionarily advanced ant farmers rear polyploid fungal crops. *Journal of Evolutionary Biology* 28: 1911–1924.

Kokolo, B., Attéké Nkoulémbéné, C., Ibrahim, B., M'Batchi, B., and Blatrix, R. (2020). Phenotypic plasticity in size of ant-domatia. *Scientific Reports* 10: 20948.

Korb, J., and Aanen, D.K. (2003). The evolution of uniparental transmission of fungal symbionts in fungus-growing termites (Macrotermitinae). *Behavioral Ecology and Sociobiology* 53: 65–71.

Krishnan, A., Ghara, M., Kasinathan, S., Pramanik, G.K., Revadi, S., and Borges, R.M. (2015). Plant reproductive traits mediate tritrophic feedback effects within an obligate brood-site pollination mutualism. *Oecologia* 179: 797–809.

Laird, R.A., and Addicott, J.F. (2007). Arbuscular mycorrhizal fungi reduce the construction of extrafloral nectaries in *Vicia faba*. *Oecologia* 152: 541–551.

Lázaro, A., Gómez-Martínez, C., Alomar, D., González-Estévez, M.A., and Traveset, A. (2020). Linking species-level network metrics to flower traits and plant fitness. *Journal of Ecology* 108: 1287–1298.

Leal, L.C., Nogueira, A., and Peixoto, P.E. (2023). Which traits optimize plant benefits? Meta-analysis on the effect of partner traits on the outcome of an ant–plant protective mutualism. *Journal of Ecology* 111: 263–275.

Ledón-Rettig, C.C., Moczek, A.P., and Ragsdale, E.J. (2018). *Diplogastrellus* nematodes are sexually transmitted mutualists that alter the bacterial and fungal communities of their beetle host. *Proceedings of the National Academy of Sciences of the United States of America* 115: 10696–10701.

Leichty, A.R., and Poethig, R.S. (2019). Development and evolution of age-dependent defenses in ant-acacias. *Proceedings of the National Academy of Sciences of the United States of America* 116: 15596–15601.

Lemaire, B., van Cauwenberghe, J., Verstraete, B., Chimphango, S., Stirton, C., Honnay, O., Smets, E., Sprent, J., James, E.K., and Muasya, A.M. (2016). Characterization of the papilionoid–*Burkholderia* interaction in the Fynbos biome: the diversity and distribution of beta-rhizobia nodulating *Podalyria calyptrata* (Fabaceae, Podalyrieae). *Systematic and Applied Microbiology* 39: 41–48.

Magalhães, D.M., Lourenção, A.L., and Bento, J.M.S. (2023). Beneath the blooms: unearthing the effect of rhizospheric bacteria on floral signals and pollinator preferences. *Plant, Cell & Environment* 2023: 1–17.

Marting, P.R., Kallman, N.M., Wcislo, W.T., and Pratt, S.C. (2018). Ant-plant sociometry in the *Azteca-Cecropia* mutualism. *Scientific Reports* 8: 17968.

Mehdiabadi, N.J., Mueller, U.G., Brady, S.G., Himler, A.G., and Schultz, T.R. (2012). Symbiont fidelity and the origin of species in fungus-growing ants. *Nature Communications* 3: 840.

Montoya, A.P., Wendlandt, C.E., Benedict, A.B., Roberts, M., Piovia-Scott, J., Griffitts, J.S., and Porter, S.S. (2023). Hosts winnow symbionts with multiple layers of absolute and conditional discrimination mechanisms. *Proceedings of the Royal Society B: Biological Sciences* 290: 20222153.

Moran, N.P., Caspers, B.A., Chakarov, N., Ernst, U.R., Fricke, C., Kurtz, J., Lilie, N.D., Lo, L.K., Müller, C., Takola, E., and Trimmer, P.C. (2022). Shifts between cooperation and antagonism driven by individual variation: a systematic synthesis review. *Oikos* 2022: e08201.

Morán-López, T., Benadi, G., Lara-Romero, C., Chacoff, N., Vitali, A., Pescador, D., Lomascolo, S.B., Morente-López, J., Vazquez, D.P., and Morales, J.M. (2022). Flexible diets enable pollinators to cope with changes in plant community composition. *Journal of Ecology* 110: 1913–1927.

Morrison, C., and Hero, J.M. (2003). Geographic variation in life-history characteristics of amphibians: a review. *Journal of Animal Ecology* 72: 270–279.

Mueller, U.G. (2002). Ant versus fungus versus mutualism: ant-cultivar conflict and the deconstruction of the attine ant-fungus symbiosis. *The American Naturalist* 160: S67–S98.

Muller, K.E., and Denison, R.F. (2018). Resource acquisition and allocation traits in symbiotic rhizobia with implications for life-history outside of legume hosts. *Royal Society Open Science* 5: 181124.

Nakazawa, T. (2020). A perspective on stage-structured mutualism and its community consequences. *Oikos* 129: 297–310.

Nelsen, M.P., Ree, R.H., and Moreau, C.S. (2018). Ant–plant interactions evolved through increasing interdependence. *Proceedings of the National Academy of Sciences of the United States of America* 115: 12253–12258.

Nobre, T., Fernandes, C., Boomsma, J.J., Korb, J., and Aanen, D.K. (2011). Farming termites determine the genetic population structure of *Termitomyces* fungal symbionts. *Molecular Ecology* 20: 2023–2033.

Noë, R., and Kiers, E.T. (2018). Mycorrhizal markets, firms, and co-ops. *Trends in Ecology & Evolution* 33: 777–789.

Noë, R., van Schaik, C.P., and van Hooff, J.A.R.A.M. (1991). The market effect: an explanation for pay-off asymmetries among collaborating animals. *Ethology* 87: 97–118.

Nyholm, S.V., and McFall-Ngai, M.J. (2021). A lasting symbiosis: how the Hawaiian bobtail squid finds and keeps its bioluminescent bacterial partner. *Nature Reviews Microbiology* 19: 666–679.

O'Brien, A.M., Jack, C.N., Friesen, M.L., and Frederickson, M.E. (2021). Whose trait is it anyways? Coevolution of joint phenotypes and genetic architecture in mutualisms. *Proceedings of the Royal Society B: Biological Sciences* 288: 20202483.

Oliveira, K.N., Coley, P.D., Kursar, T.A., Kaminski, L.A., Moreira, M.Z., and Campos, R.I. (2015). The effect of symbiotic ant colonies on plant growth: a test using an *Azteca-Cecropia* system. *PLoS ONE* 10(3): e0120351.

Paz, C., Öpik, M., Bulascoschi, L., Bueno, C.G., and Galetti, M. (2021). Dispersal of arbuscular mycorrhizal fungi: evidence and insights for ecological studies. *Microbial Ecology* 81: 283–292.

Perkovsky, E., and Wegierek, P. (2016). Aphid-*Buchnera*-ant symbiosis; or why are aphids rare in the tropics and very rare further south? *Earth and Environmental Science Transactions of the Royal Society of Edinburgh* 107: 297–310.

Pierce, N.E., and Dankowicz, E. (2022). Behavioral, ecological and evolutionary mechanisms underlying caterpillar-ant symbioses. *Current Opinion in Insect Science* 52: 100898.

Pierce, S., Negreiros, D., Cerabolini, B.E., Kattge, J., Díaz, S., Kleyer, M., Shipley, B., Wright, S.J., Soudzilovskaia, N.A., Onipchenko, V.G., van Bodegom, P.M., Frenette-Dussault, C., Weiher, E., Pinho, B.X., Cornelissen, J.H.C., Grime, G.P., Thompson, K., Hunt, R., Wilson, P.J., Buffa, G., Nyakunga, O.C., Reich, P.B., Caccianiga, M., Mangili, F., Ceriani, R.M., Luzzaro, A., Brusa, G., Siefert, A., Barbosa, N.P.U., Chapin III, F.S., Cornwell, W.K., Fang, J., Fernandes, G.W., Garnier, E., Le Stradic, S., Peñuelas, J., Melo, F.P.L., Slaviero, A., Tabarelli, M., and Tampucci, D. (2017). A global method for calculating plant CSR ecological strategies applied across biomes world-wide. *Functional Ecology* 31: 444–457.

Pion, M., Spangenberg, J.E., Simon, A., Bindschedler, S., Flury, C., Chatelain, A., Bshary, R., Job, D., and Junier, P. (2013). Bacterial farming by the fungus *Morchella crassipes*. *Proceedings of the Royal Society B: Biological Sciences* 280: 20132242.

Powell, J.A. (2001). Longest insect dormancy: yucca moth larvae (Lepidoptera: Prodoxidae) metamorphose after 20, 25, and 30 years in diapause. *Annals of the Entomological Society of America* 94: 677–680.

Primack, R.B. (1987). Relationships among flowers, fruits, and seeds. *Annual Review of Ecology and Systematics* 18: 409–430.

Primieri, S., Magnoli, S.M., Koffel, T., Stürmer, S.L., and Bever, J.D. (2022). Perennial, but not annual legumes synergistically benefit from infection with arbuscular mycorrhizal fungi and rhizobia: a meta-analysis. *New Phytologist* 233: 505–514.

Quintero, C., Barton, K.E., and Boege, K. (2013). The ontogeny of plant indirect defenses. *Perspectives in Plant Ecology, Evolution and Systematics* 15: 245–254.

Reis, F., Kirsch, R., Pauchet, Y., Bauer, E., Bilz, L.C., Fukumori, K., Fukatsu, T., Kölsch, G., and Kaltenpoth, M. (2020). Bacterial symbionts support larval sap feeding and adult folivory in (semi-) aquatic reed beetles. *Nature Communications* 11: 2964.

Reis, A.S., Sá-Neto, R.J., do Nascimento, I.C., Carneiro, M.A., Gaglioti, A.L., and Carvalho, K.S. (2022). Habitat as a conditionality factor of ant-plant mutualistic interaction in the *Cecropia-Azteca* system. *Arthropod-Plant Interactions* 16: 275–284.

Riveros, A.J., Seid, M.A., and Wcislo, W.T. (2012). Evolution of brain size in class-based societies of fungus-growing ants (Attini). *Animal Behaviour* 83: 1043–1049.

Roff, D.A. (2002). *Life history evolution*. Sinauer Associates.

Sachs, J.L., Mueller, U.G., Wilcox, T.P., and Bull, J.J. (2004). The evolution of cooperation. *The Quarterly Review of Biology* 79: 135–160.

Sachs, J.L., and Simms, E.L. (2006). Pathways to mutualism breakdown. *Trends in Ecology & Evolution* 21: 585–592.

Sadeh, A., Guterman, H., Gersani, M., and Ovadia, O. (2009). Plastic bet-hedging in an amphicarpic annual: an integrated strategy under variable conditions. *Evolutionary Ecology* 23: 373–388.

Sakai, S. (2002). A review of brood-site pollination mutualism: plants providing breeding sites for their pollinators. *Journal of Plant Research* 115: 161–168.

Salguero-Gómez, R., Jones, O.R., Jongejans, E., Blomberg, S.P., Hodgson, D.J., Mbeau-Ache, C., Zuidema, P.A., De Kroon, H., and Buckley, Y.M. (2016). Fast-slow continuum and reproductive strategies structure plant life-history variation worldwide. *Proceedings of the National Academy of Sciences of the United States of America* 113: 230–235.

Sánchez-Martín, R., Gómez, J.M., Cheptou, P.O., and Rubio de Casas, R. (2021). Differences in seed dormancy and germination in amphicarpic legumes: manifold bet-hedging in space and time. *Journal of Plant Ecology* 14: 662–672.

Satler, J.D., Herre, E.A., Jandér, K.C., Eaton, D.A., Machado, C.A., Heath, T.A., and Nason, J.D. (2019). Inferring processes of coevolutionary diversification in a community of Panamanian strangler figs and associated pollinating wasps. *Evolution* 73: 2295–2311.

Sazatornil, F.D., More, M., Benitez-Vieyra, S., Cocucci, A.A., Kitching, I.J., Schlumpberger, B.O., Oliveira, P.E., Sazima, M., and Amorim, F.W. (2016). Beyond neutral and forbidden links: morphological matches and the assembly of mutualistic hawkmoth-plant networks. *Journal of Animal Ecology* 85: 1586–1594.

Schultz, T.R. (2022). The convergent evolution of agriculture in humans and fungus-farming ants. In: *The convergent evolution of agriculture in humans and insects* (eds. Schultz, T.R., Gawne, R., and Peregrine, P.N.), 281–313. Vienna Series in Theoretical Biology, MIT Press.

Schwarz, J., Weis, V., and Potts, D. (2002). Feeding behavior and acquisition of zooxanthellae by planula larvae of the sea anemone *Anthopleura elegantissima*. *Marine Biology* 140: 471–478.

Simmons, B.I., Beckerman, A.P., Hansen, K., Maruyama, P.K., Televantos, C., Vizentin-Bugoni, J., and Dalsgaard, B. (2021). Niche and neutral processes leave distinct structural imprints on indirect interactions in mutualistic networks. *Functional Ecology* 35: 753–763.

Smith, S.A., and Donoghue, M.J. (2008). Rates of molecular evolution are linked to life history in flowering plants. *Science* 322: 86–89.

Smith, G.P., Gardner, J., Gibbs, J., Griswold, T., Hauser, M., Yanega, D., and Ponisio, L.C. (2021). Sex-associated differences in the network roles of pollinators. *Ecosphere* 12: e03863.

Sonne, J., Vizentin-Bugoni, J., Maruyama, P.K., Araujo, A.C., Chávez-González, E., Coelho, A.G., Cotton, P.A., Marín-Gómez, O.H., Lara, C., Lasprilla, L.R. and Machado, C.G., Maglianesi, M.A., Malucelli, T.S., González, A.M.M., Oliveira, G.M., Oliveira, P.E., Ortiz-Pulido, R., Rocca, M.A., Rodrigues, L.C., Sazima, I., Simmons, B.I., Tinoco, B., Varassin, I.G., Vasconcelos, M.F., O'Hara, B., Schleuning, M., Rahbek, C., Sazima, M., and Dalsgaard, B. (2020). Ecological mechanisms explaining interactions within plant-hummingbird networks: morphological matching increases towards lower latitudes. *Proceedings of the Royal Society B: Biological Sciences* 287: 20192873.

Spribille, T., Resl, P., Stanton, D.E., and Tagirdzhanova, G. (2022). Evolutionary biology of lichen symbioses. *New Phytologist* 234: 1566–1582.

Stearns, S.C. (1976). Life-history tactics: a review of the ideas. *The Quarterly Review of Biology* 51: 3–47.

Stearns, S.C. (1980). A new view of life-history evolution. *Oikos* 35: 266–281.

Stephens, P.A., Boyd, I.L., McNamara, J.M., and Houston, A.I. (2009). Capital breeding and income breeding: their meaning, measurement, and worth. *Ecology* 90: 2057–2067.

Sudakaran, S., Kost, C., and Kaltenpoth, M. (2017). Symbiont acquisition and replacement as a source of ecological innovation. *Trends in Microbiology* 25: 375–390.

Trager, M.D., Bhotika, S., Hostetler, J.A., Andrade, G.V., Rodriguez-Cabal, M.A., McKeon, C.S., Osenberg, C.W., and Bolker, B.M. (2010). Benefits for plants in ant-plant protective mutualisms: a meta-analysis. *PLoS ONE* 5(12): e14308.

Trager, M.D., and Bruna, E.M. (2006). Effects of plant age, experimental nutrient addition and ant occupancy on herbivory in a neotropical myrmecophyte. *Journal of Ecology* 94: 1156–1163.

Vázquez, D.P., Blüthgen, N., Cagnolo, L., and Chacoff, N.P. (2009). Uniting pattern and process in plant–animal mutualistic networks: a review. *Annals of Botany* 103: 1445–1457.

Venkateswaran, V., Kumble, A.L., and Borges, R.M. (2018). Resource dispersion influences dispersal evolution of highly insulated insect communities. *Biology Letters* 14: 20180111.

Venkateswaran, V., Shrivastava, A., Kumble, A.L, and Borges, R.M. (2017). Life-history strategy, resource dispersion and phylogenetic associations shape dispersal of a fig wasp community. *Movement Ecology* 5: 25.

Visick, K.L., Stabb, E.V., and Ruby, E.G. (2021). A lasting symbiosis: how *Vibrio fischeri* finds a squid partner and persists within its natural host. *Nature Reviews Microbiology* 19: 654–665.

Vizentin-Bugoni, J., Maruyama, P.K., de Souza, C.S., Ollerton, J., Rech, A.R., and Sazima, M. (2018). Plant-pollinator networks in the tropics: a review. In: *Ecological networks in the tropics: an integrative overview of species interactions from some of the most species-rich habitats on earth* (eds. Dátillo, W., and Rico-Gray, V.), 73–91. Springer.

Werner, G.D., Cornelissen, J.H., Cornwell, W.K., Soudzilovskaia, N.A., Kattge, J., West, S.A., and Kiers, E.T. (2018). Symbiont switching and alternative resource acquisition strategies drive mutualism breakdown. *Proceedings of the National Academy of Sciences of the United States of America* 115: 5229–5234.

Werner, G.D., and Kiers, E.T. (2015). Partner selection in the mycorrhizal mutualism. *New Phytologist* 205: 1437–1442.

Wisselink, M., Aanen, D.K., and van't Padje, A. (2020). The longevity of colonies of fungus-growing termites and the stability of the symbiosis. *Insects* 11: 527.

Wolfe, B.E., Husband, B.C., and Klironomos, J.N. (2005). Effects of a belowground mutualism on an aboveground mutualism. *Ecology Letters* 8: 218–223.

Wootton, J.T. (1994). The nature and consequences of indirect effects in ecological communities. *Annual Review of Ecology and Systematics* 25: 443–466.

Young, J.P.W. (1996). Phylogeny and taxonomy of rhizobia. *Plant and Soil* 186: 45–52.

Yu, D.W., Wilson, H.B., and Pierce, N.E. (2001). An empirical model of species coexistence in a spatially structured environment. *Ecology* 82: 1761–1771.

Zachariah, N., Das, A., Murthy, T.G., and Borges, R.M. (2017). Building mud castles: a perspective from brick-laying termites. *Scientific Reports* 7: 4692.

Zeng, Y., and Wiens, J.J. (2021). Do mutualistic interactions last longer than antagonistic interactions? *Proceedings of the Royal Society B: Biological Sciences* 288: 20211457.

Zera, A.J., and Harshman, L.G. (2001). The physiology of life history trade-offs in animals. *Annual Review of Ecology and Systematics* 32: 95–126.

Zhang, K., Baskin, J.M., Baskin, C.C., Cheplick, G.P., Yang, X., and Huang, Z. (2020). Amphicarpic plants: definition, ecology, geographic distribution, systematics, life history, evolution and use in agriculture. *Biological Reviews* 95: 1442–1466.

Zhang, Z., Yan, C., and Zhang, H. (2021). Mutualism between antagonists: its ecological and evolutionary implications. *Integrative Zoology* 16: 84–96.

Zhao, Y.H., Lázaro, A., Li, H.D., Tao, Z.B., Liang, H., Zhou, W., Ren, Z.X., Xu, K., Li, D.Z., and Wang, H. (2022). Morphological trait-matching in plant–Hymenoptera and plant–Diptera mutualisms across an elevational gradient. *Journal of Animal Ecology* 91: 196–209.

Part III

Applications

18

Life History and Climate Change

Juha Merilä[1,2] *and Lei Lv*[3,4]

[1] *Ecological Genetics Research Unit, Organismal and Evolutionary Biology Research Programme, Faculty Biological & Environmental Sciences, University of Helsinki, Helsinki, Finland*
[2] *Area of Ecology and Biodiversity, School of Biological Sciences, The University of Hong Kong, Hong Kong, Hong Kong SAR*
[3] *School of Environmental Science and Engineering, Southern University of Science and Technology, Shenzhen, China*
[4] *Division of Ecology and Evolution, Research School of Biology, Australian National University, Canberra, ACT, Australia*

18.1 Introduction

Life history theory seeks to explain the enormous intra- and inter-specific variation in traits that define organismal life histories. The theory is founded on two main premises. First, natural selection tends to maximize organismal fitness and the diversity of life histories we see are adaptations to current environment. Second, trade-offs arising from genetic, developmental, physiological and energetic constraints set the boundaries as to what kind of life histories are possible and optimal, given the prevailing environmental conditions (Stearns 1992, Roff 2001). Life histories can be characterized by the combination of life history (or demographic) traits such as age at maturity, lifespan, body size, number and size of offspring, investment in self-maintenance and reproduction of individuals. They interact with the environment or environmental change and consequently influence individual fitness, thereby also contributing to population dynamics and extinction risk.

The diverse impacts of anthropogenic climate change (ACC) on organismal life history traits have been documented in an increasing number of studies from terrestrial, freshwater and marine habitats (*e.g.*, Parmesan 2006, Cheung et al. 2009, Lavergne et al. 2010, Munday et al. 2013, Peñuelas et al. 2013, Scheffers et al. 2016, Sheldon 2019). By changing ecosystems' abiotic conditions both spatially and temporally, ACC can have both direct (*e.g.*, accelerated growth through thermal benefits; Bestion et al. 2015) and indirect (*e.g.*, limited growth due to reduced food availability; Ozgul et al. 2009) effects on organismal life histories. These changes in abiotic conditions can in turn lead to changes in biotic conditions as well as in biotic interactions. For instance, changes in temperature regime or rainfall patterns can lead to reduced (or increased) primary production and food availability with cascading effects to higher trophic levels (*e.g.*, Blanchard et al. 2012). Likewise, increased frequency of extreme climatic events brought along by ACC (IPCC 2021) has the potential to have strong impacts on life histories (*e.g.*, Moreno and Møller 2011, Thornton et al. 2014). Consequently, ACC can influence organismal ecology and life histories in diverse and complex ways.

Any discussion about the effects of ACC on life history traits should make an explicit distinction between genetic *vs.* plastic changes in trait means (Gienapp et al. 2008, Merilä and Hendry 2014). Namely, observing a temporal shift in the mean value of a life history trait that is correlated with some proxy of ACC could be underlaid by at least three distinct phenomena. First, it could represent an evolutionary change driven by natural selection that has acted on genetic variation in the focal trait. Second, it could be a phenotypically (adaptive) plastic response to changed environmental conditions grounded on pre-existing, and possibly genetically based, variation. Third, a shift in the mean value of a life history trait can also be simply a non-genetic (and non-adaptive or neutral) plastic change induced directly by changes in environmental conditions (Merilä 2012, Merilä and Hendry 2014). The distinction between the two latter possibilities is subtle but important: whereas the former kind of shifts could be adaptive (having a positive effect on fitness), the latter kind of plastic change can be either neutral or even non-adaptive (having a negative effect on fitness). As discussed and outlined in the review by Merilä and Hendry (2014), differentiation between the three alternative explanations for climate-driven changes in life history traits requires understanding of their genetic basis.

Life History Evolution: Traits, Interactions, and Applications, First Edition. Edited by Michal Segoli and Eric Wajnberg.
© 2025 John Wiley & Sons Ltd. Published 2025 by John Wiley & Sons Ltd.

In this chapter, we provide a brief synopsis of what is known about the effects of ACC on different life history traits of animals and outline some possible avenues for future research. Our review is not exhaustive – there have been many reviews of effects of ACC on various traits in the wild (*e.g.*, Holt 1990, Parmesan 2006, Gienapp et al. 2008, Hoffmann and Sgrò 2011, Merilä 2012, Peñuelas et al. 2013, Merilä and Hendry 2014, Piao et al. 2019, Inouye 2022), albeit few of them have focussed explicitly on life history traits (but see Isaac 2009, Lancaster et al. 2017, Iler et al. 2021, Wells et al. 2022). Our focus is on classical life history traits (*i.e.*, timing of key life history events [*e.g.*, phenology], body size, reproductive output and success, actuarial senescence) as well as in trade-offs, but we also cover what is known about effects of ACC on population dynamics and extinction risk. Finally, we discuss the relative roles of environmentally induced plastic changes *vs.* evolution in response to ACC, and what is known about populations' ability to track ACC through phenotypic plasticity and evolutionary adaptations.

18.2 Effects of ACC on Life History Strategies and Trade-Offs

ACC can be expected to lead to an increasing number of species adopting 'faster' (*i.e.*, r-selected) life history strategy (Parmesan 2007, Lancaster et al. 2017). This expectation is grounded on two observations. First, ACC has commonly resulted in earlier and shorter (sometimes also longer) growing or breeding seasons potentially favouring and selecting for advanced phenology and faster growth, maturation and reproduction (Møller et al. 2010, Lancaster et al. 2017). Second, many life history traits such as developmental and growth rates are thermal-dependent, and warming climate will automatically accelerate these processes in ectotherms. For instance, it is well established that acceleration of growth and developmental rates in ectotherms results in earlier maturation at smaller sizes, a phenomenon known as the temperature–size rule (Atkinson 1994, Angiletta and Dunham 2003).

These kinds of ACC-driven changes in organismal life history strategies and mean life history trait values can lead to trophic mismatches (asynchrony) between species (Both and Visser 2005, Menéndez 2007, Renner and Zoehner 2018) as well as changes in species competitive interactions (Lancaster et al. 2017). As for the latter, species' convergence towards faster life history strategies due to ACC can be expected to increase opportunity and intensity of competitive interactions (Lancaster et al. 2017). In the same vein, ACC-driven phenological and range shifts can cause species to experience unpredictable novel biotic interactions. For instance, warming seawater increased relative dominance of a warm-water fish species to cause a cool-water species to relocate in a less preferred habitat within the same thermal environment (Milazzo et al. 2013). At the moment, there is no general theory to predict the outcome of these novel interactions albeit general ecological theory predicts that at least one of the competing species should face a risk of competitive exclusion (Lancaster et al. 2017). Finally, it might be instructive to note here that the way a species can be expected to respond and cope with ACC is likely to depend strongly on its basic life history strategy. Namely, large, low fecund species with long generation times (K-selected species) may be more resilient to climatic perturbations in the short term, but less likely to persist and adapt to environmental changes brought by ACC than small-sized, high fecund species (r-selected species) with short generation times in the long run (*e.g.*, Isaac 2009). This is under the reasonable assumption that r-selected species can adapt to changes in environmental conditions faster than K-selected species. While the paradigms of r- and k-selection are over-simplifications and subject to criticism (*e.g.*, Reznick et al. 2002), they still serve in this context as useful heuristics to illustrate how ACC is likely to impact organisms with different life history strategies.

K- and r-selected species can also be viewed to occupy the extreme ends of the negative self-maintenance (survival)-reproduction trade-off relationship. This kind of trade-off between life history traits is ubiquitous (Roff 2001) and it is conceivable that ACC can influence trade-off allocations and their strength. By causing a reduction (resp. increase) in some critical resources to an organism, ACC can strengthen (resp. relax) trade-offs between different life history traits. For instance, it is well known that when resources are plentiful, the correlation between reproduction and survival can be positive or absent (no apparent trade-off) but turns negative (trade-off) under resource shortage (van Noordwijk and de Jong 1986).

Although studies on the effects of ACC on trade-offs are scarce, there are grounds to believe that such effects can be expected. For instance, given that global warming has been predicted to increase metabolic rates of ectotherms by 10–75% (Bickford et al. 2010, Dillon et al. 2010) and that organisms must divide their limited resources between maintenance, growth and reproduction, the increased metabolic costs can be expected to reduce growth and/or reproduction (but see Kingsolver and Huey 2008). As these increases mean increased need for food, this may also trade off with survival if increased foraging time increases exposure to predation (*e.g.*, Gotthard 2000), or with reproduction if the food intake cannot be increased to compensate for both increased metabolic demands and resources needed for reproduction. It is currently difficult to formulate general predictions regarding effects of ACC on life history trade-offs but given what is known about the effects of ACC on metabolic demands, it seems self-evident that ACC will have an impact on them.

Aestivation is an unusual but rather common life history strategy in which animals undergo periods of dormancy to avoid hot or dry periods (Storey and Storey 2012). Some aestivating animals, such as burrowing fishes, are believed to be particularly susceptible to ACC as their habitats are becoming unsuitable for aestivation (Saddlier et al. 2010, Ogston et al. 2016). ACC is also suggested to have already impacted aestivation schedules of butterflies (Birch et al. 2021) and aestivation behaviour of moths (Lownds et al. 2023). Given that a range of organisms from molluscs and arthropods to amphibians and reptiles aestivate to overcome periods of heat and drought, more studies on the effects of ACC on aestivation strategies are needed.

18.3 Phenology

The evidence that ACC has impacted the timing of key life history events of many organisms is overwhelming. A review of studies of 1700 plant, insect, amphibian and bird species revealed that 87% of them had advanced their phenology (Parmesan and Yohe 2003). A later meta-analysis of plant and animal studies estimated that blooming and breeding have advanced at the rate of *ca.* five days per decade (Root et al. 2003). While these phenological shifts can be viewed mainly as adaptive offsetting of the negative consequences of ACC (Radchuk et al. 2019), they can also have negative fitness consequences. This will happen for instance when advancing phenology leads to trophic mismatches between prey and predators (see Chapter 13, this volume), or between flowers and their pollinators (see Chapter 16, this volume), because organisms at lower trophic levels usually advance their phenology faster than those on higher trophic levels (Both and Visser 2005, Inouye 2022). However, while there is evidence to suggest the existence of significant trophic mismatches (*e.g.*, Both and Visser 2005, Both et al. 2010, Poloczanska et al. 2013, Cohen et al. 2018), several recent reviews have called to question the quality of evidence for consistent and widespread trophic mismatches from methodological grounds (Renner and Zohner 2018, Kharouba and Wolkovich 2023). In addition, there is a need for more studies on the demographic consequences of trophic mismatches. For instance, Reed et al. (2013) found that the mismatch between breeding phenology and seasonal food peak intensified directional selection for earlier breeding in warmer springs for the great tits (*Parus major*) but did not affect population growth due to relaxed competition.

Another way ACC-driven phenological changes can influence animal life histories is through increased voltinism (*i.e.*, production of multiple generations or clutches per year) by advancing emergence and reproduction, as well as by prolonging the time window suitable for reproduction. Increased voltinism in response to longer growing seasons has been reported for many terrestrial and aquatic arthropods, fish and birds (reviewed in Iler et al. 2021). However, whether the increase in voltinism has actually had any positive effect on population growth or persistence in the face of ACC is still unclear (Iler et al. 2021), and there are examples to suggest the opposite (Bestion et al. 2015, van Dyck et al. 2015).

Climate-induced phenological changes may have demographic consequences. Prolonged growth and activity seasons under climate warming have been observed to benefit some organisms, in particular those at high latitudes where populations have been constrained by long winters (Wells et al. 2022). For instance, Ozgul et al. (2010) found that prolonged growth season allowed the yellow-bellied marmots (*Marmota flaviventris*) to increase their pre-hibernation body mass which in turn reduced adult mortality, triggering a population size increase. However, for migratory animals, the lack of correlation between ACC in wintering and breeding areas may lead to a phenological mismatch on breeding grounds and consequently affect their population dynamics. In a study conducted in the Netherlands, Both et al. (2010) found that long-distance migratory birds in seasonal forests showed much stronger declines than those breeding in less seasonal marshes. This result was attributed to stronger phenological shifts in the forest than in marsh habitats, as well as to constraints in long-distance migrants' ability to advance their migration schedule. Hence, ACC-induced phenological shifts may have both direct and trait-mediated effects on organismal life histories.

A recent review (Iler et al. 2021) of 238 studies on demographic consequences of ACC-driven phenological shifts found that the majority of these studies had been conducted on animals ($n = 146$), birds in particular ($n = 86$), followed by plants ($n = 92$). There was also a strong bias towards these studies to have been conducted on terrestrial systems (89% of the studies) and only 3% came from tropical study systems. Only 14.7% of these studies had measured population-level consequences (*i.e.*, population abundance or population growth rate) of ACC-driven phenological changes. The authors concluded that 'relatively few studies can convincingly link demographic consequences of phenological shifts to consequences for population growth rate, which is striking, especially in light of the frequency with which population-level consequences are cited as a motivation to understand phenological shifts' (Iler et al. 2021). Hence, while it is obvious that ACC-driven phenological changes are common and widespread in the northern hemisphere, much less is known about their ecological and population-level consequences, particularly at the lower latitudes.

18.4 Body Size

Body size is an important life history trait as it influences many other aspects of organismal life histories (Blueweiss et al. 1978; see also Chapter 1, this volume). For instance, in many organisms, female fecundity and male mating success are positive functions of body size (Pincheira-Donoso and Hunt 2017). Across different taxa, body size is also associated with different life history strategies: large species tend to be long-lived and exhibit lower fecundity than smaller species which typically have shorter lifespans and higher fecundity (Roff 2001). Consequently, ACC-induced changes in the mean body size of organisms can have important consequences on other aspects of organismal life histories.

There is evidence that ACC has led to decreases in the mean size of many terrestrial and aquatic organisms (Daufresne et al. 2009, Sheridan and Bickford 2011, Baudron et al. 2014). In fact, based on a meta-analysis of the effects of ACC on body size of ectothermic aquatic organisms, Daufresne et al. (2009) suggested that the decrease in mean body size is 'the third universal ecological response to global warming', the two others being latitudinal and altitudinal shifts in species distribution ranges, as well as phenological shifts. However, while body size declines have been commonly observed in both endotherms and homeotherms, they are not universal in the strict sense: there are species that do not show or show even opposite trends over time (*e.g.*, Gardner et al. 2011, Sheridan and Bickford 2011, Youngflesh et al. 2022). Many of the exceptions to declining body size trends appear to be from higher latitudes where increased temperature and precipitation have supported increased net primary production (Sheridan and Bickford 2011). In addition, most of the studies not reporting body size decline, or reporting positive or equivocal change, appear to come from mammals and birds (Meiri et al. 2009, Sheridan and Bickford 2011), albeit not exclusively (*e.g.*, Solokas et al. 2023). In line with this, a meta-analysis of morphological traits in arachnids, insects, amphibians, reptiles, birds and mammals found no significant effect of warming temperatures on population trait means or consistent signal of directional selection on size (Radchuk et al. 2019; see also Baar et al. 2018). Hence, against all this background, it appears that there is a great deal of heterogeneity in how body size is responding to warming climate. While some of this heterogeneity may reflect genuine biological differences in how ACC directly (or indirectly) influences body size in different species and populations, there are also reasons to be cautious about some of the reported body size trends or the lack of them thereof. For instance, unless possible shifts in population age structure (*e.g.*, due to changes in harvesting or natural mortality patterns; Tu et al. 2018) or food availability over time are accounted for, ACC effects on body size could be masked or even reversed. It is our impression that there might be a difference as to how ACC is influencing the mean body size of endotherms and ectotherms, and also that of terrestrial and aquatic organisms. However, a formal meta-analysis testing for these possible differences is currently lacking. Hence, as pointed out by Riemer et al. (2018), there appears to be a need for integrative and data-intensive analysis of the effects of temperature on organismal size which accounts for confounding factors and their interactions.

A modelling study by Cheung et al. (2013) provides an illustration of the complexity of the factors in play influencing body size trends in both time and space. By examining how integrated effects of physiology, dispersal, distribution and population dynamics likely influence the maximum body size of over 600 marine fish species under ACC, they found that the assembly averaged maximum body size is expected to shrink by 14–24% by 2050 at the rate of 2.8–4.8% per decade (Cheung et al. 2013). They further estimated that approximately half of this shrinkage would be attributable to changes in distribution and abundance, the other half to warming-induced increase in metabolic rate and reduced oxygen supply. The predicted reductions were expected to be largest in tropical and temperate zone seas, and less in high-latitude seas. Although these predictions are derived from a model making many assumptions (Cheung et al. 2013), they illustrate the fact that many interacting factors likely influence body size trends and simple predictions-based temperature trends alone are unlikely to be informative. A corollary to this is that these complications are likely propagating in the published analyses of ACC-associated body size trends, potentially explaining some of the heterogeneity in observed trends.

In the cases where body mass declines have been observed, one can ask what is the ecological relevance of the observed changes? For instance, in analysing body mass changes of 105 North American landbird species over three decades, Youngflesh et al. (2022) quantified an average decline of 0.56% across all species, the largest decline for an individual species being 2.78%. These figures do not come across as very large effects and their implications for individual and population fitness remain opaque. Experimental studies conducted on fish and salamanders (reviewed in Sheridan and Bickford 2011) have found stronger effects: each additional degree of warming has been found to decrease body size by 6–22% in fish and 14% in salamanders. Likewise, van Rijn et al. (2017) estimated a 15% reduction in fish size in the Mediterranean in response to a 1.5 °C increase in sea surface temperatures. In the same vein, a 14–24% global reduction in fish size from 2000 to 2050 due to warming oceans was predicted by Cheung et al. (2013), a prediction which aligns with results from the analysis of historical data from North Sea commercial fisheries (Baudron et al. 2014). Given the hyperallometric scaling of reproductive output to body size in fish (*i.e.*, the tendency for reproductive output to increase disproportionally with size; Barneche et al. 2018), 15% size decrease has been predicted to lead to a 50% reduction in fecundity of Atlantic mackerel (*Scomber scombrus*; Barneche et al. 2014, 2018). Hence, decreases in projected and observed body size declines in fish can be expected to have major consequences for important organismal life history traits and thereby their population persistence. The same is likely to apply to other ectothermic animals, but the overall picture of the prevalence and proximate mechanisms driving body size trends remain unclear except for a few well-worked case studies (*e.g.*, Ozgul et al. 2010, Bestion et al. 2015).

18.5 Reproductive Output and Success

The effects of ACC on reproductive output (number of offspring produced) and success (number of offspring surviving to a certain age) can be both negative and positive as a multitude of factors related to warming climate can influence individual reproduction (Isaac 2009). The reproductive success of many organisms is determined by their breeding phenology. While many organisms have advanced their phenology, these advances have not always occurred at the same rate in different species. Thackeray et al. (2016) found that secondary consumers (*e.g.*, non-granivorous birds) showed lower climate sensitivity to rising temperature than primary consumers (*e.g.*, herbivorous insects) or primary producers (*e.g.*, plants), and, therefore, had advanced their phenology slower. This can lead to trophic mismatches between prey and predators (*e.g.*, insects and insectivorous birds) and therefore to reduced reproductive success of predators due to the mismatch between the timing of peak food availability and maximum energetic requirements of their offspring. A recent global analysis of 201 populations of 104 bird species revealed an overall reduction in reproductive success rather than in reproductive output (*i.e.*, clutch size): declining reproductive success was observed in 56.7% of populations (Halupka et al. 2023). The negative effects of ACC on reproductive success were most pronounced in migratory and large-sized species.

Apart from the negative effects of trophic mismatch on reproduction, climate-induced advancements in breeding phenology can also increase the probability of encountering extreme climatic events in the early breeding season and thereby breeding failures (*e.g.*, Dunn and Winkler 2010, Goodenough et al. 2010). For instance, tree swallows (*Tachycineta bicolor*) have advanced reproduction in response to warming springs, but their offspring are more likely to be exposed to cold spells and therefore face an increased mortality rate (Shipley et al. 2020). Likewise, the increased incidence of extreme hot and cold events in recent years has reduced the reproductive output of great tits (Regan and Sheldon 2023). Hence, the constraints imposed by increasingly frequent extreme climatic events may limit the capability of organisms to advance their breeding phenology in response to ACC.

There is also evidence for the positive effects of a warming climate on individual reproductive output and success. For instance, Halupka et al. (2023) found that rising temperatures during the chick-rearing period were associated with increased reproductive success in small and sedentary bird species. While the reasons for this remain unclear, it was suggested that smaller species may be faster in responding to environmental changes and therefore more likely to benefit from ACC due to their high fecundity, early maturation and shorter generation time (Matthews et al. 2011, Halupka et al. 2023). In addition, a warming climate led to higher reproductive success in multi-brooded bird species but did not influence the reproductive success of single-brooded species (Halupka et al. 2023). For hibernating mammals, there was predominantly positive or no effect of a warming climate on reproductive success (Wells et al. 2022). For instance, female yellow-bellied marmots are more likely to breed after warmer shorter winters (Schwartz and Armitage 2005). Likewise, warm and shorter winters lead to higher reproductive output in common hamsters (*Cricetus cricetus*; Hufnagl et al. 2011). Overall, ACC may not only affect reproduction directly but also through interacting with the life history and ecological traits of species (Halupka et al. 2023). This may explain the diverse effects of ACC on reproductive success in different species.

General life history theory predicts that under high environmental variability and temporal stochasticity, low reproductive effort may be the optimal fitness-maximising strategy because the costs of reproduction are expected to be disproportionately high during unfavourable climate conditions (Stearns 1992, Roff 2001). For instance, breeding in drought years incurred severe survival costs for eastern tiger salamanders (*Ambystoma tigrinum*; Church et al. 2007). By reducing reproductive output or even skipping reproduction in some years, individuals may reduce the negative effects of catastrophic years on their lifetime reproductive success. As the variability in climate conditions is predicted to increase under ACC (Thornton et al. 2014), this could select for 'bet-hedging' (*i.e.*, risk spreading) tactics that may maximize their lifetime reproductive success (Simons 2011) since extreme climate events can strongly affect short-term reproductive output and success (reviewed in Sergio et al. 2018). For instance, Marcelino et al. (2020) found that extreme climatic events can have more significant effects on the breeding success of lesser kestrels (*Falco naumanni*) than gradual climate changes. Overall, high environmental variability and frequent extreme climate events may reduce the reproductive output and success of organisms and therefore limit their capability of tracking ACC. However, little is known about the relative importance of bet-hedging *vs.* phenotypic plasticity and adaptive tracking as a means of coping with ACC (Simons 2011).

18.6 Survival and Senescence

The impacts of climate on organismal survival probability show typically season- and age-specific patterns (Wells et al. 2022). Organisms usually suffer most mortality under climatic conditions at the extreme ends of the climate niches to which they are adapted. For instance, for both endotherms and ectotherms living in seasonal environments in temperate regions, survival rates are typically lowest in winter (Williams et al. 2015). A recent study of Northern Bobwhite (*Colinus virginianus*) found that extreme cold temperature events reduced their survival (Tanner et al. 2017). Likewise, severe heat waves have occasionally led to catastrophic mortality via lethal hyperthermia and consequently caused major population crashes in both birds and mammals (McKechnie et al. 2012, 2021). Reviewing a small number ($n = 6$) of vertebrate studies, Moreno and Møller (2011) found that the adult survival rate in extremely dry or warm years was only around one-third of that in normal years. Thus, survival probability is usually better predicted by extreme climatic events rather than mean annual temperature or rainfall (*e.g.*, Gardner et al. 2017). As extreme climatic events are predicted to increase in frequency with advancing ACC (IPCC 2021), this is expected to reduce the mean and increase the variance of yearly survival rates which in turn are expected to translate into reduced population growth and increased extinction risk (Frederiksen et al. 2008).

The impacts of ACC can be age-specific when the same climatic conditions affect younger and older age classes differently. In general, older age classes are expected to be better buffered against unfavourable climate conditions than younger age classes due to their better ability to thermoregulate as well as due to their experience in finding suitable shelters (McKechnie et al. 2012, Williams et al. 2015, Wells et al. 2022). For instance, older age classes of Uinta ground squirrels (*Urocitellus armatus*) experienced higher survival rates than juveniles during cold winters (Falvo et al. 2019). As the survival of older age classes of many animals and plants often contributes more to population growth and is less sensitive to ACC than that of younger age classes (Hilde et al. 2020), older age classes may contribute disproportionately to buffering populations against ACC-induced fitness loss. However, empirical evidence towards this end is scarce, presumably because relevant analyses require long-term data covering the full life cycle of a given species.

Although the evidence of climate impacts on organismal survival probabilities is abundant, there is no consistent directional effect of ACC on survival probabilities. For instance, in the case of hibernating mammals, warming conditions had either no or inconsistent effects on survival probabilities (Wells et al. 2022). In smaller hibernating rodents, higher annual temperature was associated with a lower annual survival rate whereas, for larger hibernating rodents or non-hibernating rodents, there were no (or much weaker) negative impacts of climate warming on survival rates (Turbill and Prior 2016). In the case of yellow-bellied marmots where a positive effect was observed, warmer springs led to earlier emergence from hibernation, prolonged growth season resulting in increased pre-hibernation body mass and thereby higher adult survival rate (Inouye et al. 2000, Ozgul et al. 2010). A recent study of superb fairywrens (*Malurus cyaneus*; Figure 18.1a) found that ACC-driven increase in winter maximum temperatures and intensity of summer heatwaves over decades had nearly doubled their mortality rates outside the breeding season (Lv et al. 2023). As colder seasons are warming faster than warmer seasons (IPCC 2021), climate warming may reduce the survival of many other organisms in the same way as in superb fairywrens, and, consequently, contribute to population declines. Unfortunately, relevant studies that have quantified the contribution of ACC to the changes in survival rates over time are still scarce. Currently, most studies tend to focus on the climate effects on survival rate differences between seasons. However, to reveal the mechanisms of climate-induced population changes which are still poorly understood, more knowledge on the climate-induced changes in survival rates is urgently needed (McLean et al. 2016).

Figure 18.1 Examples of species subject to long-term population studies enabling investigations to how climate change influences individual life histories. (a) Superb fairywrens (*Malurus cyaneus*) breeding in Australia has been studied since 1986. (b) Red-billed gulls (*Larus novahollandie*) breeding in Kaikoura, New Zealand, have been studied since 1969. (c) A colour-banded red-billed gull in the Kaikoura colony. *Source:* Photo courtesy by Geoffrey Dabb (a) and Juha Merilä (b and c).

Rates and patterns of actuarial senescence, an increase in mortality with age, are extremely variable across the tree of life (*e.g.*, Ricklefs 2010, Jones et al. 2014; see also Chapter 2, this volume). While much of this variability is associated with phylogeny and body size, there are also reasons to believe that climatic conditions might constitute a major driver of intraspecific variation in senescence at least in ectotherms (Flouris and Piantoni 2015, Keil et al. 2015, Burraco et al. 2020). For instance, Cayuela et al. (2021) provided evidence to suggest that ACC could lead to widespread acceleration of senescence in amphibians. Since female reproductive output in continuously growing amphibians increases with increasing body size and age, the resulting increase in mortality among old and highly fecund females has potential to influence amphibian population dynamics and persistence. Likewise, there is also evidence to suggest that high water temperatures are associated with accelerated senescence in the pearl mussel (*Margaritifera margaritifera*) indicating that warming climate might negatively impact their vital rates (Hassall et al. 2017). On a related note, Dupoué et al. (2022) found that most neonates of cold-adapted common lizards (*Zootoca vivipara*) were born physiologically old with shorter telomeres if their parents had experienced a warm environment. This in turn led to lower recruitment probability and consequently increased risk of local extirpation (Dupoué et al. 2022). While there is also evidence to suggest that low temperatures are associated with increased lifespan in endotherms (Flouris and Piantoni 2015, Keil et al. 2015), studies from wild populations are currently lacking. Hence, while it is conceivable that ACC-induced stress could also increase senescence in endotherms, there is currently little data to suggest that this would be a major concern. However, since increased senescence can lead to dramatic increase in extinction risk in long-lived and slowly reproducing species (Robert et al. 2015), there is clearly an incentive for future studies to monitor effects of ACC on senescence rates. For instance, given that there is data indicating an association between lifespan and hibernation (Keil et al. 2015), as well as results showing reduced survival amongst the older age classes of Uinta ground squirrels in warm winters and springs, it might be interesting to estimate if raising temperatures have increased senescence rate also amongst hibernating mammals.

18.7 Population Demography and Extinction Risk

There is evidence that ACC can affect population demographic rates (*e.g.*, through survival and fecundity of individuals), and consequently lead to species extinctions. A recent meta-analysis of extinction risk from ACC found that global extinction risk was predicted not only to increase but also to accelerate due to increasing ambient temperatures (Urban 2015). Overall, under the unabated ACC, about one in six species will be threatened and 7.9% of species are predicted to face extinction (Urban 2015). A survey of 538 plant and animal species found that locations with local extinctions had larger changes in daily maximum temperature of the warmest month and, surprisingly, smaller changes in annual mean temperature (Román-Palacios and Wiens 2020). ACC may influence extinction risk by increasing temporal variation in demographic rates which reduces population growth rate in the long run (Pearson et al. 2014). It is also possible that ACC increases extinction risk by influencing a specific demographic trait or a combination of demographic traits. For instance, Bonnot et al. (2018) found that warming temperatures posed a significant risk of quasi-extinction on Acadian flycatchers (*Empidonax virescens*) by reducing their reproductive output.

At the intraspecific level, there are examples of experimental studies that ACC modified life histories of species and predicted extinction risk. For instance, by allocating common lizards to climatic treatments of either 'present climate' or 'warm climate (~2°C warmer)', Bestion et al. (2015) found that warmer climate led to faster growth, an earlier onset of reproduction, and an increased voltinism, but a reduced adult survival, which as a whole, increased extinction risk. These findings suggest that although species may benefit from ACC at certain life history stages, they may suffer more at others. Therefore, a full life cycle approach is needed to test the impacts of ACC on all demographic rates and to predict the extinction risk of organisms. Additionally, more experimental studies are needed to reveal the proximate mechanisms of climate-induced changes in population demography. A case in point is provided by a whole-system manipulation experiment in the rainbow trout (*Oncorhynchus mykiss*) where elevated water temperature increased metabolism, forced individuals to feed more to maintain growth rate and reduced juvenile survival due to greater exposure to predation (Biro et al. 2007).

The effects of ACC on population demography and extinction risk vary interspecifically with some species being more sensitive to ACC than others. These differences are underlined by differences in species' life histories. General life history theories predict that species with slow life histories (K-selected species) are more likely to prevail in stable environments having high competitive ability, high adult survival and low reproductive output (Stearns 1992, Roff 2001). While such species may have high resilience and be initially resistant to ACC-driven environmental changes in the short term, their capacity to recover from long-term environmental changes is reduced by their low reproductive potential. In contrast, species with fast life histories (r-selected species) having low adult survival and high reproductive output may be more likely to experience temporary changes in population size under ACC but are more likely to adapt to long-term environmental changes (Sæther et al. 2019). For instance, a recent study on the impacts of climate warming on marine fisheries production revealed that species with fast life histories were more responsive to ACC than those with slow life histories (Free et al. 2019). Additionally, an analysis of 332 Indo-Pacific fish species found that ACC tended to increase population growth for slow life history species but reduce population growth for fast life history species (Wang et al. 2020). For the long-term extinction risk, Pearson et al. (2014) estimated it for 36 amphibian and reptilian species by using a modelling approach with a set of spatial and demographic variables with and without taking ACC into account. They found that ACC is predicted to lead to a pronounced (23–28%) increase in extinction risk and that species vulnerability depends on interactions between life history traits and spatial characteristics of species distribution. The most important life history trait defining extinction risk due to ACC was generation length. Species with longer generation times (usually slow life history species) have a lower extinction risk. This may be due to the fact that long-lived species are less likely to go extinct within the projected time frame of 100 years. In the long run, species with longer generation times may be more vulnerable to climate change due to their lower potential for genetic adaptation than short-lived species.

To sum up, the contribution of ACC-induced changes in survival or reproduction to population growth differs between species with differing life histories. Life history theory predicts that population growth rates should be sensitive for survival rate in long-lived organisms producing few offspring or having long generation times, whereas population growth rates of short-lived organisms producing many offspring are sensitive to variation in reproduction (Stearns 1992, Roff 2001). For instance, in long-lived emperor penguins (*Aptenodytes forsteri*), population growth is more sensitive to changes in adult survival than reproductive output or juvenile survival (Jenouvrier et al. 2005). Furthermore, as adults are usually more experienced and efficient in foraging than juveniles, their survival can be less likely to be affected by ACC than juvenile survival (*e.g.*, Oro et al. 2010). Hence, long-lived species can be more resistant to ACC than short-lived species (Free et al. 2019). However, long-term studies are needed to test this pattern considering that short-lived species are more likely to recover from ACC due to their high reproductive output.

18.8 Genetic or Environmental Responses

As it is clear from the above, ACC has already impacted many life history traits in a diverse array of taxa from various habitats. While these phenotypic shifts in the mean trait values over time, in response to (or correlated with) ACC, are unquestionable, the challenge is to know to what extent these represent non-genetic environmentally induced changes or whether they have a genetic basis (or both; Gienapp et al. 2008, Merilä and Hendry 2014). The question about the genetic basis of the observed changes is important as it influences our understanding of the drivers of the observed shifts in population mean trait values, as well as predictions about species and populations' ability to adapt to ACC. A case in point is provided by a long-term study of red-billed gulls (*Larus novaehollandiae*; Figure 18.1b,c) from New Zealand where the mean body size in the study population declined significantly over a 47-year period and this decline was correlated with a concomitant increase in ambient temperatures at the study site (Teplitsky et al. 2008). While such trends have been earlier (*e.g.*, Millien et al. 2006) interpreted as adaptive responses to warming climate in the context of temporal interpretation of Bergmann's rule (Bergmann 1847; *i.e.*, selection in warming climate favouring small body size), quantitative genetic analyses revealed no evidence for selection favouring smaller body size and the observed body size changes had no genetic basis (Teplitsky et al. 2008). Hence, the size decline in this population was not an adaptive microevolutionary change, but rather a plastic change most likely reflecting growth-reducing deteriorating environmental conditions in the colony (Teplitsky et al. 2008).

While there are examples of studies where adaptive genetically based responses to ACC-driven selection on life history traits have been demonstrated (*e.g.*, Bradshaw and Holzapfel 2001, Franks et al. 2007, Karell et al. 2011), there are probably many more studies that have found the shifts in trait means to be underlined by phenotypic plasticity (Merilä and Hendry 2014; see also: Ozgul et al. 2009, 2010, Hoy et al. 2018). This does not necessarily mean that plastic responses outpace genetic responses to ACC because demonstrating genetic responses is more challenging than demonstrating plastic responses (Merilä and Hendry 2014). However, there are good reasons to believe that plastic responses are at least initially the predominant path for populations to respond to ACC. First, most traits and populations exhibit pre-existing plasticity that can be used to adjust phenotype to changes in environmental conditions. As long as adjustments based on pre-existing plasticity allow organisms to track changing environmental conditions without marked fitness loss, there should be little selection to drive genetic changes in either trait means or plasticity itself. Second, the expression of many life history traits is sensitive to changes in abiotic and biotic conditions caused by ACC. For instance, as discussed above, a warming climate is expected to reduce ectotherm, and perhaps also endotherm (Weeks et al. 2022), body sizes as dictated by temperature–size rule (Angilletta and Dunham 2003). Therefore, it would not be surprising if much of the already observed body size declines represent simple non- or even maladaptive environmentally induced responses to ACC. In fact, one should not expect that the direction of ACC-driven natural selection on body size is always negative: smaller size makes individuals more vulnerable to dehydration and overheating and, hence, extreme high-temperature events might actually select for larger size (McKechnie and Wolf 2010, Gardner et al. 2011, Peralta-Maraver and Rezende 2021). In the same vein, a meta-analysis found no evidence for significant selection of thermal plasticity suggesting that documented plastic responses are seldom adaptive (Arnold et al. 2019). Hence, there are good reasons to believe that much of the observed shifts in mean values of life history traits can be accounted for with non-adaptive plasticity rather than genetically based adaptations.

The relative paucity of evidence for adaptive ACC-driven shifts (Merilä and Hendry 2014) could be also explained by another perspective. Namely, although most life history traits are likely to be heritable and should therefore respond to ACC-driven selection pressures, a number of factors can constrain such responses. First, there is a tendency for life history traits closely related to fitness to have lower heritabilities than, *e.g.*, morphological traits less closely related to fitness (Merilä and Sheldon 1999). Second, ACC tends to increase environmental variance which in turn is expected to reduce heritabilities (Hoffmann and Merilä 1999, Charmantier and Garant 2005, Simons 2011). Third, negative correlations between heritability and strength of selection have often been observed, suggesting that selection responses can be weakest when the selection is strongest (Merilä 1997, Hoffmann and Merilä 1999, Wilson et al. 2006). All these factors could be in play simultaneously and at least partially explain why examples of adaptive responses to ACC-driven selection are still rare. However, a more complete understanding of the genetic responses, and consequently whether species and populations could adapt to ACC, requires more studies quantifying both genetic and plastic responses in the phenotypic shifts of the mean trait values.

We further note that the number of case studies providing firm evidence of ACC-driven evolutionary responses has not grown much since the first comprehensive review of such studies (Gienapp et al. 2008) whereas number of reviews and perspectives on this topic have propagated as mushrooms after rain (*e.g.*, Hoffmann and Willi 2008, Hoffmann and Sgrò 2011, Anderson et al. 2012, Franks and Hoffmann 2012, Merilä 2012, Alberto et al. 2013, Munday et al. 2013, Boutin and Lane 2014, Charmantier and Gienapp 2014, Crozier and Hutchings 2014, Franks et al. 2014, Merilä and Hendry

2014, Reusch 2014, Schilthuizen and Kellermann 2014, Stoks et al. 2014, Merilä and Hoffmann 2016, Gienapp and Merilä 2018, Catullo et al. 2019, Kellermann and van Heerwaarden 2019, Aguirre-Liguori et al. 2021, McGaughran et al. 2021; this chapter), probably by now exceeding the number of solid case studies demonstrating ACC-driven evolution. We believe this might be symptomatic of the gravity of the challenge demonstrating evolution in response to ACC (see Merilä and Hendry 2014) rather than actual lack of evolutionary responses to ACC. Nevertheless, the fact is that there is currently more evidence to support the view that observed changes in mean life history trait values reflect phenotypic plasticity rather than genetically based evolutionary changes. While phenotypically plastic changes can help organisms to cope with ACC, plastic responses have their limits and they are unlikely to allow species and populations to track changing climatic conditions forever (e.g., DeWitt et al. 1998, Gienapp and Merilä 2018). In fact, recent analysis of natural selection on phenological advances concluded that the observed consistent selection for earlier timing provides evidence to suggest that the observed advances are imperfect (otherwise there would not be selection), and that these imperfect responses to ACC likely threaten persistence of some of the analysed populations (Radchuk et al. 2019).

18.9 Conclusions and Outlook

To sum up, it is clear that ACC has already had diverse effects on organismal life histories, and likely will have them increasingly also in the future. The best and most widely documented responses involve various phenological traits followed by impacts on organismal size. Yet many issues regarding ACC effects on these traits remain not well understood, calling for additional case studies and meta-analyses. Effects of ACC on reproductive traits and survival have been even less studied, and, hence, ACC impacts on these important life history traits providing a basis for estimating strength of ACC-induced natural selection remain poorly understood. Consequently, how ACC impacts organismal life histories translate to population-level consequences, and thereby to extinction risk, remain even less well understood. While it is conceivable that ACC will impose selection pressures on life histories of many populations and species, it is still unclear how often such selection is actually taking place and will lead to adaptive responses allowing populations to keep up with the pace of ACC without facing extirpations or extinctions. While there are hopes that genomic and advanced modelling approaches will eventually improve our ability to predict species responses to climate change (Waldvogel et al. 2020), only time will show whether and when these hopes will be realized. Judging from the history of studies focussed on evolutionary responses to climate change, the progress in this respect might not materialize as quickly and easily as we would hope. Finally, it is worth noting that a lot of the accumulated evidence for the effects of ACC on organismal life histories comes from correlational approaches: causalities have seldom been firmly established. However, experimental approaches, long-term studies of individually marked organisms together with modelling approaches, meta-analyses and genomic tools will continue to be instrumental in improving the quality of inference and evidence for ACC effects on life histories.

Acknowledgements

During the writing of this chapter, we were supported by grants from the Research Grants Council of Hong Kong (JM) and grants from National Natural Science Foundation of China (LL).

References

Aguirre-Liguori, J.A., Ramírez-Barahona, S., and Gaut, B.S. (2021). The evolutionary genomics of species' responses to climate change. *Nature Ecology & Evolution* 5(10): 1350–1360.

Alberto, F.J., Aitken, S.N., Alía, R., González-Martínez, S.C., Hänninen, H., Kremer, A., Lefèvre, F., Lenormand, T., Yeaman, S., Whetten, R., and Savolainen, O. (2013). Potential for evolutionary responses to climate change–evidence from tree populations. *Global Change Biology* 19(6): 1645–1661.

Anderson, J.T., Panetta, A.M., and Mitchell-Olds, T. (2012). Evolutionary and ecological responses to anthropogenic climate change: update on anthropogenic climate change. *Plant Physiology* 160(4): 1728–1740.

Angilletta, Jr, M.J., and Dunham, A.E. (2003). The temperature-size rule in ectotherms: simple evolutionary explanations may not be general. *The American Naturalist* 162: 332–342.

Arnold, P.A., Nicotra, A.B., and Kruuk, L.E.B. (2019). Sparse evidence for selection on phenotypic plasticity in response to temperature. *Philosophical Transactions of the Royal Society B* 374(1768): 20180185.

Atkinson, D. (1994). Temperature and organism size – a biological law for ectotherms? *Advances in Ecological Research* 25: 1–58.

Baar, Y., Friedman, A.L.L., Meiri, S., and Scharf, I. (2018). Little effect of climate change on body size of herbivorous beetles. *Insect Science* 25(2): 309–316.

Barneche, D.R., Kulbicki, M., Floeter, S.R., Friedlander, A.M., Maina, J., and Allen, A.P. (2014). Scaling metabolism from individuals to reef-fish communities at broad spatial scales. *Ecology Letters* 17(9): 1067–1076.

Barneche, D.R., Robertson, D.R., White, C.R., and Marshall, D.J. (2018). Fish reproductive-energy output increases disproportionately with body size. *Science* 360(6389): 642–645.

Baudron, A.R., Needle, C.L., Rijnsdorp, A.D., and Marshall, C.T. (2014). Warming temperatures and smaller body sizes: synchronous changes in growth of North Sea fishes. *Global Change Biology* 20: 1023–1031.

Bergmann. C. (1847). Ueber die verhältnisse der wärme ökonomie der thiere zu ihrer grösse. *Gottinger Studien* 3:595–708.

Bestion, E., Teyssier, A., Richard, M., Clobert, J., and Cote, J. (2015). Live fast, die young: experimental evidence of population extinction risk due to climate change. *PLoS Biology* 13(10): e1002281.

Bickford, D., Howard, S.D., Ng, D.J., and Sheridan, J.A. (2010). Impacts of climate change on the amphibians and reptiles of Southeast Asia. *Biodiversity and Conservation* 19: 1043–1062.

Birch, R.J., Markl, G., and Gottschalk, T.K. (2021). Aestivation as a response to climate change: the great banded grayling *Brintesia circe* in Central Europe. *Ecological Entomology* 46(6): 1342–1352.

Biro, P.A., Post, J.R., and Booth, D.J. (2007). Mechanisms for climate-induced mortality of fish populations in whole-lake experiments. *Proceedings of the National Academy of Sciences of the United States of America* 104(23): 9715–9719.

Blanchard, J.L., Jennings, S., Holmes, R., Harle, J., Merino, G., Allen, J.I., Holt, J., Dulvy, N.K., and Barange, M. (2012). Potential consequences of climate change for primary production and fish production in large marine ecosystems. *Philosophical Transactions of the Royal Society, B: Biological Sciences* 367(1605): 2979–2989.

Blueweiss, L., Fox, H., Kudzma, V., Nakashima, D., Peters, R., and Sams, S. (1978). Relationships between body size and some life history parameters. *Oecologia* 37(2): 257–272.

Bonnot, T.W., Cox, W.A., Thompson, F.R., and Millspaugh, J.J. (2018). Threat of climate change on a songbird population through its impacts on breeding. *Nature Climate Change* 8(8): 718–722.

Both, C., van Turnhout, C.A., Bijlsma, R.G., Siepel, H., van Strien, A.J., and Foppen, R.P. (2010). Avian population consequences of climate change are most severe for long-distance migrants in seasonal habitats. *Proceedings of the Royal Society B: Biological Sciences* 277(1685): 1259–1266.

Both, C., and Visser, M.E. (2005). The effect of climate change on the correlation between avian life-history traits. *Global Change Biology* 11(10): 1606–1613.

Boutin, S., and Lane, J.E. (2014). Climate change and mammals: evolutionary versus plastic responses. *Evolutionary Applications* 7(1): 29–41.

Bradshaw, W.E., and Holzapfel, C.M. (2001). Genetic shift in photoperiodic response correlated with global warming. *Proceedings of the National Academy of Sciences of the United States of America* 98: 14509–14511.

Burraco, P., Orizaola, G., Monaghan, P., and Metcalfe, N.B. (2020). Climate change and ageing in ectotherms. *Global Change Biology* 26(10): 5371–5381.

Catullo, R.A., Llewelyn, J., Phillips, B.L., and Moritz, C.C. (2019). The potential for rapid evolution under anthropogenic climate change. *Current Biology* 29(19): R996–R1007.

Cayuela, H., Lemaître, J.F., Muths, E., McCaffery, R.M., Frétey, T., Garff, B.L., Schmidt, B.R., Grossenbacher, K., Lenzi, O., Hossack, B.R., Eby, L.A., Lambert, B.A., Elmberg, J., Merilä, J., Gippet, J.M.W., Gaillard, J.M., and Pilliod, D.S. (2021). Thermal conditions predict intraspecific variation in senescence rate in frogs and toads. *Proceedings of the National Academy of Sciences of the United States of America* 118(49): e2112235118.

Charmantier, A., and Garant, D. (2005). Environmental quality and evolutionary potential: lessons from wild populations. *Proceedings of the Royal Society B: Biological Sciences* 272:1415–1425.

Charmantier, A., and Gienapp, P. (2014). Climate change and timing of avian breeding and migration: evolutionary versus plastic changes. *Evolutionary Applications* 7(1): 15–28.

Cheung, W.W., Lam, V.W., Sarmiento, J.L., Kearney, K., Watson, R., and Pauly, D. (2009). Projecting global marine biodiversity impacts under climate change scenarios. *Fish and Fisheries* 10: 235–251.

Cheung, W.W., Sarmiento, J.L., Dunne, J., Frölicher, T.L., Lam, V.W., Palomares, M.L.D., Watson, R., and Pauly, D. (2013). Shrinking of fishes exacerbates impacts of global ocean changes on marine ecosystems. *Nature Climate Change* 3(3): 254–258.

Church, D.R., Bailey, L.L., Wilbur, H.M., Kendall, W.L., and Hines, J.E. (2007). Iteroparity in the variable environment of the salamander *Ambystoma tigrinum*. *Ecology* 88(4): 891–903.

Cohen, J.M., Lajeunesse, M.J., and Rohr, J.R. (2018). A global synthesis of animal phenological responses to climate change. *Nature Climate Change* 8(3): 224–228.

Crozier, L.G., and Hutchings, J.A. (2014). Plastic and evolutionary responses to climate change in fish. *Evolutionary Applications* 7(1): 68–87.

Daufresne, M., Lengfellner, K., and Sommer, U. (2009). Global warming benefits the small in aquatic ecosystems. *Proceedings of the National Academy of Sciences of the United States of America* 106(31): 12788–12793.

DeWitt, T.J., Sih, A., and Wilson, D.S. (1998). Costs and limits of phenotypic plasticity. *Trends in Ecology & Evolution* 13(2): 77–81.

Dillon, M.E., Wang, G., and Huey, R.B. (2010). Global metabolic impacts of recent climate warming. *Nature* 467: 704–706.

Dunn, P.O., and Winkler, D.W., (2010). Effects of climate change on timing of breeding and reproductive success in birds. In: *Effects of climate change on birds* (eds. Møller, A.P., Fiedler, W., and Berthold, P.), 113–126. Oxford University Press.

Dupoué, A., Blaimont, P., Angelier, F., Ribout, C., Rozen-Rechels, D., Richard, M., Miles, D., Villemereuil, P., Rutschmann, A., Badiane, A., Aubret, F., Lourdais, O., Meylan, S., Cote, J., Clobert, J., and Le Galliard, J.F. (2022). Lizards from warm and declining populations are born with extremely short telomeres. *Proceedings of the National Academy of Sciences of the United States of America* 119(33): e2201371119.

Falvo, C.A., Koons, D.N., and Aubry, L.M. (2019). Seasonal climate effects on the survival of a hibernating mammal. *Ecology and Evolution* 9(7): 3756–3769.

Flouris, A.D., and Piantoni, C. (2015). Links between thermoregulation and aging in endotherms and ectotherms. *Temperature* 2(1): 73–85.

Franks, S.J., and Hoffmann, A.A. (2012). Genetics of climate change adaptation. *Annual Review of Genetics* 46: 185–208.

Franks, S.J., Sim, S., and Weis, A.E. (2007). Rapid evolution of flowering time by an annual plant in response to a climate fluctuation. *Proceedings of the National Academy of Sciences of the United States of America* 104: 1278–1282.

Franks, S.J., Weber, J.J., and Aitken, S.N. (2014). Evolutionary and plastic responses to climate change in terrestrial plant populations. *Evolutionary Applications* 7(1): 123–139.

Frederiksen, M., Daunt, F., Harris, M.P., and Wanless, S. (2008). The demographic impact of extreme events: stochastic weather drives survival and population dynamics in a long-lived seabird. *Journal of Animal Ecology* 77(5): 1020–1029.

Free, C.M., Thorson, J.T., Pinsky, M.L., Oken, K.L., Wiedenmann, J., and Jensen, O.P. (2019). Impacts of historical warming on marine fisheries production. *Science* 363(6430): 979–983.

Gardner, J.L., Peters, A., Kearney, M.R., Joseph, L., and Heinsohn, R. (2011). Declining body size: a third universal response to warming? *Trends in Ecology & Evolution* 26(6): 285–291.

Gardner, J.L., Rowley, E., De Rebeira, P., De Rebeira, A., and Brouwer, L. (2017). Effects of extreme weather on two sympatric Australian passerine bird species. *Philosophical Transactions of the Royal Society, B: Biological Sciences* 372(1723): 20160148.

Gienapp, P., and Merilä, J. (2018). Evolutionary responses to climate change. *Encyclopedia of the Anthropocene* 2: 51–59.

Gienapp, P., Teplitsky, C., Alho, J.S., Mills, J.A., and Merilä, J. (2008). Climate change and evolution: disentangling environmental and genetic responses. *Molecular Ecology* 17(1): 167–178.

Goodenough, A.E., Hart A.G., and Stafford R. (2010). Is adjustment of breeding phenology keeping pace with the need for change? Linking observed response in woodland birds to changes in temperature and selection pressure. *Climate Change* 102: 687–697.

Gotthard, K. (2000). Increased risk of predation as a cost of high growth rate: an experimental test in a butterfly. *Journal of Animal Ecology* 69(5): 896–902.

Halupka, L., Arlt, D., Tolvanen, J., Millon, A., Bize, P., Adamík, P., Albert, P., Arendt, W.J., Artemyev, A.V., Baglione, V., Bańbura, J., Bańbura, M., Barba, E., Barrett, R.T., Becker, P.H., Belskii, E., Bolton, M., Bowers, E.K., Bried, J., Brouwer, L., Bukacińska, M., Bukaciński, D., Bulluck, L., Carstens, K.F., Catry, I., Charter, M., Chernomorets, A., Covas, R., Czuchra, M., Dearborn, D.C., de Lope, F., Di Giacomo, A.S., Dombrovski, V.C., Drummond, H., Dunn, M.J., Eeva, T., Emmerson, L.M., Espmark, Y., Fargallo, J.A., Gashkov, S.I., Golubova, E.Y., Griesser, M., Harris, M.P., Hoover, J.P., Jagiełło, Z., Karell, P., Kloskowski, J., Koenig, W.D., Kolunen, H., Korczak-Abshire, M., Korpimäki, E., Krams, I., Krist, M., Krüger, S.C., Kuranov, B.D., Lambin, X., Lombardo, M.P., Lyakhov, A., Marzal, A., Møller, A.P., Neves, V.C., Nielsen, J.T., Numerov, A., Orłowska, B., Oro, D., Öst, M., Phillips, R.A., Pietiäinen, H., Polo, V., Porkert, J., Potti, J., Pöysä, H., Printemps, T., Prop, J., Quillfeldt, P., Ramos, J.A., Ravussin, P.A., Rosenfield, R.N., Roulin, A., Rubenstein, D.R., Samusenko, I.E., Saunders, D.A., Schaub, M., Senar, J.C., Sergio, F., Solonen, T., Solovyeva, D.V., Stępniewski, J., Thompson, P.M., Tobolka, M., Török, J., van de Pol, M., Vernooij, L., Visser, M.E., Westneat, D.F., Wheelwright, N.T., Wiącek, J., Wiebe, K.L., Wood, A.G., Wuczyński, A., Wysocki, D., Zárybnická, M., Margalida, A., and Halupka, K., (2023). The effect of climate change on avian offspring production: a global meta-analysis. *Proceedings of the National Academy of Sciences of the United States of America* 120(19): e2208389120.

Hassall, C., Amaro, R., Ondina, P., Outeiro, A., Cordero-Rivera, A., and Miguel, E.S. (2017). Population-level variation in senescence suggests an important role for temperature in an endangered mollusc. *Journal of Zoology* 30(1): 32–40.

Hilde, C.H., Gamelon, M., Sæther, B.E., Gaillard, J.M., Yoccoz, N.G., and Pélabon, C. (2020). The demographic buffering hypothesis: evidence and challenges. *Trends in Ecology & Evolution* 35(6): 523–538.

Hoffmann, A.A., and Merilä, J. (1999). Heritable variation and evolution under favourable and unfavourable conditions. *Trends in Ecology & Evolution* 14(3): 96–101.

Hoffmann, A.A., and Sgrò, C.M. (2011). Climate change and evolutionary adaptation. *Nature* 470(7335): 479–485.

Hoffmann, A.A., and Willi, Y. (2008). Detecting genetic responses to environmental change. *Nature Reviews Genetics* 9: 421–432.

Holt, R.D. (1990). The microevolutionary consequences of climate change. *Trends in Ecology & Evolution* 5(9): 311–315.

Hoy, S.R., Peterson, R.O., and Vucetich, J.A. (2018). Climate warming is associated with smaller body size and shorter lifespans in moose near their southern range limit. *Global Change Biology* 24(6): 2488–2497.

Hufnagl, S., Franceschini-Zink, C., and Millesi, E. (2011). Seasonal constraints and reproductive performance in female common hamsters (*Cricetus cricetus*). *Mammalian Biology* 76: 124–128.

Iler, A.M., CaraDonna, P.J., Forrest, J.R., and Post, E. (2021). Demographic consequences of phenological shifts in response to climate change. *Annual Review of Ecology, Evolution, and Systematics* 52: 221–245.

Inouye, D.W. (2022). Climate change and phenology. *Climate Change* 13(3): e764.

Inouye, D.W., Barr, B., Armitage, K.B., and Inouye, B.D. (2000). Climate change is affecting altitudinal migrants and hibernating species. *Proceedings of the National Academy of Sciences of the United States of America* 97(4): 1630–1633.

IPCC (2021). Climate change 2021: the physical science basis. Contribution of Working Group I to the Sixth Assessment Report of the Intergovernmental Panel on Climate Change. https://www.ipcc.ch/report/ar6/wg1/downloads/report/IPCC_AR6_WGI_FrontMatter.pdf.

Isaac, J.L. (2009). Effects of climate change on life history: implications for extinction risk in mammals. *Endangered Species Research* 7(2): 115–123.

Jenouvrier, S., Barbraud, C., and Weimerskirch, H. (2005). Long-term contrasted responses to climate of two Antarctic seabird species. *Ecology* 86(11): 2889–2903.

Jones, O.R., Scheuerlein, A., Salguero-Gómez, R., Camarda, C.G., Schaible, R., Casper, B.B., Dahlgren, J.P., Ehrlén, J., García, M.B., Menges, E.S., Quintana-Ascencio, P.F., Caswell, H., Baudisch, A., and Vaupel, J.W. (2014). Diversity of ageing across the tree of life. *Nature* 505(7482): 169–173.

Karell, P., Ahola, K., Karstinen, T., Valkama, J., and Brommer, J.E. (2011). Climate change drives microevolution in a wild bird. *Nature Communications* 2: 208.

Keil, G., Cummings, E., and de Magalhaes, J.P. (2015). Being cool: how body temperature influences ageing and longevity. *Biogerontology* 16: 383–397.

Kellermann, V., and van Heerwaarden, B. (2019). Terrestrial insects and climate change: adaptive responses in key traits. *Physiological Entomology* 44(2): 99–115.

Kharouba, H.M., and Wolkovich, E.M. (2023). Lack of evidence for the match-mismatch hypothesis across terrestrial trophic interactions. *Ecology Letters* 26(2): 955–964.

Kingsolver, J., and Huey, R. (2008). Size, temperature, and fitness: three rules. *Evolutionary Ecology Research* 10(2): 251–268.

Lancaster, L.T., Morrison, G., and Fitt, R.N. (2017). Life history trade-offs, the intensity of competition, and coexistence in novel and evolving communities under climate change. *Philosophical Transactions of the Royal Society, B: Biological Sciences* 372(1712): 20160046.

Lavergne, S., Mouquet, N., Thuiller, W., and Ronce, O. (2010). Biodiversity and climate change: integrating evolutionary and ecological responses of species and communities. *Annual Review of Ecology, Evolution, and Systematics* 41: 321–350.

Lownds, R.M., Turbill, C., White, T.E., and Umbers, K.D. (2023). The impact of elevated aestivation temperatures on the behaviour of bogong moths (*Agrotis infusa*). *Journal of Thermal Biology* 113:103538.

Lv, L., van de Pol, M., Osmond, H.L., Liu, Y., Cockburn, A., and Kruuk, L.E.B. (2023). Winter mortality of a passerine bird increases following hotter summers and during winters with higher maximum temperatures. *Science Advances* 9(1): eabm0197.

Marcelino, J., Silva, J.P., Gameiro, J., Silva, A., Rego, F.C., Moreira, F., and Catry, I. (2020). Extreme events are more likely to affect the breeding success of lesser kestrels than average climate change. *Scientific Reports* 10(1): 7207.

Matthews, L.J., Arnold, C., Machanda, Z., and Nunn, C.L. (2011). Primate extinction risk and historical patterns of speciation and extinction in relation to body mass. *Proceedings of the Royal Society B: Biological Sciences* 278(1709): 1256–1263.

McGaughran, A., Laver, R., and Fraser, C. (2021). Evolutionary responses to warming. *Trends in Ecology & Evolution* 36(7): 591–600.

McKechnie, A.E., Hockey, P.A., and Wolf, B.O. (2012). Feeling the heat: Australian landbirds and climate change. *Emu - Austral Ornithology* 112(2): i–vii.

McKechnie, A.E., Rushworth, I.A., Myburgh, F., and Cunningham, S.J. (2021). Mortality among birds and bats during an extreme heat event in eastern South Africa. *Austral Ecology* 46(4): 687–691.

McKechnie, A.E., and Wolf B.O. (2010). Climate change increases the likelihood of catastrophic avian mortality events during extreme heat waves. *Biology Letters* 6:253–256.

McLean, N., Lawson, C.R., Leech, D.I., and van de Pol, M. (2016). Predicting when climate-driven phenotypic change affects population dynamics. *Ecology Letters* 19(6): 595–608.

Meiri, S., Guy, D., Dayan, T., and Simberloff, D. (2009). Global change and carnivore body size: data are stasis. *Global Ecology and Biogeography* 18: 240–247.

Menéndez, R. (2007). How are insects responding to global warming? *Tijdschrift voor Entomologie* 150(2): 355.

Merilä, J. (1997). Expression of genetic variation in body size of the collared flycatcher under different environmental conditions. *Evolution* 51(2): 526–536.

Merilä, J. (2012). Evolution in response to climate change: in pursuit of the missing evidence. *BioEssays* 34(9): 811–818.

Merilä, J., and Hendry, A.P. (2014). Climate change, adaptation, and phenotypic plasticity: the problem and the evidence. *Evolutionary Applications* 7(1): 1–14.

Merilä, J., and Hoffmann, A.A. (2016). Evolutionary impacts of climate change. Oxford Research Encyclopedia of Environmental Science. https://doi.org/10.1093/acrefore/9780199389414.013.136.

Merilä, J., and Sheldon, B.C. (1999). Genetic architecture of fitness and nonfitness traits: empirical patterns and development of ideas. *Heredity* 83(2): 103–109.

Milazzo, M., Mirto, S., Domenici, P., and Gristina, M. (2013). Climate change exacerbates interspecific interactions in sympatric coastal fishes. *Journal of Animal Ecology* 82(2): 468–477.

Millien, V., Kathleen Lyons, S., Olson, L., Smith, F.A., Wilson, A.B., and Yom-Tov, Y. (2006). Ecotypic variation in the context of global climate change: revisiting the rules. *Ecology Letters* 9(7): 853–869.

Møller, A.P., Flensted-Jensen, E., Klarborg, K., Mardal, W., and Nielsen, J.T. (2010). Climate change affects the duration of the reproductive season in birds. *Journal of Animal Ecology* 79(4): 777–784.

Moreno, J., and Møller, A.P. (2011). Extreme climatic events in relation to global change and their impact on life histories. *Current Zoology* 57(3): 375–389.

Munday, P.L., Warner, R.R., Monro, K., Pandolfi, J.M., and Marshall, D.J. (2013). Predicting evolutionary responses to climate change in the sea. *Ecology Letters* 16(12): 1488–1500.

Ogston, G., Beatty, S.J., Morgan, D.L., Pusey, B.J., and Lymbery, A.J. (2016) Living on burrowed time: aestivating fishes in south-western Australia face extinction due to climate change. *Biological Conservation* 195: 235–244.

Oro, D., Torres, R., Rodríguez, C., and Drummond, H. (2010). Climatic influence on demographic parameters of a tropical seabird varies with age and sex. *Ecology* 91(4): 1205–1214.

Ozgul, A., Childs, D.Z., Oli, M.K., Armitage, K.B., Blumstein, D.T., Olson, L.E., Tuljapurkar, S., and Coulson, T. (2010). Coupled dynamics of body mass and population growth in response to environmental change. *Nature* 466(7305): 482–485.

Ozgul, A., Tuljapurkar, S., Benton, T.G., Pemberton, J.M., Clutton-Brock, T.H., and Coulson, T. (2009). The dynamics of phenotypic change and the shrinking sheep of St. Kilda. *Science* 325(5939): 464–467.

Parmesan, C. (2006). Ecological and evolutionary responses to recent climate change. *Annual Review of Ecology, Evolution, and Systematics* 37: 637–669.

Parmesan, C. (2007). Influences of species, latitudes and methodologies on estimates of phenological response to global warming. *Global Change Biology* 13(9): 1860–1872.

Parmesan, C., and Yohe, G. (2003). A globally coherent fingerprint of climate change impacts across natural systems. *Nature* 421(6918): 37–42.

Pearson, R.G., Stanton, J.C., Shoemaker, K.T., Aiello-Lammens, M.E., Ersts, P.J., Horning, N., Fordham, D.A., Raxworthy, C.J., Ryu, H.Y., McNees, J., and Akçakaya, H.R. (2014). Life history and spatial traits predict extinction risk due to climate change. *Nature Climate Change* 4(3): 217–221.

Peñuelas, J., Sardans, J., Estiarte, M., Ogaya, R., Carnicer, J., Coll, M., Barbeta, A., Rivas-Ubach, A., Llusià, J., Garbulsky, M., Filella, I., and Jump, A.S. (2013). Evidence of current impact of climate change on life: a walk from genes to the biosphere. *Global Change Biology* 19(8): 2303–2338.

Peralta-Maraver, I., and Rezende, E.L. (2021). Heat tolerance in ectotherms scales predictably with body size. *Nature Climate Change* 11(1): 58–63.

Piao, S., Liu, Q., Chen, A., Janssens, I.A., Fu, Y., Dai, J., Liu, L., Lian, X., and Zhu, X. (2019). Plant phenology and global climate change: current progresses and challenges. *Global Change Biology* 25: 1922–1940.

Pincheira-Donoso, D., and Hunt, J. (2017). Fecundity selection theory: concepts and evidence. *Biological Reviews* 92(1): 341–356.

Poloczanska, E.S., Brown, C.J., Sydeman, W.J., Kiessling, W., Schoeman, D.S., Moore, P.J., Brander, K., Bruno, J.F., Buckley, L.B., Burrows, M.T., Duarte, C.M., Halpern, B.S., Holding, J., Kappel, C.V., O'Connor, M.I., Pandolfi, J.M., Parmesan, C., Schwing, F., Thompson, S.A., and Richardson, A.J. (2013). Global imprint of climate change on marine life. *Nature Climate Change* 3(10): 919–925.

Radchuk, V., Reed, T., Teplitsky, C., van de Pol, M., Charmantier, A., Hassall, C., Adamík, P., Adriaensen, F., Ahola, M.P., Arcese, P., Avilés, J.M., Balbontin, J., Berg, K.S., Borras, A., Burthe, S., Clobert, J., Dehnhard, N., de Lope, F., Dhondt, A.A., Dingemanse, N.J., Doi, H., Eeva, T., Fickel, J., Filella, I., Fossøy, F., Goodenough, A.E., Hall, S.J.G., Hansson, B., Harris, M., Hasselquist, D., Hickler, T., Joshi, J., Kharouba, H., Martínez, J.G., Mihoub, J.B., Mills, J.A., Molina-Morales, M., Moksnes, A., Ozgul, A., Parejo, D., Pilard, P., Poisbleau, M., Rousset, F., Rödel, M.O., Scott, D., Senar, J.C., Stefanescu, C., Stokke, B.G., Kusano, T., Tarka, M., Tarwater, C.E., Thonicke, K., Thorley, J., Wilting, A., Tryjanowski, P., Merilä, J., Sheldon, B.C., Møller, A.P., Matthysen, E., Janzen, F., Dobson, F.S., Visser, M.E., Beissinger, S.R., Courtiol, A., and Kramer-Schadt, S. (2019). Adaptive responses of animals to climate change are most likely insufficient. *Nature Communications* 10(1): 3109.

Reed, T.E., Grøtan, V., Jenouvrier, S., Sæther, B.E., and Visser, M.E. (2013). Population growth in a wild bird is buffered against phenological mismatch. *Science* 340(6131): 488–491.

Regan, C.E., and Sheldon, B.C. (2023). Phenotypic plasticity increases exposure to extreme climatic events that reduce individual fitness. *Global Change Biology* 29(11): 2968–2980.

Renner, S.S., and Zohner, C.M. (2018). Climate change and phenological mismatch in trophic interactions among plants, insects, and vertebrates. *Annual Review of Ecology, Evolution, and Systematics* 49: 165–182.

Reusch, T.B. (2014). Climate change in the oceans: evolutionary versus phenotypically plastic responses of marine animals and plants. *Evolutionary Applications* 7(1): 104–122.

Reznick, D., Bryant, M.J., and Bashey, F. (2002). r- and K-selection revisited: the role of population regulation in life-history evolution. *Ecology* 83(6): 1509–1520.

Ricklefs, R.E. (2010). Life-history connections to rates of aging in terrestrial vertebrates. *Proceedings of the National Academy of Sciences of the United States of America* 107(22): 10314–10319.

Riemer, K., Guralnick, R.P., and White, E.P. (2018). No general relationship between mass and temperature in endothermic species. *eLife* 7: e27166.

Robert, A., Chantepie, S., Pavard, S., Sarrazin, F., and Teplitsky, C. (2015). Actuarial senescence can increase the risk of extinction of mammal populations. *Ecological Applications* 25(1):116–124.

Roff, D. 2001. *Life history evolution*. Oxford University Press.

Román-Palacios, C., and Wiens, J.J. (2020). Recent responses to climate change reveal the drivers of species extinction and survival. *Proceedings of the National Academy of Sciences of the United States of America* 117(8): 4211–4217.

Root, T.L., Price, J.T., Hall, K.R., Schneider, S.H., Rosenzweig, C., and Pounds, J.A. (2003). Fingerprints of global warming on wild animals and plants. *Nature* 421(6918): 57–60.

Saddlier, S., Jackson, J., and Hammer, M. (2010). *National recovery plan for the dwarf galaxias:* Galaxiella Pusilla. Department of Sustainability and Environment.

Sæther, B.E., Engen, S., Gamelon, M., and Grøtan, V. (2019). Predicting the effects of climate change on bird population dynamics. In: *Effects of climate change on birds* (eds. Dunn, P.O., and Møller, A.P.), 74–90. Oxford University Press.

Scheffers, B.R., De Meester, L., Bridge, T.C., Hoffmann, A.A., Pandolfi, J.M., Corlett, R.T., Butchart, S.H., Pearce-Kelly, P., Kovacs, K.M., Dudgeon, D., Pacifici, M., Rondinini, C., Foden, W.B., Martin, T.G., Mora, C., Bickford, D., and Watson, J.E. (2016). The broad footprint of climate change from genes to biomes to people. *Science* 354(6313): aaf7671.

Schilthuizen, M., and Kellermann, V. (2014). Contemporary climate change and terrestrial invertebrates: evolutionary versus plastic changes. *Evolutionary Applications* 7(1): 56–67.

Schwartz, O.A., and Armitage, K.B. (2005). Weather influences on demography of the yellow-bellied marmot (*Marmota flaviventris*). *Journal of Zoology* 265(1): 73–79.

Sergio, F., Blas, J., and Hiraldo, F. (2018). Animal responses to natural disturbance and climate extremes: a review. *Global and Planetary Change* 161: 28–40.

Sheldon, K.S. (2019). Climate change in the tropics: ecological and evolutionary responses at low latitudes. *Annual Review of Ecology, Evolution, and Systematics* 50: 303–333.

Sheridan, J.A., and Bickford, D. (2011). Shrinking body size as an ecological response to climate change. *Nature Climate Change* 1(8): 401–406.

Shipley, J.R., Twining, C.W., Taff, C.C., Vitousek, M.N., Flack, A., and Winkler, D.W. (2020). Birds advancing lay dates with warming springs face greater risk of chick mortality. *Proceedings of the National Academy of Sciences of the United States of America* 117(41): 25590–25594.

Simons, A.M. (2011). Modes of response to environmental change and the elusive empirical evidence for bet hedging. *Proceedings of the Royal Society B: Biological Sciences* 278(1712): 1601–1609.

Solokas, M.A., Feiner, Z.S., Al-Chokhachy, R., Budy, P., DeWeber, J.T., Sarvala, J., Sass, G.G., Tolentino, S.A., Walsworth, T.E., and Jensen, O.P. (2023). Shrinking body size and climate warming: many freshwater salmonids do not follow the rule. *Global Change Biology* 29(9): 2478–2492.

Stearns, S.J. (1992). *The evolution of life histories.* Oxford University Press.

Stoks, R., Geerts, A.N., and De Meester, L. (2014). Evolutionary and plastic responses of freshwater invertebrates to climate change: realized patterns and future potential. *Evolutionary Applications* 7(1): 42–55.

Storey, K.B., and Storey, J.M. (2012). Aestivation: signaling and hypometabolism. *Journal of Experimental Biology* 215(9):1425–1433.

Tanner, E.P., Elmore, R.D., Fuhlendorf, S.D., Davis, C.A., Dahlgren, D.K., and Orange, J.P. (2017). Extreme climatic events constrain space use and survival of a ground-nesting bird. *Global Change Biology* 23(5): 1832–1846.

Teplitsky, C., Mills, J.A., Alho, J.S., Yarrall, J.W., and Merilä, J. (2008). Bergmann's rule and climate change revisited: disentangling environmental and genetic responses in a wild bird population. *Proceedings of the National Academy of Sciences of the United States of America* 105(36): 13492–13496.

Thackeray, S.J., Henrys, P.A., Hemming, D., Bell, J.R., Botham, M.S., Burthe, S., Helaouet, P., Johns, D.G., Jones, I.D., Leech, D.I., Mackay, E.B., Massimino, D., Atkinson, S., Bacon, P.J., Brereton, T.M., Carvalho, L., Clutton-Brock, T.H., Duck, C., Edwards, M., Elliott, J.M., Hall, S.J., Harrington, R., Pearce-Higgins, J.W., Høye, T.T., Kruuk, L.E.B., Pemberton, J.M., Sparks, T.H., Thompson, P.M., White, I., Winfield, I.J., and Wanless, S. (2016). Phenological sensitivity to climate across taxa and trophic levels. *Nature* 535(7611): 241–245.

Thornton, P.K., Ericksen, P.J., Herrero, M., and Challinor, A.J. (2014). Climate variability and vulnerability to climate change: a review. *Global Change Biology* 20(11): 3313–3328.

Tu, C.Y., Chen, K.T., and Hsieh, C. (2018). Fishing and temperature effects on the size structure of exploited fish stocks. *Scientific Reports* 8: 7132.

Turbill, C., and Prior, S. (2016). Thermal climate-linked variation in annual survival rate of hibernating rodents: shorter winter dormancy and lower survival in warmer climates. *Functional Ecology* 30(8): 1366–1372.

Urban, M.C. (2015). Accelerating extinction risk from climate change. *Science* 348(6234): 571–573.

van Dyck, H., Bonte, D., Puls, R., Gotthard, K., and Maes, D. (2015). The lost generation hypothesis: could climate change drive ectotherms into a developmental trap? *Oikos* 124: 54–61.

van Noordwijk, A.J., and de Jong, G. (1986). Acquisition and allocation of resources: their influence on variation in life history tactics. *The American Naturalist* 128(1): 137–142.

van Rijn, I., Buba, Y., DeLong, J., Kiflawi, M., and Belmaker, J. (2017). Large but uneven reduction in fish size across species in relation to changing sea temperatures. *Global Change Biology* 23(9): 3667–3674.

Waldvogel, A.M., Feldmeyer, B., Rolshausen, G., Exposito-Alonso, M., Rellstab, C., Kofler, R., Mock, T., Schmid, K., Schmitt, I., Bataillon, T., Savolainen, O., Bergland, A., Flatt, T., Guillaume, F., Pfenninger, M. (2020). Evolutionary genomics can improve prediction of species' responses to climate change. *Evolution Letters* 4(1): 4–18.

Wang, H.Y., Shen, S.F., Chen, Y.S., Kiang, Y.K., and Heino, M. (2020). Life histories determine divergent population trends for fishes under climate warming. *Nature Communications* 11(1): 4088.

Weeks, B.C., Klemz, M., Wada, H., Darling, R., Dias, T., O'Brien, B.K., Probst, C.M., Zhang, M., and Zimova, M. (2022). Temperature, size and developmental plasticity in birds. *Biology Letters* 18(12): 20220357.

Wells, C.P., Barbier, R., Nelson, S., Kanaziz, R., and Aubry, L.M. (2022). Life history consequences of climate change in hibernating mammals: a review. *Ecography* 2022(6): 06056.

Williams, C.M., Henry, H.A., and Sinclair, B.J. (2015). Cold truths: how winter drives responses of terrestrial organisms to climate change. *Biological Reviews* 90(1): 214–235.

Wilson, A.J., Pemberton, J.M., Pilkington, J.G., Coltman, D.W., Mifsud, D.V., Clutton-Brock, T.H., and Kruuk, L.E.B. (2006). Environmental coupling of selection and heritability limits evolution. *PLoS Biology* 4(7): e216.

Youngflesh, C., Saracco, J.F., Siegel, R.B., and Tingley, M.W. (2022). Abiotic conditions shape spatial and temporal morphological variation in North American birds. *Nature Ecology & Evolution* 6: 1860–1870.

19

Environmental Pollution Effects on Life History

Denis Réale[1], Loïc Quevarec[2], and Jean-Marc Bonzom[2]

[1] Département de sciences biologiques, Université du Québec à Montréal, Montréal, QC, Canada
[2] Institut de Radioprotection et de Sûreté Nucléaire (IRSN), PSE-ENV/SERPEN/LECO, Cadarache, Saint Paul Lez Durance, France

19.1 Introduction

Pollution is a global environmental concern that has significant impacts on various aspects of life. A pollutant is a naturally produced or human-engineered substance or energy that, when released, degrades the environment, or impacts the life of living organisms. As such, the definition of pollutants excludes the negative effects caused by living organisms, such as viruses, pathogenic bacteria or parasites. Along with the large diversity of chemicals, other sources of pollution (*e.g.*, noise, light and temperature) affect living organisms, ecosystems and biodiversity (see also Chapter 21, this volume). Pollution affects all the ecosystems worldwide (*e.g.*, soils, wetlands, rivers, oceans, mountains, polar zones), and is one of the most important causes of the decline in biodiversity, together with habitat destruction, over-exploitation, invasive species and climate change (IPBES 2019; see also Chapter 18, this volume). To date, the fragmentation of knowledge about the biological effects of pollution represents an obstacle to assessing the impacts of human activities on ecosystems and to implementing policies aimed at stopping, and even reversing, the loss of biodiversity (Sylvester et al. 2023).

Life history traits are not spared. Life history traits are the fundamental parameters of a species' life cycle. They thus are tightly linked to an individual's fitness and to the demography of the population (Roff 1992, Stearns 1992). Pollutants can reduce survival and longevity, growth, adult size and fecundity, and delay maturity (see below). Studying them is thus essential for our understanding of the health and demographic consequences of pollutants on species and ultimately on the future of biodiversity.

Pollution can impact life history traits both over the short-term, through toxic effects at the individual organism level (*e.g.*, on molecular, physiological functions), and over the longer term, by influencing evolutionary processes (*e.g.*, Sibly and Calow 1989, Calow and Sibly 1990, Holloway et al. 1990, Coutellec and Barata 2011, Dutilleul et al. 2014). Regulatory procedures are used to assess the ecotoxicological risks of a pollutant. These studies are routinely used as a quick test to evaluate the acute toxicity of chemical substances or effluents. They generally expose a single model species to a single stressor and focus primarily on survival at a single life stage (*e.g.*, egg, larva, adult), or more rarely on reproduction. Such studies do not tell us anything about the chronic effects of a pollutant on living organisms and the potential adaptive evolutionary responses of these organisms (Newman and Clements 2007). By focusing on short-term effects on survival, these experiments do not provide any information on whether the pollutant acts by constraining all the vital parameters or if individuals show an adaptive, plastic or evolutionary response to it, for instance by moving along the fast–slow life history continuum (Coutellec and Barata 2011).

Allocation trade-offs have played a central role in the development of life history theories (Williams 1966, Stearns 1989, Roff 1992, Stearns 1992), and the pioneering work of van Noordwijk and de Jong (1986) highlighted the importance of considering resource acquisition in conjunction with allocation. In this chapter, we review studies on the effects of pollutants on life history traits under the light of the acquisition/allocation principle (see below).

Numerous pollutants, with different action modes, impact living organisms. For simplicity, we have regrouped pollutants into two categories. The first comprises physical pollutants (*e.g.*, ionising radiation, fine particles, macro-, micro-, and

Life History Evolution: Traits, Interactions, and Applications, First Edition. Edited by Michal Segoli and Eric Wajnberg.
© 2025 John Wiley & Sons Ltd. Published 2025 by John Wiley & Sons Ltd.

nanoplastics, light, sonic, thermal and electromagnetic pollution). The second includes chemical pollutants (*e.g.*, trace metal elements, salt, benzene, nonylphenol, polycyclic aromatic hydrocarbons, pesticides, nitrates, endocrine disruptors, pharmaceutical residues and greenhouse gases). Some pollutants, such as plastics, can be a source of both physical (*e.g.*, Zolotova et al. 2022) and chemical (*e.g.*, Sarkar et al. 2023) pollution. Depending on their nature, pollutants act, directly and indirectly, on organisms through different pathways and action modes (*e.g.*, molecular, metabolic, physiological, neuronal, ecological). However, we are far from understanding how these effects translate into changes in life history traits at different time scales. We summarize knowledge on these effects over both the short (*i.e.*, within a generation) and long (*i.e.*, over multiple generations) terms. In closing, we discuss future directions for studying the links between pollutants and life history, how they can provide vital information on the fate of living organisms in the Anthropocene era and generate essential information on the plastic and evolutionary responses of populations' life history to rapid environmental changes.

19.2 The Role of Life History Theories in Ecotoxicology

Life history theories are based on the central notion of resource allocation trade-offs (Williams 1966, Stearns 1989, Roff 1992, Stearns 1992). An organism must find and exploit resources, and every organism is limited in its ability to acquire resources (the notion of resources goes beyond energy alone and includes essential nutrients and time). As a result, at every phase of their life cycle, organisms must trade off resources allocated to different biological functions, such as growth, maintenance, reproduction or storage (Roff 1992, Stearns 1992), or within the reproduction function, between the production of many small offspring or a few large ones (Lack 1968; see also Chapter 3, this volume).

Although allocation trade-offs should occur under many circumstances, they often go undetected in natural populations (Reznick et al. 2000). In their seminal paper, van Noordwijk and de Jong (1986) proposed, with the so-called Y model, an explanation for this paradox. According to the Y model, if the variation among individuals in their ability to acquire resources is greater than the variation in allocation, this can obscure the allocation trade-off between two biological functions (Figure 19.1). The Y model thus highlights the importance of considering both acquisition and allocation of resources in studying the response of traits to a change in the environment, such as the arrival of a new pollutant (Roff and Fairbairn 2007). The Y model makes two main predictions: (1) increasing variance in acquisition reduces our ability to detect the trade-off (*i.e.*, leading to a null or a positive correlation between the two measured life history traits); (2) for a constant mean acquisition and allocation and variance in acquisition, increasing variance in allocation increases our ability to detect a trade-off (*i.e.*, leading to a more negative correlation between the two life history traits; King et al. 2011; Figure 19.1).

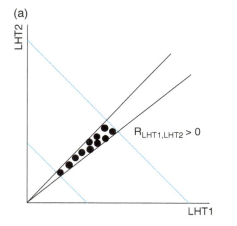 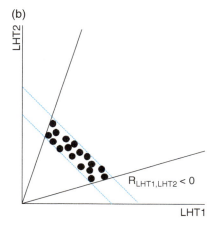

Figure 19.1 Graphical representation of the van Noordwijk and de Jong 1986 model of the effect of the relative variance in acquisition and allocation on the detection of a trade-off. (a) The variance in acquisition is greater than the variance in allocation between the two life history traits LHT1 and LHT2. In this situation, the relationship between the two traits ($R_{LHT1, LHT2}$) is positive and the trade-off is undetected. (b) The variance in acquisition is less than the variance in allocation between LHT1 and LHT2. Here the relationship between the two traits is negative indicating a trade-off. Black dots represent individuals with different levels of acquisition and allocation of resources to LHT1 and LHT2 (*Source:* Adapted from van Noordwijk and de Jong 1986).

The van Noordwijk and de Jong (1986) model assumes random variation in resource allocation to reproduction around its average (King et al. 2011). Although the Y model has provided conceptual advances in life history theories, its predictions have rarely been tested empirically (King et al. 2011).

Ecotoxicologists study the effects of pollutants on living organisms and their ecological consequences at multiple levels of organisation (*i.e.*, individuals, populations, communities and ecosystems). One of the goals of ecotoxicologists is to evaluate how a pollutant disrupts the metabolic pathways of organisms and the fate of populations (Kooijman 2000, Congdon et al. 2001). Since metabolism and resources, directly and indirectly, translate into changes in life history traits, many authors have naturally focused on life history theories to link pollution-disrupted metabolism, to individual life history traits and population growth rate (Calow and Sibly 1990, Forbes and Calow 1996). Furthermore, ecotoxicologists have embraced the concept of resource allocation trade-off and often assume that polluted organisms allocate resources to detoxification and cell repair at the expense of other biological functions (Sibly and Calow 1989, Calow and Sibly 1990, Calow and Forbes 1998, Congdon et al. 2001). They have also focused on ways to analyse the demographic consequences of the phenotypically plastic responses of individual organisms to pollutants (Schaaf et al. 1987).

More recently, some researchers have argued that independent of resource acquisition and allocation, trade-offs are also mediated by switches in signalling pathways at the molecular level (Flatt et al. 2011). The occurrence of such pathways does not reduce the importance of the Y model paradigm in life history theory (Stearns 2011). If pollutants can disturb hormones and neurosensory systems, we would also expect that pollution can lead to important changes in life history traits beyond their effects on acquisition and allocation of resources. This makes the researchers' task even more complex.

19.3 The Acquisition/Allocation Principle and the Responses of Organisms to Pollution

Using the resource acquisition/allocation principle as a framework can help us describe the possible responses of an organism to a pollutant. First, we must distinguish between responses (1) at the individual organism level, (2) at the population level over the shorter term (*i.e.*, within one generation) and (3) over the longer term (*i.e.*, multiple generations).

19.3.1 Responses at the Individual Organism Level

Within individuals, a pollutant can impact life history traits in two main ways, which result from the individual's acquisition and allocation of resources presented above. Let first focus on the phenotypically plastic responses of an individual to a pollutant. These responses may depend on many factors, such as the nature of the pollutant, its mode of action, its concentration or intensity, the species, the life stage exposed to the pollutant and other (biotic and abiotic) factors that could interact with the pollutant.

The pollutant may reduce the organism's ability to acquire resources by impacting movements and foraging rate, or assimilation efficiency (Congdon et al. 2001, Montiglio and Royauté 2014). This in turn should lower the resources the organism can invest in all its biological functions simultaneously (Figure 19.2a). As a result, we expect a systematic decrease in the values of all the life history traits (*e.g.*, survival, growth, reproduction). We will call this scenario the reduced acquisition hypothesis. This effect may seem trivial, but it can play an important role in blurring the interpretation of the results. We will see later in Tables 19.2–19.4 that this response is commonly described in the literature.

The pollutant may also indirectly affect an individual organism by altering its operative environment (*i.e.*, population density, prey or resource abundance, refuges, mates, competition, predation pressures; Congdon et al. 2001, Montiglio and Royauté 2014). A parallel decline in resource abundance caused by the pollutant may act by limiting the resource acquisition capacities of the individual. In contrast, a decrease in population density or the abundance of other species may decrease the inter- and intra-specific competition, or reduce the transmission of parasites and disease, which could result in a higher acquisition capacity for the individual. These conflicting effects may blur our ability to detect phenotypically plastic responses to the pollutants (Barata et al. 2002). Note that laboratory experiments can control for these effects to focus on the direct effects of the pollutant, but by doing so they will not provide a complete picture of the effects of pollution in the wild.

When confronted with pollution, an individual can allocate more resources to one biological function at the expense of another function (Figure 19.2b; the so-called allocation principle: Sibly and Calow 1989, Calow and Sibly 1990, Calow and Forbes 1998, Congdon et al. 2001, Janssens et al. 2021). Let us call this scenario the compensatory response hypothesis. For instance, the individual could invest more resources in its biochemical defence mechanisms, such as detoxification

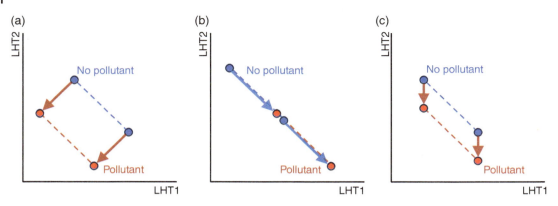

Figure 19.2 Hypothetical consequences of a pollutant on resources allocated by the individual organism to two biological functions (life history traits LHT1 and LHT2: *e.g.*, maintenance, growth, or reproduction). (a) According to the reduced acquisition hypothesis, the resources acquired by the organism decrease and lead to a systematic decline in the resources available for the two biological functions (red arrows) by each individual (from blue dots to red dots), although the trade-off (negative relationship; dashed lines) between the two life history traits is maintained. (b) With compensatory response hypothesis, allocation of resource moves along the life of trade-off: each individual increases the resources dedicated to one biological function (blue arrow to LHT1) at the expense of the other function (LHT2), without reducing their overall acquisition ability. (c) These two options can act simultaneously, for example, with decreased acquisition capacities, individuals may decrease one life history trait (LHT2) to prioritize another (LHT1). Fitness is represented here as the blue dashed and the red dashed line. With the compensatory response hypothesis, the organism maintains its fitness, but fitness decreases in scenario 19.2a,c.

and repair of cellular damage by the toxins. This allocation of resources would theoretically favour its maintenance and survival but decrease fecundity (Sibly and Calow 1989). Empirically it has shown to slow their pace of life by decreasing early fecundity (*e.g.*, salt treatment in Dutilleul et al. 2014). Alternatively, the individual could speed its reproduction, and adopt a faster pace of life, with reduced longevity (*e.g.*, uranium treatment in Dutilleul et al. 2014). Under polluted conditions, it may also produce fewer but larger offspring that could withstand pollution effects better (*e.g.*, mine effluent in Maltby 1991).

The hypotheses in Figure 19.2a, b are not mutually exclusive. Individuals may be constrained in their acquisition but still prioritize one biological function over the other (Figure 19.2c). As a result, we may observe a decrease in the value of one trait (life history trait 2), while the value of the other trait (life history trait 1) does not change, increase or decrease. Note that the allocation principle may explain the quadratic effect of a pollutant on an organism's traits, also known as hormesis (Newman and Clements 2007). Differential allocation of resources to one trait (LHT1) at the expense of another (LHT2; scenario Figure 19.2b) may first lead to an increase in LHT1 with an increase in the concentration of the pollutant. Past a certain concentration, the organism switches from scenario 19.2b to scenario 19.2c, leading to a stabilisation of LHT1 and then to scenario 19.2a with a concomitant decrease in LHT1.

What are the consequences for fitness? If fitness depends on the combination of the two life history traits as seen in Figure 19.2, scenario 19.2b maintains fitness at a similar level, whereas fitness declines in both scenarios 19.2a and 19.2c. With the reduced acquisition hypothesis, we thus expect a major decline in fitness, whereas the compensatory response hypothesis alone can predict a more stable lifetime fitness, and the shift towards a different life history strategy (*e.g.*, a faster or a slower pace of life; Table 19.1).

Table 19.1 Predictions from the two main responses of an organism to a pollutant at the population level. LTH1 and LTH2 are two life history traits (see Figure 19.2).

Scenario	Acquisition	LHT1	LHT2	Fitness	Trade-off
Reduced acquisition (Figure 19.2a)	Down	Down	Down	Down	Hidden
Compensatory response (Figure 19.2b)	Stable	Up	Down	Stable	Visible

19.3.2 Short-Term Mean Population-Level Responses

Different factors can cause changes in the mean value of a trait in a population subjected to new environmental conditions (*e.g.*, exposure to pollution). Developmental plasticity is the first factor commonly considered, with the assumption that the short-term mean response of the population reflects individual plastic responses to the new conditions. This hypothesis holds when all the individuals/genotypes show a similar response. For example, they all decrease their acquisition (Figure 19.2a). Individuals/genotypes, however, can vary in both their acquisition and allocation of resources in the original conditions (Roff 1992, Stearns 1992) and when subjected to the pollutant (Mulvey et al. 1995, Newman and Clements 2007, Montiglio and Royauté 2014, Dutilleul et al. 2015). For example, some may forage more efficiently in the polluted environment. Some may prioritize detoxification and maintenance, while others may favour growth (Holloway et al. 1990) or increase their reproductive efforts (Sibly and Calow 1989). In other words, some may exhibit reduced acquisition, others may opt for compensatory responses and they may also differ in the degree of their responses.

Variation among individuals, some of it genetic, has important consequences for the short-term responses observed at the population level. The mean population response is likely to reflect the most common individual response, but some individual responses may cancel each other, reducing the contrast between the polluted treatments and the control. Furthermore, variation in responses may lead to different survival rates. For example, depending on their acquisition level, individuals may survive differently during the study and may not be equally represented in the measure of different traits ('selective disappearance': Hayward et al. 2012). Thus, the short-term, average population response may provide confusing results that are difficult to interpret.

We synthesize the different predictions associated with each hypothesis on the average population response in Table 19.1. Stable or increased values of one life history trait in the polluted environment provide evidence for some compensatory response occurring in that population. In contrast, a decline in all the measured life history traits provides some evidence for a reduced acquisition but does not exclude the possibility that one unmeasured life history trait is not increasing with pollution.

In the absence of information on the trade-off, it will be difficult to predict the potential evolutionary responses to the pollutant. Other analyses would be necessary, such as examining among-individual or genetic covariance between the traits in the two environments.

19.3.3 Long-Term Responses of the Population

One less-studied goal of ecotoxicology is to evaluate whether populations could persist in the longer term under the new polluted conditions (Holloway et al. 1990, Medina et al. 2007, Coutellec and Barata 2011). Because of their different fitness outcomes, the different responses may be counter-selected or favoured by selection and passed on to the next generation through either genetic or epigenetic changes (*e.g.*, Dutilleul et al. 2015, Quevarec et al. 2023). We thus expect that polluted populations will not show the same responses to a pollutant as pollutant-naïve populations (Holloway et al. 1990, Maltby 1991). Note that the decline in average fitness in the population following exposure does not tell us anything about selection. The important information to consider is whether the covariance between the traits and fitness changes between the unpolluted and the polluted environment (Lande and Arnold 1983, Arnold and Wade 1984).

Assuming some non-negligible additive genetic variance for life history traits and acquisition, the population can evolve in response to pollution (Holloway et al. 1990, Dutilleul et al. 2015). If pollution directly reduces acquisition (see above), it may also lead to selection pressures improving the capacities for acquisition in the new conditions over successive generations. Such changes, however, can occur with a concomitant phenotypically plastic decline in all life history traits (*e.g.*, slow growth, delayed maturity, lower fecundity or shorter lifespan, see Cameron et al. (2013) for an example with food restriction). It is, thus, hard to predict how the polluted individuals will perform, compared to naive ones, when put back in an unpolluted environment.

To be adaptive, compensatory responses should favour the life history traits and functions that have the strongest impact on the organism's fitness. A few models have made predictions about how populations should evolve depending on the effects of the pollutant on different age classes (reviewed in Newman and Clements 2007). By decreasing adult survival, a pollutant should lead to the evolution of lower allocation to defence and maintenance (Sibly and Calow 1989) and a faster pace of life (*i.e.*, rapid growth, early maturity; Sibly and Calow 1989, Réale et al. 2010). In contrast, a pollutant that decreases survival at the juvenile stage should lead to the evolution towards a slower pace of life (*i.e.*, delayed maturity, slower growth, more spread reproductive effort; see also Chapter 1, this volume). Other models made opposite predictions. Ernande et al. (2004) found that harvesting adult stages would delay age at maturity and lead to bigger adults while harvesting juveniles or both juvenile and adult stages would decrease age and size at maturity (see Cameron et al. 2013; see also Chapter 23, this volume). The discrepancy between the predictions of the different models could come from the conditions chosen by Ernande et al. (2004), which penalize fast-growing individuals. If pollution impact is size-dependent, we expect it to impact small individuals more strongly (*e.g.*, due to the effect of the area/volume ratio on the accumulation of the pollutant). Paradoxically, the new environment (*e.g.*, pollution) could even act directly on the plastic response to growth while exerting new selection pressures favouring better growth capacity, with no phenotypic change in growth and adult size in the population (Conover and Schultz 1995). Over multiple generations, we can expect a switch from reduced acquisition, a decrease in all the life history traits and declining population during the first generations to a compensatory response and an increased population growth rate with time (Dutilleul et al. 2017, Quevarec et al. 2023).

Following a disturbance, slow-lived species often decrease their population size, whereas populations of fast-lived ones often increase it (Gamelon et al. 2015). Short- and long-term effects of pollution may thus vary depending on the species' life history. Furthermore, because they are conservative in their life histories (*i.e.*, prioritize survival over fecundity), long-lived species in a polluted environment should prioritize maintenance and adult survival over short-term reproduction. In contrast, short-lived species should prioritize short-term reproduction (*i.e.*, increased in immediate fecundity) or even faster development, at the expense of survival.

19.3.4 The Cost of Adapting to Pollutants

Adaptation to one pollutant may increase resistance to other pollutants. Oliver and Brooke (2018) report a case of the mosquito DDT-resistant strain that also displays resistance to permethrin, deltamethrin, λ-cyhalothrin and Malathion. This capacity for multiple resistance comes from a higher production of detoxification enzymes. Multiple resistance can provide a fitness advantage when the population is subjected to new pollutants. In this case, we would have to consider that the individuals are not naive and show some adaptive phenotypically plastic response to the new pollutant. As we will see below, it is thus important to ask whether pollutants have the same global effects on life history traits or not.

Based on the acquisition/allocation principle, the organism diverts resources from some biological functions, linked to performances and fitness, to allocate them to detoxification, resistance and maintenance (Jansen et al. 2011a). Adaptation to a polluted environment can thus come with some metabolic and fitness costs. Adaptation to certain conditions such as pollution can also lead to maladaptation in other conditions (Kawecki and Ebert 2004). First, if the allocation trade-off has a genetic basis (Reznick et al. 2000, Roff and Fairbairn 2007), the frequency of alleles linked to more efficient detoxification will increase in the polluted environment, since their benefits outweigh their costs. However, once back in an unpolluted environment, the benefits of these alleles will be lost while the costs will be maintained (Jansen et al. 2011a). Second, as explained above, some genotypes with a given pace of life may be able to bypass the stages of life where selection is the strongest. For example, if the pollutant mostly kills adults, fast-lived but not slow-lived individuals can reproduce before they die. Furthermore, adaptive response to a polluted environment may generally reduce genetic diversity, as shown by Athrey et al. (2007) in cadmium-resistant populations of least killifish (*Heterandria formosa*). This effect may reduce the capacity of the populations to adapt to additional stressors (Jansen et al. 2011a).

Many studies, however, do not observe any costs or trade-offs on the life history traits associated with adaptation to a pollutant. For example, Lopes et al. (2008) and Brausch and Smith (2009) observed an improvement in resistance to pesticides without any associated costs in the nematode *Caenorhabditis elegans* and the crustacean *Daphnia magna*, respectively. Several possible causes can explain the absence of observed costs. The experimental evolution approach may not effectively assess changes in resistance and detect associated costs (Kawecki et al. 2012). Indeed, the study may fail to measure the traits impacted by adaptive costs (Jansen et al. 2011b). In addition, the costs may not be apparent if they are spread over multiple traits at the same time, or if the experimental background noise reduces the statistical power to detect significant results. Identifying and integrating adaptation costs is essential for characterizing individual responses to pollution and, more generally, to environmental stressors. However, the evolution and functioning of these costs are still relatively poorly understood (Roff and Fairbairn 2007).

19.4 Literature Survey on Mechanisms Involved in the Life History Responses to Pollutants

Tables 19.2–19.4 synthesize the results of studies on the effects of different pollutants on life history traits (*e.g.*, larvae size, juvenile survival, growth, adult size, development rate, fecundity, adult survival, longevity, sex ratio) for physical, metal chemical and organic chemical pollutants, respectively. Rather than trying to be exhaustive, we aimed to highlight the general trends in the impact of these pollutants. In this review, we provide a snapshot of the state of research into the effects of pollutants and the potential gaps in our knowledge. More specifically, we can check whether pollutants, which differ in how they act on living organisms, result in a systematic decline in all life history traits or in compensatory responses by these different traits (Figure 19.2, Table 19.1).

We have considered two types of studies on the effects of pollutants on life history traits: effects of a range of concentrations of a pollutant (RoC in Tables 19.2–19.4) and common garden experiments (CG). Most often RoC studies focus on the effect of acute toxicity on survival. Authors used this approach to determine the LD_X (lethal dose x, administered to an organism) or the LC_X (lethal concentration x, in the air or water), where x is the percentage of mortality linked to a quantity or concentration of a pollutant. The results for a pollutant on different species can be aggregated to produce species sensitivity distribution curves to estimate its concentration that is hazardous to a small proportion of species and determine the non-observed effect concentrations (the highest tested concentration at which the pollutant shows no statistically significant effect). Such meta-analyses are available for different potential pollutants, for example, copper (Simpson et al. 2011), radioactive substances (Garnier-Laplace et al. 2006), herbicides (van den Brink et al. 2006) and an insecticide such as neonicotinoids (Morrissey et al. 2015). This approach is likely useful for setting safety limits for chemical concentrations to protect the structure and functioning of ecosystems (Del Signore et al. 2016). However, our goal was to examine the effects of pollutants on several life history traits that directly influence fitness and ultimately the demography of the population. We thus omit most of these studies in our tables.

By measuring several life history traits, one can highlight the consequences of exposure to a pollutant at concentrations much lower than those directly affecting survival. For example, Dutilleul et al. (2015) showed a decrease in average fecundity and growth of *C. elegans* population in a uranium-polluted environment, despite no effect on survival. The study of multiple life history traits may also allow examining the effects of pollutants on organisms as a function of life stages or sex. Numerous works have shown differential effects on life history traits according to sex or developmental stages. For example, exposure of the marine nematode, *Litoditis marina*, to 0.006% sodium dodecyl sulphate reduced male but not female survival (Oliveira et al. 2020).

In the 'systematic or compensatory' column (Tables 19.2–19.4), we compiled the evidence that the pollutant impacted all the traits systematically (Figure 19.2a), with some negative consequences for fitness. For instance, a lower growth rate, a slower developmental rate, a lower fecundity and a lower survival, would indicate that the pollutant globally reduces the resources available for growth, reproduction, and maintenance in individuals. In contrast, the combination of positive or no effect with negative effects in a study indicates a compensatory response, when at least one of the traits changes in a compensating way. For example, lower survival, along with faster development and increased fecundity, would suggest a shift towards a faster pace of life in response to the pollutant. In a few cases, the pollutant has only beneficial or neutral effects on all the traits measured. However, in those studies, the pollutant might have affected some unmeasured traits.

Unsurprisingly, most of the studies were done on invertebrates and short-lived species (see also Loria et al. 2019). Time and logistics often restrict these experimental laboratory studies. Reasonably, the authors chose an appropriate biological model, such as planktonic crustaceans or other arthropods, and very rarely vertebrates, which are harder or even impossible to breed in laboratory conditions. These studies, however, may not inform us about the effects of pollutants on long-lived species with different life histories. As a result, we have scant information on the life history consequences of pollution on long-lived organisms.

19.4.1 Studies Analysing the Effects of a Range of Concentrations of a Pollutant

Most of the reviewed studies subjected animals to a RoC, alone or with other factors (*e.g.*, temperature, other pollutants). They examine the developmental consequences of these pollutants, within one generation. Thus, we have good information on the intra-generational, plastic responses of life history traits to the effects of pollutants.

Table 19.2 Effects of physical pollutants on life history traits in animal species, grouped by type of study.

Pollution (other factors)*	Type of study	Target organism	Phylum	Larvae size or mass	Juvenile survival	Growth	Size at maturity or adult size	Developmental rate	Fecundity or Fertility	Adult survival	Longevity	Sex ratio (male/female)	Systematic or Compensatory	References
Electric fields	RoC	*Heterocypris incongruens*	Crustacea		0			0		−			Compensatory	Bieszke et al. (2020)
Flexible polyvinylchloride (diisononylphthalate + glass beads)*	RoC	*Daphnia magna*	Crustacea				0*	−	0*	0			Compensatory	Schrank et al. (2019)
Gamma radiation	RoC	*Caenorhabditis elegans*	Nematode		−	−			−				Systematic	Quevarec et al. (2023)
Magnetic fields	RoC	*Heterocypris incongruens*	Crustacea		0			+		−			Compensatory	Bieszke et al. (2020)
Nanoplastics	RoC	*Brachionus plicatilis*	Rotifera				−	−	−		−		Systematic	Wang et al. (2022)
Particulate matter	RoC	*Tigriopus japonicus*	Crustacea					−	0	0			Compensatory	Han et al. (2022)
Polyethylene microplastics	RoC	*Chironomus riparius*	Hexapoda	−		−		−					Systematic	Silva et al. (2019)
Polyethylene microplastics (deltamethrin)*	RoC	*Daphnia magna*	Crustacea				0	0*	0*		0*		Neutral	Felten et al. (2020)
Polyethylene-terephtalate/polystyrene/polyvinyl-chloride/polyamide	RoC	*Chironomus riparius*	Hexapoda	+				−		0			Compensatory	Stanković et al. (2020)
Polystyrene microplastics	RoC	*Brachionus plicatilis*	Rotifera				−	−	−	−	−		Systematic	Sun et al. (2019b)
Polystyrene microplastics	RoC	*Oryzias melastigma*	Vertebrata	−		−		+	−				Compensatory	Wang et al. (2021)
Polystyrene microplastics	RoC	*Daphnia pulex*	Crustacea				−	0	0	−			Compensatory	Zhu et al. (2022)
Star polycation nanocarrier	RoC	*Drosophila melanogaster*	Hexapoda						−	−	−		Systematic	Yan et al. (2022)
Percentage of negative impact				**67**	**50**	**100**	**60**	**55**	**56**	**63**	**67**	N.A.		
Gamma radiation	CG	*Caenorhabditis elegans*	Nematode									+	N.A.	Quevarec et al. (2022)
Nanoplastics	CG	*Brachionus plicatilis*	Rotifera				0	0	0		0		Neutral	Wang et al. (2022)

RoC: Range of concentration experiments submitting animals to a range of concentration of pollutant; CG: common CGs comparing strains or populations having lived in polluting environment with pollutant-naïve strains or populations; + sign means a positive effect; − means a negative effect; 0: mean no significant effect. N.A. = no clear evidence for either systematic impact or compensatory response (e.g., when a maximum of one trait changed). For CG studies, A '+' indicates that, in similar environmental conditions, the polluted population performs better than the unpolluted population for that trait. Percentage of negative impact represents the percentage of studies that show a negative impact of the pollutants on each trait. In bold, traits with at least five studies. Because of the small number of common CG we did not calculate the percentage of negative impact for this type of study. Other factors* = other factors that interact with the pollutant on the traits.
*: in interaction with another factor.

Table 19.3 Effects of chemical pollutants (metals) on life history traits in animal species, grouped by type of study.

Pollution (other factors)*	Type of study	Target organism	Phylum	Larvae size or mass	Juvenile survival	Growth	Size at maturity or adult size	Developmental rate	Fecundity or Fertility	Adult survival	Longevity	Sex ratio (male/female)	Systematic or Compensatory	References
Cadmium	RoC	*Chironomus riparius*	Hexapoda		–	–		–	–	–	0		Systematic	Postma et al. (1995)
Cadmium	RoC	*Daphnia magna*	Crustacea				–	–	–	–	0		Systematic	Wei et al. (2022)
Cadmium	RoC	*Daphnia magna*	Crustacea				–	–	–	–			Systematic	Wei et al. (2022)
Cadmium	RoC	*Eurytemora affinis*	Crustacea		–		0		–	–	–	0	Systematic	Kadiene et al. (2017)
Cadmium	RoC	*Helicoverpa armigera*	Hexapoda	–	–		–		–	–	–		Systematic	Zhan et al. (2017)
Cadmium	RoC	*Lymantria dispar* L.	Hexapoda	–				–		–	–	0	Systematic	Mirčić et al. (2010)
Cadmium	RoC	*Lymnaea palustris*	Mollusca		–				–				Systematic	Coeurdassier et al. (2003)
Cadmium	RoC	*Lymnaea stagnalis*	Mollusca			–			–				Systematic	Coeurdassier et al. (2003)
Cadmium	RoC	*Pseudodiaptomus annandalei*	Crustacea		–		0		–	–	–		Systematic	Kadiene et al. (2017)
Cadmium	RoC	*Pseudodiaptomus annandalei*	Crustacea					–	–	–	–	–	Systematic	Kadiene et al. (2019)
Cadmium	RoC	*Pardosa saltans*	Chelicerata		0	0		0					Neutral	Eraly et al. (2010)
Cadmium	RoC	*Pseudodiaptomus annandalei*	Crustacea						–				N.A.	Kadiene et al. (2022)
Cadmium	RoC	*Pseudodiaptomus annandalei*	Crustacea				–*		+				N.A.	Kadiene et al. (2022)
Cadmium (food)*	RoC	*Brachionus plicatilis*	Rotifera					–*	–*	–*	–*		Systematic	Sun et al. (2019a)
Cadmium chloride (malathion + deltamethrin)*	RoC	*Anopheles arabiensis*	Hexapoda				0	–		–			Systematic	Oliver and Brooke (2018)
Chromium	RoC	*Tisbe holoturiae*	Crustacea						–	–			N.A.	Miliou et al. (2000)
Cobalt	RoC	*Tisbe holoturiae*	Crustacea						–	–			N.A.	Miliou et al. (2000)
Copper	RoC	*Daphnia longispina*	Crustacea			–			–	–			Systematic	Agra et al. (2011)
Copper (temperature)*	RoC	*Protophormia terraenovae*	Hexapoda				–*	–*	–	–			Systematic	Pölkki et al. (2014)
Copper + ammonium sulfate	RoC	*Brachionus calicyflorus*	Rotifera		–			–	–	0			Systematic	Schanz et al. (2021)
Copper + zinc	RoC	*Theodoxus fluviatilis*	Mollusca			–			–	–			Systematic	Bighiu et al. (2017)

(*Continued*)

Table 19.3 (Continued)

Pollution (other factors)*	Type of study	Target organism	Phylum	Larvae size or mass	Juvenile survival	Growth	Size at maturity or adult size	Developmental rate	Fecundity or Fertility	Adult survival	Longevity	Sex ratio (male/female)	Systematic or Compensatory	References
Copper nitrate (malathion + deltamethrin)*	RoC	*Anopheles arabiensis*	Hexapoda				0	−	−	−			Systematic	Oliver and Brooke (2018)
Diverse metal: Pb, and Zn	RoC	*Chironomus riparius*	Hexapoda	−	0		−					0	Systematic	Arambourou et al. (2020)
Diverse metal: Zn, Pb, Cd, Cu	RoC	*Dendrobaena octaedra*	Annelida		0		−		−		−		Systematic	Rożen (2006)
Fluidized bed combustion ashes: silico-alumineous ashes	RoC	*Eisenia andrei*	Annelida		0	−	−		−				Systematic	Grumiaux et al. (2007)
Fluidized bed combustion ashes: sulfo-calcical ashes	RoC	*Eisenia andrei*	Annelida		−	−	−		−				Systematic	Grumiaux et al. (2007)
Lead nitrate (malathion + deltamethrin)*	RoC	*Anopheles arabiensis*	Hexapoda				0	0	−	−			N.A.	Oliver and Brooke (2018)
Mercury: mercuric chloride	RoC	*Tigriopus japonicus*	Crustacea				−	−	−	0			Systematic	Li et al. (2015)
Metal-adapted populations to Cu and Zn	RoC	*Isotoma notabili*	Hexapoda						−	−			Systematic	Tranvik et al. (1993)
Metal-adapted populations to Cu and Zn	RoC	*Onychiurus armatus*	Hexapoda			+	+		−	−			Compensatory	Tranvik et al. (1993)
Metal-adapted pop response to cadmium	RoC	*Orchesella cincta*	Hexapoda			−	−		−	−			Systematic	Posthuma et al. (1993)
Petroleum-derived substances	RoC	*Rhopalosiphum padi L.*	Hexapoda			−			−	−	−		Systematic	Rusin et al. (2017)
Salt-NaCl	RoC	*Caenorhabditis elegans*	Nematode			−			−	−			Systematic	Dutilleul et al. (2015)
Sodium dodecyl sulfate	RoC	*Diplolaimella dievengatensis*	Nematoda			+			+	0			Benefit	Oliveira et al. (2020)
Sodium dodecyl sulfate (sex)*	RoC	*Litoditis marina*	Nematoda			−			−	−*			Systematic	Oliveira et al. (2020)
Uranium	RoC	*Caenorhabditis elegans*	Nematode						−	0			Compensatory	Dutilleul et al. (2015)
Uranium + a cocktail of metals	RoC	*Daphnia magna*	Crustacea				−	0	0	−			N.A.	Reis et al. (2018)
Zinc $ZnCl_2$	RoC	*Ischnura elegans*	Hexapoda			−							Systematic	Debecker and Stocks (2019)
Percentage of negative impact				100	64	80	57	65	89	81	75	25		

Pollutant	Study type	Species	Taxon								Category	Reference
Cadmium	CG	*Pardosa saltens*	Chelicerata	0	0		0		0		Neutral	Eraly et al. (2010)
Cadmium chloride (malathion + deltamethrin)*	CG	*Anopheles arabiensis*	Hexapoda	0	0	0	0		0		Neutral	Oliver and Brooke (2018)
Chromium	CG	*Tisbe holoturiae*	Crustacea				+				N.A.	Miliou et al. (2000)
Cobalt	CG	*Tisbe holoturiae*	Crustacea				+				N.A.	Miliou et al. (2000)
Copper	CG	*Daphnia longispina*	Crustacea		+	+	+				Benefit	Agra et al. (2011)
Copper + Ammonium sulfate	CG	*Brachionus calicyflorus*	Rotifera	−	0		0		0		Compensatory	Schanz et al. (2021)
Copper nitrate (malathion + deltamethrin)*	CG	*Anopheles arabiensis*	Hexapoda		0	0	0		+		Neutral	Oliver and Brooke (2018)
Diverse metal: Zn, Pb, Cd, Cu	CG	*Dendrobaena octaedra*	Annelida	0	+	0	0		0	0	N.A.	Rożen (2006)
Diverse metal: Zn, Pb, Cd, Cu	CG	*Pterostichus oblongopunctatus*	Hexapoda	−	0	0	0		+	0	Compensatory	Lagisz and Laskowski (2008)
Lead nitrate (malathion + deltamethrin)*	CG	*Anopheles arabiensis*	Hexapoda		0	0	0		0		Neutral	Oliver and Brooke (2018)
Mercury: mercuric chloride	CG	*Tigriopus japonicus*	Crustacea			0	0		0		Neutral	Li et al. (2015)
Metal-adapted populations to Cu and Zn	CG	*Isotoma notabili*	Hexapoda			+			+		Benefit	Tranvik et al. (1993)
Metal-adapted populations to Cu and Zn	CG	*Onychiurus armatus*	Hexapoda		+	0	0		+		Benefit	Tranvik et al. (1993)
Metal-adapted pop reponse to Cadmium	CG	*Orchesella cincta*	Hexapoda		+	+	+		+		Benefit	Posthuma et al. (1993)
Metal and petrochemical contaminants	CG	*Melita plumulosa*	Crustacea			−	−				Systematic	Chung et al. (2008)
Salt-NaCl (sex)*	CG	*Caenorhabditis elegans*	Nematode		−*				−		Systematic	Dutilleul et al. (2017)
Uranium (sex)*	CG	*Caenorhabditis elegans*	Nematode		−*				−		Systematic	Dutilleul et al. (2017)
Zinc ZnCl$_2$	CG	*Ischnura elegans*	Hexapoda		−				+		Compensatory	Debecker and Stocks (2019)
Percentage of negative impact			N.A.	50	43	12	20		27	0	0	0

RoC: Range of concentration experiments submitting animals to a range of concentration of pollutant; CG: common CGs comparing strains or populations having lived in polluting environment with pollutant-naïve strains or populations; + sign means a positive effect; − means a negative effect; 0: mean no significant effect. N.A. = no clear evidence for either systematic impact or compensatory response (e.g., when a maximum of one trait changed). For CG studies, A '+' indicates that, in similar environmental conditions, the polluted population performs better than the unpolluted population for that trait. Percentage of negative impact represents the percentage of studies that show a negative impact of the pollutants on the traits. In bold, traits with at least five studies. Other factors* = other factors that interact with the pollutant on the traits.

*: in interaction with another factor.

Table 19.4 Effects of chemical pollutants (organic) on life history traits in animal species, grouped by type of study.

Pollution (other factors)*	Type of study	Target organism	Phylum	Larvae size or mass	Juvenile survival	Growth	Size at maturity or adult size	Developmental rate	Fecundity or Fertility	Adult survival	Longevity	Sex ratio (male/female)	Systematic or Compensatory	References
17α-ethinylestradiol	RoC	Daphnia magna	Crustacea			−		+	0				Compensatory	Rodrigues et al. (2021)
Atrazine (kairomones)*/***	RoC	Daphnia pulex	Crustacea			−	−*	−	−				Systematic	Qin et al. (2022)
Benzo[a]pyrene	RoC	Lymantria dispar L.	Hexapoda	−		−		−					Systematic	Ilijin et al. (2015)
Benzo[a]pyrene	RoC	Aedes aegypti	Hexapoda		−		0	+	0				Compensatory	Prud'homme et al. (2017)
Bisphenol A	RoC	Aedes aegypti	Hexapoda		0		0	0	0			0	Neutral	Prud'homme et al. (2017)
Chlorpyrifos	RoC	Ischnura elegans	Hexapoda			−				−			Systematic	Janssens et al. (2021)
Chlorpyrifos	RoC	Ischnura elegans	Hexapoda			+	+			−			Compensatory	Arambourou and Stocks (2015)
Chlorpyrifos	RoC	Tecia solanivora	Hexapoda							−			N.A.	Gutiérrez et al. (2019)
Deltamethrin (Polyethylene microplastics)	RoC	Daphnia magna	Crustacea				0	+*	0*		0*		Benefit	Felten et al. (2020)
Diazinon and chlorpyrifos	RoC	Xenopus laevis	Vertebrata	−		−							Systematic	Boualit et al. (2022)
Endocrine disrupting chemicals	RoC	Gammarus pulex	Crustacea				+		+			−	Benefit	Schneider et al. (2015)
Esfenvalerate + polystyrene microplastics (food)*	RoC	Chironomus riparius	Hexapoda		−		0	0		−*		0	Systematic	Varg et al. (2021)
EstRoCenic wastewater	RoC	Gammarus pulex	Crustacea				+		+	0		−	Benefit	Schneider et al. (2015)
Fipronil and 2,4-D (kairomones)*/****	RoC	Ceriodaphnia silvestrii	Crustacea				−*	0	−	−			Systematic	Moreira et al. (2020)
Fluoxetine	RoC	Daphnia magna	Crustacea			+	+	0	+				Benefit	Aulsebrook et al. (2022)
Ibuprofen	RoC	Aedes aegypti	Hexapoda		0		0	0	0				Neutral	Prud'homme et al. (2017)
Imidacloprid	RoC	Culex pipiens	Hexapoda		0		0	0					Neutral	Pigeault et al. (2021)
Imidacloprid (by soil or food)*	RoC	Hypogastrura viatica	Hexapoda		−		+	−	−				Compensatory	Kristiansen et al. (2021)
Imidacloprid (temperature)*	RoC	Brachionus calicifloras	Rotifera		−				−				Systematic	Wen et al. (2022)
Malathion (temperature)*	RoC	Culex restuans	Hexapoda										N.A.	Muturi et al. (2011)
Malathion (temperature)*	RoC	Aedes albopictus	Hexapoda		−		0	0			0		N.A.	Muturi et al. (2011)

Pollutant	Study type	Species	Taxon	T1	T2	T3	T4	T5	T6	T7	Percentage of negative impact	Response	Reference
Phenanthrene (mild heat shocks)*	RoC	*Folsomia candida*	Hexapoda	+							0	Compensatory	Dai et al. (2023)
Polychlorinated biphenyls + polybrominated diphenyl ethers	RoC	*Danio rerio*	Vertebrata		–		–	–	–	–*		Systematic	Horri et al. (2018)
Pyrimethanil	RoC	*Hyla intermedia*	Vertebrata	–		+	–				0	Compensatory	Bernabò et al. (2016)
Sertraline	RoC	*Daphnia magna*	Crustacea		0	0	–	+				Benefit	Minguez et al. (2015)
Tebuconazole	RoC	*Hyla intermedia*	Vertebrata	–		+	+				0	Compensatory	Bernabò et al. (2016)
Temephos-organophosphate (sex)*	RoC	*Culex quinquefasciatus*	Hexapoda		0		0			+*	–	N.A.	Mpho et al. (2001)
Tributyltin	RoC	*Chironomus riparius*	Hexapoda	–				–				Systematic	Nowak et al. (2009)
Tributyltin	RoC	*Chironomus riparius*	Hexapoda	–		–		–	+			Compensatory	Lilley et al. (2012)
Venlafaxine	RoC	*Daphnia magna*	Crustacea		0		0	–			0	Compensatory	Minguez et al. (2015)
Percentage of negative impact				**67**	**75**	**55**	**20**	**35**	**41**	**64**	**40**		
Benzo[a]pyrene	CG	*Aedes aegypti*	Hexapoda	–	0	+	0	0	–			Compensatory	Prud'homme et al. (2017)
Bisphenol A	CG	*Aedes aegypti*	Hexapoda	0	0	0	0	0	0			Neutral	Prud'homme et al. (2017)
Carbofuran and chlorpyrifos	CG	*Tecia solanivora*	Hexapoda	–		+	0**			0		Compensatory	Gutiérrez et al. (2019)
Chlorpyrifos	CG	*Ischnura elegans*	Hexapoda		+				+			Benefit	Janssens et al. (2021)
Chlorpyrifos	CG	*Ischnura elegans*	Hexapoda		0				0			Neutral	Arambourou and Stocks (2015)
Tributyltin	CG	*Chironomus riparius*	Hexapoda	0	0		–	0				N.A.	Lilley et al. (2012)
Ibuprofen	CG	*Aedes aegypti*	Hexapoda	–			0	–		–	67	Compensatory	Prud'homme et al. (2017)
Percentage of negative impact			N.A.	**60**	**40**	**0**	**0**	**0**	**0**	**67**			

RoC: Range of concentration experiments submitting animals to a range of concentrations of pollutant; CG: common CGs comparing strains or populations having lived in polluting environment with pollutant-naïve strains or populations; + sign means a positive effect; – means a negative effect; 0: mean no significant effect. N.A. = no clear evidence for either systematic impact or compensatory response (*e.g.*, when a maximum of one trait changed). For CG studies, A '+' indicates that, in similar environmental conditions, the polluted population performs better than the unpolluted population for that trait. Percentage of negative impact represents the percentage of studies that show a negative impact of the pollutants on each trait. In bold, traits with at least five studies. Other factors* = other factors that interact with the pollutant on the traits.

*: in interaction with another factor.
**: the resistant strain showed delayed fecundity.
***: kairomones of rosy bitterling (*Rhodeus ocellatus*).
****: kairomones of serpea tetra (*Hyphessobrycon eques*).

Most RoC studies have found a systematic decline in life history trait values in response to physical (Table 19.2) and metal (Table 19.3) pollution. The results are more variable for organic pollutants (Table 19.4). Metal and physical pollution seem to have more similar general effects on different life history traits, while we observed more frequent evidence for compensatory responses to organic pollution. We have no clear explanation for such contrasted effects of the different families of pollutants.

We have also calculated the percentage of studies that showed a negative impact of the pollutants on each trait (Tables 19.2–19.4). Between 55% and 63% of the RoC studies on physical pollution have a negative impact on the life history traits in question, and traits do not seem to vary a lot in how they were impacted (Table 19.2; here we only considered traits with at least five studies). These percentages vary between 57% and 89% for metallic pollution, with adult survival, fecundity and growth being the most impacted traits (Table 19.3). They vary between 20% and 75% for organic pollution, juvenile and adult survival being the most impacted traits (Table 19.4). These results suggest that different pollutants may disturb life history traits in different ways, and thus may lead to different evolution patterns of polluted populations. Heavy metals may be selected for delayed maturity and slower pace of life, but organic pollutants do not always have such consequences. Further studies are needed to examine the degree to which different pollutants affect the evolution of life history traits.

19.4.2 Multigenerational Studies, Common Garden Experiments and Long-Term Evolutionary Responses of Populations

RoC studies do not evaluate the long-term responses of populations to pollutants. To fill this gap, a second experimental approach compares phenotypic changes of populations subjected to a pollutant with control populations over multiple generations (*i.e.*, more than three). With multigenerational experiments, it is possible to study the evolutionary processes occurring within a population facing a polluted environment (Shirley and Sibly 1999, Nowak et al. 2009, Bickham 2011, Kadiene et al. 2022, Quevarec et al. 2023). This approach tests whether populations quickly adapt to these new conditions. For example, *D. magna* exposed to waterborne uranium showed a decline in their life history traits (survival, growth and fecundity) between the first and the third generation (Massarin et al. 2010). In contrast, Dutilleul et al. (2017) found an increase in the values of life history traits (fertility or growth) of *C. elegans* exposed to uranium or salt for 22 generations.

A few studies have also been conducted in the field to compare life history traits of individuals in populations with different histories of pollution (*e.g.*, Chung et al. 2008, Benejam et al. 2010, Dechamps et al. 2011, Otero et al. 2018). This approach is essential for assessing the impact of pollutants on natural populations, but because it often tests mixtures of pollutants, it cannot establish cause–effect relationships specific to each pollutant. Because organisms are directly collected in the field, it is not possible to separate the different factors (*e.g.*, plasticity, genetic or epigenetic differentiation) responsible for phenotypic differences between populations. Furthermore, in these studies, organisms are rarely monitored throughout their life cycle, and few life history traits are measured (Chung et al. 2008, Otero et al. 2018).

To overcome these limitations some studies have used CG experiments to detect genetic differentiation between populations, or reciprocal transplant-like experiments (regrouped under the same denomination: CG, in Tables 19.2–19.4), to test whether such differentiation is adaptive or not (Chung et al. 2008, Dutilleul et al. 2015, 2017, Zweerus et al. 2017, Quevarec et al. 2023). Differences between populations could also be caused by transgenerational epigenetic effects if they persist over a few generations (Dutilleul et al. 2017). No difference in the traits between the polluted and unpolluted populations, once they are both raised in an unpolluted environment, indicates that the changes observed in the polluted conditions were caused by phenotypic plasticity only. For example, Wang et al. (2022) showed that nanoplastics reduced the fecundity of a marine rotifers *Brachionus plicatilis* but found no difference in fecundity once polluted and unpolluted individuals were placed back in clean seawater. Compared to RoC studies, fewer CG experiments have been published. We thus have less knowledge of the long-term effects of pollutants (see Tables 19.2–19.4). This was also noted by Loria et al. (2019) in their meta-analysis of 258 studies. It prevents us from making any inference about the effects of physical pollutants (Table 19.2).

Studies showing compensatory responses in polluted populations compared to unpolluted ones, either in the unpolluted or polluted environment, indicate that evolution in polluted conditions has led to a shift in life history strategies (Tables 19.2–19.4). A few CG experiments demonstrated a systematic decline of the values of all the life history traits in response to pollution by metals (Table 19.3), and no common CG experiments have found a systematic decline of life history traits in response to chemical pollutants (Table 19.4). Chemical or organic pollution, thus, seems to lead to the evolution of compensatory responses more often than a reduction of acquisition.

The general pattern becomes less clear when considering interactions with other stressors (Pölkki et al. 2014, Oliver and Brooke 2018). Many stressors simultaneously impact wild organisms over their life, and it seems essential to look at the combined effects of several pollutants. For example, in their RoC experiment, Pölkki et al. (2014) showed that copper reduced adult weight in the northern blowfly *Protophormia terraenovae* and that increasing temperature amplified this effect. On the other hand, in the same study, the authors show an antagonistic effect of copper and temperature on the development time of individuals, with the former increasing and the latter decreasing it. Other studies have also observed such interactive effects (see '*' in Tables 19.2–19.4), which shows the importance of considering multiple stressors when studying the impact of pollutants on life history traits.

Neutral results indicated in the tables ('0' in the tables) show that the polluted population does not differ from the unpolluted one. This can happen for studies run over too few generations to permit a significant evolution, or on traits with low heritability. Life history traits usually show low heritability, but high additive genetic variance and thus should evolve in response to natural selection (Price and Schluter 1991, Stirling et al. 2002). Trade-offs (*i.e.*, genetic correlations) between life history traits should also constrain their independent evolution (Roff 1992, Stearns 1992). This is most likely what we see with compensatory response to a pollutant when some values for some traits increase at the expense of others. CG experiments on heavy metals often found genetic differentiation between polluted and unpolluted populations (Table 19.3). Seven out of 18 studies show at least one trait with an increased value because of evolution in polluted conditions. However, these results did not display any general trend, which indicates no evidence for convergent evolution of populations subjected to a chemical pollutant towards a slower or faster pace of life. This supports the findings of Newman and Clements (2007).

According to what is explained in this chapter, we could predict that pollutants would increase the efficiency of acquisition or shift life history towards a faster or slower pace of life (although we cannot predict the direction of this shift). Few empirical studies have formally tested predictions from life history theories on the evolutionary consequences of pollutants. Maltby (1991) observed decreased age-specific survival and juvenile growth rate caused by pollution in a population of the crustacean *Aselus aquaticus*, living below a mine effluent. Following a model by Sibly and Calow (1989), Maltby (1991) predicted that the polluted population should have evolved towards a lower allocation of resources to reproduction and the production of fewer, larger offspring, and confirmed these predictions with a common garden experiment. In contrast, Zweerus et al. (2017) did not find any evidence for adaptive changes in the rotifer *Brachionus calyciflorus*, subject to copper pollution. In their meta-analysis, Loria et al. (2019) could not find clear evidence for adaptive responses to pollution.

19.5 Case Studies

In this section, we present two experimental studies which have evaluated short- and long-term (evolutionary) responses of *C. elegans* populations to uranium (a radioactive heavy metal), or sodium chloride or to gamma ionising radiation emitted by a ^{137}Cs source and assessed their potential costs of adaptation.

19.5.1 Responses of *Caenorhabditis elegans* to Environmental Stressors

Dutilleul et al. (2017) studied the microevolutionary response of *C. elegans* populations exposed to environmental stressors and measured their costs of adaptation. Negative cross-environment genetic correlations caused by antagonistic pleiotropic effects are assumed to be at the origin of adaptation costs (Fry 1993, Falconer and Mackay 1996). as mentioned above, genes with a positive effect on fitness in the polluted environment could be counter-selected in the unpolluted environment. Dutilleul et al. (2017) experimentally exposed populations for 22 generations to a high concentration of uranium (U populations), sodium chloride, an extreme hypertonic stress (NaCl populations), alternating uranium/salt treatments (U/NaCl populations) and to an unpolluted control environment (C populations). They then measured survival, early and late fertility, total fertility, fitness and body length as an index of growth from age 0 to 96 h.

C. elegans populations adapted rapidly to both the uranium (U populations) and salt environments (NaCl populations). Populations responded to salt by reducing early and total fertility, thereby producing individuals with slower life cycles and lower fertility. Moreover, NaCl populations showed reduced survival compared to both uranium and control treatments. Here we could suggest the evolution through both reduced acquisition and compensatory response to lead to a slower pace of life because resources are diverted from reproduction towards water regulation. In contrast, uranium selected for fast growth, high early fertility and thus a short generation time. Growing faster and becoming larger may allow individuals to detoxify their bodies, prevent internalisation of the pollutant and reduce internal pollutant concentrations (Sibly and Calow 1989, Guedes et al. 2006). U/NaCl (*i.e.*, alternating) populations seemed also to have evolved towards faster life histories. Hence, the different pollution regimes led to the evolution of life histories in different directions. Pollutants may thus have strong consequences on the evolution of populations along the fast–slow life history continuum (Stearns 1983, Promislow and Harvey 1990) with potentially strong implications for their dynamics.

After 22 generations of exposure, once transferred in the uranium, NaCl populations showed both lower fertility and slower male growth than U or C populations. These results indicate that adaptation to salt bears a fitness cost in terms of tolerance to uranium. Furthermore, when they were returned to the control environment, U, NaCl and U/NaCl populations showed lower fertility than the control populations, indicating adaptation costs incurred by the adaptive evolution in a polluted environment. Such costs limit the ability of polluted populations to deal with their new environmental conditions once the environment is depolluted.

Such opposite selection pressures between NaCl and U may lead to maladaptation when exposed to the other environment. A lower total fertility of NaCl populations in uranium and of U populations in NaCl would suggest such a maladaptation. Populations experiencing alternating conditions, however, did not seem to show such maladaptive outcomes, which indicates that evolving in a more heterogeneous environment may help populations deal with future environmental changes.

19.5.2 Responses of *Caenorhabditis elegans* Exposed to Ionising Radiation

Quevarec et al. (2022, 2023) exposed *C. elegans* populations to three dose rates of gamma radiation (0, 1.4 and 50 mGy h^{-1}) for 60 days (corresponding to around 20 generations). During this multigenerational experiment, the authors measured population growth, male proportion, hatching success, individual fecundity (*i.e.*, survival of juveniles), delayed egg laying and female fecundity. They showed that, at the highest dose rate, population growth, reproduction, hatching success and fitness decreased compared to the control condition (Quevarec et al. 2023). At the lowest dose rate, the authors observed a decrease in females' fecundity. However, they observed a slight increase in population growth and male proportion compared to the control condition. With a reciprocal transplant experiment they showed adaptive responses of populations exposed to low doses of ionising radiation: populations that evolved in the low-dose environment showed improved embryo survival and a slower pace of life compared to the control populations transferred in that environment. At high dose rates, individuals' fitness improved within a few generations. These changes came with some adaptive costs related to gamma radiation: back in the control environment, the irradiated individuals showed a lower hatching success and late-life brood size compared to control populations.

Quevarec et al. (2022) observed an increase in the proportion of males in the radiated populations (Figure 19.3a). Using a theoretical approach, Stewart and Phillips (2002) found that the increased male proportion could be attributed to genetic changes (Figure 19.3b). The model showed that an increase in male fertilisation success or a decrease in hermaphrodite self-fertilisation could explain this increase in the frequency of males (Figure 19.3c). This approach allowed the authors to establish that irradiation favoured reproduction by outcrossing to the detriment of self-fertilisation in hermaphrodites.

Finally, Quevarec et al. (2024) exposed the different populations to the pathogenic bacteria *Serratia marcescens* and found that populations evolving under ionising radiation survived less in the presence of the bacteria, suggesting some possible adaptive costs or evolutionary trade-offs.

The combined results demonstrate that ionising radiation can modify the evolutionary trajectory of a population by having a significant impact on life history and demographic traits. It therefore seems essential to consider the long-term survival of these organisms and underlines the need to integrate the evolutionary approach in the study of the effects of pollutants on organisms (*i.e.*, as part of ecological risk assessment).

19.6 Conclusion and Future Directions

Our current state of knowledge is still limited. This review shows that organisms respond primarily by systematically reducing the values of their life history traits (*e.g.*, growth, adult size, fecundity or longevity), indicating that pollutants affect the acquisition of resources by organisms. Pollutants, however, can impact many facets of the biology and ecology of an organism, which can act additively, antagonistically or synergistically on an organism's fitness and its resulting demography (Congdon et al. 2001). Furthermore, we could not find evidence for a general evolutionary response to pollutants common to all pollutants and all species, or for more specific responses to certain families of pollutants or even to each pollutant, target species and conditions in which contamination occurs. Based on these observations, we propose to combine approaches used in evolutionary biology with those used in ecotoxicology studies to deepen our understanding of the acquisition–allocation principle and the plasticity and evolution of life history traits.

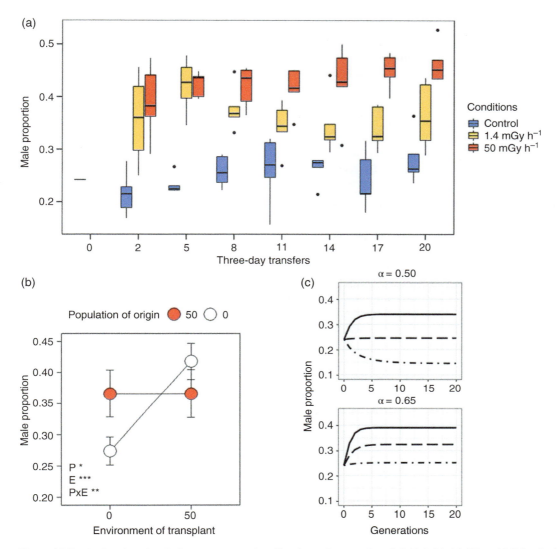

Figure 19.3 (a) Boxplot of male frequency over time (*i.e.*, three-day transfers: 0, 2, 5, 8, 11, 14, 17, and 20) for *C. elegans* populations living in different gamma radiation environments. Blue: Control; yellow: Low radiation (1.4 mGy h^{-1}); red: High radiation (50 mGy h^{-1}). Each boxplot shows the distribution of data in each condition, displaying from the centre outwards: the median (thick horizontal line in the box), the interquartile range where 50% of the data are located (box), the upper and lower values outside the interquartile range (two vertical lines outside the box) and the potential outliers (dots). (b) Mean ± SE male frequency of *C. elegans* populations after four transfers of reciprocal transplant between control and high (50 mGy h^{-1}) radiation treatments. The colour of the dot represents the populations' treatment during the multigenerational experiment. White: Control; red: High radiation (50 mGy h^{-1}). The significance of each main effect [population of origin (P), environment of transplant (e), and their interaction (P × E)] is indicated at the bottom left of each graph. *$p < 0.05$; **$p < 0.01$; ***$p < 0.001$. (c) Model results showing changes in *C. elegans* male frequency over 20 generations as a function of male fertilisation success (α) and the proportion of eggs not fertilized by males that are self-fertilized (β). This model is adapted from Stewart and Phillips (2002). α was set to 0.50 or 0.65. The three lines correspond to three values of β: 0.10 (solid line), 0.15 (long dashed line) and 0.20 (dot dashed line). (*Source:* Adapted from Quevarec et al. (2022)).

We first need more studies estimating the shape and the strength of selection pressures on life history and other traits in unpolluted and polluted environments. We can, for instance, estimate the covariance between the phenotype of the trait and fitness (*e.g.*, ultimately lifetime reproductive success; Lande and Arnold 1983, Arnold and Wade 1984, Dingemanse et al. 2021) in each environment. Alternatively, quantitative genetic breeding designs allowing the estimation of the genetic covariance between traits and fitness would provide stronger results by estimating the direct effects of selection on the genetic component of the traits (Morrissey et al. 2010). However, the large samples needed for these experiments may not be compatible with the strong reduction of populations in the polluted environment. Isogenic or inbred lines may provide a neat alternative to breeding designs to study pollution-induced selection on life history traits (Dutilleul et al. 2015).

Isogenic lines allow the use of numerous replicates of the same genotypes in different environments to estimate genetic correlations between traits and fitness.

Second, we need to go beyond average-population-effect studies and run more experiments that allow evaluating the variance among individuals in their responses to the pollutant. This can be done by doing experiments that look at the genetic basis of differences in plastic response to the pollutant (Newman and Clements 2007). Again, given the constraints imposed by studies on pollution, breeding designs could be the most convenient approach to estimate genotype-by-environment interaction or cross-environment genetic correlations (Roff 1997, Lynch and Walsh 1998) for different life histories, behaviours, and physiological traits. These results would help us understand the potential for variation in acquisition and allocation responses within populations. Here again, isogenic lines could be a great candidate approach for such studies. These studies will provide some information on the potential for populations to evolve towards new life history strategies in polluted conditions. Changes in the genetic (co)variance of life history traits in the new polluted compared to the control environment, and their effects on the evolutionary potential of the population, would improve our knowledge of the evolutionary constrains imposed by trade-offs (Holloway et al. 1990, Hendrickx et al. 2008, Dutilleul et al. 2015). Common garden or reciprocal transplant experiments would help us to test how frequently populations evolve in response to one or multiple pollutants and stressors, and/or whether responses and the potential costs of adaptation to new environments differ according to the pollutant and/or to the life history strategy used by the species (Maltby 1991, Agra et al. 2011, Dutilleul et al. 2015, 2017, Quevarec et al. 2023). Additionally, experiments on how pollutants are involved in switches in the signalling pathways that could change life history traits beyond their effects on the acquisition and allocation of resources might provide new insight into the complexity of life history changes under polluted environments.

Studying the effects of pollutants on life histories of wild populations could improve our knowledge of the consequences of pollution on biodiversity. Since pollutants affect both the acquisition and allocation of resources, they could also provide knowledge for theorists to refine their analyses of the mechanisms by which life history traits evolve.

References

Agra, A.R., Soares, A.M., and Barata, C. (2011). Life-history consequences of adaptation to pollution: *Daphnia longispina* clones historically exposed to copper. *Ecotoxicology* 20: 552–562.

Arambourou, H., Llorente, L., Moreno-Ocio, I., Herrero, Ó., Barata, C., Fuertes, I., Delorme, N., Méndez-Fernández, L., and Planelló, R. (2020). Exposure to heavy metal-contaminated sediments disrupts gene expression, lipid profile, and life history traits in the midge *Chironomus riparius*. *Water Research* 168: 115165.

Arambourou, H., and Stoks, R. (2015). Warmer winters modulate life history and energy storage but do not affect sensitivity to a widespread pesticide in an aquatic insect. *Aquatic Toxicology* 167: 38–45.

Arnold, S.J., and Wade, M.J. (1984). On the measurement of natural and sexual selection: applications. *Evolution* 38(4): 720–734.

Athrey, N.G., Leberg, P.L., and Klerks, P.L. (2007). Laboratory culturing and selection for increased resistance to cadmium reduce genetic variation in the least killifish, *Heterandria formosa*. *Environmental Toxicology and Chemistry* 26(9): 1916–1921.

Aulsebrook, L.C., Wong, B.B., and Hall, M.D. (2022). Warmer temperatures limit the effects of antidepressant pollution on life-history traits. *Proceedings of the Royal Society B: Biological Sciences* 289: 20212701.

Barata, C., Baird, D.J., and Soares, A.M.V.M. (2002). Demographic responses of a tropical cladoceran to cadmium: effects of food supply and density. *Ecological Applications* 12(2): 552–564.

Benejam, L., Benito, J., and García-Berthou, E. (2010). Decreases in condition and fecundity of freshwater fishes in a highly polluted reservoir. *Water, Air, & Soil Pollution* 210: 231–242.

Bernabò, I., Guardia, A., Macirella, R., Sesti, S., Crescente, A., and Brunelli, E. (2016). Effects of long-term exposure to two fungicides, pyrimethanil and tebuconazole, on survival and life history traits of Italian tree frog (*Hyla intermedia*). *Aquatic Toxicology* 172: 56–66.

Bickham, J.W. (2011). The four cornerstones of evolutionary toxicology. *Ecotoxicology* 20: 497–502.

Bieszke, B., Namiotko, L., and Namiotko, T. (2020). Life history traits of a temporary water ostracod *Heterocypris incongruens* (Crustacea, Ostracoda) are affected by power frequency (50 Hz) electromagnetic environmental pollution. *The European Zoological Journal* 87(1): 148–155.

Bighiu, M.A., Gorokhova, E., Carney Almroth, B., and Eriksson Wiklund, A.K. (2017). Metal contamination in harbours impacts life-history traits and metallothionein levels in snails. *PLoS ONE* 12(7): e0180157.

Boualit, L., Cayuela, H., Cattin, L., and Chèvre, N. (2022). The amphibian short-term assay: evaluation of a new ecotoxicological method for amphibians using two organophosphate pesticides commonly found in nature – assessment of biochemical, morphological, and life-history traits. *Environmental Toxicology and Chemistry* 41(11): 2688–2699.

Brausch, J.M., and Smith, P.N. (2009). Development of resistance to cyfluthrin and naphthalene among *Daphnia magna*. *Ecotoxicology* 18: 600–609.

Calow, P., and Forbes, V.E. (1998). How do physiological responses to stress translate into ecological and evolutionary processes? *Comparative Biochemistry and Physiology A: Molecular Integrative Physiology* 120: 11–16.

Calow, P., and Sibly, R.M. (1990). A physiological basis of population processes: ecotoxicological implications. *Functional Ecology* 4(3): 283–288.

Cameron, T.C., O'Sullivan, D., Reynolds, A., Piertney, S.B., and Benton, T.G. (2013). Eco-evolutionary dynamics in response to selection on life history. *Ecology Letters* 16: 754–763.

Chung, P.P., Hyne, R.V., Mann, R.M., and Ballard, J.W.O. (2008). Genetic and life-history trait variation of the amphipod *Melita plumulosa* from polluted and unpolluted waterways in eastern Australia. *Science of the Total Environment* 403(1–3): 222–229.

Coeurdassier, M., De Vaufleury, A., and Badot, P.M. (2003). Bioconcentration of cadmium and toxic effects on life-history traits of pond snails (*Lymnaea palustris* and *Lymnaea stagnalis*) in laboratory bioassays. *Archives of Environmental Contamination and Toxicology* 45: 102–109.

Congdon, J.D., Dunham, A.E., Hopkins, W.A., Rowe, C.L., and Hinton, T.G. (2001) Resource allocation-based life histories: a conceptual basis for studies of ecological toxicology. *Environmental Toxicology* 20: 1698–1703.

Conover, D.O., and Schultz, E.T. (1995). Phenotypic similarity and the evolutionary significance of countergradient variation. *Trends in Ecology & Evolution* 10: 248–252.

Coutellec, M.-A., and Barata, C. (2011). An introduction to evolutionary processes in ecotoxicology. *Ecotoxicology* 20: 493–496.

Dai, W., Slotsbo, S., Holmstrup, M., and van Gestel, C.A. (2023). Evaluation of life-history traits in *Folsomia candida* exposed to combined repeated mild heat shocks with phenanthrene. *Environmental Science and Pollution Research* 30(19): 55132–55142.

Debecker, S., and Stocks, R. (2019). Pace of life syndrome under warming and pollution: integrating life history, behavior, and physiology across latitudes. *Ecological Monographs* 89(1): e01332.

Dechamps, C., Elvinger, N., Meerts, P., Lefèbvre, C., Escarré, J., Colling, G., and Noret, N. (2011). Life history traits of the pseudometallophyte *Thlaspi caerulescens* in natural populations from Northern Europe. *Plant Biology* 13: 125–135.

Del Signore, A., Hendriks, A.J., Lenders, H.R., Leuven, R.S., and Breure, A.M. (2016). Development and application of the SSD approach in scientific case studies for ecological risk assessment. *Environmental Toxicology and Chemistry* 35(9): 2149–2161.

Dingemanse, N.J., Araya-Ajoy, Y.G., and Westneat, D.F. (2021). Most published selection gradients are underestimated: why this is and how to fix it. *Evolution* 75: 806–818.

Dutilleul, M., Bonzom, J.-M., Lecomte, C., Goussen, B., Daian, F., Galas, S., and Réale, D. (2014). Rapid evolutionary responses of life history traits to different experimentally-induced pollutions in *Caenorhabditis elegans*. *BMC Evolutionary Biology* 14: 252.

Dutilleul, M., Goussen, B., Bonzom, J.M., Galas, S., and Réale, D. (2015). Pollution breaks down the genetic architecture of life history traits in *Caenorhabditis elegans*. *PLoS ONE* 10(2): e0116214.

Dutilleul, M., Réale, D., Goussen, B., Lecomte, C., Galas, S., and Bonzom, J.M. (2017). Adaptation costs to constant and alternating polluted environments. *Evolutionary Applications* 10(8): 839–851.

Eraly, D., Hendrickx, F., Bervoets, L., and Lens, L. (2010). Experimental exposure to cadmium affects metallothionein-like protein levels but not survival and growth in wolf spiders from polluted and reference populations. *Environmental Pollution* 158(6): 2124–2131.

Ernande, B., Dieckmann, U., and Heino, M. (2004). Adaptive changes in harvested populations: plasticity and evolution of age and size at maturation. *Proceedings of the Royal Society B: Biological Sciences* 271: 415–423.

Falconer, D.S., and Mackay, T.F.C. (1996). *Introduction to quantitative genetics*. Prentice Hall.

Felten, V., Toumi, H., Masfaraud, J.F., Billoir, E., Camara, B.I., and Férard, J.F. (2020). Microplastics enhance *Daphnia magna* sensitivity to the pyrethroid insecticide deltamethrin: effects on life history traits. *Science of the Total Environment* 714: 136567.

Flatt, T., Heyland, A., and Stearns, S.C. (2011). What mechanistic insights can or cannot contribute to life history evolution: an exchange between Stearns, Heyland and Flatt. In: *Mechanism of life history evolution: the genetics and physiology of life history traits and trade-offs* (eds. Flatt, T., and Heyland, A.), 365–379. Oxford University Press.

Forbes, V.E., and Calow, P. (1996). Population growth rate as a basis for ecological risk assessment of toxic chemicals. In: *Wildlife population growth rates* (eds. Sibly, R.M., Hone, J., and Clutton-Brock, T.H.), 269–283. Cambridge University Press.

Fry, J.D. (1993). The "general vigor" problem: can antagonistic pleiotropy be detected when genetic covariances are positive? *Evolution* 47: 327–333.

Gamelon, M., Gaillard, J.-M., Gimenez, O., Coulson, T., Tuljapurkar, S., and Baubet, E. (2015). Linking demographic responses and life history tactics from longitudinal data in mammals. *Oikos* 125(3): 395–404.

Garnier-Laplace, J., Della-Vedova, C., Gilbin, R., Copplestone, D., Hingston, J., and Ciffroy, P. (2006). First derivation of predicted-no-effect values for freshwater and terrestrial ecosystems exposed to radioactive substances. *Environmental Science & Technology* 40(20): 6498–6505.

Grumiaux, F., Demuynck, S., Schikorski, D., Lemière, S., Vandenbulcke, F., and Leprêtre, A. (2007). Effect of fluidized bed combustion ashes used in metal polluted soil remediation on life history traits of the *oligochaeta Eisenia andrei*. *European Journal of Soil Biology*, 43: S256–S260.

Guedes, R.N.C., Oliveira, E.E., Guedes, N.M.P., Ribeiro, B., and Serrão, J.E. (2006). Cost and mitigation of insecticide resistance in the maize weevil, *Sitophilus zeamais*. *Physiological Entomology* 31: 30–38.

Gutiérrez, Y., Bacca, T., Zambrano, L.S., Pineda, M., and Guedes, R.N. (2019). Trade-off and adaptive cost in a multiple-resistant strain of the invasive potato tuber moth *Tecia solanivora*. *Pest Management Science* 75(6): 1655–1662.

Han, J., Park, Y., Jeong, H., and Park, J.C. (2022). Effects of particulate matter ($PM_{2.5}$) on life history traits, oxidative stress, and defensome system in the marine copepod *Tigriopus japonicus*. *Marine Pollution Bulletin* 178: 113588.

Hayward, A.D., Wilson, A.J., Pilkington, J.G., Clutton-Brock, T.H., Pemberton, J.M., and Kruuk, L.E.B. (2012). Reproductive senescence in female Soay sheep: variation across traits and contributions of individual ageing and selective disappearance. *Functional Ecology* 27: 184–195.

Hendrickx, F., Maelfait, J.-P., and Lens, L. (2008). Effect of metal stress on life history divergence and quantitative genetic architecture in a wolf spider. *Journal of Evolutionary Biology* 21: 183–193.

Holloway, G.J., Sibly, R.M., and Povey, S.R. (1990). Evolution in toxin-stressed environments. *Functional Ecology* 4: 289–294.

Horri, K., Alfonso, S., Cousin, X., Munschy, C., Loizeau, V., Aroua, S., Bégout, M.L., and Ernande, B. (2018). Fish life-history traits are affected after chronic dietary exposure to an environmentally realistic marine mixture of PCBs and PBDEs. *Science of the Total Environment* 610: 531–545.

Ilijin, L., Mrdaković, M., Todorović, D., Vlahović, M., Gavrilović, A., Mrkonja, A., and Perić-Mataruga, V. (2015). Life history traits and the activity of antioxidative enzymes in *Lymantria dispar* L. (Lepidoptera, Lymantriidae) larvae exposed to benzo[a]pyrene. *Environmental Toxicology and Chemistry* 34(11): 2618–2624.

IPBES (2019). *Global assessment report on biodiversity and ecosystem services of the Intergovernmental Science-Policy Platform on Biodiversity and Ecosystem Services* (eds. Brondizio, E.S., Settele, J., Díaz, S., and Ngo, H.T.). IPBES Secretariat. https://doi.org/10.5281/zenodo.3831673.

Jansen, M., Coors, A., Stoks, R., and De Meester, L. (2011a). Evolutionary ecotoxicology of pesticide resistance: a case study in *Daphnia*. *Ecotoxicology* 20: 543–551.

Jansen, M., Stoks, R., Coors, A., van Doorslaer, W., and De Meester, L. (2011b). Collateral damage: rapid exposure-induced evolution of pesticide resistance leads to increased susceptibility to parasites. *Evolution* 65(9): 2681–2691.

Janssens, L., Verberk, W., and Stoks, R. (2021). The pace-of-life explains whether gills improve or exacerbate pesticide sensitivity in a damselfly larva. *Environmental Pollution* 282: 117019.

Kadiene, E.U., Bialais, C., Ouddane, B., Hwang, J.S., and Souissi, S. (2017). Differences in lethal response between male and female calanoid copepods and life cycle traits to cadmium toxicity. *Ecotoxicology* 26: 1227–1239.

Kadiene, E.U., Meng, P.J., Hwang, J.S., and Souissi, S. (2019). Acute and chronic toxicity of cadmium on the copepod *Pseudodiaptomus annandalei*: a life history traits approach. *Chemosphere* 233: 396–404.

Kadiene, E.U., Ouddane, B., Gong, H.Y., Hwang, J.S., and Souissi, S. (2022). Multigenerational study of life history traits, bioaccumulation, and molecular responses of *Pseudodiaptomus annandalei* to cadmium. *Ecotoxicology and Environmental Safety* 230: 113171.

Kawecki, T.J., and Ebert, D. (2004). Conceptual issues in local adaptation. *Ecology Letters* 7(12): 1225–1241.

Kawecki, T.J., Lenski, R.E., Ebert, D., Hollis, B., Olivieri, I., and Whitlock, M.C. (2012). Experimental evolution. *Trends in Ecology & Evolution* 27(10): 547–560.

King, E.G., Roff, D.A., and Fairbairn, D.J. (2011). Trade-off acquisition and allocation in *Gryllus firmus*: a test of the Y model. *Journal of Evolutionary Biology* 24: 256–264.

Kooijman, S.A.L.M. (2000). *Dynamic energy and mass budgets in biological systems*. Cambridge University Press.

Kristiansen, S.M., Borgå, K., Rundberget, J.T., and Leinaas, H.P. (2021). Effects on life-history traits of *Hypogastrura viatica* (Collembola) exposed to imidacloprid through soil or diet. *Environmental Toxicology and Chemistry* 40(11): 3111–3122.

Lack, D. (1968). *Ecological adaptations for breeding in birds*. Methuen & Co., Ltd.

Lagisz, M., and Laskowski, R. (2008) Evidence for between-generation effects in carabids exposed to heavy metals pollution. *Ecotoxicology* 17: 59–66.

Lande, R., and Arnold, S.J. (1983). The measurement of selection on correlated characters. *Evolution* 37: 1210–1226.

Li, H., Shi, L., Wang, D., and Wang, M. (2015) Impacts of mercury exposure on life history traits of *Tigriopus japonicus*: multigeneration effects and recovery from pollution. *Aquatic Toxicology* 166: 42–49.

Lilley, T.M., Ruokolainen, L., Pikkarainen, A., Laine, V.N., Kilpimaa, J., Rantala, M.J., and Nikinmaa, M. (2012). Impact of tributyltin on immune response and life history traits of *Chironomus riparius*: single and multigeneration effects and recovery from pollution. *Environmental Science & Technology* 46(13): 7382–7389.

Lopes, P.C., Sucena, É., Santos, M.E., and Magalhães, S. (2008). Rapid experimental evolution of pesticide resistance in *C. elegans* entails no costs and affects the mating system. *PLoS ONE* 3(11): e3741.

Loria, A., Cristescu, M.E., and Gonzalez, A. (2019). Mixed evidence for adaptation to environmental pollution. *Evolutionary Applications* 12: 1259–1273.

Lynch, M., and Walsh, B. (1998). *Genetics and analysis of quantitative traits*. Sinaeur.

Maltby, L. (1991). Pollution as a probe of life-history adaptation in *Asellus aquaticus* (Isopoda). *Oikos* 61: 11–18.

Massarin, S., Alonzo, F., Garcia-Sanchez, L., Gilbin, R., Garnier-Laplace, J., and Poggiale, J.C. (2010). Effects of chronic uranium exposure on life history and physiology of *Daphnia magna* over three successive generations. *Aquatic Toxicology* 99(3): 309–319.

Medina, M.H., Correa, J.A., and Barata, C. (2007). Micro-evolution due to pollution: possible consequences for ecosystem responses to toxic stress. *Chemosphere* 67: 2105–2114.

Miliou, H., Verriopoulos, G., Maroulis, D., Bouloukos, D., and Moraitou-Apostolopoulou, M. (2000). Influence of life-history adaptations on the fidelity of laboratory bioassays for the impact of heavy metals (Co_{2+} and Cr_{6+}) on tolerance and population dynamics of *Tisbe holothuriae*. *Marine Pollution Bulletin* 40(4): 352–359.

Minguez, L., Ballandonne, C., Rakotomalala, C., Dubreule, C., Kientz-Bouchart, V., and Halm-Lemeille, M.P. (2015). Transgenerational effects of two antidepressants (sertraline and venlafaxine) on *Daphnia magna* life history traits. *Environmental Science & Technology* 49(2): 1148–1155.

Mirčić, D., Janković Tomanić, M., Nenadović, V., Franeta, F., and Lazarević, J. (2010). The effects of cadmium on the life history traits of *Lymantria dispar* L. *Archives of Biological Sciences* 62(4): 1013–1020.

Montiglio, P.-O., and Royauté, R. (2014). Contaminants as a neglected source of behavioural variation. *Animal Behaviour* 88: 29–35.

Moreira, R.A., Rocha, O., Pinto, T.J.D.S., da Silva, L.C.M., Goulart, B.V., Montagner, C.C., and Espindola, E.L.G. (2020). Life-history traits response to effects of fish predation (kairomones), fipronil and 2,4-D on neotropical cladoceran *Ceriodaphnia silvestrii*. *Archives of Environmental Contamination and Toxicology* 79: 298–309.

Morrissey, M.B., Kruuk, L.E.B., and Wilson, A.J. (2010). The danger of applying the breeder's equation in observational studies of natural populations. *Journal of Evolutionary Biology* 23: 2277–2288.

Morrissey, C.A., Mineau, P., Devries, J.H., Sanchez-Bayo, F., Liess, M., Cavallaro, M.C., and Liber, K. (2015). Neonicotinoid contamination of global surface waters and associated risk to aquatic invertebrates: a review. *Environment International* 74: 291–303.

Mpho, M., Holloway, G.J., and Callaghan, A. (2001). A comparison of the effects of organophosphate insecticide exposure and temperature stress on fluctuating asymmetry and life history traits in *Culex quinquefasciatus*. *Chemosphere* 45(6–7): 713–720.

Mulvey M., Newman M.C., Chazal A., Keklak M.M., Heagler M.G., and Hales L.S. (1995). Genetic and demographic responses of mosquitofish (*Gambusia holbrooki* Girard 1859) populations stressed by mercury. *Environmental Toxicology & Chemistry* 14: 1411–1418.

Muturi, E.J., Lampman, R., Costanzo, K., and Alto, B.W. (2011). Effect of temperature and insecticide stress on life-history traits of *Culex restuans* and *Aedes albopictus* (Diptera: Culicidae). *Journal of Medical Entomology* 48(2): 243–250.

Newman, M.C., and Clements, W.H. (2007). *Ecotoxicology, a comprehensive treatment*. Routledge.

Nowak, C., Vogt, C., Pfenninger, M., Schwenk, K., Oehlmann, J., Streit, B., and Oetken, M. (2009). Rapid genetic erosion in pollutant-exposed experimental chironomid populations. *Environmental Pollution* 157(3): 881–886.

Oliveira, N.R., Moens, T., Fonseca, G., Nagata, R.M., Custódio, M.R., and Gallucci, F. (2020). Response of life-history traits of estuarine nematodes to the surfactant sodium dodecyl sulfate. *Aquatic Toxicology* 227: 105609.

Oliver, S.V., and Brooke, B.D. (2018). The effect of metal pollution on the life history and insecticide resistance phenotype of the major malaria vector *Anopheles arabiensis* (Diptera: Culicidae). *PLoS ONE* 13(2): e0192551.

Otero, M.A., Pollo, F.E., Grenat, P.R., Salas, N.E., and Martino, A.L. (2018). Differential effects on life history traits and body size of two anuran species inhabiting an environment related to fluorite mine. *Ecological Indicators* 93: 36–44.

Pigeault, R., Bataillard, D., Glaizot, O., and Christe, P. (2021). Effect of neonicotinoid exposure on the life history traits and susceptibility to Plasmodium infection on the major avian malaria vector *Culex pipiens* (Diptera: Culicidae). *Parasitologia* 1(1): 20–33.

Pölkki, M., Kangassalo, K., and Rantala, M.J. (2014). Effects of interaction between temperature conditions and copper exposure on immune defense and other life-history traits of the blow fly *Protophormia terraenovae*. *Environmental Science & Technology* 48(15): 8793–8799.

Posthuma, L., Verweij, R.A., Widianarko, B., and Zonneveld, C. (1993). Life-history patterns in metal-adapted Collembola. *Oikos* 67(2): 235–249.

Postma, J.F., van Kleunen, A., and Admiraal, W. (1995). Alterations in life-history traits of *Chironomus riparius* (Diptera) obtained from metal contaminated rivers. *Archives of Environmental Contamination and Toxicology* 29: 469–475.

Price, T., and Schluter, D. (1991). On the low heritability of life-history traits. *Evolution* 45: 853–861.

Promislow, D.E., and Harvey, P.H. (1990). Living fast and dying young: a comparative analysis of life-history variation among mammals. *Journal of Zoology* 220: 417–437.

Prud'homme, S.M., Chaumot, A., Cassar, E., David, J.P., and Reynaud, S. (2017). Impact of micropollutants on the life-history traits of the mosquito *Aedes aegypti*: on the relevance of transgenerational studies. *Environmental Pollution* 220: 242–254.

Qin, S., Yang, T., Yu, B., Zhang, L., Gu, L., Sun, Y., and Yang, Z. (2022). The stress effect of atrazine on the inducible defense traits of *Daphnia pulex* in response to fish predation risk: evidences from morphology, life history traits, and expression of the defense-related genes. *Environmental Pollution* 311: 119965.

Quevarec, L., Réale, D., Dufourcq-Sekatcheff, E., Armant, O., Adam-Guillermin, C., and Bonzom, J.M. (2023). Ionising radiation affects the demography and the evolution of *Caenorhabditis elegans* populations. *Ecotoxicology and Environmental Safety* 249: 114353.

Quevarec, L., Réale, D., Dufourcq-Sekatcheff, E., Car, C., Armant, O., Dubourg, N., Adam-Guillermin, C., and Bonzom, J.M. (2022). Male frequency in *Caenorhabditis elegans* increases in response to chronic irradiation. *Evolutionary Applications* 15(9): 1331–1343.

Quevarec, L., Morran, L.T., Dufourcq-Sekatcheff, E., Armant, O., Adam-Guillermin, C., Bonzom, J.M., and Réale, D. (2024). Host defense alteration in Caenorhabditis elegans after evolution under ionizing radiation. *BMC Ecology and Evolution* 24(1): 95.

Réale, D., Garant, D., Humphries, M.M., Bergeron, P., Careau, V., and Montiglio, P.-O. (2010). Personality and the emergence of the pace-of-life syndrome concept at the population level. *Philosophical Transactions Royal Society B: Biological Sciences* 365: 4051–4063.

Reis, P., Pereira, R., Carvalho, F.P., Oliveira, J., Malta, M., Mendo, S., and Lourenço, J. (2018). Life history traits and genotoxic effects on *Daphnia magna* exposed to waterborne uranium and to a uranium mine effluent – a transgenerational study. *Aquatic Toxicology* 202: 16–25.

Reznick, D., Nunney, L., and Tessier, A. (2000). Big houses, big cars, superfleas and the costs of reproduction. *Trends in Ecology & Evolution* 15(10): 421–425.

Rodrigues, S., Silva, A.M., and Antunes, S.C. (2021) Assessment of 17 α-ethinylestradiol effects in *Daphnia magna*: life-history traits, biochemical and genotoxic parameters. *Environmental Science and Pollution Research* 28: 23160–23173.

Roff, D.A. (1992). *The evolution of life histories*. Chapman & Hall.

Roff, D.A. (1997). *Evolutionary quantitative genetics*. Chapman & Hall.

Roff, D.A., and Fairbairn, D.J. (2007). The evolution of trade-offs: where are we? *Journal of Evolutionary Biology* 20(2): 433–447.

Rożen, A. (2006). Effect of cadmium on life-history parameters in *Dendrobaena octaedra* (Lumbricidae: Oligochaeta) populations originating from forests differently polluted with heavy metals. *Soil Biology and Biochemistry* 38(3): 489–503.

Rusin, M., Gospodarek, J., Nadgórska-Socha, A., and Barczyk, G. (2017). Effect of petroleum-derived substances on life history traits of black bean aphid (*Aphis fabae* Scop.) and on the growth and chemical composition of broad bean. *Ecotoxicology* 26: 308–319.

Sarkar, S., Diab, H., and Thompson, J. (2023). Microplastic pollution: chemical characterization and impact on wildlife. *International Journal of Environmental Research and Public Health* 20(3): 1745.

Schaaf, W.E., Peters, D.S., Vaughan, D.S., Coston-Clements, L., and Krouse, C.W. (1987). Fish population responses to chronic and acute pollution: the influence of life history strategies. *Estuaries* 10: 267–275.

Schanz, F.R., Sommer, S., Lami, A., Fontaneto, D., and Ozgul, A. (2021). Life-history responses of a freshwater rotifer to copper pollution. *Ecology and Evolution* 11(16): 10947–10955.

Schneider, I., Oehlmann, J., and Oetken, M. (2015). Impact of an estrogenic sewage treatment plant effluent on life-history traits of the freshwater amphipod *Gammarus pulex*. *Journal of Environmental Science and Health A: Toxic/Hazardous Substances and Environmental Engineering* 50(3): 272–281.

Schrank, I., Trotter, B., Dummert, J., Scholz-Böttcher, B.M., Löder, M.G., and Laforsch, C. (2019). Effects of microplastic particles and leaching additive on the life history and morphology of *Daphnia magna*. *Environmental Pollution* 255: 113233.

Shirley, M.D., and Sibly, R.M. (1999) Genetic basis of a between-environment trade-off involving resistance to cadmium in *Drosophila melanogaster*. *Evolution* 53(3): 826–836.

Sibly, R.M., and Calow, P. (1989). A life-cycle theory of responses to stress. *Biological Journal of the Linnean Society* 37: 101–116.

Silva, C.J., Silva, A.L.P., Gravato, C., and Pestana, J.L. (2019). Ingestion of small-sized and irregularly shaped polyethylene microplastics affect *Chironomus riparius* life-history traits. *Science of the Total Environment* 672: 862–868.

Simpson, S.L., Batley, G.E., Hamilton, I.L., and Spadaro, D.A. (2011). Guidelines for copper in sediments with varying properties. *Chemosphere* 85(9): 1487–1495.

Stanković, J., Milošević, D., Savić-Zdraković, D., Yalçın, G., Yildiz, D., Beklioğlu, M., and Jovanović, B. (2020). Exposure to a microplastic mixture is altering the life traits and is causing deformities in the non-biting midge *Chironomus riparius* Meigen (1804). *Environmental Pollution* 262: 114248.

Stearns, S.C. (1983). The influence of size and phylogeny on patterns of covariation among life-history traits in the mammals. *Oikos* 41: 173–187.

Stearns, S.C. (1989). Trade-offs in life-history evolution. *Functional Ecology* 3: 259–268.

Stearns, S.C. (1992). *The evolution of life histories*. Oxford University Press.

Stearns, S.C. (2011). Does impressive progress on understanding mechanisms advance life history theory? In: *Mechanism of life history evolution: the genetics and physiology of life history traits and trade-offs* (eds. Flatt, T., and Heyland, A.), 365–374. Oxford University Press.

Stewart, A.D., and Phillips, P.C. (2002). Selection and maintenance of androdioecy in *Caenorhabditis elegans*. *Genetics* 160(3): 975–982.

Stirling, D.G., Réale, D., and Roff, D.A. (2002). Selection, structure and the heritability of behaviour. *Journal of Evolutionary Biology* 15: 277–289.

Sun, Y., Lei, J., Wang, Y., Cheng, J., Zhou, Q., Wang, Z., Zhang, L., Gu, L., Huang, Y., and Yang, Z. (2019a). High concentration of *Phaeocystis globosa* reduces the sensitivity of rotifer *Brachionus plicatilis* to cadmium: based on an exponential approach fitting the changes in some key life-history traits. *Environmental Pollution* 246: 535–543.

Sun, Y., Xu, W., Gu, Q., Chen, Y., Zhou, Q., Zhang, L., Gu, L., Huang, Y., Luy, K., and Yang, Z. (2019b). Small-sized microplastics negatively affect rotifers: changes in the key life-history traits and rotifer – *Phaeocystis* population dynamics. *Environmental Science & Technology* 53(15): 9241–9251.

Sylvester, F., Weichert, F.G., Lozano, V.L., Groh, K.J., Bálint, M., Baumann, L., Bässler, C., Brack, W., Brandl, B., Curtius, J., Dierkes, P., Döll, P., Ebersberger, I., Fragkostefanakis, S., Helfrich, E.J.N., Hickler, T., Johann, S., Jourdan, J., Klimpel, S., Kminek, H., Liquin, F., Möllendorf, D., Mueller, T., Oehlmann, J., Ottermanns, R., Pauls, S.U., Piepenbring, M., Pfefferle, J., Schenk, G.J., Scheepens, J.F., Scheringer, M., Schiwy, S., Schlottmann, A., Schneider, F., Schulte, L.M., Schulze-Sylvester, M., Stelzer, E., Strobl, F., Sundermann, A., Tockner, K., Tröger, T., Vilcinskas, A., Völker, C., Winkelmann, R., and Hollert, H. (2023). Better integration of chemical pollution research will further our understanding of biodiversity loss. *Nature Ecology & Evolution* 7: 1552–1555.

Tranvik, L., Bengtsson, G., and Rundgren, S. (1993). Relative abundance and resistance traits of two Collembola species under metal stress. *Journal of Applied Ecology* 30(1): 43–52.

van den Brink, P.J., Blake, N., Brock, T.C., and Maltby, L. (2006). Predictive value of species sensitivity distributions for effects of herbicides in freshwater ecosystems. *Human and Ecological Risk Assessment* 12(4): 645–674.

van Noordwijk, A.J., and de Jong, G. (1986). Acquisition and allocation of resources: their influence on variation in life history tactics. *The American Naturalist* 128: 137–142.

Varg, J.E., Kunce, W., Outomuro, D., Svanbäck, R., and Johansson, F. (2021). Single and combined effects of microplastics, pyrethroid and food resources on the life-history traits and microbiome of *Chironomus riparius*. *Environmental Pollution* 289: 117848.

Wang, D., Ru, S., Zhang, W., Zhang, Z., Li, Y., Zhao, L., Li, L., and Wang, J. (2022). Impacts of nanoplastics on life-history traits of marine rotifer (*Brachionus plicatilis*) are recovered after being transferred to clean seawater. *Environmental Science and Pollution Research* 29(28): 42780–42791.

Wang, J., Zheng, M., Lu, L., Li, X., Zhang, Z., and Ru, S. (2021). Adaptation of life-history traits and trade-offs in marine medaka (*Oryzias melastigma*) after whole life-cycle exposure to polystyrene microplastics. *Journal of Hazardous Materials* 414: 125537.

Wei, X., Li, X., Liu, H., Lei, H., Sun, W., Li, D., Dong, W., Chen, H., and Xie, L. (2022). Altered life history traits and transcripts of molting- and reproduction-related genes by cadmium in *Daphnia magna*. *Ecotoxicology* 31(5): 735–745.

Wen, Y., Cao, M.M., Huang, Z.Y., and Xi, Y.L. (2022) Combined effects of warming and imidacloprid on survival, reproduction and population growth of *Brachionus calyciflorus* (Rotifera). *Bulletin of Environmental Contamination and Toxicology* 109: 990–995.

Williams, G.C. (1966) Natural selection, the costs of reproduction, and a refinement of Lack's principle. *The American Naturalist* 100: 687–690.

Yan, S., Li, N., Guo, Y., Chen, Y., Ji, C., Yin, M., Shen, J., and Zhang, J. (2022). Chronic exposure to the star polycation (SPc) nanocarrier in the larval stage adversely impairs life history traits in *Drosophila melanogaster*. *Journal of Nanobiotechnology* 20(1): 515.

Zhan, H., Zhang, J., Chen, Z., Huang, Y., Ruuhola, T., and Yang, S. (2017). Effects of Cd_{2+} exposure on key life history traits and activities of four metabolic enzymes in *Helicoverpa armigera* (Lepidopteran: Noctuidae). *Chemistry and Ecology* 33(4): 325–338.

Zhu, C., Zhang, T., Liu, X., Gu, X., Li, D., Yin, J., Jiang, Q., and Zhang, W. (2022). Changes in life-history traits, antioxidant defense, energy metabolism and molecular outcomes in the cladoceran *Daphnia pulex* after exposure to polystyrene microplastics. *Chemosphere* 308: 136066.

Zolotova, N., Kosyreva, A., Dzhalilova, D., Fokichev, N., and Makarova, O. (2022). Harmful effects of the microplastic pollution on animal health: a literature review. *PeerJ* 10: e13503.

Zweerus, N.L., Sommer, S., Fontaneto, D., and Ozgul, A. (2017). Life-history responses to environmental change revealed by resurrected rotifers from a historically polluted lake. *Hydrobiologia* 796: 121–130.

20

Life History Evolution on Expansion Fronts

Elodie Vercken[1] and Ben L. Phillips[2]

[1] *Biologie des Populations Introduites, UMR 1355 INRAE-CNRS-UniCA, Institut Sophia Agrobiotech, Sophia Antipolis, France*
[2] *School of Molecular and Life Sciences, Curtin University, Perth, Australia*

20.1 What Are Expansion Fronts and Why Are They Hotspots for Rapid Evolution

20.1.1 A Quick Introduction to Expansion Theory

Population expansions are an astonishing dynamic phenomenon. From a small number of initial founders in a small area, expanding populations can ultimately spread over thousands of kilometres. Dramatic though they often are, expansions are a spontaneous process, like a chemical reaction, or the spread of a bushfire. Under the right conditions, invasions simply happen. Those conditions are simply that (1) individuals disperse and (2) the population increases in density at its leading edge. Under these circumstances, any given population will spread into new areas.

An expansion front is the leading edge of an expanding population. Expansion fronts emerge from the process of expansion: they are a place where the density of the population forms a cline down to zero, as we move from the part of space already occupied to the part of space that is yet to be occupied (Figure 20.1). The ecological processes occurring on the expansion front determine whether the population continues to propagate forward or not. If individuals on the front disperse, and if they find themselves in an environment where population growth rate is positive, then the expansion front will tend to move forward. This idea was captured beautifully in early expansion theory (Fisher 1937, Kolmogorov et al. 1937, Skellam 1951), where, if a population is growing logistically (with growth rate r, and carrying capacity K) and diffusing through space (at a rate determined by D), then the spreading speed of the population is given as $v = 2\sqrt{rD}$. This elegant theoretical prediction shows the clear importance of population growth and dispersal to expansion. If either the growth rate, r, or dispersal, D, increases, then the rate of expansion will increase.

This insight into the importance of r and D to the expansion process is a well-established theoretical result (Giometto et al. 2014, Gandhi et al. 2016). In recent years, however, it has become clear that evolutionary processes are also playing out on expansion fronts. These evolutionary processes are a natural consequence of the dynamics of expansion, and they naturally cause r and D to change. In many circumstances, it appears that evolutionary processes on the expansion front act to maximize rD (Perkins et al. 2013, Korolev 2015, Deforet et al. 2019). These pronumerals – r and D – are masterpieces of minimalism. They capture all the aspects of phenotype and life history that lead to population growth and dispersal. Thus, when, for example, the growth rate, r, is under selection we can expect to see complex selection acting on all the component parts of life history underlying r, like fecundity, development time, or size. When r or D changes, we can expect this to be affected by some observable changes in life history. However, while this theoretical basis is simple, the actual responses of life history traits to these broad selection pressures might be surprisingly complex.

Below, we detail the evolutionary processes at play on expansion fronts, and we investigate how these processes may affect the evolution of life histories. While these processes will often be transitory – springing into existence on expansion fronts – they may nonetheless have much longer-term consequences for life history evolution. This is because the phenotypes that emerge on expansion fronts may be very different from those that emerge in equilibrium populations (Burton et al. 2010), because these phenotypes often go on to colonize large areas (Klopfstein et al. 2006), and because expansion processes are a common phenomenon.

Life History Evolution: Traits, Interactions, and Applications, First Edition. Edited by Michal Segoli and Eric Wajnberg.
© 2025 John Wiley & Sons Ltd. Published 2025 by John Wiley & Sons Ltd.

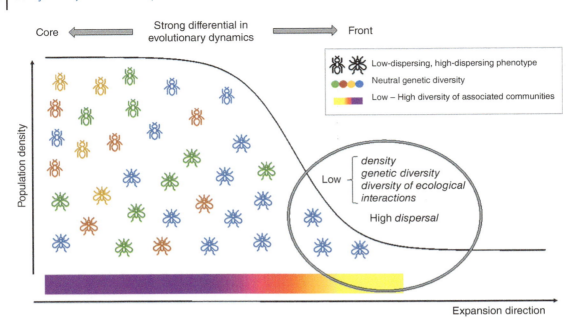

Figure 20.1 Evolutionary dynamics on a pulled expansion front. The front is defined by a decreasing cline in density along the direction of the expansion. Deterministic (spatial sorting and *r*-selection) and stochastic processes (loss of neutral diversity, loss of specialized interactions with the associated communities) interact on the front to drive evolutionary dynamics that are qualitatively different from those in the population core.

20.1.2 Evolutionary Processes

20.1.2.1 Spatial Sorting

The first and maybe most intuitive life history shift often observed on expansion fronts is an increased rate of dispersal. An expansion front is defined as the colonisation of a new habitat by dispersing individuals at the edge of the population range. The leading edge of the expansion front, thus, usually contains individuals who are good at dispersal (one potential exception being expansions where external forces, like human transportation, are the main dispersal mode). Space acts therefore like a selective sieve that preferentially retains dispersive individuals on the front of the expansion where they will then mate with each other (Figure 20.1). This assortative mating by dispersal ability was initially referred to as the 'Olympic village effect' (Phillips et al. 2010a) but is now called 'spatial sorting'. It can be considered as a spatial analogue to natural selection, in which selection is driven by a differential in dispersal ability rather than a differential in fitness (Shine et al. 2011, Phillips and Perkins 2019). The evolutionary process of spatial sorting has been the subject of extensive theoretical work with different modelling approaches (Travis and Dytham 2002, Burton et al. 2010, Phillips et al. 2010a, Shine et al. 2011, Bouin et al. 2012, Phillips and Perkins 2019), and increased dispersal abilities on expansion fronts have been observed repeatedly in range-shifting species in nature (see, for instance, Phillips et al. 2006, Lindström et al. 2013, for the famous example of the invasive cane toad in Australia; Figure 20.2, Narimanov et al. 2022, Mowery et al. 2022, in spiders; Hughes et al. 2007, Lombaert et al. 2014, in insects; and Huang et al. 2015, Tabassum and Leishman 2018, in plants).

Figure 20.2 The cane toad *Rhinella marina*, a species that has become a textbook example of invasion biology, with more than 100 articles over the last 20 years dissecting the eco-evolutionary dynamics of its expansion in Australia. *Source:* Photo by B. Phillips.

While increased dispersal abilities are regularly observed on expansion fronts, it must be noted that dispersal is a complex trait that is often condition-dependent (dependent on either individual or environmental conditions), so its response to selection on the expansion front must be measured under conditions that are representative of the specific environment in which this is observed. In particular, expansion fronts are typically low-density environments, so spatial sorting will specifically select for higher dispersal at low density. Because dispersal is frequently positively density-dependent in animals (with increased dispersal at high density to avoid intraspecific competition), spatial sorting is expected to flatten or even reverse the dispersal reaction norm on expansion fronts (Travis et al. 2009, De Bona et al. 2019). Such increased dispersal from low densities has in fact been observed in several empirical systems (Fronhofer et al. 2017, Weiss-Lehman et al. 2017, Clark et al. 2022). It is also worth noting that other selective forces acting on dispersal can act in synergy with spatial sorting to produce an amplified evolutionary response. A leading contender here is kin competition, which is also expected to select for increased dispersal during expansions in conditions that promote the emergence of strong spatial structure in relatedness (Kubisch et al. 2013, van Petegem et al. 2018).

A major consequence of a general increase in dispersal on expansion fronts is that, all else being equal, the expansion speed should also increase, because we have an increased dispersal rate, D. The coupled effect of spatial sorting on increased dispersal capacities and the acceleration of expansion has been demonstrated in several independent systems, *in natura* and in the laboratory (Perkins et al. 2013, Ochocki and Miller 2017, Narimanov et al. 2022). In a recent meta-analysis of controlled evolution experiments, however, Miller et al. (2020) found that, if evolution consistently results in an increase in mean expansion speed, the traits involved are not always the same and not always related to dispersal (*e.g.*, sometimes they are related to reproductive rate or competitive ability). Additionally, an expansion front is an inherently transient environment. If we fix our view on one location, the expansion front is rapidly replaced by a high-density population similar to that in the population core. When the selective environment changes again once the core of the population has caught up in space, the evolutionary patterns may reverse (Perkins et al. 2016). For instance, historical analysis of museum collections shows that initial continental colonisation events in monarch butterflies were associated with highly dispersing phenotypes, but the establishment was then followed by a reduction in wing size and a loss of migration over the next 1000 generations (Freedman et al. 2020).

20.1.2.2 *r*-Selection

While the evolution of dispersal on an expansion front is mostly driven by spatial sorting rather than classical natural selection, other life history traits will evolve in response to classical selective pressures. The density of conspecific has long been recognized as a major influence in life history evolutionary dynamics (Reznick et al. 2002). At low density, competition does not constrain population growth, and individuals that have the highest reproductive rate also have higher fitness (MacArthur and Wilson 1967). In contrast, individuals in high-density environments experience higher intraspecific competition, so their fitness ultimately depends not only on their reproductive rate but also on their competitive ability. Such contrasting life history strategies were first described as *r*- or *K*-strategies in reference to the dominance of exponential or logistic population dynamics (Charlesworth 1971, Roff 1993). This concept has since been extended to the pace-of-life hypothesis, which states that life history, physiological, or behavioural traits co-vary on a continuum of fast–slow life history strategies (Ricklefs and Wikelski 2002). The density of conspecific thus directly shapes the spatial structure of the selective environment for expanding populations: as expanding populations are characterized by a more or less steep cline in conspecific density between the core and the front (Figure 20.1), they also experience a gradient between *K/r* or slow/fast selective environments. Because of this, different life history strategies should be selected at the core and in front of the expanding population (Phillips 2009, Burton et al. 2010).

This general prediction is largely supported by empirical evidence from naturally expanding populations (Chuang and Peterson 2016). Cane toad populations at the front of the expansion have both faster growth rates and higher feeding rates (Phillips 2009, Lindström et al. 2013). In insects, females from edge populations have been found to be larger and to reproduce earlier (Roy et al. 2016, Clark et al. 2022), and in several species of plants, individuals from invasive populations grow or germinate faster than individuals from the native range (Kilkenny and Galloway 2013, Tabassum and Leishman 2018). In the vendace fish, individuals from invasive populations are characterized by faster growth rate (Bøhn et al. 2004), earlier sexual maturation, greater reproductive input and lower longevity (Amundsen et al. 2012), all of which are consistently indicative of a shift towards a faster life history. Some behavioural traits – exploratory behaviour or lower neophobia for instance – can also be related to pace-of-life syndromes (Hall et al. 2015) and are found to be selected for on expansion fronts because they provide a competitive advantage in novel environments (see for instance Liebl and Martin 2012, 2014). However, this pattern is somehow less obvious in experimental expansions, where traits related to reproductive rate or fast life history were often not found to evolve on edge populations relative to core populations (no evolution of reproductive

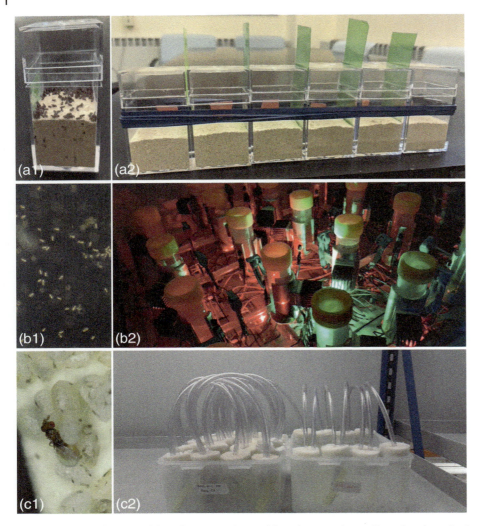

Figure 20.3 Unlike natural invasions, experimental invasions can be replicated and manipulated, allowing us to explore the generality of our results and to directly test key hypotheses. Three examples of experimental systems that have greatly improved our understanding of life history evolution during invasion, from the demonstration of spatial sorting with flour beetles (a, Weiss-Lehman et al. 2017, 2019) to complex eco-evolutionary feedbacks with protists (b, Fronhofer and Altermatt 2015, Zilio et al. 2023) and pushed expansion dynamics with micro-wasps (c, Dahirel et al. 2021a, 2021b). (a): *Tribolium castaneum* flour beetles (a1, *Source:* Photo by C. Weiss-Lehman.) expanding through an array of growth medium connected through dispersal holes (a2, *Source:* Photo by C. Weiss-Lehman.); (b): *Tetrahymena* protists (b1, *Source:* Photo by C. Gougat-Barbera) grown across an array of liquid medium habitat patches linked by tubing (b2, *Source:* Photo by E. Fronhofer.); (c): *Trichogramma* micro-wasps (c1, *Source:* Photo by G. Groussier) grown in tubes connected through tubing to form a meta population (c2, *Source:* Photo by C. Guicharnaud).

traits during expansion: Ochocki and Miller 2017; similar evolution in core and edge populations: Szúcs et al. 2017, Dahirel et al. 2021a) or evolve in the direction opposite to what was expected (lower growth rate in edge populations: Fronhofer and Altermatt 2015, Williams et al. 2016, Weiss-Lehman et al. 2017). Experimental expansions are typically done in simplified laboratory microcosms (see examples in Figure 20.3), with little or no environmental variation along the expansion, no interactions with other species and for a limited number of generations. While these conditions allow separating the influence of ecological and evolutionary factors with a resolution that cannot be achieved in natural populations, they have limitations in the degree of complexity and the time/space scales they can embrace. It is thus possible that selective pressures for faster life histories are exacerbated when the competition occurs not only at the intraspecific but also at the interspecific level, and/or that these life history shifts involving multiple traits take longer than the increase in dispersal rate, and so have not been easily captured in experimental systems so far. Furthermore, when edge populations in nature do display higher reproductive rates or faster development, it might be in some cases more directly related to, for instance, relaxed competition at low density (Therry et al. 2016, Jan et al. 2019) or adaptation to warmer climatic conditions (Swaegers et al. 2022) than to increased *r*-selection. To address these potential limitations, we recommend that studies that aim at investigating the evolutionary processes driving life history evolution on expanding range edges should try to cover a large range of traits that are components of *r* (*e.g.*, morphology, reproduction, survival and behaviour), representative of the different life stages, and to include common garden or reciprocal transplant experiments to compare the performance of individuals from core and edge populations in similar environments.

20.1.2.3 Stochasticity

In addition to these directional selective processes, stochastic processes based on sampling effects are known to impact the evolution of life history traits on expansion fronts and to increase the variability of evolutionary trajectories along range expansions. If we consider that the leading edge of an expansion is caused by a small number of dispersers that have moved outwards from the core of the range, then it is clear that there will be sampling effects. The leading edge in each generation is a small sample of the genetic diversity of the larger population immediately interior to it. And this sampling process happens in each generation, hence, expansion fronts are subject to serial founder events. These serial founder effects amount to a spatial analogue of genetic drift (Slatkin and Excoffier 2012) and we will refer to the process as 'spatial drift' hereafter. Expansion fronts are also of course subject to standard genetic drift because of their relatively small population size. We will refer to genetic drift as 'temporal drift' hereafter to make it clearly distinct from its spatial analogue. The combination of these two processes means that expansion fronts are a place where considerable stochasticity is at play.

High stochasticity, coupled with expansion dynamics, can lead to surprising outcomes. The first is that expansion fronts can be expected to rapidly lose genetic diversity as a consequence of the repeated subsampling (through time and space) that is characteristic of spatial and temporal drift (Figure 20.1). Genetic diversity may be restored later, once the more diverse population core catches up in space or due to multiple founding events, but the recovery capacity should be constrained by the accumulation of bottlenecks. For example, in a controlled introduction experiment on bush crickets, Kaňuch et al. (2021) found that even populations started from a single pair of individuals recovered levels of genetic diversity similar to the large source population after only 15 generations, probably under the combined influence of an exceptionally rapid genome-wide mutation rate and a polyandrous mating system. In contrast, when analysing the natural range expansion of a lice species subject to repeated bottleneck events, Demastes et al. (2019) found that the post-bottleneck genetic recovery is much slower, with only 30% recovery of genetic diversity over 225 generations. A decrease in genetic diversity along a geographical range is thus classically interpreted as a signature of a range expansion in population genetics studies (*e.g.*, Hewitt 1999, Estoup et al. 2004).

In addition to this intense loss of genetic diversity, low effective population size makes selection less efficient on the front, so front populations might accumulate mildly deleterious mutations over time and space, creating an 'expansion load' (Travis et al. 2007, Peischl et al. 2013, Bosshard et al. 2017). This expansion load should result in a decreasing cline in mean fitness between the front and the core of the population, whose steepness increases over time as deleterious mutations continue to accumulate on the front. Although this mutation load will be progressively eliminated by selection once the core has caught up with the expansion or once enough time has passed for new, fitter variants to appear, this process is expected to be quite slow and expansion load can remain for thousands of generations (Peischl et al. 2013).

Compounding the issue of expansion load is the issue of mutation surfing. Genetic variants that happen to arrive on the front (either from standing variation or from *de novo* mutations), even if they are rare in the core, can quickly fix and go on to fund all future invasion fronts. In this way, even deleterious alleles can spread over large spatial scales (Klopfstein et al. 2006, Hallatschek and Nelson 2008). This phenomenon has been verified empirically in several laboratory studies considering the spread patterns of bacteria on Petri dishes: initially, mixed colonies grow into several adjacent monomorphic radial sectors, and even rare, deleterious mutants can overcome whole sectors of the expansion (Hallatschek et al. 2007, Korolev et al. 2011, Gralka et al. 2016). Combining the idea of mutation surfing and spatial drift, Möbius et al. (2015, 2021) show that environmental heterogeneity can amplify the process of spatial drift. Here, sectors of the expansion front that encounter less favourable environments slow down or stop, and the expansion beyond is overtaken by adjacent parts of the expansion front. In this way, mutants that dominate particular segments of the invasion front can be left behind if their part of the front encounters a poor environment. Thus, environmental variation works to compound the effect of spatial drift by speeding up the fixation of alleles on expansion fronts.

Stochastic and directional processes can also interact to either reinforce or counterbalance their respective effects. The phenomenon of mutation surfing may of course impact beneficial mutations and strongly increase their fixation rate on the front, thus enhancing adaptation during expansion (Lehe et al. 2012). The evolution of dispersal through spatial sorting may also generate a rescue effect of expansion load at the front: in a stepping-stone model (a model of discrete one-dimensional space with nearest-neighbour dispersal), when dispersal increases along the expansion front, founder events and spatial drift become less intense until selection eventually manages to reduce or even reverse the expansion load (Peischl and Gilbert 2020). Drift and loss of diversity also result in increased relatedness on the front and increased kin competition, which in turn select for higher dispersal (van Petegem et al. 2018; see the section above on spatial sorting). Furthermore, the balance between selective and stochastic processes is also expected to vary across life history traits. For instance, in an analysis of life history shifts on the expansion fronts of three plant species, Latron et al. (2023) found that while the evolution of dispersal-related traits displayed a common trend compatible with directional selection at range edges, other life history traits were characterized by inconsistent evolutionary changes, suggesting also a strong influence of stochastic evolutionary processes or varying environmental conditions.

Recent work on experimental expansions in laboratory systems confirmed that both selective and stochastic processes drive the evolutionary dynamics of expansion and that they might sometimes be difficult to disentangle. In an experiment on red flour beetles (*Tribolium castaneum*; Figure 20.3a), Weiss-Lehman et al. (2017) found that even if evolution did increase expansion speed on average as predicted under the influence of spatial sorting and natural selection, expansion rates were highly variable between replicates. A similar pattern was observed repeatedly in other evolution experiments (Fronhofer and Altermatt 2015, Williams et al. 2016, Ochocki and Miller 2017). Indeed, as stochastic evolutionary processes can affect both the neutral genetic diversity and also non-neutral variants, populations are expected to display variable expansion patterns, depending on the beneficial or deleterious effects on individual performance of the mutations that happen to surf on the front (Phillips 2015). In that same *Tribolium* experiment, Weiss-Lehman et al. (2019) observed significant differentiation at the genome scale between front populations from independent replicates after eight generations of expansion only, thus confirming a strong influence of stochasticity. Another key result was that front populations had reduced reproductive fitness on average, which suggests the consistent presence of an expansion load across replicates (Weiss-Lehman et al. 2017). Within a single expansion boot, expansion load can also generate fluctuations in speed over time if the accumulation of deleterious mutations at the front slows its progression enough for higher fitness genotypes from the core to catch up with the edge (Peischl et al. 2015). Ultimately, it is the balance between selective and stochastic processes at edges that determines outcomes. Factors such as population size and mating system are predicted to play a crucial role in this balance (Williams et al. 2019).

20.1.3 Loss of Ecological Interactions

Another selective process specific to expanding populations is related to the loss of specialized interactions on the front. Because only a few individuals from the core population manage to reach the front during each founder event, there is a high probability that by chance, these individuals will not harbour all the diversity of the parasites, pathogens or symbionts present in the core (Phillips et al. 2010b). This general outcome occurs through a confluence of several processes. First is serial foundering: parasite or symbiont diversity on an invasion front is a sample of the diversity present on the front in the previous generation. So, just as we expect loss of genetic diversity through temporal and spatial drift, we can also expect a loss of parasite diversity. Second, there may also be deterministic processes at play: a threshold density of host/prey is typically required to support specialized pathogens/predators, and, below this density, the growth rate of pathogens and predators is negative. Thus, the low density on expansion fronts may make the persistence of parasites/pathogens difficult (*e.g.*, Shigesada and Kawasaki 1997, Perkins 2012). For example, many microparasites exhibit density-dependent transmission (*e.g.*, for most horizontal transmissions; Anderson and May 1979, Hu et al. 2013) and so require a threshold density of hosts before the number of infections can be expected to grow positively (see also Chapter 24, this volume). Third, we might often expect individuals harbouring a parasite or a pathogen to be sick and so disperse less or over smaller distances (Fellous et al. 2011, Zilio et al. 2023), thus spatial sorting will selectively remove infected individuals from the expansion front. Finally, predators or mutualists can also be lost during an expansion due to dispersal capacities insufficient to follow the expanding species. Overall, this confluence of processes should cause a cline in the abundance and diversity of specialized interactions along an expanding front (Phillips et al. 2010b). This gradient is expected to have different impacts on the evolutionary dynamics of life history traits depending on the nature of the ecological interactions at play.

The loss of natural enemies, be they parasites, pathogens or predators, is classically predicted to relax selective pressures related to immunity and defence, with a potential reallocation of resources towards other life history traits, like reproductive effort, which might provide a competitive advantage over the native populations within the expansion area (evolution of increased competitive ability, *i.e.*, evolution of increased competitive ability [EICA] hypothesis; Blossey and Notzold 1995).

The first part of this prediction is that populations at the front of an expansion should have reduced diversity or abundance of natural enemies. This prediction has been tested many times with heterogeneous results (*e.g.*, Keane and Crawley 2002, Phillips et al. 2010b, Mlynarek 2015, Edwards and Edwards 2023), yet it is globally supported in reviews and meta-analyses (mostly in plant-herbivore systems, Liu and Stiling 2006, Kirichenko et al. 2013, Meijer et al. 2016; but see Torchin et al. 2003, in animals). However, it can be argued that a decline in the diversity of natural enemies does not automatically translate into a relaxed selective pressure on defence, which may further depend on the relative impact and specificity of the enemies (Colautti et al. 2004, Brown et al. 2015, Lucero et al. 2020). A robust confirmation of the EICA hypothesis should thus require also that investment in immunity or defence be lower on expansion fronts, to the benefit of reproduction. This prediction relies on two assumptions: (1) that there is a strong trade-off between defence and reproduction and (2) that the gradient in the selective pressure from natural enemies between the core and the front of the expansion is steep enough to shift the cost–benefit ratio towards alternative life history strategies. The need for both of these assumptions to be met might make the evolutionary conditions favouring EICA rare in practice (Honor and Colautti 2020). Indeed, examples can be found either in support (Felker-Quinn et al. 2013(or in disagreement (Rotter and Holeski 2018) with this second assumption (based on meta-analysis of the EICA hypothesis in plants), suggesting that the evolutionary dynamics of trade-offs between immunity and reproduction might be more complex than expected.

Regarding mutualists, the loss or limitation of interactions can select compensatory mechanisms in the remaining partner, but these are likely to be costly to some extent. For instance, pollinator limitation can select increased plant self-crossing in small, isolated populations on the front of an expansion, but this will further increase the amount of drift and genetic load (Mullarkey et al. 2013, Zhang et al. 2021, Busch et al. 2022). Another example comes from invasive populations of St. John's wort in North America, which were found to display a modified root architecture to compensate for weaker association with mycorrhizal fungi, but this change came with a transition towards a weedier life history strategy (earlier reproduction, lower investment in roots; Seifert et al. 2009).

Finally, facultative symbionts have been identified as major mediators of individual phenotype and fitness (*e.g.*, with effects on lifespan, fecundity, immunity or metabolism, Carthey et al. 2020; see also Chapter 14, this volume), so selection processes can be considered to act at the collective scale of the hosts and their symbionts (the 'holobiont'; Zilber-Rosenberg and Rosenberg 2008). Therefore, when host populations undergo repeated bottlenecks that deplete their genetic diversity, they may also experience parallel 'hologenomic bottlenecks' (Arnaud-Haond et al. 2017, Goddard-Dwyer et al. 2021), with potential consequences on survival, reproduction and adaptive capacities (Arnaud-Haond et al. 2017, Le Roux 2022, Ørsted et al. 2022). These integrated effects at the holobiont scale are just beginning to be uncovered and may emerge as a major component of the eco-evolutionary dynamics of life history traits on expansion fronts.

20.2 Trade-Offs Matter

It is clear that the evolutionary processes playing out on expansion fronts will cause phenotypes to arise that may be markedly different from the phenotypes in the core of a population. While the expansion front favours life histories with high rD values, the core of a range will typically favour high values of r or K (Burton et al. 2010, Deforet et al. 2019, Phillips and Perkins 2019), K referring here to competitive ability at high density (*i.e.*, a property of the organism) and not the carrying capacity of the environment. Thus, we might imagine the optimal phenotype moving within a phenotype space defined by axes of r, K and D as we travel from the long-established populations out to the invasion front.

It is a central tenet of life history theory that trade-offs between fitness-relevant traits impose constraints on how population might evolve (Stearns 1992). No organism can be good at everything: there are fundamental constraints of time, energy and biology that limit how far any population can move within the r, K and D space as well as the possible pathways it might take (Figure 20.4). As a simple example, imagine a population in which all individuals have an identical energy budget but there is genetic variation as to how this budget is allocated between r, K and D within an individual. On an expansion front, we would expect genotypes that invest heavily in r and D (at the expense of K) to become dominant, whereas, in the core of the range, we might expect high investment in K (at the expense of D). This is precisely what is demonstrated in simulation models incorporating such a trade-off (Burton et al. 2010) and is also evident in other models of evolutionary trade-offs on expansion fronts (*e.g.*, Gallaher et al. 2019, Ochocki et al. 2019).

With regard to evolution on an expansion front – where populations tend to maximize rD – it is clear that the trade-off between r and D will be of great importance. Indeed, if there is, for example, a negative trade-off such that increasing r leads

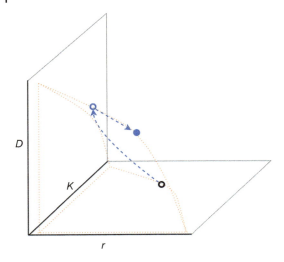

Figure 20.4 A hypothetical *r, K and D* space showing a possible evolutionary trajectory through this space. The orange dotted line defines the space of possible phenotypes: it is impossible, for reasons of fundamental constraint or fitness, for phenotypes outside of this space. The edges of this space are called Pareto fronts. We imagine here the outer Pareto fronts to be defined by curvilinear trade-offs between *r, K* and *D*. We show the evolutionary trajectory of a hypothetical population starting with a range core phenotype (open black circle), which sits along the *r, K* trade-off. During expansion, the front will evolve to maximize *rD*. Because movement within the Pareto space is relatively unconstrained, the population may rapidly increase *D* until it hits the Pareto front defined by the *r, D* trade-off (open blue circle). The population will then evolve along this trade-off until *rD* is maximized (closed blue circle).

to a proportional decline in D, then there is no evolutionary advantage to increasing r on the invasion front because the value of rD remains the same regardless of what value we choose for r. Of course, such a perfect trade-off is unlikely to manifest in reality, but the point is made that the nature of the trade-off will be important to the evolutionary outcome and to the subsequent invasion dynamics. Perhaps, more realistically, it is easy to imagine trade-offs in which D increases at some (proportionally smaller) cost to r. Under these circumstances, we might expect a population to decrease growth rate in order to increase dispersal rate. Such a trade-off may be an additional explanation for the inconsistent evolutionary response shown by growth rate in empirical invasions (Chuang and Peterson 2016).

Because of the likely large shift in optimum phenotype between the core and the invasion front, we might expect some time to elapse before trade-offs become evident. That is, there are initially likely to be 'easy wins' where both r and D might increase (see Figure 20.4). Eventually, however, the phenotype will hit a Pareto front – a line in phenotype space between possible and impossible phenotypes (Shoval et al. 2012). The shape of this line through the phenotype space then defines the trade-off to be optimized. The shape of this trade-off curve, and where the population first hits it, will then go on to determine the maximum value of rD that is attainable. This maximum value of rD also places an evolutionary limit on the maximum speed of the invasion.

Linking these theoretical ideas to empirical work is challenging, however. The primary obstacle is that measuring r, K and D directly is difficult in most systems. As mentioned above, we typically measure phenotypic components of these population-level parameters and hope that all else is equal. For example, we might measure fecundity as a proxy for r, hoping that reproductive intervals and death rates are approximately the same between the groups we are comparing. Such *ceteris paribus* assumptions rarely hold, of course, and when we consider age-structured populations, additional complexities also arise that may confound our intuition (Reznick et al. 2002).

20.3 Other Types of Expansions, How Our Expectations Might Change

20.3.1 Pushed *vs.* Pulled Expansions

Interestingly, all of the above predictions arise from the interplay of demographic and evolutionary processes operating at low density on the front, and so are related to only one specific type of expansion dynamic called a 'pulled' wave. Pulled waves occur when a population expands by sending a few pioneer individuals ahead, who eventually manage to reproduce and send a few descendants even further, and so forth. In this process, the expansion of the whole population is thus 'pulled' by a relatively few individuals at the leading edge of the invasion. However, in the presence of positive density-dependence in reproduction or dispersal (*i.e.*, higher reproductive success or higher dispersal rate when density is high), individual performance might be so impaired at low density that individuals from the front cannot contribute efficiently to expansion. In this case, the expansion is called 'pushed', meaning that the dynamics are driven mostly by the individuals from the core of the population, where density is high (Figure 20.5).

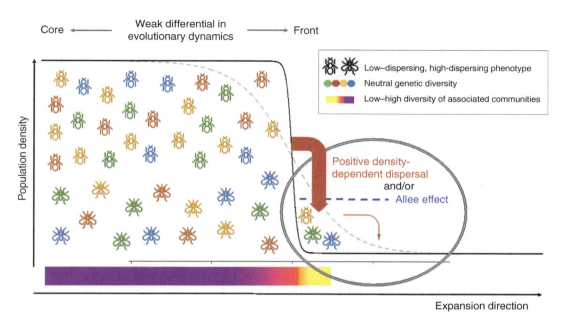

Figure 20.5 Evolutionary dynamics on a pushed front. The density profile of a pulled front appears as a dashed grey line for reference (see Figure 20.1). Colonisation ahead of the front is impaired because low-density populations do not establish (in the presence of an Allee effect) or because they do not produce enough dispersers (in the presence of positive density-dependent dispersal). The expansion progresses only when the front has reached enough density to overcome one of the other of these limitations.

Historically, the distinction between pulled and pushed dynamics derives from the analysis of reaction–diffusion equations used to describe the spatial expansion of populations. The classical reaction–diffusion model (mentioned in the introduction of this chapter; Kolmogorov et al. 1937, Fisher 1937) yields a pulled expansion. Here, the formula for the expansion speed of the population (*i.e.*, $v = 2\sqrt{rD}$) is derived from a simplification of the reaction–diffusion equation around zero density, which captures the dominant influence of low-density conditions on the front on the spatial dynamics of the whole population. In many other models, however, the equation cannot be simplified to calculate the spreading speed. The expansion is then considered to be driven by the dynamics over the whole range of the population (Stokes 1976, Mollison 1991) and is said to be pushed. However, this distinction between pulled and pushed dynamics is restricted to certain types of reaction–diffusion models where the spreading speed of the travelling waves can be computed analytically, and thus may not be generalized to other contexts (*e.g.*, discrete space or time, empirical data). Recently, pushed expansions have been the object of a renewed interest from the community of theoretical and empirical ecologists, and a quantitative criterion for determining the nature of an expansion has been proposed by Birzu et al. (2019). This criterion is based on the ratio of the observed spreading speed to the spreading speed expected under a pulled model (*i.e.*, based on the values of reproductive and dispersal rate at density 'zero'). When this velocity ratio is equal to one, the expansion is pulled, *i.e.*, the expansion is behaving as a classic Fisher (1937) travelling wave. In contrast, for pushed waves, the population's growth and/or dispersal rate are at a maximum deeper into the expansion front. The growth and/or dispersal rate at the leading tip of the invasion is less than this maximum. Because of this, if we put these low-density growth/dispersal rates into the formula for expansion speed, we get an expected speed that is lower than it would be if the maximum rates of growth and dispersal occurred on the leading tip. Thus, for pushed expansions, the actual expansion speed is higher than that expected under the Fisher (1937) model, and so the ratio of observed to expected velocity is higher than one. Pushed expansions can be further classified into semi-pushed or fully pushed depending on the value of the velocity ratio (Birzu et al. 2018), but we will focus here only on the case of fully pushed expansions.

Pushed expansions emerge whenever colonisation happens to be more difficult at low density, either because population growth or dispersal is reduced. The first developments related to pushed expansions were done when comparing populations with or without an Allee effect (Roques et al. 2012, Wittmann et al. 2014, Gandhi et al. 2016, Lewis et al. 2016), and most of these results were confirmed later in the case of positive density-dependence in dispersal (Birzu et al. 2019, Haond et al. 2021, Morel-Journel et al. 2022). An Allee effect is a phenomenon in which *per capita* population growth rate is highest at some intermediate population density. Allee effects arise naturally from a range of phenomena including cooperation and mate limitation, at low density (Courchamp et al. 1999). The common characteristic of pushed waves is that low-density populations on the front either do not establish (in case of an Allee effect) or do not disperse to colonize further (in case of density-dependent dispersal). In both cases, because the leading edge is now driven forward by movement from the bulk of population, the expanding population does not experience the intense drift that is characteristic of pulled invasions. Thus, in pushed expansions, neutral genetic diversity tends to be conserved on the front (Figure 20.5), a result that has been consistently supported by theoretical and empirical work (Roques et al. 2012, Birzu et al. 2019, Gandhi et al. 2019). In addition, because there is no repeated selective process at low density, the evolutionary response of dispersal and life history traits is likely to be muted (Travis and Dytham 2002; Figure 20.5), and in some cases even reversed (Korolev 2015). This prediction still lacks thorough investigation, but some elements of evidence are nevertheless available. First, in some earlier theoretical work, Burton et al. (2010) explored the evolution of life history traits along an expansion front. While they did not address the issue of pulled or pushed dynamics explicitly, they investigated a case in which expansion occurs in the presence of a competitor (a situation which can generate pushed waves: Roques et al. 2015), and they did find that trait evolution is greatly reduced in this case. More recently, however, several empirical studies have investigated the evolution of traits during pushed expansions, and the patterns appear more equivocal. For instance, while Dahirel et al. (2021a, 2021b) managed to generate *Trichogramma* expansions (Figure 20.3c) with different dynamics of neutral diversity loss (*i.e.*, more or less pushed), they did not find any clear evidence for corresponding different evolutionary dynamics of phenotypic traits.

In addition to these distinct evolutionary patterns emerging from pulled or pushed dynamics, the mechanisms driving one type or the other of expansion can evolve during the course of an invasion. Indeed, positive density-dependent dispersal or an Allee effect is expected to be strongly counter-selected on expansion fronts, so that pushed expansions are expected to evolve into pulled ones in the longer term, either by the evolution of resistance to the Allee effect (Kanarek and Webb 2010, Erm and Phillips 2020) or of density-independent dispersal (Travis et al. 2009). This last prediction was confirmed experimentally in the laboratory, with the evolution of density-independent movement in replicated populations of the ciliate *Tetrahymena* (Figure 20.3b) and associated acceleration of spread (Fronhofer et al. 2017), and further supported by the observation of increased social attraction in several naturally expanding populations (Rodríguez et al. 2010, Gruber et al. 2017), a mechanism for increasing cooperation and decreasing mate limitation in low-density populations. Recent modelling results, however, point out that density-dependent dispersal is actually a complex trait, of which different components may evolve independently (Dahirel et al. 2022). When this complexity is taken into account, it appears that pushed expansions do not always evolve into pulled ones and that the opposite trajectory can also occur.

20.3.2 Interaction with the Environment

All previous processes were described assuming a homogeneous environment. The spatial structure of environmental variation is, however, expected to impact eco-evolutionary dynamics on expansion fronts. It has long been known that for populations expanding along an environmental gradient, adaptation and dispersal were subjected to conflicting selective pressures. When encountering a novel environment, populations may adapt by evolving locally advantageous traits, but dispersal brings maladapted individuals from other environments that counteract this adaptive process (Kirkpatrick and Barton 1997, García-Ramos and Rodríguez 2002). At the same time, dispersal itself comes under strong selection for reduced dispersal rates (Balkau and Feldman 1973, Hastings 1983). The reason for this is that, on an environmental gradient, dispersing individuals on average tend to arrive in places that they are maladapted to, thus dispersal itself is selected against (as an aside, this can be framed either as spatial sorting or natural selection). The presence of an environmental gradient thus imposes a limit on the evolution of increased dispersal rate on the expansion front, and in some cases selects for reduced dispersal (García-Ramos and Rodríguez 2002, Kubisch et al. 2014). A theoretical work concluded that the results of an interaction between dispersal evolution and local adaptation depend on the initial dispersal rate of the population (Andrade-Restrepo et al. 2019). Below a certain threshold (the clustering threshold), the expansion will proceed in steps, characterized by strong temporal variation in expansion rate and dispersal selection, with short phases of high dispersal, followed by expansion pauses during which the front population adapts to the newly colonized habitat. For populations above this initial clustering threshold, the expansion will take a wave-like, continuous form until selection against dispersal becomes too strong and generates an abrupt transition into the clustering regime.

Fragmentation of the environment is also expected to influence the evolutionary dynamics of dispersal, with heterogeneous effects depending on the process driving habitat fragmentation (Legrand et al. 2017), and on the ecological characteristics of the dispersing species (Bonte et al. 2018, Miller et al. 2020). Increased isolation between patches, coupled with decreasing patch size, is expected to increase dispersal costs and thus to select against dispersal, although this effect can be compensated partly in temporally variable environments that select for increased dispersal as a bet-hedging strategy (*i.e.*, a risk-spreading strategy where a given genotype can give rise to different phenotypes, each with an advantage in alternative environmental conditions; McPeek and Holt 1992, Kubisch et al. 2014). Furthermore, trade-offs between dispersal and fecundity interact with fragmentation: when the trade-off is strong and the landscape is highly fragmented, fecundity is expected to be favoured at the expense of dispersal, and variability in expansion speed increases (Urquhart and Williams 2021). As mentioned earlier, environmental heterogeneity can also directly affect the selection/drift balance (Möbius et al. 2015, 2021). Using experimental expansions with bacteria, Gralka and Hallatschek (2019) indeed showed that surface irregularities creating physical obstacles to expansion could strongly reduce the efficacy of selection so that the local characteristics of the environment where a mutation occurs have more influence in its following fixation or loss than its actual effect on fitness.

Finally, the characteristics of the environment may also modulate the pulled or pushed nature of an expansion, and so the evolutionary dynamics on the front. For instance, the presence of a competing species in the invaded area was found to reduce both expansion speed (Burton et al. 2010, Legault et al. 2020) and evolutionary speed (for life history parameters, Burton et al. 2010; for neutral diversity, Roques et al. 2015), thus suggesting a shift towards pushed dynamics. There are other ways that an expansion might be pushed. For example, a population expanding its geographical range in response to a climatic shift will be pushed if the speed at which suitable habitat becomes available is slower than the expansion potential of the focus species (Garnier and Lewis 2016). In addition to their effect on dispersal evolution and drift/selection balance described above, high fragmentation and reduced connectivity can also interact with density-dependence to generate pushed expansions, with more fragmented landscapes constraining low-density colonisation dynamics (Dahirel et al. 2021b).

20.4   An Applied Case Study: The Cane Toad

Cane toads (Figure 20.2) were introduced into Australia in 1935 and have spread to occupy more than 1.6 million km^2 of the continent (Urban et al. 2007). This is a biological invasion writ large, and an excellent example of how we might put to use an understanding of the ecological and evolutionary dynamics of expansion.

In toads, spatial sorting and natural selection on the invasion front have caused the expansion to accelerate: an invasion that once spread at 10 km/year now spreads around five times faster (Phillips et al. 2006, Perkins et al. 2013). However, this acceleration appears to have reached its limits. There are clear signs of trade-offs – particularly a trade-off between reproductive rate and dispersal rate – that likely work to limit the maximum speed the toad invasion will achieve (Hudson et al. 2015, Perkins et al. 2016).

Around 2026, the toad expansion will be due to enter a narrow corridor of suitable habitat, through which it must pass in order to colonize a much larger area further to the south (Florance et al. 2011). Plans are underway to create a 'waterless barrier' across this corridor – a length of the corridor around 70 km wide in which all surface water is managed to prevent access by toads. Modelling indicates that such an action would stop the toad invasion, containing further spread and preventing toads from invading 270,000 km^2 of Australia (an area larger than Great Britain; Tingley et al. 2013).

The barrier needs to be around 70 km wide because it needs to stop invasion front toads that have, through spatial sorting, evolved to be highly dispersive. One way to make the barrier substantially more effective, then, may be to implement a 'genetic backburn' (Phillips et al. 2016). Here, animals from the core of the toads' range are brought to the nearside of the barrier and released in order to spread back towards the oncoming invasion front. The idea is that these less dispersive (but more competitive) genotypes from the core of the range will outcompete the highly dispersive (but less competitive) genotypes characteristic of the invasion front (Perkins et al. 2016). The result is, in theory, highly dispersive genotypes then never make it to the barrier. In this way, the barrier can be made more effective simply by choosing the genotype that it is being asked to stop. The absorbing nature of the barrier also sets up ongoing selection against dispersal, so the barrier should become more effective over time, simply because the dispersal rate of the toad population abutting the barrier should decline over time (Phillips et al. 2016).

20.5 Summary and Future Directions

It is clear that expansion fronts offer evolutionary conditions that are dramatically different from those in the stable core of a species range. This shift in evolutionary conditions will often cause life histories to diverge between range core and expansion front, and the dynamics of expansion can cause expansion front phenotypes to be spread widely across space. The process of expansion, then, can be seen as a (currently underappreciated) source of life history diversity within a species range.

Expansion fronts set both deterministic and stochastic evolutionary processes into motion. When expansions are pulled, we have a strong deterministic expectation that life histories will evolve (within the constraint of trade-offs) to maximize the product of growth and dispersal rates. But at the same time, these pulled expansions are strongly stochastic, such that maladaptive phenotypes can emerge to dominate the expansion front. Thus, pulled expansions can be highly variable and difficult to predict. When expansions are pushed, by contrast, the evolutionary response often appears muted, due to strong mixing of genotypes across the invasion front. This mixing across the front also means that pushed expansions manifest substantially less stochasticity relative to pulled ones. Pushed expansions are, as a result, slower and more predictable. These results have emerged over the last 20 years, through a mix of theoretical and empirical work. While the work on pulled expansions is now well established, the cutting edge is currently focused on pushed expansions. Pushed expansions are a more complex set of circumstances than pulled expansions – they arise through a wide range of processes and are mathematically challenging – and because of this, we know substantially less about their evolutionary dynamics.

For both pushed and pulled expansions we also know that trade-offs are critical to evolutionary trajectories. Here, there is strong scope for an improved interplay between life history theory and spatial dynamics. A well-informed and general theory of life history trade-offs applied to expansions offers the possibility that we might predict evolutionary constraints and so define the ultimate limits of expansion speed (see Figure 20.4). Careful consideration of life history is also necessary as we attempt to map theoretical predictions to empirical circumstances. Given the myriad life history components underlying growth rate, dispersal, competitive ability and their responses to density, a nuanced and sophisticated understanding of life history will be needed if we are to make the connection between empirical and theoretical understanding without getting confused by complex empirical patterns.

Overall, an evolutionary perspective gives us a much richer understanding of population expansions. The process of population expansion sets evolutionary forces in play that can cause rapid shifts in life history. These life history shifts in turn affect the speed and dynamics of the expansion. The basic ideas are now in place, but there is much still to be learned, and prediction remains a formidable challenge. Meeting these challenges is a worthy enterprise because expansions are common and are very often consequential for biodiversity, food production and human health. By unearthing the fundamental processes that apply to population expansions of all types – from invasive species, to the spread of disease or the growth of a tumour – we place ourselves in a much stronger position to make clear predictions, acknowledge our uncertainty and develop new methods of containment and control.

References

Amundsen, P.-A., Salonen, E., Niva, T., Gjelland, K.O., Praebel, K., Sandlund, O.T., Knudsen, R., and Bøhn, T. (2012). Invader population speeds up life history during colonization. *Biological Invasions* 14(7): 1501–1513.

Anderson, R.M., and May, R.M. (1979). Population biology of infectious diseases: part I. *Nature* 280(5721): 361–367.

Andrade-Restrepo, M., Champagnat, N., and Ferrière, R. (2019). Local adaptation, dispersal evolution, and the spatial eco-evolutionary dynamics of invasion. *Ecology Letters* 22(5): 767–777.

Arnaud-Haond, S., Aires, T., Candeias, R., Teixeira, S.J.L., Duarte, C.M., Valero, M., and Serrão, E.A. (2017). Entangled fates of holobiont genomes during invasion: nested bacterial and host diversities in *Caulerpa taxifolia*. *Molecular Ecology* 26(8): 2379–2391.

Balkau, B.J., and Feldman, M.W. (1973). Selection for migration modification. *Genetics* 74(1): 171–174.

Birzu, G., Hallatschek, O., and Korolev, K.S. (2018). Fluctuations uncover a distinct class of traveling waves. *Proceedings of the National Academy of Sciences of the United States of America* 115(16): E3645–E3654.

Birzu, G., Matin, S., Hallatschek, O., and Korolev, K.S. (2019). Genetic drift in range expansions is very sensitive to density dependence in dispersal and growth. *Ecology Letters* 22(11): 1817–1827.

Blossey, B., and Notzold, R. (1995). Evolution of increased competitive ability in invasive nonindigenous plants: a hypothesis. *Journal of Ecology* 83(5): 887–889.

Bøhn, T., Terje Sandlund, O., Amundsen, P.-A., and Primicerio, R. (2004). Rapidly changing life history during invasion. *Oikos* 106(1): 138–150.

Bonte, D., Masier, S., and Mortier, F. (2018). Eco-evolutionary feedbacks following changes in spatial connectedness. *Current Opinion in Insect Science* 29: 64–70.

Bosshard, L., Dupanloup, I., Tenaillon, O., Bruggmann, R., Ackermann, M., Peischl, S., and Excoffier, L. (2017). Accumulation of deleterious mutations during bacterial range expansions. *Genetics* 207(2): 669–684.

Bouin, E., Calvez, V., Meunier, N., Mirrahimi, S., Perthame, B., Raoul, G., and Voiturier, R. (2012). Invasion fronts with variable motility: phenotype selection, spatial sorting and wave acceleration. *Comptes Rendus Mathematique* 350(15): 761–766.

Brown, G., Phillips, B.L., Dubey, S., and Shine, R. (2015). Invader immunology: invasion history alters immune-system function in cane toads (*Rhinella marina*) in tropical Australia. *Ecology Letters* 18: 57–65.

Burton, O.J., Phillips, B.L., and Travis, J.M.J. (2010). Trade-offs and the evolution of life-histories during range expansion. *Ecology Letters* 13(10): 1210–1220.

Busch, J.W., Bodbyl-Roels, S., Tusuubira, S., and Kelly, J.K. (2022). Pollinator loss causes rapid adaptive evolution of selfing and dramatically reduces genome-wide genetic variability. *Evolution* 76(9): 2130–2144.

Carthey, A.J.R., Blumstein, D.T., Gallagher, R.V., Tetu, S.G., and Gillings, M.R. (2020). Conserving the holobiont. *Functional Ecology* 34(4): 764–776.

Charlesworth, B. (1971). Selection in density-regulated populations. *Ecology* 52(3): 469–474.

Chuang, A., and Peterson, C.R. (2016). Expanding population edges: theories, traits, and trade-offs. *Global Change Biology* 22(2): 494–512.

Clark, E.I., Bitume, E.V., Bean, D.W., Stahlke, A.R., Hohenlohe, P.A., and Hufbauer, R.A. (2022). Evolution of reproductive life-history and dispersal traits during the range expansion of a biological control agent. *Evolutionary Applications* 15(12): 2089–2099.

Colautti, R.I., Ricciardi, A., Grigorovich, I.A., and MacIsaac, H.J. (2004). Is invasion success explained by the enemy release hypothesis? *Ecology Letters* 7(8): 721–733.

Courchamp, F., Clutton-Brock, T., and Grenfell, B. (1999). Inverse density-dependence and the Allee effect. *Trends in Ecology & Evolution* 14(10): 405–410.

Dahirel, M., Bertin, A., Calcagno, V., Duraj, C., Fellous, S., Groussier, G., Lombaert, E., Mailleret, L., Marchand, A., and Vercken, E. (2021a). Landscape connectivity alters the evolution of density-dependent dispersal during pushed range expansions. bioRxiv 2021.03.03.433752, ver. 4 peer-reviewed and recommended by Peer Community in Evolutionary Biology. https://doi.org/10.1101/2021.03.03.433752.

Dahirel, M., Bertin, A., Haond, M., Blin, A., Lombaert, E., Calcagno, V., Fellous, S., Mailleret, L., Malausa, T., and Vercken, E. (2021b). Shifts from pulled to pushed range expansions caused by reduction of landscape connectivity. *Oikos* 130(5): 708–724.

Dahirel, M., Guicharnaud, C., and Vercken, E. (2022). Individual variation in dispersal, and its sources, shape the fate of pushed *vs.* pulled range expansions. bioRxiv 2022.01.12.476009. https://doi.org/10.1101/2022.01.12.476009.

De Bona, S., Bruneaux, M., Lee, A.E.G, Reznick, D.N., Bentzen, P., and López-Sepulcre, A. (2019). Spatio-temporal dynamics of density-dependent dispersal during a population colonisation. *Ecology Letters* 22(4): 634–644.

Deforet, M., Carmona-Fontaine, C., Korolev, K.S., and Xavier, J.B. (2019). Evolution at the edge of expanding populations. *The American Naturalist* 194(3): 291–305.

Demastes, J.W., Hafner, D.J., Hafner, M.S., Light, J.E., and Spradling, T.A. (2019). Loss of genetic diversity, recovery and allele surfing in a colonizing parasite, *Geomydoecus aurei*. *Molecular Ecology* 28(4): 703–720.

Edwards, D.D., and Edwards, O.M. (2023). Range expansion of green treefrogs (*Hyla cinerea*) in Southern Illinois: no evidence of parasite release. *Journal of Parasitology* 109(2): 51–55.

Erm, P., and Phillips, B.L. (2020). Evolution transforms pushed waves into pulled waves. *The American Naturalist* 195(3): E87–E99.

Estoup, A., Beaumont, M., Sennedot, F., Moritz, C., and Cornuet, J.-M. (2004). Genetic analysis of complex demographic scenarios: spatially expanding populations of the cane toad, *Bufo marinus*. *Evolution* 58(9): 2021–2036.

Felker-Quinn, E., Schweitzer, J.A., and Bailey, J.K. (2013). Meta-analysis reveals evolution in invasive plant species but little support for evolution of increased competitive ability (EICA). *Ecology and Evolution* 3(3): 739–751.

Fellous, S., Quillery, E., Duncan, A.B., and Kaltz, O. (2011). Parasitic infection reduces dispersal of ciliate host. *Biology Letters* 7(3): 327–329.

Fisher, R.A. (1937). The wave of advance of advantageous genes. *Annals of Eugenics* 7(4): 355–369.

Florance, D., Webb, J.K., Dempster, T., Kearney, M.R., Worthing, A., and Letnic, M. (2011). Excluding access to invasion hubs can contain the spread of an invasive vertebrate. *Proceedings of the Royal Society B: Biological Sciences* 278(1720): 21345870.

Freedman, M.G., Dingle, H., Strauss, S.Y, and Ramírez, S.R. (2020). Two centuries of monarch butterfly collections reveal contrasting effects of range expansion and migration loss on wing traits. *Proceedings of the National Academy of Sciences of the United States of America* 117(46): 28887–28893.

Fronhofer, E.A., and Altermatt, F. (2015). Eco-evolutionary feedbacks during experimental range expansions. *Nature Communications* 6: 6844.

Fronhofer, E.A., Gut, S., and Altermatt, F. (2017). Evolution of density-dependent movement during experimental range expansions. *Journal of Evolutionary Biology* 30(12): 2165–2176.

Gallaher, J.A., Brown, J.S., and Anderson, A.R.A. (2019). The impact of proliferation-migration tradeoffs on phenotypic evolution in cancer. *Scientific Reports* 9(1): 2425.

Gandhi, S.R., Korolev, K.S., and Gore, J. (2019). Cooperation mitigates diversity loss in a spatially expanding microbial population. *Proceedings of the National Academy of Sciences of the United States of America* 116(47): 23582–23587.

Gandhi, S.R., Yurtsev, E.A., Korolev, K.S., and Gore, J. (2016). Range expansions transition from pulled to pushed waves as growth becomes more cooperative in an experimental microbial population. *Proceedings of the National Academy of Sciences of the United States of America* 113(25): 6922–6927.

García-Ramos, G., and Rodríguez, D. (2002). Evolutionary speed of species invasions. *Evolution* 56(4): 661–668.

Garnier, J., and Lewis, M.A. (2016). Expansion under climate change: the genetic consequences. *Bulletin of Mathematical Biology* 78(11): 2165–2185.

Giometto, A., Rinaldo, A., Carrara, F., and Altermatt, F. (2014). Emerging predictable features of replicated biological invasion fronts. *Proceedings of the National Academy of Sciences of the United States of America* 111(1): 297–301.

Goddard-Dwyer, M., López-Legentil, S., and Erwin, P.M. (2021). Microbiome variability across the native and invasive ranges of the ascidian *Clavelina oblonga*. *Applied and Environmental Microbiology* 87(2): e02233–20.

Gralka, M., and Hallatschek, O. (2019). Environmental heterogeneity can tip the population genetics of range expansions. *eLife* 8: e44359.

Gralka, M., Stiewe, F., Farrell, F., Möbius, W., Waclaw, B., and Hallatschek, O. (2016). Allele surfing promotes microbial adaptation from standing variation. *Ecology Letters* 19(8): 889–898.

Gruber, J., Whiting, M.J., Brown, G., and Shine, R. (2017). The loneliness of the long-distance toad: invasion history and social attraction in cane toads (*Rhinella marina*). *Biology Letters* 13(11): 20170445.

Hall, M.L., van Asten, T., Katsis, A.C., Dingemanse, N.J., Magrath, M.J.L., and Mulder, R.A. (2015). Animal personality and pace-of-life syndromes: do fast-exploring fairy-wrens die young? *Frontiers in Ecology and Evolution* 3: 28.

Hallatschek, O., Hersen, P., Ramanathan, S., and Nelson, D.R. (2007). Genetic drift at expanding frontiers promotes gene segregation. *Proceedings of the National Academy of Sciences of the United States of America* 104(50): 19926–19930.

Hallatschek, O., and Nelson, D.R. (2008). Gene surfing in expanding populations. *Theoretical Population Biology* 73(1): 158–170.

Haond, M., Morel-Journel, T., Lombaert, E., Vercken, E., Mailleret, L., and Roques, L. (2021). When higher carrying capacities lead to faster propagation. *Peer Community Journal* 1: e57.

Hastings, A. (1983). Can spatial variation alone lead to selection for dispersal? *Theoretical Population Biology* 24(3): 244–251.

Hewitt, G.M. (1999). Post-glacial re-colonization of European biota. *Biological Journal of the Linnean Society* 68(1–2): 87–112.

Honor, R., and Colautti, R. (2020). EICA 2.0: a general model of enemy release and defence in plant and animal invasions. In: *Plant invasions: the role of biotic interactions* (eds. Traveset, A., and Richardson, D.M.), 192–207. CABI.

Hu, H., Nigmatulina, K., and Eckhoff, P. (2013). The scaling of contact rates with population density for the infectious disease models. *Mathematical Biosciences* 244(2): 125–134.

Huang, F., Peng, S., Chen, B., Liao, H., Huang, Q., Lin, Z., and Liu, G. (2015). Rapid evolution of dispersal-related traits during range expansion of an invasive vine *Mikania micrantha*. *Oikos* 124(8): 1023–1030.

Hudson, C.M., Phillips, B.M., Brown, G.P., and Shine, R. (2015). Virgins in the vanguard: low reproductive frequency in invasion front toads. *Biological Journal of the Linnean Society* 116(4): 743–747.

Hughes, C.L., Dytham, C., and Hill, J.K. (2007). Modelling and analysing evolution of dispersal in populations at expanding range boundaries. *Ecological Entomology* 32(5): 437–445.

Jan, P.-L., Lehnen, L., Besnard, A.-L., Kerth, G., Biedermann, M., Schorcht, W., Petit, E.J., Le Gouar, P., and Puechmaille, S.J. (2019). Range expansion is associated with increased survival and fecundity in a long-lived bat species. *Proceedings of the Royal Society B: Biological Sciences* 286(1906): 20190384.

Kanarek, A., and Webb, C. (2010). Allee effects, adaptive evolution, and invasion success. *Evolutionary Applications* 3(2): 122–135.

Kaňuch, P., Berggren, Å., and Cassel-Lundhagen, A. (2021). A clue to invasion success: genetic diversity quickly rebounds after introduction bottlenecks. *Biological Invasions* 23(4): 1141–1156.

Keane, R.M., and Crawley, M.J. (2002). Exotic plant invasions and the enemy release hypothesis. *Trends in Ecology & Evolution* 17(4): 164–170.

Kilkenny, F.F., and Galloway, L.F. (2013). Adaptive divergence at the margin of an invaded range. *Evolution* 67(3): 722–731.

Kirichenko, N., Péré, C., Baranchikov, Y., Schaffner, U., and Kenis, M. (2013). Do alien plants escape from natural enemies of congeneric residents? Yes but not from all. *Biological Invasions* 15(9): 2105–2113.

Kirkpatrick, M., and Barton, N.H. (1997). Evolution of a species' range. *The American Naturalist* 150(1): 1–23.

Klopfstein, S., Currat, M., and Excoffier, L. (2006). The fate of mutations surfing on the wave of range expansion. *Molecular Biology and Evolution* 23(3): 482–490.

Kolmogorov, A.N., Petrovsky, I.G., and Piskunov, N.S. (1937). Etude de l'équation de la diffusion avec croissance de la quantité de matière et son application à un problème biologique. *Bulletin de l'Université d'État de Moscou, Série Internationale Secteur A* 1: 1–26.

Korolev, K.S. (2015). Evolution arrests invasions of cooperative populations. *Physical Review Letters* 115(20): 208104.

Korolev, K.S., Xavier, J.B., Nelson, D.R., and Foster, K.R. (2011). A quantitative test of population genetics using spatiogenetic patterns in bacterial colonies. *The American Naturalist* 178(4): 538–552.

Kubisch, A., Fronhofer, E.A., Poethke, H.J., and Hovestadt, T. (2013). Kin competition as a major driving force for invasions. *The American Naturalist* 181(5): 700–706.

Kubisch, A., Holt, R.D., Poethke, H.-J., and Fronhofer, E.A. (2014). Where am I and why? Synthesizing range biology and the eco-evolutionary dynamics of dispersal. *Oikos* 123(1): 5–22.

Latron, M., Arnaud, J.-F., Schmitt, E., and Duputié, A. (2023). Idiosyncratic shifts in life-history traits at species' geographic range edges. *Oikos* 2023(2): e09098.

Le Roux, J. (2022). *The evolutionary ecology of invasive species*. Elsevier.

Legault, G., Bitters, M.E., Hastings, A., and Melbourne, B.A. (2020). Interspecific competition slows range expansion and shapes range boundaries. *Proceedings of the National Academy of Sciences of the United States of America* 117(43): 26854–26860.

Legrand, D., Cote, J., Fronhofer, E.A., Holt, R.D., Ronce, O., Schtickzelle, N., Travis, J.M.J., and Clobert, J. (2017). Eco-evolutionary dynamics in fragmented landscapes. *Ecography* 40(1): 9–25.

Lehe, R., Hallatschek, O., and Peliti, L. (2012). The rate of beneficial mutations surfing on the wave of a range expansion. *PLoS Computational Biology* 8(3): e1002447.

Lewis, M.A., Petrovskii, S.V., and Potts, J.R. (2016). *The mathematics behind biological invasions*. Cham: Springer International Publishing.

Liebl, A.L., and Martin, L.B. (2012). Exploratory behaviour and stressor hyper-responsiveness facilitate range expansion of an introduced songbird. *Proceedings of the Royal Society B: Biological Sciences* 279(1746): 4375–4381.

Liebl, A.L., and Martin, L.B. (2014). Living on the edge: range edge birds consume novel foods sooner than established ones. *Behavioral Ecology* 25(5): 1089–1096.

Lindström, T., Brown, G.P., Sisson, S.A., Phillips, B.L., and Shine, R. (2013). Rapid shifts in dispersal behavior on an expanding range edge. *Proceedings of the National Academy of Sciences of the United States of America* 110(33): 13452–13456.

Liu, H., and Stiling, P. (2006). Testing the enemy release hypothesis: a review and meta-analysis. *Biological Invasions* 8(7): 1535–1545.

Lombaert, E., Estoup, A., Facon, B., Joubard, B., Grégoire, J.-C., Jannin, A., Blin, A., and Guillemaud, T. (2014). Rapid increase in dispersal during range expansion in the invasive ladybird *Harmonia axyridis*. *Journal of Evolutionary Biology* 27(3): 508–517.

Lucero, J.E., Arab, N.M., Meyer, S.T., Pal, R.W., Fletcher, R.A., Nagy, D.U., Callaway, R.M., and Weisser, W.W. (2020). Escape from natural enemies depends on the enemies, the invader, and competition. *Ecology and Evolution* 10(19): 10818–10828.

MacArthur, R.H., and Wilson, E.O. (1967). *The theory of island biogeography*. Princeton University Press.

McPeek, M.A., and Holt, R.D. (1992). The evolution of dispersal in spatially and temporally varying environments. *The American Naturalist* 140(6): 1010–1027.

Meijer, K., Schilthuizen, M., Beukeboom, L., and Smit, C. (2016). A review and meta-analysis of the enemy release hypothesis in plant–herbivorous insect systems. *PeerJ* 4: e2778.

Miller, T.E.X., Angert, A.L., Brown, C.D., Lee-Yaw, J.A., Lewis, M., Lutscher, F., Marculis, N.G., Melbourne, B.A., Shaw, A.K., Szúcs, M., Tabares, O., Usui, T., Weiss-Lehman, C., and Williams, J.L. (2020). Eco-evolutionary dynamics of range expansion. *Ecology* 101(10): e03139.

Mlynarek, J.J. (2015). Testing the enemy release hypothesis in a native insect species with an expanding range. *PeerJ* 3: e1415.

Möbius, W., Murray, A.W., and Nelson, D.R. (2015). How obstacles perturb population fronts and alter their genetic structure. *PLoS Computational Biology* 11(12): e1004615.

Möbius, W., Tesser, F., Alards, K.M.J., Benzi, R., Nelson, D.R., and Toschi, F. (2021). The collective effect of finite-sized inhomogeneities on the spatial spread of populations in two dimensions. *Journal of the Royal Society Interface* 18(183): 20210579.

Mollison, D. (1991). Dependence of epidemic and population velocities on basic parameters. *Mathematical Biosciences* 107(2): 255–287.

Morel-Journel, T., Haond, M., Lamy, L., Muru, D., Roques, L., Mailleret, L., and Vercken, E. (2022). When expansion stalls: an extension to the concept of range pinning in ecology. *Ecography* 2022: e06018.

Mowery, M.A., Lubin, Y., Harari, A., Mason, A.C., and Andrade, M.C.B. (2022). Dispersal and life history of brown widow spiders in dated invasive populations on two continents. *Animal Behaviour* 186: 207–217.

Mullarkey, A.A., Byers, D.L., and Anderson, R.C. (2013). Inbreeding depression and partitioning of genetic load in the invasive biennial *Alliaria petiolata* (Brassicaceae). *American Journal of Botany* 100(3): 509–518.

Narimanov, N., Bauer, T., Bonte, D., Fahse, L., and Entling, M.H. (2022). Accelerated invasion through the evolution of dispersal behaviour. *Global Ecology and Biogeography* 31(12): 2423–2436.

Ochocki, B.M., and Miller, T.E.X. (2017). Rapid evolution of dispersal ability makes biological invasions faster and more variable. *Nature Communications* 8(1): 14315.

Ochocki, B.M., Saltz, J.B., and Miller, T.E.X. (2019). Demography-dispersal trait correlations modify the eco-evolutionary dynamics of range expansion. *The American Naturalist* 195(2): 231–246.

Ørsted, M., Yashiro, E., Hoffmann, A.A., and Kristensen, T.N. (2022). Population bottlenecks constrain host microbiome diversity and genetic variation impeding fitness. *PLoS Genetics* 18(5): e1010206.

Peischl, S., Dupanloup, I., Kirkpatrick, M., and Excoffier, L. (2013). On the accumulation of deleterious mutations during range expansions. *Molecular Ecology* 22(24): 5972–5982.

Peischl, S., and Gilbert, K.J. (2020). Evolution of dispersal can rescue populations from expansion load. *The American Naturalist* 195(2): 349–360.

Peischl, S., Kirkpatrick, M., and Excoffier, L. (2015). Expansion load and the evolutionary dynamics of a species range. *The American Naturalist* 185(4): E81–93.

Perkins, T.A. (2012). Evolutionarily labile species interactions and spatial spread of invasive species. *The American Naturalist* 179(2): E37–E54.

Perkins, S.E., Boettiger, C., and Phillips, B.L. (2016). After the games are over: life-history trade-offs drive dispersal attenuation following range expansion. *Ecology and Evolution* 6(18): 6425–6434.

Perkins, T.A., Phillips, B.L., Baskett, M.L., and Hastings, A. (2013). Evolution of dispersal and life history interact to drive accelerating spread of an invasive species. *Ecology Letters* 16(8): 1079–1087.

Phillips, B.L. (2009). The evolution of growth rates on an expanding range edge. *Biology Letters* 5(6): 802–804.

Phillips, B.L. (2015). Evolutionary processes make invasion speed difficult to predict. *Biological Invasions* 17: 1949–1960.

Phillips, B.L., Brown, G.P., and Shine, R. (2010a). Life-history evolution in range-shifting populations. *Ecology* 91(6): 1617–1627.

Phillips, B.L., Brown, G.P., Webb, J.K., and Shine, R. (2006). Invasion and the evolution of speed in toads. *Nature* 439(7078): 803–803.

Phillips, B.L., Kelehear, C., Pizzatto, L., Brown, G.P., Barton, D., and Shine, R. (2010b). Parasites and pathogens lag behind their host during periods of host range advance. *Ecology* 91(3): 872–881.

Phillips, B.L., and Perkins, T.A. (2019). Spatial sorting as the spatial analogue of natural selection. *Theoretical Ecology* 12(2): 155–163.

Phillips, B.L., Shine, R., and Tingley, R. (2016). The genetic backburn: using evolution to halt invasions. *Proceedings of the Royal Society B: Biological Sciences* 283(1825): 20153037.

Reznick, D., Bryant, M.J., and Bashey, F. (2002). r- and K-selection revisited: the role of population regulation in life-history evolution. *Ecology* 83(6): 1509–1520.

Ricklefs, R.E., and Wikelski, M. (2002). The physiology/life-history nexus. *Trends in Ecology & Evolution* 17(10): 462–468.

Rodriguez, A., Hausberger, M., and Clergeau, P. (2010). Flexibility in European starlings' use of social information: experiments with decoys in different populations. *Animal Behaviour* 80(6): 965–973.

Roff, D. (1993). *Evolution of life histories: theory and analysis*. Springer.

Roques, L., Garnier, J., Hamel, F., and Klein, E.K. (2012). Allee effect promotes diversity in traveling waves of colonization. *Proceedings of the National Academy of Sciences of the United States of America* 109(23): 8828–8833.

Roques, L., Hosono, Y., Bonnefon, O., and Boivin, T. (2015). The effect of competition on the neutral intraspecific diversity of invasive species. *Journal of Mathematical Biology* 71(2): 465–489.

Rotter, M.C., and Holeski, L.M. (2018). A meta-analysis of the evolution of increased competitive ability hypothesis: genetic-based trait variation and herbivory resistance trade-offs. *Biological Invasions* 20(9): 2647–2660.

Roy, H.E., Brown, P.M.J., Adriaens, T., Berkvens, N., Borges, I., Clusella-Trullas, S., Comont, R.F., de Clercq, P., Eschen, R., and Estoup, A. (2016). The harlequin ladybird, *Harmonia axyridis*: global perspectives on invasion history and ecology. *Biological Invasions* 18(4): 997–1044.

Seifert, E.K., Bever, J.D., and Maron, J.L. (2009). Evidence for the evolution of reduced mycorrhizal dependence during plant invasion. *Ecology* 90(4): 1055–1062.

Shigesada, N., and Kawasaki, K. (1997). *Biological invasions: theory and practice*. Oxford University Press.

Shine, R., Brown, G.P., and Phillips, B.L. (2011). An evolutionary process that assembles phenotypes through space rather than through time. *Proceedings of the National Academy of Sciences of the United States of America* 108(14): 5708–5711.

Shoval, O., Sheftel, H., Shinar, G., Hart, Y., Ramote, O., Mayo, A., Dekel, E., Kavanagh, K., and Alon, U. (2012). Evolutionary trade-offs, Pareto optimality, and the geometry of phenotype space. *Science* 336(6085): 1157–1160.

Skellam, J.G. (1951). Random dispersal in theoretical populations. *Biometrika* 38(1–2): 196–218.

Slatkin, M., and Excoffier, L. (2012). Serial founder effects during range expansion: a spatial analog of genetic drift. *Genetics* 191(1): 171–81.

Stearns, S.C. (1992). *The evolution of life histories*. Oxford University Press.

Stokes, A.N. (1976). On two types of moving front in quasilinear diffusion. *Mathematical Biosciences* 31(3): 307–315.

Swaegers, J., Sánchez-Guillén, R.A., Carbonell, J.A., and Stoks, R. (2022). Convergence of life history and physiology during range expansion toward the phenotype of the native sister species. *Science of the Total Environment* 816: 151530.

Szúcs, M., Vahsen, M.L., Melbourne, B.A., Hoover, C., Weiss-Lehman, C., and Hufbauer, R.A. (2017). Rapid adaptive evolution in novel environments acts as an architect of population range expansion. *Proceedings of the National Academy of Sciences of the United States of America* 114(51): 13501–13506.

Tabassum, S., and Leishman, M.R. (2018). Have your cake and eat it too: greater dispersal ability and faster germination towards range edges of an invasive plant species in eastern Australia. *Biological Invasions* 20(5): 1199–1210.

Therry, L., Swaegers, J., van Dinh, K., Bonte, D., and Stoks, R. (2016). Low larval densities in northern populations reinforce range expansion by a Mediterranean damselfly. *Freshwater Biology* 61(9): 1430–1441.

Tingley, R., Phillips, B.L., Letnic, M., Brown, G.P., Shine, R., and Baird, S.J.E. (2013). Identifying optimal barriers to halt the invasion of cane toads *Rhinella marina* in arid Australia. *Journal of Applied Ecology* 50: 129–137.

Torchin, M.E., Lafferty, K.D., Dobson, A.P., McKenzie, V.J., and Kuris, A.M. (2003). Introduced species and their missing parasites. *Nature* 421(6923): 628–630.

Travis, J.M.J., and Dytham, C. (2002). Dispersal evolution during invasions. *Evolutionary Ecology Research* 4(8): 1119–1129.

Travis, J.M.J., Münkemüller, T., Burton, O.J., Best, A., Dytham, C., and Johst, K. (2007). Deleterious mutations can surf to high densities on the wave front of an expanding population. *Molecular Biology and Evolution* 24(10): 2334–2343.

Travis, J.M.J., Mustin, K., Benton, T.G., and Dytham, C. (2009). Accelerating invasion rates result from the evolution of density-dependent dispersal. *Journal of Theoretical Biology* 259(1): 151–158.

Urban, M.C., Phillips, B.L., Skelly, D.K., and Shine, R. (2007). The cane toad's (*Chaunus* [*Bufo*] *marinus*) increasing ability to invade Australia is revealed by a dynamically updated range model. *Proceedings of the Royal Society B: Biological Sciences* 274(1616): 1413–1419.

Urquhart, C.A., and Williams, J.L. (2021). Trait correlations and landscape fragmentation jointly alter expansion speed via evolution at the leading edge in simulated range expansions. *Theoretical Ecology* 14(3): 381–394.

van Petegem, K., Moerman, F., Dahirel, M., Fronhofer, E.A., Vandegehuchte, M.L., van Leeuwen, T., Wybouw, N., Stoks, R., and Bonte, D. (2018). Kin competition accelerates experimental range expansion in an arthropod herbivore. *Ecology Letters* 21(2): 225–234.

Weiss-Lehman, C., Hufbauer, R.A., and Melbourne, B.A. (2017). Rapid trait evolution drives increased speed and variance in experimental range expansions. *Nature Communications* 8(1): 14303.

Weiss-Lehman, C., Tittes, S., Kane, N.C., Hufbauer, R.A., and Melbourne, B.A. (2019). Stochastic processes drive rapid genomic divergence during experimental range expansions. *Proceedings of the Royal Society B: Biological Sciences* 286(1900): 20190231.

Williams, J.L., Hufbauer, R.A., and Miller, T.E.X. (2019). How evolution modifies the variability of range expansion. *Trends in Ecology & Evolution* 34(10): 903–913.

Williams, J.L., Kendall, B.E., and Levine, J.M. (2016). Rapid evolution accelerates plant population spread in fragmented experimental landscapes. *Science* 353(6298): 482–485.

Wittmann, M.J., Gabriel, W., and Metzler, D. (2014). Genetic diversity in introduced populations with an Allee effect. *Genetics* 198(1): 299–310.

Zhang, W., Hu, Y.F., He, X., Zhou, W., and Shao, J.W. (2021). Evolution of autonomous selfing in marginal habitats: spatiotemporal variation in the floral traits of the distylous *Primula wannanensis*. *Frontiers in Plant Science* 12: 781281.

Zilber-Rosenberg, I., and Rosenberg, E. (2008). Role of microorganisms in the evolution of animals and plants: the hologenome theory of evolution. *FEMS Microbiology Reviews* 32(5): 723–735.

Zilio, G., Nørgaard, L.S., Gougat-Barbera, C., Hall, M.D., Fronhofer, E.A., and Kaltz, O. (2023). Travelling with a parasite: the evolution of resistance and dispersal syndromes during experimental range expansion. *Proceedings of the Royal Society B: Biological Sciences* 290(1990): 20221966.

21

Adaptive Evolution of Life History Traits in Urban Environments

Yuval Itescu[1,2,3*], *Maud Bernard-Verdier*[1,2,4*], *and Jonathan M. Jeschke*[1,2]

[1] *Leibniz Institute of Freshwater Ecology and Inland Fisheries (IGB), Berlin, Germany*
[2] *Institute of Biology, Freie Universität Berlin, Berlin, Germany*
[3] *Department of Evolutionary and Environmental Biology, University of Haifa, Haifa, Israel*
[4] *Centre d'Ecologie et des Sciences de la Conservation (CESCO), Sorbonne Université, Paris, France*

21.1 Introduction

Urbanization, the conversion of land into urban landscapes, has become a defining feature of the modern era (Pickett et al. 2001, Grimm et al. 2008, Alberti et al. 2020). Urban land-use and human populations are rapidly expanding and are projected to continue doing so (Seto et al. 2012). Urbanization operates a rapid and dramatic transformation of the environment, creating conditions not typically found elsewhere (Shochat et al. 2006). These novel environmental filters modify species distribution and abundance, reorganize ecological networks, affect species diversity (Kowarik 2011, Alberti et al. 2020) and may drive local extinctions (Roy et al. 2005, Morelli et al. 2023).

Studying life history evolution in urban areas is crucial and fascinating. Urban environments act as natural experiments, introducing diverse selective pressures that shape species' life history strategies (Grimm et al. 2008). Cities also serve as proxies for climate change in evolutionary experiments due to their warmer microclimates (Ziska et al. 2003, Jochner and Menzel 2015, Lambrecht et al. 2016). Investigating species' responses to urban environments enhances our understanding of rapid adaptation and contemporary evolutionary processes. Urban areas offer easy accessibility and long-term monitoring potential, facilitating extensive data collection and the study of life history evolution over time. Altered temporal cycles in urban environments, disrupting natural biotic rhythms, offer opportunities to investigate adaptation to changes in crucial life history events like reproduction and migration.

This chapter first presents the defining characteristics of urban environments, their potential implications on evolution of urban biota and the diverse methodological approaches applied to study evolution in urban areas. We then focus on the life history traits of urban organisms and review evolutionary patterns and underlying mechanisms, particularly highlighting evidence of adaptive evolution. Finally, we discuss the main generalities, biases and gaps in current knowledge, practical implications of life-history evolution in cities and future research directions.

21.2 Urban Drivers of Selection on Life History Traits

Urban ecological conditions create a set of distinctive characteristics that can affect every aspect of the life cycle of urban organisms. While urban drivers can be classified in various ways (*e.g.*, Rivkin et al. 2019, Diamond and Martin 2021), we identify five key dimensions (Figure 21.1) relevant to selection pressures shaping life histories in cities: (1) habitat change and fragmentation (spatial dimension), (2) altered temporal cycles (temporal dimension), (3) novel abiotic stressors (abiotic dimension), (4) novel organisms and biotic interactions (biotic dimension) and (5) anthropogenic resources and habitats (anthropogenic inputs dimension). Finally, these different dimensions combine and interact to create pan-urban selection pressures affecting multiple aspects of urban life history.

---

[*]Equal contribution

Life History Evolution: Traits, Interactions, and Applications, First Edition. Edited by Michal Segoli and Eric Wajnberg.
© 2025 John Wiley & Sons Ltd. Published 2025 by John Wiley & Sons Ltd.

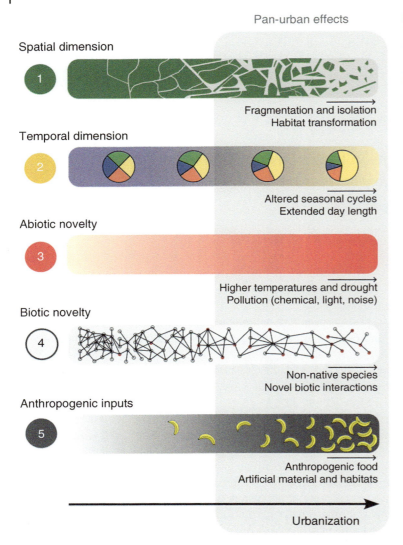

Figure 21.1 Urban drivers of life history evolution. Five main dimensions of environmental change characterize the urban landscape: (1) spatial dimension, (2) temporal dimension, (3) abiotic dimension, (4) biotic dimension and (5) anthropogenic input dimension. Finally, pan-urban effects result from the combination of multiple factors.

21.2.1 The Spatial Dimension: Habitat Change and Spatial Fragmentation

Urban landscapes possess unique structural components that distinguish them from natural or rural environments. These components include human-made structures like buildings, roads, bridges and other infrastructures that shape physical urban layouts. Urban areas also incorporate green spaces, parks and gardens that introduce natural elements into the built environment. The combination of human-made structures and natural elements creates a dynamic and heterogeneous landscape with abrupt transitions at small spatial scales, influencing ecological processes, species distribution and human-nature interactions. These characteristics of urban landscapes have significant impacts on individuals and population dynamics, affecting, in particular, the spatial dimensions of life history, *i.e.*, foraging, mate finding and dispersal (Lokatis et al. 2023).

Habitat fragmentation resulting from the division of natural habitats into smaller, isolated patches by human infrastructure is a highly significant environmental pressure in urban areas (Fahrig 2003; Fischer and Lindenmayer 2007; Figure 21.1). This process severely contributes to the loss of both biodiversity and key ecosystem functions (Haddad et al. 2015). Fragmentation negatively affects dispersal and gene flow, leading to decreased genetic diversity within populations and increased genetic differentiation between populations (Templeton et al. 1990). It also reduces the quality and availability of suitable habitats (Rahel et al. 1996, Arroyo-Rodríguez and Mandujano 2006), potentially forcing organisms to occupy suboptimal environments. Habitat change and fragmentation in urban areas, directly and indirectly, impact the availability and quality of natural resources, affecting food sources, water access, nesting sites, shelters and vital mutualistic services (Cheptou et al. 2017, Theodorou 2022).

21.2.2 The Temporal Dimension: Altered Temporal Cycles

Urban environments are characterized by altered temporal cycles compared to their rural surroundings, with microclimatic conditions and light pollution altering the rhythm of seasons and day-night cycles. Modifications of urban temporal cycles in temperature, light and noise, and their consequences on the timing and availability of natural resources, can profoundly shape the performance and life history of urban organisms. Such temporal changes impact the phenology of development, reproduction, seasonal dormancy, migration and dispersal, which may lead to changes in reproductive investment and resource allocation (Neil and Wu 2006). Such phenological shifts are typically not coordinated across species (Miller-Rushing et al. 2010) and may lead to a temporal mismatch between interaction partners, affecting, for instance, prey availability during the peak demand by offspring (Branston et al. 2021), plant-pollinator interactions (Fisogni et al. 2020; see also Chapter 16, this volume) or host-parasite interactions (Meineke et al. 2014; see also Chapter 13, this volume).

The 'urban heat island' (UHI) effect creates a microclimatic 'bubble' over cities, which imposes thermal and drought-related stresses and disrupts seasonal and diurnal cycles (Oke 1982). Urbanized areas, both terrestrial and aquatic, experience elevated temperatures throughout the year, which disrupts the natural course of seasons and results in changes like earlier springtime, extended summers and even complete disappearance of deep winter temperatures, particularly in temperate regions (Rizwan et al. 2008, Somers et al. 2013). The UHI also modifies temperature fluctuations between day and night due to heat-absorbing surfaces and human activities: urban surface temperatures may show dramatic peaks during the day, while air temperatures tend to show reduced cooling at night due to heat-absorbing surfaces releasing heat (Youngsteadt and Terando 2020). Climate change can exacerbate these changes in cities (Villalobos-Jiménez and Hassall 2017; see also Chapter 18, this volume).

Light and noise pollution in urban areas have profound impacts on the circadian rhythms of organisms, particularly for nocturnal species. Light pollution is caused by excessive and misdirected artificial lighting in cities, emanating from streetlights, indoor illumination, cars and neon signs. It eliminates natural darkness and induces skyglow, affecting urban areas and their surroundings (Gaston and Sánchez de Miguel 2022). Noise pollution, the excessive and often uncontrolled sound of human activities (*e.g.*, transportation, construction and industrial activities), is highly heterogeneous in space and time, and reinforced by vertical reflective surfaces (Warren et al. 2006). In animals, the consequences of disrupted diel cycles are diverse and include altered activity patterns, reduced foraging efficiency, disrupted sleep cycles and changes in reproductive strategies (Russart and Nelson 2018, Boyes et al. 2021).

21.2.3 The Abiotic Dimension: A Diversity of Novel Abiotic Stressors

Cities are hot spots for a diversity of pollution types. High concentrations of chemical pollutants in the air, water and soil, including smoke, heavy metals, pesticides, fertilizers and pharmaceuticals, are common in cities. This chemical pollution stems from urban runoff, industrial activities, transportation and agricultural activities, and significantly threatens both terrestrial and aquatic ecosystems (Diamond and Hodge 2007). Exposure to diverse chemical pollutants in urban areas often reduces reproductive success, accelerates senescence, increases mortality rates and induces changes in behaviour, morphology and physiology (Croteau et al. 2008, Honour et al. 2009, Bauerová et al. 2017; see also Chapter 19, this volume). For instance, Eurasian dippers *Cinclus cinclus* in urban and rural streams in Wales showed similar clutch sizes and nest success, but urban nestlings were lighter, showed male-biased brood sex ratios and their thyroid hormone profile reflected exposure to pollutants (polychlorinated biphenyls (PCBs) and polybrominated diphenyl ethers (PBDEs)) (Morrissey et al. 2014). Urban contaminants, such as heavy metals from car exhausts, microplastics and pesticides, which are widely used in urban landscaping, agriculture and pest control, can also negatively affect organisms' life history traits and fitness (*e.g.*, in *Bufo bufo* toads: Bókony et al. 2018, Adams et al. 2021b). However, urban pollution can have indirect positive effects on some organisms. Munich air pollution positively affected the growth rates of an aphid in experimental conditions (*e.g.*, Dohmen 1985), highlighting the importance of cascading effects of pollution-driven plant stress in urban areas (Butler and Trumble 2008, Raupp et al. 2010).

Other forms of urban pollution, like the abovementioned noise and light pollution, are more unique to cities and impact life history traits beyond altering the diel cycle (Hölker et al. 2010, Gaston and Sánchez de Miguel 2022). Noise pollution interferes with a range of life history traits (Halfwerk et al. 2011, Duquette et al. 2021). Chronic noise has been shown to disturb communication between parents and offspring in house sparrows *Passer domesticus* (Schroeder et al. 2012), or cause delayed development in field crickets *Teleogryllus oceanicus* (Gurule-Small and Tinghitella 2019).

Nocturnal species are particularly affected by artificial lights that interfere with their navigation, foraging, mating and other essential activities (Owens and Lewis 2018, Adams et al. 2021a). The effect of supplemental light at night can differ across related species, for instance, affecting the growth rate and flower production of some plant species but not others (reviewed in Bennie et al. 2016). Light pollution impacts biotic interactions, such as predator-prey and plant-pollinator interactions, by disrupting both the probability of partner encounters in space and time and the outcome of interactions (Macgregor et al. 2015). For instance, light sources at night can provide novel urban food resources by creating hotspots of easy insect prey, which has cascading effects on a diversity of predators (*e.g.*, bats: Rydell 2006; reptiles: Perry et al. 2008; birds: Robertson et al. 2010).

Finally, the UHI effect, via higher heat stress and reduced winter frost, directly affects organism physiology, reproduction, embryonic development and survival rates, potentially exerting selection pressures on the thermal thresholds and timing of the life history cycle (Angilletta et al. 2007, Chick et al. 2019, see also the discussion above in the context of the temporal dimension).

21.2.4 The Biotic Dimension: Novel Communities and Biotic Interactions

Community assembly in urban areas is influenced not only by the unique abovementioned abiotic conditions but also by high levels of biological invasions (Aronson et al. 2016). Urbanization thus generates novel communities of species that rarely coexist and interact in other ecosystems (Heger et al. 2019; Figure 21.1). Interactions between native and novel organisms (non-native species, domestic species, new hybrids or genetically modified organisms; Jeschke et al. 2013) and between different novel organisms can be described as eco-evolutionarily novel (*sensu* Saul et al. 2013), in that neither partner has encountered the other during their evolutionary history. Such novel interactions may create new selection pressures on both partners.

As trade and migration routes converge in urban areas, along with a high concentration of pets, zoos and gardens, the likelihood of non-native species introductions and successful invasions increases dramatically (Gaertner et al. 2017, and references therein). While the heterogeneity of urban landscapes filters multiple functional strategies, a few widespread successful urban species, often globally invasive species, have been shown to drive urban biotic homogenization worldwide (McKinney 2006, Lokatis and Jeschke 2022). Certain life histories and functional traits typical of invasive species appear particularly well suited for urban lifestyles, *e.g.*, higher niche flexibility, dispersal ability and tolerance to environmental conditions and disturbances (Callaghan et al. 2021, Wolf et al. 2022, Neate-Clegg et al. 2023).

Human presence itself can be perceived as a threat by a diversity of species (Landscape of fear hypothesis; Laundre et al. 2010, Lokatis et al. 2023), altering their activity patterns to avoid encounters (George and Crooks 2006, Stillfried et al. 2017, Green et al. 2023). Other species, often non-natives accustomed to living close to humans (Human commensalism hypothesis; Jeschke and Strayer 2006), benefit from human presence and even adjust their activity patterns accordingly (Spelt et al. 2021).

Non-native species introduced to a novel biotic environment are more likely to undergo rapid evolution, in particular in cities (Borden and Flory 2021). Urban environments, with reduced biodiversity, additional anthropogenic resources and disturbed interaction networks, are likely to provide ecological opportunities or new empty niches, which invaders may evolve to occupy (Stroud and Losos 2016). Rapid evolution and adaptation of novel organisms in urban environments are, in return, likely to increase the invasion potential of species, selecting, for instance, for increased reproductive output or dispersal capacities (Borden and Flory 2021; see also Chapter 10, this volume).

21.2.5 The Anthropogenic Inputs Dimension: Human Introduction of Artificial and Supplemental Resources

A vast range of novel anthropogenic resources is available in urban areas. Human presence and its material input directly affect resource abundance and distribution in cities (Pickett and Cadenasso 2009; Figure 21.1), including food (Contesse et al. 2004, Tryjanowski et al. 2015), water (Sun and Lockaby 2012), shelter (Ben-Moshe and Iwamura 2020) and nesting sites (Charter et al. 2016). Such new opportunities can impose selection for new behaviours and life history traits. For instance, increased and stable food availability from a novel source (garden bird feeders) and elevated urban temperature contributed to the loss of migration in populations of blackcaps, *Sylvia atricapilla*, in the United Kingdom (Plummer et al. 2015). Urban organisms exploit anthropogenic food, garbage, feeding stations, ornamental vegetation, wastewater, artificial nesting houses, built structures, indoor spaces, artificial water bodies, excessive irrigation, fertilizers, etc., with both beneficial and adverse consequences. A study on red-winged starling *Onychognathus morio* breeding in Cape Town, South Africa, revealed that the consumption of human-provided food sources contributed to adult condition and survival on the one hand due to high-calorie nutrition but hindered nestling growth on the other hand due to imbalances in offspring nutrient intake (Catto et al. 2021).

The availability, diversity and composition of anthropogenic resources can create spatial and temporal mismatches between different types of vital resources (Meyrier et al. 2017). They also alter ecological communities, leading to changes in species abundance, distribution, competition and predation (Faeth et al. 2005, Wilson and Jamieson 2019). In one example, additional light at night, acting as a new resource for plants able to exploit its particular wave frequencies for photosynthesis, gave a competitive advantage to an invasive plant in urban areas (Murphy et al. 2022). Resource heterogeneity in cities can also induce cascading effects on biodiversity (LaDeau et al. 2013).

21.2.6 Pan-urban Effects

Urbanization is a complex process from which 'pan-urban' effects can emerge, where multiple drivers interact, synergistically or antagonistically, to shape evolutionary trajectories (Diamond and Martin 2021). Urban organisms face a multitude of interacting environmental conditions in addition to individual selection pressures. Factors like light and noise pollution, habitat fragmentation, UHI and chemical pollution create a complex set of selection pressures with unpredictable, non-additive effects on biodiversity (Dominoni et al. 2020, Fenoglio et al. 2021). In such a multifactorial context, understanding the individual drivers of evolution in urban areas is challenging, as their specific influences are often difficult or even impossible to disentangle (Sprau et al. 2017, Rivkin et al. 2019).

Urbanization, as an integrative driver of selection, can lead to shifts in overall life history strategies such as mating systems and social structures, and in particular stages of the life cycle. For instance, unusual social polygyny has been documented in urban populations of the San Joaquin kit fox *Vulpes macrotis mutica* in California, United States (Westall et al. 2019) and in odorous house ants across North America (Blumenfeld et al. 2022). Decreases in reproductive output and growth due to multifactorial urban effects have been documented in a variety of species that are otherwise successful urban dwellers, possibly reflecting shifts in complex selection gradients, though heritability is yet to be shown. For instance, studies conducted on blue tits and great tits have unveiled decreased reproductive success in urban settings. In great tits, this was not only evident between urban and rural populations but also along an urbanization gradient within urban areas (Charmantier et al. 2017). Additionally, in the case of urban blue tits, the strength of natural selection favouring early breeding and larger clutch sizes appeared to be weaker when compared to non-urban environments, possibly due to the reduced and unpredictable availability of natural insect food in urban surroundings (Branston et al. 2021). Similarly, there is evidence that black widow spiders, while reaching exceptionally high densities in desert cities of Arizona, USA, are displaying reduced reproductive success and body size, possibly associated with both changes in prey quality and heat stress (Johnson et al. 2012). The multifaceted nature of urban environments presents a challenging and exciting research area for evolutionary ecologists, studying the persistence and adaptation of organisms at the intersection of multiple global changes (Grimm et al. 2008).

21.3 Studying Evolution in Urban Areas

21.3.1 The Field of Urban Evolution

Urban evolution, a relatively young research field, examines how organisms evolve in response to urban environments and human activities, with implications for urban infrastructure and socio-economic factors (Szulkin et al. 2020b). Recent reviews and books highlight concerns about the evolutionary impact of urbanization on organisms, ecosystems and human health (Johnson and Munshi-South 2017, Rivkin et al. 2019, Szulkin et al. 2020b, Diamond and Martin 2021, Lambert et al. 2021, Alberti 2023). Urbanization drives rapid and interconnected evolutionary changes. Studies across different ecological systems reveal how urbanization influences, both adaptively and non-adaptively, the genetic diversity, phenotypic means and variations within and between populations (Donihue and Lambert 2015, Alberti et al. 2017b, Johnson and Munshi-South 2017, Alter et al. 2021, Diamond and Martin 2021, Thompson et al. 2022). This underscores the significance of understanding evolutionary dynamics in urban areas for urban ecosystem functioning and the well-being of urban inhabitants. Urban population fragmentation leads to reduced gene flow and increased genetic drift for many species (Johnson and

Munshi-South 2017). This may explain why *ca.* 60% of evolutionary studies in cities to date have focussed on documenting genetic structure and drift rather than selection (Diamond and Martin 2021). To advance the field, researchers are currently developing and refining methods for observing phenotypic divergence, measuring fitness consequences of genetically based phenotypes and experimentally manipulating potential drivers of urban adaptation.

21.3.2 Methodological Approaches to Study Evolution in Urban Areas

Studying urban evolution presents unique challenges due to the complexity and heterogeneity of urban environments (Winchell et al. 2022). Diamond and Martin (2021) identified three research axes for urban evolutionary studies: investigating the relative role of different mechanisms (*e.g.*, selection, gene flow, mutation, drift) in shaping evolution in urban areas, addressing basic *vs.* applied research questions and examining pan-urban effects *vs.* isolating specific variables. They suggested that, as the specificity of selected study sites in urban evolutionary studies increases (*i.e.*, selecting sites that represent variation in a specific factor), the generalizability of inferences shifts from the broad impact of urbanization as a complex of unique factors to impacts of single variables, which can be also generalized to non-urban landscapes.

To investigate how urbanization shapes species traits, researchers have developed various approaches. One approach compares urban to rural or other non-urban populations, examining trait differences. Another approach views urbanization as a gradient, examining trait responses to varying levels of urbanization. Such an urbanization gradient is often quantified by the percentage of impervious built surfaces, but numerous other indices exist (Szulkin et al. 2020a). A third approach considers the heterogeneity of urban areas, treating them as mosaics of habitats with different urbanization levels and adaptive clines within cities (Szulkin et al. 2020a, Santangelo et al. 2022b).

Urban evolution can be studied at different spatial scales. Local studies focus on specific urban habitats to understand trait variations and local evolutionary drivers (*e.g.*, Jackson et al. 2022). Regional comparisons of different urban areas identify general patterns of urban adaptation and assess the influence of both local and regional factors (*e.g.*, Salmón et al. 2021). Global comparative analyses explore evolution across regions, searching for convergences and identifying underlying factors (*e.g.*, Santangelo et al. 2022a).

21.4 Available Evidence of Adaptive Life History Evolution in Urban Areas

Studies that robustly demonstrate the adaptive evolution of life history traits in response to selection pressures in urban environments, substantiated by compelling evidence of both fitness advantages and heritability, are still relatively rare. In this section, we present such studies alongside cases that demonstrate at least one of these components together with some indications for the other component (Table 21.1).

21.4.1 The Spatial Dimension

21.4.1.1 Dispersal in Fragmented Landscapes

Fragmentation, by reducing population size and gene flow, is expected to generate stochastic evolutionary processes such as genetic drift, but recent studies have shown that it can also drive natural selection and adaptation (Cheptou et al. 2017). Fragmented landscapes can be selected either for increased dispersal ability, if reaching distant, suitable habitats offers a high colonization advantage, or for reduced dispersal to avoid dispersal costs. The selected strategy depends largely on a balance between dispersal costs and chances of colonizing empty patches, which depend on both landscape characteristics and species dispersal abilities (Cheptou et al. 2017). A seminal study of the weed *Crepis sancta* demonstrated selection for reduced seed dispersal in less than 12 generations in highly fragmented urban populations surrounded by a harsh concrete matrix (Cheptou et al. 2008; see Box 21.1). Alternatively, selection for increased dispersal is also observed in fragmented natural and anthropogenic habitats, particularly in already highly mobile animals, although these are often more related to foraging than dispersal (reviewed in Cheptou et al. 2017). For instance, dispersal-enhancing adaptations such as longer femurs and wings were found in an urban herbivore insect, the grasshopper *Chorthippus brunneus* (San Martin Y Gomez and van Dyck 2012).

Mutualist dispersers (see also Chapter 17, this volume) may help overcome fragmentation in cities. While there is evidence that selection by urban frugivore birds may affect plant reproductive traits (Palacio and Ordano 2023) and potential costs from mismatches between urban dispersal partners (Stanley and Arceo-Gómez 2020), robust proof of adaptive plant evolution in response to urban animal-mediated dispersal is still lacking.

Table 21.1 Selected case studies of adaptive evolution of life history traits in urban areas, providing strong evidence for either both selection and heritability or at least one of them and suggestive evidence for the other. All studies measured some proxy for life history trait selection in the field or in controlled conditions. Our assessment of the evidence level for fitness differences and heritability follows principles from Lambert et al. (2021). Certain studies suggest rather maladaptive evolution (marked with *). Adapted from Lambert et al. (2021).

Taxon	Species	LH	Selection pressure	Evolutionary change in urban areas	Spatial context	Selection evidence	Heritability evidence	References
(1) Spatial dimension: habitat change and fragmentation
Plant	*Commelina communis*	M	Limited pollination	Lower proportion of staminate flowers increased reproductive success	Urbanization gradient	Strong	Suggestive	Ushimaru et al. (2014)
Plant	*Crepis sancta*	D	Higher dispersal cost	Higher proportion of non-dispersing seeds increased reproductive success (see Box 21.1)	Urban contrast (urban *vs.* rural)	Strong	Strong	Cheptou et al. (2008), Dubois and Cheptou (2017)
Plant	*C. sancta*	M	Limited pollination	Lower flower attractiveness and delayed flowering reduced fitness cost of investment in pollinator attraction	Urban contrast (urban *vs.* rural)	Strong	Suggestive	Dubois and Cheptou (2017)
Plant	*Gelsemium sempervirens*	M	Limited pollination	Larger floral display increased reproductive success	Urban contrast (suburban *vs.* rural)	Strong	Suggestive	Irwin et al. (2014, 2018)
Insect	*Chorthippus brunneus*	D	Dispersal barrier, isolation	Ecotypic differentiation in dispersal-related traits of urban grasshoppers, urban adults having longer wings and femurs and lower abdomen mass relative to body size	Urban contrast (urban *vs.* rural)	Suggestive	Strong	San Martin Y Gomez and van Dyck (2012)
Crustacean	*Ceriodaphnia cornuta*	G	isolation (leading to decreased toxic cyanobacteria and increased food quality)	Loss of tolerance to poor food quality. Decelerated growth rate and delayed maturity reduced survival under low food quality treatment*	Urbanization gradient	Strong	Strong	Zhang et al. (2022)
(2) Temporal dimension: altered temporal cycles
Bird	*Cyanistes caeruleus*	P	Light pollution	Earlier dawn singing and egg-laying increased reproductive success	Within population	Strong	Suggestive	Kempenaers et al. (2010)
Bird	*Turdus merula*	P	Urban heat island and food availability	Reduced migratory propensity and earlier timing of gonadal development in males increased survival (in winter) and reproductive success	Urban contrast (urban *vs.* rural)	Strong	Strong	Partecke and Gwinner (2007)

(Continued)

Table 21.1 (Continued)

Taxon	Species	LH	Selection pressure	Evolutionary change in urban areas	Spatial context	Selection evidence	Heritability evidence	References
Insect	Pieris napi, Chiasmia clathrata	P	Urban heat island and light pollution	Elongated flight season and diminished photoperiod threshold for direct development due to urban heat island (but not light pollution) increased probabilities for survival and reproductive success	Urban contrast (urban vs. rural)	Suggestive	Strong	Merckx et al. (2021)
Plant	Ambrosia artemisiifolia	P	Earlier summer drought	Earlier flowering of urban genotypes compensates for plastic responses delaying flowering in urban areas (counter-gradient variation compared to plastic response)	Urban contrast (urban vs. rural)	Strong	Strong	Gorton et al. (2018)
Plant	Lepidium virginicum	P, G, R	Urban heat island	Earlier bolting, longer interval between bolting and flowering, larger size and higher seed production increased reproductive success	Urban contrast (multi-city urban vs. rural)	Strong	Strong	Yakub and Tiffin (2017)

(3) Abiotic dimension: novel abiotic stressors

Taxon	Species	LH	Selection pressure	Evolutionary change in urban areas	Spatial context	Selection evidence	Heritability evidence	References
Bird	Parus major	M	Noise pollution	Lower minimal song frequency increased reproductive success	Within population	Strong	Suggestive	Slabbekoorn and Peet (2003)
Crustacean	Daphnia magna	G, R	Elevated temperatures	Accelerated pace of life increased reproductive success and population growth (see Box 21.2)	Urbanization gradient	Strong	Strong	Brans et al. (2017b), Brans and De Meester. (2018)
Fungi	Chrysosporium pannorum Penicillium bilaii Trichoderma koningii Torulomyces lagena	G	Elevated temperatures	Accelerated growth in urban populations, though only at high temperatures for some species (C. pannorum and T. koningii)	Urban contrast (urban vs. rural)	Strong	Strong	McLean et al. (2005)
Insect	C. brunneus	G, R	Elevated temperatures	Longer larval and adult stages, heavier eggs, larger clutch size, accelerated growth and larger adults (in females) increased reproductive success	Urban contrast (urban vs. rural)	Strong	Suggestive	San Martin Y Gomez and van Dyck (2012)
Insect	Coenagrion puella	G	Elevated temperatures (heat waves)	Decelerated growth increased survival	Urban contrast (urban vs. rural)	Strong	Strong	Tüzün et al. (2017), Tüzün and Stoks (2021)

Insect	*Dianemobius nigrofasciatus*	P, G, R	Noise and light pollution	Delayed development and prevented egg diapause, particularly by light pollution, but no change of survival rates (reduced in rural areas)	Urban contrast (urban vs. rural)	Strong	Strong	Ichikawa and Kuriwada (2023)
Plant	*C. sancta*	P, G	Elevated abiotic stress	Earlier flowering and senescence and larger size increased reproductive success	Urban contrast (urban vs. rural)	Strong	Strong	Lambrecht et al. (2016)
(4) Biotic dimension: novel organisms and novel biotic interactions								
Insect	*Cytisus scoparius*	M	Novel pollinator availability	Selection by urban pollinators, but not rural ones, for larger flowers associated with niche expansion to new pollinators in this invasive plant	Urban contrast (urban vs. rural)	Strong	Suggestive	Bode and Tong (2018)
(5) Anthropogenic input dimension (none)								
—	—	—	—	—	—	—	—	—
(6) Pan-urban effects								
Fish	*Poecilia reticulata*	G, R	Urbanization *per se* (implied: food availability)	Larger adult size and offspring production increased reproductive success and population growth	Urban contrast (urban vs. rural)	Strong	Suggestive	Santana Marques et al. (2020)
Plant	*Taraxacum officinale*	R	Urbanization and lower herbivory risk	Reduced early production of seeds (reproductive fitness) in urban populations in response to experimental herbivory*	Urbanization gradient	Strong	Strong	Pisman et al. (2020)
Plant	*Trifolium repens*	P, G, R	Urbanization, shifts in pollinator community and increased pollinator visits	Multivariate shift towards delayed germination and flowering, larger biomass, larger flowers and higher fecundity increased reproductive success (seed set) in urban areas	Urbanization gradient	Strong	Suggestive	Santangelo et al. (2020b)
Reptile	*Anolis cristatellus*	P, G, R	Urbanization *per se*	Earlier oviposition, accelerated growth, larger hatchling and adult size and higher fecundity (number of eggs produced) increased reproductive success	Urban contrast (urban vs. rural)	Strong	Suggestive	Hall and Warner (2017)

D: Dispersal, G: Development and growth, LH: Life history stage, M: Mating system and group composition, P: Phenology, R: Reproduction effort.

> **Box 21.1 *Crepis sancta*: fragmentation selects for reduced dispersal**
>
> Organism: Holy hawksbeard *C. sancta* (Asteraceae)
> Trait: Proportion of dispersing seeds
> Drivers: Habitat fragmentation
> Habitat: Mediterranean ruderal grasslands, sidewalks
> Methods: Common garden, quantitative genetics
> Evidence: Strong evidence of selection
>
> *Crepis sancta* is an annual weed characterized by composite flowers producing two types of seeds: many small seeds equipped with a pappus for wind dispersal and fewer large, non-dispersing seeds. The relative proportion of these two types of seeds per flower captures the dispersal strategy of the plant, with higher proportions of non-dispersing seeds reducing chances of long-distance dispersal. A series of common garden and transplantation studies from Cheptou et al. (2008), Lambrecht et al. (2016) and Dubois and Cheptou (2017) in the south of France have shown that populations from highly fragmented (*i.e.*, sidewalk tree bases) urban habitats have evolved a high proportion of non-dispersing seeds compared to rural unfragmented populations in under 12 generations (Figure 21.2). Quantitative genetic analyses across more or less fragmented urban and rural habitats further suggest that this pattern is the result of direct selection by fragmentation itself, not urbanization in general. Theory and field measurements of dispersal cost indicate that this selection is likely driven by the high cost of dispersal when dispersing seeds are likely to land on an inhospitable concrete urban matrix.
>
>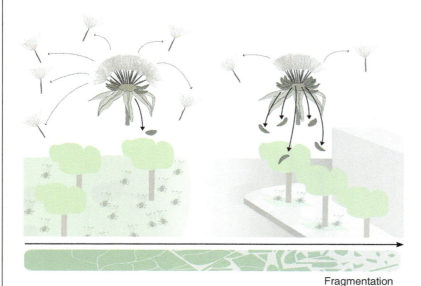
>
> **Figure 21.2** Evolution of reduced dispersal in the Mediterranean weed *Crepis sancta* in highly fragmented urban landscapes. Fragmented populations produced higher proportions of non-dispersing seeds, likely to avoid high dispersal costs in an inhospitable concrete matrix.

21.4.1.2 Finding a Mate in a Fragmented Urban Landscape

By reducing the likelihood of encountering a mate, or a pollinator, fragmentation may have a selective effect on mating systems. Pollinator limitations are expected in urban areas due to pollinator population fragmentation and associated Allee effects (*i.e.*, positive density-dependence as fitness increases with population size; Stephens et al. 1999), in which pollinator density positively correlates with plant density, reinforcing pollinator limitations in small isolated patches (Cheptou and Avendano 2006). This may select for increased self-reliance for reproduction *vs.* outcrossing, with reduced investments in attracting pollinators, such as reduced flower size (*e.g.*, *C. sancta*; Dubois and Cheptou 2017) or reduced proportion of staminate flowers (*e.g.*, *Commelina communis*; Ushimaru et al. 2014). Strategies avoiding self-fertilization such as herkogamy (*i.e.*, the spatial separation of stigmas and anthers in flowers that limits self-pollination in a hermaphroditic flower) have been shown to decrease in some urban populations (*e.g., C. communis*; Ushimaru et al. 2014), and higher selfing (*i.e.*, self-pollination) rates are observed (Cheptou and Avendano 2006). Urban plant populations may even shift from sexual to predominantly asexual reproduction, as observed for the small forb *Linaria vulgaris,* which displayed more clonal reproduction, less clonal diversity and lower seed set in urban populations (Bartlewicz et al. 2015). It is not clear whether these changes in mating systems are the result of selection for selfing and asexual reproduction or rather a consequence of isolation and a potential evolutionary trap. Nevertheless, strong evidence for rapid evolution of increased selfing under experimental pollinator limitation (Brys and Jacquemyn 2012) supports the hypothesis that these patterns could be selected and locally adaptive.

By contrast, increased investment in pollinator attraction may occur in urban areas, with, for instance, a selection for larger flower displays, as demonstrated in the vine *Gelsemium sempervirens* in the suburbs of North Carolina, USA (Irwin et al. 2018). Selection for increased pollinator attraction may occur particularly in larger suburban vegetation patches where pollinator limitation is moderate, with a diversity of habitats for arthropods, while mate availability and pollinator competition remain high (Irwin et al. 2014). In some cases, pollinator visits can actually increase with urbanization: urban populations of *Trifolium repens* received more frequent pollinator visits and had high reproductive outputs (Santangelo et al. 2020b).

While adaptive shifts in social and mating systems in animals due to urban fragmentation have not been demonstrated to date, some observations indicate the potential for such shifts. For example, urban fragmentation in West Lafayette, Indiana, USA, has diminished tree density in these areas and resulted in increased competition for nesting sites among predominantly polydomous (*i.e.,* occupying multiple nests) black carpenter ants *Camponotus pennsylvanicus* compared to suburban areas. This has led to simplified colony structure, smaller colony size and increased monodomy (*i.e.*, the use of a single nest; Buczkowski 2011).

21.4.1.3 Patch Isolation and the Island Syndrome Hypothesis

Increased isolation of urban fragmented populations can impact life history traits, potentially aligning with the island syndrome hypothesis, which suggests relaxed biotic selection pressures in isolated habitats (*e.g.*, extrinsic mortality agents) that lead to predictable changes in life history traits (*e.g.*, reduced dispersal and 'slower' life history, along morphological, physiological and behavioural traits) compared to non-isolated (*i.e.*, connected, continuous or mainland) populations (Adler and Levins 1994). In Shanghai, three fragmented populations of the water flea *Ceriodaphnia cornuta* experienced different degrees of isolation and ecological conditions. The urban pond population lost tolerance to toxic cyanobacteria blooms due to isolation and release from toxicity effects, diverging genetically (lower genetic diversity and higher genetic differentiation) and phenotypically in life history traits (lower growth rates and later maturity), from the urban river and rural lake populations (Zhang et al. 2022). The adaptive nature of other observations of life history trait shifts in urban isolated populations has not been tested. For example, an isolated population of the freshwater crab *Potamon fluviatile* in Rome, Italy, showed reduced body growth rate, larger body size and longer lifespan compared to non-urban populations (Scalici et al. 2008). While these differences seem to follow the island syndrome hypothesis, further investigation is still needed to determine their underlying mechanisms.

21.4.1.4 Reduced Access to High-Quality Resources and Food Limitation

Loss of access to high-quality resources in a fragmented urban landscape can have indirect consequences on nutrition and breeding success. Studies from urban areas support the food limitation hypothesis, which associates scarcity of food with poor breeding success (Martin 1987). Limited high-quality natural food resources in urban areas led to poor-quality eggs and nestlings, high nestling mortality and reduced breeding success in several bird species, including western jackdaw (*Corvus monedula*; Meyrier et al. 2017), blue tits (*Cyanistes caeruleus*; Jarrett et al. 2020), great tits (*Parus major*; Seress et al. 2020) and common blackbird (*Turdus merula*; Ibáñez-Álamo and Soler 2010). Organisms can successfully adjust nesting behaviours in modified and fragmented urban landscapes. An example from Melbourne shows that tawny frogmouths *Podargus strigoides* altered their nesting site choices in urban areas in response to diminished availability of preferred nesting sites with increased road density and urban land-use and, consequently, exhibited higher reproductive success (Weaving et al. 2016). However, while clear fitness consequences were observed, heritability has not been demonstrated in any of these cases.

21.4.2 The Temporal Dimension

21.4.2.1 Adaptation to Changes in Seasonal Cycles

The UHI effect plays a significant role in altering the timing, length and quality of seasons, which has been shown to trigger not just phenotypic shifts in the phenology of urban organisms but also adaptive evolutionary changes (Jochner and Menzel 2015). Beyond the opportunities offered by shorter winters and longer growing seasons, especially in cold and temperate regions, a key aspect of survival in warmer cities is to avoid the adverse effects of longer and more severe summer heat and drought, in particular in tropical areas. This seasonal constraint is expected to select evolutionary strategies to avoid or tolerate summer heat, with, for instance, an earlier reproduction of short-lived species to allow successful life cycle completion before drought hits (Neil and Wu 2006).

Urban shifts in plant phenology have long been observed (Neil and Wu 2006), but evidence for evolutionary adaptive responses is still scarce. Evidence for urban evolutionary shifts towards earlier (*e.g.*, *Lepidium virginicum*: Yakub and Tiffin 2017; *Ambrosia artemisiifolia*: Gorton et al. 2018) and delayed (*e.g.*, *T. repens*: Santangelo et al. 2020b; *C. sancta*: Lambrecht et al. 2016) phenology have both been reported. Phenological shifts are often associated with changes in other life history traits, suggesting a selection for whole-life history syndromes rather than the timing of the life cycle alone. For instance, in the examples above, urban *T. repens* or *C. sancta* individuals showed not only delayed phenology but also grew larger and produced more seeds, possibly due to an extended growing period or to other factors such as increased resource availability. The direction of selection on phenology may depend on functional types, with, for instance, earlier phenology being selected for spring flowers and delayed phenology for late flowering species, as they tend to follow different environmental cues (*e.g.*, photoperiod; Neil and Wu 2006, Christmann et al. 2023), but the low number of studies does not yet allow for synthesis across functional groups.

In animals, similar trends have been observed towards earlier reproductive phenology. For example, reproduction in urban populations of acorn-dwelling ants *Temnothorax curvispinosus* in American cities happens a month earlier than in rural populations, which likely leads to reproductive isolation between them (Chick et al. 2019). A few meta-analyses showed that, generally, urban bird populations tend to breed earlier and have smaller broods, and that these changes may be either plastic or heritable (Chamberlain et al. 2009, Sepp et al. 2018, Capilla-Lasheras et al. 2022). Such temporal shifts, together with spatial isolation by fragmentation, further reduce gene flow between populations, especially for those species with a short reproduction window, such as winter annual plants (Jochner and Menzel 2015, Chick et al. 2019).

Altered temperature regimes, and in particular warmer winters, not only select for changes in phenology but can also transform or even eliminate whole stages of the life cycle, such as migration or dormancy, enabling organisms to exploit urban resources more effectively. Urban European blackbirds, particularly males, exhibit adaptively reduced migratory propensity, leading to earlier gonadal development and longer reproductive seasons (Partecke and Gwinner 2007). In the butterfly *Pieris napi* and the moth *Chiasmia clathrate*, UHI effects extend the flight seasons and inhibit diapause induction (Merckx et al. 2021).

21.4.2.2 Effects of Changes to the Diel Cycle

In animals, fitness consequences of disrupted diel cycles have been documented. For example, artificial light at night has introduced altered cues that influence the timing of singing behaviour in blue tits, driving males to initiate singing earlier in the morning and thereby increasing their extra-pair success and overall fitness (Kempenaers et al. 2010). Many plant species using photoperiod as a seasonal cue are also susceptible to light pollution, and impacts of diel cycle changes as a new potential evolutionary urban selection agent are now well recognized (Gaston et al. 2013, Gaston and Sánchez de Miguel 2022). However, there is still a lack of studies testing for adaptive evolutionary responses to altered circadian rhythms and photoperiods in cities (Hopkins et al. 2018).

21.4.3 The Abiotic Dimension

21.4.3.1 Light Pollution

The evolutionary potential of light pollution as a selection agent of life history traits, even outside possible consequences on diel cycles, has not been thoroughly investigated (Hopkins et al. 2018), but implicit indications have been reported. In some urban carabid beetles and moths, a trade-off between attraction to artificial light sources and dispersal ability has been detected (Altermatt and Ebert 2016, Kaunath and Eccard 2022). In túngara frogs, stronger illumination, through its effect on predation risk perception, may relax sexual selection on male call complexity, although contrasting findings have been demonstrated in this regard (Rand et al. 1997, Halfwerk et al. 2019).

21.4.3.2 Noise Pollution

Excessive noise from traffic, construction and other urban activities can mask communication signals, such as animal courtship vocalizations, and affect sexual selection, as shown for the great tit, whose males that sing at higher notes have a fitness advantage in noisy urban areas (Slabbekoorn and Peet 2003). However, we found no robust example of demonstrated selection and adaptive evolution in response to excessive noise pollution in cities.

21.4.3.3 Chemical Pollution

While the first (and seminal) evidence for rapid adaptation in urban areas involved chemical pollution (Peppered moths; Kettlewell 1955) and urban evolution of tolerance to pollutants has been documented in multiple taxa (Alberti et al. 2017a), studies on adaptive evolution in life history traits in response to pollution are scarce. The direct and indirect evolutionary consequences of chemical selection pressure on life history strategies remain to be further explored in the urban context (see also Chapter 19, this volume).

21.4.3.4 Elevated Temperatures and Thermal Stress

Many populations appear to adaptively evolve higher growth rates and faster pace of life under the higher temperatures of urban areas, including the water flea *Daphnia magna* (Box 21.2) and the chitinolytic fungi *Chrysosporium pannorum* and *Trichoderma koningii* (McLean et al. 2005). By contrast, urban populations of damselflies *Coenagrion puella* in Belgium experienced reduced larvae growth rate under heat wave conditions as an adaptive trade-off, improving their coping abilities and immune response to parasites (Tüzün et al. 2017, Tüzün and Stoks 2021).

Box 21.2 *Daphnia magna*: faster pace of life in urban ponds

Organism: Water flea *Daphnia magna* (Crustacea)
Traits: Traits related to pace of life, particularly age of maturity, size at maturity and fecundity
Drivers: Elevated temperatures and other pan-urban effects
Habitat: Ponds in Belgium
Methods: Common-garden experiment
Evidence: Strong evidence of genetic differentiation along the urbanization gradient

Water fleas (*Daphnia* spp.) are keystone species in freshwater habitats and can also be found in cities. A large member of this genus is *D. magna*, which served as a model organism for studies in Belgium that investigated evolutionary responses to urbanization (Brans et al. 2017b, Brans and De Meester 2018). Using a common-garden design, these studies found that water fleas living in more urban, and thus warmer (Brans et al. 2018), areas reach maturity more quickly than those living in less urban areas, that they are smaller at maturity, have more offspring (*i.e.*, a higher fecundity), and can overall achieve a higher population growth rate (Figure 21.3). Thus, urban water fleas are closer to the 'fast' end of the fast-slow continuum of life histories (Jeschke et al. 2019) than rural water fleas.

Figure 21.3 Water fleas of the species *Daphnia magna* have a faster 'pace of life' in urban than in rural environments. In particular, they are younger, smaller at maturity and more fecund.

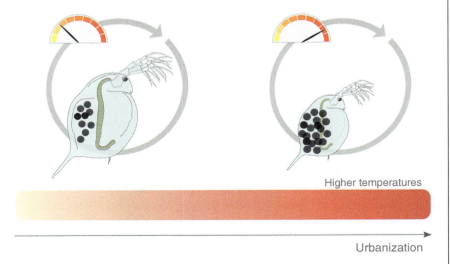

Evolutionary shifts in thermal tolerance, a well-documented example of global urban adaptation in both plants (*e.g.*, *T. repens*: Santangelo et al. 2022b) and animals (*e.g.*, the ant *T. curvispinosus*: Diamond et al. 2018), sometimes rely on changes in life history traits. For instance, urban populations of the grasshopper *C. brunneus* lay heavier (and therefore more heat-resilient) eggs than rural ones (San Martin Y Gomez and van Dyck 2012). Adaptation to urban heat in plants may be, for instance, associated with drought-resistance strategies taking the form of temporal avoidance strategies via earlier reproduction and seasonal senescence (Yakub and Tiffin 2017). Interestingly, tests of adaptation of lizard life history to elevated heat in cities in different species have failed to provide supporting evidence (Angilletta et al. 2013, Tiatragul et al. 2017, Hall and Warner 2018).

21.4.4 The Biotic Dimension

Novel organisms (including humans and our domesticated animals and plants) may become selection agents for other species, acting as new enemies or resources (Mooney and Cleland 2001, Shine 2012). Adaptation of native species to invasive species is predicted to be less common in response to small and isolated invader populations (Stotz et al. 2016), while invaders, that are widespread and well connected across the urban landscape are more likely to drive evolution. One example of this comes from Mérida, Mexico, where 'kissing bugs' *Triatoma dimidiata* show adaptive dispersal-related morphological simplification (smaller thorax and simplified antennas; Montes de Oca-Aguilar et al. 2022) in highly urban areas in response to the stable availability of novel hosts (domicile vertebrates, particularly humans, chickens and dogs; Ordóñez-Krasnowski et al. 2020). However, such responses are still understudied in urban areas.

Introduced species can have different evolutionary impacts in adjacent urban and rural areas. In Ohio, United States, an evolutionary trap laid by an invasive plant was reversed in urban environments: in rural habitats, the introduced Amur honeysuckle *Lonicera maackii* is a favourite plant for males of the Northern Cardinal *Cardinalis cardinalis*, as it provides carotenoids for their red colouration. However, the plant occupies territories that have higher predation risks and limit reproductive success, creating an evolutionary trap. In urban areas, males have access to various other carotenoid sources and territories, relaxing sexual selection for colouration and allowing avoidance of this evolutionary trap (Rodewald et al. 2011).

Non-native species not only flourish in urban areas and impose selection pressures on other species but also evolve themselves, including in life history traits. For example, abundant food resources (chironomid larvae) in urban streams have relaxed trade-offs in life history traits and thereby boosted the invasive capacity of guppies *Poecilia reticulata* through increased reproduction and somatic investment (Santana Marques et al. 2020). The diverse novel pollinator community encountered by the invasive scotch broom *Cytisus scoparius* in urban areas of Washington State, USA, created a selection gradient for larger flowers, which was not detected in the surrounding rural areas, allowing for pollinator niche expansion by the invader (Bode and Tong 2018). Repeated introductions of scotch broom in the area may have increased genetic and phenotypic diversity compared to its native range, providing ample variation for selection on this trait. In other cases, reduced gene flow and low genetic diversity in urban settings are associated with invasion success. Indeed, for the odorous house ant *Tepinoma sessile*, urban habitats with reduced gene flow are the exclusive settings for extreme polygyny and polydomous colonies, characteristics identified with successful invasive ant species and indicating that the species can only achieve supercoloniality within urbanized areas. The convergent pattern across several cities of a peculiar breeding structure confined to urban populations implies that, while social structure is often plastic in this species, adaptive evolution may play a role in divergence between urban and non-urban populations (Blumenfeld et al. 2022).

21.4.5 The Anthropogenic Inputs Dimension

In a landscape of unfamiliar or human-altered resources, innovative behavioural changes allowing a shift to novel resources such as anthropogenic food or artificial breeding sites and materials are likely to be adaptive, but we did not find strong evidence for heritable adaptive changes in such behaviours. There is some support for urban selection for innovative breeding behaviours, with, for instance, dark-eyed junco *Junco hyemalis* populations in urban areas of California, USA, gaining a fitness advantage by nesting in ecologically novel locations, such as off-ground and on artificial surfaces (Bressler et al. 2020).

However, novel urban behaviours can also become an evolutionary trap. The preference of urban western jackdaw *Corvus modedula* for nesting in cavities in buildings is tightly associated with reduced breeding success, as buildings are often located in areas lacking high-quality natural food (Meyrier et al. 2017). Some evolutionary traps due to novel anthropogenic resources likely result from plastic behaviour. For example, urban-dwelling house finches *Carpodacus mexicanus* in Mexico

City use cigarette butts in nest construction. This plastic behavioural change is beneficial against increasing ectoparasite levels but may have potential long-term disadvantages due to genotoxic damage inflicted on offspring (Suárez-Rodríguez and Garcia 2017).

21.4.6 Pan-urban Effects

In many studies of evolution in cities, the perceived selection pressure is an 'urbanization level' *per se*, or a general 'pan-urban' effect on life history trait selection. Some of these pan-urban trait changes may be adaptive. For example, urban females of *Anolis cristatellus*, a non-native lizard in South Florida, USA, are larger and lay eggs earlier than rural females under laboratory conditions, and their hatchlings grow faster, resulting in higher reproductive success, which was linked to the urban environment as a whole rather than a specific factor (Hall and Warner 2017). A study on band-legged ground crickets *Dianemobius nigrofasciatus* in Kagoshima City, Japan, revealed, through a common-garden experiment, that light and noise pollution had detrimental synergistic effects on life history traits such as growth rate and diapause induction in individuals originating from rural, but not urban, areas, indicating potential adaptation within urban environments (Ichikawa and Kuriwada 2023).

Evolution in response to urbanization can sometimes be difficult to detect and may only be revealed in interaction with another factor. For instance, a heritable phenotypic change in urban populations of the plant *Taraxacum officinale* was only detectable in interaction with an herbivory treatment: early seed production decreased with experimental herbivory only for urban plants, but no other differences were detected between urban and rural populations (Pisman et al. 2020). In this case, it is the plastic reaction norm of the organisms to predation that has evolved, possibly as a consequence of enemy release in urban conditions.

21.5 Synthesis and Perspectives

21.5.1 Patterns, Biases and Gaps in Research on Urban Life History Evolution

Our literature review, though not systematic, revealed a set of studies across different cities and species that provide some evidence of life history trait changes in response to urban selection pressures. However, only a minority of these studies give clear support for adaptive evolution by demonstrating both a change in selection levels and evidence for heritability. Heritability is often assumed or suggested in many urban selection studies but not formally tested (Table 21.1). Thus, despite the progress made, we are still in the early stages of consolidating evidence (Johnson and Munshi-South 2017, Diamond and Martin 2021). While some urban selection pressures on the life cycle of animals and plants have now started to be well explored, in particular the spatial and temporal dimensions, the full extent of urban selection pressures remains to be studied (Figure 21.4). In particular, we still lack studies exploring the effects of some urban abiotic stressors such as noise and light, novel anthropogenic resources and novel biotic interactions in cities.

The current research body on life history evolution in cities is subject to biases and knowledge gaps, which are not unique to life history investigations (see Diamond and Martin 2021 for a review of research patterns in urban evolution in general). These biases and gaps, together with the paucity of evidence, limit our ability to draw generalizations but also offer interesting opportunities for future research. We identified three major biases related to taxonomic groups, geography and habitat.

First, most studies have focussed on birds, flowering plants and insects (particularly pollinators), taxa that are generally more studied by urban biologists (*e.g.*, Aronson et al. 2014) and for which reviews on urban evolution already exist (birds: Chamberlain et al. 2009, Sepp et al. 2018; plants: Gorton et al. 2020; pollinating insects: Irwin et al. 2020). For many other taxa, including mammals, reptiles, molluscs, annelids, ferns and mosses, studies on this topic are rare.

Second, there is a geographic bias with a disproportionate focus on cities in temperate regions, particularly in Europe and North America. This bias misrepresents global climatic and biodiversity gradients and hinders the ability to draw general conclusions, particularly regarding climate-related urban selection pressures. It also neglects the most rapidly urbanizing regions and biomes as well as diverse urban landscapes worldwide (Elmqvist et al. 2013). Consequently, the current evidence predominantly reflects the impacts of evolutionary drivers characterizing western-world urbanization.

Third, aquatic habitats are underrepresented in studies of urban life history evolution. This is true for freshwater and particularly coastal and marine habitats, despite their merits for evolutionary studies (Touchard et al. 2023). Critically, these

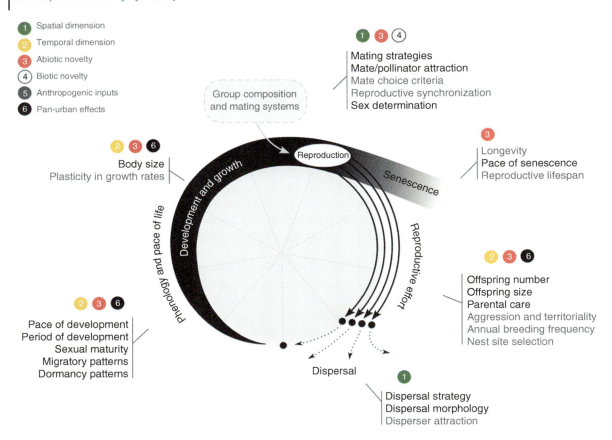

Figure 21.4 Synthetic summary of urban dimensions driving evolution across life history traits. General dimensions of life history are represented here schematically and do not necessarily offer a true representation of the chronology or ontology of life history phases for all organisms. Life history traits mentioned in this chapter are listed for each stage (black for evidence of adaptive evolution – see Table 21.1; grey for only suggested evidence of selection). The urban dimensions most likely to be causing these selection pressures are marked as numbers (see Figure 21.1): (1) habitat modification and spatial fragmentation (spatial dimension), (2) altered seasonal and diel cycles (temporal dimension), (3) novel abiotic stressors (abiotic dimension), (4) novel organisms and biotic interactions (biotic dimension) and (5) new anthropogenic resources and habitats (anthropogenic inputs dimension).

habitats are rich in biodiversity, and the global urbanization of coasts is rapidly advancing (Pelling and Blackburn 2013, Scherner et al. 2013). Thus, more efforts are needed to explore the potential impact of urbanization on life history evolution in aquatic environments and their rich biodiversity.

21.5.1.1 Common Patterns and Observed Trade-Offs

Despite the general paucity of robust evidence for adaptive life history evolution in cities, existing studies demonstrate that urbanization imposes novel selection pressures on all life cycle stages (Figure 21.4), with certain life history traits appearing frequently under selection.

Urban-dwelling species, known for their flexibility and generalist tendencies, thrive in the dynamic and heterogeneous conditions of urban areas (Ducatez et al. 2018). This flexibility suggests that their phenotypes are more likely to respond plastically to selection by urban conditions, possibly explaining the relative scarcity of studies that robustly demonstrate adaptive evolution in cities (Lambert et al. 2021). A combination of adaptive evolution and phenotypic plasticity can drive parallel or synergistic shifts in life history strategies among species in urban areas (Brans and De Meester 2018, Branston et al. 2021), although, in some cases, counter-gradient effects can happen where trait selection seems to be in an opposite direction from plastic responses to urbanization (Gorton et al. 2018, 2020). Additionally, it is important to acknowledge that plasticity itself can undergo natural selection (Ghalambor et al. 2007), as seen in invertebrates' thermal tolerance responses (Brans et al. 2017a, Diamond et al. 2018, Johnson et al. 2020). Exploring the role of selection on plasticity and its adaptive responses in urban environments remains a major challenge for urban evolutionary ecology.

While trade-offs between traits or fitness components in urban organisms have received limited attention (Gorton et al. 2020), some examples have been observed. These include trade-offs between adult size and development span in blowflies due to UHI (Hwang and Turner 2009), between lifetime fitness (survival and reproduction) and mobility (affecting foraging and dispersal) in moths influenced by flight-to-light behaviour in cities (Altermatt and Ebert 2016) and between survival and reproductive success in white-throated dippers affected by chemical pollution and food limitation (Morrissey et al. 2014). Trade-offs can also occur between traits of adults and their offspring (Catto et al. 2021), and even between different stages of offspring development (Charmantier et al. 2017).

21.5.1.2 Variation in Selection Pressures in Urban Areas

Urbanization imposes diverse selection pressures on species, leading to varying selection gradients between urban and non-urban areas, which can intensify or relax selection, sometimes in opposite directions even in sympatric and ecologically related species (Kaiser et al. 2016, Branston et al. 2021, Jagiello et al. 2022). High spatial heterogeneity means that even conspecifics within a city can experience contrasting selection pressures (Byrne and Nichols 1999, Hwang and Turner 2009), potentially driving the emergence of ecotypes, reproductive isolation and the evolution of novel life history strategies (Charmantier et al. 2017, Chick et al. 2019). Studying life history traits presents challenges in distinguishing direct selection from their role as fitness indicators for other traits, particularly in cases of indirect evolutionary impact (San Martin Y Gomez and van Dyck 2012, Kolonin et al. 2022).

21.5.1.3 Cascading Effects and Indirect Selection via Biotic Interactions

Urban evolution driven by abiotic factors often involves indirect biotic effects through disruption of intra- or inter-specific interactions or shifts in community composition. Temporal mismatches, in particular due to species responding differently to the same selection pressures, may have cascading effects on competition, mutualisms and food–web interactions (Závodská et al. 2012, Meineke et al. 2014, Fisogni et al. 2020). Habitat change and fragmentation create physical barriers and alter selection landscapes driven by prey/host availability or mutualisms (*e.g.*, pollination, seed dispersal; Cheptou et al. 2017). Urbanization may lead to hyper-abundance of certain species, which in turn drives selection on others (*e.g.*, Zhang et al. 2022). Such cascading effects make it challenging to predict evolutionary trends in urban ecosystems.

21.5.2 The Future of Life History Evolution in Cities: Research and Practice

21.5.2.1 Island Biology in the City

The application of principles from island biology to urban biology holds great potential to help understand how urban fragmentation affects life history trait evolution, but it is currently underutilized. Although true islands and habitat fragments are not fully ecologically equivalent, the shared characteristics of urban areas with islands can stimulate intriguing comparisons (Itescu 2019). Island biogeography theory and concepts such as species-area relationships, colonization-extinction dynamics and meta population theory have already been adapted to investigate urban habitats (Davis and Glick 1978, Marzluff 2005, Fattorini 2016). In particular, island biology offers insights into the role of dispersal and gene flow in life history evolution (Cheptou et al. 2017). It also predicts the evolution of an 'island syndrome' (Adler and Levins 1994). Such parallels can provide theoretical frameworks needed to understand urban population dynamics and life history selection and may help predict future evolutionary trajectories (Dunn et al. 2022).

21.5.2.2 Urban Trait Syndrome and Convergent Evolution

If conditions of urban life are comparable across cities and continents, we may expect the emergence of an urban trait syndrome through parallel convergent evolution, contributing to urban biotic homogenization (Santangelo et al. 2020a, Lokatis and Jeschke 2022). The concept of an 'urban syndrome' in life history traits, akin to the 'island syndrome', offers a promising research avenue (Gorton et al. 2020). By studying and comparing multiple traits across diverse cities, a general eco-evolutionary pattern associated with urbanization might be identified, encompassing characteristics such as accelerated growth and development, changes in body size, modified reproductive strategies, advanced and prolonged seasonal activity, increased tolerance to environmental stressors or shifts in resource allocation (Santangelo et al. 2020a). A few studies have discussed the idea of a 'typical' urban life history strategy shaped by multivariate trait selection (*e.g.*, Cheptou et al. 2017, Santangelo et al. 2020a). However, current research on urban life history evolution still tends to overlook the co-variation among multiple traits (see, *e.g.*, Scalici et al. 2008, Ibáñez-Álamo and Soler 2010, Brans and De Meester 2018). This narrow focus limits our understanding, and future research may benefit from adopting a multivariate approach to detect such life history syndromes.

Global urbanization constitutes a large-scale replicated evolutionary experiment, providing opportunities to test for such parallel evolution. Several recent studies detected convergent selection within species in various plant and animal traits in response to urbanization (Diamond et al. 2018, Johnson et al. 2018, Campbell-Staton et al. 2020, Blumenfeld et al. 2022, Cosentino and Gibbs 2022). Nevertheless, few of these studies have compared selection across continents (Santangelo et al. 2022a), and only a few have specifically focussed on life history traits (Yakub and Tiffin 2017). Although convergent selection contributes to homogenization, facilitated by increased human-mediated dispersal among cities, local factors such as climate and vegetation cover can lead to divergent influences on urban evolution, even within the same species (Santangelo et al. 2022a). There is yet little evidence of convergent urban selection across species, with opposite trends found for closely related or ecologically similar species. Large-scale factors, such as climatic zones, interact with local characteristics and species' evolutionary potential, resulting in different types of 'urban syndromes'. Thus, future studies should consider environmental factors at different scales and taxonomic differences to account for the complexity and variability of urban evolutionary patterns.

21.5.2.3 Understanding Life History Evolution in Cities Is of High Practical Relevance

The study of life history evolution in urban species, particularly their adaptive responses, has practical implications for effectively addressing the challenges and opportunities of urbanization. Urban areas are home to diverse species, including pests and disease vectors that impact human well-being (Neiderud 2015). The economic sustainability of urban areas is also influenced by species' life history evolution, as urban ecosystems provide economic benefits such as pollination, pest control and recreation (Haase et al. 2014). Recognizing and preserving the adaptive capacities of urban species helps maintain these ecosystem services and enhances economic sustainability. Understanding the drivers of success for invasive species in urban environments allows for effective pest management strategies, reducing economic costs associated with biological invasions (Diagne et al. 2021). By studying adaptations in urban environments and identifying underlying mechanisms of success, such as changes in reproduction rates, dispersal patterns or habitat preferences, targeted management strategies can be developed to minimize human–wildlife conflicts, mitigate disease risks and address pest-related problems in urban infrastructures (Szczepaniec et al. 2011, Meineke et al. 2014).

Such studies also support nature conservation. Urbanization causes habitat loss, fragmentation and changes in ecological processes. Identifying key life history traits associated with successful colonization and persistence in urban areas informs habitat restoration initiatives, urban policies, planning strategies and infrastructure development. Considering ecological requirements and species' life histories enables the design of green corridors, connectivity networks and green infrastructure, facilitating the movement and persistence of urban wildlife (Snäll et al. 2016). Incorporating ecological and evolutionary principles into urban planning supports the preservation of biodiversity and ecological processes within and beyond urban areas. By integrating the adaptive responses of urban species into decision-making processes, we can create more resilient, sustainable and healthy cities and ensure human–nature coexistence in an era of expanding human-driven global changes.

21.5.2.4 Integrative Approaches for Studying Life History Evolution in Cities Are Essential

Recognizing the significance of urbanization-driven eco-evolutionary changes requires a comprehensive consideration of multiple factors. As urban areas expand, it becomes increasingly important to go beyond characterizing ecological conditions and understand the influences of human dimensions like urban policy, socio-economic drivers, social–ecological networks, urban nature's contributions to human well-being and systems of racial oppression (Schell et al. 2020, Kinnunen et al. 2022). These social dynamics shape environmental conditions, drive biological changes and influence the adaptive evolution of life history traits in urban organisms (*e.g.*, Lowe et al. 2014). Urban policies prioritizing green and blue spaces (*i.e.*, vegetated areas and water bodies) and ecological networks can shape life histories by providing dispersal opportunities, resource access and suitable habitats. Social drivers, such as human behaviour and cultural practices, can impact selection pressures through habitat modification, pollution and resource availability. Socio-ecological networks can create feedback loops shaping organisms' adaptive responses to urban environments. Systems of racial oppression and social class segregation can affect resource distribution and environmental quality, leading to selection gradients on life history traits within cities (Schell et al. 2020). The diversity of cultural, religious and social perspectives on human–nature relationships may also generate selection gradients within and between cities and, compared to non-urban areas, driving divergent life history strategies.

To comprehensively study urban evolution, an interdisciplinary approach is essential. Incorporating diverse disciplines from biology and social sciences, varied knowledge domains and research methods can provide a deeper understanding of the mechanisms and unique anthropogenic drivers shaping urban evolution. It will enable effective assessment of urbanization impacts as well as predict organisms' responses to these novel and unfamiliar environments (Des Roches et al. 2021). Recognizing the interconnectedness of human activities, urban design and ecological processes will facilitate the development of more sustainable and resilient cities supporting human well-being and biodiversity.

Acknowledgements

We would like to thank Kristien Brans and Luc De Meester for their helpful input on parts of this chapter. YI was funded by a Humboldt Stiftung post-doctoral fellowship.

References

Adams, C.A., Fernández-Juricic, E., Bayne, E.M., and St. Clair, C.C. (2021a). Effects of artificial light on bird movement and distribution: a systematic map. *Environmental Evidence* 10(1): 37.

Adams, E., Leeb, C., and Brühl, C.A. (2021b). Pesticide exposure affects reproductive capacity of common toads (*Bufo bufo*) in a viticultural landscape. *Ecotoxicology* 30(2): 213–223.

Adler, G.H., and Levins, R. (1994). The island syndrome in rodent populations. *The Quarterly Review of Biology* 69(4): 473–490.

Alberti, M. (2023). Cities of the Anthropocene: urban sustainability in an eco-evolutionary perspective. *Philosophical Transactions of the Royal Society, B: Biological Sciences* 379(1893): 20220264.

Alberti, M., Correa, C., Marzluff, J.M., Hendry, A.P., Palkovacs, E.P., Gotanda, K.M., Hunt, V.M., Apgar, T.M., and Zhou, Y. (2017a). Global urban signatures of phenotypic change in animal and plant populations. *Proceedings of the National Academy of Sciences of the United States of America* 114(34): 8951–8956.

Alberti, M., Marzluff, J., and Hunt, V.M. (2017b). Urban driven phenotypic changes: empirical observations and theoretical implications for eco-evolutionary feedback. *Philosophical Transactions of the Royal Society, B: Biological Sciences* 372(1712): 20160029.

Alberti, M., Palkovacs, E.P., Roches, S.D., Meester, L.D., Brans, K.I., Govaert, L., Grimm, N.B., Harris, N.C., Hendry, A.P., Schell, C.J., Szulkin, M., Munshi-South, J., Urban, M.C., and Verrelli, B.C. (2020). The complexity of urban eco-evolutionary dynamics. *BioScience* 70(9): 772–793.

Alter, S.E., Tariq, L., Creed, J.K., and Megafu, E. (2021). Evolutionary responses of marine organisms to urbanized seascapes. *Evolutionary Applications* 14(1): 210–232.

Altermatt, F., and Ebert, D. (2016). Reduced flight-to-light behaviour of moth populations exposed to long-term urban light pollution. *Biology Letters* 12(4): 20160111.

Angilletta, M.J. Jr, Wilson, R.S., Niehaus, A.C., Sears, M.W., Navas, C.A., and Ribeiro, P.L. (2007). Urban physiology: city ants possess high heat tolerance. *PLoS ONE* 2(2): e258.

Angilletta, M.J., Zelic, M.H., Adrian, G.J., Hurliman, A.M., and Smith, C.D. (2013). Heat tolerance during embryonic development has not diverged among populations of a widespread species (*Sceloporus undulatus*). *Conservation Physiology* 1(1): cot018.

Aronson, M.F.J., La Sorte, F.A., Nilon, C.H., Katti, M., Goddard, M.A., Lepczyk, C.A., Warren, P.S., Williams, N.S.G., Cilliers, S., Clarkson, B., Dobbs, C., Dolan, R., Hedblom, M., Klotz, S., Kooijmans, J.L., Macgregor-fors, I., Mcdonnell, M., Mörtberg, U., Pyšek, P., Siebert, S., Sushinsky, J., Werner, P., and Winter, M. (2014). A global analysis of the impacts of urbanization on bird and plant diversity reveals key anthropogenic drivers. *Proceedings of the Royal Society B: Biological Sciences* 281: 20133330.

Aronson, M.F.J., Nilon, C.H., Lepczyk, C.A., Parker, T.S., Warren, P.S., Cilliers, S.S., Goddard, M.A., Hahs, A.K., Herzog, C., Katti, M., La Sorte, F.A., Williams, N.S.G., and Zipperer, W. (2016). Hierarchical filters determine community assembly of urban species pools. *Ecology* 97(11): 2952–2963.

Arroyo-Rodríguez, V., and Mandujano, S. (2006). Forest fragmentation modifies habitat quality for *Alouatta palliata*. *International Journal of Primatology* 27(4): 1079–1096.

Bartlewicz, J., Vandepitte, K., Jacquemyn, H., and Honnay, O. (2015). Population genetic diversity of the clonal self-incompatible herbaceous plant *Linaria vulgaris* along an urbanization gradient. *Biological Journal of the Linnean Society* 116(3): 603–613.

Bauerová, P., Vinklerová, J., Hraníček, J., Čorba, V., Vojtek, L., Svobodová, J., and Vinkler, M. (2017). Associations of urban environmental pollution with health-related physiological traits in a free-living bird species. *Science of the Total Environment* 601–602: 1556–1565.

Ben-Moshe, N., and Iwamura, T. (2020). Shelter availability and human attitudes as drivers of rock hyrax (*Procavia capensis*) expansion along a rural–urban gradient. *Ecology and Evolution* 10(9): 4044–4065.

Bennie, J., Davies, T.W., Cruse, D., and Gaston, K.J. (2016). Ecological effects of artificial light at night on wild plants. *Journal of Ecology* 104(3): 611–620.

Blumenfeld, A.J., Eyer, P.-A., Helms, A.M., Buczkowski, G., and Vargo, E.L. (2022). Consistent signatures of urban adaptation in a native, urban invader ant *Tapinoma sessile*. *Molecular Ecology* 31(18): 4832–4850.

Bode, R.F., and Tong, R. (2018). Pollinators exert positive selection on flower size on urban, but not on rural Scotch broom (*Cytisus scoparius* L. Link). *Journal of Plant Ecology* 11(3): 493–501.

Bókony, V., Üveges, B., Ujhegyi, N., Verebélyi, V., Nemesházi, E., Csíkvári, O., and Hettyey, A. (2018). Endocrine disruptors in breeding ponds and reproductive health of toads in agricultural, urban and natural landscapes. *Science of the Total Environment* 634: 1335–1345.

Borden, J.B., and Flory, S.L. (2021). Urban evolution of invasive species. *Frontiers in Ecology and the Environment* 19(3): 184–191.

Boyes, D.H., Evans, D.M., Fox, R., Parsons, M.S., and Pocock, M.J.O. (2021). Street lighting has detrimental impacts on local insect populations. *Science Advances* 7(35): eabi8322.

Brans, K.I., and De Meester. L. (2018). City life on fast lanes: urbanization induces an evolutionary shift towards a faster lifestyle in the water flea *Daphnia*. *Functional Ecology* 32: 2225–2240.

Brans, K.I., Engelen, J.M.T., Souffreau, C., and De Meester, L. (2018). Urban hot-tubs: local urbanization has profound effects on average and extreme temperature in ponds. *Landscape and Urban Planning* 176: 22–29.

Brans, K.I., Govaert, L., Engelen, J.M.T., Gianuca, A.T., Souffreau, C., and De Meester, L. (2017a). Eco-evolutionary dynamics in urbanized landscapes: evolution, species sorting and the change in zooplankton body size along urbanization gradients. *Philosophical Transactions of the Royal Society, B: Biological Sciences* 372(1712): 20160030.

Brans, K.I., Jansen, M., Vanoverbeke, J., Tüzün, N., Stoks, R., and De Meester, L. (2017b). The heat is on: genetic adaptation to urbanization mediated by thermal tolerance and body size. *Global Change Biology* 23(12): 5218–5227.

Branston, C.J., Capilla-Lasheras, P., Pollock, C.J., Griffiths, K., White, S., and Dominoni, D.M. (2021). Urbanisation weakens selection on the timing of breeding and clutch size in blue tits but not in great tits. *Behavioral Ecology and Sociobiology* 75(11): 155.

Bressler, S.A., Diamant, E.S., Tingley, M.W., and Yeh, P.J. (2020). Nests in the cities: adaptive and non-adaptive phenotypic plasticity and convergence in an urban bird. *Proceedings of the Royal Society B: Biological Sciences* 287(1941): 20202122.

Brys, R., and Jacquemyn, H. (2012). Effects of human-mediated pollinator impoverishment on floral traits and mating patterns in a short-lived herb: an experimental approach. *Functional Ecology* 26(1): 189–197.

Buczkowski, G. (2011). Suburban sprawl: environmental features affect colony social and spatial structure in the black carpenter ant, *Camponotus pennsylvanicus*. *Ecological Entomology* 36(1): 62–71.

Butler, C.D., and Trumble, J.T. (2008). Effects of pollutants on bottom-up and top-down processes in insect–plant interactions. *Environmental Pollution* 156(1): 1–10.

Byrne, K., and Nichols, R.A. (1999). *Culex pipiens* in London underground tunnels: differentiation between surface and subterranean populations. *Heredity* 82(1): 7–15.

Callaghan, C.T., Bowler, D.E., and Pereira, H.M. (2021). Thermal flexibility and a generalist life history promote urban affinity in butterflies. *Global Change Biology* 27(15): 3532–3546.

Campbell-Staton, S.C., Winchell, K.M., Rochette, N.C., Fredette, J., Maayan, I., Schweizer, R.M., and Catchen, J. (2020). Parallel selection on thermal physiology facilitates repeated adaptation of city lizards to urban heat islands. *Nature Ecology & Evolution* 4(4): 652–658.

Capilla-Lasheras, P., Thompson, M.J., Sánchez-Tójar, A., Haddou, Y., Branston, C.J., Réale, D., Charmantier, A., and Dominoni, D.M. (2022). A global meta-analysis reveals higher variation in breeding phenology in urban birds than in their non-urban neighbours. *Ecology Letters* 25(11): 2552–2570.

Catto, S., Sumasgutner, P., Amar, A., Thomson, R.L., and Cunningham, S.J. (2021). Pulses of anthropogenic food availability appear to benefit parents, but compromise nestling growth in urban red-winged starlings. *Oecologia* 197(3): 565–576.

Chamberlain, D.E., Cannon, A.R., Toms, M.P., Leech, D.I., Hatchwell, B.J., and Gaston, K.J. (2009). Avian productivity in urban landscapes: a review and meta-analysis. *Ibis* 151(1): 1–18.

Charmantier, A., Demeyrier, V., Lambrechts, M., Perret, S., and Grégoire, A. (2017). Urbanization is associated with divergence in pace-of-life in great tits. *Frontiers in Ecology and Evolution* 5: 53.

Charter, M., Izhaki, I., Ben Mocha, Y., and Kark, S. (2016). Nest-site competition between invasive and native cavity nesting birds and its implication for conservation. *Journal of Environmental Management* 181: 129–134.

Cheptou, P.-O., and Avendano, L.G. (2006). Pollination processes and the Allee effect in highly fragmented populations: consequences for the mating system in urban environments. *New Phytologist* 172(4): 774–783.

Cheptou, P.-O., Carrue, O., Rouifed, S., and Cantarel, A. (2008). Rapid evolution of seed dispersal in an urban environment in the weed *Crepis sancta*. *Proceedings of the National Academy of Sciences of the United States of America* 105(10): 3796–3799.

Cheptou, P.-O., Hargreaves, A.L., Bonte, D., and Jacquemyn, H. (2017). Adaptation to fragmentation: evolutionary dynamics driven by human influences. *Philosophical Transactions of the Royal Society, B: Biological Sciences* 372(1712): 20160037.

Chick, L.D., Strickler, S.A., Perez, A., Martin, R.A., and Diamond, S.E. (2019). Urban heat islands advance the timing of reproduction in a social insect. *Journal of Thermal Biology* 80: 119–125.

Christmann, T., Kowarik, I., Bernard-Verdier, M., Buchholz, S., Hiller, A., Seitz, B., and von der Lippe, M. (2023). Phenology of grassland plants responds to urbanization. *Urban Ecosystems* 26(1): 261–275.

Contesse, P., Hegglin, D., Gloor, S., Bontadina, F., and Deplazes, P. (2004). The diet of urban foxes (*Vulpes vulpes*) and the availability of anthropogenic food in the city of Zurich, Switzerland. *Mammalian Biology* 69(2): 81–95.

Cosentino, B.J., and Gibbs, J.P. (2022). Parallel evolution of urban–rural clines in melanism in a widespread mammal. *Scientific Reports* 12(1): 1752.

Croteau, M.C., Hogan, N., Gibson, J.C., Lean, D., and Trudeau, V.L. (2008). Toxicological threats to amphibians and reptiles in urban environments. In: *Urban herpetology* (eds. Mitchell, J.C., Brown, R.E.J., and Bartholomew, B.), 197–209. Society for the Study of Amphibians and Reptiles.

Davis, A.M., and Glick, T.F. (1978). Urban ecosystems and island biogeography. *Environmental Conservation* 5(4): 299–304.

Des Roches, S., Brans, K.I., Lambert, M.R., Rivkin, L.R., Savage, A.M., Schell, C.J., Correa, C., De Meester, L., Diamond, S.E., Grimm, N.B., Harris, N.C., Govaert, L., Hendry, A.P., Johnson, M.T.J., Munshi-South, J., Palkovacs, E.P., Szulkin, M., Urban, M.C., Verrelli, B.C., and Alberti, M. (2021). Socio-eco-evolutionary dynamics in cities. *Evolutionary Applications* 14(1): 248–267.

Diagne, C., Leroy, B., Vaissière, A.-C., Gozlan, R.E., Roiz, D., Jarić, I., Salles, J.-M., Bradshaw, C.J.A., and Courchamp, F. (2021). High and rising economic costs of biological invasions worldwide. *Nature* 592(7855): 571–576.

Diamond, S.E., Chick, L.D., Perez, A., Strickler, S.A., and Martin, R.A. (2018). Evolution of thermal tolerance and its fitness consequences: parallel and non-parallel responses to urban heat islands across three cities. *Proceedings of the Royal Society B: Biological Sciences* 285(1882): 20180036.

Diamond, M.L., and Hodge, E. (2007). Urban contaminant dynamics: from source to effect. *Environmental Science & Technology* 41(11): 3796–3800.

Diamond, S.E., and Martin, R.A. (2021). Evolution in cities. *Annual Review of Ecology, Evolution, and Systematics* 52(1): 519–540.

Dohmen, G.P. (1985). Secondary effects of air pollution: Enhanced aphid growth. *Environmental Pollution Series A, Ecological and Biological* 39(3): 227–234.

Dominoni, D., Smit, J.A.H., Visser, M.E., and Halfwerk, W. (2020). Multisensory pollution: artificial light at night and anthropogenic noise have interactive effects on activity patterns of great tits (*Parus major*). *Environmental Pollution* 256: 113314.

Donihue, C.M., and Lambert, M.R. (2015). Adaptive evolution in urban ecosystems. *Ambio* 44(3): 194–203.

Dubois, J., and Cheptou, P.-O. (2017). Effects of fragmentation on plant adaptation to urban environments. *Philosophical Transactions of the Royal Society, B: Biological Sciences* 372(1712): 20160038.

Ducatez, S., Sayol, F., Sol, D., and Lefebvre, L. (2018). Are urban vertebrates city specialists, artificial habitat exploiters, or environmental generalists? *Integrative and Comparative Biology* 58(5): 929–938.

Dunn, R.R., Burger, J.R., Carlen, E.J., Koltz, A.M., Light, J.E., Martin, R.A., Munshi-South, J., Nichols, L.M., Vargo, E.L., Yitbarek, S., Zhao, Y., and Cibrián-Jaramillo, A. (2022). A theory of city biogeography and the origin of urban species. *Frontiers in Conservation Science* 3: 761449.

Duquette, C.A., Loss, S.R., and Hovick, T.J. (2021). A meta-analysis of the influence of anthropogenic noise on terrestrial wildlife communication strategies. *Journal of Applied Ecology* 58(6): 1112–1121.

Elmqvist, T., Fragkias, M., Goodness, J., Güneralp, B., Marcotullio, P.J., McDonald, R.I., Parnell, S., Schewenius, M., Sendstad, M., Seto, K.C., and Wilkinson, C. (2013). *Urbanization, biodiversity and ecosystem services: challenges and opportunities*. Springer.

Faeth, S.H., Warren, P.S., Shochat, E., and Marussich, W.A. (2005). Trophic dynamics in urban communities. *BioScience* 55(5): 399–407.

Fahrig, L. (2003). Effects of habitat fragmentation on biodiversity. *Annual Review of Ecology, Evolution, and Systematics* 34(1): 487–515.

Fattorini, S. (2016). Insects and the city: what island biogeography tells us about insect conservation in urban areas. *Web Ecology* 16(1): 41–45.

Fenoglio, M.S., Calviño, A., González, E., Salvo, A., and Videla, M. (2021). Urbanisation drivers and underlying mechanisms of terrestrial insect diversity loss in cities. *Ecological Entomology* 46(4): 757–771.

Fischer, J., and Lindenmayer, D.B. (2007). Landscape modification and habitat fragmentation: a synthesis. *Global Ecology and Biogeography* 16(3): 265–280.

Fisogni, A., Hautekèete, N., Piquot, Y., Brun, M., Vanappelghem, C., Michez, D., and Massol, F. (2020). Urbanization drives an early spring for plants but not for pollinators. *Oikos* 129(11): 1681–1691.

Gaertner, M., Wilson, J.R.U., Cadotte, M.W., MacIvor, J.S., Zenni, R.D., and Richardson, D.M. (2017). Non-native species in urban environments: patterns, processes, impacts and challenges. *Biological Invasions* 19(12): 3461–3469.

Gaston, K.J., Bennie, J., Davies, T.W., and Hopkins, J. (2013). The ecological impacts of nighttime light pollution: a mechanistic appraisal: nighttime light pollution. *Biological Reviews* 88(4): 912–927.

Gaston, K.J., and Sánchez de Miguel, A. (2022). Environmental impacts of artificial light at night. *Annual Review of Environment and Resources* 47(1): 373–398.

George, S.L., and Crooks, K.R. (2006). Recreation and large mammal activity in an urban nature reserve. *Biological Conservation* 133(1): 107–117.

Ghalambor, C.K., McKay, J.K., Carroll, S.P., and Reznick, D.N. (2007). Adaptive versus non-adaptive phenotypic plasticity and the potential for contemporary adaptation in new environments. *Functional Ecology* 21(3): 394–407.

Gorton, A.J., Burghardt, L.T., and Tiffin, P. (2020). Adaptive evolution of plant life history in urban environments. In: *Urban evolutionary biology* (eds. Szulkin, M., Munshi-South, J., and Charmantier, A.), 142–156. Oxford University Press.

Gorton, A.J., Moeller, D.A., and Tiffin, P. (2018). Little plant, big city: a test of adaptation to urban environments in common ragweed (*Ambrosia artemisiifolia*). *Proceedings of the Royal Society B: Biological Sciences* 285(1881): 20180968.

Green, L., Faust, E., Hinchcliffe, J., Brijs, J., Holmes, A., Orn, F.E., Svensson, O., Roques, J.A.C., Leder, E.H., Sandblom, E., and Kvarnemo, C. (2023). Invader at the edge – genomic origins and physiological differences of round gobies across a steep urban salinity gradient. *Evolutionary Applications* 16(2): 321–337.

Grimm, N.B., Faeth, S.H., Golubiewski, N.E., Redman, C.L., Wu, J., Bai, X., and Briggs, J.M. (2008). Global change and the ecology of cities. *Science* 319(5864): 756–760.

Gurule-Small, G.A., and Tinghitella, R.M. (2019). Life history consequences of developing in anthropogenic noise. *Global Change Biology* 25(6): 1957–1966.

Haase, D., Larondelle, N., Andersson, E., Artmann, M., Borgström, S., Breuste, J., Gomez-Baggethun, E., Gren, Å., Hamstead, Z., and Hansen, R. (2014). A quantitative review of urban ecosystem service assessments: concepts, models, and implementation. *Ambio* 43: 413–433.

Haddad, N.M., Brudvig, L.A., Clobert, J., Davies, K.F., Gonzalez, A., Holt, R.D., Lovejoy, T.E., Sexton, J.O., Austin, M.P., Collins, C.D., Cook, W.M., Damschen, E.I., Ewers, R.M., Foster, B.L., Jenkins, C.N., King, A.J., Laurance, W.F., Levey, D.J., Margules, C.R., Melbourne, B.A., Nicholls, A.O., Orrock, J.L., Song, D.-X., and Townshend, J.R. (2015). Habitat fragmentation and its lasting impact on Earth's ecosystems. *Science Advances* 1(2): e1500052.

Halfwerk, W., Blaas, M., Kramer, L., Hijner, N., Trillo, P.A., Bernal, X.E., Page, R.A., Goutte, S., Ryan, M.J., and Ellers, J. (2019). Adaptive changes in sexual signalling in response to urbanization. *Nature Ecology & Evolution* 3: 374–380.

Halfwerk, W., Bot, S., Buikx, J., van der Velde, M., Komdeur, J., ten Cate, C., and Slabbekoorn, H. (2011). Low-frequency songs lose their potency in noisy urban conditions. *Proceedings of the National Academy of Sciences of the United States of America* 108(35): 14549–14554.

Hall, J.M., and Warner, D.A. (2017). Body size and reproduction of a non-native lizard are enhanced in an urban environment. *Biological Journal of the Linnean Society* 122(4): 860–871.

Hall, J.M., and Warner, D.A. (2018). Thermal spikes from the urban heat island increase mortality and alter physiology of lizard embryos. *Journal of Experimental Biology* 221(14): jeb181552.

Heger, T., Bernard-Verdier, M., Gessler, A., Greenwood, A.D., Grossart, H.-P., Hilker, M., Keinath, S., Kowarik, I., Kueffer, C., Marquard, E., Müller, J., Niemeier, S., Onandia, G., Petermann, J.S., Rillig, M.C., Rödel, M.-O., Saul, W.-C., Schittko, C., Tockner, K., Joshi, J., and Jeschke, J.M. (2019). Towards an integrative, eco-evolutionary understanding of ecological novelty: studying and communicating interlinked effects of global change. *BioScience* 69(11): 888–899.

Hölker, F., Wolter, C., Perkin, E.K., and Tockner, K. (2010). Light pollution as a biodiversity threat. *Trends in Ecology & Evolution* 25(12): 681–682.

Honour, S.L., Bell, J.N.B., Ashenden, T.W., Cape, J.N., and Power, S.A. (2009). Responses of herbaceous plants to urban air pollution: effects on growth, phenology and leaf surface characteristics. *Environmental Pollution* 157(4): 1279–1286.

Hopkins, G.R., Gaston, K.J., Visser, M.E., Elgar, M.A., and Jones, T.M. (2018). Artificial light at night as a driver of evolution across urban–rural landscapes. *Frontiers in Ecology and the Environment* 16(8): 472–479.

Hwang, C.C., and Turner, B.D. (2009). Small-scaled geographical variation in life-history traits of the blowfly *Calliphora vicina* between rural and urban populations. *Entomologia Experimentalis et Applicata* 132(3): 218–224.

Ibáñez-Álamo, J.D., and Soler, M. (2010). Does urbanization affect selective pressures and life-history strategies in the common blackbird (*Turdus merula* L.)? *Biological Journal of the Linnean Society* 101(4): 759–766.

Ichikawa, I., and Kuriwada, T. (2023). The combined effects of artificial light at night and anthropogenic noise on life history traits in ground crickets. *Ecological Research* 38(3): 446–454.

Irwin, R.E., Warren, P.S., and Adler, L.S. (2018). Phenotypic selection on floral traits in an urban landscape. *Proceedings of the Royal Society B: Biological Sciences* 285(1884): 20181239.

Irwin, R.E., Warren, P.S., Carper, A.L., and Adler, L.S. (2014). Plant–animal interactions in suburban environments: implications for floral evolution. *Oecologia* 174(3): 803–815.

Irwin, R.E., Youngsteadt, E., Warren, P.S., and Bronstein, J.L. (2020). The evolutionary ecology of mutualisms in urban landscapes. In: *Urban evolutionary biology* (eds. Szulkin, M., Munshi-South, J., and Charmantier, A.), 111–129. Oxford University Press.

Itescu, Y. (2019). Are island-like systems biologically similar to islands? A review of the evidence. *Ecography* 42(7): 1298–1314.

Jackson, N., Littleford-Colquhoun, B.L., Strickland, K., Class, B., and Frere, C.H. (2022). Selection in the city: rapid and fine-scale evolution of urban eastern water dragons. *Evolution* 76(10): 2302–2314.

Jagiello, Z., Corsini, M., Dylewski, Ł., Ibáñez-Álamo, J.D., and Szulkin, M. (2022). The extended avian urban phenotype: anthropogenic solid waste pollution, nest design, and fitness. *Science of the Total Environment* 838: 156034.

Jarrett, C., Powell, L.L., McDevitt, H., Helm, B., and Welch, A.J. (2020). Bitter fruits of hard labour: diet metabarcoding and telemetry reveal that urban songbirds travel further for lower-quality food. *Oecologia* 193(2): 377–388.

Jeschke, J.M., Gabriel, W., and Kokko, H. (2019). *r*-strategists/*K*-strategists. *Encyclopedia of Ecology* 3: 193–201.

Jeschke, J.M., Keesing, F., and Ostfeld, R.S. (2013). Novel organisms: comparing invasive species, GMOs, and emerging pathogens. *Ambio* 42(5): 541–548.

Jeschke, J.M., and Strayer, D.L. (2006). Determinants of vertebrate invasion success in Europe and North America. *Global Change Biology* 12(9): 1608–1619.

Jochner, S., and Menzel, A. (2015). Urban phenological studies – Past, present, future. *Environmental Pollution* 203: 250–261.

Johnson, J.C., Garver, E., and Martin, T. (2020). Black widows on an urban heat island: extreme heat affects spider development and behaviour from egg to adulthood. *Animal Behaviour* 167: 77–84.

Johnson, M.T.J., and Munshi-South, J. (2017). Evolution of life in urban environments. *Science* 358(6363): eaam8327.

Johnson, M.T.J., Prashad, C.M., Lavoignat, M., and Saini, H.S. (2018). Contrasting the effects of natural selection, genetic drift and gene flow on urban evolution in white clover (*Trifolium repens*). *Proceedings of the Royal Society B: Biological Sciences* 285: 20181019.

Johnson, J.C., Trubl, P.J., and Miles, L.S. (2012). Black widows in an urban desert: city-living compromises spider fecundity and egg investment despite urban prey abundance. *The American Midland Naturalist* 168(2): 333–340.

Kaiser, A., Merckx, T., and van Dyck, H. (2016). The urban heat island and its spatial scale dependent impact on survival and development in butterflies of different thermal sensitivity. *Ecology and Evolution* 6(12): 4129–4140.

Kaunath, V., and Eccard, J.A. (2022). Light attraction in carabid beetles: comparison among animals from the inner city and a dark sky reserve. *Frontiers in Ecology and Evolution* 10: 75128.

Kempenaers, B., Borgström, P., Loës, P., Schlicht, E., and Valcu, M. (2010). Artificial night lighting affects dawn song, extra-pair siring success, and lay date in songbirds. *Current Biology* 20(19): 1735–1739.

Kettlewell, H.B.D. (1955). Selection experiments on industrial melanism in the Lepidoptera. *Heredity* 9(3): 323–342.

Kinnunen, R.P., Fraser, K.C., Schmidt, C., and Garroway, C.J. (2022). The socioeconomic status of cities covaries with avian life-history strategies. *Ecosphere* 13(2): e3918.

Kolonin, A.M., Bókony, V., Bonner, T.H., Zúñiga-Vega, J.J., Aspbury, A.S., Guzman, A., Molina, R., Calvillo, P., and Gabor, C.R. (2022). Coping with urban habitats via glucocorticoid regulation: physiology, behavior, and life history in stream fishes. *Integrative and Comparative Biology* 62(1): 90–103.

Kowarik, I. (2011). Novel urban ecosystems, biodiversity, and conservation. *Environmental Pollution* 159(8–9): 1974–1983.

LaDeau, S., Leisnham, P., Biehler, D. and Bodner, D. (2013). Higher mosquito production in low-income neighborhoods of Baltimore and Washington, DC: understanding ecological drivers and mosquito-borne disease risk in temperate cities. *International Journal of Environmental Research and Public Health* 10(4): 1505–1526.

Lambert, M.R., Brans, K.I., Des Roches, S., Donihue, C.M., and Diamond, S.E. (2021). Adaptive evolution in cities: progress and misconceptions. *Trends in Ecology & Evolution* 36(3): 239–257.

Lambrecht, S.C., Mahieu, S., and Cheptou, P.-O. (2016). Natural selection on plant physiological traits in an urban environment. *Acta Oecologica* 77(2016): 67–74.

Laundre, J.W., Hernandez, L., and Ripple, W.J. (2010). The landscape of fear: ecological implications of being afraid. *The Open Ecology Journal* 3: 1–7.

Lokatis, S., and Jeschke, J.M. (2022). Urban biotic homogenization: approaches and knowledge gaps. *Ecological Applications* 32(8): e2703.

Lokatis, S., Jeschke, J.M., Bernard-Verdier, M., Buchholz, S., Grossart, H.-P., Havemann, F., Hölker, F., Itescu, Y., Kowarik, I., Kramer-Schadt, S., Mietchen, D., Musseau, C.L., Planillo, A., Schittko, C., Straka, T.M., and Heger, T. (2023). Hypotheses in urban ecology: building a common knowledge base. *Biological Reviews* 98(5): 1530–1547.

Lowe, E.C., Wilder, S.M., and Hochuli, D.F. (2014). Urbanisation at multiple scales is associated with larger size and higher fecundity of an orb-weaving spider. *PLoS ONE* 9(8): e105480.

Macgregor, C.J., Pocock, M.J.O., Fox, R., and Evans, D.M. (2015). Pollination by nocturnal Lepidoptera, and the effects of light pollution: a review. *Ecological Entomology* 40(3): 187–198.

Martin, T.E. (1987). Food as a limit on breeding birds: a life-history perspective. *Annual Review of Ecology and Systematics* 18(1): 453–487.

Marzluff, J.M. (2005). Island biogeography for an urbanizing world: how extinction and colonization may determine biological diversity in human-dominated landscapes. *Urban Ecosystems* 8(2): 157–177.

McKinney, M.L. (2006). Urbanization as a major cause of biotic homogenization. *Biological Conservation* 127(3): 247–260.

McLean, M.A., Angilletta, M.J., and Williams, K.S. (2005). If you can't stand the heat, stay out of the city: thermal reaction norms of chitinolytic fungi in an urban heat island. *Journal of Thermal Biology* 30(5): 384–391.

Meineke, E.K., Dunn, R.R., and Frank, S.D. (2014). Early pest development and loss of biological control are associated with urban warming. *Biology Letters* 10(11): 20140586.

Merckx, T., Nielsen, M.E., Heliölä, J., Kuussaari, M., Pettersson, L.B., Pöyry, J., Tiainen, J., Gotthard, K., and Kivelä, S.M. (2021). Urbanization extends flight phenology and leads to local adaptation of seasonal plasticity in Lepidoptera. *Proceedings of the National Academy of Sciences of the United States of America* 118(40): e2106006118.

Meyrier, E., Jenni, L., Bötsch, Y., Strebel, S., Erne, B., and Tablado, Z. (2017). Happy to breed in the city? Urban food resources limit reproductive output in Western Jackdaws. *Ecology and Evolution* 7(5): 1363–1374.

Miller-Rushing, A.J., Høye, T.T., Inouye, D.W., and Post, E. (2010). The effects of phenological mismatches on demography. *Philosophical Transactions of the Royal Society, B: Biological Sciences* 365(1555): 3177–3186.

Montes de Oca-Aguilar, A.C., González-Martínez, A., Chan-González, R., Ibarra-López, P., Smith-Ávila, S., Córdoba-Aguilar, A., and Ibarra-Cerdeña, C.N. (2022). Signs of urban evolution? Morpho-functional traits co-variation along a nature-urban gradient in a chagas disease vector. *Frontiers in Ecology and Evolution* 10: 805040.

Mooney, H.A. and Cleland, E.E. (2001). The evolutionary impact of invasive species. *Proceedings of the National Academy of Sciences of the United States of America* 98(10): 5446–5451.

Morelli, F., Tryjanowski, P., Ibáñez-Álamo, J.D., Díaz, M., Suhonen, J., Møller, A.P., Prosek, J., Moravec, D., Bussière, R., Mägi, M., Kominos, T., Galanaki, A., Bukas, N., Markó, G., Pruscini, F., Reif, J., and Benedetti, Y. (2023). Effects of light and noise pollution on avian communities of European cities are correlated with the species' diet. *Scientific Reports* 13(1): 4361.

Morrissey, C.A., Stanton, D.W.G., Tyler, C.R., Pereira, M.G., Newton, J., Durance, I., and Ormerod, S.J. (2014). Developmental impairment in Eurasian dipper nestlings exposed to urban stream pollutants: effects of urban stream pollutants on dippers. *Environmental Toxicology and Chemistry* 33(6): 1315–1323.

Murphy, S.M., Vyas, D.K., Sher, A.A., and Grenis, K. (2022). Light pollution affects invasive and native plant traits important to plant competition and herbivorous insects. *Biological Invasions* 24(3): 599–602.

Neate-Clegg, M.H.C., Tonelli, B.A., Youngflesh, C., Wu, J.X., Montgomery, G.A., Şekercioğlu, Ç.H., and Tingley, M.W. (2023). Traits shaping urban tolerance in birds differ around the world. *Current Biology* 33(9): 1677–1688.

Neiderud, C.-J. (2015). How urbanization affects the epidemiology of emerging infectious diseases. *Infection Ecology & Epidemiology* 5(1): 27060.

Neil, K., and Wu, J. (2006). Effects of urbanization on plant flowering phenology: a review. *Urban Ecosystems* 9(3): 243–257.

Oke, T.R. (1982). The energetic basis of the urban heat island. *Quarterly Journal of the Royal Meteorological Society* 108(455): 1–24.

Ordóñez-Krasnowski, P.C., Lanati, L.A., Gaspe, M.S., Cardinal, M.V., Ceballos, L.A., and Gürtler, R.E. (2020). Domestic host availability modifies human-triatomine contact and host shifts of the Chagas disease vector *Triatoma infestans* in the humid Argentine Chaco. *Medical and Veterinary Entomology* 34(4): 459–469.

Owens, A.C.S., and Lewis, S.M. (2018). The impact of artificial light at night on nocturnal insects: a review and synthesis. *Ecology and Evolution* 8(22): 11337–11358.

Palacio, F.X., and Ordano, M. (2023). Urbanization shapes phenotypic selection of fruit traits in a seed-dispersal mutualism. *Evolution* 77(8): 1769–1779.

Partecke, J., and Gwinner, E. (2007). Increased sedentariness in European blackbirds following urbanization: a consequence of local adaptation? *Ecology* 88(4): 882–890.

Pelling, M., and Blackburn, S. (2013). *Megacities and the coast: risk, resilience and transformation*. Routledge, Taylor & Francis Group.

Perry, G., Buchanan, B.W., Fisher, R.N., Salmon, M., and Wise, S.E. (2008). Effects of artificial night lighting on amphibians and reptiles in urban environments. *Urban Herpetology* 3: 239–256.

Pickett, S.T.A., and Cadenasso, M.L. (2009). Altered resources, disturbance, and heterogeneity: a framework for comparing urban and non-urban soils. *Urban Ecosystems* 12(1): 23–44.

Pickett, S.T.A., Cadenasso, M.L., Grove, M.J., Nilon, C.H., Pouyat, R.V., Zipperer, W.C., and Costanza, R. (2001). Urban ecological systems: linking terrestrial ecological, physical, and socioeconomic components of metropolitan areas. *Annual Review of Ecological Systems* 32: 127–157.

Pisman, M., Bonte, D., and de la Peña, E. (2020). Urbanization alters plastic responses in the common dandelion *Taraxacum officinale*. *Ecology and Evolution* 10(9): 4082–4090.

Plummer, K.E., Siriwardena, G.M., Conway, G.J., Risely, K., and Toms, M.P. (2015). Is supplementary feeding in gardens a driver of evolutionary change in a migratory bird species? *Global Change Biology* 21(12): 4353–4363.

Rahel, F.J., Keleher, C.J., and Anderson, J.L. (1996). Potential habitat loss and population fragmentation for cold water fish in the North Platte River drainage of the Rocky Mountains: response to climate warming. *Limnology and Oceanography* 41(5): 1116–1123.

Rand, A.S., Bridarolli, M.E., Dries, L., and Ryan, M.J. (1997). Light levels influence female choice in túngara frogs: predation risk assessment? *Copeia* 2: 447–450.

Raupp, M.J., Shrewsbury, P.M., and Herms, D.A. (2010). Ecology of herbivorous arthropods in urban landscapes. *Annual Review of Entomology* 55(1): 19–38.

Rivkin, L.R., Santangelo, J.S., Alberti, M., Aronson, M.F.J., de Keyzer, C.W., Diamond, S.E., Fortin, M.J., Frazee, L.J., Gorton, A.J., Hendry, A.P., Liu, Y., Losos, J.B., MacIvor, J.S., Martin, R.A., McDonnell, M.J., Miles, L.S., Munshi-South, J., Ness, R.W., Newman, A.E.M., Stothart, M.R., Theodorou, P., Thompson, K.A., Verrelli, B.C., Whitehead, A., Winchell, K.M., and Johnson, M.T.J. (2019). A roadmap for urban evolutionary ecology. *Evolutionary Applications* 12(3): 384–398.

Rizwan, A.M., Dennis, L.Y.C., and Liu, C. (2008). A review on the generation, determination and mitigation of Urban Heat Island. *Journal of Environmental Sciences* 20(1): 120–128.

Robertson, B., Kriska, G., Horvath, V., and Horvath, G. (2010). Glass buildings as bird feeders: urban birds exploit insects trapped by polarized light pollution. *Acta Zoologica Academiae Scientiarum Hungaricae* 56(3): 283–293.

Rodewald, A.D., Shustack, D.P., and Jones, T.M. (2011). Dynamic selective environments and evolutionary traps in human-dominated landscapes. *Ecology* 92(9): 1781–1788.

Roy, A.H., Freeman, M.C., Freeman, B.J., Wenger, S.J., Ensign, W.E., and Meyer, J.L. (2005). Investigating hydrologic alteration as a mechanism of fish assemblage shifts in urbanizing streams. *Journal of the North American Benthological Society* 24(3): 656–678.

Russart, K.L.G., and Nelson, R.J. (2018). Artificial light at night alters behavior in laboratory and wild animals. *Journal of Experimental Zoology Part A, Ecological and Integrative Physiology* 329(8–9): 401–408.

Rydell, J. (2006). Bats and their insect prey at streetlights. In: *Ecological consequences of artificial night lighting* (eds. Rich, C., and Longcore, T.), 43–60. Island Press.

Salmón, P., Jacobs, A., Ahrén, D., Biard, C., Dingemanse, N.J., Dominoni, D.M., Helm, B., Lundberg, M., Senar, J.C., Sprau, P., Visser, M.E., and Isaksson, C. (2021). Continent-wide genomic signatures of adaptation to urbanisation in a songbird across Europe. *Nature Communications* 12(1): 2983.

San Martin Y Gomez, G., and van Dyck, H. (2012). Ecotypic differentiation between urban and rural populations of the grasshopper *Chorthippus brunneus* relative to climate and habitat fragmentation. *Oecologia* 169(1): 125–133.

Santana Marques, P., Resende Manna, L., Clara Frauendorf, T., Zandonà, E., Mazzoni, R., and El-Sabaawi, R. (2020). Urbanization can increase the invasive potential of alien species. *Journal of Animal Ecology* 89(10): 2345–2355.

Santangelo, J.S., Miles, L.S., Breitbart, S.T., Murray-Stoker, D., Rivkin, L.R., Johnson, M.T.J. and Ness, R.W. (2020a). Urban environments as a framework to study parallel evolution. In: *Urban evolutionary biology*, (eds. Szulkin, M., Munshi-South, J., and Charmantier, A.), 36–53. Oxford University Press.

Santangelo, J.S., Ness, R.W., Cohan, B., Fitzpatrick, C.R., Innes, S.G., Koch, S., Miles, L.S., Munim, S., Peres-Neto, P.R., Prashad, C., Tong, A.T., Aguirre, W.E., Akinwole, P.O., Alberti, M., Álvarez, J., Anderson, J.T., Anderson, J.J., Ando, Y., Andrew, N.R., Angeoletto, F., Anstett, D.N., Anstett, J., Aoki-Gonçalves, F., Arietta, A.Z.A., Arroyo, M.T.K., Austen, E.J., Baena-Díaz, F., Barker, C.A., Baylis, H.A., Beliz, J.M., Benitez-Mora, A., Bickford, D., Biedebach, G., Blackburn, G.S., Boehm, M.M.A., Bonser, S.P., Bonte, D., Bragger, J.R., Branquinho, C., Brans, K.I., Bresciano, J.C., Brom, P.D., Bucharova, A., Burt, B., Cahill, J.F., Campbell, K.D., Carlen, E.J., Carmona, D., Castellanos, M.C., Centenaro, G., Chalen, I., Chaves, J.A., Chávez-Pesqueira, M., Chen, X.-Y., Chilton, A.M., Chomiak, K.M., Cisneros-Heredia, D.F., Cisse, I.K., Classen, A.T., Comerford, M.S., Fradinger, C.C., Corney, H., Crawford, A.J., Crawford, K.M., Dahirel, M., David, S., De Haan, R., Deacon, N.J., Dean, C., del-Val, E., Deligiannis, E.K., Denney, D., Dettlaff, M.A., DiLeo, M.F., Ding, Y.-Y., Domínguez-López, M.E., Dominoni, D.M., Draud, S.L., Dyson, K., Ellers, J., Espinosa, C.I., Essi, L., Falahati-Anbaran, M., Falcão, J.C.F., Fargo, H.T., Fellowes, M.D.E., Fitzpatrick, R.M., Flaherty, L.E., Flood, P.J., Flores, M.F., Fornoni, J., Foster, A.G., Frost, C.J., Fuentes, T.L., Fulkerson, J.R., Gagnon, E., Garbsch, F., Garroway, C.J., Gerstein, A.C., Giasson, M.M., Girdler, E.B., Gkelis, S., Godsoe, W., Golemiec, A.M., Golemiec, M., González-Lagos, C., Gorton, A.J., Gotanda, K.M., Granath, G., Greiner, S., Griffiths, J.S., Grilo, F., Gundel, P.E., Hamilton, B., Hardin, J.M., He, T., Heard, S.B., Henriques, A.F., Hernández-Poveda, M., Hetherington-Rauth, M.C., Hill, S.J., Hochuli, D.F., Hodgins, K.A., Hood, G.R., Hopkins, G.R., Hovanes, K.A., Howard, A.R., Hubbard, S.C., Ibarra-Cerdeña, C.N., Iñiguez-Armijos, C., Jara-Arancio, P., Jarrett, B.J.M., Jeannot, M., Jiménez-Lobato, V., Johnson, M., Johnson, O., Johnson, P.P., Johnson, R., Josephson, M.P., Jung, M.C., Just, M.G., Kahilainen, A., Kailing, O.S., Kariñho-Betancourt, E., Karousou, R., Kirn, L.A., Kirschbaum, A., Laine, A.-L., LaMontagne, J.M., Lampei, C., Lara, C., Larson, E.L., Lázaro-Lobo, A., Le, J.H., Leandro, D.S., Lee, C., Lei, Y., León, C.A., Lequerica Tamara, M.E., Levesque, D.C., Liao, W.-J., Ljubotina, M., Locke, H., Lockett, M.T., Longo, T.C., Lundholm, J.T., MacGillavry, T., Mackin, C.R., Mahmoud, A.R., Manju, I.A., Mariën, J., Martínez, D.N., Martínez-Bartolomé, M., Meineke, E.K., Mendoza-Arroyo, W., Merritt, T.J.S., Merritt, L.E.L., Migiani, G., Minor, E.S., Mitchell, N., Mohammadi Bazargani, M., Moles, A.T., Monk, J.D., Moore, C.M., Morales-Morales, P.A., Moyers, B.T., Muñoz-Rojas, M., Munshi-South, J., Murphy, S.M., Murúa, M.M., Neila, M., Nikolaidis, O., Njunjić, I., Nosko, P., Núñez-Farfán, J., Ohgushi, T., Olsen, K.M., Opedal, Ø.H., Ornelas, C., Parachnowitsch, A.L., Paratore, A.S., Parody-Merino, A.M., Paule, J., Paulo, O.S., Pena, J.C., Pfeiffer, V.W., Pinho, P., Piot, A., Porth, I.M., Poulos, N., Puentes, A., Qu, J., Quintero-Vallejo, E., Raciti, S.M., Raeymaekers, J.A.M., Raveala, K.M., Rennison, D.J., Ribeiro, M.C., Richardson, J.L., Rivas-Torres, G., Rivera, B.J., Roddy, A.B., Rodriguez-Muñoz, E., Román, J.R., Rossi, L.S., Rowntree, J.K., Ryan, T.J., Salinas, S., Sanders, N.J., Santiago-Rosario, L.Y., Savage, A.M., Scheepens, J.F., Schilthuizen, M., Schneider, A.C., Scholier, T., Scott, J.L., Shaheed, S.A., Shefferson, R.P., Shepard, C.A., Shykoff, J.A., Silveira, G., Smith, A.D., Solis-Gabriel, L., Soro, A., Spellman, K.V., Whitney, K.S., Starke-Ottich, I., Stephan, J.G., Stephens, J.D., Szulc, J., Szulkin, M., Tack, A.J.M., Tamburrino, Í., Tate, T.D., Tergemina, E., Theodorou, P., Thompson, K.A., Threlfall, C.G., Tinghitella, R.M., Toledo-Chelala, L., Tong, X., Uroy, L., Utsumi, S., Vandegehuchte, M.L., VanWallendael, A., Vidal, P.M., Wadgymar, S.M., Wang, A.-Y., Wang, N., Warbrick, M.L., Whitney, K.D., Wiesmeier, M., Wiles, J.T., Wu, J., Xirocostas, Z.A., Yan, Z., Yao, J., Yoder, J.B., Yoshida, O., Zhang, J., Zhao, Z., Ziter, C.D., Zuellig, M.P., Zufall, R.A., Zurita, J.E., Zytynska, S.E., and Johnson, M.T.J. (2022a). Global urban environmental change drives adaptation in white clover. *Science* 375(6586): 1275–1281.

Santangelo, J.S., Rivkin, L.R., Advenard, C., and Thompson, K.A. (2020b). Multivariate phenotypic divergence along an urbanization gradient. *Biology Letters* 16(9): 9–14.

Santangelo, J.S., Roux, C., and Johnson, M.T.J. (2022b). The effects of environmental heterogeneity within a city on the evolution of clines. *Journal of Ecology* 110(12): 2950–2959.

Saul, W.-C., Jeschke, J.M., and Heger, T. (2013). The role of eco-evolutionary experience in invasion success. *NeoBiota* 17(2013): 57–74.

Scalici, M., Macale, D., Schiavone, F., Gherardi, F., and Gibertini, G. (2008). Effect of urban isolation on the dynamics of river crabs. *Fundamental and Applied Limnology* 172(2): 167–174.

Schell, C.J., Dyson, K., Fuentes, T.L., Des Roches, S., Harris, N.C., Miller, D.S., Woelfle-Erskine, C.A., and Lambert, M.R. (2020). The ecological and evolutionary consequences of systemic racism in urban environments. *Science* 369(6510): eaay4497.

Scherner, F., Horta, P.A., de Oliveira, E.C., Simonassi, J.C., Hall-Spencer, J.M., Chow, F., Nunes, J.M.C., and Pereira, S.M.B. (2013). Coastal urbanization leads to remarkable seaweed species loss and community shifts along the SW Atlantic. *Marine Pollution Bulletin* 76(1–2): 106–115.

Schroeder, J., Nakagawa, S., Cleasby, I.R., and Burke, T. (2012). Passerine birds breeding under chronic noise experience reduced fitness. *PLoS ONE* 7(7): e39200.

Sepp, T., McGraw, K.J., Kaasik, A., and Giraudeau, M. (2018). A review of urban impacts on avian life-history evolution: does city living lead to slower pace of life? *Global Change Biology* 24(4): 1452–1469.

Seress, G., Sándor, K., Evans, K.L., and Liker, A. (2020). Food availability limits avian reproduction in the city: an experimental study on great tits *Parus major*. *Journal of Animal Ecology* 89(7): 1570–1580.

Seto, K.C., Güneralp, B., Hutyra, L.R., Guneralp, B., and Hutyra, L.R. (2012). Global forecasts of urban expansion to 2030 and direct impacts on biodiversity and carbon pools. *Proceedings of the National Academy of Sciences of the United States of America* 109(40): 16083–16088.

Shine, R. (2012). Invasive species as drivers of evolutionary change: cane toads in tropical Australia. *Evolutionary Applications* 5(2): 107–116.

Shochat, E., Warren, P.S., Faeth, S.H., McIntyre, N.E., and Hope, D. (2006). From patterns to emerging processes in mechanistic urban ecology. *Trends in Ecology & Evolution* 21(4): 186–191.

Slabbekoorn, H., and Peet, M. (2003). Birds sing at a higher pitch in urban noise. *Nature* 424(6946): 267–267.

Snäll, T., Lehtomäki, J., Arponen, A., Elith, J., and Moilanen, A. (2016). Green infrastructure design based on spatial conservation prioritization and modeling of biodiversity features and ecosystem services. *Environmental Management* 57(2): 251–256.

Somers, K.A., Bernhardt, E.S., Grace, J.B., Hassett, B.A., Sudduth, E.B., Wang, S., and Urban, D.L. (2013). Streams in the urban heat island: spatial and temporal variability in temperature. *Freshwater Science* 32(1): 309–326.

Spelt, A., Soutar, O., Williamson, C., Memmott, J., Shamoun-Baranes, J., Rock, P., and Windsor, S. (2021). Urban gulls adapt foraging schedule to human-activity patterns. *Ibis* 163(1): 274–282.

Sprau, P., Mouchet, A., and Dingemanse, N.J. (2017). Multidimensional environmental predictors of variation in avian forest and city life histories. *Behavioral Ecology* 28(1): 59–68.

Stephens, P.A., Sutherland, W.J., and Freckleton, R.P. (1999). What is the Allee effect? *Oikos* 87(1): 185–190.

Stanley, A., and Arceo-Gómez, G. (2020). Urbanization increases seed dispersal interaction diversity but decreases dispersal success in *Toxicodendron radicans*. *Global Ecology and Conservation* 22: e01019.

Stillfried, M., Gras, P., Börner, K., Göritz, F., Painer, J., Röllig, K., Wenzler, M., Hofer, H., Ortmann, S., and Kramer-Schadt, S. (2017). Secrets of success in a landscape of fear: urban wild boar adjust risk perception and tolerate disturbance. *Frontiers in Ecology and Evolution* 5: 157.

Stotz, G.C., Gianoli, E., and Cahill, J.F. (2016). Spatial pattern of invasion and the evolutionary responses of native plant species. *Evolutionary Applications* 9(8): 939–951.

Stroud, J.T., and Losos, J.B. (2016). Ecological opportunity and adaptive radiation. *Annual Review of Ecology, Evolution, and Systematics* 47(1): 507–532.

Suárez-Rodríguez, M., and Garcia, C.M. (2017). An experimental demonstration that house finches add cigarette butts in response to ectoparasites. *Journal of Avian Biology* 48(10): 1316–1321.

Sun, G., and Lockaby, B.G. (2012). Water quantity and quality at the urban-rural interface. In: *Urban-rural interfaces* (eds. Laband, D.N., Lockaby, B.G., and Zipperer, W.C.), 29–48. American Society of Agronomy, Soil Science Society of America, Crop Science Society of America, Inc.

Szczepaniec, A., Creary, S.F., Laskowski, K.L., Nyrop, J.P., and Raupp, M.J. (2011). Neonicotinoid insecticide imidacloprid causes outbreaks of spider mites on elm trees in urban landscapes. *PLoS ONE* 6(5): e20018.

Szulkin, M., Garroway, C.J., Corsini, M., Kotarba, A.Z., and Dominoni, D. (2020a). How to quantify urbanization when testing for urban evolution? In: *Urban evolutionary biology*, (eds. Szulkin, M., Munshi-South, J., and Charmantier, A.), 13–35. Oxford University Press.

Szulkin, M., Munshi-South, J., and Charmantier, A. (2020b). *Urban evolutionary biology*. Oxford University Press.

Templeton, A.R., Shaw, K., Routman, E., and Davis, S.K. (1990). The genetic consequences of habitat fragmentation. *Annals of the Missouri Botanical Garden* 77(1): 13–27.

Theodorou, P. (2022). The effects of urbanisation on ecological interactions. *Current Opinion in Insect Science* 52: 100922.

Thompson, M.J., Capilla-Lasheras, P., Dominoni, D.M., Réale, D., and Charmantier, A. (2022). Phenotypic variation in urban environments: mechanisms and implications. *Trends in Ecology & Evolution* 37(2): 171–182.

Tiatragul, S., Kurniawan, A., Kolbe, J.J., and Warner, D.A. (2017). Embryos of non-native anoles are robust to urban thermal environments. *Journal of Thermal Biology* 65: 119–124.

Touchard, F., Simon, A., Bierne, N., and Viard, F. (2023). Urban rendezvous along the seashore: ports as Darwinian field labs for studying marine evolution in the Anthropocene. *Evolutionary Applications* 16(2): 560–579.

Tryjanowski, P., Skórka, P., Sparks, T.H., Biaduń, W., Brauze, T., Hetmański, T., Martyka, R., Indykiewicz, P., Myczko, Ł., Kunysz, P., Kawa, P., Czyż, S., Czechowski, P., Polakowski, M., Zduniak, P., Jerzak, L., Janiszewski, T., Goławski, A., Duduś, L., Nowakowski, J.J., Wuczyński, A., and Wysocki, D. (2015). Urban and rural habitats differ in number and type of bird feeders and in bird species consuming supplementary food. *Environmental Science and Pollution Research* 22(19): 15097–15103.

Tüzün, N., Op De Beeck, L., Brans, K.I., Janssens, L., and Stoks, R. (2017). Microgeographic differentiation in thermal performance curves between rural and urban populations of an aquatic insect. *Evolutionary Applications* 10(10): 1067–1075.

Tüzün, N., and Stoks, R. (2021). Lower bioenergetic costs but similar immune responsiveness under a heat wave in urban compared to rural damselflies. *Evolutionary Applications* 14(1): 24–35.

Ushimaru, A., Kobayashi, A., and Dohzono, I. (2014). Does urbanization promote floral diversification? Implications from changes in herkogamy with pollinator availability in an urban-rural area. *The American Naturalist* 184(2): 258–267.

Villalobos-Jiménez, G., and Hassall, C. (2017). Effects of the urban heat island on the phenology of Odonata in London, UK. *International Journal of Biometeorology* 61(7): 1337–1346.

Warren, P.S., Katti, M., Ermann, M., and Brazel, A. (2006). Urban bioacoustics: It's not just noise. *Animal Behaviour* 71(3): 491–502.

Weaving, M.J., White, J.G., Isaac, B., Rendall, A.R., and Cooke, R. (2016). Adaptation to urban environments promotes high reproductive success in the tawny frogmouth (*Podargus strigoides*), an endemic nocturnal bird species. *Landscape and Urban Planning* 150: 87–95.

Westall, T.L., Cypher, B.L., Ralls, K., and Wilbert, T. (2019). Observations of social polygyny, allonursing, extrapair copulation, and inbreeding in urban San Joaquin kit foxes (*Vulpes macrotis mutica*). *The Southwestern Naturalist* 63(4): 271–276.

Wilson, C.J., and Jamieson, M.A. (2019). The effects of urbanization on bee communities depends on floral resource availability and bee functional traits. *PLoS ONE* 14(12): e0225852.

Winchell, K.M., Aviles-Rodriguez, K.J., Carlen, E.J., Miles, L.S., Charmantier, A., De León, L.F., Gotanda, K.M., Rivkin, L.R., Szulkin, M., and Verrelli, B.C. (2022). Moving past the challenges and misconceptions in urban adaptation research. *Ecology and Evolution* 12: e9552.

Wolf, J.M., Jeschke, J.M., Voigt, C.C., and Itescu, Y. (2022). Urban affinity and its associated traits: a global analysis of bats. *Global Change Biology* 28(19): 5667–5682.

Yakub, M., and Tiffin, P. (2017). Living in the city: urban environments shape the evolution of a native annual plant. *Global Change Biology* 23(5): 2082–2089.

Youngsteadt, E., and Terando, A.J. (2020). Ecology of urban climates: the need for landscape biophysics in cities. In: *Urban ecology: its nature and challenges*, (ed. Barbosa, P.), 144–159. CABI.

Závodská, R., Fexová, S., von Wowern, G., Han, G.-B., Dolezel, D., and Sauman, I. (2012). Is the sex communication of two pyralid moths, *Plodia interpunctella* and *Ephestia kuehniella*, under circadian clock regulation? *Journal of Biological Rhythms* 27(3): 206–216.

Zhang, H., He, Y., Yang, J., Mao, H., and Jiang, X. (2022). Contemporary adaptive evolution in fragmenting river landscapes: evidence from the native waterflea *Ceriodaphnia cornuta*. *Journal of Plankton Research* 44(1): 88–98.

Ziska, L.H., Gebhard, D.E., Frenz, D.A., Faulkner, S., Singer, B.D., and Straka, J.G. (2003). Cities as harbingers of climate change: common ragweed, urbanization, and public health. *Journal of Allergy and Clinical Immunology* 111(2): 290–295.

22

Life History and Biological Control

Paul K. Abram[1] and Ryan L. Paul[2]

[1] Agriculture and Agri-Food Canada, Agassiz, BC, Canada
[2] Department of Horticulture, Oregon State University, Corvalis, OR, USA

22.1   Introduction

The life histories of consumers can determine their potential to reduce the populations of their hosts or prey. This is one of the main hypotheses that links life history theory to the practice of biological control. Biological control includes the management of populations of pest organisms (mostly insects, plants or microbes) with their natural enemies (usually insects or microbes) (DeBach 1964, DeBach and Rosen 1991, Heimpel and Mills 2017), which are referred to as biological control 'agents'. Throughout its centuries long history, biological control has been used to manage hundreds of species of pests (*e.g.*, insects, weedy plants) in a wide variety of contexts, including forestry, field and greenhouse food crop production, livestock grazing systems, and the protection of species of conservation concern in natural environments. In general, biological control is practised in one of three ways: augmentation, importation or conservation (Figure 22.1). Outcomes of successful biological control programmes have included reduced pesticide usage, improved crop yield, natural habitat protection and restoration and improvements to human welfare (Heimpel and Mills 2017, Mason 2021). One of the questions often asked by biological control researchers is, 'What are the characteristics of an effective biological control agent?' (Murdoch et al. 1985, Segoli et al. 2023). Life history traits, because they determine how long a biological control agent lives and how much and when it reproduces, are obvious candidates for potentially important characteristics. The study of life history traits in a biological control context could help identify some of the factors responsible for past biological control successes and failures and generate accurate predictions about future biological control programmes. At the same time, because the life histories of biological control agents are subjected to a variety of environments over the course of a biological control agent programme (Figure 22.2), and different life history traits may be important at these different stages of biological control programmes, biological control research provides fertile ground for understanding the evolution of animal life histories.

The basic idea that life history traits could be important for biological control was present from the early days of applying ecological theory to biological control systems. During the 20th century, there was an effort to understand what factors influence how much a pest population is suppressed by natural enemies and how stable the suppression is, which were largely investigated with consumer/resource population dynamics modelling (Nicholson and Bailey 1935, Hassell and May 1973, Beddington et al. 1978; reviewed in Mills and Getz 1996). Many of these population dynamics models included aspects of natural enemy and prey life histories that influenced their population growth rates, such as their reproductive capacities, development rates and mortality rates. However, while these models contained life history-related parameters, they did not explicitly incorporate or test predictions from life history theory.

Subsequently, some groups of organisms used as biological control agents also became important study subjects for the development of life history theory. For example, insect parasitoids are among the most commonly used biological control agents for insect pests (Greathead and Greathead 1992, van Lenteren 2012, Brodeur et al. 2018) and are also a rich testing ground for life history theory (Blackburn 1991a, 1991b, Godfray 1994, Heimpel and Rosenheim 1998, Jervis et al. 2001, Jervis and Ferns 2011, Mayhew 2016). This work discovered a diverse array of life history strategies in these organisms, as well as a

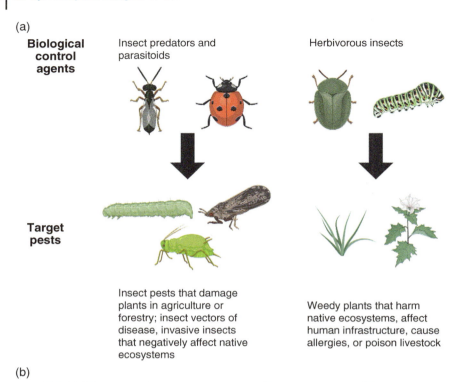

Figure 22.1 Types of biological control agents, target species and approaches are usually studied in relation to life history traits of biological control agents. (a) Common types of natural enemies ('biological control agents') and the pests they attack ('target pests') that are the focus of this chapter. (b) The three main types of biological control. References: 1 – van Hezewijk et al. (2010); 2 – Hoddle et al. (1998); 3 – Brennan (2013). Created with BioRender.com

tendency for these traits to be arranged as 'suites' of related traits that cluster in phylogenetically, biologically and ecologically similar species, as well as a variety of trade-offs (and lack thereof) among different traits that could hypothetically be considered desirable for biological control efficacy (*e.g.*, longevity and reproductive capacity). These findings prompted the question of how variation in life history traits, their arrangements in suites and trade-offs among them relate to observed variation in biological control agents' ability to control their target pests' populations.

However, this question turned out not to be easily answered (Segoli et al. 2023). For example, it has become apparent that there is seldom a clear relationship between the efficacy of biological control agents and any single life history trait (Kimberling 2004, Stiling and Cornelissen 2005, Jarrett and Szűcs 2022). However, it does appear that when focussing on particular target pest taxa, clearer relationships can emerge, such as a positive relationship between parasitoid fecundity

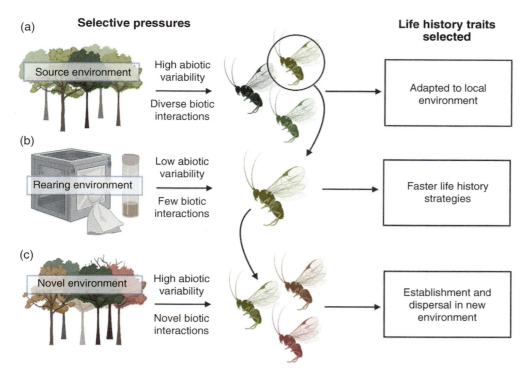

Figure 22.2 Biological control agents are often collected from a natural environment (a), reared for several generations in an indoor facility (b), and then released into a novel environment (c). Agents are selected from a subset of life history trait variation in the source environment (a). This source population then experiences selective pressures during rearing that may alter the life history parameters of the agent in a way that increases its fitness under artificial rearing conditions (b). This agent, bearing the legacy of selection imposed by laboratory rearing, is released into a different context with novel environmental conditions and heterogeneity, where new selection pressures and environmental conditions can further shape life history traits (c). Created with BioRender.com

and the ability to suppress host populations, but only for moth and butterfly pests (Lane et al. 1999, Mills 2001). Nonetheless, it is often the case that very different suites of agent traits can lead to successful biological control (*e.g.*, Murdoch et al. 1985, Janssen and Sabelis 1992). Thus, it appears that a more comprehensive understanding of the link between life history traits and biological control efficacy, and its environmental context-dependency, still needs to be developed if life history theory is going to be usefully applied to improve biological control outcomes.

The foundation for this more comprehensive approach is already beginning to be laid. Recent developments in life history theory that are beginning to be applied to understanding the processes involved in determining outcomes of biological control programmes include: (1) an increased focus on suites, or 'axes' of life history traits, rather than single traits in isolation (Tayeh et al. 2015, Mayhew 2016, Mills and Heimpel 2018, Clark et al. 2022, Guicharnaud et al. 2023); (2) focussing on the relevance of not only inter- but also intraspecific life history trait variation (Lommen et al. 2017, Leung et al. 2020, Clark et al. 2022) and (3) consideration of how environmental context influences life history trade-offs and the optimality of an organism's life history traits for biological control (Segoli and Rosenheim 2013a, Plouvier and Wajnberg 2018, Segoli et al. 2018, Segoli and Wajnberg 2020).

In this chapter, we outline the current understanding of how insect life histories relate to the application of biological control, focussing mostly on importation and augmentative biological control, but with some discussion of conservation biological control as well. We adopt a definition of 'life history traits' that focusses specifically on traits that directly affect the timing and amount of biological control agent survival and reproduction (rates of development, mortality and reproduction over their lifetimes). We do not explicitly focus on other traits such as voracity, attack rates, virulence, dispersal, host specificity or diapause that are sometimes referred to as life history traits in the biological control literature but that we do not consider to be life history traits in the traditional sense. In addition, we centre our discussion on parasitoids, predators and herbivore insects used in biological control of pest arthropods and weeds. Although microbes (*e.g.*, bacteria, viruses, nematodes, fungi) are important groups of biological control agents that have been extensively studied and applied (Brodeur et al. 2018, van Lenteren et al. 2018), we do not include them in this overview.

In addition, we focus on the life histories of biological control agents, rather than those of their target pests. Especially for weed biological control agents, the link between biological control success and the life history of the target plants has been the subject of extensive discussion, and some patterns, albeit inconsistent, between the traits of pest plants and their suitability for biological control have been identified (Crawley 1988, Charudattan 2005). For arthropods, the traits of the target pest have less often been considered as potential determinants of biological control success (Heimpel and Mills 2017), although it has been hypothesized that they have the potential to interact with life history traits of the agent to determine the effectiveness of biological control (*e.g.,* Jarrett and Szűcs 2022). Most biological control research related to life histories, in practice, focusses on selecting among, managing and enhancing the life histories of the biological control agents themselves. This can involve selecting among or managing inter- and intraspecific life history variation in biological control agents, manipulating environments, or adjusting release strategies in a way that considers the life histories of biological control agents.

22.2 Selecting Among Interspecific Life History Variation

In the process of developing a biological control programme against a given pest, researchers would ideally be able to identify the most effective natural enemies from among the pool of candidate species as early as possible. Throughout the history of biological control, several life history traits have been suggested as primary criteria for predicting the success of introduced natural enemies. Indeed, the measurement of basic life history parameters of candidate biological control agents, such as development time and lifetime fecundity has become a routine part of biological control research. Individual traits such as high fecundity and attack rates (Beddington et al. 1978, van Lenteren et al. 2019) as well as aggregate life history metrics that are determined from life history trait performance in a given environment (*e.g.,* intrinsic rate of increase (r_m) and killing potential; Bigler 1989, Tommasini et al. 2004, van Lenteren et al. 2019) have been proposed as key predictors of biological control agent success. However, the diversity of environmental contexts, evolutionary histories and hosts of biological control agents have made it challenging to identify life history traits predictive of success from interspecific variation, and a rather diverse set of natural enemy traits have been associated with both successful and unsuccessful biological control attempts (Segoli et al. 2023). In this section, we explore how life history traits are currently used to evaluate interspecific variation in biological control agents and the potential areas for future progress. First, however, it is important to note a few important differences between the two main types of biological control (*i.e.,* importation, augmentative) that often compare the life histories of different species of agents in order to evaluate their biological control potential.

In general, when evaluating natural enemies for a specific target pest, there is less opportunity to make use of interspecific comparisons among life history traits of candidate importation biological control agents than among augmentative biological control candidates. This is because, for importation biological control, host range (specificity) has become the primary selection criteria among exotic natural enemies due to the ecological risk to non-target species associated with lower specificity (Heimpel and Cock 2018). While there are some loose associations between life history strategies and specificity, these are generally not predictive, so evaluation of individual life history traits of each candidate species is still necessary (Heimpel and Mills 2017). Thus, comparisons of life history traits would be limited to a subset of candidate agents already deemed to have high specificity. Or, in cases where there is only one sufficiently specific candidate agent, life history comparisons with natural enemy species that have been successful in past programmes have the potential to provide *a priori* predictions of whether that agent would be effective or not when released (van Lenteren et al. 2019). However, a more complete understanding of the associations between life history traits and the efficacy of importation biological control agents is still needed to make useful predictions (Segoli et al. 2023).

For augmentative biological control, where specificity is less constraining than for importation biological control because a new species is not typically being irreversibly introduced to the ecosystem, the practical pool of candidate species is usually larger. This can create greater potential for direct interspecific comparisons of life history traits. Two other features of augmentative biological control may also make interspecific comparisons of life history traits more practical as criteria for selecting among agents. First, augmentative agents must be amenable to mass-rearing methods in order to be produced in sufficiently large numbers for release, and this usually means that they must have relatively fast development and high fecundity (van Lenteren et al. 2021). Second, the release environment may often be less heterogeneous compared to areas in which importation biological control is used. This is because augmentation targets a restricted area (*e.g.,* crop field or greenhouse) rather than a wide geographic region (*e.g.,* a country or continent). In addition, the aim is for a rapid reduction in pest populations rather than long-term establishment of the agent. This leaves fewer variables that may impact efficacy and helps create clearer criteria for the success of the biological control releases that can be evaluated on the scale of days to months, rather than years, as is the case for importation biological control. As a result, life history traits of many augmentative biological control candidates can more easily be compared to those of biological control agents already being used for biological control (often closely related species or natural enemies of the same pest), which have been successful in pest control and/or mass rearing (*e.g.*, Albuquerque et al. 1994, Nomikou et al. 2001, Abad-Moyano et al. 2009, Stahl et al. 2019). One of the life history metrics most often used to compare biological control agents in this way is the intrinsic rate of increase (Bigler 1989, Tommasini et al. 2004, van Lenteren et al. 2019).

22.2.1 Intrinsic Rate of Increase as an Aggregate Metric for Comparing the Life Histories of Biological Control Agents

The intrinsic rate of increase, r_m, encompasses information on an insect's development time, longevity and age-specific fecundity, offering a better interspecific comparison of the 'speed' of an insect's life history than the net reproductive rate, which does not account for generation times (Birch 1948). Higher r_m values correspond to faster population growth, which should improve establishment and more rapid suppression of pest populations in the absence of other trade-offs (Birch 1948). Furthermore, natural enemies with r_m values higher than that of the target pest should be capable of suppressing pest populations as their populations should grow faster than the pest (Gassman 1996, Fournet et al. 2000, Nomikou et al. 2001, Abad-Moyano et al. 2009, Mills 2018, Bouvet et al. 2019), but empirical evidence that higher r_m is consistently associated with better biological control efficacy is still lacking. Retrospective stage-structured modelling has supported components of r_m, such as short generation time, as being key factors leading to suppression from exotic natural enemies in importation biological control as well (Mills 2005, Murdoch et al. 2006), though explicit field tests of the impacts of different biological control agents with different r_m values are needed to better support this hypothesis as other trade-offs may be present.

Despite the common use of r_m as part of the evaluation of biological control agents, there are limitations to using it to make predictive comparisons of the relative efficacy of different species. The relationship between life history traits and biological control efficacy can vary based on environmental context, and so defining efficacy of a biological control agent solely based on r_m is often insufficient, especially given that its components are often measured under conditions (*i.e.*, in microcosms in the laboratory) that are not representative of where the biological control agents are used (Janssen and Sabelis 1992). At a minimum, the r_m of a biological control agent should be considered as a ratio in relation to the target pest (which we describe as relative r_m), which might provide a more clear and consistent criteria for natural enemy success (Mills 2018). Relative r_m can be expressed as r_{m_agent}/r_{m_pest} where relative $r_m > 1$ indicates agents that can reproduce more quickly than their host/prey and should be able to drive their target pest's populations to low levels, at least at the scale of local populations with zero or minimal immigration (Janssen and Sabelis 1992). Additionally, even some biological control agents with low relative r_m have the potential to be effective if they have sufficiently high pest attack rates (Yu et al. 2013, van Lenteren et al. 2019, 2021). Lower r_m can also be compensated by higher release rates or a high ability of the biological control agent to persist in times of host scarcity due to either a long lifespan or use of alternative food resources (Janssen et al. 2022, Segoli et al. 2023). However, perhaps the most important criticism of using r_m is that it does not reflect the actual impact of biological control organisms on pest populations because most natural enemy species do not produce offspring for every pest they kill (Tommasini et al. 2004, van Lenteren et al. 2019). For example, a parasitoid species may produce one to several offspring from a single host killed, whereas predators typically consume many prey items in order to produce offspring. As a result, another metric, *i.e.*, kill rate (k_m), has been proposed as an alternative estimate of natural enemy efficacy as it directly quantifies the number of hosts or prey killed per unit time (Tommasini et al., 2004, van Lenteren et al. 2019). Although not yet widely used, high kill rate appears to correspond well with efficacy of biological control agents of *Tuta absoluta* (van Lenteren et al. 2021).

22.2.2 Using Life History Trait Syndromes to Explain Interspecific Patterns

Trade-offs and co-dependence among life history traits likely prevent biological control agents from possessing a combination of all the traits that would be considered 'ideal' for efficacy (Murdoch and Briggs 1996, Leung et al. 2020). This may partially explain the diverse range of individual traits found in successful biological control agents (Segoli et al. 2023). It may be useful to consider life history strategies in terms of 'suites' of traits rather than individual life history traits, and this thinking is beginning to be applied to groups of organisms often used as biological control agents (Mayhew 2016,

Guicharnaud et al. 2023). For insect parasitoids in particular, correlations between developmental strategies and life history traits (sometimes also referred to as 'axes of life history variation') have been long recognized and have been useful in understanding trait co-variation and predicting ecological outcomes (Godfray 1994, Mayhew 2016).

The application of trait syndromes (phenotypic covariance in sets of life history traits) to organisms used in biological control has so far been done mostly in relation to the pace-of-life-hypothesis, which places species along a 'fast–slow continuum' based on reproductive age, frequency of reproduction and fecundity (Stearns 1983, Blackburn 1991b, Tayeh et al. 2015, Guicharnaud et al. 2023). Although similar to the categorization of *K* vs. *r* selected species, the pace-of-life hypothesis is not based on density-dependence but rather on the relative mortality of juveniles *vs.* adults (Blackburn 1991b). 'Fast' species tend to have earlier reproduction and higher fecundity with shorter lifespans compared to 'slow' species (Blackburn 1991b, Tayeh et al. 2015, Guicharnaud et al. 2023). As a result, species on the faster end of the continuum should generally have higher potential for biological control (especially in an augmentative context), although to date only a few studies have explicitly explored the fast–slow continuum in relation to biological control (Tayeh et al. 2015, Guicharnaud et al. 2023). It is notable that placing biological control candidates along the fast–slow continuum is very similar to comparing their r_m (see above), and so the added value of applying pace-of-life hypothesis to evaluating life history variation in biological control agents remains somewhat unclear. While laboratory environments may promote faster life histories, this may not be consistently favourable across biological control systems. For example, environments where high disturbance keeps populations low would likely favour fast strategies where few individuals can contribute quickly to population growth (Fowler 1981, Wright et al. 2019). In higher-density environments facing heavier competition, in contrast, slow types may be more favourable (Wright et al. 2019). In order to better incorporate life history comparisons into the process of selecting among candidate species of biological control agents, we need to improve our understanding of the role of these traits and syndromes in pest control efficacy under different environmental contexts.

There may be some potential to improve comparisons among life history syndromes of different species of biological control agents by relating them to other axes of trait variation that are important to biological control, such as dispersal (Mayhew 2016; see also Chapter 10, this volume) or patch exploitation behaviour (Plouvier and Wajnberg 2018). For example, there is strong support for trade-offs between dispersal and life history traits (*e.g.*, fecundity, age at maturity and survival). However, these trade-offs are sometimes not evident (Guicharnaud et al. 2023) and dispersal ability can vary widely among species with similar life history strategies (Clobert 2012). The lack of support for general relationships between currently used life history metrics and biological control efficacy could be due in part to interactions with these other traits.

22.2.3 Case Study: Life Histories of Some of the Most Commonly Used Augmentative Biological Control Agents

More than 200 species of commercially available arthropods have been released for augmentative biological control programmes around the world (van Lenteren et al. 2021). A select few of these natural enemies hold a large share of the market value and are used against common greenhouse pests such as whiteflies, thrips, aphids and mites (van Lenteren 2012). The extensive use of these select augmentative biological control agents could, in theory, indicate that they possess life history characteristics conducive to biological control efficacy, in particular, high relative r_m (see above). This is often the criteria used in selecting among related biological control agents in augmentative biological control.

The r_m values of frequently used augmentative biological control agents span a range from higher than the target pest to considerably lower. Some augmentative agents have estimated r_m values at least equal to their primary targets. For example, the parasitoid wasp *Aphidius colemani*, a biological control agent used against aphids, has a similar r_m to several of its target pest species (relative $r_m \approx 1$) (Frazer 1972, van Steenis 1993, Torres et al. 2007). *Encarsia formosa*, an important whitefly parasitoid (Figure 22.3), has a considerably higher r_m (relative $r_m > 1$) than whiteflies under equivalent conditions (Vis and van Lenteren 2002). Interestingly, the predatory midge *Aphidoletes aphidimyza* (Figure 22.3) has an intrinsic rate of increase much less than its target pest (relative $r_m < 1$) (Havelka and Zemek 1999, Madahi et al. 2013). Nonetheless, *A. aphidimyza* is one of the most widely used augmentative biological control agents for aphids and has demonstrated high efficacy against several different target species (Barriault et al. 2019, Boulanger et al. 2019). High predation rates could account for the high success despite low relative r_m of some predatory species (van Lenteren et al. 2019), although *A. aphidimyza* may have relatively low predation rates compared to some predators (Farhadi et al. 2011). Alternatively, other behavioural traits such as the avoidance of intraguild predation (predation of other natural enemies) could explain the ability of *A. aphidimyza* to successfully control the pest despite low relative r_m (Barriault et al. 2019).

Figure 22.3 The parasitoid wasp *Encarsia formosa* (adult ovipositing into a whitefly crawler, pictured on the left), a widely used augmentative biological control agent against greenhouse pests, has a higher intrinsic rate of population increase (r_m) than the whitefly pest it is released against. The predatory midge *Aphidoletes aphidimyza* (larvae predating on an aphid pictured on the right), also a popular biological control agent in greenhouses, has a lower r_m than the aphid pests, it is released against. *Source:* Courtesy of Dave Gillespie.

Comparisons of intrinsic rates of increase among these well-studied biological control agents exemplify some of the issues with r_m comparisons across biological control agents and pests. First, many individual species of natural enemies and pests have a wide range of published r_m values as a result of differences in diet or hosts (Park et al. 2011, Riahi et al. 2017) or temperature (van Steenis 1993, Tsueda and Tsuchida 2011). In order to properly compare r_m between species, measurements need to be taken under standardized biotic and abiotic rearing conditions and they should be calculated relative to the pest under equivalent conditions. Without such standardization, r_m measurements can vary greatly, making proper interspecific comparisons unreliable. Thus, despite the widespread adoption of r_m as a composite life history metric for comparison of biological control agents, making use of the metric in integrative analyses of how life history affects biological control performance remains difficult, as most experimental calculations of r_m are performed on individual natural enemy species under specific environmental conditions. Nevertheless, existing evidence suggests that biological control agents with a range of relative r_m values can receive widespread adoption for augmentative biological control programmes.

22.3 Managing or Manipulating Intraspecific Life History Variation

In addition to the large variation in life history traits that exist among insect species used as biological control agents, there is considerable genetically determined life history trait variation within species (Wajnberg 2004, 2010, Lommen et al. 2017, Leung et al. 2020). There is also evidence that this intraspecific variation in life history can rapidly evolve in response to novel selective environments. For example, the same species of natural enemy can evolve considerably different life histories (*e.g.*, reproductive schedules, mortality rates) in different populations through laboratory rearing or introduction into new geographic areas (McEvoy et al. 2012, Tayeh et al. 2012, 2015, Hoffmann and Ross 2018).

Thus, there is potential to 'fine-tune' the selection and evaluation of biological control agents across a variety of contexts through the management and selection of intraspecific life history variants, with the potential to observe and exploit trait 'syndromes' and trade-offs (or lack thereof) among traits. In fact, future selection of biological control agents might be increasingly forced to focus on exploiting intraspecific variation because other factors (*e.g.*, cross-border regulations on movement of non-native species, strict host specificity requirements) limit the pool of available candidate biological control species (and thus the amount of interspecific trait variation) that can be considered (van Lenteren et al. 2011, Moffat et al. 2021). There are at least four main contexts under which measuring, manipulating and/or managing intraspecific variation in life history traits may be considered, which we discuss in the sections below.

22.3.1 Selecting Biological Control Agents from Among Naturally Occurring Intraspecific Life History Variation

When a natural enemy species is being evaluated as an importation or augmentative biological control agent, it may be possible to choose among intraspecific variants to select the ones with life history traits best suited for rearing and release programmes. In importation biological control, it is relatively common to consider multiple biological control agent strains (also referred to as 'biotypes' or 'ecotypes'), often from different geographic source regions, for importation and release (reviewed in Stahlke et al. 2022). However, at least to our knowledge, life history traits have not been the primary driver of the selection of particular biotypes. Rather, considerations such as differences in climatic tolerance, genetic compatibility with pre-existing populations and host specificity (*i.e.*, the degree to which the agent does not attack species other than the target pest) are the main factors underlying such decisions (reviewed in Heimpel and Mills 2017). Host specificity of agents, in particular, has become a dominant factor in decision-making because of the lower ecological risk posed by more host-specific agents (Heimpel and Cock 2018). Therefore, the selection of an agent biotype based on desirable life history traits (*e.g.*, those thought to be related to pest control efficacy) would need to consider how those traits co-vary with the relative host specificity of that biotype.

In augmentative biological control, intraspecific variation in life history traits is not typically taken into account when selecting seed populations for mass rearing in commercial facilities – and, in any case, any intraspecific variation that may be present when a laboratory colony is established is likely to be altered by unintentional selection and drift within a few generations (see below). However, intraspecific variation in such life history traits is known to exist, particularly for agents that are easily reared and kept in multiple lines, such as *Trichogramma* parasitoid wasps (Benvenuto et al. 2012, Guicharnaud et al. 2023; see Smith 1996 and Wajnberg 2004 for a review). Unfortunately, there is seldom data available to link intraspecific differences in life history traits to differences in biological control efficacy in the field (Smith 1996, Lommen et al. 2017; but see Kazmer and Luck 1995 for one example).

While much of the intraspecific variation in life history traits among 'strains' or 'biotypes' of biological control agents has a genetic basis (Leung et al. 2020), it is likely that some of this variation is associated with infection by microbial symbionts (see also Chapter 14, this volume), and this can be relevant for how efficiently agents can be mass-produced for augmentative biological control. For example, infection with symbiotic bacteria such as *Wolbachia* can cause parthenogenesis (production of all-female offspring) in *Trichogramma* parasitoid wasps (Stouthamer et al. 1990, 1999), which is beneficial for the yield of mass rearing because in sexually reproducing strains, male wasps do not produce offspring or kill hosts (Stouthamer 1993). However, *Wolbachia* infection is also associated with longer development times and lower fecundity, making *Wolbachia*-infected strains of *Trichogramma* less efficient to mass produce and potentially less effective when released (*e.g.*, Stouthamer and Luck 1993, Wang and Smith 1996). In addition to the effect of microbial symbionts, variation in microbiomes more generally may explain significant intraspecific variation in life history traits of biological control agents, but this remains mostly unexplored (see Leung et al. 2020).

We currently lack a deep enough understanding of the relationship between biological control agent success and variation in intraspecific traits to effectively design biological control programmes based on biotype-level differences. Past analyses of life history traits relating to the efficacy of importation biocontrol agents to control pests have focussed on variation at the interspecific, rather than intraspecific, level (Segoli et al. 2023; see above). More studies are clearly needed to understand the relationship between intraspecific trait variation and biological control efficacy under realistic settings.

22.3.2 Leveraging Intraspecific Variation in Life History Traits for Breeding Programmes to Improve Biological Control Agent Performance

Selective breeding is an intuitively appealing way to shape the life history traits of biological control agents in order to improve their efficacy. The possibility of selectively breeding biological control agents to improve their efficacy has been explored for decades, mostly for commercially produced augmentative biological control agents, and is reviewed extensively elsewhere (Wilkes 1947, Hoy 1986, Wajnberg 2004, 2010, Lommen et al. 2017, Lirakis and Magalhães 2019, Leung et al. 2020). Some attributes of commercially available biological control agents, notably pesticide resistance, have been improved through selection (Lirakis and Magalhães 2019), but, to our knowledge, are no longer commercially available.

As of the time of the review done by Lirakis and Magalhães (2019), a surprising minority of artificial selection/experimental evolution studies on biological control agents (6 out of more than 150) have targeted life history traits – in the sense that we define them here (faster development time, higher fecundity, more female-biased sex ratios, etc.) – for artificial selection. These studies did sometimes observe responses to artificial selection in these life history traits, as well as correlated responses in other traits in some instances (*e.g.*, Wilkes 1947, Wajnberg 2004, Siddiqui et al. 2015). For example, in one species of ladybird beetle, selection for shorter development time correlated with increased overall prey/host consumption and shorter handling times but with lower attack rates (compared to slow developers) (Siddiqui et al. 2015). However, most of these experimental evolution studies did not test how changes in the selected life history traits translated to differences in biological control performance. More recent studies artificially selected laboratory populations of native parasitoid wasps to increase their performance on an invasive pest (Jarrett et al. 2022, Linder et al. 2022), but did not observe any correlated changes in development time or body size.

While promise remains for using selective breeding to shape the life histories of biological control agents, there is still a pressing need to link differences in life histories to the actual performance of the agents in the context under which they are used. If future studies could produce intraspecific variants of biological control agents with different life histories through experimental evolution, it would provide a unique opportunity to explore intraspecific associations between different suites of life history traits and biological control efficacy. An alternative (albeit challenging) approach would be, reciprocally, to select lines of biological control agents based on their ability to suppress host populations and then explore what life history trait changes are associated with strains that are more effective at host population suppression (Wajnberg 1991).

22.3.3 Preventing Unintentional Artificial Selection in Captive Rearing

In both importation and augmentative biological control, biological control agents that are collected from their natural environment are usually held in captive rearing under artificial conditions for several generations – sometimes for decades – before being released. Captive rearing usually provides constant and enemy-free conditions with relatively high conspecific densities and high resource availability and unintentionally selects for short generation times due to the necessity of rearing in discrete generations (Hoffmann and Ross 2018). There has long been concern that artificial selection, genetic drift and inbreeding depression imposed by laboratory rearing could compromise the establishment ability and efficacy of biological control agents after their release (reviewed in Leung et al. 2020), due in part to alteration of their life history traits. It is also possible that these factors could also make maintenance of intentionally selected traits more challenging (see previous section). Indeed, there is widespread evidence that laboratory rearing affects the life history traits of insects and usually promotes faster life histories with shorter development times, reduced lifespans and higher fecundity in the laboratory environment (Tayeh et al. 2012, Hoffmann and Ross 2018). Some guidance has been developed for how to manage genetic (and consequent phenotypic) variation in colonies of insects used in biological control (*e.g.*, Hopper et al. 1993). For example, colonies can be maintained in isofemale lines and then combined shortly before release. In the absence of testing for phenotypic differences among lines, this maintains genetic variation but does not select for any particular traits. However, these recommendations are not necessarily widely implemented, and the genetics of most commercial and laboratory populations of biological control agents are not intentionally managed, and so are highly likely to bear the genetic and life history signatures of being kept in captivity. However, empirical evidence showing that differences in life history imposed by captivity (relative to a reference wild-type population) actually affect the performance (efficacy) of biological control agents in the environmental context that they are released in is still lacking. This is part of a more general problem that the field performance of biological control agents is still rarely tested as part of their quality control (van Lenteren and Bigler 2010). Future studies on this topic should explore not only how the life histories of biological control agents evolve over time in laboratory colonies but also how these differences in life history traits translate into differences (or lack thereof) in their biological control efficacy.

22.3.4 Monitoring for Evolution of Life History Traits After Biological Control Releases

After biological control agents are released into a new environment, there is the potential for their life histories to evolve over time and space. This is most relevant for importation biological control, where the released agents establish self-sustaining populations. Populations of augmentative biological control agents, in contrast, probably do not persist in the environment long enough to pass through enough generations after their release for life history evolution to take place (Figure 22.1). For importation biological control, because life history variation is hypothesized to be linked to biological control efficacy (see above), it is important to document and understand how life histories of biological control agents evolve after their introduction, as this could lead to changes in biological control efficacy over time (*e.g.*, failures of previously effective biological control agents) and/or to significant regional (*i.e.*, spatial) variation in biological control efficacy. However, it is only relatively recently that rapid life history evolution has begun to be documented in populations of previously released biological control agents (Abram and Moffat 2018).

Life histories of introduced biological control agents may evolve after their introduction due to selection imposed by their new environment (McEvoy et al. 2012). Often, for invasive species, it is argued that successfully establishing and spreading in a new area should be associated with 'faster' life histories (short lifespan, earlier reproduction; Sakai et al. 2001, Davis 2005, Allen et al. 2017). However, a variety of other life history strategies can also be associated with successful invasions (Sakai et al. 2001), depending on the group of organisms being considered. For example, in birds, invasion success is associated with 'hedged' life history strategies, wherein reproduction is delayed and spread out over a longer lifespan (Sol et al. 2012). Few studies have investigated how the life history strategies of intentionally introduced biological control agents have changed (relative to their source population or the laboratory colony from which they were sourced) after their introduction to a new environment (reviewed in McEvoy et al. 2012). In a classic example, McEvoy et al. (2012) demonstrated that mountain populations of an importation biological control agent of weeds, the cinnabar moth, had evolved shorter development times than their valley-bottom counterparts. Although the life history trait examined (development time) did not have a clear relationship with biological control safety or efficacy in this study system, it was suggested that testing for the potential for post-release trait evolution could become a part of pre-release importation biological control evaluations. Another series of studies of accidentally introduced (invasive) North American populations of the ladybird beetle *Harmonia axyridis* showed that insects introduced into a new environment can rapidly evolve life histories that are inconsistent with the fast–slow continuum of life history variation (*i.e.*, both early and sustained reproduction and long lifespan), and that this life history strategy is markedly different both from their native populations (later reproduction and intermediate lifespan) and laboratory-held populations used in biological control (early reproduction and short lifespan) (Tayeh et al. 2012, 2015). The factors involved in this unexpected combination of life history traits may include the purging of deleterious mutations during the invasion process, differences in resource availability between native and invasive ranges or trade-offs with other life history traits that have not been quantified (Tayeh et al. 2015).

After a biological control agent is established in a new area, differences in life history may evolve at the range limits of the population relative to those at the core of the population (Clark et al. 2022). This can occur due to different selective pressures operating at the range edge *vs.* the range core, resulting in faster life histories at the range edge and slower, more competitive life histories at the range core (Phillips et al. 2010). Alternatively, non-adaptive processes such as genetic drift can reduce the fitness of individuals at the range edge, or higher-dispersing range-edge populations created through assortative mating can have lower fecundity due to dispersal-fecundity life history trade-offs (Phillips et al. 2010; see also Chapter 20, this volume). To some extent, these dynamics can be modified by environmental factors, such as environmental gradients, that constrain an agent's expansion (Clark et al. 2022). In biological control systems, where this phenomenon has only just begun to be investigated (Clark et al. 2022), it is not known whether or not these range-core evolutionary dynamics of life histories have any impact on the efficacy of biological control over space and time.

When multiple distinct strains, or closely related species, of a biological control agent are introduced to the same area, it can create opportunities for admixture or hybridization (Stahlke et al. 2022). The life history consequences of admixture among strains can include hybrid vigour (*e.g.*, reduced development time and increased fecundity), hybrid breakdown (*e.g.*, increased mortality or complete reproductive breakdown after the first generation of hybridization) or may have no detectable effect on life history traits (*e.g.*, Benvenuto et al. 2012, Szűcs et al. 2012, Bitume et al. 2017, Clark et al. 2023). In some biological control agents of weeds, changes in life history parameters can also be associated with shifts in preferences for non-target plant species (relative to the target plant species; *e.g.*, Hoffmann et al. 2002, Bitume et al. 2017, Clark et al. 2023). However, hybridization between different biological control agent strains or species has not been shown to cause a complete host shift, wherein hybrid offspring use non-target host plants that neither parent was able to use (Szűcs et al. 2021).

22.3.5 Case Study: Range-Core Evolutionary Dynamics and Hybridization in Biological Control Agents of Tamarisk Shrubs

Six ecotypes, representing a complex of four cryptic species of leaf beetles (*Diorhabda* spp.), were introduced to western North America from source populations across Eurasia beginning in 2001 (reviewed in Bean and Dudley 2018, Stahlke et al. 2022). The goal was to help control invasive woody shrubs in the genus *Tamarisk*, which were causing a broad range of economic and environmental issues including increased fire risk and negative impacts on native flora and fauna. The biological control programme, although it showed some early measures of success, became controversial in large part because of its perceived negative impact on a bird that used *Tamarisk* shrubs as nesting sites (Bean and Dudley 2018).

Figure 22.4 An adult (left) and larva (right) of the northern tamarisk beetle, *Diorhabda carinulata*, on *Tamarisk* shrubs. This species provides the best-studied example of how life histories of biological control agents can evolve after their introduction to a new area. *Sources:* Roman Jashenko / Wikimedia / Public Domain (left); Judy Gallagher / Flickr / CC BY 2.0 (right).

This motivated a series of studies to understand the population biology and life history of rapidly spreading and evolving populations of this biological control agent (Stahlke et al. 2022).

Focussing on one out of the four established species (*Diorhabda carinulata*; see Figure 22.4), Clark et al. (2022) found that, relative to individuals in the population's core (*i.e.*, in locations where it had been established for longer), populations at the range edge had 'faster' life histories (earlier reproduction and higher fecundity), suggesting that natural selection has shaped the life histories at the population's leading edge. Edge populations also had larger body sizes and higher dispersal under certain conditions of mating status and density. Although these life history changes at the range edge have the potential to affect the beetle's biological control efficacy, it is not yet clear whether this is the case.

There was also evidence in this system that hybridization was occurring among three out of the four leaf beetle species to varying extents (Stahlke et al. 2022), validating previous laboratory studies showing the potential for hybridization (Bitume et al. 2017). Collecting naturally occurring hybrids and comparing their life histories in common garden experiments, relative to pure species, revealed that hybridization usually did not have meaningful positive or negative impacts on fecundity or body size (Clark et al. 2022). In addition, although there was an increase in preference for a non-target plant species in some hybrids, their host use patterns were overall similar to pure species (Clark et al. 2023). These findings suggest that, at least for *Tamarisk* biological control, hybridization is unlikely to substantially negatively affect an agent's life history or associated host specificity.

22.4 Using Life History to Inform Environmental Management and Agent Release Strategies

A common theme in this chapter has been the role of environmental context in influencing the strength of the relationships between a biological control agent's efficacy and its life history traits. It follows that environmental conditions could potentially be altered to favour or improve natural enemy life history parameters, often as part of conservation biological control programmes but also in augmentative biological control (Figure 22.1). While environmental conditions impact the fitness of biological control agents (often measured through changes in longevity or reproductive output), this is not the focus of this chapter. In this section, we will focus on the long-term consequences of environmental variation for life history traits and how life history strategies might influence biological control agent success in different environments.

22.4.1 Field Resource Supplementation

Many biological control agents, mostly in their adult life stage, consume food resources other than their target pest and supplementing environments with these resources could boost the efficacy of biological control (however, see Schuldiner-Harpaz et al. 2022). Floral resources, in particular, have received particular research attention because a range of arthropod biological control agents (*e.g.*, Chrysopidae, Syrphidae, mites, parasitoids) use floral nectar and/or pollen as alternative food sources (Lee and Heimpel 2008, Hogg et al. 2011, Delisle et al. 2015, Villa et al. 2016, Riahi et al. 2017, He et al. 2021). Many laboratory experiments have provided support for these alternative foods benefiting life history traits such as fecundity and longevity (Villa et al. 2016, Riahi et al. 2017, Araj et al. 2019, Herz et al. 2021). However, there is mixed support for the benefits of floral resources on foraging natural enemies and the efficacy of biological control agents under more realistic field conditions (Lee and Heimpel 2008, Hogg et al. 2011, Lu et al. 2014, Gurr et al. 2016, Miall et al. 2021, Schuldiner-Harpaz et al. 2022). There is a general assumption that life history characteristics benefiting from these environmental inputs, such as shorter development time, increased longevity and higher fecundity, are beneficial for natural enemies and biological control, but they may not be favourable in all environmental contexts (Blackburn 1991a, Wright et al. 2019, Segoli and Wajnberg, 2020, Schuldiner-Harpaz et al. 2022). For example, longer development times may be favourable under intense intraspecific competition (Wright et al. 2019, Guicharnaud et al. 2023), which could result from low host availability.

The benefits of food resources for biological control agents may especially depend on their egg production and factors limiting reproduction. Female insects are generally expected to be time-limited or egg-limited in their reproductive potential (see also Chapter 4, this volume). Time-limited individuals die before being able to exhaust their egg supply, while egg-limited individuals exhaust their egg supply, either absolutely if they cannot mature additional eggs in adulthood (pro-ovigenic) or temporarily if they can mature additional eggs (synovigenic) (Heimpel and Rosenheim 1998, Rosenheim et al. 2008; see also Chapter 4, this volume). Increasing longevity benefits natural enemies that are more prone to time limitation (*e.g.*, synovigenic species with low host densities), while increased fecundity has the reverse outcome benefiting those facing higher risk of egg limitation (*e.g.*, for pro-ovigenic species or in very high host densities) (Heimpel and Rosenheim 1998, Zhang et al. 2014). For example, for pro-ovigenic insects that cannot mature additional eggs after emergence, increasing longevity may have little effect on population growth, given that the number of hosts attacked is still limited by pre-determined egg production (Jervis et al. 2001, 2007, 2008). Similarly, increases in fecundity alone may not benefit individuals facing high risk of time limitation (*i.e.*, short lifespans or high mortality) as they may perish before using even the normal egg supply, and empirical data suggests time limitation even for pro-ovigenic species may be common (Rosenheim et al. 2008, Segoli and Rosenheim 2013a). Thus, the benefits of supplemental food resources will likely depend on the life histories of the biological control agents involved. More studies on the relationship between these fitness and life history characteristics and biological control efficacy, especially in the field, are needed to understand the impact that these alternative food resources have.

22.4.2 Using Life History Information to Inform Release Strategies

Natural enemy population density is a key factor affecting whether a biological control agent establishes and provides effective control of the target pest. Generally, establishment is positively correlated with release numbers (Fauvergue et al. 2012, Sinclair et al. 2019, Williams et al. 2021; but see Bellows et al. 2006), but this is dependent on the interaction with environmental context and life history (Grevstad 1999). At least some studies have suggested that demographic effects (including life history traits) alone are unlikely to lead to population extinction, but higher population growth rates should increase the likelihood of establishment (Hopper and Roush 1993, Grevstad 1999). High fecundity of the individual organisms released for biological control is especially important at low densities (such as following initial release) where there are few individuals to contribute to population growth (Wright et al. 2019). Additionally, high fecundity could help prevent stochastic effects and environmental variability leading to extinction, which likely play a dominant role in determining establishment success (Grevstad 1999, Shea and Possingham 2000, Fauvergue et al. 2012). Natural enemies with higher population growth rates (*i.e.*, r_m) thus should perform better at lower release rates than species with lower growth rates (Plouvier and Wajnberg 2018, Wright et al. 2019), but, as stated earlier in this chapter, releasing higher numbers may still allow slower growth rate species to be effective in biological control (Janssen et al. 2022). Lastly, since prey/host availability *per capita* increases with lower release rates, the realized fecundity of natural enemies may also increase, offering some explanation for the equivalent success of some agents in augmentative biological control at both low and high release rates (Hoddle et al. 1998, Alomar et al. 2006, Crowder 2007).

22.4.3 Life History Traits and the Resilience of Biological Control to Climate Change

Climatic conditions can affect the establishment and efficacy of biological control agents due to effects on their life histories (see Chapter 18, this volume). For example, temperature has a well-known impact on the development of insects, and studying the thermal threshold for biological control agents to ensure they match with intended release climates is common practice (Torres et al. 2007, Carrasco et al. 2015, Cowie et al. 2016, Falla et al. 2019, Fischbein et al. 2019, Wang et al. 2022). While life history traits are not known to have any broad relationships to temperature sensitivity, certain life history strategies are likely to be more resilient to temperature stress, including heat waves, which are expected to increase in frequency as a result of climate change (Harvey et al. 2020). Empirical evidence suggests that immature stages are likely more susceptible than adults to extreme heat stress (Zhang et al. 2019, Walzer et al. 2022); thus, natural enemies with shorter development times (and thus shorter periods of exposure to susceptible stages) may perform better under frequent heat waves (Zhang et al., 2019). However, reductions in pest development times also reduce exposure time to biological control agents that attack immature stages (Benrey and Denno 1997, Walzer et al. 2022). In some mites, shortening development time is a plastic response to heat stress (Walzer et al. 2022). Life history theory predicts that this reduction in development time will have a stronger overall positive effect on growth rates for species with higher reproductive output (Caswell 1982, Walzer et al. 2022), thus generally favouring faster life history strategies under heat stress. However, to our knowledge, explicit comparisons of which life history syndromes of biological control agents perform best under heat stress have not been performed but could provide valuable insights into the resilience of biological control agents in a changing climate.

22.4.4 Life History Traits and the Resilience of Biological Control to Insecticides

An apparent impediment to the establishment and survival of some biological control agents is the use of other pest management strategies within the same system. Perhaps the most obvious of these is the use of insecticides in agricultural systems, which can kill not only pests but also their natural enemies, potentially to differing degrees. In general, insect species with higher growth rates, shorter development times and fewer development stages (essentially 'fast' species) may be more resilient to insecticides (Stark et al. 2004), although individuals with less exposed environments as immatures could sustain greater variation in life history traits in the face of insecticide applications. Insecticide sprays may act in a similar way to other stochastic mortality effects that greatly reduce population sizes, favouring individuals that reproduce rapidly (Grevstad 1999, Plouvier and Wajnberg 2018, Wright et al. 2019). Thus, biological control agents with r_m values greater than their target pest, all else being equal, should also be more resilient to insecticide sprays. However, while high-r_m natural enemies are potentially more resilient to insecticide mortality over the longer term, initial knockdown of their populations may be proportionally larger than lower-r_m species (Stark et al. 2007). Indeed, because lost reproduction per individual is greater for higher-r_m species, a 50% reduction in the population of a high-r_m species will cause a proportionally larger reduction in population size and growth rate than an initial 50% reduction in a low-r_m species with a smaller population size (Stark et al. 2007). These interactions between sudden mortality and population growth rates demonstrate the importance of including life history strategies in our understanding of the impacts of insecticides in tandem with biological control. In addition to the other aspects of life history trait selection in biological control, relative r_m may be valuable in determining the relative effect of insecticides on biological control agents and their target pests, assuming the mode of action of the pesticide affects the pest and biological control agent in a similar way. Furthermore, the use of life table response experiments that incorporate r_m (or other life history traits/metrics) into evaluations of pesticide impacts along with matrix models can aid in understanding interactions between biological control and insecticide applications (Stark et al. 2007).

22.4.5 Case Study: The Interaction Between Environmental Conditions and Life Histories of *Anagrus* Parasitoids of Leafhopper Eggs

Life history strategies, including adaptive responses in life history traits to resource availability, can greatly affect the success of biological control agents in the field. Insects should evolve life history strategies that reduce the risk of resource limitation (Heimpel and Rosenheim 1998, Segoli and Rosenheim 2013a). For example, insect parasitoids have long served as models for understanding the dynamics of life history theory and resource limitation (Blackburn 1991a, Godfray, 1994, Rosenheim et al. 2008, Jervis and Ferns 2011, Segoli and Wajnberg 2020) and provide valuable insight for these theories in biological control. Parasitoid reproductive success is considered to be limited by either time or egg limitation (see above). Thus, the availability of hosts as well as resources in the environment that can improve egg availability or longevity (*e.g.*, food for the adult stage) will be critical to the success of parasitoids (and other insects) in biological control.

Important progress in understanding these dynamics in field populations has been made with *Anagrus* parasitoids of *Erythroneura* leafhopper eggs (Segoli and Rosenheim 2013a, 2013b, Segoli 2016, Segoli et al. 2018) (Figure 22.5). Populations of *A. daanei* (a pro-ovigenic species) from agricultural systems, where hosts are abundant, were predicted to emerge with larger egg loads than populations from natural areas where hosts are more scarce (Segoli and Rosenheim 2013a). Field studies generally supported these predictions: parasitoids from vineyards (high host availability) had higher egg loads than those

Figure 22.5 The parasitoid wasp *Anagrus daanei* is a biological control agent of leafhopper (*Erythroneura* spp.) eggs in grape vineyards. The wasp lays its eggs inside the leafhoppers' eggs (left, middle), killing them as a result of its offspring's development. The adult parasitoids can consume sugary liquids as food to boost their longevity (right). *Source:* Courtesy of Houston Wilson.

from natural areas (low host availability) (Segoli and Rosenheim 2013a, Segoli et al. 2018). Given that egg loads of pro-ovigenic species are determined before emergence rather than being plastic responses to host availability, this suggests that local adaptation could be responsible for differences in egg loads and fecundity (Segoli et al. 2018). However, there was poor support for the expected negative correlation between host density (where egg loads were higher) and longevity, possibly due to low food availability leading to selection for higher initial resource allocation (fecundity) instead of longevity (Segoli and Rosenheim 2013b, Segoli and Wajnberg 2020; see also Chapter 4, this volume). Under these conditions, egg load resources may trade-off with egg size rather than energy reserves (fecundity) (Segoli and Wajnberg 2020). This has potentially important implications for biological control, indicating a strong effect of host density on life history traits and the strategies that will be successful (Segoli and Rosenheim 2013a, Segoli 2016), as well as supporting the addition of floral resources for field parasitoid populations (Segoli and Rosenheim 2013b).

22.5 Future Directions and Conclusions

The idea that there is a single set of life history characteristics that consistently makes a biological control organism effective has not been borne out by empirical or theoretical research (Murdoch et al. 1985, Janssen and Sabelis 1992, Segoli et al. 2023). The picture that now seems to be emerging is that, for a given pest problem, there is probably more than one set of life history traits possessed by a biological control organism that could lead to effective control. In addition, the relationship between the suitability of a biological control organism and its life history depends heavily on the type of biological control being practised (Figure 22.1) and its associated environmental context (Figure 22.2). The biggest challenge that remains for using life history traits to select and evaluate biological control agents, in our view, is a lack of knowledge as to what the ideal 'biological control life histories' actually are in specific contexts: that is, a lack of quantitative empirical analyses of what suites of life history traits (and how they interact with other important traits and environmental variables) are correlated with the amount of pest suppression provided by a biological control agent, for particular pest systems.

What is needed to rectify this issue? The first step would be to develop and validate hypotheses for what an ideal set of life history traits would be for a biological control agent for a particular pest problem and environmental context (Plouvier and Wajnberg 2018). Here, the metric for biological control efficacy could be economic (*e.g.*, predicted increase in crop yield) and/or related to the amount of pest suppression (*e.g.*, how much lower the pest population is in the presence *vs.* absence of the agent; see Beddington et al. 1978). The second step would be to quantify the inter- and intraspecific life history variation actually present in natural enemies of the target pest, which could then be expressed along axes of life history variation (*e.g.*, the slow–fast continuum) or developed empirically using dimension reduction techniques (*e.g.*, Ellers et al. 2018). In addition to life history traits, other axes of trait variation relevant to the biological control context would also need to be quantified. This could be, for example, variation in important behavioural traits such as dispersal or patch exploitation strategies, or variation in ecological traits such as host specificity that are important practical filters on biological control candidates. The third step would be to plot the observed life history variation against the other relevant axis (or axes) of trait variation and select from among or further manipulate the inter- and intraspecific variation present in candidate biological control agents in the ways we have described in this chapter. Lastly, methods of release could be improved in ways that compensate for suboptimality of life history traits by employing environmental modifications, such as supplementation of food resources, that could make environmental conditions more suitable for the life history traits found in the selected agent (see Figure 22.6 for a summary). In addition to identifying agents (and intraspecific variants) that have life histories that are close to predicted pest control optima, this approach could help identify whether there are multiple candidate biological control agents for a given pest that may have different, but potentially complementary, life histories.

The relative importance of axes of life history variation and the other axes of trait variation will depend on the type of biological control (Figure 22.1). For example, for importation biological control, practitioners would be limited to selecting among life history variations within species that have high specificity. However, practitioners of augmentative biological control would be able to select among organisms with a broader range of specificity. In fact, in some cases, more generalist agents would actually be preferable because the market for their sale would be larger. However, selection of insects for use in augmentative biological control would typically be more restricted by axes of life history trait variation, particularly the pace-of-life axis, because commercially produced biological control agents typically need to have relatively short development time, and laboratory rearing tends to select for faster life histories (Figure 22.2).

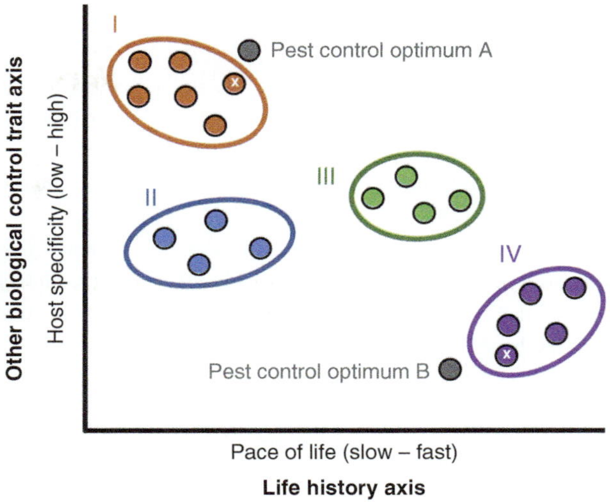

The approach described above is highly challenging and data-hungry. In particular, it will require improvements and extensions to existing datasets that compile life history information about biological control agents (Segoli et al. 2023), continued theoretical work to generate predictions about life history optima for biological control (Plouvier and Wajnberg 2018) and empirical studies to measure trait variation of many species (both inter- and intraspecifically) and quantify pest suppression in realistic or at least semi-realistic settings, to validate these predictions. High-throughput phenotyping approaches are beginning to be developed and applied to biological control systems, particularly for behavioural traits (*e.g.*, Burte et al. 2023, Cointe et al. 2023), which could make this task more feasible in the future if these approaches are creatively extended to facilitate the measurement of life history traits and the quantification of pest population suppression by natural enemies. Quantifying the variation in life history strategies of biological control agents, how they evolve in different environments, and how they relate to successful pest control will no doubt continue to occupy researchers. We anticipate that this research will contribute not only to improving the application of life history to sustainable pest management but will also contribute to empirical and theoretical advances in life history theory more broadly.

Figure 22.6 An example of how understanding the role of life history variation in biological control efficacy could aid in the selection of biological control agents from among a pool of available life histories, represented in 'trait space' along an axis of life history variation (here, the pace of life as one example) and another trait axis relevant to biological control (here, host specificity). Each coloured oval represents the trait space occupied by each of four natural enemy species (I, II, III, IV), with each point within the ovals representing an intraspecific variant or 'biotype' of that species. The grey points represent two hypothetical experimentally or theoretically determined trait space optima (A, B) for control of the target pest in a given environmental context. Selecting the highest performing biological control agents based on their traits could, for example, involve: (1) selecting among interspecific variation (I, IV) for traits that are closest to the optima; (2) choosing intraspecific variants with traits that are closest to the optima (labelled with white 'x'); (3) selectively breeding natural enemies so their trait spaces come closer to or encompass the biological control optima; and/or (4) manipulating the environment to reduce the distance in trait-space between the species' traits and the trait optima.

Acknowledgements

We are grateful to the editors, Michal Segoli and Eric Wajnberg, for inviting us to contribute to this volume. We thank Jacques Brodeur and Yonathan Uriel for their helpful comments on an earlier version of the chapter. Thanks to Houston Wilson and Dave Gillespie for providing images, and Briana Price for help with designing figures.

References

Abad-Moyano, R., Pina, T., Ferragut, F., and Urbaneja, A. (2009). Comparative life-history traits of three phytoseiid mites associated with *Tetranychus urticae* (Acari: Tetranychidae) colonies in clementine orchards in eastern Spain: implications for biological control. *Experimental and Applied Acarology* 47: 121–132.

Abram, P.K., and Moffat, C.E. (2018). Rethinking biological control programs as planned invasions. *Current Opinion in Insect Science* 27: 9–15.

Albuquerque, G.S., Tauber, C.A., and Tauber, M.J. (1994). *Chrysoperla externa* (Neuroptera: Chrysopidae): life history and potential for biological control in central and South America. *Biological Control* 4: 8–13.

Allen, W.L., Street, S.E.and Capellini, I. (2017). Fast life history traits promote invasion success in amphibians and reptiles. *Ecology Letters* 20: 222–230.

Alomar, O., Riudavets, J., and Castañe, C. (2006). *Macrolophus caliginosus* in the biological control of *Bemisia tabaci* on greenhouse melons. *Biological Control* 36: 154–162.

Araj, S.-E., Shields, M.W., and Wratten, S.D. (2019). Weed floral resources and commonly used insectary plants to increase the efficacy of a whitefly parasitoid. *BioControl* 64: 553–561.

Barriault, S., Fournier, M., Soares, A.O., and Lucas, E. (2019). *Leucopis glyphinivora*, a potential aphidophagous biocontrol agent? Predation and comparison with the commercial agent *Aphidoletes aphidimyza*. *BioControl* 64: 21–31.

Bean, D., and Dudley, T. (2018). A synoptic review of *Tamarix* biocontrol in North America: tracking success in the midst of controversy. *BioControl* 63: 361–376.

Beddington, J.R., Free, C.A., and Lawton, J.H. (1978). Characteristics of successful natural enemies in models of biological control of insect pests. *Nature* 273: 513–519.

Bellows, T.S., Paine, T.D., Bezark, L.G., and Ball, J. (2006). Optimizing natural enemy release rates, and associated pest population decline rates, for *Encarsia inaron* walker (Hymenoptera: Aphelinidae) and *Siphoninus phillyreae* (Haliday) (Homoptera: Aleyrodidae). *Biological Control* 37: 25–31.

Benrey, B., and Denno, R.F. (1997). The slow-growth-high-mortality hypothesis: a test using the cabbage butterfly. *Ecology* 78: 987–999.

Benvenuto, C., Tabone, E., Vercken, E., Sorbier, N., Colombel, E., Warot, S., Fauvergue, X., and Ris, N. (2012). Intraspecific variability in the parasitoid wasp *Trichogramma chilonis*: can we predict the outcome of hybridization? *Evolutionary Applications* 5: 498–510.

Bigler, F. (1989). Quality assessment and control in entomophagous insects used for biological control. *Journal of Applied Entomology* 108: 390–400.

Birch, L.C. (1948). The intrinsic rate of natural increase of an insect population. *Journal of Animal Ecology* 17: 15–26.

Bitume, E.V., Bean, D., Stahlke, A.R., and Hufbauer, R.A. (2017). Hybridization affects life-history traits and host specificity in *Diorhabda* spp. *Biological Control* 111: 45–52.

Blackburn, T.M. (1991a). A comparative examination of life-span and fecundity in parasitoid Hymenoptera. *Journal of Animal Ecology* 60: 151–164.

Blackburn, T.M. (1991b). Evidence for a 'fast-slow' continuum of life-history traits among parasitoid Hymenoptera. *Functional Ecology* 5: 65–74.

Boulanger, F.-X., Jandricic, S., Bolckmans, K., Wäckers, F.L., and Pekas, A. (2019). Optimizing aphid biocontrol with the predator *Aphidoletes aphidimyza*, based on biology and ecology. *Pest Management Science* 75: 1479–1493.

Bouvet, J.P.R., Urbaneja, A., Perez-Hedo, M., and Monzo, C. (2019). Contribution of predation to the biological control of a key herbivorous pest in citrus agroecosystems. *Journal of Animal Ecology* 88: 915–926.

Brennan, E.B. (2013). Agronomic aspects of strip intercropping lettuce with alyssum for biological control of aphids. *Biological Control* 65: 302–311.

Brodeur, J., Abram, P.K., Heimpel, G.E., and Messing, R.H. (2018). Trends in biological control: public interest, international networking and research direction. *BioControl* 63: 11–26.

Burte, V., Cointe, M., Perez, G., Mailleret, L., and Calcagno, V. (2023). When complex movement yields simple dispersal: behavioural heterogeneity, spatial spread and parasitism in groups of micro-wasps. *Movement Ecology* 11: 13.

Carrasco, D., Larsson, M.C., and Anderson, P. (2015). Insect host plant selection in complex environments. *Current Opinion in Insect Science* 8: 1–7.

Caswell, H. (1982). Life history theory and the equilibrium status of populations. *The American Naturalist* 120: 317–339.

Charudattan, R (2005). Ecological, practical, and political inputs into selection of weed targets: what makes a good biological control target? *Biological Control* 35: 183–196.

Clark, E.I., Bitume, E.V., Bean, D.W., Stahlke, A.R., Hohenlohe, P.A., and Hufbauer, R.A. (2022). Evolution of reproductive life-history and dispersal traits during the range expansion of a biological control agent. *Evolutionary Applications* 15: 2089–2099.

Clark, E.I., Stahlke, A.R., Gaskin, J.F., Bean, D.W., Hohenlohe, P.A., Hufbauer, R.A., and Bitume, E.V. (2023). Fitness and host use remain stable in a biological control agent after many years of hybridization. *Biological Control* 177: 105102.

Clobert, J. (2012). *Dispersal ecology and evolution*. Oxford University Press.

Cointe, M., Burte, V., Perez, G., Mailleret, L., and Calcagno, V. (2023). A double-spiral maze and hi-resolution tracking pipeline to study dispersal by groups of minute insects. *Scientific Reports* 13: 5200.

Cowie, B.W., Venturi, G., Witkowski, E.T.F., and Byrne, M.J. (2016). Does climate constrain the spread of *Anthonomus santacruzi*, a biological control agent of *Solanum mauritianum*, in South Africa? *Biological Control* 101: 1–7.

Crawley, M. (1988). Plant life-history and the success of weed biological control projects. Proceedings of the VII international symposium on the biological control of weeds, pp. 17–26.

Crowder, D.W. (2007). Impact of release rates on the effectiveness of augmentative biological control agents. *Journal of Insect Science* 7: 15.

Davis, H.G. (2005). r-Selected traits in an invasive population. *Evolutionary Ecology* 19: 255–274.

DeBach, P. (1964). *Biological control of insect pests and weeds*. Chapman and Hall.

DeBach, P., and Rosen, D. (1991). *Biological control by natural enemies*, 2nd ed. Cambridge University Press.

Delisle, J.F., Shipp, L., and Brodeur, J. (2015). Apple pollen as a supplemental food source for the control of western flower thrips by two predatory mites, *Amblyseius swirskii* and *Neoseiulus cucumeris* (Acari: Phytoseiidae), on potted chrysanthemum. *Experimental and Applied Acarology* 65: 495–509.

Ellers, J., Berg, M.P., Dias, A.T.C., Fontana, S., Ooms, A., and Moretti, M. (2018). Diversity in form and function: vertical distribution of soil fauna mediates multidimensional trait variation. *Journal of Animal Ecology* 87: 933–944.

Falla, C., Najar-Rodriguez, A., Minor, M., Harrington, K., Paynter, Q., and Wang, Q. (2019). Effects of temperature, photoperiod and humidity on the life history of *Gargaphia decoris*. *BioControl* 64: 633–643.

Farhadi, R., Gholizadeh, M., Chi, H., Mou, D.-F., Allahyari, H., Yu, J.-Z., Huang, Y.-B., and Yang, T.-C. (2011). Finite predation rate: a novel parameter for the quantitative measurement of predation potential of predator at population level. *Nature Precedings*. https://doi.org/10.1038/npre.2011.6651.1.

Fauvergue, X., Vercken, E., Malausa, T., and Hufbauer, R.A. (2012). The biology of small, introduced populations, with special reference to biological control. *Evolutionary Applications* 5: 424–443.

Fischbein, D., Lantschner, M.V., and Corley, J.C. (2019). Modelling the distribution of forest pest natural enemies across invaded areas: towards understanding the influence of climate on parasitoid establishment success. *Biological Control* 132: 177–188.

Fournet, S., Stapel, J.O., Kacem, N., Nenon, J.-P., and Brunel, E. (2000). Life history comparison between two competitive *Aleochara* species in the cabbage root fly, *Delia radicum*: implications for their use in biological control. *Entomologia Experimentalis et Applicata* 96: 205–211.

Fowler, C.W. (1981). Density dependence as related to life history strategy. *Ecology* 62: 602–610.

Frazer, B.D. (1972). Life tables and intrinsic rates of increase of apterous black bean aphids and pea aphids, on broad bean (Homoptera: Aphididae). *The Canadian Entomologist* 104: 1717–1722.

Gassman, A. (1996). Classical biological control of weeds with insects: a case for emphasizing agent demography. Proceedings of the IX International symposium on biological control, pp. 19–26.

Godfray, H.C.J. (1994). *Parasitoids: behavioral and evolutionary ecology*. Princeton University Press.

Greathead, D.J., and Greathead, A.H. (1992). Biological control of insect pests by insect parasitoids and predators: the BIOCAT database. *Biocontrol News and Information* 13: 61N–68N.

Grevstad, F.S. (1999). Factors influencing the chance of population establishment: implications for release strategies in biocontrol. *Ecological Applications* 9: 1439–1447.

Guicharnaud, C., Groussier, G., Beranger, E., Lamy, L., Vercken, E., and Dahirel, M. (2023). Life history traits, pace-of-life and dispersal among and within five species of *Trichogramma* wasps: a comparative analysis. *Peer Community Journal* 3: e57.

Gurr, G.M., Lu, Z., Zheng, X., Xu, H., Zhu, P., Chen, G., Yao, X., Cheng, J., Zhu, Z., Catindig, J.L., Villareal, S., van Chien, H., Cuong, L.Q., Channoo, C., Chengwattana, N., Lan, L.P., Hai, L.H., Chaiwong, J., Nicol, H.I., Perovic, D.J., Wratten, S.D., and Heong, K.L. (2016). Multi-country evidence that crop diversification promotes ecological intensification of agriculture. *Nature Plants* 2: 16014.

Harvey, J.A., Heinen, R., Gols, R., and Thakur, M.P. (2020). Climate change-mediated temperature extremes and insects: from outbreaks to breakdowns. *Global Change Biology* 26: 6685–6701.

Hassell, M.P., and May, R.M. (1973). Stability in insect host-parasite models. *Journal of Animal Ecology* 42: 693–726.

Havelka, J., and Zemek, R. (1999). Life table parameters and oviposition dynamics of various populations of the predacious gall-midge *Aphidoletes aphidimyza*. *Entomologia Experimentalis et Applicata* 91: 483–486.

He, X., Kiær, L.P., Jensen, P.M., and Sigsgaard, L. (2021). The effect of floral resources on predator longevity and fecundity: a systematic review and meta-analysis. *Biological Control* 153: 104476.

Heimpel, G.E., and Cock, M.J.W. (2018). Shifting paradigms in the history of classical biological control. *BioControl* 63: 27–37.

Heimpel, G.E., and Mills, N.J. (2017). *Biological control: ecology and applications*. Cambridge University Press.

Heimpel, G.E., and Rosenheim, J.A. (1998). Egg limitation in parasitoids: a review of the evidence and a case study. *Biological Control* 11: 160–168.

Herz, A., Dingeldey, E., and Englert, C. (2021). More power with flower for the pupal parasitoid *Trichopria drosophilae*: a candidate for biological control of the spotted wing drosophila. *Insects* 12: 628.

Hoddle, M.S., van Driesche, R.G., and Sanderson, J.P. (1998). Biology and use of the whitefly parasitoid *Encarsia formosa*. *Annual Review of Entomology* 43: 645–669.

Hoffmann, J.H., Impson, F.A.C., and Volchansky, C.R. (2002). Biological control of cactus weeds: implications of hybridization between control agent biotypes. *Journal of Applied Ecology* 39: 900–908.

Hoffmann, A.A., and Ross, P.A. (2018). Rates and patterns of laboratory adaptation in (mostly) insects. *Journal of Economic Entomology* 111: 501–509.

Hogg, B.N., Bugg, R.L., and Daane, K.M. (2011). Attractiveness of common insectary and harvestable floral resources to beneficial insects. *Biological Control* 56: 76–84.

Hopper, K.R., and Roush, R.T. (1993). Mate finding, dispersal, number released, and the success of biological control introductions. *Ecological Entomology* 18: 321–331.

Hopper, K.R., Roush, R.T., and Powell, W. (1993). Management of genetics of biological-control introductions. *Annual Review of Entomology* 38: 27–51.

Hoy, M.A. (1986). Use of genetic improvement in biological control. *Agriculture, Ecosystems & Environment* 15: 109–119.

Janssen, A., Fonseca, M.M., Marcossi, I., Kalile, M.O., Cardoso, A.C., Walerius, A.H., Hanel, A., Marques, V., Ferla, J.J., Farias, V., Carbajal, P.A.F., Pallini, A., and Nachman, G. (2022). Estimating intrinsic growth rates of arthropods from partial life tables using predatory mites as examples. *Experimental and Applied Acarology* 86: 327–342.

Janssen, A., and Sabelis, M.W. (1992). Phytoseiid life-histories, local predator-prey dynamics, and strategies for control of tetranychid mites. *Experimental & Applied Acarology* 14: 233–250.

Jarrett, B.J.M., Linder, S., Fanning, P.D., Isaacs, R., and Szűcs, M. (2022). Experimental adaptation of native parasitoids to the invasive insect pest, *Drosophila suzukii*. *Biological Control* 167: 104843.

Jarrett, B.J.M., and Szűcs, M. (2022). Traits across trophic levels interact to influence parasitoid establishment in biological control releases. *Ecology and Evolution* 12: e8654.

Jervis, M.A., Boggs, C.L., and Ferns, P.N. (2007). Egg maturation strategy and survival trade-offs in holometabolous insects: a comparative approach: life-history strategies in insects. *Biological Journal of the Linnean Society* 90: 293–302.

Jervis, M.A., Ellers, J., and Harvey, J.A. (2008). Resource acquisition, allocation, and utilization in parasitoid reproductive strategies. *Annual Review of Entomology* 53: 361–385.

Jervis, M.A., and Ferns, P.N. (2011). Towards a general perspective on life-history evolution and diversification in parasitoid wasps: parasitoid wasp life-history predictors. *Biological Journal of the Linnean Society* 104: 443–461.

Jervis, M.A., Heimpel, G.E., Ferns, P.N., Harvey, J.A., and Kidd, N.A.C. (2001). Life-history strategies in parasitoid wasps: a comparative analysis of 'ovigeny'. *Journal of Animal Ecology* 70: 442–458.

Kazmer, D.J., and Luck, R.F. (1995). Field tests of the size-fitness hypothesis in the egg parasitoid *Trichogramma pretiosum*. *Ecology* 76: 412–425.

Kimberling, D.N. (2004). Lessons from history: predicting successes and risks of intentional introductions for arthropod biological control. *Biological Invasions* 6: 301–318.

Lane, S.D., Mills, N.J., and Getz, W.M. (1999). The effects of parasitoid fecundity and host taxon on the biological control of insect pests: the relationship between theory and data. *Ecological Entomology* 24: 181–190.

Lee, J.C., and Heimpel, G.E. (2008). Floral resources impact longevity and oviposition rate of a parasitoid in the field. *Journal of Animal Ecology* 77: 565–572.

Leung, K., Ras, E., Ferguson, K.B., Ariëns, S., Babendreier, D., Bijma, P., Bourtzis, K., Brodeur, J., Bruins, M.A., Centurión, A., Chattington, S.R., Chinchilla-Ramírez, M., Dicke, M., Fatouros, N.E., González-Cabrera, J., Groot, T.V.M., Haye, T., Knapp, M.,

Koskinioti, P., Le Hesran, S., Lyrakis, M., Paspati, A., Pérez-Hedo, M., Plouvier, W.N., Schlötterer, C., Stahl, J.M., Thiel, A., Urbaneja, A., van de Zande, L., Verhulst, E.C., Vet, L.E.M., Visser, S., Werren, J.H., Xia, S., Zwaan, B.J., Magalhães, S., Beukeboom, L.W., and Pannebakker, B.A. (2020). Next-generation biological control: the need for integrating genetics and genomics. *Biological Reviews* 95: 1838–1854.

Linder, S., Jarrett, B.J.M., Fanning, P., Isaacs, R., and Szűcs, M. (2022). Limited gains in native parasitoid performance on an invasive host beyond three generations of selection. *Evolutionary Applications* 15: 2113–2124.

Lirakis, M., and Magalhães, S. (2019). Does experimental evolution produce better biological control agents? A critical review of the evidence. *Entomologia Experimentalis et Applicata* 167: 584–597.

Lommen, S.T.E., de Jong, P.W., and Pannebakker, B.A. (2017). It is time to bridge the gap between exploring and exploiting: prospects for utilizing intraspecific genetic variation to optimize arthropods for augmentative pest control – a review. *Entomologia Experimentalis et Applicata* 162: 108–123.

Lu, Z., Zhu, P., Gurr, G., Zheng, X.-S., Read, D., Heong, K., Yang, Y., and Xu, H.-X. (2014). Mechanisms for flowering plants to benefit arthropod natural enemies of insect pests: prospects for enhanced use in agriculture. *Insect Science* 21: 1–12.

Madahi, K., Sahragard, A., and Hosseini, R. (2013). Influence of *Aphis gossypii* Glover (Hemiptera: Aphididae) density on life table parameters of *Aphidoletes aphidimyza* Rondani (Diptera: Cecidomyiidae) under laboratory conditions. *Journal of Crop Protection* 2: 355–368.

Mason, P.G. (2021). *Biological control: global impacts, challenges and future directions of pest management*. CSIRO Publishing.

Mayhew, P.J. (2016). Comparing parasitoid life histories. *Entomologia Experimentalis et Applicata* 159: 147–162.

McEvoy, P.B., Higgs, K.M., Coombs, E.M., Karaçetin, E., and Starcevich, L.A. (2012). Evolving while invading: rapid adaptive evolution in juvenile development time for a biological control organism colonizing a high-elevation environment. *Evolutionary Applications* 5: 524–536.

Miall, J.H., Abram, P.K., Cappuccino, N., Bennett, A.M.R., Fernández-Triana, J.L., Gibson, G.A.P., and Mason, P.G. (2021). Addition of nectar sources affects a parasitoid community without improving pest suppression. *Journal of Pest Science* 94: 335–347.

Mills, N. (2005). Selecting effective parasitoids for biological control introductions: codling moth as a case study. *Biological Control* 34: 274–282.

Mills, N.J. (2001). Factors influencing top-down control of insect pest populations in biological control systems. *Basic and Applied Ecology* 2: 323–332.

Mills, N. (2018). An alternative perspective for the theory of biological control. *Insects* 9: 131.

Mills, N.J., and Getz, W.M. (1996). Modelling the biological control of insect pests: a review of host-parasitoid models. *Ecological Modelling* 92, 121–143.

Mills, N.J., and Heimpel, G.E. (2018). Could increased understanding of foraging behavior help to predict the success of biological control? *Current Opinion in Insect Science* 27: 26–31.

Moffat, C., Abram, P., and Ensing, D. (2021). An evolutionary ecology synthesis for biological control. In: *Biological control: a global endeavor* (ed. Mason, P.G.), 584–615. CSIRO Publishing.

Murdoch, W.W., and Briggs, C.J. (1996). Theory for biological control: recent developments. *Ecology* 77: 2001–2013.

Murdoch, W.W., Chesson, J., and Chesson, P.L. (1985). Biological control in theory and practice. *The American Naturalist* 125: 344–366.

Murdoch, W.W., Swarbrick, S.L., and Briggs, C.J. (2006). Biological control: lessons from a study of California red scale. *Population Ecology* 48: 297–305.

Nicholson, A.J., and Bailey, V.A. (1935). The balance of animal populations – part I. *Proceedings of the Zoological Society of London* 105: 551–598.

Nomikou, M., Janssen, A., Schraag, R., and Sabelis, M.W. (2001). Phytoseiid predators as potential biological control agents for *Bemisia tabaci*. *Experimental & Applied Acarology* 25: 271–291.

Park, H.-H., Shipp, L., Buitenhuis, R., and Ahn, J.J. (2011). Life history parameters of a commercially available *Amblyseius swirskii* (Acari: Phytoseiidae) fed on cattail (*Typha latifolia*) pollen and tomato russet mite (*Aculops lycopersici*). *Journal of Asia-Pacific Entomology* 14: 497–501.

Phillips, B.L., Brown, G.P., and Shine, R. (2010). Life-history evolution in range-shifting populations. *Ecology* 91: 1617–1627.

Plouvier, W.N., and Wajnberg, E. (2018). Improving the efficiency of augmentative biological control with arthropod natural enemies: a modeling approach. *Biological Control* 125: 121–130.

Riahi, E., Fathipour, Y., Talebi, A.A., and Mehrabadi, M. (2017). Linking life table and consumption rate of *Amblyseius swirskii* (Acari: Phytoseiidae) in presence and absence of different pollens. *Annals of the Entomological Society of America* 110: 244–253.

Rosenheim, J.A., Jepsen, S.J., Matthews, C.E., Smith, D.S., and Rosenheim, M.R. (2008). Time limitation, egg limitation, the cost of oviposition, and lifetime reproduction by an insect in nature. *The American Naturalist* 172: 486–496.

Sakai, A.K., Allendorf, F.W., Holt, J.S., Lodge, D.M., Molofsky, J., With, K.A., Baughman, S., Cabin, R.J., Cohen, J.E., Ellstrand, N. C., McCauley, D.E., O'Neil, P., Parker, I.M., Thompson, J.N., and Weller, S.G. (2001). The population biology of invasive species. *Annual Review of Ecology and Systematics* 32: 305–332.

Schuldiner-Harpaz, T., Coll, M., and Wajnberg, E. (2022). Optimal foraging strategy to balance mixed diet by generalist consumers: a simulation model. *Behaviour* 159: 1263–1284.

Segoli, M. (2016). Effects of habitat type and spatial scale on density dependent parasitism in *Anagrus* parasitoids of leafhopper eggs. *Biological Control* 92: 139–144.

Segoli, M., Abram, P.K., Ellers, J., Hardy, I.C.W., Greenbaum, G., Heimpel, G.E., Keasar, T., Ode, P.J., Sadeh, A., and Wajnberg, E. (2023). Trait-based approaches to predicting biological control success: challenges and prospects. *Trends in Ecology & Evolution* 38: 802–811.

Segoli, M., and Rosenheim, J.A. (2013a). The link between host density and egg production in a parasitoid insect: comparison between agricultural and natural habitats. *Functional Ecology* 27: 1224–1232.

Segoli, M., and Rosenheim, J.A. (2013b). Spatial and temporal variation in sugar availability for insect parasitoids in agricultural fields and consequences for reproductive success. *Biological Control* 67: 163–169.

Segoli, M., Sun, S., Nava, D.E., and Rosenheim, J.A. (2018). Factors shaping life history traits of two proovigenic parasitoids. *Integrative Zoology* 13: 297–306.

Segoli, M., and Wajnberg, E. (2020). The combined effect of host and food availability on optimized parasitoid life-history traits based on a three-dimensional trade-off surface. *Journal of Evolutionary Biology* 33: 850–857.

Shea, K., and Possingham, H.P. (2000). Optimal release strategies for biological control agents: an application of stochastic dynamic programming to population management. *Journal of Applied Ecology* 37: 77–86.

Siddiqui, A., Omkar, Paul, S.C., and Mishra, G. (2015). Predatory responses of selected lines of developmental variants of ladybird, *Propylea dissecta* (Coleoptera: Coccinellidae) in relation to increasing prey and predator densities. *Biocontrol Science and Technology* 25: 992–1010.

Sinclair, J.S., Arnott, S.E., Millette, K.L., and Cristescu, M.E. (2019). Benefits of increased colonist quantity and genetic diversity for colonization depend on colonist identity. *Oikos* 128: 1761–1771.

Smith, S.M. (1996). Biological control with *Trichogramma*: advances, successes, and potential of their use. *Annual Review of Entomology* 41: 375–406.

Sol, D., Maspons, J., Vall-llosera, M., Bartomeus, I., García-Peña, G.E., Piñol, J., and Freckleton, R.P. (2012). Unraveling the life history of successful invaders. *Science* 337: 580–583.

Stahl, J.M., Babendreier, D., and Haye, T. (2019). Life history of *Anastatus bifasciatus*, a potential biological control agent of the brown marmorated stink bug in Europe. *Biological Control* 129: 178–186.

Stahlke, A.R., Bitume, E.V., Özsoy, Z.A., Bean, D.W., Veillet, A., Clark, M.I., Clark, E.I., Moran, P., Hufbauer, R.A., and Hohenlohe, P.A. (2022). Hybridization and range expansion in tamarisk beetles (*Diorhabda* spp.) introduced to North America for classical biological control. *Evolutionary Applications* 15: 60–77.

Stark, J.D., Banks, J.E., and Acheampong, S. (2004). Estimating susceptibility of biological control agents to pesticides: influence of life history strategies and population structure. *Biological Control* 29: 392–398.

Stark, J.D., Sugayama, R.L., and Kovaleski, A. (2007). Why demographic and modeling approaches should be adopted for estimating the effects of pesticides on biocontrol agents. *BioControl* 52: 365–374.

Stearns, S.C. (1983). The influence of size and phylogeny on patterns of covariation among life-history traits in the mammals. *Oikos* 41: 173–187.

Stiling, P., and Cornelissen, T. (2005). What makes a successful biocontrol agent? A meta-analysis of biological control agent performance. *Biological Control* 34: 236–246.

Stouthamer, R. (1993). The use of sexual versus asexual wasps in biological control. *Entomophaga* 38: 3–6.

Stouthamer, R., Breeuwer, J.A., and Hurst, G.D. (1999). *Wolbachia pipientis*: microbial manipulator of arthropod reproduction. *Annual Review of Microbiology* 53: 71–102.

Stouthamer, R., and Luck, R.F. (1993). Influence of microbe-associated parthenogenesis on the fecundity of *Trichogramma deion* and *T. pretiosum*. *Entomologia Experimentalis et Applicata* 67: 183–192.

Stouthamer, R., Luck, R.F., and Hamilton, W.D. (1990). Antibiotics cause parthenogenetic *Trichogramma* (Hymenoptera/Trichogrammatidae) to revert to sex. *Proceedings of the National Academy of Sciences of the United States of America* 87: 2424–2427.

Szűcs, M., Clark, E.I., Schaffner, U., Littlefield, J.L., Hoover, C., and Hufbauer, R.A. (2021). The effects of intraspecific hybridization on the host specificity of a weed biocontrol agent. *Biological Control* 157: 104585.

Szűcs, M., Eigenbrode, S.D., Schwarzländer, M., and Schaffner, U. (2012). Hybrid vigor in the biological control agent, *Longitarsus jacobaeae*. *Evolutionary Applications* 5: 489–497.

Tayeh, A., Estoup, A., Laugier, G., Loiseau, A., Turgeon, J., Toepfer, S., and Facon, B. (2012). Evolution in biocontrol strains: insight from the harlequin ladybird *Harmonia axyridis*. *Evolutionary Applications* 5: 481–488.

Tayeh, A., Hufbauer, R.A., Estoup, A., Ravigné, V., Frachon, L., and Facon, B. (2015). Biological invasion and biological control select for different life histories. *Nature Communications* 6: 7268.

Tommasini, M.G., van Lenteren, J.C., and Burgio, G. (2004). Biological traits and predation capacity of four *Orius* species on two prey species. *Bulletin of Insectology* 57: 79–93.

Torres, A.D.F, Bueno, V.H.P., Sampaio, M.V., and De Conti, B.F. (2007). Fertility life table of *Aphidius colemani* Viereck (Hymenoptera: Braconidae, Aphidiinae) on *Aphis gossypii* Glover (Hemiptera: Aphididae). *Neotropical Entomology* 36: 532–536.

Tsueda, H., and Tsuchida, K. (2011). Reproductive differences between Q and B whiteflies, *Bemisia tabaci*, on three host plants and negative interactions in mixed cohorts. *Entomologia Experimentalis et Applicata* 141: 197–207.

van Hezewijk, B.H., Bourchier, R.S., and De Clerck-Floate, R.A. (2010). Regional-scale impact of the weed biocontrol agent *Mecinus janthinus* on Dalmatian toadflax (*Linaria dalmatica*). *Biological Control* 55: 197–202.

van Lenteren, J.C. (2012). The state of commercial augmentative biological control: plenty of natural enemies, but a frustrating lack of uptake. *BioControl* 57: 1–20.

van Lenteren, J.C., and Bigler, F. (2010). Quality control of mass reared egg parasitoids. In: *Egg parasitoids in agroecosystems with emphasis on* Trichogramma (eds. Cônsoli, F.L., Parra, J.R.P, and Zucchi, R.A.), 315–340. Springer.

van Lenteren, J.C., Bolckmans, K., Köhl, J., Ravensberg, W.J., and Urbaneja, A. (2018). Biological control using invertebrates and microorganisms: plenty of new opportunities. *BioControl* 63: 39–59.

van Lenteren, J.C., Bueno, V.H.P., Burgio, G., Lanzoni, A., Montes, F.C., Silva, D.B., de Jong, P.W., and Hemerik, L. (2019). Pest kill rate as aggregate evaluation criterion to rank biological control agents: a case study with Neotropical predators of *Tuta absoluta* on tomato. *Bulletin of Entomological Research* 109: 812–820.

van Lenteren, J.C., Cock, M.J.W., Brodeur, J., Barratt, B.I.P., Bigler, F., Bolckmans, K., Haas, F., Mason, P.G., and Parra, J.R.P. (2011). Will the convention on biological diversity put an end to biological control? *Revista Brasileira de Entomologia* 55: 1–5.

van Lenteren, J.C., Lanzoni, A., Hemerik, L., Bueno, V.H.P., Bajonero Cuervo, J.G., Biondi, A., Burgio, G., Calvo, F.J., de Jong, P.W., López, S.N., Luna, M.G., Montes, F.C., Nieves, E.L., Aigbedion-Atalor, P.O., Riquelme Virgala, M.B., Sánchez, N.E., and Urbaneja, A. (2021). The pest kill rate of thirteen natural enemies as aggregate evaluation criterion of their biological control potential of *Tuta absoluta*. *Scientific Reports* 11: 10756.

van Steenis, M.J. (1993). Intrinsic rate of increase of *Aphidius colemani* Vier. (Hym., Braconidae), a parasitoid of *Aphis gossypii* Glov. (Hom., Aphididae), at different temperatures. *Journal of Applied Entomology* 116: 192–198.

Villa, M., Santos, S.A.P., Benhadi-Marín, J., Mexia, A., Bento, A., qnd Pereira, J.A. (2016). Life-history parameters of *Chrysoperla carnea s.l.* fed on spontaneous plant species and insect honeydews: importance for conservation biological control. *BioControl* 61: 533–543.

Vis, R.M.J., and van Lenteren, J. (2002). Longevity, fecundity, oviposition frequency and intrinsic rate of increase of the greenhouse whitefly, *Trialeurodes vaporariorum* on greenhouse tomato in Colombia. *Bulletin of Insectology* 55: 3–8.

Wajnberg E. (1991). Quality control of mass-reared arthropods: a genetical and statistical approach. Proceedings of the 5th workshop on quality control of mass-reared arthropods, pp. 15–25.

Wajnberg, E. (2004). Measuring genetic variation in natural enemies used for biological control: why and how? In: *Genetics, evolution and biological control* (ed. Ehler, T., Sforza, L.E., and Mateille, R.), 19–37. CABI International.

Wajnberg, E. (2010). Genetics of the behavioral ecology of egg parasitoids. In: *Egg parasitoids in agroecosystems with emphasis on* Trichogramma (eds. Cônsoli, F.L., Parra, J.R.P., and Zucchi, R.A.), 149–165. Springer.

Walzer, A., Nachman, G., Spangl, B., Stijak, M., and Tscholl, T. (2022). Trans- and within-generational developmental plasticity may benefit the prey but not its predator during heat waves. *Biology* 11: 1123.

Wang, L., Etebari, K., Zhao, Z., Walter, G.H., and Furlong, M.J. (2022). Differential temperature responses between *Plutella xylostella* and its specialist endo-larval parasitoid *Diadegma semiclausum* – implications for biological control. *Insect Science* 29: 855–864.

Wang, Z., and Smith, S.M. (1996). Phenotypic differences between thelytokous and arrhenotokous *Trichogramma minutum* from *Zeiraphera canadensis*. *Entomologia Experimentalis et Applicata* 78: 315–323.

Wilkes, A. (1947). The effects of selective breeding on the laboratory propagation of insect parasites. *Proceedings of the Royal Society B: Biological Sciences* 134: 227–245.

Williams, H.E., Brockerhoff, E.G., Liebhold, A.M., and Ward, D.F. (2021). Probing the role of propagule pressure, stochasticity, and Allee effects on invasion success using experimental introductions of a biological control agent. *Ecological Entomology* 46: 383–393.

Wright, J., Bolstad, G.H., Araya-Ajoy, Y.G., and Dingemanse, N.J. (2019). Life-history evolution under fluctuating density-dependent selection and the adaptive alignment of pace-of-life syndromes. *Biological Reviews* 94: 230–247.

Yu, J.-Z., Chi, H., and Chen, B.-H. (2013). Comparison of the life tables and predation rates of *Harmonia dimidiata* (F.) (Coleoptera: Coccinellidae) fed on *Aphis gossypii* Glover (Hemiptera: Aphididae) at different temperatures. *Biological Control* 64: 1–9.

Zhang, Y.-B., Zhang, G.-F., Liu, W.-X., and Wan, F.-H. (2019). Continuous heat waves change the life history of a host-feeding parasitoid. *Biological Control* 135: 57–65.

Zhang, Y., Zhang, H., Yang, N., Wang, J., and Wan, F. (2014). Income resources and reproductive opportunities change life history traits and the egg/time limitation trade-off in a synovigenic parasitoid. *Ecological Entomology* 39: 723–731.

23

Life History and Exploitative Management of Fish and Wildlife

Marco Festa-Bianchet

Département de biologie, Université de Sherbrooke, Sherbrooke, Québec, Canada

23.1 Introduction

Many book chapters and scientific articles about wild animal harvest begin by mentioning that humans have hunted and fished since they first evolved (Bichel and Hart 2023). That is true, as is the observation that for thousands of years, unsustainable harvests by humans have resulted in local and sometimes planetary extinctions (Johnson 2002). How can life history theory help ensure that harvests are sustainable? Two main ecological characteristics make some species particularly prone to unsustainable harvests. One is weak density-dependence in population dynamics. If harvests increase mortality without eliciting a compensatory population response, the population will decline to extinction if harvests continue. The other is low recruitment rate. Species that evolved with low or non-existent predation on the sex–age classes that are targets of human harvests may only sustain very limited harvest rates. For example, mammals that went extinct in Australia, after humans arrived, likely had lower reproductive rates than those that survived (Johnson 2002). Life history theory provides a biological explanation for why some species have low recruitment rate: they likely evolved under selective pressures where ensuring adult survival was more important than maximizing reproduction (Gaillard and Yoccoz 2003). Therefore, a consideration of life history strategies is essential for sustainable harvest because recruitment of new individuals must compensate for the mortality induced by harvest. Density-dependence is required in nearly all sustainable harvest programmes. The fundamental principle here is that harvest mortality will reduce intraspecific competition and allow increased survival or recruitment by individuals that are not harvested (Boyce et al. 1999). A species with high, mostly density-independent adult survival and low recruitment rate is less likely to respond to the increased resource availability when density is reduced by harvest than a species with variable, resource-dependent adult survival and flexible recruitment. This chapter will explore how life history affects sustainable yield of fish and wildlife and how some life histories make certain species vulnerable to local extinction under human harvests. It will include a consideration of how, in turn, human harvests may affect the evolution of life histories.

23.2 Life History Traits, Density-Dependence and Sustainable Harvest

The sustainable yield of an exploited population is the proportion of individuals that can be taken each year while maintaining the size of the population at a desired level (Caughley and Sinclair 1994). The concept of sustainable yield relies on the assumption of density-dependent population growth. The proportion of individuals that can be taken annually without causing a continuing decline in population size will depend on how a population's life history traits provide it with a capacity to increase in size. That capacity is often referred to as Rmax, or the maximum growth rate of the population when resources are unlimited. While the concept of Rmax is often used in fisheries science as a species-specific attribute (Hutchings 2015), it is less frequently encountered in the literature about wildlife harvesting. The abundance of an exploited population is lower than the abundance of an unexploited population. It is often assumed that non-harvested populations are in long-term equilibrium with resource availability in their environment. At that equilibrium population size, there is no net population growth. Any level of harvest can be defined as sustainable if new recruits to the population replace individuals that are harvested. The definition of 'recruitment' is important here. For example, if sexually mature adults are

Life History Evolution: Traits, Interactions, and Applications, First Edition. Edited by Michal Segoli and Eric Wajnberg.
© 2025 John Wiley & Sons Ltd. Published 2025 by John Wiley & Sons Ltd.

harvested, then recruitment must replace those adults in the population. We will see later how the need to replace individuals of the sex–age classes that are harvested can complicate harvest sustainability in species where several years are required to reach adult age.

Most harvest models assume that recruitment rate increases as density decreases because of greater *per capita* availability of resources. Harvesters, however, are more interested in the size of the harvest than in the rate of exploitation. Sustainable harvest size is the product of population size and population growth rate. If the population growth rate decreases with increasing population density, harvest will be maximized at an intermediate population size or density. Above this size, there will be more individuals but a lower reproductive rate, and below this size, there will be a greater reproductive rate but fewer individuals. Many fishery and wildlife management textbooks illustrate a population growth rate that decreases linearly from a very low density to the environmental carrying capacity K, at which the population has a growth rate of zero. If the population growth rate decreases linearly as population density increases, the maximum sustainable yield will be when the population is at 50% of K. Indeed, at that population size, the product of the growth rate and the number of harvestable individuals is maximized. Such a model, however, is based on one untested assumption and one unlikely shortcut. The untested assumption is a linear relationship between population size and population growth rate over the entire range of population sizes from very small to the carrying capacity. Several populations, however, show Allee effects so that at very low density, the growth rate decreases as numbers decline (Hutchings 2015, Nagel et al. 2021). The unlikely shortcut is that carrying capacity is fixed over time and that we actually know what it is. A more realistic scenario is illustrated in Figure 23.1. The population growth rate only decreases after some resource scarcity is encountered, as population size increases from a very low density. Below the level when all individuals have access to enough resources to express the maximum reproductive and survival rate given a population's life history traits (Rmax), there is no change in population growth rate. For simplicity, Figure 23.1 is based on the unlikely shortcut that, at some point, population growth rate will drop to zero strictly because of resource scarcity. In reality, however, population size and carrying capacity will vary as a function of many ecological variables, such as the prevalence and severity of disease and parasites, resource seasonality, interspecific competition and both occurrence and type of predation on different sex–age classes. The important point, shared with the more simplistic model of linear density-dependence at any density, is that sustained yield is maximized at an intermediate population size.

Of course, the sustainable harvest of a given population will vary each year according to many ecological and environmental variables. Its assessment is complicated by uncertainties in estimation of population size, changes in resource availability and the reliability of harvest data. Consequently, it is not advisable to seek the maximum sustained yield because errors could lead to over-harvest and population declines, which would be compounded over time if the decline is not immediately detected. When maximum sustainable yield is over-estimated, as the population declines, the rate of over-harvest increases every year (Caughley and Sinclair 1994). By definition, sustainable harvest cannot exceed the rate of recruitment to the population. Therefore, the opportunity for sustainable harvest is intimately tied to a population's life history traits, especially when harvest is indiscriminate in relation to sex–age class or targets adults of both sexes.

Assuming some level of density-dependence, one can calculate an approximate sustainable harvest rate for any species if life history parameters are known, once it is acknowledged that the rate is neither fixed over time nor known with certainty (Eberhardt 2002). In a real-world management situation, it would be important to monitor population size, sex ratio, age structure and other ecological parameters, but here we will focus on how life history differences may affect sustainable harvests. Generally, species that are at the 'fast' end of the life history continuum, with early maturation, high fecundity, short life expectancy and short generation time (Gaillard et al. 2005), can sustain a greater rate of harvest than species at the 'slow' end of the continuum. For example,

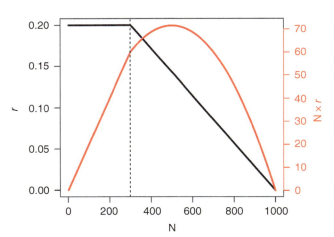

Figure 23.1 The relationship between population size and growth rate (*r*) when growth is independent of density from 0 to 300 individuals, with a growth rate fixed at 0.2. As population size increases past 300 individuals, intraspecific competition leads to a linear decline in growth rate until carrying capacity is reached at 1000 individuals. At any population size, yield is given by $r \times N$ and corresponds to the number of individuals recruited yearly into the population. Yield increases linearly until 300 (dotted vertical line), then follows a curve peaking at about 550 and then declining.

Reynolds et al. (2005) examined vulnerability to over-exploitation in marine fishes and identified late age of first reproduction and large body size as strong correlates of vulnerability to over-fishing. Large body size is likely correlated with reduced risk of predation, simply because large animals face fewer predators than small animals. Larger fish likely evolved with low predation rate on adults. Persistence of species that evolved with low predation likely relies on high adult survival and is easily affected by fishery mortality that targets adults. Slow-growing fish species with an exponential relationship between body size and fecundity (Barneche et al. 2018) are especially vulnerable, likely because females must reach an advanced age and a large size before they become very fecund. Reynolds et al. (2005) also point out that often fisheries managers over-emphasize how the high fecundity of some fish species may affect recruitment. High fecundity prompted the assumption that a fish stock was nearly inexhaustible simply because a single female could lay millions of eggs. That fallacious assumption persists as a 'zombie' idea: it keeps resurfacing despite demonstrations that it is invalid (Kindsvater et al. 2016). Although faith in large numbers of eggs leading to rapid population growth and recovery from harvest may appear sensible, life history theory predicts that if a fish species lays a very large number of small eggs, it is likely that most of those eggs will not lead to a reproducing adult. Some of the most sought-after fish species are large predatory fish with late age of first reproduction and high natural survival of adults, such as Atlantic cod (*Gadus morhua*) or rockfish (*Sebastes* spp.). In these species, the greatest contributions to recruitment are provided by large and older females (Ahrens et al. 2020), and under heavy fishery pressure, essentially no female will reach that age and size. The unjustified reliance on high fecundity rates as insurance against harvest-induced declines also underlines the importance of how recruitment is measured and how it is identified as a life history trait. Fish with 'slow' life history, long-lived adults, late age of first reproduction and often episodic recruitment have low tolerance for harvest, even if they produce large numbers of eggs. For example, some species of rockfish do not start to reproduce until they are 20 years of age or older and, as adults, have natural yearly survival rates over 95% (COSEWIC 2008). Although an adult female can lay more than two million eggs in a year, recruitment is episodic, juvenile mortality is extremely high and the population largely persists because of high adult survival. These species, are therefore, highly sensitive to human harvests. Similarly, large elasmobranchs (sharks and rays) are at risk of local extinction at much lower rates of fishing than similar-sized teleosts (bony fishes) because elasmobranchs tend to have lower recruitment rates and higher natural adult survival than teleosts. Therefore, they are much more sensitive than teleost populations to changes in adult mortality (Le Quesne and Jennings 2012).

Life histories that suggest that population persistence relies mostly on adult survival are found in many taxa. When adults are harvested by humans, in many cases, populations of these taxa undergo drastic declines or are locally extirpated. In the fossil record, there is often a temporal correlation between the arrival of human hunter-gatherers and the extinction of large taxa, apparently because of over-exploitation (Smith et al. 2018). Johnson (2002) underlined how, regardless of body size, extinction through apparent human over-exploitation appeared linked to low recruitment rates. More recent examples include both freshwater and marine turtles, where population persistence is heavily dependent on high adult survival, while egg and juvenile survival are low and highly variable from year to year. Bycatch and directed harvests of these species can lead to severe and rapid population declines (Spotila et al. 2000, Midwood et al. 2015). Of course, the high sensitivity to declines in adult survival does not mean that juvenile harvests or bycatch can be ignored; eventually, populations need new recruits to persist. Recent apparent increases in populations of several species of marine turtles have been attributed to better protection of nesting beaches, which have improved hatching success and juvenile survival, coupled with measures to reduce bycatch and direct harvest of adults (Mazaris et al. 2017).

23.3 Contrasting Life Histories and Harvest Potential

A useful illustration of the importance of life history for sustainable harvests is a comparison of two long-lived reptiles, turtles and pythons. In some freshwater turtles, simulations suggested that removals of as few as 1% of adult females each year would lead to high probabilities of extirpation of populations of a few hundred individuals (Midwood et al. 2015). Indeed, several species of turtles are threatened by the harvest of adults for the pet trade (Ceballos and Fitzgerald 2004). In contrast, the harvest of wild reticulated pythons (*Python reticulatus*) appears sustainable. Even though older and larger snakes produce more eggs and are more valuable to hunters, harvest surveys conducted 20 years apart suggested no declines in size, proportion of very large snakes or changes in other life history traits (Natusch et al. 2016). Both turtles and snakes likely have high natural survival of adults, but tropical pythons have an earlier age of first reproduction and may have higher juvenile survival compared to turtles because of maternal defence (Shine et al. 1999, Natusch et al. 2016). Shine et al. (1999)

Table 23.1 Key life history traits of native populations of mountain goats (*Oreamnos americanus*) and white-tail deer (*Odocoileus virginianus*).

	Mountain goats	White-tailed deer
Age of primiparity (years)	4–5	1–2
Prime-aged adult female survival (excluding harvests)	90–95%	80–93%
Litter size	1	2–3
Proportion of prime-aged females not reproducing in a given year	20%	0–5%

Mountain goat data from Festa-Bianchet and Côté (2008), deer data from DelGiudice et al. (2006), Webb and Gee (2014) and Michel et al. (2019).

underline how the apparent sustainability of the harvest of these pythons surprises herpetologists familiar with large snakes in temperate areas that, like turtles, are highly sensitive to adult mortality. Instead, key differences in life history are responsible for contrasting sustainable levels of harvest. The contrast between tropical pythons and other large, long-lived reptiles underlines the importance of considering all life history attributes to assess the sustainability of harvest. Table 23.1 contrasts some life history attributes of two similar-sized North American ungulates: mountain goats (*Oreamnos americanus*) and white-tailed deer (*Odocoileus virginianus*; see Figure 23.2). Compared to white-tailed deer, mountain goats have a later age of first reproduction, smaller litter size and a greater probability of not reproducing in any given year. Not surprisingly, white-tailed deer in productive habitats can withstand annual harvests of up to 20% (Nagy-Reis et al. 2022), but harvests of more than 2% of native mountain goat populations often lead to local declines (Hamel et al. 2006). Even within the same species, there can be important differences in life history attributes among populations or over time, sometimes related to resource availability. In ungulates, the age of first reproduction and juvenile survival can vary substantially over time as resource availability changes, while age-specific adult survival is mostly independent of resource availability (Bonenfant et al. 2009). A comparison of native and introduced populations of mountain goats further illustrates how intraspecific variability in life history traits can affect sustainable harvest. While native populations only tolerate minimal harvest, introduced populations can sustain harvests of 10% (Williams 1999). In introduced populations, females can first reproduce at two to three years of age and occasionally have twins (Bailey 1991), while in native populations, most females first reproduce at four to five years and twins are extremely rare (Festa-Bianchet and Côté 2008). These differences in life history among populations of the same species lead to a somewhat paradoxical situation. While wildlife managers now recognize that native populations of mountain goats must be managed very conservatively, mountain goats can be difficult to control in areas where they have been artificially introduced, largely because of differences in reproductive rates (Festa-Bianchet and Côté 2008).

Figure 23.2 Mountain goats and white-tailed deer are similar-sized ungulates with different life history traits (see Table 23.1) that lead to an order of magnitude in sustainable harvest rates. *Source:* photo credits: mountain goats by Édouard Bélanger, deer by Steeve Côté.

23.4 Ecological Plasticity and Evolutionary Sources of Variability: A Few Ungulate Examples

Much of the spatial and temporal variability in reproductive parameters among ungulate populations, including age of primiparity, fecundity, litter size and weaning success, can be attributed to environmental variability, particularly food availability but also climatic conditions, disease and predation (Gaillard et al. 2000). The comparison of white-tailed deer and mountain goats (Table 23.1), however, illustrates how species-specific life history attributes may affect harvest. In this example, deer appear to 'be in a hurry' and goats seem to be 'taking their time', so they may be considered at opposite ends of the slow-fast continuum that appears to fascinate many students of life history. Why is that? I suggest that deer likely evolved under greater predation pressure than mountain goats, leading to higher mortality for all age classes and higher reproductive rate. Note that higher mortality for adult deer during their evolution could well include thousands of years of harvests by humans. Mountain goats have an effective antipredator strategy against cursorial predators: when they see wolves (*Canis lupus*) or bears, they rapidly move to escape terrain where predators cannot follow and can also defend themselves against wolves (Côté et al. 1997). Perhaps mountain goats evolved with low adult mortality because of limited predation. Low adult mortality allows a species to persist with lower recruitment, so mountain goat females can afford to allocate all metabolic resources to growth and maintenance until four to five years of age before they start to reproduce. They can also skip reproduction about once every five years, which is likely to reduce their fitness costs (Hamel et al. 2010). Such life history strategy leads to a population that mostly persists through high adult survival, with limited recruitment. These traits may explain the high vulnerability of mountain goats to human harvests (Hamel et al. 2006). Mountain goats, however, are very vulnerable to cougar (*Puma concolor*) predation. Cougars use an ambush hunting technique that foils the antipredator behaviour of mountain goats. There are no cougars in much of the current distribution of mountain goats, and goats may have evolved under predation pressure mostly from cursorial predators (Dulude-de-Broin et al. 2020). Another ungulate example of how life history traits affect sustainability of harvest is provided by moose (*Alces alces*). Possibly as an adaptation to an unpredictable environment where forest fires provide local flushes of resources, unexpectedly for a very large ungulate, moose can have a very high reproductive rate, with early primiparity and frequent twinning (Gaillard 2007). In the absence of predation, moose can sustain a high harvest rate (Solberg et al. 2000) because adults that are shot by hunters are quickly replaced through recruitment.

These examples illustrate a fundamental point: life history strategies can vary, and the key traits for harvest sustainability are not always the same. Inevitably, however, some life history traits must show certain correlation structures, or the species or population would go extinct. If a species has low recruitment, it must have high adult survival. When humans harvest wild species and remove adults, the consequences of that harvest will be strongly affected by whether or not the harvested species evolved under high adult mortality.

23.5 How Can Knowledge of Life History Traits Improve Harvest Management?

When information on life history traits is unavailable for harvested species, managers must rely on knowledge from similar species. For fish, taxonomic group and asymptotic body length are useful indices of life history attributes and, therefore, sensitivity to harvest (Le Quesne and Jennings 2012). Generation length, the average age of parents in a population, is an important correlate of life history traits, particularly adult survival (Gaillard et al. 2005) and is often a good index of vulnerability to over-exploitation. Species with long generation length, be they rockfish, sharks, turtles or elephants, are more vulnerable to over-exploitation than species with short generation length, such as herring, quail or white-tailed deer. Generation length is an important variable in many of the quantitative criteria to assess risk of extinction developed by the International Union for the Conservation of Nature. A given decline in population size over a fixed time is a greater concern for species of long rather than of short generation length (Akçakaya et al. 2000). By extension, a given harvest rate is less likely to be sustainable in species with long than in species with short generation length. Generation length should be estimated in the absence of human harvests, as harvested populations often have shorter generation lengths than unharvested populations. That is because most human harvests selectively remove adult individuals so that the average age of reproducing individuals declines. For example, just 5% of male moose shot by hunters in Norway are aged five years or older (Mysterud et al. 2005) and over 70% of female brown bears (*Ursus arctos*) in Sweden are shot by hunters before their second reproductive opportunity (Zedrosser et al. 2013). In heavily harvested populations, animals have a reduced life expectancy and very few survive to old age.

When harvest can be targeted at specific sex–age classes, a knowledge of life history can help maximize harvest while minimizing or maximizing impact on population growth, depending on management objectives. The elasticity of each life history trait to population growth becomes important in these considerations. Elasticity refers to the proportional impact of a change in a given trait, such as adult survival or age of first reproduction, upon population growth rate. Elasticity is estimated based on a life-table analysis. It quantifies what would happen to population growth rate if a vital rate changed by a given proportion. A vital rate with high elasticity may have little impact on population dynamics if it did not vary much. Knowledge of elasticity of different vital rates can be used to modulate the impact of harvest. If the goal is to increase harvest while preventing population declines, harvest should be directed at sex–age classes with low elasticity, because their removal will have a smaller effect on population growth than the removal of an equal number of individuals from sex–age classes with high elasticity (Gaillard et al. 1998). For example, in long-lived mammals, harvest of adult males and juveniles has a lower impact on population dynamics than harvest of adult females (Milner et al. 2011). This is because the population growth rate is much more elastic to adult female survival than to juvenile survival, and, in many polygynous species, a high proportion of males can be removed without affecting fertility rates. This approach has its limits: although adult females have a much higher reproductive value than juveniles in large mammals, in most populations of large mammals, there are many more adult females than juveniles. Consequently, heavy harvest of juveniles will eventually lead to population declines (Kokko et al. 2001). The key issue here is that, given a choice between removing a juvenile or an adult female, removal of the young will typically have a much weaker impact on population dynamics. This concept is not easy to communicate to some hunters. Intuitively, people may think that an adult female has already 'used up' some of her years, while her young has all her life ahead of her. In reality, the mother has a much greater chance to survive and reproduce again than her dependent offspring. Because of life history differences, killing the mother will have a much greater impact on population dynamics than shooting the young, even without considering the possible detrimental effects of orphaning on juvenile survival (Rughetti et al. 2017). Age-specific survival and inter-annual variability in survival are important concepts in life history evolution, but not always easy ones, to communicate to harvester groups. Yet, social preferences are an important issue in human harvests and must be taken into account. Typically, hunters prefer not to harvest lactating females if they can recognize them, as reported for moose (Markussen et al. 2018). Some hunting regulations also discourage the harvest of lactating females. Those regulations are based on ethical considerations and presumably on the assumption that, if a lactating female is killed, her young will die as well. In chamois (*Rupicapra rupicapra*), such regulations can be counterproductive. Where there are heavy penalties for harvesting female with an offspring, hunters harvest many two-year-old females (Rughetti and Festa-Bianchet 2014). These young females are pre-reproductive but have gone through the filter of juvenile survival. They can look forward to many years of high survival and reproduction. Considering residual reproductive potential over the lifetime, they are the most productive age class in the population. A random harvest of females of any age regardless of reproductive status would have a weaker impact on population dynamics, unless survival rate of orphaned kids was extremely low (Rughetti and Festa-Bianchet 2014). In this case, a regulation meant to limit the impact of harvests on population growth has the opposite effect, partly because of a failure to consider all life history traits.

A better knowledge of life history variables and hunter preferences could help avoid counter-productive regulations. Hunting regulations in Sweden that protect brown bear family groups direct much of the harvest to pre-reproductive subadults. Subadult females, aged two to three years, that are yet to reproduce, are usually alone and therefore legal to harvest. These young females have high reproductive value because they have high natural survival and many years of reproduction ahead of them. Consequently, their harvest will have a strong negative effect on population dynamics. This regulation may also select for a lower female reproductive rate. Given that females accompanied by cubs are illegal to hunt, they have higher survival rate than females without cubs. When they wean their cubs in the spring, females are therefore at risk of harvesting the following autumn. Over time, a greater proportion of females in the hunted populations have been keeping their cubs for an extra year. This gives females one more year of protection from sport hunting, but decreases their reproductive rate (van de Walle et al. 2018).

Life history considerations are very important when harvests seek to reduce population size, for example, in the case of pest control. Many urban deer and kangaroo populations are considered over-abundant, leading to human–wildlife conflict, removing preferred forage species and altering habitat for other species (Côté et al. 2004, Gordon et al. 2021). Often, the public advocates the use of contraceptives to reduce their numbers, but removal of adult females is more efficient and less costly than fertility control when the goal is population reduction (Walker et al. 2021). That is because adult female survival in these species has the highest elasticity on population growth, while juvenile production has a very low elasticity (Gaillard et al. 2000).

An insightful example of the importance of life history traits for sustainable harvest comes from comparisons between human harvests and predation. Wright et al. (2006) found that wolves mostly killed senescent or juvenile female elk

(*Cervus canadensis*), while hunters took adult females in proportion to their age structure but avoided juveniles. The average reproductive value of females taken by hunters was therefore higher than that of females killed by wolves. Consequently, for the same number of females taken by hunters and wolves, human harvests had a greater effect on elk population dynamics compared to wolf kills. When life histories include senescence in survival and reproduction, as is the case for ungulates (Gaillard et al. 2000), directing the harvest to older animals may decrease its impact on population growth. On the other hand, for many fishes, the oldest females are the most productive. In this case, to limit impacts on population growth, harvest could be regulated by slot sizes to ensure that enough fish reach older age, large size and therefore greater productivity (Ahrens et al. 2020).

23.6 Life History and Trophy Hunting

Life history considerations are important also in the case of trophy hunting, a type of harvest that can have major economic and conservation implications if the large amounts of money it generates are used partly for conservation activities (Di Minin et al. 2016). Trophy hunting is controversial, partly because its links with conservation activities are not always clear (Bichel and Hart 2023). This activity has attracted the fund-raising interests of animal rights groups and is generally detested by much of the general public, including people who support other forms of consumptive wildlife use (Bichel and Hart 2023). One common narrative in support of trophy hunting is that removal of post-reproductive males would have minimal impact on either the demography or the evolution of the hunted species. Although in theory, that argument is correct, in many cases, trophy animals are shot in their prime reproductive years. If 'trophy' size is strongly age-related, the males with the largest antlers, horns or tusks will be of advanced age. They would likely have already disseminated their genes in the population and would be close to their natural deaths. In some species, it is possible to identify older males by physical traits such as colour or horn size. For example, in male ibex (*Capra ibex*), horns continue to grow substantially until 10–11 years of age. Swiss hunting regulations specify quotas for 5 age classes, from yearlings and 2-year-olds to males aged 11 years and older (Büntgen et al. 2018). In many other cases, however, the 'trophy' harvest is directed at younger males who may not yet have had a chance to breed (Festa-Bianchet et al. 2014, Loveridge et al. 2023). In these cases, intense selective removal of 'trophy' males can lead to undesirable genetic changes such as the evolution of smaller horns (Pigeon et al. 2016).

There are three ways in which life history data could improve management for trophy hunting. One is the relationship between the traits that make an animal a desirable trophy and its age-specific reproductive success. That relationship can vary substantially across species. For example, horn size is a major determinant of male reproductive success in wild sheep but not in mountain goats (Mainguy et al. 2009, Martin et al. 2016). Another is the skew in male reproductive success according to both trait size and age. When siring success is highly skewed, as, for example, in fallow deer (*Dama dama*) (Apollonio et al. 1989), removal of a few highly successful males may have substantial genetic consequences in the population because of the redistribution of many matings among surviving males. This situation likely increases the speed at which possible evolution of traits affected by selective hunting may occur. In species where skew is limited, such as eastern grey kangaroos (*Macropus giganteus*; Rioux-Paquette et al. 2015), a similar removal of a few large and dominant males would not lead to a strong redistribution of siring success. Kangaroos are not trophy-hunted, but the commercial harvest does select large males, which provide more meat and other products than other sex–age classes (Tenhumberg et al. 2004). Finally, knowledge of age-specific survival would allow managers to better assess the availability of 'trophy' males at different ages. Unless adult survival is high, redirecting the harvest towards older age classes to reduce possible evolutionary impacts of selective hunting will reduce the number of males available for harvest. For example, adult male survival is much lower in bighorn sheep than in Alpine ibex, so if the 'target age' at harvest was changed from, say, 6–10 years, there would be about 40–50% fewer bighorn rams available for harvest, but only about 10% fewer ibex (Jorgenson et al. 1997, Toïgo et al. 2007). Unfortunately, we currently know very little about the distribution of siring success among males in most wild species, or about the relative effects on siring success of different male traits, such as size, dominance, age or weapon size (Festa-Bianchet 2012).

23.7 Life History and Compensatory Population Responses to Harvest

The sustainability of nearly all fish and wildlife harvests is critically based on the assumption of density-dependent compensatory population dynamics. Animals that are harvested must be replaced over a year by increased recruitment or survival (Caughley and Sinclair 1994). The strength of compensation, however, depends partly on the proportion of a population made up of sex–age classes that show density-dependence (Sinclair and Pech 1996). Species with slow life histories, where population persistence mostly depends on adult survival, are likely to show weaker density-dependence than species with fast life histories, where population persistence mostly depends on recruitment (Gaillard and Yoccoz 2003). In long-lived vertebrates, density-dependence typically affects vital rates according to a predictable order (Eberhardt 2002): as density increases, juvenile survival is the first vital rate to decline, followed by an increase in age of primiparity, then a decrease in fertility and finally a decrease in adult survival. Evidence for density-dependence in adult survival of long-lived species is rare (Bonenfant et al. 2009). In ducks, species with high natural survival show a weaker compensatory response to harvest (Nichols 1991), while those with high reproductive rates and lower adult survival generally rebound quickly from harvest (Riecke et al. 2022). In many long-lived species, natural adult survival is often above 90% on an annual basis (Eberhardt 2002). The reasoning here is somewhat tautological: a species cannot be long-lived unless it has very high adult survival. The relevant issue, however, is that when adult survival is high and varies little from year to year (Gaillard et al. 2000), there is limited opportunity for compensation to harvest to be generated by changes in adult survival. It is unlikely that natural survival of adults will increase and make up for the extra mortality caused by harvests. Sustainable harvests will then depend on compensation generated from other life history traits, such as fertility, juvenile survival and earlier age of first reproduction. If only juvenile survival is density-dependent, and most individuals in a population are adults, then any compensatory response to harvest will be weakened compared to populations with a greater proportion of juveniles. A 'slow' life history with naturally high adult survival makes species more difficult to harvest sustainably. A more cautious approach is required compared to species with 'fast' life histories that could quickly respond to the reduced density brought about by harvests. Species with short generation times sometimes show compensatory changes in adult survival following harvest, but for species with long generation times, the effect of harvest on adult survival is typically additive (Péron 2013).

23.8 The Special Case of Sexually Selected Infanticide

For species whose life history is affected by social factors, the intensity and selectivity of harvest can present additional challenges because they may affect the social structure. That is the case, for example, when sexually selected infanticide affects population dynamics. Sexually selected infanticide can be considered a life history trait because it appears to be an adaptation of males to increase their reproductive success. Within a wildlife management context, it has mostly been studied in large predators such as brown bears and lions (*Panthera leo*). In both species, harvest of dominant males can lead surviving males to kill juveniles. If nursing juveniles are killed, their mother will stop lactating and reach her next estrus sooner than if she had continued lactation until her young were weaned. The infanticidal male can then mate with the female and increase his rate of reproduction. In brown bears in Sweden, the frequency of infanticide following hunting of mature males led researchers to estimate that killing one adult male would have population-dynamic consequences comparable to those expected if an adult female was shot (Swenson et al. 1997). In lions, the high frequency of infanticide when pride-controlling males are shot has led to management recommendations to restrict harvest to males aged seven to eight years and older to limit pride male turnover and the frequency of infanticide (Creel et al. 2016). The important role of sexually selected infanticide in some species and populations underlines the importance of knowledge of life history traits for consumptive wildlife management. These polygynous species would normally be expected to tolerate a high rate of harvest of mature males, but that is not the case when harvest of a mature male increases juvenile mortality through sexually selected infanticide. Prevalence of infanticide can also vary among populations, presenting additional challenges for management. For example, there appears to be no evidence for sexually selected infanticide in North American brown bears (McLellan 2005), supporting management strategies aimed at directing the harvest to mature males (McLellan et al. 2017).

23.9 Can Harvest Affect the Evolution of Life History Strategies?

So far, this chapter has examined the importance of knowledge of life history traits for sustainable harvests. Life history strategies likely evolved partly in response to differences in sex- and age-specific mortality patterns. The biological principle underlying this expectation is simple: if an age-specific mortality rate increases, for whatever reason, unless other vital rates change and compensate for the additional mortality, the population will go extinct. Human harvests usually change the mortality patterns of exploited populations, typically by increasing adult mortality. Consequently, harvests could affect the evolution of life history traits (Pelletier and Coltman 2018). There is substantial evidence that selective harvests can affect the evolution of certain morphological traits, such as horn size, in wild sheep under intense trophy hunting (Festa-Bianchet 2017). Evidence of harvest-induced evolutionary changes in life history traits in wild species, however, is limited. This is despite modelling approaches (Ayllon et al. 2018), early observational studies (Boyce 1981), and numerous experimental simulations of size-selective harvests (Uusi-Heikkila et al. 2016) suggesting that evolutionary changes should occur.

There are several reasons why evidence for harvest-induced changes in life history traits of wild animals is limited. Life history traits tend to have lower heritability than morphological traits (Mousseau and Roff 1987), so their evolution may be slower and could be detected only through long-term monitoring over many generations. Immigration from no-harvest refuges, or periods with curtailed harvests, may dampen any selective effects (Sørdalen et al. 2022). For selection to lead to evolution, the selective pressure from harvests must be consistent over multiple generations and over very large areas (Festa-Bianchet 2017). Changes in harvest regulations, harvest pressure and harvester preferences could modify the strength and direction of selection over time. Information on what genetic changes may underline life history adaptations to harvest is available from laboratory studies (Uusi-Heikkila et al. 2015) but is very limited for wild populations (Therkildsen et al. 2019). Consequently, studies of wild populations mostly rely on phenotypic changes in life history traits. Many of the changes in life history traits that one could expect from artificially increased adult mortality, however, are also predicted as plastic responses to the greater resource abundance that should follow reductions in population density induced by harvest (Dunlop et al. 2018). These include an earlier age of first reproduction and greater allocation to reproduction. These changes could simply originate from a plastic response to lowered intraspecific competition, making it difficult to distinguish the relative contribution, if any, of evolutionary responses (Pelletier and Coltman 2018).

The potential for fishery-induced evolution in life history traits has been the subject of considerable interest (Hutchings and Kuparinen 2020), partly because, by driving evolutionary change in a direction opposed to natural selection, fishery-induced evolution could slow the recovery of over-exploited fish populations (Swain et al. 2007). Extremely high fishing mortality over several generations may select for fish that spawn at a young age and at a small size because their chances of reproducing later in life are drastically curtailed. This may be the best option under very high fishery pressure, but under natural selective pressures, a strategy of delaying reproduction in favour of body growth may be selected for and lead to faster population growth. Consequently, intensely over-fished populations could be maladapted and take longer to recover if fishing pressure was relaxed. Several experimental studies of fish in captivity have established that intense, selective harvest can lead to changes in life history traits (review in Heino et al. 2015). For example, experimental selective harvest of small or large Atlantic silverside (*Menidia menidia*) led to substantial shifts in fish size over four generations (Conover and Munch 2002), and selection against large fish led to a decrease in both mass and yield. In the wild, however, selective fishing of larger individuals can have complex consequences and may not necessarily simply favour evolution of smaller size and early maturation, as suggested by long-term monitoring of pike (*Esox lucius*) in Lake Windermere (Edeline et al. 2009). While the experimental fishery that only removed large pike did select against larger fish, it also selected for a faster growth rate, leading to increased variance in growth and age-specific size. In laboratory experiments, conditions are controlled. In the wild, size-selective harvests may interact with several uncontrolled and often unknown variables (Thambiturai and Kuparinen 2023), possibly leading to unexpected results (see also Chapter 12, this volume). Hutchings and Kuparinen (2020) argued that fishery-induced evolution is likely a minor concern for the recovery of most over-exploited fish populations. This does not imply that fish harvest, when intense and age- or size-selective, cannot cause evolutionary changes in life history. It suggests, instead, that those changes likely will not strongly affect population growth after fishing pressure is relaxed. That is partly because, in many over-exploited fish populations, adult mortality is extremely high. The cessation of fishing would have an overwhelming positive effect on adult survival. In comparison, the possible additional mortality of young adults induced by evolution of a suboptimal, harvest-induced reproductive strategy would likely have a minor effect on population dynamics (Hutchings 2009).

There are fewer studies of harvest-related changes in life history traits for terrestrial vertebrates than for fishes. That taxonomic difference may simply arise because there are more data and longer time series for fish than for terrestrial mammals. Perhaps, fisheries biologists are more interested than wildlife managers in the possible effects of harvest on life history traits. Alternatively, however, the greater flexibility in life history traits in fish than, for example, in mammals and birds may make fish more amenable to studies of life history variation induced by harvests. The two ungulates presented in Table 23.1 were chosen because they show substantial interspecific differences in life history traits. Still, variability is limited: traits vary by factors of two to four only. In contrast, some species of fish can vary their size at maturity and their fecundity by two orders of magnitude (Hutchings et al. 2019). In comparison, plasticity and inter-population variability in life history traits in mammals and birds are more limited (Kuparinen and Festa-Bianchet 2017). The few examples of possible harvest-induced life history evolution in terrestrial mammals remain questionable. In bighorn sheep (*Ovis canadensis*), breeding values in horn size, which decrease under intense selective hunting, are genetically correlated with breeding values of several life history traits in both sexes (Coltman et al. 2005). Given that horn size in females appears linked to reproductive success after accounting for body mass (Deakin et al. 2022), harvest-induced evolution of smaller horns in males may eventually lead to changes in life history traits in both sexes. There is currently, however, no clear evidence that harvest has affected the evolution of life history traits in wild mammals or birds. Most monitoring programmes of terrestrial vertebrates have documented changes over a few generations. Over the longer term, however, there are suggestions that heavily harvested populations of brown bears may have evolved life history traits that can be interpreted as adaptations to the high adult mortality induced by hunting over centuries or even millennia. Those changes include an earlier age of primiparity and larger litter size after accounting for female mass (Zedrosser et al. 2011). Other studies that have attributed life history changes to selection from hunting include delayed birth date in male moose (Kvalnes et al. 2016) and earlier birth date in wild boar (*Sus scrofa*) (Gamelon et al. 2011). These contrasting effects originate because early born male moose are more likely to be shot, possibly because they are more visible, while early born female wild boars have a greater probability of reproducing before being shot.

Such research raises an important question: we are concerned about the effects of harvest on contemporary evolution of life history traits, but harvest by humans over millennia may also have affected the life history evolution of exploited species. Given the evidence that human harvest has likely led to extinction of some prey species during the Pleistocene (Johnson 2002), the possibility that surviving species adapted to human harvest by changing some of their life history traits seems plausible and worthy of additional research.

23.10 Conclusion and Future Directions

The importance of life history traits for sustainable harvests is unquestionable, simply because individuals that are removed must be replaced. What life history traits are most important and how they may change in harvested populations, however, is not always clear. Life histories can be complicated, and so is their relationship with harvest of wild species. Harvest strategies must consider all life history traits of the target species, account for the variability in those traits and consider how some traits, but not others, may change when humans harvest populations. Fundamentally, sustainable harvest depends on changes in one or more vital rates to compensate for individuals that are harvested. Compensation is most likely to arise from changes in recruitment, but this is not always the case. Because harvest typically targets adults, compensation must involve the recruitment of more adults to the population. This chapter examined some cases where incomplete knowledge of life histories led to unsustainable harvest. Indeed, fish may lay millions of eggs but cannot compensate for over-fishing, and well-meaning protection of lactating females in chamois and bears can lead to detrimental effects on population dynamics.

The chapter also identified some gaps in knowledge. The importance of harvest-induced evolutionary changes in life history traits, now or historically, remains mostly unresolved. Collaboration with wildlife and fisheries managers provides ample opportunities for long-term, experimental manipulations of harvest strategies to study possible evolutionary consequences. Recent advances in genomics will allow us to identify specific genes that may be affected by harvests and contribute to harvest strategies that are both ecologically and evolutionarily sustainable. These are complex systems because both harvest dynamics and life history traits may change in response to each other. Fruitful areas of future research include the impacts of networks of no-take areas as a potential source of immigrants not exposed to harvest selection and better quantification of plastic and genetic contributions to changes in life history traits under different levels of harvest. For many harvested species, however, we still need basic life history data such as age- and sex-specific reproductive value and parental attributes that affect recruitment potential, with recruitment measured as the number of new potentially harvestable individuals in the population, not just the number of eggs or newborns. Despite the major development of theory, the effectiveness of harvest strategies is often limited by lack of data on harvested species.

Acknowledgements

The Natural Sciences and Engineering Research Council of Canada provided long-term support for my research programme. I am grateful for constructive criticism of an earlier draft of this chapter by John Reynolds, Eric Wajnberg and Michel Segoli. I thank Guillaume Blanchet for drawing Figure 23.1.

References

Ahrens, R.N.M., Allen, M.S., Walters, C., and Arlinghaus, R. (2020). Saving large fish through harvest slots outperforms the classical minimum-length limit when the aim is to achieve multiple harvest and catch-related fisheries objectives. *Fish and Fisheries* 21: 483–510.

Akçakaya, H.R., Ferson, S., Burgman, M.A., Keith, D.A., Mace, G.M., and Todd, C.R. (2000). Making consistent IUCN classifications under uncertainty. *Conservation Biology* 14: 1001–1013.

Apollonio, M., Festa-Bianchet, M., and Mari, F. (1989). Correlates of copulatory success in a fallow deer lek. *Behavioral Ecology and Sociobiology* 25: 89–97.

Ayllon, D., Railsback, S.F., Almodovar, A., Nicola, G.G., Vincenzi, S., Elvira, B., and Grimm V. (2018). Eco-evolutionary responses to recreational fishing under different harvest regulations. *Ecology and Evolution* 8: 9600–9613.

Bailey, J.A. (1991). Reproductive success in female mountain goats. *Canadian Journal of Zoology* 69: 2956–2961.

Barneche, D.R., Robertson, D.R., White, C.R., and Marshall, D.J. (2018). Fish reproductive-energy output increases disproportionately with body size. *Science* 360: 642–645.

Bichel, N., and Hart, A. (2023). *Trophy hunting*. Springer.

Bonenfant, C., Gaillard, J.M., Coulson, T., Festa-Bianchet, M., Loison, A., Garel, M., Loe, L.E., Blanchard, P., Pettorelli, N., Owen-Smith, N., Du Toit, J., and Duncan, P. (2009). Empirical evidence of density-dependence in populations of large herbivores. *Advances in Ecological Research* 41: 313–357.

Boyce, M.S. (1981). Beaver life-history responses to exploitation. *Journal of Applied Ecology* 18: 749–753.

Boyce, M.S., Sinclair, A.R.E., and White, G.C. (1999). Seasonal compensation of predation and harvesting. *Oikos* 87: 419–426.

Büntgen, U., Galvan, J.D., Mysterud, A., Krusic, P.J., Hülsmann, L., Jenny, H., Senn, J., and Bollmann, K. (2018). Horn growth variation and hunting selection of the Alpine ibex. *Journal of Animal Ecology* 87: 1069–1079.

Caughley, G., and Sinclair, A.R.E. (1994). *Wildlife ecology and management*. Blackwell Scientific Publications.

Ceballos, C., and Fitzgerald, L.A. (2004). The trade in native and exotic turtles in Texas. *Wildlife Society Bulletin* 32: 881–892.

Coltman, D.W., O'Donoghue, P., Jorgenson, J.T., Hogg, J.T., and Festa-Bianchet, M. (2005). Selection and genetic (co)variance in bighorn sheep. *Evolution* 59: 1372–1382.

Conover, D.O., and Munch, S.B. (2002). Sustaining fisheries yields over evolutionary time scales. *Science* 297: 94–96.

COSEWIC (2008). COSEWIC assessment and status report on the Yelloweye Rockfish *Sebastes ruberrimus*, Pacific Ocean inside waters population and Pacific Ocean outside waters population, in Canada.

Côté, S.D., Peracino, A., and Simard, G. (1997). Wolf (*Canis lupus*) predation and maternal defensive behavior in mountain goat (*Oreamnos americanus*). *Canadian Field-Naturalist* 111: 389–392.

Côté, S.D., Rooney, T.P., Tremblay, J.P., Dussault, C., and Waller, D.M. (2004). Ecological impacts of deer overabundance. *Annual Review of Ecology and Systematics* 35: 113–147.

Creel, S., M'Soka, J., Dröge, E., Rosenblatt, E., Becker, M., Matadinko, W., and Simpamba, T. (2016). Assessing the sustainability of African lion trophy hunting, with recommendations for policy. *Ecological Applications* 26: 2347–2357.

Deakin, S., Festa-Bianchet, M., Miller, J.M., Pelletier, F., and Coltman, D.W. (2022). Ewe are what ewe wear: bigger horns, better ewes and the potential consequence of trophy hunting on female fitness in bighorn sheep. *Proceedings of the Royal Society B: Biological Sciences* 289: 2021534.

DelGiudice, G.D., Fieberg, J., Riggs, M.R., Carstensen Powell, M., and Pan, W. (2006). A long-term age-specific survival analysis of female white-tailed deer. *Journal of Wildlife Management* 70: 1556–1568.

Di Minin, E., Leader-Williams, N., and Bradshaw, C.J.A. (2016). Banning trophy hunting will exacerbate biodiversity loss. *Trends in Ecology & Evolution* 39: 99–102.

Dulude-de-Broin, F., Hamel, S., Mastromonaco, G.F., and Côté, S.D. (2020). Predation risk and mountain goat reproduction: evidence for stress-induced breeding suppression in a wild ungulate. *Functional Ecology* 34: 1003–1014.

Dunlop, E.S., Feiner, Z.S., and Höök, T.O. (2018). Potential for fisheries-induced evolution in the Laurentian Great Lakes. *Journal of Great Lakes Research* 44: 734–748.

Eberhardt, L.L. (2002). A paradigm for population analysis of long-lived vertebrates. *Ecology* 83: 2841–2854.

Edeline, E., Le Rouzic, A., Winfield, I.J., Fletcher, J.M., James, J.B., Stenseth, N.C., and Vøllestad, L.A. (2009). Harvest-induced disruptive selection increases variance in fitness-related traits. *Proceedings of the Royal Society B: Biological Sciences* 276: 4163–4171.

Festa-Bianchet, M. (2012). The cost of trying: weak interspecific correlations among life-history components in male ungulates. *Canadian Journal of Zoology* 90: 1072–1085.

Festa-Bianchet, M. (2017). When does selective hunting select, how can we tell, and what should we do about it? *Mammal Review* 47: 76–81.

Festa-Bianchet, M., and Côté, S.D. (2008). *Mountain goats: ecology, behavior and conservation of a mountain ungulate*. Island Press.

Festa-Bianchet, M., Pelletier, F., Jorgenson, J.T., Feder, C., and Hubbs, A. (2014). Decrease in horn size and increase in age of trophy sheep in Alberta over 37 years. *Journal of Wildlife Management* 78: 133–141.

Gaillard, J.M. (2007). Are moose only a larger deer? Some life-history considerations. *Alces* 43: 1–11.

Gaillard, J.-M., Festa-Bianchet, M., and Yoccoz, N.G. (1998). Population dynamics of large herbivores: variable recruitment with constant adult survival. *Trends in Ecology & Evolution* 13: 58–63.

Gaillard, J.-M., Festa-Bianchet, M., Yoccoz, N.G., Loison, A., and Toïgo, C. (2000). Temporal variation in fitness components and population dynamics of large herbivores. *Annual Review of Ecology and Systematics* 31: 367–393.

Gaillard, J.-M., and Yoccoz, N.G. (2003). Temporal variation in survival of mammals: a case of environmental canalization? *Ecology* 84: 3294–3306.

Gaillard, J.M., Yoccoz, N.G., Lebreton, J.-D., Bonenfant, C., Devillard, S., Loison, A., Pontier, D., and Allainé, D. (2005). Generation time: a reliable metric to measure life-history variation among mammalian populations. *The American Naturalist* 166: 119–123.

Gamelon, M., Besnard, A., Gaillard, J.M., Servanty, S., Baubet, E., Brandt, S., and Gimenez, O. (2011). High hunting pressure selects for earlier birth date: wild boar as a case study. *Evolution* 65: 3100–3112.

Gordon, I.J., Snape, M., Fletcher, D., Howland, B., Coulson, G., Festa-Bianchet, M., Caley, P., McIntyre, S., Pople, A., Wimpenny, C., Baines, G., and Alcock, D. (2021). Herbivore management for biodiversity conservation: a case study of kangaroos in the Australian Capital Territory (ACT). *Ecological Management and Restoration* 22: 124–137.

Hamel, S., Côté, S.D., and Festa-Bianchet, M. (2010). Maternal characteristics and environment affect the costs of reproduction in female mountain goats. *Ecology* 91: 2034–2043.

Hamel, S., Côté, S.D., Smith, K.G., and Festa-Bianchet, M. (2006). Population dynamics and harvest potential of mountain goat herds in Alberta. *Journal of Wildlife Management* 70: 1044–1053.

Heino, M., Pauli, B.D., and Dieckmann, U. (2015). Fisheries-induced evolution. *Annual Review of Ecology and Systematics* 46: 461–480.

Hutchings, J.A. (2009). Avoidance of fisheries-induced evolution: management implications for catch selectivity and limit reference points. *Evolutionary Applications* 2: 324–334.

Hutchings, J.A. (2015). Thresholds for impaired species recovery. *Proceedings of the Royal Society B: Biological Sciences* 282: 20150654.

Hutchings, J.A., Ardren, W., Barlaup, B.T., Bergman, E.J., Clarke, K.D., Greenberg, L.A., Lake, C., Piironen, J., Sirois, P., Sundt-Hansen, L.E., and Fraser, D.J. (2019). Life-history variability and conservation status of landlocked Atlantic salmon: an overview. *Canadian Journal of Fisheries and Aquatic Sciences* 76: 1697–1708.

Hutchings, J.A., and Kuparinen, A. (2020). Implications of fisheries-induced evolution for population recovery: refocusing the science and refining its communication. *Fish and Fisheries* 21: 453–464.

Johnson, C.N. (2002). Determinants of loss of mammal species during the Late Quaternary 'megafauna' extinctions: life history and ecology, but not body size. *Proceedings of the Royal Society B: Biological Sciences* 269: 2221–2227.

Jorgenson, J.T., Festa-Bianchet, M., Gaillard, J.-M., and Wishart, W.D. (1997). Effects of age, sex, disease, and density on survival of bighorn sheep. *Ecology* 78: 1019–1032.

Kindsvater, H.K., Mangel, M., Reynolds, J.D., and Dulvy, N.K. (2016). Ten principles from evolutionary ecology essential for effective marine conservation. *Ecology and Evolution* 6: 2125–2138.

Kokko, H., Lindström, J., and Ranta, E. (2001). Life histories and sustainable harvesting. In: *Conservation of exploited species* (eds. Reynolds, J.D., Mace, G.M., Redford, K.H., and Robinson, J.G.), 301–322. Cambridge University Press.

Kuparinen, A., and Festa-Bianchet, M. (2017). Harvest-induced evolution: insights from aquatic and terrestrial systems. *Philosophical Transactions of the Royal Society B* 372: 20160036.

Kvalnes, T., Saether, B.-E., Haanes, H., Røed, K.H., Engen, S., and Solberg, E.J. (2016). Harvest-induced phenotypic selection in an island population of moose, *Alces alces*. *Evolution* 70: 1486–1500.

Le Quesne, W.J.F., and Jennings, S. (2012). Predicting species vulnerability with minimal data to support rapid risk assessment of fishing impacts on biodiversity. *Journal of Applied Ecology* 49: 20–28.

Loveridge, A.J., Wijers, M., Mandisodza-Xhikerema, R., Macdonald, D.W., and Chapron, G. (2023). Anthropogenic edge effects and aging errors by hunters can affect the sustainability of lion trophy hunting. *Scientific Reports* 13: 95.

Mainguy, J., Côté, S.D., Festa-Bianchet, M., and Coltman, D.W. (2009). Father–offspring phenotypic correlations suggest intralocus sexual conflict for a fitness-linked trait in a wild sexually dimorphic mammal. *Proceedings of the Royal Society B: Biological Sciences* 276: 4067–4075.

Markussen, S.S., Loison, A., Herfindal, I., Solberg, E., Haanes, H., Røed, K.H., Heim, M., and Saether, B.-E. (2018). Fitness correlates of age at primiparity in a hunted moose population. *Oecologia* 186: 447–458.

Martin, A.M., Festa-Bianchet, M., Coltman, D.W., and Pelletier, F. (2016). Demographic drivers of age-dependent sexual selection. *Journal of Evolutionary Biology* 29: 1437–1446.

Mazaris, A.D., Schofield, G., Gkazinou, C., Almpanidou, V., and Hays, G.C. (2017). Global sea turtle conservation successes. *Science Advances* 3: e1600730.

McLellan, B.N. (2005). Sexually selected infanticide in grizzly bears: the effects of hunting on cub survival. *Ursus* 16: 141–156.

McLellan, B.N., Mowat, G., Hamilton, T., and Hatter, I.W. (2017). Sustainability of the grizzly bear hunt in British Columbia, Canada. *Journal of Wildlife Management* 81: 218–229.

Michel, E.S., Demarais, S., Strickland, B.K., Belant, J.L., and Castle, L.E. (2019). Body mass influences maternal allocation more than parity status for a long-lived cervid mother. *Journal of Mammalogy* 100: 1459–1465.

Midwood, J.D., Cairns, N.A., Stoot, L.J., Cooke, S.J., and Blouin-Demers, G. (2015). Bycatch mortality can cause extirpation in four freshwater turtle species. *Aquatic Conservation: Marine and Freswater Ecosystems* 27: 71–80.

Milner, J.M., Bonenfant, C., and Mysterud, A. (2011). Hunting Bambi – evaluating the basis for selective harvesting of juveniles. *European Journal of Wildlife Research* 57: 565–574.

Mousseau, T.A., and Roff, D.A. (1987). Natural selection and the heritability of fitness components. *Heredity* 59: 181–197.

Mysterud, A., Solberg, E.J., and Yoccoz, N.G. (2005). Ageing and reproductive effort in male moose under variable levels of intrasexual competition. *Journal of Animal Ecology* 74: 742–754.

Nagel, R., Stainfield, C., Fox-Clarke, C., Toscani, C., Forcada, J., and Hoffman, J.I. (2021). Evidence for an Allee effect in a declining fur seal population. *Proceedings of the Royal Society B: Biological Sciences* 288: 20202882.

Nagy-Reis, M., Reimer, J.R., Lewis, M.A., Jensen, W.F., and Boyce, M.S. (2022). Aligning population models with data: adaptive management for big game harvests. *Global Ecology and Conservation* 26: e01501.

Natusch, D.J.D., Lyons, J.A., Mumpuni, Riyanto, A., and Shine, R. (2016). Jungle giants: assessing sustainable harvesting in a difficult-to-survey species (*Python reticulatus*). *PLoS ONE* 11(7) e0158397.

Nichols, J.D. (1991). Responses of North American duck populations to exploitation. In: *Bird population studies: relevance to conservation and management* (eds. Perrins, C., Lebreton, J.-D., and Hirons, G.), 498–525. Oxford University Press.

Pelletier, F., and Coltman, D.W. (2018). Will human influences on evolutionary dynamics in the wild pervade the Anthropocene? *BMC Biology* 16: 7.

Péron, G. (2013). Compensation and additivity of anthropogenic mortality: life-history effects and review of methods. *Journal of Animal Ecology* 82: 408–417.

Pigeon, G., Festa-Bianchet, M., Coltman, D.W., and Pelletier, F. (2016). Intense selective hunting leads to artificial evolution in horn size. *Evolutionary Applications* 9: 521–530.

Reynolds, J.D., Dulvy, N.K., Goodwin, N.B., and Hutchings, J.A. (2005). Biology of extinction risk in marine fishes. *Proceedings of the Royal Society B: Biological Sciences* 272: 2337–2344.

Riecke, T.V., Lohman, M.G., Sedinger, B.S., Arnold, T.W., Feldheim, C.L., Koons, D.N., Rohwer, F.C., Schaub, M., Williams, P.J., and Sedinger, J.S. (2022). Density-dependence produces spurious relationships among demographic parameters in a harvested species. *Journal of Animal Ecology* 91: 2261–2272.

Rioux-Paquette, E., Garant, D., Martin, A.M., Coulson, G.C., and Festa-Bianchet, M. (2015). Paternity in eastern grey kangaroos: moderate skew despite strong sexual dimorphism. *Behavioral Ecology* 26: 1147–1155.

Rughetti, M., and Festa-Bianchet, M. (2014). Effects of selective harvest of non-lactating females on chamois population dynamics. *Journal of Applied Ecology* 51: 1075–1084.

Rughetti, M., Festa-Bianchet, M., Côté, S.D., and Hamel, S. (2017). Ecological and evolutionary effects of selective harvest of non-lactating female ungulates. *Journal of Applied Ecology* 54:) 1571–1580.

Shine, R., Ambariyanto, Harlow, P.S., and Mumpuni (1999). Reticulated pythons in Sumatra: biology, harvesting and sustainability. *Biological Conservation* 87: 349–357.

Sinclair, A.R.E., and Pech, R.P. (1996). Density dependence, stochasticity, compensation and predator regulation. *Oikos* 75: 164–173.

Smith, F.A., Elliott Smith, R.E., Lyons, S.K., and Payne, J.L. (2018). Body size downgrading of mammals over the late Quaternary. *Science* 360: 310–313.

Solberg, E.J., Loison, A., Saether, B.E., and Strand, O. (2000). Age-specific harvest mortality in a Norwegian moose *Alces alces* population. *Wildlife Biology* 6: 41–52.

Sørdalen, T.K., Halvorsen, K.T., and Olsen, E.M. (2022). Protection from fishing improves body growth of an exploited species. *Proceedings of the Royal Society B: Biological Sciences* 289: 20221718.

Spotila, J.R., Reina, R.D., Steyermark, A.C., Plotkin, P.T., and Paladino, F.V. (2000). Pacific leatherback turtles face extinction. *Nature* 405: 529–530.

Swain, D.P., Sinclair, A.F., and Hanson, J.M. (2007). Evolutionary response to size-selective mortality in an exploited fish population. *Proceedings of the Royal Society B: Biological Sciences* 274: 1015–1022.

Swenson, J.E., Sandegren, F., Soderberg, A., Bjarvall, A., Franzen, R., and Wabakken, P. (1997). Infanticide caused by hunting of male bears. *Nature* 386: 450–451.

Tenhumberg, B., Tyre, A.J., Pople, A.R., and Possingham, H.P. (2004). Do harvest refuges buffer kangaroos against evolutionary responses to selective harvesting? *Ecology* 85: 2003–2017.

Thambiturai, D., and Kuparinen, A. (2023). Environmental forcing alters fisheries selection. *Trends in Ecology & Evolution* 39: 131–140. https://doi.org/10.1016/j.tree.2023.08.015.

Therkildsen, N.O., Wilder, A.P., Conover, D.O., Munch, S.B., Baumann, H., and Palumbi, S.R. (2019). Contrasting genomic shifts underlie parallel phenotypic evolution in response to fishing. *Science* 365: 487–490.

Toïgo, C., Gaillard, J.-M., Festa-Bianchet, M., Largo, E., Michallet, J., and Maillard, D. (2007). Sex- and age-specific survival of the highly dimorphic Alpine ibex: evidence for a conservative life-history tactic. *Journal of Animal Ecology* 76: 679–686.

Uusi-Heikkila, S., Lindstrom, K., Parre, N., Arlinghaus, R., Alos, J., and Kuparinen, A. (2016). Altered trait variability in response to size-selective mortality. *Biology Letters* 12: 20160584.

Uusi-Heikkila, S., Whiteley, A R., Kuparinen, A., Matsumura, S., Venturelli, P.A., Wolter, C., Slate, J., Primmer, C.R., Meinelt, T., Killen, S.S., Bierbach, D., Polverino, G., Ludwig, A., and Arlinghaus, R. (2015). The evolutionary legacy of size-selective harvesting extends from genes to populations. *Evolutionary Applications* 8: 597–620.

van de Walle, J., Pigeon, G., Zedrosser, A., Swenson, J.E., and Pelletier, F. (2018). Hunting regulation favors slow life histories in a large carnivore. *Nature Ecology & Evolution* 9: 1100.

Walker, M.J., Shank, G.C., Stoskopf, M.K., Minter, L.J., and DePerno, C.S. (2021). Efficacy and cost of GonaCon™ for population control in a free-ranging white-tailed deer population. *Wildlife Society Bulletin* 45: 589–596.

Webb, S.L., and Gee, K.L. (2014). Annual survival and site fidelity of free-ranging white-tailed deer (*Odocoileus virginianus*): comparative demography before (1983–1992) and after (1993–2005) spatial confinement. *Integrative Zoology* 9: 24–33.

Williams, J.S. (1999). Compensatory reproduction and dispersal in an introduced mountain goat population in central Montana. *Wildlife Society Bulletin* 27: 1019–1024.

Wright, G.J., Peterson, R.O., Smith, D.W., and Lemke, T.O. (2006). Selection of Northern Yellowstone elk by gray wolves and hunters. *Journal of Wildlife Management* 70: 1070–1078.

Zedrosser, A., Pelletier, F., Bischof, R., Festa-Bianchet, M., and Swenson, J.E. (2013). Determinants of lifetime reproduction in female brown bears: early body mass, longevity, and hunting regulations. *Ecology* 94: 231–240.

Zedrosser, A., Steyaert, S.M.J.G., Gossow, H., and Swenson, J.E. (2011). Brown bear conservation and the ghost of persecution past. *Biological Conservation* 144: 2163–2170.

24

Life History and the Control of Diseases

Jessica E. Metcalf and Justin K. Sheen

Department of Ecology and Evolutionary Biology, Princeton University, Princeton, NJ, USA

24.1 Introduction

Pathogens exhibit a vast spectrum of life histories, from mode of transmission between hosts (direct, environmental, vector-borne), to duration of infection (acute, chronic) and host species range (one, or multiple; sometimes in sequence). Since pathogens must grow and/or reproduce on or within hosts and transmit between hosts, the detail of their life histories is determined via an interplay of their own characteristics, those of their hosts, and potentially the environment (Figure 24.1).

Control and management efforts are widely deployed to reduce the impact of pathogens on human populations, as well as wild and domestic animals and plants. Such interventions can be organized into three main categories:

1) Therapeutic treatments (drugs). Drugs are typically provided to individuals experiencing symptoms of disease. They can be administered orally, intramuscularly, intravenously, *etc.*, and have the intended effect of reducing disease symptoms. To understand selection pressures imposed by drugs, it is useful to note that 'disease' and 'infection' are not necessarily synonymous: some infections can be asymptomatic, while symptoms of disease may persist after infection has cleared. Drugs may act directly by killing pathogens and even clearing the infection (*e.g.*, antibiotics), or by modulating the indirect disease outcomes of infection, for example, reducing pathological reactions of the immune system (*e.g.*, steroids).
2) Vaccinations. Vaccination is typically provided prophylactically to prevent infection or disease associated with infection and can also be administered via a range of modes (orally, intramuscularly, *etc.*). Vaccines typically act by exposing hosts to 'antigens', *i.e.*, foreign substances that induce an immune response and trigger the formation of immune memory. This immune memory will leave the host prepared to respond rapidly to pathogens that bear similar antigens. The effect may be complete prevention of infection (the case of the measles vaccine) or reduction or prevention of disease following infection (the case of SARS-CoV-2 vaccines). Partial or short-duration immunity to either infection or disease can also occur.
3) Environmental interventions. A range of interventions have the effect of reducing the environment's suitability for transmission of pathogens. Examples include sanitation (*i.e.*, reduction or elimination of transmission of waterborne pathogens via chlorination), vector control (*i.e.*, efforts such as indoor residual spraying or deployment of larvicides that reduce vector abundance and thus transmission of vector-borne diseases) and non-pharmaceutical interventions (*e.g.*, travel bans, or school closures that reduce contact between individuals and thus prevent spread of directly transmitted pathogens).

Each intervention may drive selection for changes in pathogen life histories. Where pathogens can respond to this selection, the result may be altered outcomes at the scale of a single patient, or at the scale of populations, and may ultimately require a change in the intervention.

Predicting the evolutionary response to selection pressures resulting from control efforts requires knowledge of the trade-offs that govern pathogen life history. The only evolutionary prediction that can be made without information on trade-offs is that survival and reproduction will be maximized (Metcalf 2016). For example, without information on trade-offs, the existence of 'reproductive restraint', a term used to describe the less-than-total commitment to production of gametes in

Life History Evolution: Traits, Interactions, and Applications, First Edition. Edited by Michal Segoli and Eric Wajnberg.
© 2025 John Wiley & Sons Ltd. Published 2025 by John Wiley & Sons Ltd.

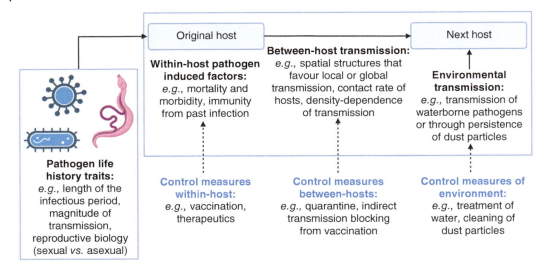

Figure 24.1 Many pathogen life history traits (left box) are shaped by within-host, between-host and environmental factors (right box), which can be impacted by control measures (blue text, dashed arrows). For example, pathogen-induced host mortality will reduce the duration of the transmitting life stage. Pathogen transmission (indicated by solid arrows) may be either direct (from host to host and thus depend on host contact networks) or indirect via a vector or a reservoir host, or via the environment (*e.g.*, waterborne).

blood-stage malaria, is evolutionarily puzzling. This observation can be reconciled with evolutionary expectations if there are costs of reproductive commitment, *e.g.*, to survival, or reproduction of other life stages (Mideo and Day 2008). However, quantifying trade-offs in pathogen systems can require tracking pathogen survival and reproduction within and between hosts. These processes involve phenomena that are hard to measure directly, such as within-host replication. Available measures such as viral titres in the blood or culturable bacteria in the stool may not capture replication in specific organs or organs most relevant to transmission. Such processes may also be conditional on individual features of the host, such as immune status, that may be highly heterogeneous and hard to quantify. Coevolution between hosts and pathogens (*e.g.*, 'Red Queen' dynamics, where pathogens and hosts evolve counter-adaptations to each other in sequence) and other eco-evolutionary feedback can intensify and complicate these dependencies.

Here, we first introduce one of the simplest and well-established quantitative models of infectious disease dynamics (Anderson and May 1992) and use it to illustrate expectations for how pathogen life history should respond to selection, assuming initially that each life history trait can be considered in isolation. We then introduce complexities associated with levels of selection and variance and covariance across life history traits, with particular emphasis on trade-offs. Throughout, we consider how each of the three types of interventions might drive selection based on pathogen life history traits, and how this shapes options for control, encompassing increasing levels of complexity. We conclude by introducing some of the current frontiers in the analysis of pathogen life history responses to interventions.

24.2 Life History Outcomes: A Classic Theoretical Scaffold to Illustrate Predictions

Early work investigating when pathogens should evolve to harm their hosts (Anderson and May 1982) led to the development of a rich array of predictions on pathogen life history trait responses to selection (Kamo and Boots 2006, Mideo and Day 2008, Ashby et al. 2019). Such predictions hinge on how each focal life history trait fits within the pathogen's life cycle (Figure 24.1) and on the existence of direct and indirect trade-offs between different life history traits. For example, the fitness benefit to pathogens of a lengthy infectious period (*e.g.*, a chronic rather than acute infection) will be contingent on the rate at which transmission occurs, *etc*. Theoretical models have been an important tool in the development of such predictions. We start by introducing a classic model (Anderson and May 1992) in the field of epidemiology (the susceptible-infected-recovered model; Figure 24.2) and use it to illustrate some of these predictions, first focussing on the case where different life history characteristics can be considered in isolation, *i.e.*, initially ignoring trade-offs or covariance between different life history traits.

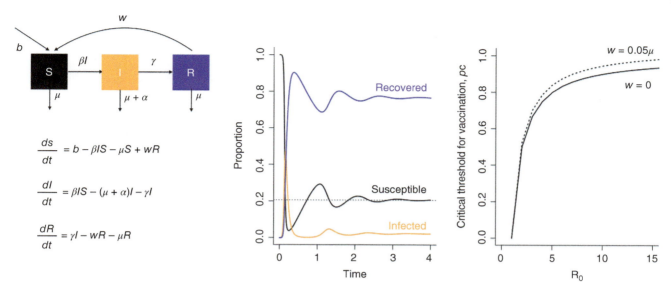

Figure 24.2 The susceptible-infected-recovered model with waning of immunity (schematic and associated equations, left panel). Individuals are born into the 'susceptible' stage at birth rate b, and then may acquire infection at a rate determined by a transmission term β and density of other infected individuals I. They recover at rate γ, and the immunity they have acquired may wane at rate w. Mortality μ occurs in all stages, and infected individuals may incur additional mortality α. If the birth rate b is equal to the mortality rate μ, i.e., $b = \mu$, and additional mortality attributable to infection, $\alpha = 0$, then population size ($N = S + I + R$) is constant, as the rate of flow into the population (birth) is equal to the rate of flow out of the population (death). Setting $N = 1$ corresponds to framing the dynamics in proportions. Following the introduction of an infected individual into a completely susceptible population, the proportion of infected individuals first increases, then declines and oscillates towards an equilibrium as susceptibles are depleted by infection and mortality, and replenished by birth (middle panel, parameters are $\beta = 87, \gamma = 17, w = 0.001, b = \mu = 0.5, \alpha = 0$). Setting $dI/dt = 0$ reveals that the equilibrium proportion of susceptible individuals, S^*, is equal to the inverse of the net reproduction number, $S^* = 1/R_0$. Since R_0 measures pathogen fitness, this indicates that the pathogen strain or species that reduces the susceptible population to the lowest level will outcompete all others. This echoes the concept of the 'pessimization principle' of ecology (Lion and Metz 2018). This framework can also be used to explore the impact of interventions: if vaccination occurs at birth and prevents infection, the rate at which susceptible (unvaccinated) births enter the population will be $b(1 - p)$ with p the proportion of births that are vaccinated. We can then identify the critical proportion of births that need to be vaccinated to ensure pathogen elimination (p_c, right panel) and consider implications of different life histories. The vaccination coverage required to obtain an equilibrium proportion infected $I^* = 0$, or p_c, depends on the magnitude of R_0 and the degree of waning, $p_c = \left(1 - \frac{1}{R_0}\right)\left(1 + \frac{w}{\mu}\right)$.

Although there are many pathogens for which the burden across hosts is highly variable (*e.g.*, many pathogenic worm species), for others, including viral infections like measles or bacterial infections like whooping cough, infection can be considered in binary terms: host individuals either are infected or they are not. For pathogens within this category, although individuals may experience more or less severe infections or lower or higher pathogen loads, these differences have a negligible effect on transmission and thus on pathogen fitness.

Such pathogens may also be directly transmitted and confined to a single host species, so that only contact patterns between individuals of the focal host species need to be considered to understand the spread of infection, and thus pathogen fitness, which will determine the course of pathogen life history evolution. The result is perhaps the simplest of possible pathogen life cycles (Figure 24.2). Hosts are born susceptible (entering the category S; ignoring maternal immunity here for convenience), acquire infection via encounter with an infected individual (entering the category I) and then recover into a state that is associated with immunity from reinfection (entering the category R) immunity may wane, returning individuals to the category S. We can use this simple framing to develop insights into selection pressures on pathogen life histories.

Assuming that the interval during which individuals have been exposed and are infected but not that infectious is short (an assumption that is readily relaxed), one metric that can be used to reflect pathogen fitness is R_0, the basic reproductive number, or the number of new infections per infected individual in a completely susceptible population:

$$R_0 = \frac{\beta}{\gamma + \mu + \alpha} \tag{24.1}$$

where β is the rate of transmission, γ is the rate of recovery, μ is the baseline rate of mortality and α is the additional mortality in infected individuals. Intuitively, R_0 is the balance of loss (denominator) and gain (numerator) of infected individuals in a completely susceptible population, *i.e.*, when the proportion of susceptible individuals $S = 1$ (assuming frequency-dependent transmission) (McCallum et al. 2001).

The expression of R_0 in Eq. (24.1) indicates some of the ways that control interventions may unintentionally select for greater pathogen fitness, in turn eroding the benefits of control measures for hosts. For example, interventions that reduce transmission, β, such as mask wearing or indoor ventilation for airborne pathogens, will also diminish pathogen fitness as measured by R_0, and thus favour the spread of pathogen mutations that reduce the impact of such interventions. For example, one hypothesis for the higher transmissibility of the alpha variant of SARS-CoV-2 that emerged in late 2020, compared to ancestral strains, was that this reflected a response to selection imposed by the reduction in transmission as a consequence of non-pharmaceutical interventions. However, it is difficult to disentangle whether this trait emerged in response to selection imposed by interventions that reduced transmission (of which there were so many over the course of 2020, including movement restrictions, mask wearing and school closures), or selection imposed by a build-up of immunity within the population, or simply reflected that sufficient time had elapsed to allow the right combination of mutations to appear.

Interventions that increase recovery rates γ will similarly reduce pathogen fitness by increasing the denominator in the expression for R_0 in Eq. (24.1), biologically corresponding to reducing the time that hosts are infectious. Thus, pathogen mutations that reduce the impact of such interventions will spread within the pathogen population. Examples include the evolution of drug resistance (*e.g.*, anti-malarials or antibiotics), which prolongs the duration of infectiousness and thus increases the fitness of drug-resistant variants by providing them with more time to spread to new hosts.

In contrast to this, interventions that target reductions of infection-associated mortality α (*e.g.*, therapeutics, toxin-blocking vaccines) but that do not affect pathogen burden *per se*, should not see their benefits to hosts eroded by selection (Allen et al. 2014), since reducing infection-associated mortality equates to reducing the magnitude in the denominator of the expression for R_0, and therefore increasing fitness, again by providing pathogens with more time to spread to new hosts. It is interesting to note that many pathogens induce high host mortality or morbidity (also termed virulence), even though this should reduce their fitness. This can be explained by covariation between other life history traits, such as transmission, with virulence, resulting in trade-offs between the benefits (*e.g.*, higher transmission) and disadvantages of increasing virulence. We introduce core concepts around trade-offs in life history evolution here, using the example of virulence evolution, to illustrate the potential need to expand beyond the focus on single traits (transmission, recovery) deployed to this point.

It is often suggested that increases in transmission may be tied to increased virulence (Anderson and May 1982). Mechanistically, this might occur because, for example, higher viral titres of a respiratory pathogen in the lungs might not only result in higher transmission but also higher risk of mortality for the host (Figure 24.3). This would mean that reduced virulence (captured by the term α in the denominator of R_0) can only be achieved at the cost of reduced transmission (captured by the term β in the numerator of R_0), as depicted in Figure 24.3a, so that pathogens might be unlikely to evolve towards no virulence, $\alpha = 0$, even though virulence levels above 0 result in a reduction in transmission time and thus fitness. The shape of the relationship between these two traits (transmission β and virulence, α) will determine what magnitude of virulence maximizes fitness for the pathogen (Figure 24.3c), corresponding to the optimal strategy and the expected outcome of selection, all else equal and genetic variation permitting. Importantly, this shape may be difficult to quantify (for these or any other pairs of pathogen traits), yet it will be essential to predicting pathogen responses to selection in both the absence and presence of interventions, further detailed below.

Beyond covariance between different traits, the details of the overall life history will be important in shaping selection pressure on traits – for example, microbes that transmit not only horizontally (*i.e.*, between different hosts) but also vertically (where offspring inherit microbes from their mothers) may evolve reduced virulence (Ewald 1987), since host reproduction now aligns with pathogen transmission (Lipsitch et al. 1996), a feature not reflected in the model depicted in Figure 24.2.

While R_0 in Eq. (1) provides a useful starting point for considering how interventions alter pathogen fitness, it also has limited scope. For example, control measures such as vaccines may alter the pool of available susceptibles or reduce the rate of waning of immunity, w (shown in Figure 24.2). Intuitively, it is clear that there will be selection for pathogens to escape from immunity generated by either vaccination or prior infection since access to more susceptibles will allow pathogens to spread faster, as exemplified by the higher critical threshold of vaccination for elimination in the presence of waning (Figure 24.2, right-hand panel). However, the expression of R_0 provided in Eq. (24.1) contains neither w, nor the size of the susceptible population – and, therefore, cannot capture the role of competition for susceptibles beyond the phase of early growth in determining pathogen fitness. More nuanced expressions of fitness will be required to capture the role of competition for susceptibles.

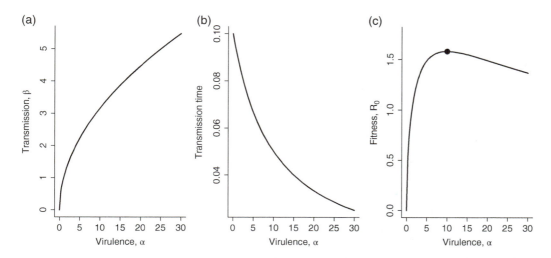

Figure 24.3 Evolution of virulence. The persistence of virulence or the tendency for pathogens to kill their hosts is initially puzzling because infected host mortality reduces the time available for pathogens to transmit, reducing pathogen fitness (indicated by the presence of α in the denominator of R_0). This paradox is resolved if virulence and transmission are correlated, as depicted in (a), which might occur if, for example, higher parasite burden translated into greater transmission potential but also greater host mortality (empirical evidence is reviewed in Acevedo et al. 2019). The benefits to the pathogen of high transmission corresponding to high virulence (a) must then be balanced against a shorter time available for transmission as a result of infected host mortality associated with virulence (b). Combining these two components results in an intermediate level of virulence mapping to the highest fitness as measured by R_0 (c), which is also the expected outcome of selection. Very low virulence results in low levels of transmission and thus lower fitness, while very high virulence results in a short transmission time as a result of rapid mortality and thus also lower fitness. Intermediate virulence might thus evolve.

There are a range of further situations where R_0 maximization might not adequately reflect evolutionary outcomes. Over short time scales and at the start of an outbreak, returns on immediate growth will outweigh benefits to the pathogen of equilibrium performance and the per infected growth rate of pathogens described in the following equation:

$$\frac{dI}{Idt} = \beta S - \gamma - \mu - \alpha \tag{24.2}$$

This might provide a more appropriate measure of fitness than R_0 (Day and Proulx 2004, Day and Gandon 2007, Bull and Ebert 2008).

Over the longer term, pathogen competition may require methods for modelling evolution that allow for density- and/or frequency-dependent feedback to reflect situations where the success of a particular pathogen mutation will depend on the density or frequency of pathogen variants currently circulating in the population. One approach is to use game theory and separation of time scales, also termed 'adaptive dynamics', to explore evolutionary outcomes (Geritz et al. 2013). Under this framing, ecological processes are assumed to occur on more rapid time scales than evolutionary processes so that evolutionary dynamics can be considered to occur only once ecological equilibria have been reached. Evolutionary outcomes can then be explored by a sequence of 'invasions' of different strategies into an environment set by a 'resident strategy' – if the invader can successfully invade, it becomes the new resident strategy, and the process is repeated until a strategy that cannot be invaded by any other strategy is identified. This strategy is termed the Evolutionarily Stable Strategy, or ESS. The different approaches may coincide, such that the strategy identified under fitness maximization (e.g., the maximal R_0) also indicates the strategy that cannot be invaded by any other strategy (ESS), but this is not always the case (Lion and Metz 2018). For example, if mortality is density-dependent (e.g., $\mu = \mu_0 + kN$, where k captures the additional impact of population size N on background host mortality μ_0), the context into which a novel mutant pathogen must invade is defined by two features set by the current resident pathogen strategy: (1) the landscape of host susceptibility to infection (a more transmissible resident pathogen will deplete susceptibles to a lower level, reducing invading pathogen spread) and (2) the total population density of hosts, which defines total host mortality (a more virulent or more transmissible resident pathogen will reduce the total host population size, thereby reducing the background mortality rate and potentially increasing spread of an invader). The related fitness proxy is:

$$R = \frac{\beta \hat{S}}{\gamma + \mu_0 + k\hat{N} + \alpha} \tag{24.3}$$

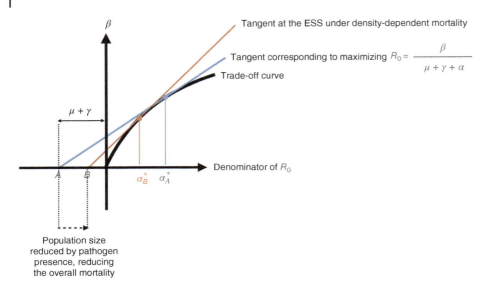

Figure 24.4 Evolution of virulence: beyond R_0 maximization. Where higher pathogen burdens result in both higher infected host mortality and higher rates of transmission to the next host (curved black line), this is termed a 'virulence-transmission trade-off' curve. An interesting life history question is then the evolutionarily optimal pathogen level of virulence (see also Figure 24.3). Another way to consider the optimization problem is to note that the largest feasible slope on this landscape will yield the highest R_0 by yielding the greatest increase in the numerator of R_0 (transmission, β, y axis) per unit increase in the denominator of R_0 (x axis, including mortality and recovery, $\mu + \gamma$, shown left of the vertical axis starting at point A, and virulence, α starting at the vertical axis). Mortality and recovery have no effect on transmission, which is accordingly zero across their span, i.e., to the left of the vertical axis. The question is then how much to increase α to maximize fitness. The line passing through A that is tangent to the trade-off curve indicates the greatest possible slope, and thus value of virulence α_A^* that corresponds to the highest R_0 (i.e., maps to the α value corresponding to the peak in Figure 24.3c). If we can ignore the effect of the pathogen on host population density, this is also the ESS, or strategy that cannot be invaded by any other strategy. However, if some other form of feedback is operating, e.g., density-dependent mortality, then the start of the x axis may be altered by reduced host mortality because pathogen presence might, for example, reduce population size (depicted in point B). In this case, the ESS virulence α_B^* may no longer correspond to the quantity that maximizes R_0 if the resident strategy's effect on fitness via its effect on host mortality is not considered, as illustrated by the altered tangent (Adapted from Lion et al, 2018.).

where the "*" symbol indicates the two environmental features (susceptibles and total population size) set by the resident pathogen strategy. We can illustrate the difference in evolutionary optima and how it emerges graphically. In Figure 24.4, point A reflects a situation ignoring the effects of pathogen strategy on host population density so that the optimal is only defined by susceptible availability and host mortality in the absence of the pathogen. In point B, the existence of density-dependent mortality means that the current resident pathogen population alters the background mortality μ in the host population into which mutant pathogen invasion is occurring (since $\mu = \mu_0 + kN$). This alters the tangent and thus the optimal virulence (Figure 24.4, orange line). Since the resident affects the environment in two different ways (e.g., first via susceptible availability and second via the mortality pressure resulting from density-dependence), this expression does not lend itself to one-dimensional optimization (Lion and Metz 2018, Ashby et al. 2019), and thus maximization of R_0 is no longer adequate to capture fitness outcomes. A wider lens on possible infectious disease dynamics illustrates that ESS outcomes, but also branching points and cycling scenarios, are possible (Dieckmann et al. 2004).

Separation of time scales or sequential evaluation of ecological and evolutionary outcomes has proved to be a useful tool for analysis of many evolutionary questions but may not be sufficient for some aspects of pathogen evolution (Luo and Koelle 2013). Pathogen population dynamics are often far from endemic equilibrium (Day and Gandon 2007, Sasaki et al. 2022). An array of models with overlapping time scales of ecology and evolution have been developed to address this issue, including the evolution of pathogen ability to evade immunity (also termed immune escape) by mutations that alter pathogen features recognized by immunity (termed 'antigenic drift'). Early work established the requirement of immunity that could recognize a spectrum of mutations (termed 'strain-transcending immunity') to produce evolutionary trajectories in pathogens that would reflect the empirically characterized 'ladder-like phylogenies' of pathogens such as influenza (Ferguson et al. 2003), where ancestral variants repeatedly go extinct (Figure 24.5). Population genetic approaches addressing the same issue were subsequently formalized (Day and Gandon 2007), and an analytical framework encompassing both

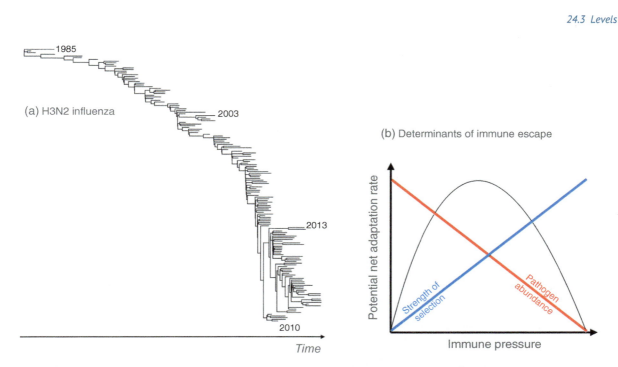

Figure 24.5 Immune escape. (a) The phylogeny of the influenza A H3N2 viruses, shows a 'ladder-like' structure (from Volz and Frost 2017), suggesting the footprint of immune adaptation where each year, the circulating strain goes extinct and is replaced by a new strain that can escape the updated circulating immunity. Position along the x axis indicates the passage of time, and numbers at label tips indicate years of circulation of a few of the strains. The ladder-like phylogeny reflects repeated selection and spread of 'immune escape' variants, raising the question of what contexts are likely to most favour immune escape. (b) Potential for immune escape should peak in contexts of intermediate immune pressure (Grenfell et al. 2004) since the strength of selection will be zero in the absence of immune pressure and increase as immune pressure increases (blue line), while, conversely, pathogen population sizes will be greatest when immune pressure is close to zero and close to zero when immune pressure is at its highest (red line), and pathogen population size is a key determinant of the occurrence of escape mutants. Since pathogen adaptation will reflect a combination of selection pressure and the occurrence of escape mutants, adaptation to escape immunity will peak at intermediate levels of immune pressure, noting that little is known about the details of the shape of this landscape in any real system.

immune escapes via antigenic drift and virulence was recently developed and revealed that 'antigenic escape' selects for evolution of higher pathogen transmission and virulence (Sasaki et al. 2022).

The most appropriate model framework to capture eco-evolutionary dynamics will be dictated by details of both the ecology of infection and immunity in the focal system (mode of transmission, acute *vs.* chronic infection, mode of competition between pathogens, *etc.*), as well as details of the determinants of the emergence of novel pathogen variants (mutations, recombination), which might have higher fitness. In some cases, additional complexity emerges from the fact that pathogen life history is shaped by different levels of selection.

24.3 Levels of Selection

The model underlying the fitness expression captured by Eq. (24.1) assumes that selection occurs at the scale of the individual host within the host population. In fact, pathogen biology features multiple levels of selection (Figure 24.6). Alleles that maximize fitness on one scale may not translate into fitness on another. For example, although alleles that improve persistence or replication of pathogens within a host will come to dominate within-host–pathogen populations (and this may occur following principles laid out in the previous section, but at the scale of, *e.g.*, movement between host cells, rather than movement between hosts), these alleles may or may not map to traits that enhance between-host spread (*e.g.*, via impacts on transmission, recovery, virulence or waning, pathogen life history traits indicated on Figure 24.2). Mutations in the SARS-CoV-2 spike protein seem to have increased both within- and between-host transmission, making the virus able to bind more tightly to host cell receptors, thus increasing within-host growth and enhancing immune escape, and increasing spread to previously immunized hosts. However, there is no general reason for mutations that are advantageous

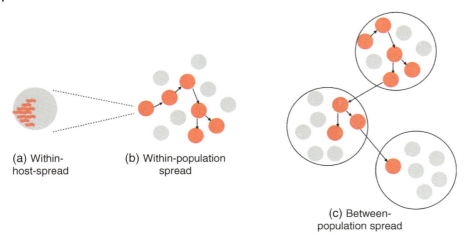

Figure 24.6 Levels of selection (a) over the course of an infection, pathogens (shown in red) replicate and spread within a host (grey-filled circle). Mutations that enhance replication and evasion of adaptive and innate immunity may reach high prevalence. Decline in pathogen prevalence overall may be dictated by immunity or target cell depletion, or, in the extremes, host mortality. (b) Each infected host's probability of transmission per unit time (which may be dictated by within-host–pathogen prevalence) but also duration of infection (dictated by immune control, target cell depletion, or virulence) will shape movement between hosts (grey and red indicating susceptible and infected hosts, respectively), as could the existence of an environmental transmission phase. (c) The intersection between immunity or mortality and spread within populations may affect spread between populations; if susceptible depletion occurs faster than movement of infected individuals between populations, then pathogens may fail to spread beyond their initial population and go extinct.

for within-host transmission or persistence to translate into enhanced between-host transmission. Faster within-host transmission (Figure 24.6a) may map to greater virulence, with the effect that alleles that rapidly dominate within-host–pathogen populations may also result in rapid host mortality, meaning that there is no opportunity for pathogen transmission to the next host (Figure 24.6b). For viral pathogens, the potential for opposing directions of selection at different scales, paired with narrow bottlenecks occurring during transmission (*i.e.*, only a very small number of virions may move between hosts, estimated, *e.g.*, at 2–8 in SARS-CoV-2), may dramatically slow pathogen responses to selection (Sobel Leonard et al. 2017, Braun et al. 2021).

Shifting up a scale, pathogens that spread effectively within a population (Figure 24.6c) may deplete susceptible hosts (by immunity or mortality) too rapidly to ensure persistence for sufficiently long to be transmitted to another geographically separated population (Restif and Grenfell 2006). Relatedly, in small populations, demographic stochasticity may weaken postulated relationships between virulence and transmission (Figure 24.3a), such that chance events give advantage to less variable strategies (Parsons et al. 2018). Stochastically varying the population sizes will also favour slower strategies with long infectious periods and less transmission compared to faster strategies with short infectious periods and more transmission, simply because slower strategies will not have as many transmission events per unit time, and thus will encounter fewer 'bad luck' population fluctuations. (Parsons et al. 2018). However, although predominantly local, short-distance contact between individuals can select for reduced virulence, even relatively rare occurrences of transmission between distant populations have been shown to lead to more global epidemics, eliminating this effect (Watts and Strogatz 1998, Boots and Sasaki 1999). Reduced importance of local contact patterns may be an inevitable feature of our increasingly connected world. It may also emerge when the pathogen itself is able to evolve to disperse over greater distances (Kamo and Boots 2006). However, in some circumstances, predominantly short-distance contacts can have the reverse effect on selection on virulence: if new (susceptible) hosts can only recruit following host mortality, low virulence strategies might block their own spread by infecting, but then immunizing, all their contacts, leaving them with no susceptible hosts to spread to, while highly virulent strategies might kill hosts, freeing up area for recruitment of new susceptible hosts that would allow them to spread (Boots et al. 2004).

How pathogens respond to selection imposed by control efforts will be shaped by how control efforts intersect with these different levels of selection. Therapeutics often aim to reduce symptoms as much as to cure infections. For example, aspirin can be taken to reduce fever, antibiotics can be taken to reduce the painful tonsils associated with strep throat, *etc.* Symptoms are likely to be in part determined by the within-host–pathogen burden (Figure 24.6a), which could be associated with within-host fitness, but, importantly, the relationship may be slight or fleeting. Possible reasons for the disconnect include,

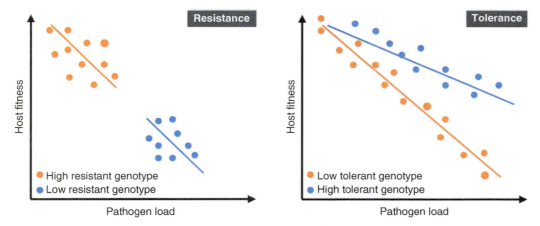

Figure 24.7 Host heterogeneity in pathogen responses. Distinction can be made between-host genetic variation in resistance (left), where some hosts can more effectively diminish the pathogen load, which results in an increase in fitness (both lines are negative, since pathogen load always reduces fitness, but the 'orange host' more effectively reduces burden and obtains higher fitness than the 'blue host'), and host genetic variation in tolerance (right), where at the same pathogen load, some hosts suffer a much-diminished effect on fitness (and likely health) than others, with much interesting work around how such variance can be maintained (Best et al. 2008). The underlying biology can drive varied effects on pathogen transmission, with, *e.g.*, higher tolerance resulting in more 'competent' hosts as they are capable of carrying the pathogen and transmitting it for longer (Ruden and Adelman 2021).

Vaccination can change the trajectory of the within-host–pathogen burden (Figure 24.6a) in vaccinated hosts; in the best case, it largely blocks infection (*e.g.*, the measles vaccine), but in some cases, it prevents disease outcomes, for example, by preventing the action of toxins (*e.g.*, the pertussis vaccine; Miller and Metcalf 2019). Covariance between pathogen traits can mean that such symptom-blocking vaccine impacts can drive selection for greater virulence in unvaccinated hosts. This occurs because higher virulence correlates with higher transmission, and in vaccinated hosts, the costs of virulence associated with host mortality are reduced or eliminated (Gandon et al. 2001). Conversely, transmission-blocking vaccines that elicit immunity that prevents pathogen colonization will drive selection for pathogen variants that are no longer detectable by host immunity (*i.e.*, they can 'escape' immunity). This selection pressure plays out at both within- and between-host levels (Figure 24.6a,c), as the success of a variant that can escape one host's immunity will depend on the degree to which other such hosts within the population are accessible to it. Covariance between traits will further modulate trajectories.

first, host adaptations associated with host 'tolerance' of pathogens (Figure 24.7), such that high burden has little effect on health. A second possible reason is the dangers posed by immunity (Graham et al. 2022): in the extremes, feedback loops of signalling molecules ('cytokine storms') may drive excessive immune responses that cause catastrophic mortality of healthy cells in such a way that even low burdens of infection could have a large effect on health. Therapeutics that only reduce symptoms (*e.g.*, steroids that only attack inflammation) might exert minimal selection pressures on pathogens by minimally affecting pathogen fitness (at the within- and between-host level, Figure 24.6a,b), noting that they might also increase fitness indirectly by reducing host mortality. However, in most cases, therapeutics are likely to also affect pathogens directly, and resistance to therapeutics has evolved repeatedly.

Environmental interventions can reduce within-population spread (Figure 24.6b), *e.g.*, by reducing persistence of environmental transmission stages, vector populations, or contact between individuals. Non-pharmaceutical interventions, such as travel bans implemented during the COVID-19 pandemic, may also affect mobility between populations (Figure 24.6c). All of these processes may differentially affect selection pressures on variants, *e.g.*, slowing down or speeding up the rates of spread (Otto et al. 2021).

Shifting from directly transmitted infections to vector-borne infections (*e.g.*, malaria, dengue, lyme disease), the question of levels of selection emerges again, with the effective addition of an extra host (*i.e.*, the vector). Evolution of resistance to insecticides is a recurrent problem. To counter this, insecticides that target senescent vector individuals have been proposed. The advantage is that such treatments will exert minor selection pressures on vectors, so minimal evolution of resistance to the insecticide is expected. Since it takes pathogens like malaria a long time to develop within the mosquito, such insecticides may considerably reduce pathogen burden (Read et al. 2009). Control efforts that combine treatments that spatially repel vectors from indoor spaces with insecticide treatment of these spaces have been indicated as another potentially profitable path to reduce the burden of malaria (Lynch and Boots 2016). In this case, vectors may be strongly selected to respond to the spatial repellent, which will keep them out of indoor spaces where they would otherwise die. Response to this dual selection pressure will have the beneficial effect of reducing human biting in indoor spaces.

24.4 The Complexities of Variance and Covariation in Empirical Systems

Even for pathogens whose life cycle is broadly reflected by the basic S-I-R model (Figure 24.2) and without complex feedback between levels of selection (Figure 24.6), developing predictions about the evolutionary outcome of control efforts is difficult. This is largely because of the challenge of characterizing the rich degree of variation and covariation across host individuals in the manifestation of pathogen life history traits. The latter is of particular importance in reflecting key trade-offs, as introduced in Figure 24.3a. We address variance and covariance in an array of different pathogen life history traits and their implications for selection as applied by control efforts in turn.

24.4.1 Variance

Transmission often varies between hosts, even for largely identical pathogens. At the most extreme, a small fraction of host individuals are responsible for most of the onwards transmission (Lloyd-Smith et al. 2005), implying that the number of new infections per infected individual, R_0 is both variable across individuals and highly skewed (Figure 24.8). Examples include SARS-CoV-2, where, early during the pandemic, the average number of new infections per infected individual was around

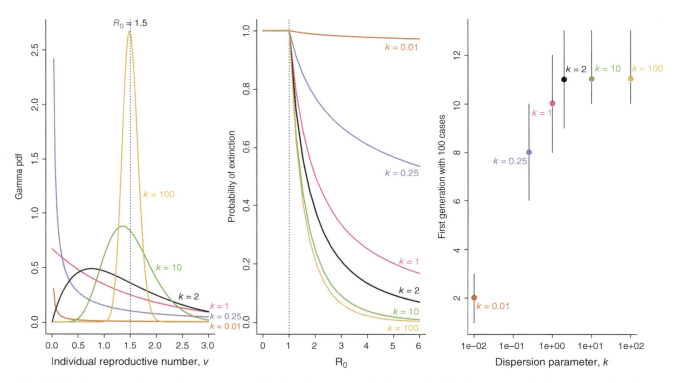

Figure 24.8 Superspreading and the dynamics of infection. A spectrum of pathogen life histories with an identical average $R_0 = 1.5$ but ranging from limited individual variability (yellow line centred around 1.5) to extreme variability and superspreading (long-tailed orange line) can be captured by modelling individual reproductive numbers, denoted v (left-hand panel, x axis), as following the probability density function (pdf) of a Gamma distribution with mean 1.5 (y axis). Setting the dispersion parameter $k = 100$ corresponds to very little variability (yellow line, tight peak around 1.5), while a small value of the dispersion parameter, e.g., $k = 0.01$ results in extreme variability (orange line, distributions with a long tail to the right). Other colours reflect magnitudes of k shown in the matching coloured labels. The shape of these distributions has consequences for the probability of stochastic extinction of the pathogen following introduction into a completely susceptible population (middle panel, y axis), which also depends on R_0 (middle panel, x axis), as can be derived using a branching process model (Lloyd-Smith et al. 2005). The shape of the distribution of individual reproductive numbers v also determines how explosive an outbreak will be following the introduction of the pathogen into a completely susceptible population, here captured by simulating the introduction of a pathogen with a mean $R_0 = 1.5$ into a susceptible population and tracking the first generation at which 100 cases are observed (right panel, y axis), for different values of k (x axis, colours as before). Pathogens characterized by more superspreading events (small values of k) reach 100 cases on average much faster, shown here across 8000 simulations, excluding simulations where the pathogen went extinct. Points indicate the median time to the first generation with 100 cases, while vertical lines indicate the interquartiles of the time to the first generation with 100 cases (asymmetries may emerge as the distributions can be long-tailed).

two, but there were clear (although relatively rare) examples of situations where one individual had infected more than 40 others (Althouse et al. 2020). Ebola is another example: during the 2014 outbreak, around 70% of infected individuals generated no further cases, while some generated more than 40 (Faye et al. 2015). This variation may be the result of biological differences among hosts (more or less social individuals with smaller or larger number of contacts, more or less resistant individuals or individuals with lower immunocompetence who transmit more per time unit or for longer; Illingworth et al. 2021, *etc.*) or even ecological differences (*e.g.*, positioning next to an air vent in a restaurant and living in higher density subpopulations).

Outbreaks for pathogens characterized by superspreading (*i.e.*, large variance in transmission across individuals, Figure 24.8) are both more extinction-prone (middle panel in Figure 24.8 indicates that transmission with higher variance results in greater extinction probability for equivalent R_0) but also more explosive if they do take off (right-hand panel on Figure 24.8 indicates a shorter time on average to 100 cases for equivalent mean transmission but higher variance). Control measures that target individuals responsible for the bulk of the spread will clearly be most effective, but this is often hard to achieve, in part because it is hard to identify the host individuals that drive spread sufficiently early to reduce their impact. Therapies that can transmit from the treated individual to their contacts (*e.g.*, live-attenuated vaccines, like the polio vaccine or 'viral-vectored vaccines' where a sequence that encodes the immunizing element required to block the pathogen is introduced into an asymptomatic existing or synthetic virus that transmits within the target population), and that use the same transmission networks as the pathogen might be particularly powerful in these situations (Metzger et al. 2011): such 'transmissible vaccines' might rapidly reach highly connected points in the transmission networks. However, such therapies will suffer the same extinction risk as the pathogens themselves (Figure 24.8, middle panel), and repeated reintroduction may be necessary. In general, the utility of transmissible therapies will depend on how their replication compares with and intersects with that of the pathogen, but also, in the case of engineered viral vectors, on the degree to which the introduced sequence required for the host to develop immunity to the target pathogen is lost. The sequence is likely to provide the viral vector with no fitness benefits and might incur costs, so the vaccine viral vector will be under selection to lose this sequence. Simulation of the transmission of a candidate vectored rabies vaccine and the rabies pathogen showed that inoculating a single bat could immunize >80% of a bat population, reducing the size, frequency and duration of rabies outbreaks by 50–95% (Griffiths et al. 2023), even accounting for vaccine reversion (*i.e.*, loss of the introduced sequence), making this a promising direction for control.

Beyond the examples above, three additional drivers of variance around transmission in largely identical pathogens should be noted: (1) chance events; (2) pathogen adaptive plasticity (Cornet et al. 2014) and (3) pathogen adaptive intra-genotypic variance (Bruijning et al. 2020). For the first, stochastic processes at the within-host level (Figure 24.6a) may drive highly variable transmission outcomes. Indeed, the simple fact that pathogens are generally invading networks of cells can yield vast differences in the duration of the incubation period (Ottino-Loffler et al. 2017), which could translate into variation in transmission time. For the second, pathogens may adapt to adjust behaviour to maximize transmission in context (adaptive plasticity). For example, malaria parasites may respond to the presence of vectors to increase density in the blood (Cornet et al. 2014). For the third, in a context of variable and unpredictable environments, pathogens may evolve bet-hedging strategies, where, for example, even genetically identical offspring display diverse life histories. By chance, a subset of the offspring will happen onto the optimal strategy for the environment experienced. For example, herpes viruses may enter a period of latency (no transmission) of variable and unpredictable duration allowing them to evade immunity, and potentially await build-up of susceptible individuals or average across uncertain environments (Stumpf et al. 2002). Pathogen strategies involving latency often complicate therapeutic treatment (it is often hard to attack metabolically inert pathogens) and may also complicate surveillance and monitoring of pathogen control efforts. A better understanding of the cues that pathogens use for plastic switching, or their absence, in the case of pure bet-hedging, could valuably inform public health strategies.

The degree of immunity to pathogens is often variable across hosts. This will both shape and be shaped by evolutionary dynamics. Pathogens with rare phenotypes (antigens) are likely to evade recognition by the immune systems of most hosts within the population and therefore will be at an advantage (a phenomenon termed 'negative frequency-dependence'). In the absence of adaptive immunity (*i.e.*, an immune system capable of immune memory), hosts with genotypes encoding the ability to recognize and resist the most abundant pathogen antigens might then be selected for and increase in frequency in the population. This will drive selection for novel rare pathogen variants, which will select in turn for a rise in frequency of the hosts capable of recognizing these new pathogens, driving a process of cycling between host and pathogen traits termed 'Red Queen Dynamics'. Where adaptive immunity is an important line of host defence, a host's history of exposure shapes their vulnerability to infection, as the presence of adaptive immunity means that the immune system can 'learn' to defend hosts against infections seen on past exposures. This again means that rare pathogens are most likely to escape the net of established immunity, and cycles of immune-associated variation in hosts and pathogens are again expected, but host cycles might now be embedded within a host generation rather than between generations (*e.g.*, amplification of host B-cell clones that produce receptors that best match the highest frequency pathogen antigens will occur following exposure).

Either within- or between-generation pursuit by host immunity of recognized pathogen features (antigens) might result in a 'ladder-like' phylogeny in pathogens (Figure 24.5a), reflecting regular extinction of 'common' pathogen ancestors that were identified by hosts' immunity and thus driven to extinction. Vast efforts are deployed to anticipate this process, given the importance of vaccine development (Łuksza and Lässig 2014), but considerable uncertainty remains. The ideal workaround would be 'variant proof vaccines' that train immunity to target some highly conserved (and thus likely tightly constrained) part of the pathogen (Pica and Palese 2013). Theory has indicated that such vaccines could dampen transmission and reduce prospects of pathogen evolution of escape from immunity (Subramanian et al. 2016), but the practice has proved technically very difficult. 'Mosaic vaccination', where different individuals are vaccinated with different antigens, making a complex immune landscape for the pathogen to evolve across, might achieve similar effects (McLeod et al. 2021). Both strategies hinge on modulating the effect of selection pressures on the pathogen applied by immune memory within hosts. Some pathogens have evolved to evade such selection pressure. For example, malaria deploys transcriptional switches among their *var* gene family such that new sequences are periodically expressed by the pathogen over the course of an infection (Otto et al. 2019), such that the development of immune memory to any one variant will be ineffectual. This may in part underpin the vast challenge of developing vaccines against malaria.

Pathogen adaptation rate towards immune escape is expected to peak in hosts with intermediate scales of immunity (Figure 24.5b), between extremes of no selection in the absence of immune pressure, and no pathogens to evolve where immune pressure is of sufficient magnitude as to achieve pathogen elimination. How vaccinal or natural immunity, or immunity after a second vaccine dose, or boosting or waning shape both selection pressure and pathogen clearance (Figure 24.5b, x axis) emerges as an important question. The issue had particular salience when delays between the first and the second dose deployed for SARS-CoV-2 were considered in 2020 (Saad-Roy et al. 2021). Complexities in the timing of viral diversification and host immune responsiveness may further complicate selection pressure on immune escape and how it manifests across scales (Morris et al. 2020).

Individual variation among hosts in rates of recovery, indicating clearance of infection (captured by the parameter γ, shown in Figure 24.2), might be rooted in host genetic differences in what is termed 'resistance' (Figure 24.7) (Råberg et al. 2007), or in stochastic processes happening at the within-host scale (Metcalf et al. 2020), or, as for waning (captured by the parameter w, shown in Figure 24.2), on the history of immunity that different hosts have experienced. Variable host immuno-competence may also be driven by coinfections with immuno-modulatory pathogens like HIV and worms. Considering these various sources of heterogeneity is important, since, for example, it could modulate how selection acts on pathogen resistance to therapeutics (Birger et al. 2015).

Virulence also varies from host to host, here expanding the definition from purely an impact on additional mortality of infected individuals (captured by the parameter α on Figure 24.2) to consider a range of health impacts that may link to fitness. For many pathogens, some hosts may experience almost no symptoms (in the extreme, they may be asymptomatic), while others experience extremely severe outcomes and even death. Some of this variability may be attributed to stochastic within-host processes (Ottino-Loffler et al. 2017). The surprisingly large role that the immune system itself plays in adverse health outcomes and its sensitivity to initial conditions may also contribute to this variability among hosts in health outcomes (Graham et al. 2022). As mentioned above, pathogens may not experience any selection to escape the control of therapeutics that result in recovery from symptoms alone, if symptoms and infection states (pathogen abundance, transmission, *etc.*) are uncoupled. Therapeutics that increase the period of transmission by improving patient health could even increase pathogen fitness. However, as the example of the repeated evolution of antibiotic resistance readily illustrates, there is often considerable correlation between pathogen abundance within the host and host health/symptoms so that evolution of resistance to therapeutics designed to prevent symptoms readily evolves. This, and other forms of covariance between pathogen life history traits, will be a critical element in untangling evolutionary trajectories, as discussed below.

24.4.2 Covariance

Correlations among pathogen life history traits are widespread for two reasons. First, the underlying within-host dynamics of infections drive correlations among virulence, transmission and recovery (Mideo et al. 2008). For example, rapid recovery may often translate into lower virulence purely mechanistically (the rapid absence of pathogens translating into a minimal health burden if pathogens are the key source of disease), or both transmission and virulence may be associated with higher viral load, forcing a correlation between them (Bull and Antia 2022), *etc*. Second, all else being equal, selection pressures on such traits will be inter-dependent. For example, selection on virulence is expected to depend on the rate of recovery: higher virulence might be tolerated if recovery is rapid (van Baalen 1998) or in the context of rapid immune escape (Sasaki et al. 2022).

This complicated web of inter-dependence makes characterizing how pathogen life history will respond to control efforts extremely tricky. Predicting an evolutionary trajectory requires knowledge of core trade-offs (Metcalf 2016), yet these trade-offs are dependent on characteristics of the host, environment, pathogen itself, and the dynamic, path-dependent transmission of infection. (Bull and Antia 2022), path-dependent, and involving characteristics of the host, the environment and the pathogen itself.

All of this has implications for the selection pressures that control strategies place on pathogens. To take a previously introduced example, vast amounts of theoretical work have developed around the implications of a relationship between level of virulence and magnitude of transmission. The basic intuition is that a feature like viral titre might amplify transmission potential but could lead to more rapid host mortality or reduced mobility, both of which would reduce transmission time by shortening the period of infectiousness of the pathogen (Anderson and May 1982), opening the way to evolution of intermediate virulence (*e.g.*, see Figure 24.3). However, the empirical evidence for this relationship remains variable (Alizon et al. 2009, Cressler et al. 2016, Acevedo et al. 2019), with little evidence for it in vector-borne plant pathogens, for example (Froissart et al. 2010). Where such a relationship does exist, the details will matter, leading to calls for further work to frame this relationship within the context of within-host dynamics (Bull and Antia 2022). The broader life history context also needs careful quantification. For example, theory indicates that, for pathogens with an environmental transmission stage, introducing incomplete cleaning, that only partially removes infected particles in agricultural settings can select for more virulent pathogens, as cleaning generally reduces pathogen lifespan, selecting for strains that can survive the cleaning period through increased transmission (Rozins and Day 2017), but this outcome will be sensitive to the magnitude of cleaning effects. Vaccination might also lead to unintended evolutionary consequences for the pathogen when immune escape and virulence also interact indirectly via the population scale, implying not just evolutionary but eco-evolutionary dynamics. Repeated escape and invasion lead to selection of variants that maximize the population rate of increase, r, rather than the effective reproduction number R_0, *i.e.*, a 'live fast, die young' strategy (Sasaki et al. 2022), with higher virulence.

The nuance of overlap between symptoms and transmission and when they occur (*e.g.*, asymptomatic transmission early in the infection) and how they link to the rest of the pathogen life history (recovery, *etc*.) can result in a diversity of evolutionary outcomes (Saad-Roy et al. 2020). If asymptomatic infection is associated with lower transmission, control measures that reduce transmission via a reduction in contact rates may force the pathogen to evolve towards reducing the length of an early, latent asymptomatic stage since the benefits to the pathogen of transmission occurring as a result of infection being undetected will be lost. As asymptomatic transmission can be a root cause of high infection burden in a population due to its ability to evade detection, driving evolution to reduce the early asymptomatic stage would be an additional benefit to hosts (us) conferred by control measures. Thus, control measures may not only mitigate transmission in the short-term but have long-term evolutionarily useful benefits to us (the hosts) by shaping the future life history traits of the pathogen.

Despite remaining uncertainties, there are some relatively well-understood examples of response to selection applied by an intervention to a pathogen life history trait in the context of covariation between life history traits. One particularly crisp one is that of the effect of 'imperfect vaccines' (*i.e.*, vaccines that reduce pathogen virulence but not the magnitude of pathogen transmission in vaccinated hosts) on the evolution of pathogen virulence. In the presence of covariation between virulence and transmission (*e.g.*, Figure 24.3a), theory indicates that deploying such vaccines would allow pathogens to evolve greater virulence (Gandon et al. 2001), because pathogens would no longer pay the costs of virulence in vaccinated hosts (*i.e.*, reduced time for transmission) but would gain the benefits of increased transmission. The evolutionary trajectory of a poultry pathogen, Marek's disease, following deployment of a toxin-blocking vaccine, has been empirically demonstrated to follow this theoretically predicted trajectory (Read et al. 2015).

24.5 Frontiers in Life History Evolution and Pathogen Control

24.5.1 Emerging Pathogens

The SARS-CoV-2 virus will not be the last zoonotic spillover. Identifying host and pathogen life history characteristics that define pathogens most likely to spill over into human populations and cause serious disruption would open the way to developing targets for therapeutics or surveillance (Gandon et al. 2013, Glennon et al. 2021). A range of comparative framings have begun to shed light on the question, *e.g.*, considering both what might constrain the diversity of existing endemic pathogens (Rice et al. 2021), or why, for some pathogens like the coronavirus Middle-Eastern-Respiratory-Syndrome, even repeated spillovers do not lead to emergence of widespread transmission within the human population (Rice et al. 2022), or indeed why particular host species (*e.g.*, bats) might be more likely to harbour pathogens that are most virulent in humans (Guth et al. 2022). Nevertheless, obtaining a predictive map of the sources of pathogen spillover is a challenging and perhaps remote goal. Making progress will require a combination and expansion of such efforts including further empirical measurement (*e.g.*, obtaining data on the landscape of immunity towards different pathogens; Mina et al. 2020), characterizing distributions of candidate host receptors across critical organs like the lung that will define virulence, *etc.*). These insights would shed light on core trade-offs that affect the ability for pathogen emergence (Visher et al. 2021) and, when combined with estimates of fitness across temporal and spatial scales (Figure 24.6), could help us understand their relative importance.

24.5.2 Biological Control Agents

Biological control agents are a fascinating frontier in the field of pathogen control. Life history considerations come very much to the fore, as such agents may both drive pathogen evolution and also evolve themselves. Transmissible vaccines are one promising example of biological control agents, with examples including live-attenuated vaccines (*e.g.,* polio), or existing asymptomatic viruses edited to include an antigen that will trigger the desired host immunity (engineered viral vectors, introduced in the context of heterogeneity in transmission, see above). However, expanding the utility of 'transmissible vaccines' requires ensuring that these agents will neither evolve virulence themselves nor lose the antigen that drives the desired immunization of hosts (Layman et al. 2021). The latter may be a particular risk as the introduced epitope will generally be foreign to the virus being engineered and thus likely costly, such that variants without it are more fit.

The use of bacteriophage to control bacterial infections is another exciting direction of research into biological control agents. Their existence opens the way to avoiding use of antibiotics and thus the risk of evolution of antibiotic resistance. Bacteria will, of course, evolve resistance to such therapeutically deployed phages, and efforts to make such control agents maximally effective include working to identify phages for which bacterial escape mutants will be more vulnerable to other available sources of treatment (*e.g.*, antibiotics) or will have reduced virulence, *i.e.*, limited impacts on host health (Kortright et al. 2019). The mechanisms that underpin both pathogen and control agent life histories across scales are increasingly accessible, which will help refine efforts to both isolate (or generate) and make available phages with such desirable properties.

24.5.3 Pathogen Control Efforts in the Light of Host-Pathogen Coevolution

One final interesting intersection between life history evolution and pathogen control emerges from the fact that we ourselves have co-evolved in the context of a range of microbes. Pathogen control efforts have led to a relatively microbe-depauperate world (Blaser 2012). This paucity may lead to mismatch between our environment and immune functioning: early life exposure to microbes has been suggested to be an important component of developing immune tolerance, and an absence of such exposure could result in later life immunopathology, as our immune system 'over-reacts'. A range of interesting life history questions around immune ontogeny emerge in considering the impacts of a microbe-depleted world, for example, what is the optimal duration of a critical window during which the immune system learns to tolerate commensal or neutral microbes (Metcalf et al. 2022).? The better our understanding of past transitions in the relationship between the immune system and microbes, the better our predictive capability of what control efforts can alter these relationships in the future, positively or negatively.

24.6 Conclusions

The intersection between pathogen life history and pathogen control efforts can lead to a wide range of evolutionary outcomes. An array of conceptual and theoretical tools have been developed to tackle this question, and data is increasingly available at a variety of scales. Progress in this area hinges on an ever-refined understanding of the trade-offs that underpin both pathogen and control agent features – likely to require an ever-deeper understanding of the underlying dynamical mechanisms and an area where there is much more work to be done.

References

Acevedo, M.A., Dillemuth, F.P., Flick, A.J., Faldyn, M.J., and Elderd, B.D. (2019). Virulence-driven trade-offs in disease transmission: a meta-analysis. *Evolution* 73(4): 636–647.

Alizon, S., Hurford, A., Mideo, N., and van Baalen, M. (2009). Virulence evolution and the trade-off hypothesis: history, current state of affairs and the future. *Journal of Evolutionary Biology* 22(2): 245–259.

Allen, R.C., Popat, R., Diggle, S.P., and Brown, S.P. (2014). Targeting virulence: can we make evolution-proof drugs? *Nature Reviews Microbiology* 12(4): 300–308.

Althouse, B.M., Wenger, E.A., Miller, J.C., Scarpino, S.V., Allard, A., Hébert-Dufresne, L., and Hu, H. (2020). Superspreading events in the transmission dynamics of SARS-CoV-2: opportunities for interventions and control. *PLoS Biology* 18(11): e3000897.

Anderson, R.M., and May, R.M. (1982). Coevolution of hosts and parasites. *Parasitology* 85: 411–426.

Anderson, R.M., and May, R.M. (1992). *Infectious diseases of humans: dynamics and control*. Oxford University Press.

Ashby, B., Iritani, R., Best, A., White, A., and Boots, M. (2019). Understanding the role of eco-evolutionary feedbacks in host-parasite coevolution. *Journal of Theoretical Biology* 464: 115–125.

Best, A., White, A., and Boots, M. (2008). Maintenance of host variation in tolerance to pathogens and parasites. *Proceedings of the National Academy of Sciences of the United States of America* 105(52): 20786–20791.

Birger, R.B., Kouyos, R.D., Cohen, T., Griffiths, E.C., Huijben, S., Mina, M.J., Volkova, V., Grenfell, B., and Metcalf, C.J.E. (2015). The potential impact of coinfection on antimicrobial chemotherapy and drug resistance. *Trends in Microbiology* 23(9): 537–544.

Blaser, M.J. (2012). Equilibria of humans and our indigenous microbiota affecting asthma. *Proceedings of the American Thoracic Society* 9(2): 69–71.

Boots, M., Hudson, P.J., and Sasaki, A. (2004). Large shifts in pathogen virulence relate to host population structure. *Science* 303(5659): 842–844.

Boots, M., and Sasaki, A. (1999). 'Small worlds' and the evolution of virulence: infection occurs locally and at a distance. *Proceedings of the Royal Society B: Biological Sciences* 266(1432): 1933–1938.

Braun, K.M., Moreno, G.K., Wagner, C., Accola, M.A., Rehrauer, W.M., Baker, D.A., Koelle, K., O'Connor, D.H., Bedford, T., Friedrich, T.C., and Moncla, L.H. (2021). Acute SARS-CoV-2 infections harbor limited within-host diversity and transmit via tight transmission bottlenecks. *PLoS Pathogen*, 17(8): e1009849.

Bruijning, M., Metcalf, C.J.E., Jongejans, E., and Ayroles, J.F. (2020). The evolution of variance control. *Trends in Ecology & Evolution* 35(1): 22–33.

Bull, J.J., and Antia, R. (2022). Which 'imperfect vaccines' encourage the evolution of higher virulence? *Evolution, Medicine, and Public Health* 10(1): 202–213.

Bull, J.J., and Ebert, D. (2008). Invasion thresholds and the evolution of nonequilibrium virulence. *Evolutionary Applications* 1(1): 172–182.

Cornet, S., Nicot, A., Rivero, A., and Gandon, S. (2014). Evolution of plastic transmission strategies in avian malaria. *PLoS Pathogens* 10(9): e1004308.

Cressler, C.E., McLeod, D.V., Rozins, C., van den Hoogen, J., and Day, T. (2016). The adaptive evolution of virulence: a review of theoretical predictions and empirical tests. *Parasitology* 143(7): 915–930.

Day, T., and Gandon, S. (2007). Applying population-genetic models in theoretical evolutionary epidemiology. *Ecology Letters* 10(10): 876–888.

Day, T., and Proulx, S.R. (2004). A general theory for the evolutionary dynamics of virulence. *The American Naturalist* 163(4): 40–63.

Dieckmann, U., Metz, J.A.J., Sabelis, M.W., and Sigmund, K. (2004). *Adaptive dynamics of infectious diseases: in pursuit of virulence management*. Cambridge University Press.

Ewald, P.W. (1987). Transmission modes and evolution of the parasitism-mutualism continuum. *Annals of the New York Academy of Sciences* 503: 295–306.

Faye, O., Boëlle, P.-Y., Heleze, E., Faye, O., Loucoubar, C., Magassouba, N., Soropogui, B., Keita, S., Gakou, T., Bah, E.H.I., Koivogui, L., Sall, A.A., and Cauchemez, S. (2015). Chains of transmission and control of Ebola virus disease in Conakry, Guinea, in 2014: an observational study. *The Lancet Infectious Diseases* 15(3): 320–326.

Ferguson, N.M., Galvani, A.P., and Bush, R.M. (2003). Ecological and immunological determinants of influenza evolution. *Nature* 422(6930): 428–433.

Froissart, R., Doumayrou, J., Vuillaume, F., Alizon, S., and Michalakis, Y. (2010). The virulence-transmission trade-off in vector-borne plant viruses: a review of (non-)existing studies. *Philosophical Transactions of the Royal Society, B: Biological Sciences* 365(1548): 1907–1918.

Gandon, S., Hochberg, M.E., Holt, R.D., and Day, T. (2013). What limits the evolutionary emergence of pathogens? *Philosophical Transactions of the Royal Society, B: Biological Sciences* 368(1610): 20120086.

Gandon, S., Mackinnon, M.J., Nee, S., and Read, A.F. (2001). Imperfect vaccines and the evolution of pathogen virulence. *Nature* 414(6865): 751–756.

Geritz, S.A.H., Kisdi, E., Meszéna, G., and Metz, J.A.J., (2013). Evolutionarily singular strategies and the adaptive growth and branching of the evolutionary tree. *Evolutionary Ecology* 12(1): 35–57.

Glennon, E.E., Bruijning, M., Lessler, J., Miller, I.F., Rice, B.L., Thompson, R.N., Wells, K., and Metcalf, C.J.E. (2021). Challenges in modeling the emergence of novel pathogens. *Epidemics* 37: 100516.

Graham, A.L., Schrom, E.C., II, and Metcalf, C.J.E. (2022). The evolution of powerful yet perilous immune systems. *Trends in Immunology* 43(2): 117–131.

Grenfell, B.T., Pybus, O.G., Gog, J.R., Wood, J.L.N., Daly, J.M., Mumford, J.A., and Holmes, E.C. (2004). Unifying the epidemiological and evolutionary dynamics of pathogens. *Science* 303(5656): 327–332.

Griffiths, M.E., Meza, D.K., Haydon, D.T., and Streicker, D.G. (2023). Inferring the disruption of rabies circulation in vampire bat populations using a betaherpesvirus-vectored transmissible vaccine. *Proceedings of the National Academy of Sciences of the United States of America* 120(11): e2216667120.

Guth, S., Mollentze, N., Renault, K., Streicker, D.G., Visher, E., Boots, M., and Brook, C.E. (2022). Bats host the most virulent-but not the most dangerous-zoonotic viruses. *Proceedings of the National Academy of Sciences of the United States of America* 119(14): e2113628119.

Illingworth, C., Jr, Hamilton, W.L., Warne, B., Routledge, M., Popay, A., Jackson, C., Fieldman, T., Meredith, L.W., Houldcroft, C.J., Hosmillo, M., Jahun, A.S., Caller, L.G., Caddy, S.L., Yakovleva, A., Hall, G., Khokhar, F.A., Feltwell, T., Pinckert, M.L., Georgana, I., Chaudhry, Y., Curran, M.D., Parmar, S., Sparkes, D., Rivett, L., Jones, N.K., Sridhar, S., Forrest, S., Dymond, T., Grainger, K., Workman, C., Ferris, M., Gkrania-Klotsas, E., Brown, N.M., Weekes, M.P., Baker, S., Peacock, S.J., Goodfellow, I.G., Gouliouris, T., de Angelis, D., and Török, M.E. (2021). Superspreaders drive the largest outbreaks of hospital onset COVID-19 infections. *eLife* 10: e67308.

Kamo, M., and Boots, M. (2006). The evolution of parasite dispersal, transmission, and virulence in spatial host populations. *Evolutionary Ecology Research* 8(7): 1333–1347.

Kortright, K.E., Chan, B.K., Koff, J.L., and Turner, P.E. (2019). Phage therapy: a renewed approach to combat antibiotic-resistant bacteria. *Cell Host & Microbe* 25(2): 219–232.

Layman, N.C., Tuschhoff, B.M., and Nuismer, S.L. (2021). Designing transmissible viral vaccines for evolutionary robustness and maximum efficiency. *Virus Evolution* 7(1): veab002.

Lion, S., and Metz, J.A.J. (2018). Beyond R_0 maximisation: on pathogen evolution and environmental dimensions. *Trends in Ecology & Evolution* 33(6): 458–473.

Lipsitch, M., Siller, S., and Nowak, M.A. (1996). The evolution of virulence in pathogens with vertical and horizontal transmission. *Evolution* 50(5): 1729–1741.

Lloyd-Smith, J.O., Schreiber, S.J., Kopp, P.E., and Getz, W.M. (2005). Superspreading and the effect of individual variation on disease emergence. *Nature* 438(7066): 355–359.

Łuksza, M., and Lässig, M. (2014). A predictive fitness model for influenza. *Nature.* 507(7490): 57–61.

Luo, S., and Koelle, K. (2013). Navigating the devious course of evolution: the importance of mechanistic models for identifying eco-evolutionary dynamics in nature. *The American Naturalist* 181(Suppl 1): 58–75.

Lynch, P.A., and Boots, M. (2016). Using evolution to generate sustainable malaria control with spatial repellents. *eLife* 5: e15416.

McCallum, H., Barlow, N., and Hone, J. (2001). How should pathogen transmission be modelled? *Trends in Ecology & Evolution* 16(6): 295–300.

McLeod, D.V., Wahl, L.M., and Mideo, N. (2021). Mosaic vaccination: how distributing different vaccines across a population could improve epidemic control. *Evolution Letters* 5(5): 458–471.

Metcalf, C.J.E., (2016). Invisible trade-offs: van Noordwijk and de Jong and life-history evolution. *The American Naturalist* 187(4): iii–v.

Metcalf, C.J.E., Grenfell, B.T., and Graham, A.L. (2020). Disentangling the dynamical underpinnings of differences in SARS-CoV-2 pathology using within-host ecological models. *PLoS Pathogens* 16(12): e1009105.

Metcalf, C.J.E., Tepekule, B., Bruijning, M., and Koskella, B. (2022). Hosts, microbiomes, and the evolution of critical windows. *Evolution Letters* 6(6): 412–425.

Metzger, V.T., Lloyd-Smith, J.O., and Weinberger, L.S. (2011). Autonomous targeting of infectious superspreaders using engineered transmissible therapies. *PLoS Computational Biology* 7(3): e1002015.

Mideo, N., Alizon, S., and Day, T. (2008). Linking within- and between-host dynamics in the evolutionary epidemiology of infectious diseases. *Trends in Ecology & Evolution* 23(9): 511–517.

Mideo, N., and Day, T. (2008). On the evolution of reproductive restraint in malaria. *Proceedings of the Royal Society B: Biological Sciences* 275(1639): 1217–1224.

Miller, I.F., and Metcalf, C.J. (2019). Vaccine-driven virulence evolution: consequences of unbalanced reductions in mortality and transmission and implications for pertussis vaccines. *Journal of the Royal Society, Interface* 16(161): 20190642.

Mina, M.J., Metcalf, C.J.E., McDermott, A.B., Douek, D.C., Farrar, J., and Grenfell, B.T. (2020). A global immunological observatory to meet a time of pandemics. *eLife* 9: e58989.

Morris, D.H., Petrova, V.N., Rossine, F.W., Parker, E., Grenfell, B.T., Neher, R.A., Levin, S.A., and Russell, C.A. (2020). Asynchrony between virus diversity and antibody selection limits influenza virus evolution. *eLife* 9: e62105.

Ottino-Loffler, B., Scott, J.G., and Strogatz, S.H. (2017). Evolutionary dynamics of incubation periods. *eLife* 6: e30212.

Otto, T.D., Assefa, S.A., Böhme, U., Sanders, M.J., Kwiatkowski, D., Berriman, M., and Newbold, C. (2019). Evolutionary analysis of the most polymorphic gene family in *falciparum* malaria. *Wellcome Open Research* 4: 193.

Otto, S.P., Day, T., Arino, J., Colijn, C., Dushoff, J., Li, M., Mechai, S., van Domselaar, G., Wu, J., Earn, D.J.D., and Ogden, N.H. (2021). The origins and potential future of SARS-CoV-2 variants of concern in the evolving COVID-19 pandemic. *Current Biology* 31(14): R918–R929.

Parsons, T.L., Lambert, A., Day, T., and Gandon, S. (2018). Pathogen evolution in finite populations: slow and steady spreads the best. *Journal of the Royal Society, Interface* 15: 20180135.

Pica, N., and Palese, P. (2013). Toward a universal influenza virus vaccine: prospects and challenges. *Annual Review of Medicine* 64: 189–202.

Råberg, L., Sim, D., and Read, A.F. (2007). Disentangling genetic variation for resistance and tolerance to infectious diseases in animals. *Science* 318(5851): 812–814.

Read, A.F., Baigent, S.J., Powers, C., Kgosana, L.B., Blackwell, L., Smith, L.P., Kennedy, D.A., Walkden-Brown, S.W., and Nair, V.K. (2015). Imperfect vaccination can enhance the transmission of highly virulent pathogens. *PLoS Biology* 13(7): e1002198.

Read, A.F., Lynch, P.A., and Thomas, M.B. (2009). How to make evolution-proof insecticides for malaria control. *PLoS Biology* 7(4): e1000058.

Restif, O., and Grenfell, B.T. (2006). Integrating life history and cross-immunity into the evolutionary dynamics of pathogens. *Proceedings of the Royal Society B: Biological Sciences*, 273(1585): 409–416.

Rice, B.L., Douek, D.C., McDermott, A.B., Grenfell, B.T., and Metcalf, C.J.E. (2021). Why are there so few (or so many) circulating coronaviruses? *Trends in Immunology* 42(9): 751–763.

Rice, B.L., Lessler, J., McKee, C., and Metcalf, C.J.E. (2022). Why do some coronaviruses become pandemic threats when others do not? *PLoS Biology* 20(5): e3001652.

Rozins, C., and Day, T. (2017). The industrialization of farming may be driving virulence evolution. *Evolutionary Applications* 10(2): 189–198.

Ruden, R.M., and Adelman, J.S. (2021). Disease tolerance alters host competence in a wild songbird. *Biology Letters* 17(10): 20210362.

Saad-Roy, C.M., Morris, S.E., Metcalf, C.J.E., Mina, M.J., Baker, R.E., Farrar, J., Holmes, E.C., Pybus, O.G., Graham, A.L., Levin, S.A., Grenfell, B.T., and Wagner, C.E. (2021). Epidemiological and evolutionary considerations of SARS-CoV-2 vaccine dosing regimes. *Science* 372(6540): 363–370.

Saad-Roy, C.M., Wingreen, N.S., Levin, S.A., and Grenfell, B.T. (2020). Dynamics in a simple evolutionary-epidemiological model for the evolution of an initial asymptomatic infection stage. *Proceedings of the National Academy of Sciences of the United States of America* 117(21): 11541–11550.

Sasaki, A., Lion, S., and Boots, M. (2022). Antigenic escape selects for the evolution of higher pathogen transmission and virulence. *Nature Ecology & Evolution* 6(1): 51–62.

Sobel Leonard, A., Weissman, D.B., Greenbaum, B., Ghedin, E., and Koelle, K. (2017). Transmission bottleneck size estimation from pathogen deep-sequencing data, with an application to human influenza A virus. *Journal of Virology* 91(14): e00171-17.

Stumpf, M.P.H., Laidlaw, Z., and Jansen, V.A.A. (2002). Herpes viruses hedge their bets. *Proceedings of the National Academy of Sciences of the United States of America*, 99(23): 15234–15237.

Subramanian, R., Graham, A.L., Grenfell, B.T., and Arinaminpathy, N. (2016). Universal or specific? A modeling-based comparison of broad-spectrum influenza vaccines against conventional, strain-matched vaccines. *PLoS Computational Biology* 12(12): e1005204.

van Baalen, M., (1998). Coevolution of recovery ability and virulence. *Proceedings of the Royal Society B: Biological Sciences* 265(1393): 317–325.

Visher, E., Evensen, C., Guth, S., Lai, E., Norfolk, M., Rozins, C., Sokolov, N.A., Sui, M., and Boots, M. (2021). The three Ts of virulence evolution during zoonotic emergence. *Proceedings of the Royal Society B: Biological Sciences* 288(1956): 20210900.

Volz, E.M., and Frost, S.D.W. (2017). Scalable relaxed clock phylogenetic dating. *Virus Evolution* 3(2): vex025.

Watts, D.J., and Strogatz, S.H. (1998). Collective dynamics of 'small-world' networks. *Nature* 393(6684): 440–442.

Index

Note: *Italicized* and **bold** page numbers refer to figures and tables, respectively.

a

abiotic dimension *376*
  urban areas, adaptive life history evolution in
    chemical pollution 387
    elevated temperatures and thermal stress 387–388
    light pollution 386
    noise pollution 387
  urban drivers of selection on life history traits 377–378
abiotic environment 107
abiotic stressors, diversity of 377–378
Acadian flycatchers (*Empidonax virescens*) 324
Acanthochromis 138
ACC *see* anthropogenic climate change (ACC)
Aché hunter-gatherers 201
acorn-dwelling ants (*Temnothorax curvispinosus*) 386, 388
acquisition/allocation principle and the responses of organisms to pollution 335
  cost of adapting to pollutants 338
  individual organism level, responses at 335–336
  long-term responses of the population 337–338
  short-term mean population-level responses 337
Acrocephalus sechellensis 119, *120*
Acroclisoides 62
actuarial senescence 34. 318, 323
adaptive dynamics 443
adaptive life history evolution in urban areas 380, **381–383**
  abiotic dimension *376*
    chemical pollution 387
    elevated temperatures and thermal stress 387–388
    light pollution 386
    noise pollution 387
  anthropogenic inputs dimension *376*, 388–389
  biotic dimension *376*, 388
  pan-urban effects 389
  spatial dimension *376*
    fragmented landscapes, dispersal in 380
    fragmented urban landscape, finding a mate in 384–385
    patch isolation and island syndrome hypothesis 385
    reduced access to high-quality resources and food limitation 385
  temporal dimension
    diel cycle, effects of changes to 386
    seasonal cycles, adaptation to changes in 386
adaptive plasticity
  in growth rates 15
  in life histories 222–223
adaptive sex allocation 114, 123, 126
adult mortality 42, 99, 101, 218
adult sex ratio (ASR) 88, 103
Aedes aegypti 234, 243, 244
African clawed frog (*Xenopus laevis*) 135
age-1 mutants 33
age and size at maturation
  classical models of 7–8
  reaction norms for 8–13
  trade-off between 15–16
ageing 29
  antagonistic pleiotropy (AP) theory of 35
  developmental theory of ageing (DTA) 31, *31*
    'disposable soma' and 34–38, *37*
  disposable soma' theory (DST) of 30–31
  evolutionary theory of ageing (ETA) 30
    asynchronous ageing 38
    mortality, age, density and condition-dependence of 42–43
    mutation accumulation and antagonistic pleiotropy 31–34
    sex differences in ageing 40–42
  sex differences in 40–42
aggression 81–83
Alcaligenes 140
Alces alces 429
Allee effect 365–366, 426
alleles 32–33, 445
allocarers 201
allocation trade-offs 34, 275, 333, 334
Allomerus ant 301
alloparenting 195, 204
Alpine ibex 431
Altrichthys 138
Ambrosia artemisiifolia 386
Ambystoma 137
Ambystoma maculatum 221
Ambystoma mexicanum 137
Ambystoma opacum 221
Ambystomatid salamanders 137
Ambystoma tigrinum 322
ametaboly 138
AMF *see* arbuscular mycorrhizal fungi (AMF)
amphicarpic plants 304

Life History Evolution: Traits, Interactions, and Applications, First Edition. Edited by Michal Segoli and Eric Wajnberg.
© 2025 John Wiley & Sons Ltd. Published 2025 by John Wiley & Sons Ltd.

amphidromous gobies   136
Amphiprion percula   125
Amphiuma   137
Amur honeysuckle (*Lonicera maackii*)   388
Anagrus   63
Anagrus daanei   64, 65, 415, *416*
Anagrus erythroneurae   64
Anagrus parasitoids of leafhopper eggs   415–416
Anastrepha   64
Andrias   137
Anemone coronaria   287, *288*, 289
   biology and reproduction of   288
   flowering phenology   290
   insect visitors   290–291
   phenology regulation in   289
   pollination biology   287–289
anemones   288, 290
animal diversity, complex life cycles across   131
   anthropogenic environmental impacts and global climate change   144–145
   metamorphic development
      integration of, within the life cycle   131, *132*
      regulation of, by hormones   131–132, *133*
   metamorphic mechanisms   132, *134*
      insect metamorphosis   138–140
      invertebrates, regulation of bentho-planktonic life cycles in   140–144
      vertebrate metamorphosis   134–138
animal movement, phoretic   305–306
anisogamy   102
Anolis cristatellus   389
Anopheles gambiae   240
antagonistic pleiotropy (AP)   30, *31*, 32, 35
   mutation accumulation and   31–34, 199
anthropogenic climate change (ACC)   317–318
   body size   320–321
   genetic/environmental responses   325–326
   on life history strategies and trade-offs   318–319
   population demography and extinction risk   324
   on reproductive output and success   321–322
   survival and senescence   322–323
anthropogenic environmental impacts and global climate change   144–145
anthropogenic inputs dimension   *376*
   adaptive life history evolution in urban areas   388–389
   artificial and supplemental resources, human introduction of   378–379
   urban drivers of selection on life history traits   378–379
antigenic drift   444–445
anti-oxidant genes   166
anti-Williams   42–43
ant–plant protective interactions   299–301
ants, fungus-farming   303
Aphia   138
aphid–*Buchnera* nutritional mutualism   307
Aphidius colemani   408
Aphidoletes aphidimyza   408
aphids   260
Aphytis aonidiae   63, 64

Aphytis lingnanensis   64
Aphytis melinus   65
apogonids   138
AP *see* antagonistic pleiotropy (AP)
Aptenodytes forsteri   324
aquatic model system   *236*
Arabidopsis thaliana   271
arbuscular mycorrhizal fungi (AMF)   302
Arctia plantaginis   139
Arsenophonus   259
Arsenophonus nasoniae   259
artificial and supplemental resources, human introduction of   378–379
Aselus aquaticus   347
Asobara tabida   64, 297
ASR *see* adult sex ratio (ASR)
Astyanax bimaculatus   216
asymmetric genetic inheritance   42
asymmetry in partner interaction   298
asynchronous ageing   38
Atlantic cod (*Gadus morhua*)   427
Atlantic mackerel (*Scomber scombrus*)   321
Atlantic silverside (*Menidia menidia*)   433
augmentative biological control agents, life histories of   408–409
Aurelia aurita   142
autumn-born females   123, *123*
Azteca ant   299, 301

b

Bacillus thuringiensis   244
bacterial co-constructed countershading mechanism in squids   299
bacterioids   302
bacteriophage   454
Bactrocera   64
band-legged ground crickets (*Dianemobius nigrofasciatus*)   389
barley (*Hordeum vulgare*)   242
Bateman gradient   85
bats   165–166
'behavioural reaction norm' approach   106
Belostoma lutarium   *102*
bentho-planktonic life cycles regulation in invertebrates   140
   hormonal systems as regulators of metamorphic development   142
   immune system-related signals   141
   marine invertebrates
      environmental stressors affecting metamorphic development and settlement among   143
      metamorphosis evolution among   143–144
   neuro-endocrine system   141–142
   settlement cues and signals   140
Berlese hypothesis   139
bet-hedging strategies   107
bighorn sheep (*Ovis canadensis*)   431, 433
Binodoxys communis   63
biological control, life history traits and the resilience of
   to climate change   414–415
   to insecticides   415
biological control agents   403, 452

augmentative 408–409
   breeding programmes to improve performance of 410–411
   intrinsic rate of increase as an aggregate metric for comparing the life histories of 407
   monitoring for evolution of life history traits after biological control releases 411–412
   range-core evolutionary dynamics and hybridization in 412–413
   selecting of, from among naturally occurring intraspecific life history variation 410
bioluminescent *Vibrio* bacteria 299
Biomphalaria glabrata 230
biotic dimension 376
   adaptive life history evolution in urban areas 388
   urban drivers of selection on life history traits 378
biotic interactions
   cascading effects and indirect selection via 391
   novel communities and 378
bird-dispersed tropical shrubs 276
blackcaps (*Sylvia atricapilla*) 378
black carpenter ants (*Camponotus pennsylvanicus*) 385
black garden ant (*Lasius niger*) 35
black guillemot (*Cepphus grylle*) 103
black poplar (*Populus nigra*) 271
blue tits (*Cyanistes caeruleus*) 385
body size
   evolutionary stability of 6
   phylogeny-based studies on 7
body size, large
   benefits of 3–6
      in females 3–5
      in males 5–6
   challenges 6
   costs of 6–7
body size and timing of maturation 3
   absolute timing of maturation 7
   classical models of age and size at maturation 7–8
   evolution, rates of 16–18
   large body size
      benefits of 3–6
      challenges 6
      costs of 6–7
   reaction norms for age and size at maturation 8, 9
      classical models, reaction norms in 9–10
      developmental thresholds, introducing 10–11
      physiological realism, integrating 12–13
      plastic growth rates 10
      probabilistic reaction norms 11–12
   reaction norms with positive slope 13–15
   size–fecundity relationship 4
   trade-off between age and size at maturation 15–16
Boltenia villosa 142
Boonekamp's (2020) study 88
Brachionus calyciflorus 347
Brachionus plicatilis 346
Branchiostoma floridae 142
Brassica oleracea 271
BR-C *see* broad complex (BR-C)
breeding programmes 410–411
breeding values 433
Breviolum minutum 257

broad complex (BR-C) 140
brood site pollination mutualisms 305
brown bears 432
Buchnera 307
Buchnera aphidicola 260
Bugula neritina 51, 53, *53*
burying beetles (*Nicrophorus vespilloides*) 101, 102
butterfly (*Pieris napi*) 386

c

cabbage root fly (*Delia radicum*) 259
Caenorhabditis elegans 33, 35, *36*, 38, 163, 237, 244, 245, 338, 339, 347
   responses of
      to environmental stressors 347–348
      exposed to ionising radiation 348
Caenorhabditis remanei 40, *41*, 43
Camponotus pennsylvanicus 385
Canary Islands 285
Candidatus 260
cane toad (*Rhinella marina*) *358*, 367
capital-breeding insects 4, 6
   females of 4
   and income-breeding insects 5
capital-breeding lepidopterans 5
Capra ibex 431
captive rearing, preventing unintentional artificial selection in 411
Carcinus sp. 216
Cardinalis cardinalis 388
Cardinium 259
Cardiocondyla obscurior 160, 162
Carinascincus ocellatus 124
Carpodacus mexicanus 388
CA *see* corpora allata (CA)
caveats 88–89
Cecropia plants 299
Cepphus grylle 103
Ceriodaphnia cornuta 385
Cervus elaphus 121
cGMP *see* guanosine 3′,5′-cyclic monophosphate (cGMP)
chamois (*Rupicapra rupicapra*) 430
Chaoborus americanus 220, 222, 223
Chaoborus flavicans 223
Charadrius spp. 105
Chiasmia clathrate 386
childhood in humans 194
Chironomus ramosus 139
chitinolytic fungi 387
Chorthippus brunneus 380, 388
chronic noise 377
Chrysosporium pannorum 387
cichlid fishes 105
Cinclus cinclus 377
Ciona intestinalis 142
CI *see* cytoplasmic incompatibility (CI)
cities, life history evolution in
   integrative approaches for studying 392–393
   island biology in the city 391
   understanding 392
   urban trait syndrome and convergent evolution 391–392

CL390 protein   142
classical models, reaction norms in   9–10
Claytonia virginica   287
climate change   317
  anthropogenic climate change (ACC)
    body size   320–321
      genetic/environmental responses   325–326
      on life history strategies and trade-offs   318–319
      phenology   319–320
      population demography and extinction risk   324
      reproductive output and success   321–322
      survival and senescence   322–323
  anthropogenic environmental impacts and   144–145
  life history traits and the resilience of biological control to   414–415
climate-induced phenological changes   319
clownfish (*Amphiprion percula*)   125
clustered regularly interspaced short palindromic repeats system (CRISPR)   235
cold-adapted common lizards (*Zootoca vivipara*)   323
Colinus virginianus   322
Collared Flycatchers (*Ficedula albicollis*)   34
Coloeus monedula   34
colonisation–competition trade-offs   300
colony life history in obligatory eusocial insects   162–163
Colorado Rocky Mountains   285
Commelina communis   384–385
common blackbird (*Turdus merula*)   385
common garden experiments   182, 221, 222, 347, 413
common hamsters (*Cricetus cricetus*)   321
communal breeding in humans   195
community-level phenology and pollination specialisation   291–292
comparative phylogenetic analysis   125
compensatory population responses to harvest, life history and   431–432
competitiveness   85, *86*
conditional sex allocation theory   121–123, 124, 126
condition-dependence and within-population variation   183
constraints   3, 18
  ontogenetic   3
  physiological   13
  plant-based allocation constraints   270–272
conventional wisdom   238
cooperative breeders   155, 158–160
cooperative breeding systems   105
cooperative female-biased sex ratio   114
Cope's rule   6
Copidosoma   69
coral reef fish   144
  environmental regulation of metamorphosis in   135–136
  metamorphosis of   133
corals, zooxanthellae and   301
Cordia nodosa   301
Cordia populations   301
corpora allata (CA)   139
Corvus modedula   388
Corvus monedula   385
Corvusmonedula   88
Corydalis ambigua   287
Coryne uchidae   140

'cost of reproduction'   34
cost type of dispersal   **176**
cotton bollworm (*Helicoverpa armigera*)   139
covariance in empirical systems   450–451
COVID-19 pandemic   417
  see also SARS-CoV-2 virus
coyote tobacco (*Nicotiana attenuata*)   272
crabs (*Carcinus* sp.)   216
Crassostrea gigas   141
Crenicichla alta   *216* 223
Crepis sancta   380, 384, 386
Cricetus cricetus   321
CRISPR *see* clustered regularly interspaced short palindromic repeats system (CRISPR)
Cryptobranchus   137
Cryptotermes secundus   161
CThTV *see Curvularia* thermal tolerance virus (CThTV)
Culex pipiens   262
Curvularia protuberata   260
Curvularia thermal tolerance virus (CThTV)   260
Cyanistes caeruleus   385
Cyptocercus wood roaches   160
Cytisus scoparius   388
cytokine storms   447
cytoplasmic incompatibility (CI)   259

d
DA *see* dopamine (DA)
daf-2 gene   33, 35, *36*
Dalechampia   275
Dama dama   431
damselfish   138
Danionella   138
Daphnia   222, 223, 239
Daphnia galeata   222
Daphnia hyalina   223
Daphnia magna   220, 223, 235, 240, 338, 346, 387
Daphnia pulex   220, 223
dark-eyed junco (*Junco hyemalis*)   388
Darwinian demons   79
Darwinian fitness   29, 30
Darwin's theory of natural selection   215
deciduous species   273
defences and life history traits in plants *see* plant defences: and life history traits
defensive microbial endosymbionts   259–260
defensive symbioses between eukaryote hosts and microbes   **261**
defensive symbiosis   259, 260
delayed breeding   158, 161, 166
Delia radicum   259
density-dependent processes   100–101, 425–427
desiccation   135
developmental theory of ageing (DTA)   31, *31*
  'disposable soma' and   34–38, *37*
developmental thresholds   10–11
Diadegma insulare   63
diadromous fish, environmental regulation of metamorphosis in   136
Dianemobius nigrofasciatus   389
diapause and microbial symbionts   262

Dichanthelium lanuginosum var. *thermale* 260
dietary restriction 34
differential predation 221
Diorhabda carinulata 413, *413*
Diorhabda spp. 412
diseases, control of 439
  empirical systems 448
    covariance in 450–451
    variance in 448–450
  frontiers in life history evolution and pathogen control
    biological control agents 452
    emerging pathogens 451–452
    host–pathogen coevolution, pathogen control efforts in the light of 452
  predictions, classic theoretical scaffold to illustrate 440–445
  selection, levels of 445–447
dispersal 175, 183
  and colonization processes 184
  cost type of **176**
  evolution in response to other life history attributes 178–179
  evolution of 243–244
  joint evolution of, and other life history attributes 179
  as part of the life history 176–177, *177*
  theory of 177–179
dispersal–fecundity trade-off 301
dispersal heterogeneity 182
dispersal-life history co-variation
  condition-dependence and within-population variation 183
  micro-geographic variation among populations 181–183
dispersal-life history reaction norms, evolution of 179
dispersal syndromes 176
  empirical evidence of 180–181
  evolution of 179, *180*
  phylogenetic signals in 181
  strength and direction of *180*
'disposable soma' and developmental theory of ageing 34–38
'disposable soma' theory (DST) of ageing 30–31, 34
dome-shaped reaction norms 10
dopamine (DA) 141
Drosophila 234
Drosophila larvae 63
Drosophila mauritiana 234
Drosophila melanogaster 32–33, 34, 41–42, 139, 163, 220, 234, 236
Drosophila model 8
Drosophila neotestacea 237
Drosophila parasitoids 63, 64
Drosophila simulans 234
DST of ageing *see* disposable soma' theory (DST) of ageing
DTA *see* developmental theory of ageing (DTA)

e

E93 gene 140
early-life inertia *see* developmental theory of ageing (DTA)
eastern grey kangaroos (*Macropus giganteus*) 431
eastern tiger salamanders (*Ambystoma tigrinum*) 322
Ebola virus 449
ecdysone receptor (EcR) 142
ecdysteroids 138
echinoderms 143

eco-evolutionary dynamics 176, 246, 363, 366
ecological interactions, loss of 362–363
ecological plasticity and evolutionary sources of variability 429
ecotoxicology, role of life history theories in 334–335
Edhazardia aedis 243
egg
  additional life history strategies to overcome the risk of egg limitation 68–69
  cost of producing an egg 67, *68*
  environmental stochasticity and the cost of producing an egg, interaction between 68
  time *vs.* egg limitation in insects 61–62
egg and time limitation, relative importance of 62–64
egg load 61, 62
egg maturation patterns 62
egg maturation rate 62
EICA hypothesis *see* evolution of increased competitive ability (EICA) hypothesis
Eisenia andrei 258
elasticity 430
Eleutherodactylus coqui 137
Empidonax virescens 324
empirical systems 448
  covariance in 450–451
  variance in 448–450
Encarsia formosa 408, *409*
energetic costs **176**
English oak (*Quercus robur*) 273
Enterococcus faecalis 237
environment, interaction with 366–367
environmental interventions 439
environmental management and agent release strategies, using life history to inform 413
  Anagrus parasitoids of leafhopper eggs 415–416
  field resource supplementation 413–414
  life history traits and resilience of biological control
    to climate change 414–415
    to insecticides 415
  using life history information to inform release strategies 414
environmental pollution effects on life history 333
  acquisition/allocation principle and the responses of organisms to pollution 335
    cost of adapting to pollutants 338
    individual organism level, responses at 335–336
    long-term responses of the population 337–338
    short-term mean population-level responses 337
  Caenorhabditis elegans
    responses of, exposed to ionising radiation 348
    responses of, to environmental stressors 347–348
  ecotoxicology, role of life history theories in 334–335
  future directions 348–350
  pollutants, life history responses to 339, **340–345**
    common garden experiments 347
    long-term evolutionary responses of populations 346
    multigenerational studies 346
    studies analysing the effects of a range of concentrations of a pollutant 339–346
environmental sex determination 124
environmental stochasticity 66–67

environmental stochasticity (*cont'd*)
  and the cost of producing an egg 68
environmental stressors, responses of *Caenorhabditis elegans* to 347–348
environmental variability, adding 8
  classical models, reaction norms in 9–10
  developmental thresholds, introducing 10–11
  physiological realism, integrating 12–13
  plastic growth rates 10
  probabilistic reaction norms 11–12
Epicephala seed-consuming moths 305
epinephrine (EPI) 141
EPI *see* epinephrine (EPI)
Erythroneura spp. 415, *416*
Erythrura gouldiae 81, *81*
Escherichia coli 242
Esox lucius 433
ESS *see* evolutionarily stable strategy (ESS)
ETA *see* evolutionary theory of ageing (ETA)
Euptoieta hegesia 273
European Jackdaws (*Coloeus monedula*) 34
eusocial animals 159, *164*
eusocial insect
  colony 155
  obligatory 162–163
eusociality 159
eusocial organisms 35, 138
evergreen species 273
evolution, rates of 16–18
evolutionarily stable strategy (ESS) 8, 217, 443, *444*
evolutionary ecology 12
evolutionary explanation of adult body size 3
evolutionary processes
  r-selection 359–361
  spatial sorting 358–359
  stochasticity 361–362
evolutionary stability of body size 6
evolutionary theory 40
evolutionary theory of ageing (ETA) 30
  age, density and condition-dependence of mortality 42–43
  asynchronous ageing 38
  'disposable soma' and the developmental theory of ageing 34–38
  mutation accumulation and antagonistic pleiotropy 31–34
  sex differences in ageing 40–42
evolution of increased competitive ability (EICA) hypothesis 362–363
Exaiptasia pallida 257
expansion fronts 357
  cane toads 367
  ecological interactions, loss of 362–363
  environment, interaction with 366–367
  evolutionary processes
    r-selection 359–361
    spatial sorting 358–359
    stochasticity 361–362
  future directions 368
  pushed *vs.* pulled expansions 364–366
  trade-offs 363–364
expansion theory 357

experimental evolution 41, 89, 117, *117*, 236, *236*, 240, 244, 245
extended genotype 237
extended phenotype 162
extrinsic mortality 31–32, *41*, 42, 43, 84

f

facultative sex ratio adjustment 121–123
Falco naumanni 322
Falco tinnunculus 88
fallow deer (*Dama dama*) 431
farming 302
  fungus-farming ants 303
  fungus-farming termites 303
fast-strategists 192
fa'afafine 195
fecundity 4–5, *4*, *5*, 11, 52, 53, 55, 161, 163, *164*, 165, 218, *233*, 245, 348, 367
fecundity advantage 3–4, 5, 8
fecundity tolerance 235, 237
female-biased population 113–114
female-biased sex ratio 114, 115, 119
female-biased sexual size dimorphism 5
female brown bears (*Ursus arctos*) 429
'femaleness' syndrome 77
female parasitoids, reproductive success of 61
females, egg-limited 63
females, larger-bodied 3–4
females of capital-breeding insects 4
female *vs.* male production *see* sex allocation
Ficedula albicollis 34
fig–fig wasp brood site pollination mutualism 305
fig wasp pollinators 305
fig wasps 64, 116
fish and wildlife, exploitative management of 425
  contrasting life histories and harvest potential 427–428
  density-dependence 425–427
  ecological plasticity and evolutionary sources of variability 429
  future directions 434
  harvest, life history and compensatory population responses to 431–432
  harvest and evolution of life history strategies 432–434
  harvest management, knowledge of life history traits and 429–431
  sexually selected infanticide, special case of 432
  sustainable harvest 425–427
  trophy hunting, life history and 431
fishery-induced evolution 433
Fisher's (1930) theory 113–116, 120
fitness consequences 56, 113, **114**
flagfish (*Jordanella floridae*) 102
flatworms 118
floral resources 413–414
flounders 134–135
flower–butterfly interactions 292
flowering phenology 286, 287
flower–insect interactions 292
folivorous lepidopteran larvae 6
Forkhead box O (FOXO) transcription factor 138, 139
fragmented landscapes, dispersal in 380

fragmented urban landscape, finding a mate in   384–385
Frankia   257
freshwater crab (*Potamon fluviatile*)   385
frogs, environmental regulation of metamorphosis in   135
frugivory   275
Fukomys species   159, 165
fungus-farming ants   303
fungus-farming termites   303

g

Gadus morhua   427
Gambusia holbrooki   123
game-theoretical approach   8
gastropod molluscs   143
Gelsemium sempervirens   385
generalized linear model (GLM)   12
generation length   429
generations, link between   53–54
genetic backburn   367
genome-wide association studies (GWAS)   33
geroscience   33
giant water bug (*Belostoma lutarium*)   *102*
GLM *see* generalized linear model (GLM)
global climate change, anthropogenic environmental impacts and   144–145
Gnatocerus cornutus   89
Gonatocerus ashmeadi   63
gongylidia   303
gonochorism   125
Gorilla beringei   **193**
Gorilla gorilla   **193**
Gouldian finch (*Erythrura gouldiae*)   81, *81*, *82*, 88
grain beetle (*Oryzaephilus surinamensis*)   258
Grandmother Hypothesis   201–202, 203
grandmothering   199, 202
grasshopper (*Chorthippus brunneus*)   380, 388
great apes, life histories of   192–195
great tits (*Parus major*)   319, 385
gregarious parasitoids   68
Grime's C-S-R (competition-stress tolerance-ruderal) life history strategy   301
growth-related traits, defences and
    herbivore pressure, role of   272–274
    plant-based allocation constraints   270–272
growth-reproduction trade-offs   277
Gryllodes sigillatus   41
guanosine 3′,5′-cyclic monophosphate (cGMP)   141
guppies (*Poecilia reticulata*)   216, 388
gut microbiome   262
GWAS *see* genome-wide association studies (GWAS)
Gymnogobius isaza   136

h

habitat fragmentation   376
Habropoda laboriosa   287
Hakea   275
Haliotis asinina   141
Hamiltonella defensa   260
Harpegnathos saltator   159
harvest
    and evolution of life history strategies   432–434
    life history and compensatory population responses to   431–432
harvest management, knowledge of life history traits and   429–431
harvest potential, contrasting life histories and   427–428
HA *see* histamine (HA)
Hawk–Dove game   81, 82
Helicoverpa armigera   139
hemimetaboly   138
hemocytes   141
herbivore pressure, role of   272–274
herbivores, evolutionary ecology of plant defences against   269–270
Herdmania momus   142
Heterandria formosa   338
Heterocephalus glaber   159, 165
Heterocephalus glaberrodents   29
heterochromatic repetitive DNA   41
Hinton hypothesis   139
histamine (HA)   141
holobiont   255
hologenomic bottlenecks   363
holometabolous insects   138
Holospora undulata   236, 239
Homalodisca vitripennis   63
Homo floresiensis   17
Homo sapiens   **193**
honeybee   274
Hoplias malabaricus   216
Hordeum vulgare   242
horizontal *vs.* vertical transmission   241–243
hormonally mediated traits   144
hormonal systems as regulators of metamorphic development   142
hormones, regulation of metamorphic development by   131–132, *133*
host life history evolution in response to parasites   230, **231–232**
    costs of resistance and brief perspective on tolerance   234–235
    evolutionary shifts in reproductive effort and phenology   230–234
    importance of eco-evolutionary feedbacks and multispecies interactions   237
    parasite effects on host dispersal evolution   235–236
    symbionts mediating host life history evolution   236–237
    see also parasite life history evolution: in response to hosts
host mortality and infection clearance   239–241
host-pathogen coevolution, pathogen control efforts in the light of   452
host phenology   233
house finches (*Carpodacus mexicanus*)   388
house sparrows (*Passer domesticus*)   378
human life histories   191
    great apes, life histories of   192–195
    menopause and post-reproductive lifespan (PRLS)   197
        as an adaptation   200–204
        as a by-product   198–200
        phylogenetic patterning of   197
        study of evolution of menopause   204–205
    trade-offs   191–192

human life histories (cont'd)
    variation in   195–197
Humboldtia brunonis   300
hunter-gatherers   192, **193**, 194, *194*, 195, 199, 201, 204
Huntington's disease   30
Hyalinobatrachium orientale   106
Hydractinia planula settlement   140
Hylobius abietis   272
hyperfunction   *37*, 38
hyperfunction theory of ageing   38
hypofunction   *37*, 38

i

IGF-2 *see* insulin-like growth factor 2 (IGF-2)
IGF signalling *see* insulin-like growth factor (IGF) signalling
IIS pathway *see* insulin/insulin-like signalling (IIS) pathway
ILPs *see* insulin-like peptides (ILPs)
immune system-related signals   141
immunopathology   234
inbreeding depression   32
income-breeding insects   5
'infectivity' genes   241
insect egg loads, evolution of   61
    additional life history strategies to overcome the risk of egg limitation   68–69
    cost of producing an egg   67, *68*
    egg and time limitation, relative importance of   62–64
    egg maturation patterns   62
    environmental stochasticity   66–67
        and cost of producing an egg   68
    future directions   69
    oviposition opportunities   64–65
        and sugar availability   65–66
    sugar availability   65
    time *vs.* egg limitation in insects   61–62
    trade-offs between early and late components of reproduction   61
insecticides, life history traits and the resilience of biological control to   415
insect metamorphosis   138
    environmental control of   138–139
    evolution of   139–140
insect–plant phenological synchronisation   286–287
insects, time *vs.* egg limitation in   61–62
insulin/insulin-like signalling (IIS) pathway   33, 138
insulin-like growth factor (IGF) signalling   139
insulin-like growth factor 2 (IGF-2)   139
insulin-like peptides (ILPs)   138
intersexual conflict   90
interspecific life history variation, selecting among   406
    augmentative biological control agents, life histories of   408–409
    intrinsic rate of increase as an aggregate metric   407
    life history trait syndromes, using   407–408
intra-sexual conflict   90
intraspecific life history variation   409
    leveraging intraspecific variation in life history traits   410–411
    monitoring for evolution of life history traits after biological control releases   411–412
    preventing unintentional artificial selection in captive rearing   411

selecting biological control agents   410
Tamarisk shrubs, biological control agents of   412–413
intrinsic ageing   29
intrinsic rate of increase as an aggregate metric   407
invertebrates, bentho-planktonic life cycles regulation in   140
    hormonal systems as regulators of metamorphic development   142
    immune system-related signals   141
    marine invertebrates
        environmental stressors affecting metamorphic development and settlement among   143
        metamorphosis evolution among   143–144
    neuro-endocrine system   141–142
    settlement cues and signals   140
ionising radiation, responses of *Caenorhabditis elegans* exposed to   348
island biology in the city   391
island syndrome hypothesis   385, 391
iteroparity   104, 162

j

jackdaws   88
JH *see* juvenile hormone (JH)
Jordanella floridae   102
Junco hyemalis   388
juvenile hormone (JH)   138–139, 163

k

kangaroos   431
killer whales (*Orcinus orca*)   77, 202
killifish
    Heterandria formosa   338
    Rivulus hartii   216
kissing bugs (*Triatoma dimidiata*)   389
knowledge gaps, embracing   84–88
Krüppel homolog1 (Kr-h1)   140
K-selected species   79–80
K-species   175

l

labrid fish   125
Lamprotornis superbus   105
Lansing effect   201
large body size
    benefits of   3
        in females   3–5
        in males   5–6
    challenges   6
    costs of   6–7
Larus novaehollandiae   325, *325*
larval-to-polyp transformation   140
Lasius niger   35
leaf beetles (*Diorhabda* spp.)   412
leaf-cutting ants, colony size in   303
leaf habit   273
leafhopper eggs, *Anagrus* parasitoids of
    environmental conditions and life histories of   415–416
leafhopper (*Erythroneura* spp.) eggs   415, *416*
leaf–stem allometry   300
leguminous tree   300
lekking birds   88

lekking manakins   88
Lepidium virginicum   386
lepidopterans, capital-breeding   5
Leptopilina boulardi   64, 234
Leptopilina clavipes   63
Leptopilina heterotoma   64
lesser kestrels (*Falco naumanni*)   322
Leucoagaricus gongylophorus   303
Liebig's law   67
'life fast, die young' paradigm   196
life history trait syndromes, using   407–408
LiFe-part   166
lifespan and fecundity (LiFe)   164
lifespan artefact hypothesis   198–199
Linaria vulgaris   385
lions (*Panthera leo*)   432
lipopolysaccharides (LPS)   140
Litoditis marina   338
'live fast, die young' strategy   87
LMC *see* local mate competition (LMC)
loblolly pine (*Pinus taeda*)   271
local mate competition (LMC)   115–117
local resource competition (LRC)   115
local resource enhancement (LRE)   115, 119–120
local sperm competition in simultaneous hermaphrodites   118–119, *119*
longevity mutants   33
long-lived emperor penguins (*Aptenodytes forsteri*)   324
long-living organisms, absolute timing of the maturation in   7
long reproductive lifespans, ultimate causes of   156, *157*
   empirical evidence
      social arthropods   159–161
         ultimate causes   161
      vertebrates   158–159
   sociality, longevity and other life history traits   161–162
   theoretical models   156–158
long-term evolutionary responses of populations   346
Lonicera maackii   388
Lotka–Euler equation   7
LPS *see* lipopolysaccharides (LPS)
LRC *see* local resource competition (LRC)
LRE *see* local resource enhancement (LRE)
L-shaped reaction norms   10–11
Lymantria dispar   239
Lytechinus pictus   141

m

Macropus giganteus   431
Macrotermes bellicosus   161
male-biased sex ratios   115, 119
male fitness   7, 8, 86, 89
male ibex (*Capra ibex*)   431
male-killing endosymbiont manipulators   178
male-male competition   88
male mate choice   199–200
male mitochondrial load   42
'maleness' syndrome   77
male-to-male competition   8
male *vs.* female production *see* sex allocation
Malthusian parameter   30
Malurus cyaneus   322, *323*

mammal-like female-biased adult sex ratios   88
Manduca sexta   139
Margaritifera margaritifera   323
marine invertebrates
   environmental stressors affecting metamorphic development and settlement among   143
   metamorphosis evolution among   143–144
marine rotifers   346
maritime pine (*Pinus pinaster*)   272
Marmota flaviventris   319
MA *see* mutation accumulation (MA)
mate competition, local   115–117
maternal grandmothers   201–202
mathematical theory for the evolution of life histories   215
mating market   194
maturation
   absolute timing of   7
   age and size at   7–8
      trade-off between   15–16
   reaction norms for age and size at   8, *9*
      classical models, reaction norms in   9–10
      developmental thresholds, introducing   10–11
      physiological realism, integrating   12–13
      plastic growth rates   10
      probabilistic reaction norms   11–12
mDNA *see* mitochondrial DNA (mDNA)
MD *see* metamorphic development (MD)
Melittobia   69, 117
Melittobia australica   117
Menidia menidia   433
menopause and post-reproductive lifespan   197
   as an adaptation   200
      Grandmother Hypothesis   201–202
      Mother Hypothesis   200–201
      Reproductive Conflict Hypothesis   203–204
   as a by-product
      antagonistic pleiotropy and mutation accumulation   199
      lifespan artefact hypothesis   198–199
      male mate choice   199–200
   phylogenetic patterning of   197
   study of evolution of menopause   204–205
menstruation   194
metamorphic development (MD)   138
   integration of, within the life cycle   131, *132*
   regulation of, by hormones   131–132, *133*
metamorphic mechanisms   132, *134*
   insect metamorphosis   138
      environmental control of   138–139
      evolution of   139–140
   invertebrates, regulation of bentho-planktonic life cycles in   140
      hormonal systems as regulators of metamorphic development   142
      immune system-related signals   141
      marine invertebrates   143–144
      neuro-endocrine system   141–142
      settlement cues and signals   140
   vertebrate metamorphosis   134
      environmental regulation of   135–136
      evolutionary patterns of life history diversity within and across vertebrate taxa   137–138
metamorphosis   134–138

metaphorical 'germline' 35
methionine 34
microbial symbionts 255
    categories of 256
    defensive 259–260
    diapause and 262
    in essential biological functions of their hosts 256–257
    nutritional 257–258
    reproductive 258–259
Microbotryum spp. 233
Microctonus hyperodae 64
micro-geographic variation among populations 181–183
Middle-Eastern-Respiratory-Syndrome 452
MIPs *see* molluscan insulin-related peptides (MIPs)
mitochondrial DNA (mDNA) 42
molluscan insulin-related peptides (MIPs) 142
Monarda fistulosa 272
monogamy 83
Monte Carlo simulations 65
moose (*Alces alces*) 429
mortality
    age, density and condition-dependence of 42–43
    tolerance 235
mosaic vaccination 450
mosquitofish (*Gambusia holbrooki*) 123
moth (*Chiasmia clathrate*) 386
Mother Hypothesis 200–201
Mother's Curse 78–79
mountain goats (*Oreamnos americanus*) 428, *428*, **428**, 429
movement
    in animals 305–306
    in plants 304–305
multicellular organisms, microbial symbionts in 255
    categories of microbial symbiosis 256
    defensive microbial endosymbionts 259–260
    diapause and microbial symbionts 262
    in essential biological functions of their hosts 256–257
    nutritional microbial symbionts 257–258
    reproductive microbial symbionts 258–259
multigenerational studies 346
multiple infections, parasite life history evolution in response to 244–245
multi-queen colonies 163
multi-trait 'dispersal syndromes' 243
mutation accumulation (MA) 30, 199
    and antagonistic pleiotropy 31–34
mutualism 297
    asymmetry in partner interaction and consequences for 298
    benefits 298–306
    examples of *304*
    farming 302
        fungus-farming ants 303
        fungus-farming termites 303
    future directions 306–307
    life history and other traits that predispose organisms to enter into 298
    movement 303
        in animals 305–306
        in plants 304–305
    nutrition 301
        plants, rhizobia and mycorrhizae 301–302
        zooxanthellae and corals 301
    protection 299
        ant–plant protective interactions 299–301
        bacterial co-constructed countershading mechanism in squids 299
Mycoplasma gallisepticum 243
mycorrhizae and rhizobia 301–302
myrmecophytic plants 300, 301
Mytilus edulis 141

n

Nasonia vitripennis 116, 262
natural death 30
natural selection 100, 113, 114
naïve group-selection approaches 30
Necturus 137
negative correlations 14
neoteny 137
    in salamanders 137
NE *see* norepinephrine (NE)
neuro-endocrine system 141–142
Nicotiana attenuata 272
Nicrophorus vespilloides 101, 102
nitric oxide 141–142
nitric oxide synthase (NOS)-expressing neurons 141
nitrogen-fixing bacteria 257
noise pollution 377
norepinephrine (NE) 141
Northern Bobwhite (*Colinus virginianus*) 322
Northern Cardinal (*Cardinalis cardinalis*) 388
northern tamarisk beetle (*Diorhab carinulata*) 413
NOS-expressing neurons *see* nitric oxide synthase (NOS)-expressing neurons
Notophthalmus viridescens 221
novel abiotic stressors, diversity of 377–378
novel communities and biotic interactions 378
null models 90
nutritional microbial symbionts 257–258
nutritional mutualisms 301
    plants, rhizobia and mycorrhizae 301–302
    zooxanthellae and corals 301

o

obligate polyembryony 68
obligatory eusocial insects, colony life history in 162–163
Odocoileus virginianus 428, *428*, **428**
odorous house ant (*Tepinoma sessile*) 388
offspring needs 99, 105
offspring sex ratio 116, *116*, 117, *117*, *118*, 119
offspring size
    affecting fitness 56
    defined 49
    knowns of 54–55
    maternal control 56
    and number 55
    and theory 51–52, *51*
    transgenerational plasticity in 53–54
    unknowns in 56
offspring size effects 49–51
offspring size–fitness relationship 51–52
offspring size–performance relationship 52
offspring traits 101–102
Olympic village effect 338

Oncorhynchus mykiss 324
ontogenetic constraints 3
ontogenetic growth 12
ontogeny 273
operational sex ratio (OSR) 103
Ophryocystis elektroscirrha 243
opioid processing 141
opportunity costs 175, **176**
optimal reaction norms 9
Orchinus orca 202
Orcinus orca 77
Oreamnos americanus 428, *428*, **428**
Orgyia antiqua, 5
Oryzaephilus surinamensis 258
OSR *see* operational sex ratio (OSR)
Otolemur crassicaudatus 115
overall productivity in the population 114
overhead threshold model 11
ovigeny index 62
oviposition opportunities 64–65
 and sugar availability 65–66
Ovis aries 38, *39*
Ovis canadensis 433

p

pace-of-life syndromes (POLS) 80–81, 83
Paedocypris 138
paedomorphosis 137
Pan paniscus **193**
Panthera leo 432
Pan troglodytes **193**
pan-urban effects 379
 adaptive life history evolution in urban areas 389
 urban drivers of selection on life history traits 379
Paramecium 236
Paramecium caudatum 236, 239
Paramecium–Holospora system 242, 243
parasite life history evolution
 in response to hosts 238, *238*
  Russian Doll effect 241
  shortened infection lifespan, sources of 239–241
  vertical *vs.* horizontal transmission 241–243
  virulence and dispersal, evolution of 243–244
  virulence evolution 239
 in response to multiple infections 244–245
 see also host life history evolution in response to parasites
parasite-mediated selection 234
parasite virulence 238
parasitic *Allomerus* ant 301
parasitoid female wasp 62
parasitoid wasp (*Aphidius colemani*) 119, 297, 408, 410
parental care
 co-evolution with offspring traits and parental traits 101–102
 future directions 108
 life history and the maintenance of 100–101
 life history and the origin of 98–100, **99**
 origin and maintenance of 98
 plasticity and the evolution of 106
  abiotic environment 107
  social environment 107–108
 relationship to life history 97–98

sexual selection, life history and sex differences in 102–104
 in stochastic environments 104
  to empirical patterns and analysis 105–106
parent and offspring traits 101–102
parent-offspring conflicts 105
parthenogenesis 410
Parus major 319, 385
Passer domesticus 378
Pasteuria ramosa 235, 243
patch isolation and island syndrome hypothesis 385
pathogen adaptation rate 450
pathogen control, life history evolution and
 biological control agents 452
 emerging pathogens 451–452
 in the light of host–pathogen coevolution 452
pathogens 439, 449
PBDEs *see* polybrominated diphenyl ethers (PBDEs)
PCBs *see* polychlorinated biphenyls (PCBs)
pearl mussel (*Margaritifera margaritifera*) 323
pelagic larval phase (PLP) 135, 143
perennial herb 272
PG *see* prothoracic gland (PG)
PHB granules *see* polyhydroxybutyrate (PHB) granules
phenoloxidase (PO) 141
phenotypic plasticity 106, 196, 222
phenotypic variation 191
phoretic animal movement 305–306
phosphatidylinositol 3-kinase (PI3K) signalling cascade 138
phospholipids 140
phylogeny-based studies on body size 7
Physalis 277
Physella virgata 223
physiological constraints 13
physiological realism, integrating 12–13
PI3K signalling cascade *see* phosphatidylinositol 3-kinase (PI3K) signalling cascade
Pieris napi 386
pike (*Esox lucius*) 433
pike cichlid (*Crenicichla alta*), 216
Pinus pinaster 272
Pinus taeda 271
Piper 276
Plantago lanceolata 244
Plant Apparency Theory 273
plant-based allocation constraints 270–272
plant defences
 against herbivores, evolutionary ecology of 269–270
 and growth-related traits
  herbivore pressure, role of 272–274
  plant-based allocation constraints 270–272
 and life history traits 269
  correlated evolution of 270–276
  future research, challenges for 277–278
 and reproductive traits 274
  pollination-related traits 274–275
  seed dispersal-related traits 275–276
plant–insect phenological synchronisation 286–287
plant life history traits, evolution of 276–277
plant–pollinator assemblages 291
plant–pollinator phenological synchrony 285
plants, movement in 304–305
plants, rhizobia and mycorrhizae 301–302

Plasmodium 240
plastic growth rates 10
plasticity
    and evolution of parental care 106
        abiotic environment 107
        social environment 107–108
    social context of 108
Platythyrea punctata 159
pleiotropy 199, 234
Plodia interpunctella 234, 235
plovers (*Charadrius* spp.) 105
PLP *see* pelagic larval phase (PLP)
Plutella xylostella 63
PMRN *see* probabilistic maturation reaction norms (PMRN)
Podosphaera plantaginis 244
Poecilia reticulata 43, 216, *216*, 221, 388
pollination networks
    community-level phenology and 291–292
    generalisation in 285–286
pollination-related traits, plant defences and 274–275
pollinator fig wasps 305
pollinator phenology 286
pollutants, life history responses to 339
    common garden experiments 347
    long-term evolutionary responses of populations 346
    multigenerational studies 346
    studies analysing the effects of a range of concentrations of a pollutant 339–346
pollution 333
    acquisition/allocation principle and the responses of organisms to 335
        cost of adapting to pollutants 338
        individual organism level, responses at 335–336
        long-term responses of the population 337–338
        short-term mean population-level responses 337
POLS *see* pace-of-life syndromes (POLS)
polyandry 83
polybrominated diphenyl ethers (PBDEs) 377
polychlorinated biphenyls (PCBs) 377
polydnaviruses 260
polydomous ant colonies 300
polyembryony 68
polygamous mating systems 178
polygynous species 125
polygyny 83
polyhydroxybutyrate (PHB) granules 299, 302
polyploidy of fungal crop 303
polyp-to-jellyfish transition 142
polysaccharides 140
Pomatoschistus minutus, *102*, 103
POMC system *see* pro-opiomelanocortin (POMC) system
Pongo abelii **193**, 194
Pongo pygmaeus **193**
population density fluctuations 106
population sex ratio 113–114
Populus 271
Populus nigra 271
Populus tremuloides 273, 277
PO *see* phenoloxidase (PO)
positive correlations 14
positive slope, reaction norms with 13–15
post-menopausal women 195

post-reproductive lifespan (PRLS) 191
    see also menopause and post-reproductive lifespan
post-reproductive representation (PrR) 197
Potamon fluviatile 385
predation, types of 216–217
predator-driven life history evolution, theory for 217–218
    predictions from theory 218–219
predator–prey interactions 215
    adaptive plasticity in life histories 222–223
    empirical evidence 220–222
    future directions 224
    theory for predator-driven life history evolution 217–218
        predictions from theory 218–219
predatory midge (*Aphidoletes aphidimyza*) 408
predictions, classic theoretical scaffold to illustrate 440–445
prime age 88
PRLS *see* post-reproductive lifespan (PRLS)
probabilistic maturation reaction norms (PMRN) 12
probabilistic reaction norms 11–12, *12*
promiscuous species 125
pro-opiomelanocortin (POMC) system 141
pro-ovigenic insects 414
pro-phenoloxidase (pro-PO) system 141
protandrous species 126
protective symbionts 236, 237
protein:carbohydrate (P:C) ratio 40–41
Proteus 137
prothoracic gland (PG) 139
prothoracicotropic hormone (PTTH) neurons 139
protogyny 124, 125
Protophormia terraenovae 347
Pseudoalteromonas 140
Pseudobranchus 137
Pseudomonas aeruginosa 235, 236
Pseudomonas fluorescens 235, 245
psychometric tools 196
Pterapogon kauderni 138
PTTH neurons *see* prothoracicotropic hormone (PTTH) neurons
puberty 194
Puerto Rican tree frog (*Eleutherodactylus coqui*) 137, 138
pushed *vs.* pulled expansions 364–366
Python reticulatus 427

q

quaking aspen (*Populus tremuloides*) 273, 277
quasi-sociality 119
queen fecundity 161
queen-to-worker lifespan divergence 158
queen-to-worker longevity ratio 159
Quercus robur 273
Quinca mirifica 138

r

Rafflesia 298
rainbow trout (*Oncorhynchus mykiss*) 324
range of concentrations of a pollutant (RoC) 339, 346–347
RA signalling *see* retinoic acid (RA) signalling
reaction norms with positive slope 13–15
reactive oxygen species (ROS) 163
recruitment rate 425–426

red-billed gulls (*Larus novaehollandiae*)   325, *325*
red-breasted nuthatch (*Sitta canadensis*)   101
red deer (*Cervus elaphus*)   121
red flour beetles (*Tribolium castaneum*)   362
Red King effect   307
Red Queen Dynamics   449
Regiella insecticola, 260
reproduction, trade-offs between early and late components of   61
Reproductive Conflict Hypothesis   203–204
reproductive effort   241–243
reproductive manipulators   259
reproductive microbial symbionts   258–259
reproductive parasites   259
reproductive restraint   439
reproductives, long lifespans of   156, *157*
   empirical evidence
      social arthropods   159–161
         ultimate causes   161
      vertebrates   158–159
   sociality, longevity and other life history traits   161–162
   theoretical models   156–158
reproductive success   114, **114**, 115
reproductive traits, defences and   274
   plant defences and pollination-related traits   274–275
   plant defences and seed dispersal-related traits   275–276
'resistance is futile' effect   237
resource competition, local   115
retinoic acid (RA) signalling   142
Retinoid X Receptor (RXR)   142
retinol dehydrogenase   142
retrospective stage-structured modelling   407
Rhabdias pseudosphaerocephala   236
Rhabdomys pumilio   29
Rhinella marina   236, *358*, 367
Rhinogobius sp.   136
rhizobia   257, 301–302
Rhopalomyia californica   63
Rhytidoponera purpurea   159
ribbon worms   143
Rickettsia spp.   259
Rickettsiella   259
Riptortus pedestris   259
risk costs   **176**
Rivulus hartii   216, 223
r-K framework   79–80, 84
Rmax   425
RNA interference (RNAi)   35
rockfish (*Sebastes* spp.)   427
ROS *see* reactive oxygen species (ROS)
r-selection   359–361
Ruellia nudiflora   274
Rupicapra rupicapra   430
Russian Doll effect   241
RXR *see* Retinoid X Receptor (RXR)

S

safe-stage hypothesis   107
salamander (*Ambystoma maculatum*)   221
Salix spp.   272
Salmo salar   89
sand goby (*Pomatoschistus minutus*)   *102*, 103

SARS-CoV-2 virus   442, 445–446, 448, 450, 451
Schindleria   138
Schistosoma mansoni   230
Sclerodermus   69
Sclerodermus harmandi   119, 120
Scomber scombrus   321
scotch broom (*Cytisus scoparius*)   388
Scotopteryx chenopodiata   64
Scytodes pallida   101
sea anemone (*Exaiptasia pallida*)   257
sea urchin
   Lytechinus pictus   141
   Strongylocentrotus purpuratus   141
Sebastes spp.   427
seed dispersal-related traits, plant defences and   275–276
selection, levels of   445–447
selection shadow   30
selective breeding   410
selective disappearance   337
self-fertilisation   288
self-shading   243
semelparous organisms   104
senescence   29
sequential hermaphrodites   120
   sex change in   124–126
sequential hermaphroditism   125
serotonin (5-HT)   141
sex allocation   113
   environmental condition   120
      environmental sex determination   124
      facultative sex ratio adjustment   121–123
      sequential hermaphrodites, sex change in   124–126
   Fisher's theory   113–115
   future directions   126
   relatives, interaction with   115
      local mate competition   115–117
      local resource competition   115
      local resource enhancement   119–120
      local sperm competition in simultaneous hermaphrodites   118–119, *119*
sex-changing organisms   126
sex chromosomes   41, 77
sex differences in ageing   40–42
sex-specific life histories   77
   Mother's Curse   78–79
   toxic Y   78–79
   trait co-variation   79
      aggression   81–83
         caveats   88–89
         knowledge gaps, embracing   84–88
         one genome, two sexes   89–90
         sex-specific predictions   83–84
   unguarded X   78–79
sex-specific predictions   83–84
sex-specific selective sieve   42
sexual conflict   89
sexually antagonistic mtDNA alleles   42
sexually selected infanticide, special case of   432
sexual selection, life history and sex differences in parental care   102–104
sexual size dimorphism (SSD)   5, 15
Seychelles warblers   158

Seychelles warblers (cont'd)
  Acrocephalus sechellensis 119, *120*
shared trait 238
Shine's safe-stage hypothesis 107
shorebirds 105
shortened infection lifespan, sources of 239–241
short-lived organisms, absolute timing of the maturation in 7
Sicyopterus lagocephalus 136
simultaneous hermaphrodites, local sperm competition in 118–119, *119*
single nucleotide polymorphisms (SNPs) 32
sirens 137
SIR models *see* susceptible-infected-recovered (SIR) models
Sitta canadensis 101
Sitta carolinensis 101
size-advantage model 124, 125
size–fecundity relationship 4, 8
size-limited predation on juveniles 219
size-selective predation 6
slow-strategists 192
snow skink (*Carinascincus ocellatus*) 124
SNPs *see* single nucleotide polymorphisms (SNPs)
Soay sheep (*Ovis aries*) 38, *39*
social arthropods 165
  LiFe-part 166
  long lifespans of 159–161
  TI-J-Part 165
social environment 107–108
social evolution 156, 161
social insect reproductives 162
sociality 155, 158
  longevity and other life history traits 161–162
social living and life history evolution 155
  colony life history in obligatory eusocial insects 162–163
  long reproductive lifespans, ultimate causes of 156, *157*
    social arthropods 159–161
    sociality, longevity and other life history traits 161–162
    theoretical models 156–158
    ultimate causes 161
    vertebrates 158–159
  proximate mechanisms 166
    social arthropods 165
      LiFe-part 166
      TI-J-Part 165
    social vertebrates 164–165
  social vertebrates 164–165
Solanum sp., 276
somatic cells 34
spatial dimension 376
  adaptive life history evolution in urban areas
    fragmented landscapes, dispersal in 380
    fragmented urban landscape, finding a mate in 384–385
    patch isolation and island syndrome hypothesis 385
    reduced access to high-quality resources and food limitation 385
  urban drivers of selection on life history traits 376–377
spatial drift 361
spatial fragmentation, habitat change and 376–377
spatial sorting 338, 358–359
spatial *vs.* temporal stochasticity 67

species-level phenological asynchrony and generalized pollination 287
  Anemone coronaria
    flowering phenology 290
    insect visitors 290–291
    pollination biology 287–289
species-poor pollination networks 285
sperm competition, local 118–119, *119*
spider mite (*Tetranychus urticae*) 117
Spiroplasma spp. 259
spitting spider (*Scytodes pallida*) 101
spring-born males and females 123
squids, bacterial co-constructed countershading mechanism in 299
SSD *see* sexual size dimorphism (SSD)
Staphylococcus aureus 237, 244
stochastic dynamic programming models 195
stochastic environments, parental care in 104
  to empirical patterns and analysis 105–106
stochasticity 66–67, 361–362
strain-transcending immunity 444
striped field mice (*Rhabdomys pumilio*) 29
Strongylocentrotus purpuratus 141
Styela clava 142
sugar availability 65
sugar-deprived *Aphytis melinus* 65
Sula granti 88
Sundadanio 138
superb fairywrens (*Malurus cyaneus*) 322, *323*
super-infectors 244
superorganismality 156, 159, 160
superorganisms 35, 155
  sensu stricto 156, 159, 162, 167
susceptible-infected-recovered (SIR) models 238, *441*
sustainable harvest 425–427
Sylvia atricapilla 378
Symbiodinium 257
Symbiodinium minutum, *257*
symbiont-mediated protection 237
symbionts–aphids–aphid parasitoids 260
symbiotic bacteria 410
synchrony 38

t

Tachycineta bicolor 321
tadpoles 135
Tamarisk shrubs 412–413
Taraxacum officinale 389
target of rapamycin (TOR) pathways 38, 138, 139, 163
Teleogryllus oceanicus 378
teleost fish 134
Temnothorax curvispinosus 386, 388
temperature–size rule 15
temporal cycles, altered 377
temporal dimension
  adaptive life history evolution in urban areas
    diel cycle, effects of changes to 386
    seasonal cycles, adaptation to changes in 386
  urban drivers of selection on life history traits 377
temporal drift 361
Tepinoma sessile 388

terminal investment   230
termites   160
   fungus-farming   303
Tetranychus truncates   259
Tetranychus urticae   117
thale cress (*Arabidopsis thaliana*)   271
the Mother's Curse   42
therapeutic treatments   439
thermal stress, elevated temperatures and   387–388
thyroid hormone (TH)   135–138, 144
thyroid hormone receptors (TRs)   137, 142
thyroid-like hormone signalling   142
TI-J-LiFe network   163, *164*, 166
TI-J-Part   165
time costs   **176**
time *vs.* egg limitation in insects   61–62
tobacco hornworm (*Manduca sexta*)   139
tolerance, defined   269
tolerating an infection   235
TOR pathways *see* target of rapamycin (TOR) pathways
toxic Y hypothesis   41, 78–79
trade-offs   363–364
   allocation   34, 275, 333, 334
   between age and size at maturation   15–16
   between early and late components of reproduction   61
   colonisation–competition   300
   common patterns and observed trade-offs   390–391
   dispersal–fecundity   301
   expansion fronts   363–364
   growth-reproduction   277
   human life histories   191–192
   virulence-dispersal   243
   virulence-transmission   238–241
trait co-variation   79
   aggression   81–83
   caveats   88–89
   knowledge gaps, embracing   84–88
   one genome, two sexes   89–90
   sex-specific predictions   83–84
trait-specific ageing   38
trait syndromes   391–392, 407–408
transformer^F (*tra^F*) gene   42
transgenerational plasticity in offspring size   53–54
transmissible vaccines   449
tree swallows (*Tachycineta bicolor*)   321
trematode parasites   230
TRIAC *see* Triiodothyroacetic acid (TRIAC)
Triatoma dimidiata   389
Tribolium castaneum   362
Trichoderma koningii   387
Trichogramma   366, 410
Trifolium repens   385, 386
Trifolium sp., 258
Triiodothyroacetic acid (TRIAC)   142
Trinidadian guppies (*Poecilia reticulata*)   43, *216*, 221
Triturus alpestris   137
Trivers and Willard hypothesis   120–122
trophy hunting, life history and   431
tropical coastal shrub   273
TRs *see* thyroid hormone receptors (TRs)
tunicates   142

Turdus merula   385
Turnera velutina   273
two-spot sardine (*Astyanax bimaculatus*)   *216*

u

UHI effect *see* 'urban heat island' (UHI) effect
Uinta ground squirrels (*Urocitellus armatus*)   322, 323
unguarded X hypothesis   41, 78–79
unpredictable environments   104
urban areas, adaptive life history evolution in   380
   abiotic dimension   *376*
     chemical pollution   387
     elevated temperatures and thermal stress   387–388
     light pollution   386
     noise pollution   387
   anthropogenic inputs dimension   *376*, 388–389
   biotic dimension   *376*, 388
   pan-urban effects   389
   spatial dimension   *376*
     fragmented landscapes, dispersal in   380
     fragmented urban landscape, finding a mate in   384–385
     patch isolation and island syndrome hypothesis   385
     reduced access to high-quality resources and food limitation   385
   temporal dimension
     diel cycle, effects of changes to   386
     seasonal cycles, adaptation to changes in   386
urban areas, studying evolution in
   field of urban evolution   379–380
   methodological approaches   380
urban drivers of selection on life history traits   375, *376*
   abiotic dimension   *376*, 377–378
   anthropogenic inputs dimension   378–379
   biotic dimension   *376*, 378
   pan-urban effects   379
   spatial dimension   376–377, *376*
   temporal dimension   *376*, 377
urban-dwelling species   390
'urban heat island' (UHI) effect   377–378
urban life history evolution
   patterns, biases and gaps in research on   389
     cascading effects and indirect selection via biotic interactions   391
     common patterns and observed trade-offs   390–391
     selection pressures, variation in   391
urban trait syndrome and convergent evolution   391–392
urban western jackdaw (*Corvus modedula*)   388
urochordates   143
Urocitellus armatus   322, 323
Ursus arctos   429

v

vaccinations   439
Vaccinium   287
variability, evolutionary sources of   429
variance in empirical systems   448–450
variant proof vaccines   450
Vavraia culicis   244, 245
Venturia canescens   65
vertebrate-dispersed seeds   277

vertebrate metamorphosis 134
    environmental regulation of 135
        in coral reef fish 135–136
        in diadromous fish 136
        in frogs 135
    evolutionary patterns of life history diversity within and across vertebrate taxa 137–138
vertebrates, long lifespans of 158–159
vertical vs. horizontal transmission 241–243
Vibrio bacteria 299
Vibrio fischeri 299
vine 385
viral-vectored vaccines 449
virulence and dispersal, evolution of 243–244
virulence-dispersal trade-off 243
virulence evolution 229, 230, 238, *238*, 239, 241, 244, 442, *443*
virulence-transmission trade-off 238
    breaking down of 241
vital substances, circulation of 3
vitellogenin (Vg) expression 165
von Willebrand factor domains 141

W

water flea
    Ceriodaphnia cornuta 385
    Daphnia magna 387
western jackdaw (*Corvus monedula*) 385
white-breasted nuthatch (*Sitta carolinensis*) 101
white-tailed deer (*Odocoileus virginianus*) 428, *428*, **428**
wild cabbage (*Brassica oleracea*) 271
wildlife and fish, exploitative management of 425
    contrasting life histories and harvest potential 427–428
    density-dependence 425–427
    ecological plasticity and evolutionary sources of variability 429
    future directions 434
    harvest, life history and compensatory population responses to 431–432
    harvest and evolution of life history strategies 432–434
    harvest management, knowledge of life history traits and 429–431
    sexually selected infanticide, special case of 432
    sustainable harvest 425–427
    trophy hunting, life history and 431
wild reticulated pythons (*Python reticulatus*) 427
Williams and anti-Williams 42–43
within-host–pathogen populations 445–446, *446*, *447*
within-population variation 183
Wolbachia 259, 410
Wolbachia-free host populations 237
Wolbachia-induced cytoplasmic incompatibility 259
Wolbachia pipientis 259
wolf fish (*Hoplias malabaricus*) *216*
wood tiger moth (*Arctia plantaginis*) 139

x

Xenopus 137
Xenopus leavis 135

y

Y chromosome 79
yellow-bellied marmots (*Marmota flaviventris*) 319
Y model 334

z

Z chromosome 81
Zootoca vivipara 323
zooxanthellae and corals 301